Pro Silverlight 4 in VB

Third Edition

Matthew MacDonald

Pro Silverlight 4 in VB, Third Edition

ISBN 978-1-4302-3548-4

ISBN 978-1-4302-3549-1 (eBook)

President and Publisher: Paul Manning
Lead Editor: Ewan Buckingham
Technical Reviewer: Fabio Claudio Ferracchiati
Editorial Board: Steve Anglin, Mark Beckner, Ewan Buckingham, Gary Cornell, Jonathan Gennick, Jonathan Hassell, Michelle Lowman, Matthew Moodie, Duncan Parkes, Jeffrey Pepper, Frank Pohlmann, Douglas Pundick, Ben Renow-Clarke, Dominic Shakeshaft, Matt Wade, Tom Welsh
Coordinating Editors: Anne Collett and Debra Kelly
Copy Editor: Kim Wimpsett
Compositor: Mary Sudul
Indexer: BIM Indexing & Proofreading Services
Artist: April Milne
Cover Designer: Anna Ishchenko

Distributed to the book trade worldwide by Springer-Verlag New York, Inc., 233 Spring Street, 6th Floor, New York, NY 10013. Phone 1-800-SPRINGER, fax 201-348-4505, e-mail orders-ny@springer-sbm.com, or visit www.springeronline.com.

For information on translations, please e-mail rights@apress.com, or visit www.apress.com.

Apress and friends of ED books may be purchased in bulk for academic, corporate, or promotional use. eBook versions and licenses are also available for most titles. For more information, reference our Special Bulk Sales–eBook Licensing web page at www.apress.com/info/bulksales.

The source code for this book is available to readers at www.apress.com. You will need to answer questions pertaining to this book in order to successfully download the code.

For my family

Contents at a Glance

Contents

About the Author

 **Matthew MacDonald** is an author, educator, and Microsoft MVP for Silverlight. He's the author of more than a dozen books about .NET programming, including Pro WPF in VB 2010, Pro ASP.NET 4 in C# 2010, and Beginning ASP.NET in VB 2010. He's also the author of *Your Brain: The Missing Manual* (Pogue Press, 2008), a popular look at getting the most from your squishy gray matter. Matthew lives in Toronto with his wife and two daughters.

About the Technical Reviewer

■ **Fabio Claudio Ferracchiati** is a prolific writer on cutting-edge technologies. Fabio has contributed to more than a dozen books on .NET, C#, Visual Basic, and ASP.NET. He is a .NET Microsoft Certified Solution Developer (MCSD) and lives in Rome, Italy.

Acknowledgments

No author can complete a book without a small army of helpful individuals. I'm deeply indebted to the whole Apress team, including Ewan Buckingham, Anne Collet, and Debra Kelly, who shepherded the book through its various stages; Fabio Ferracchiati and Damien Foggon, who hunted down stray errors; and Kim Wimpsett, who copy edited the text. Finally, I'd never write any book without the support of my wife and these special individuals: Nora, Razia, Paul, and Hamid. Thanks everyone!

Introduction

Silverlight is a framework for building rich, browser-hosted applications that run on a variety of operating systems. Silverlight works its magic through a *browser plug-in*. When you surf to a web page that includes some Silverlight content, this browser plug-in runs, executes the code, and renders that content in a specifically designated region of the page. The important part is that the Silverlight plug-in provides a far richer environment than the traditional blend of HTML and JavaScript that powers ordinary web pages. Used carefully and artfully, you can create Silverlight pages that have interactive graphics, use vector animations, and play video and sound files.

If this all sounds eerily familiar, it's because the same trick has been tried before. Several other technologies use a plug-in to stretch the bounds of the browser, including Java, ActiveX, Shockwave, and (most successfully) Adobe Flash. Although all these alternatives are still in use, none of them has become the single, dominant platform for rich web development. Many of them suffer from a number of problems, including installation headaches, poor development tools, and insufficient compatibility with the full range of browsers and operating systems. The only technology that's been able to avoid these pitfalls is Flash, which boasts excellent cross-platform support and widespread adoption. However, Flash has only recently evolved from a spunky multimedia player into a set of dynamic programming tools. It still offers less than a modern programming environment like .NET.

That's where Silverlight fits into the picture. Silverlight aims to combine the raw power and cross-platform support of Flash with a first-class programming platform that incorporates the fundamental concepts of .NET. At the moment, Flash has the edge over Silverlight because of its widespread adoption and its maturity. However, Silverlight boasts a few architectural features that Flash can't match—most importantly, the fact that it's based on a scaled-down version of .NET's common language runtime (CLR) and allows developers to write client-side code using pure VB.

Understanding Silverlight

Silverlight uses a familiar technique to go beyond the capabilities of standard web pages: a lightweight browser plug-in.

The advantage of the plug-in model is that the user needs to install just a single component to see content created by a range of different people and companies. Installing the plug-in requires a small download and forces the user to confirm the operation in at least one security dialog box. It takes a short but definite amount of time, and it's an obvious inconvenience. However, once the plug-in is installed, the browser can process any content that uses the plug-in seamlessly, with no further prompting.

Figure 1 shows two views of a page with Silverlight content. At the top is the page you'll see if you *don't* have the Silverlight plug-in installed. At this point, you can click the Get Microsoft Silverlight picture to be taken to Microsoft's website, where you'll be prompted to install the plug-in and then sent back to the original page. On the bottom is the page you'll see once the Silverlight plug-in is installed.

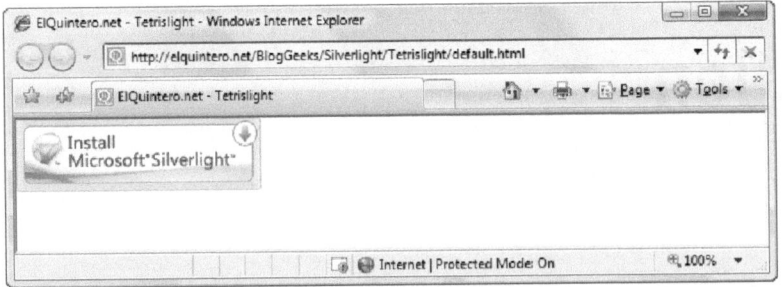

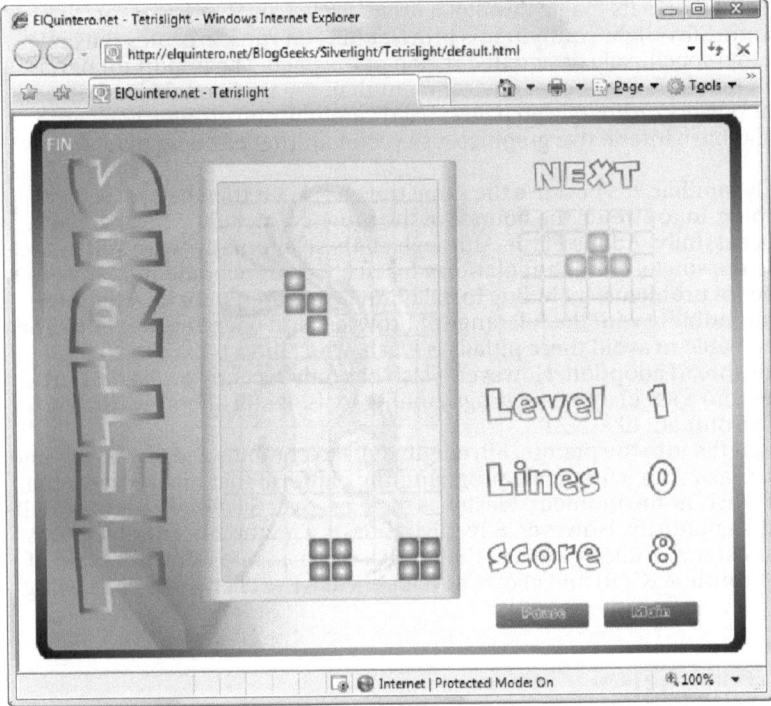

Figure 1. Installing the Silverlight plug-in

■ **Note** At the time of this writing, Silverlight 4 is installed on an estimated 53% of Internet-connected computers. (The share is higher if you consider only Windows operating systems or Interent Explorer browsers.) If you write an application that targets Silverlight 3, you can expand your reach to 62% of all web surfers. Although this is impressive for a relatively new technology, it pales in comparison to Flash, which has version 10 or 10.1 installed on a staggering 93% of all Internet-connected computers. (To get up-to-date statistics, refer to www.riastats.com.)

Silverlight System Requirements

With any Web-centric technology, it's keenly important to have compatibility with the widest possible range of computers and devices. Although Silverlight is still evolving, it's clearly stated mandate is to "support all major browsers on Mac OS X and Windows."

Currently, Silverlight's cross-browser compatibility stacks up fairly well:

- *Windows computers:* Silverlight works on PCs with Windows 7, Windows Vista, and Windows XP. The minimum browser versions that Silverlight supports are Internet Explorer 6, Firefox 1.5, and Google Chrome 4.0. Silverlight will also work in Windows 2000, but only with Internet Explorer 6. Other browsers, such as Opera and Safari (for Windows), aren't currently supported.
- *Mac computers:* Silverlight works on Mac computers with OS X 10.4.8 or later, provided they have Intel hardware (as opposed to the older PowerPC hardware). The minimum browser versions that Silverlight supports are Firefox 2 and Safari 3.
- *Linux computers:* Although Silverlight 4 doesn't currently work on Linux, the Mono team has created an open-source Linux implementation of Silverlight 1 and Silverlight 2, and implementations of Silverlight 3 and Silverlight 4 are planned for the future. This project is known as Moonlight, and it's being developed with key support from Microsoft. To learn more, visit www.mono-project.com/Moonlight.

■ **Note** The system requirements for Silverlight may change as Microsoft releases plug-ins for other browsers. For example, the Opera browser currently works on PCs through an unsupported hack, but better supported is planned in the future. To see the latest system requirements, check www.microsoft.com/silverlight/resources/install.aspx.

Installing Silverlight requires a small-sized setup (around 6MB) that's easy to download. That allows it to provide an all-important "frictionless" setup experience, much like Flash (but quite different from Java).

Silverlight vs. Flash

The most successful browser plug-in is Adobe Flash, which is installed on over 90 percent of the world's web browsers. Flash has a long history that spans more than ten years, beginning as a straightforward tool for adding animated graphics and gradually evolving into a platform for developing interactive content.

It's perfectly reasonable for .NET developers to create websites that use Flash content. However, doing so requires a separate design tool, and a completely different programming language (ActionScript) and programming environment (Flex). Furthermore, there's no straightforward way to integrate Flash content with server-side .NET code. For example, creating Flash applications that call .NET components is awkward at best. Using server-side .NET code to render Flash content (for example, a custom ASP.NET control that spits out a Flash content region) is far more difficult.

Silverlight aims to give .NET developers a better option for creating rich web content. Silverlight provides a browser plug-in with many similar features to Flash, but one that's designed from the ground up for .NET. Silverlight natively supports the VB language and embraces a range of .NET concepts. As a result, developers can write client-side code for Silverlight in the same language they use for server-side code (such as C# and VB), and use many of the same abstractions (including streams, controls, collections, generics, and LINQ).

The Silverlight plug-in has an impressive list of features, some of which are shared in common with Flash, and a few of which are entirely new and even revolutionary. Here are some highlights:

- *2D Drawing:* Silverlight provides a rich model for 2D drawing. Best of all, the content you draw is defined as shapes and paths, so you can manipulate this content on the client side. You can even respond to events (like a mouse click on a portion of a graphic), which makes it easy to add interactivity to anything you draw.

- *Controls:* Developers don't want to reinvent the wheel, so Silverlight is stocked with a few essentials, including buttons, text boxes, lists, and even a DataGrid. Best of all, these basic building blocks can be restyled with custom visuals if you want all of the functionality but none of the stock look.

- *Animation:* Silverlight has a time-based animation model that lets you define what should happen and how long it should take. The Silverlight plug-in handles the sticky details, like interpolating intermediary values and calculating the frame rate.

- *Media:* Silverlight provides playback of a range of video standards, including high-definition H.264 video and AAC audio. Silverlight doesn't use the Windows Media Player ActiveX control or browser plug-in—instead, you can create any front-end you want, and you can even show video in full-screen mode.

- *The Common Language Runtime:* Most impressively, Silverlight includes a scaled-down version of the CLR, complete with an essential set of core classes, a garbage collector, a JIT (just-in-time) compiler, support for generics, threading, and so on. In many cases, developers can take code written for the full .NET CLR and use it in a Silverlight application with only moderate changes.

- *Networking:* Silverlight applications can call old-style ASP.NET web services (.asmx) or WCF (Windows Communication Foundation) web services. They can also send manually created XML requests over HTTP and even open direct socket connections for fast two-way communication. This gives developers a great way to combine rich client-side code with secure server-side routines.

- *Data binding:* Although it's not as capable as in its big brother (WPF), Silverlight data binding provides a convenient way to display large amounts of data with minimal code. You can pull your data from XML or in-memory objects, giving you the ability to call a web service, receive a collection of objects, and display their data in a web page—often with just a couple of lines of code.

- *Multithreading:* A Silverlight application can take advantage of the multithreading capability of modern operating systems. You can easily run background code to perform time-consuming tasks (like contacting a web service or refreshing a calculation).

- *Out-of-browser execution:* Although every Silverlight application needs the Silverlight plug-in, you can configure your application to launch in a stand-alone window, like a normal desktop program. The advantages include desktop presence (for example, you can add a Start menu shortcut to the local computer) and offline support (so your application can run even without an Internet connection). You also have the option of requesting *elevated trust*, which gives trusted applications a range of capabilities that are more common to desktop applications, like document file access, main window customization, and even COM interop.

Silverlight and WPF

One of the most interesting aspects of Silverlight is the fact that it borrows the model WPF uses for rich, client-side user interfaces.

WPF is the modern toolkit for building rich Windows applications. It was introduced in .NET 3.0 as the successor to Windows Forms. WPF is notable because it not only simplifies development with a powerful set of high-level features, it also increases performance by rendering everything through the DirectX pipeline. To learn about WPF, you can refer to *Pro WPF in VB 2010* (Apress).

Silverlight obviously can't duplicate the features of WPF, because many of them rely deeply on the capabilities of the operating system, including Windows-specific display drivers and DirectX technology. However, rather than invent an entirely new set of controls and classes for client-side development, Silverlight uses a subset of the WPF model. If you've had any experience with WPF, you'll be surprised to see how closely Silverlight resembles its big brother. Here are a few common details:

- To define a Silverlight user interface (the collection of elements that makes up a Silverlight content region), you use XAML markup, just as you do with WPF. You can even map data to your display using the same data-binding syntax.
- Silverlight borrows many of the same basic controls from WPF, along with the same styling system (for standardizing and reusing formatting), and a similar templating mechanism (for changing the appearance of standard controls).
- To draw 2D graphics in Silverlight, you use shapes, paths, transforms, geometries, and brushes, all of which closely match their WPF equivalents.
- Silverlight provides a declarative animation model that's based on storyboards, and works in the same way as WPF's animation system.
- To show video or play audio files, you use the MediaElement class, as you do in WPF.

Microsoft has made no secret about its intention to continue to expand the capabilities of Silverlight by drawing from the full WPF model. With each Silverlight release, Silverlight borrows more and more features from WPF. This trend is likely to continue in the future.

■ **Note** WPF is not completely cut off from the easy deployment world of the Web. WPF allows developers to create browser-hosted applications called XBAPs (XAML Browser Applications). These applications are downloaded seamlessly, cached locally, and run directly inside the browser window, all without security prompts. However, although XBAPs run in Internet Explorer and Firefox, they are still a Windows-only technology, unlike Silverlight.

The Evolution of Silverlight

Silverlight 1 was a relatively modest technology. It included 2D drawing features and media playback support. However, it didn't include the CLR engine or support for .NET languages, so developers were forced to code in JavaScript.

Silverlight 2 was a dramatic change. It added the CLR, a subset of .NET Framework classes, and a user interface model based on WPF (as described in the next section, "Silverlight and WPF"). As a result, Silverlight 2 was one of the most hotly anticipated releases in Microsoft's history.

Silverlight 3 and Silverlight 4 haven't been as ambitious. Both keep the same development model as Silverlight 2, but adds a carefully selected group of features and performance enhancements. In Silverlight 3, these features included new controls, more powerful data binding, better graphical effects, a navigation system, and support for out-of-browser applications. In Silverlight 4, the highlights are as follows (in the order they appear in this book):

- *Basic command support:* It's still far short of the WPF model, but Silverlight now has just enough command support for diehard MVVM (Model-View-ViewModel) designers. Chapter 4 has more.

- *The RichTextBox:* This surprisingly sophisticated control allows users to edit richly formatted text, and change font attributes (like bold, italics, and underline) on a specific selection. It can even hold images and interactive controls. You'll see it in action in Chapter 5.

- *Content loaders:* If you need to extend Silverlight's navigation system (for example, to incorporate authentication or download-on-demand behavior), you can now develop your own custom content loader. Chapter 7 shows you how.

- *Printing:* Silverlight 4 adds a bitmap-based printing model that makes sending content to the printer as easy as drawing shapes. You'll learn all the details in Chapter 9.

- *Webcam and microphone support:* Silverlight 4 adds basic support for capturing video and audio input. However, you'll need to do a lot more work on your own if you want to store or transmit chunks of video or audio content. For the full story, refer to Chapter 11.

- *Automatic styles:* It's now possible to apply styles automatically to specific element types (for example, to all the buttons in a window). This convenient feature makes it almost effortless to reskin an application. Chapter 12 explains how automatic styles work.

- *Data-binding refinments:* Silverligh 4 adds data-binding properties that allow you to handle null values and apply simple formatting with a minimum of work. You'll learn more in Chapter 16.

- *File drag-and-drop:* You can now drag files from your desktop (or a Windows Explorer file listing) and drop them into a Silverlight application. Chapter 18 demonstrates this technique.

- *Elevated trust:* One of the most dramatic enhancements in Silverlight 4 is the support for elevated trust. Using it, you can configure an out-of-browser application to request greater security privileges when it's installed. If these prileges are granted, the application can then perform a number of tasks that ordinary Silverlight applications cannot, such as accessing files in the My Documents area and interacting with system components and Windows applications through COM. Chapter 21 describes these new abilities.

■ **Note** This book contains everything you need to master Silverlight 4. You don't need any experience with previous versions of Silverlight. However, if you *have* developed with Silverlight 3, you'll appreciate the "What's New" tip boxes that follow the introduction in each chapter. They point out features that are new to Silverlight 4, so you can home in on its changes and enhancements.

BACKWARD COMPATIBILITY IN SILVERLIGHT 4

At this point, you might be wondering if existing Silverlight 2 and Silverlight 3 applications can run on a computer that has only the latest version of the Silverlight plugin (version 4) installed. It's a reasonable question, as Silverlight 4 introduces some subtle changes and bug fixes that can influence the way applications work—and even change its behavior.

However, Silverlight 4 prevents these differences from causing problems by using its *quirks mode* feature. When the Silverlight 4 plugin loads an application that was compiled for an earlier version of Silverlight, it automatically switches into a quirks mode that attempts to emulate the behavior of the appropriate Silverlight runtime environment.

For more detailed information about breaking changes between Silverlight 4 and Silverlight 3, you can refer to http://msdn.microsoft.com/library/cc645049.aspx.

About This Book

This book is an in-depth exploration of Silverlight for professional developers. You don't need any experience with WPF or previous versions of Silverlight, but you do need to know the .NET platform, the VB language, and the Visual Studio development environment.

What You Need to Use This Book

In order to *run* Silverlight applications, you simply need the Silverlight browser plug-in, which is available at http://silverlight.net. In order to *create* Silverlight applications (and open the sample projects included with this book), you need Visual Studio 2010 and the Silverlight 4 Tools for Visual Studio 2010, which you can download from http://go.microsoft.com/fwlink/?LinkID=177428.

Alternatively, you can use Expression Blend 4—a graphically oriented design tool—to create, build, and test Silverlight applications. Overall, Expression Blend is intended for graphic designers who spend their time creating serious eye candy, while Visual Studio is ideal for code-heavy application programmers. This book assumes you're using Visual Studio. If you'd like to learn more about Expression Blend, you can consult one of many dedicated books on the subject.

■ **Tip** Can't wait to see what Silverlight can do? For a go-to list of the most impressive Silverlight demos (including a few mind-blowers) surf to http://adamkinney.com/blog/Showcase-Silverlight-Apps-for-Talks-and-Demos.

The Silverlight Toolkit

To keep in touch with Silverlight's latest developments, you should also download Microsoft's impressive Silverlight Toolkit, which provides a set of controls and components that extend the features of Silverlight. You can use them in your Silverlight applications simply by adding an assembly reference.

The Silverlight Toolkit isn't just a package of useful tools. It's also a development process that gradually brings new controls into the Silverlight platform. Many new controls appear first in the Silverlight Toolkit, are gradually refined, and then migrate to the core platform. Examples of controls that have made the jump from the Silverlight Toolkit to the core Silverlight plugin include the AutoCompleteBox, TreeView, and Viewbox.

To understand how this process works, you need to understand a bit more about the Silverlight Toolkit's *quality bands*—groups of controls at a particular evolutionary stage. The Silverlight Toolkit divides its features into four quality bands:

- *Mature:* The mature band has controls that are unlikely to change. Usually, these are controls that are already included with the core Silverlight plugin. However, the Silverlight Toolkit gives you access to their complete source code, which opens up customization possibilities.
- *Stable:* The stable band includes controls that are ready for inclusion in just about any application—however, there may be further tweaks and fixes in the future that subtly change behavior. This book describes many of the stable controls, including the DockPanel, WrapPanel, and Expander.
- *Preview:* The preview band includes controls that are reliable enough for most applications, but are likely to change in response to developer comments, so you expect to change your code before using newer versions.

- *Experimental:* The experimental band includes new controls that are intended to solicit developer feedback. Feel free to play with these, but include them in an application at your own risk.

To learn more about the different quality bands, try out the controls with live demos, or download the Silverlight Toolkit for yourself, go to http://silverlight.codeplex.com.

Code Samples

It's a good idea to check the Apress website or www.prosetech.com to download the up-to-date code samples. You'll need to do this to test most of the more sophisticated code examples described in this book because the less significant details are usually left out. This book focuses on the most important sections so that you don't need to wade through needless extra pages to understand a concept.

To download the source code, surf to www.prosetech.com and look for the page for this book.

Feedback

This book has the ambitious goal of being the best tutorial and reference for programming Silverlight. Toward that end, your comments and suggestions are extremely helpful. You can send complaints, adulation, and everything in between directly to apress@prosetech.com. I can't solve your Silverlight problems or critique your code, but I will benefit from information about what this book did right and wrong (or what it may have done in an utterly confusing way).

The Last Word

As you've seen, Silverlight is a .NET-based Flash competitor. Unlike the Flash development model, which is limited in several ways due to how it has evolved over the years, Silverlight is a starting-from-scratch attempt that's thoroughly based on .NET and WPF, and will therefore allow .NET developers to be far more productive. In many ways, Silverlight is the culmination of two trends: the drive to extend web pages to incorporate more and more rich-client features, and the drive to give the .NET Framework a broader reach. It's also a new direction that will only get more interesting in the months ahead.

CHAPTER 1

■ ■ ■

Introducing Silverlight

In the introduction, you learned about the design philosophy that underpins Silverlight. Now, you're ready to get your hands dirty and create your first Silverlight application.

The best starting point for coding a Silverlight application is Visual Studio, Microsoft's premiere development tool. In this chapter, you'll see how to create, compile, and deploy a Silverlight application using Visual Studio. Along the way, you'll get a quick look at how Silverlight controls respond to events, you'll see how Silverlight applications are compiled and packaged for the Web, and you'll consider the two options for hosting Silverlight content: either in an ordinary HTML web page or in an ASP.NET web form.

Silverlight Design Tools

Although it's technically possible to create the files you need for a Silverlight application by hand, professional developers always use a development tool. If you're a graphic designer, that tool is likely to be Microsoft Expression Blend 4, which provides a full complement of features for designing visually rich user interfaces. If you're a developer, you'll probably use Visual Studio 2010, which includes well-rounded tools for coding, testing, and debugging.

Because both tools are equally at home with the Silverlight application model, you can easily create a workflow that incorporates both of them. For example, a developer could create a basic user interface with Visual Studio and then hand it off to a crack design team, which would polish it up with custom graphics in Expression Blend. When the face-lift is finished, the designers deliver the project to the developers, who continue writing and refining its code in Visual Studio.

Many developers go a step further: they install both applications on their computer, load them simultaneously, and switch between them as they go. They use Visual Studio for core programming tasks such as code-writing and debugging and switch to Expression Blend to enhance the user interface—for example, to edit control templates, pick colors, refine animations, and draw simple vector art. (This back-and-forth process works because once you save the changes in one program, the other program notices. When you switch back, it will prompt you to perform a quick refresh that loads the new version. The only trick is that you need to remember to save before switching.) Whether you use this approach is up to you—but even if you do, Visual Studio will be the starting point and central hub for your development.

Visual Studio vs. Expression Blend

If you're still trying to understand how Visual Studio and Expression Blend stack up, here's a quick overview:

1

- *Visual Studio 2010*: It has everything you need to develop Silverlight applications, with a visual designer for Silverlight pages. Using this designer, you can drag, drop, and draw your user interface into existence (which isn't always the best idea), and you can get a live preview of what it looks like (which is terrifically useful).

- *Expression Blend 4*: It provides the rich support for creating Silverlight user interface, with visual tools that surpass Visual Studio. For certain types of user interface grunt work (for example, creating a nice gradient fill), it's a tremendous help. Expression Blend also supports a fun application prototyping tool called SketchFlow and includes a decent coding editor that's designed to look like Visual Studio. However, it lacks many advanced and important development tools, such as debugging, code refactoring, and project source control.

■ **Note** Visual Studio 2010 includes full support for creating Silverlight 3 projects. But to create Silverlight 4 projects, you need the Silverlight 4 Tools for Visual Studio 2010, which you can download from http://go.microsoft.com/fwlink/?LinkID=177428.

This book assumes you're working primarily with Visual Studio. You'll get occasional tips for Expression Blend (and other Expression products that work with Silverlight, including the Expression Design drawing tool and Expression Encoder video encoding tool). But if you really want to master Expression Blend, you should consider a dedicated book on the subject, spend an afternoon experimenting, or take a look through Microsoft's Expression Blend training videos at http://expression.microsoft.com/cc136535.aspx.

Understanding Silverlight Websites

You can create two types of Silverlight websites in Visual Studio or Expression Blend:

- *An ordinary website with HTML pages*: In this case, the entry point to your Silverlight application is a basic HTML file that includes a Silverlight content region.

- *ASP.NET website*: In this case, Visual Studio creates two projects—one to contain the Silverlight application files and one to hold the server-side ASP.NET website that will be deployed alongside your Silverlight files. The entry point to your Silverlight application can be an ordinary HTML file, or it can be an ASP.NET web page that includes server-generated content.

So, which approach is best? No matter which option you choose, your Silverlight application will run the same way—the client browser will receive an HTML document, which will include a Silverlight content region, and the Silverlight code will run on the local computer, *not* the web server. However, the ASP.NET web approach makes it easier to mix ASP.NET and Silverlight content. This is usually a better approach in the following cases:

- You want to create a website that contains both ASP.NET web pages and Silverlight-enhanced pages.

- You want to create a Silverlight application that calls a web service, and you want to design the web service at the same time (and deploy it to the same web server).
- You want to generate Silverlight content indirectly, using specialized ASP.NET web controls.

On the other hand, if you don't need to write any server-side code, there's little point in creating a full-fledged ASP.NET website. Many of the Silverlight applications you'll see in this book use basic HTML-only websites. The examples only include ASP.NET websites when they need specific server-side features. For example, the examples in Chapter 16 use an ASP.NET website that includes a web service. This web service allows the Silverlight application to retrieve data from a database on the web server, a feat that would be impossible without server-side code. You'll learn how to design an ASP.NET web service for Silverlight in Chapter 15.

ADDING SILVERLIGHT CONTENT TO AN EXISTING WEBSITE

A key point to keep in mind when considering the Silverlight development model is that in many cases you'll use Silverlight to *augment* the existing content of your website, which will still include generous amounts of HTML, CSS, and JavaScript. For example, you might add a Silverlight content region that shows an advertisement or allows an enhanced experience for a portion of a website (such as playing a game, completing a survey, interacting with a product, or taking a virtual tour). You may use Silverlight-enhanced pages to present content that's already available in your website in a more engaging way or to provide a value-added feature for users who have the Silverlight plug-in.

Of course, it's also possible to create a Silverlight-only website, which is a somewhat more daring approach. The key drawback is that Silverlight is still relatively new, isn't installed as widely as other web technologies such as Flash, and doesn't support legacy clients such as those running the Windows ME or Windows 2000 operating system. As a result, Silverlight doesn't have nearly the same reach as ordinary HTML. Many businesses that are adopting Silverlight are using it to distinguish themselves from other online competitors with cutting-edge content, but they aren't abandoning their traditional websites.

Creating a Stand-Alone Silverlight Project

The easiest way to start using Silverlight is to create an ordinary website with HTML pages and no server-side code. Here's how:

1. Select File ➤ New ➤ Project in Visual Studio, choose the Visual Basic ➤ Silverlight group of project types, and then select the Silverlight Application template. As usual, you need to pick a project name and a location on your hard drive before clicking OK to create the project.
2. At this point, Visual Studio will prompt you to choose whether you want to create a full-fledged ASP.NET website that can run server-side code along with your Silverlight project (see Figure 1-1). Uncheck the "Host the Silverlight application in a new Web site" option to keep things simple.

3. Underneath, choose the version of Silverlight application that you want to create. If you aren't using any of the new features in Silverlight 4, you'll get slightly more reach with Silverlight 3 (which, at the time of this writing, is still installed on more computers). If you haven't installed the Silverlight 4 Tools for Visual Studio 2010, you won't get an option for creating Silverlight 4 applications.

■ **Tip** You can change the version of Silverlight that you're targeting at any point after you've created it. To do so, just double-click the My Project node in the Solution Explorer, and change the selection in the Target Silverlight Version list.

---

4. Click OK to continue and create the project.

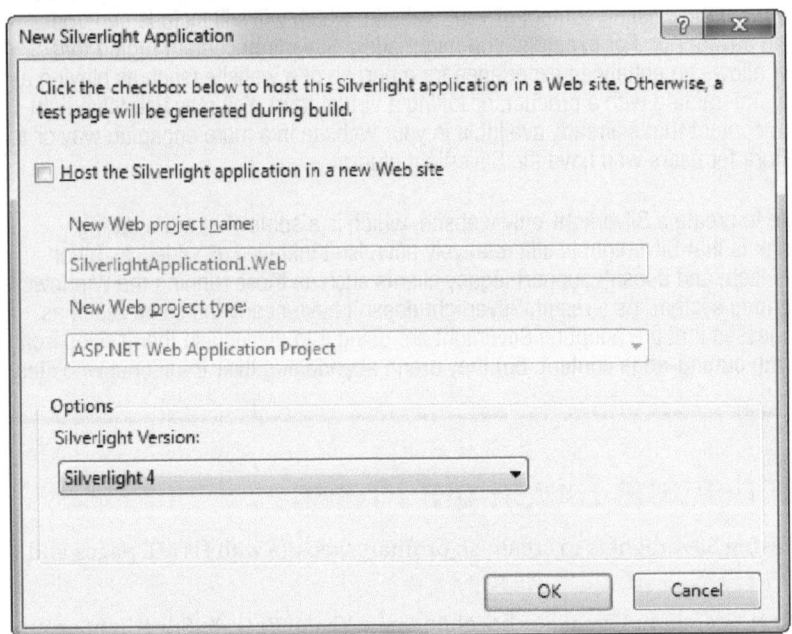

Figure 1-1. Choosing not to include an ASP.NET website

Every Silverlight project starts with a small set of essential files, as shown in Figure 1-2. All the files that end with the extension .xaml use a flexible markup standard called XAML, which you'll dissect in the next chapter. All the files that end with the extension .vb hold the VB source code that powers your application.

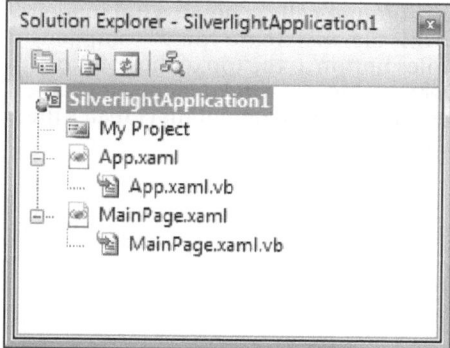

Figure 1-2. A Silverlight project

Here's a rundown of the files shown in Figure 1-2:

- *App.xaml and App.xaml.vb*: These files configure your Silverlight application. They allow you to define resources that will be made available to all the pages in your application (see Chapter 2), and they allow you react to application events such as startup, shutdown, and error conditions (see Chapter 6). In a newly generated project, the startup code in the App.xaml.vb file specifies that your application should begin by showing MainPage.xaml.
- *MainPage.xaml*: This file defines the user interface (the collection of controls, images, and text) that will be shown for your first page. Technically, Silverlight pages are *user controls*—custom classes that derive from UserControl. A Silverlight application can contain as many pages as you need—to add more, simply choose Project ➤ Add New Item, pick the Silverlight User Control template, choose a file name, and click Add.
- *MainPage.xaml.vb*: This file includes the code that underpins your first page, including the event handlers that react to user actions.

■ **Note** For the first few chapters of this book, you'll create applications that have just a single page. In Chapter 6, you'll take a closer look at the application logic that sets your initial page. In Chapter 7, you'll break free of this constraint altogether and learn the techniques you need to combine pages and navigate from one to another.

Along with these four essential files, there are a few more ingredients that you'll find only if you dig around. To see these files, click the Show All Files button at the top of the Solution Explorer (or choose Project ➤ Show All Files from the menu). Under the My Project node in the Solution Explorer, you'll find a file named AppManifest.xml, which lists the assemblies that your application uses. You'll also find a file named AssemblyInfo.vb, which contains information about your project (such as its name, version, and publisher) that's embedded into your Silverlight assembly when it's compiled. Neither of these files should be edited by hand—instead, they're modified by Visual Studio when you add references or set project properties.

Last, the gateway to your Silverlight application is an automatically generated but hidden HTML file named TestPage.html (see Figure 1-3). To see this file, make sure you've compiled your application at least once. Then, click the Show All Files button at the top of the Solution Explorer (if you haven't already), and expand the Bin\Debug folder (which is where your application is compiled). The TestPage.html file includes an <object> element that creates the Silverlight content area. You'll take a closer look at it later in this chapter.

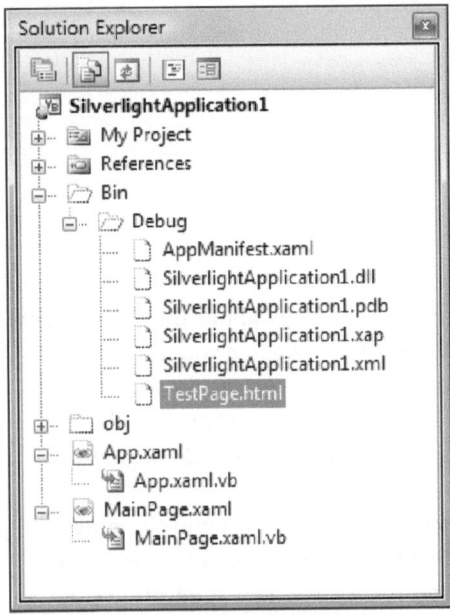

Figure 1-3. *The HTML test page*

Creating a Simple Silverlight Page

As you've already learned, every Silverlight page includes a markup portion that defines the visual appearance (the XAML file) and a source code file that contains event handlers. To customize your first Silverlight application, you simply need to open the MainPage.xaml file and begin adding markup.

Visual Studio gives you two ways to look at every XAML file—as a visual preview (known as the *design surface*) or the underlying markup (known as the *XAML view*). By default, Visual Studio shows both parts, stacked one on the other. Figure 1-4 shows this view and points out the buttons you can use to change your vantage point.

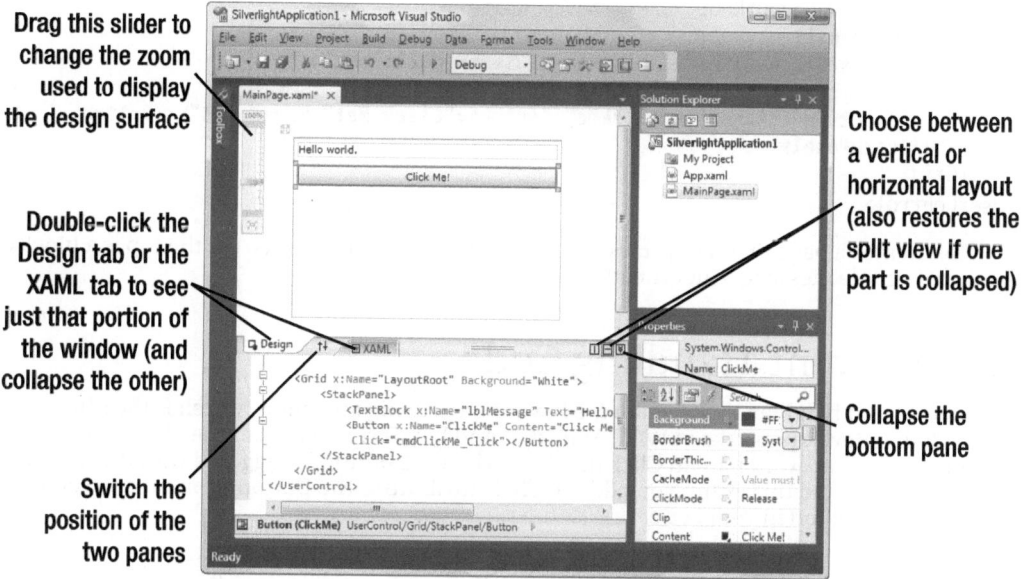

Figure 1-4. *Viewing XAML pages*

You can start designing a XAML page by selecting a control in the Toolbox and then "drawing" it onto the design surface. However, this convenience won't save you from learning the full intricacies of XAML. To organize your elements into the right layout containers, change their properties, wire up event handlers, and use Silverlight features such as animation, styles, templates, and data binding, you'll need to edit the XAML markup by hand. In fact, in many cases, you'll find that the markup Visual Studio generates when you drag and drop a page into existence might not be what you really want.

■ **Note** In Silverlight terminology, each graphical widget that meets these criteria (appears in a window and is represented by a .NET class) is called an *element*. The term *control* is generally reserved for elements that receive focus and allow user interaction. For example, a TextBox is a control, but the TextBlock is not.

To get started, you can try creating the page shown in the following example, which defines a block of text and a button. The portions in bold have been added to the basic page template that Visual Studio generated when you created the project.

```
<UserControl x:Class="SilverlightApplication1.MainPage"
  xmlns="http://schemas.microsoft.com/winfx/2006/xaml/presentation"
  xmlns:x="http://schemas.microsoft.com/winfx/2006/xaml"
  xmlns:d="http://schemas.microsoft.com/expression/blend/2008"
  xmlns:mc="http://schemas.openxmlformats.org/markup-compatibility/2006"
  mc:Ignorable="d" d:DesignWidth="300" d:DesignHeight="400">
```

```
<Grid x:Name="LayoutRoot" Background="White">
    <StackPanel>
        <TextBlock x:Name="lblMessage" Text="Hello world."
         Margin="5"></TextBlock>
        <Button x:Name="cmdClickMe" Content="Click Me!" Margin="5"></Button>
    </StackPanel>
</Grid>
</UserControl>
```

This creates a page that has a stacked arrangement of two elements. On the top is a block of text with a simple message. Underneath it is a button.

Adding Event-Handling Code

You attach event handlers to the elements in your page using attributes, which is the same approach that developers take in WPF, ASP.NET, and JavaScript. For example, the Button element exposes an event named Click that fires when the button is triggered with the mouse or keyboard. To react to this event, you add the Click attribute to the Button element and set it to the name of a method in your code:

```
<Button x:Name="cmdClickMe" Click="cmdClickMe_Click" Content="Click Me!"
 Margin="5"></Button>
```

■ **Tip** Although it's not required, it's a common convention to name event handler methods in the form *ElementName_EventName*. If the element doesn't have a defined name (presumably because you don't need to interact with it in any other place in your code), consider using the name it *would* have.

This example assumes that you've created an event-handling method named cmdClickMe_Click. Here's what it looks like in the MainPage.xaml.vb file:

```
Private Sub cmdClickMe_Click(ByVal sender As Object, ByVal e As RoutedEventArgs)
    lblMessage.Text = "Goodbye, cruel world."
End Sub
```

You can add an event handler by double-clicking an element on the design surface or by clicking the Events button in the Properties window and then double-clicking the appropriate event.

If you've already coded the event handler you need, you can use IntelliSense to quickly attach it to the right event. Begin by typing in the attribute name, followed by the equals sign. At this point, Visual Studio will pop up a menu that lists all the methods that have the right syntax to handle this event and currently exist in your code-behind class, as shown in Figure 1-5. Simply choose the right event-handling method.

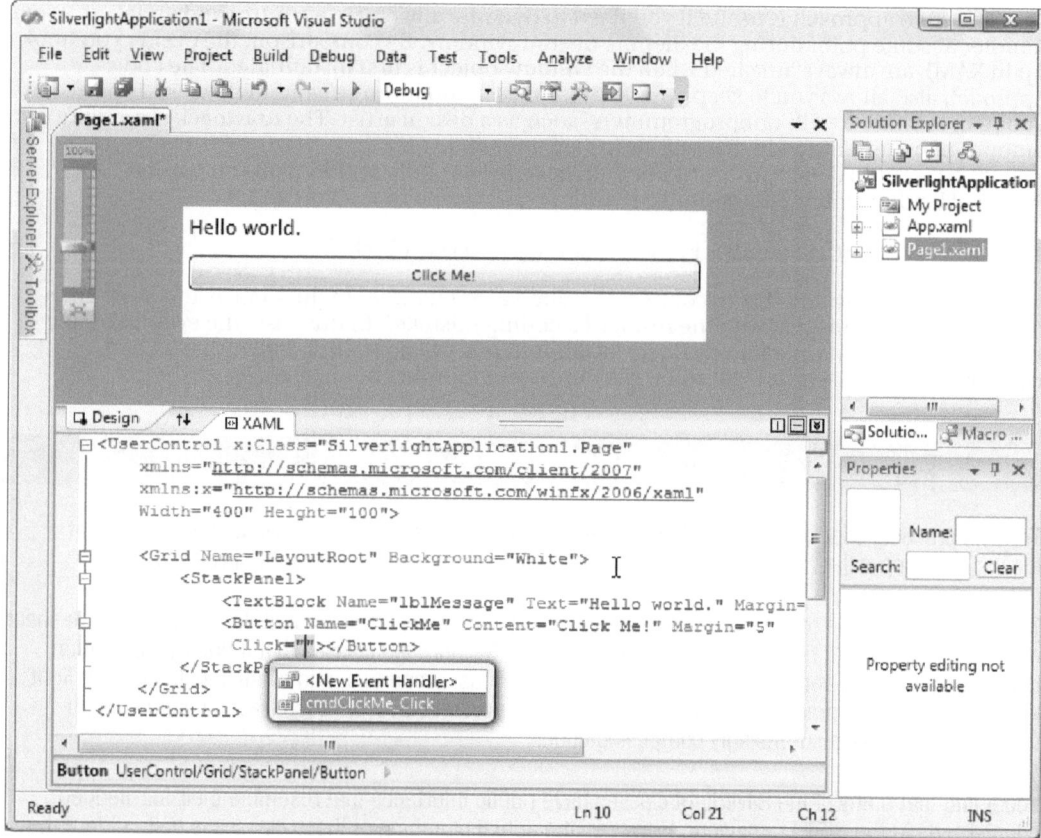

Figure 1-5. *Attaching an event handler*

It's possible to use Visual Studio (either version) to create and assign an event handler in one step by adding an event attribute and choosing the <New Event Handler> option in the menu.

■ **Tip** To jump quickly from the XAML to your event-handling code, right-click the appropriate event attribute in your markup and choose Navigate to Event Handler.

You can also connect an event with code. The place to do it is the constructor for your page, after the call to InitializeComponent(), which initializes all your controls. Here's the code equivalent of the XAML markup shown previously:

```
Public Sub New()
    InitializeComponent()
    AddHandler cmdClickMe.Click, AddressOf cmdClickMe_Click
End Sub
```

The code approach is useful if you need to dynamically create a control and attach an event handler at some point during the lifetime of your window. By comparison, the events you hook up in XAML are always attached when the window object is first instantiated. The code approach also allows you to keep your XAML simpler and more streamlined, which is perfect if you plan to share it with nonprogrammers, such as a design artist. The drawback is a significant amount of boilerplate code that will clutter up your code files.

If you want to detach an event handler, code is your only option. You can use the RemoveHandler statement, as shown here:

```
RemoveHandler cmdClickMe.Click, AddressOf cmdClickMe_Click
```

It is technically possible to connect the same event handler to the same event more than once, but this is almost always the result of a coding mistake. (In this case, the event handler will be triggered multiple times.) If you attempt to remove an event handler that's been connected twice, the event will still trigger the event handler, but just once.

THE SILVERLIGHT CLASS LIBRARIES

To write practical code, you need to know quite a bit about the classes you have to work with. That means acquiring a thorough knowledge of the core class libraries that ship with Silverlight.

The Silverlight version of the .NET Framework is simplified in two ways. First, it doesn't provide the sheer number of types you'll find in the full .NET Framework. Second, the classes that it does include often don't provide the full complement of constructors, methods, properties, and events. Instead, Silverlight keeps only the most practical members of the most important classes, which leaves it with enough functionality to create surprisingly compelling code.

You'll find that many of the Silverlight classes have public interfaces that resemble their full-fledged counterparts in the .NET Framework. However, the actual plumbing of these classes is quite different. All the Silverlight classes have been rewritten from the ground up to be as streamlined and efficient as possible.

Testing a Silverlight Application

You now have enough to test your Silverlight project. When you run a Silverlight application, Visual Studio launches your default web browser and navigates to the hidden browser test page, named TestPage.html. The test page creates a new Silverlight control and initializes it using the markup in MainPage.xaml.

■ **Note** Visual Studio sets TestPage.html to be the start page for your project. As a result, when you launch your project, this page will be loaded in the browser. You can choose a different start page by right-clicking another HTML file in the Solution Explorer and choosing Set As Start Page.

Figure 1-6 shows the previous example at work. When you click the button, the event-handling code runs, and the text changes. This process happens entirely on the client—there is no need to contact the server or post back the page, as there is in a server-side programming framework such as ASP.NET. All the Silverlight code is executed on the client side by the scaled-down version of .NET that's embedded in the Silverlight plug-in.

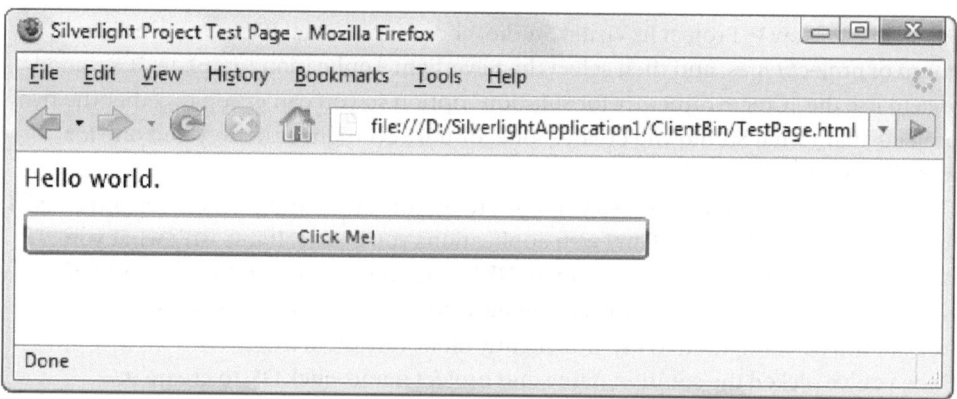

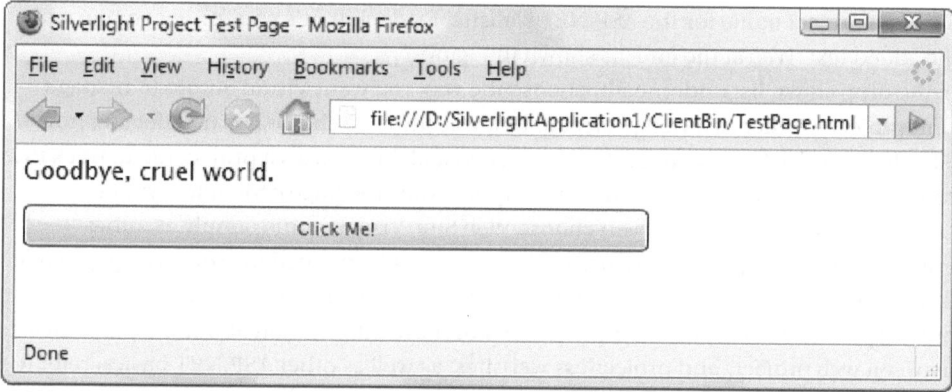

Figure 1-6. Running a Silverlight application (in Firefox)

If you're hosting your host Silverlight content in an ordinary website (with no server-side ASP.NET), Visual Studio won't use its integrated web server during the testing process. Instead, it simply opens the HTML test page directly from the file system. (You can see this in the address bar in Figure 1-6.)

In some situations, this behavior could cause discrepancies between your test environment and your deployed environment, which will use a full-fledged web server that serves pages over HTTP. The most obvious difference is the security context—in other words, you could configure your web browser to allow local web pages to perform actions that remote web content can't. In practice, this isn't often a problem, because Silverlight always executes in a stripped-down security context and doesn't include any extra functionality for trusted locations. This simplifies the Silverlight development model and ensures that features won't work in certain environments and break in others. However, when production testing a Silverlight application, it's a good idea to create an ASP.NET test website (as described in the next section) or—even better—deploy your Silverlight application to a test web server.

Creating an ASP.NET-Hosted Silverlight Project

Although Silverlight does perfectly well on its own, you can also develop, test, and deploy it as part of an ASP.NET website. Here's how to create a Silverlight project and an ASP.NET website that uses it in the same solution:

1. Select File ➤ New ➤ Project in Visual Studio, choose the Visual Basic ➤ Silverlight group of project types, and then select the Silverlight Application template. It's a good idea to use the "Create directory for solution" option so you can group together the two projects that Visual Studio will create—one for the Silverlight assembly and one for ASP.NET website.

2. Ordinarily, Visual Studio assumes you want to use the latest and greatest version of .NET for the server-side portion of any web applications you create. If this isn't what you want, you can pick a different version of .NET from the drop-down list at the top of the New Project window. For example, if you pick .NET Framework 3.5, your ASP.NET website will be configured to use this slightly older version of .NET.

3. Once you've picked the solution name and project name, click OK to create it.

4. Make sure the option "Host the Silverlight application in a new website" is checked.

5. Supply a project name for the ASP.NET website. By default, it's your project name with the added text *.Web* at the end, as shown in Figure 1-7.

6. In the drop-down list underneath, choose the way you want Visual Studio to manage your project—either as a web project or as a website. The choice has no effect on how Silverlight works. If you choose Web Project, Visual Studio uses a project file to track the contents of your web application and compiles your web page code into a single assembly before you run it. If you choose Web Site, Visual Studio simply assumes everything in the application folder is part of your web application. Your web page code will be compiled the first time a user requests a page (or when you use the precompilation tool aspnet_compiler.exe). For more information about the difference between web projects and projectless websites, as well as other ASP.NET basics, refer to *Pro ASP.NET 4 in VB 2010*.

7. Choose whether you want to create a Silverlight 3 or Silverlight 4 application in the Silverlight Version list.

8. You can also choose to enable WCF RIA services, which are a set of web services that help you bridge the gap between the client-side world of Silverlight and the server-side world of ASP.NET. For a basic Silverlight website, leave this option unchecked.

■ **Note** WCF RIA Services require a separate download (www.silverlight.net/getstarted/riaservices) and aren't discussed in this book. However, you'll get an introduction to web services, their foundational technology, in Chapter 15. For more information, check out the download site or read *Pro Business Applications with Silverlight 4*.

---

9. Finally, click OK to create the solution.

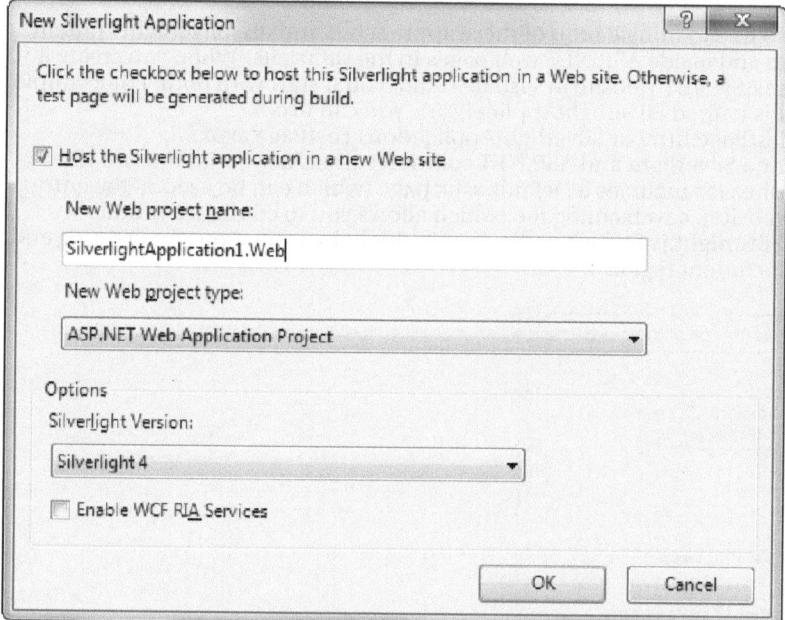

Figure 1-7. Creating an ASP.NET website to host Silverlight content

---

■ **Note** If you create an ordinary HTML-only website, you can host it on any web server. In this scenario, the web server has an easy job—it simply needs to send along your HTML files when a browser requests them. If you decide to create an ASP.NET website, your application's requirements change. Although the Silverlight portion of your application will still run on the client, any ASP.NET content you include will run on the web server, which must have the ASP.NET engine installed.

There are two ways to integrate Silverlight content into an ASP.NET application:

*   *Create HTML files with Silverlight content*: You place these files in your ASP.NET website folder, just as you would with any other ordinary HTML file. The only limitation of this approach is that your HTML file obviously can't include ASP.NET controls, because it won't be processed on the server.
*   *Place Silverlight content inside an ASP.NET web form*: In this case, the <object> element that loads the Silverlight plug-in is inserted into a dynamic .aspx page. You can add other ASP.NET controls to different regions of this page. The only disadvantage to this approach is that the page is always processed on the server. If you aren't actually using

any server-side ASP.NET content, this creates an extra bit of overhead that you don't need when the page is first requested.

Of course, you're also free to mingle both of these approaches and use Silverlight content in dedicated HTML pages and inside ASP.NET web pages in the same site. When you create a Silverlight project with an ASP.NET website in Visual Studio, you'll start with both. For example, if your Silverlight project is named SilverlightApplication1, you can use SilverlightApplication1TestPage.html or SilverlightApplication1TestPage.aspx.

Figure 1-8 shows how a Silverlight and ASP.NET solution starts. Along with the two test pages, the ASP.NET website also includes a Default.aspx page (which can be used as the entry point to your ASP.NET website), a web.config file (which allows you to configure various website settings), and a Silverlight.js file (which has JavaScript helper functions for creating and initializing the Silverlight content region).

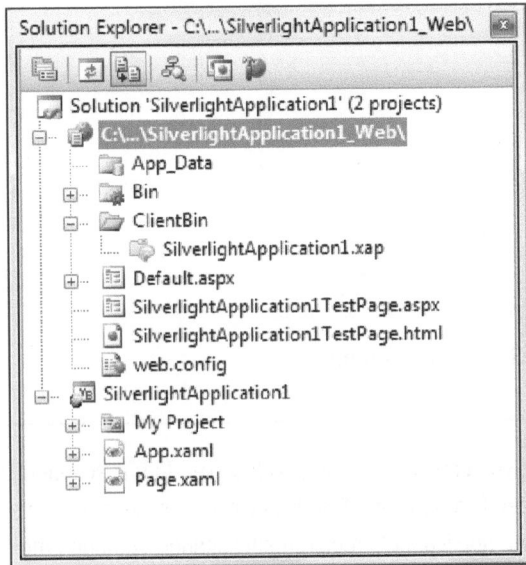

Figure 1-8. *Creating an ASP.NET website to host Silverlight content*

The Silverlight and ASP.NET option provides essentially the same debugging experience as a Silverlight-only solution. When you run the solution, Visual Studio compiles both projects and copies the Silverlight assembly to the ClientBin folder in the ASP.NET website. (This is similar to assembly references—if an ASP.NET website references a private DLL, Visual Studio automatically copies this DLL to the Bin folder.)

Once both projects are compiled, Visual Studio looks to the startup project (which is the ASP.NET website) and looks for the currently selected page. It then launches the default browser and navigates to that page. The difference is that it doesn't request the start page directly from the file system. Instead, it communicates with its built-in test web server. This web server automatically loads on a randomly chosen port. It acts like a scaled-down version of IIS but accepts requests only from the local computer. This gives you the ease of debugging without needing to configure IIS virtual directories. Figure 1-9 shows the same Silverlight application you considered earlier but hosted by ASP.NET.

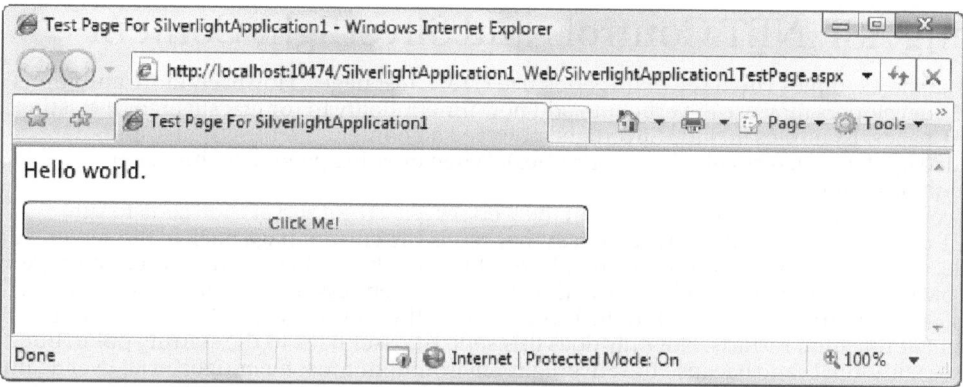

Figure 1-9. An ASP.NET page with Silverlight content

To navigate to a different page from the ASP.NET project, you can type in the address bar of the browser.

■ **Note** Remember, when building a Silverlight and ASP.NET solution, you add all your Silverlight files and code to the Silverlight project. The ASP.NET website consumes the final, compiled Silverlight assembly and makes it available through one or more of its web pages.

ASP.NET Controls That Render Silverlight Content

In the past, ASP.NET developers who wanted to incorporate Silverlight content often relied on a specially designed ASP.NET web control named Silverlight. Like all ASP.NET controls, the Silverlight control is processed on the server. When the ASP.NET engine renders the page into HTML, the Silverlight control emits the <object> element that defines the Silverlight content region. The end result is that the client gets the same content as in a normal, non-ASP.NET-hosted Silverlight application. However, the server-side programming model is a bit different.

The advantage of using a web control to generate the Silverlight content region is that it opens up possibilities for server-side interaction. For example, server-side code can dynamically set the Source property of the Silverlight control to point to a different application. However, the ASP.NET Silverlight control provided few openings for real interaction with server code. In the end, it was rarely more than a glorified wrapper for the <object> element.

Microsoft no longer promotes the use of the Silverlight control, and the Silverlight SDK no longer includes it. If you migrate an existing Silverlight 2 project that includes an ASP.NET website to Silverlight 4, your project will continue to use the ASP.NET Silverlight control. However, whenever you build a new project, Visual Studio will use the more straightforward <object> element approach. If you do still want to use the Silverlight and MediaPlayer controls in new projects, you can download them from http://code.msdn.microsoft.com/aspnetprojects.

Mixing ASP.NET Controls and Silverlight Content

Almost all the examples you'll see in this book use HTML test pages. However, more ambitious ASP.NET developers may use Silverlight to add new functionality to (or just sugarcoat) existing ASP.NET pages. Examples include Silverlight-powered ad content, menu systems, and embedded applets (such as calculators or games). When creating pages like this, a few considerations apply.

As you know, all ASP.NET code runs on the web server. To get server-side code to run, ASP.NET controls use a postback mechanism that sends the current page back to the server. For example, this happens when you click an ASP.NET button. The problem is that when the page is posted back, the current Silverlight application ends. The web server code runs, a new version of the page is sent to the browser, and the browser loads this new page, at which point your Silverlight application restarts. Not only does this send the user back to the starting point, but it also takes additional time because the Silverlight environment must be initialized all over again.

If you want to avoid this disruption, you can use ASP.NET AJAX techniques. A particularly useful tool is the UpdatePanel. The basic technique is to wrap the controls that would ordinarily trigger a postback and any other controls that they modify into one or more UpdatePanel controls. Then, when the user clicks a button, an asynchronous request is sent to the web server instead of a full postback. When the browser receives the reply, it updates the corresponding portions of the page without disrupting the Silverlight content.

■ **Tip** For a much more detailed exploration of the UpdatePanel control, refer to *Pro ASP.NET 4 in VB 2010*.

Silverlight Compilation and Deployment

Now that you've seen how to create a basic Silverlight project, add a page with elements and code, and run your application, it's time to dig a bit deeper. In this section, you'll see how your Silverlight is transformed from a collection of XAML files and source code into a rich browser-based application.

Compiling a Silverlight Application

When you compile a Silverlight project, Visual Studio uses the same vbc.exe compiler that you use for full-fledged .NET applications. However, it references a different set of assemblies, and it passes in the command-line argument *nostdlib*, which prevents the VB compiler from using the standard library (the core parts of the .NET Framework that are defined in mscorlib.dll). In other words, Silverlight applications can be compiled like normal .NET applications written in standard VB, just with a more limited set of class libraries to draw on. The Silverlight compilation model has a number of advantages, including easy deployment and vastly improved performance when compared to ordinary JavaScript.

Your compiled Silverlight assembly includes the compiled code *and* the XAML documents for every page in your application, which are embedded in the assembly as resources. This ensures that there's no way for your event-handling code to become separated from the user interface markup it needs. Incidentally, the XAML is not compiled in any way (unlike WPF, which converts it into a more optimized format called BAML).

Your Silverlight project is compiled into a DLL file named after your project. For example, if you have a project named SilverlightApplication1, the vbc.exe compiler will create the file SilverlightApplication1.dll. The project assembly is dumped into a Bin\Debug folder in your project directory, along with a few other important files:

- *A PDB file*: This file contains information required for Visual Studio debugging. It's named after your project assembly (for example, SilverlightApplication1.pdb).
- *AppManifest.xaml*: This file lists assembly dependencies.
- *Dependent assemblies*: The Bin\Debug folder contains the assemblies that your Silverlight project uses, provided these assemblies have the Copy Local property set to True. Assemblies that are a core part of Silverlight have Copy Local set to False, because they don't need to be deployed with your application (you can change the Copy Local setting by expanding the References node in the Solution Explorer, selecting the assembly, and using the Properties window).
- *TestPage.html*: This is the entry page that the user requests to start your Silverlight application. Visual Studio only generates this file for stand-alone Silverlight projects, not for ASP.NET-hosted Silverlight projects.
- *A XAP file*: This is a Silverlight package that contains everything you need to deploy your Silverlight application, including the application manifest, the project assembly, and any other assemblies that your application uses. If you're developing an ASP.NET-hosted Silverlight application, Visual Studio will also copy the XAP file to the ClientBin folder in the test website.

Of course, you can change the assembly name, the default namespace (which is used when you add new code files), and the XAP file name using the Visual Studio project properties (Figure 1-10). Just double-click the My Project node in the Solution Explorer.

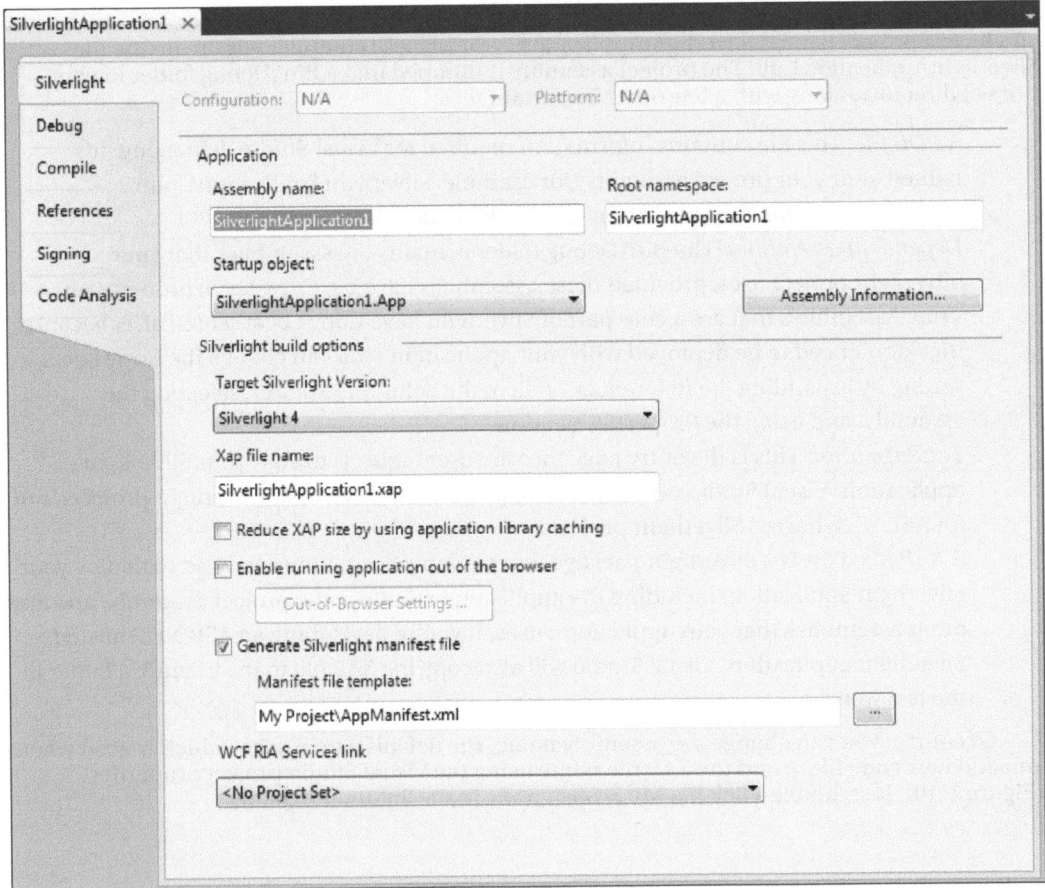

Figure 1-10. *Silverlight project properties*

Deploying a Silverlight Application

Once you understand the Silverlight compilation model, it's a short step to understanding the deployment model. The XAP file is the key piece. It wraps the units of your application (the application manifest and the assemblies) into one neat container.

Technically, the XAP file is a ZIP archive. To verify this, rename a XAP file like SilverlightApplication1.xap to SilverlightApplication1.xap.zip. You can then open the archive and view the files inside. Figure 1-11 shows the contents of the XAP file for the simple example shown earlier in this chapter. Currently, it includes the application manifest and the application assembly. If your application uses add-on assemblies such as System.Windows.Controls.dll, you'll find them in the XAP file as well.

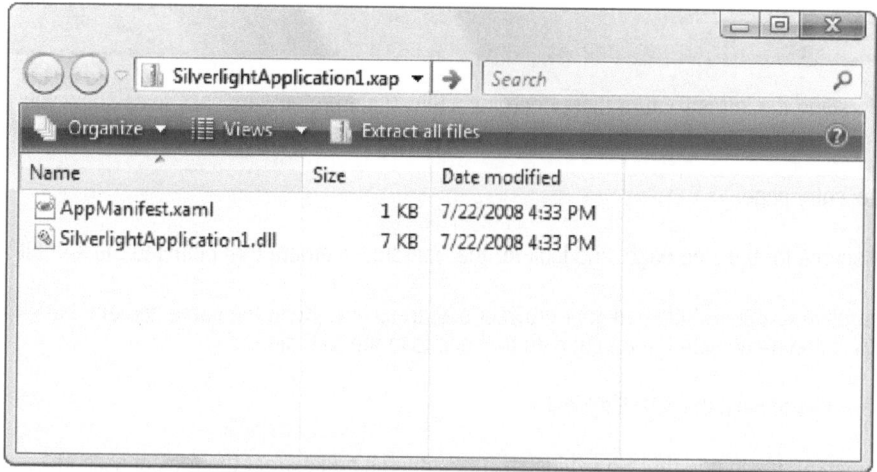

Figure 1-11. *The contents of a XAP file*

The XAP file system has two obvious benefits:

- *It compresses your content*: Because this content isn't decompressed until it reaches the client, it reduces the time required to download your application. This is particularly important if your application contains large static resources (see Chapter 6), such as images or blocks of text.
- *It simplifies deployment*: When you're ready to take your Silverlight application live, you simply need to copy the XAP file to the web server, along with TestPage.html or a similar HTML file (or ASP.NET web form) that includes a Silverlight content region. You don't need to worry about keeping track of the assemblies and resources.

Thanks to the XAP model, there's not much to think about when deploying a simple Silverlight application. Hosting a Silverlight application simply involves making the appropriate XAP file available so the clients can download it through the browser and run it on their local machines.

However, there's one potential stumbling block. When hosting a Silverlight application, your web server must be configured to allow requests for the XAP file type. This file type is included by default in IIS 7, provided you're using Windows Server 2008, Windows 7, or Windows Vista with Service Pack 1. If you have Windows Vista without Service Pack 1, if you have an earlier version of IIS, or if you have another type of web server, you'll need to add a file type that maps the .xap extension to the MIME type application/x-silverlight-app. For IIS instructions, see http://learn.iis.net/page.aspx/262/silverlight.

SILVERLIGHT DECOMPILATION

Now that you understand the infrastructure that underpins a Silverlight project, it's easy to see how you can decompile any existing application to learn more about how it works. Here's how:

1. Surf to the entry page.

2. View the source for the web page, and look for the <param> element that points to the XAP file.

3. Type a request for the XAP file into your browser's address bar. (Keep the same domain, but replace the page name with the partial path that points to the XAP file.)

4. Choose Save As to save the XAP file locally.

5. Rename the XAP file to add the .zip extension. Then, open it and extract the project assembly. This assembly is essentially the same as the assemblies you build for ordinary .NET applications. Like ordinary .NET assemblies, it contains Intermediate Language (IL) code.

6. Open the project assembly in a tool such as Reflector (www.red-gate.com/products/reflector) to view the IL and embedded resources. Using the right plug-in, you can even decompile the IL to VB syntax.

Of course, many Silverlight developers don't condone this sort of behavior (much as many .NET developers don't encourage end users to decompile their rich client applications). However, it's an unavoidable side effect of the Silverlight compilation model.

Because IL code can be easily decompiled or reverse engineered, it's not an appropriate place to store secrets (such as encryption keys, proprietary algorithms, and so on). If you need to perform a task that uses sensitive code, consider calling a web service from your Silverlight application. If you just want to prevent other hotshots from reading your code and copying your style, you may be interested in raising the bar with an *obfuscation* tool, which uses a number of tricks to scramble the structure and names in your compiled code without changing its behavior. Visual Studio ships with a scaled-down obfuscation tool named Dotfuscator, and many more are available commercially.

Silverlight Core Assemblies

Silverlight includes a subset of the classes from the full .NET Framework. Although it would be impossible to cram the entire .NET Framework into Silverlight—after all, it's a 5MB download that needs to support a variety of browsers and operating systems—Silverlight includes a remarkable amount of functionality.

Every Silverlight project starts with references to the following assemblies. All of these assemblies are part of the Silverlight runtime, so they don't need to be deployed with your application.

- *mscorlib.dll*: This assembly is the Silverlight equivalent of the mscorlib.dll assembly that includes the most fundamental parts of the .NET Framework. The Silverlight version includes core data types, exceptions, and interfaces in the System namespace; ordinary and generic collections; file management classes; and support for globalization, reflection, resources, debugging, and multithreading.
- *System.dll*: This assembly contains additional generic collections, classes for dealing with URIs, and classes for dealing with regular expressions.
- *System.Core.dll*: This assembly contains support for LINQ. The name of the assembly matches the full .NET Framework.
- *System.Net.dll*: This assembly contains classes that support networking, allowing you to download web pages and create socket-based connections.
- *System.Windows.dll*: This assembly includes many of the classes for building Silverlight user interfaces, including basic elements, shapes and brushes, classes that support animation and data binding, and a version of the OpenFileDialog that works with isolated storage.
- *System.Windows.Browser.dll*: This assembly contains classes for interacting with HTML elements.
- *System.Xml.dll*: This assembly includes the bare minimum classes you need for XML processing: XmlReader and XmlWriter.

■ **Note** Some of the members in the Silverlight assemblies are available only to .NET Framework code and aren't callable from your code. These members are marked with the SecurityCritical attribute. However, this attribute does not appear in the Object Browser, so you won't be able to determine whether a specific feature is usable in a Silverlight application until you try to use it. (If you attempt to use a member that has the SecurityCritical attribute, you'll get a SecurityException.) For example, Silverlight applications are allowed to access the file system only through the isolated storage API or the OpenFileDialog class. For that reason, the constructor for the FileStream class is decorated with the SecurityCritical attribute.

Silverlight Add-on Assemblies

The architects of Silverlight have set out to keep the core framework as small as possible. This design makes the initial Silverlight plug-in small to download and quick to install—an obvious selling point to web surfers everywhere.

To achieve this lean-and-mean goal, the Silverlight designers have removed some functionality from the core Silverlight runtime and placed it in separate add-on assemblies. These assemblies are still considered to be part of the Silverlight platform, but if you want to use them, you'll need to package them with your application. This is an obvious trade-off, because it will increase the download size of your application. (The effect is mitigated by Silverlight's built-in compression, which you'll learn about later in this chapter.)

You'll learn about Silverlight's add-on assemblies throughout this book. The most commonly used ones follow:

- *System.Windows.Controls.dll*: This assembly contains many valuable but more specialized controls, including a TreeView, a TabControl, two date controls (the DatePicker and Calendar), and the GridSplitter.
- *System.Windows.Controls.Data.dll*: This assembly has Silverlight's built-from-scratch DataGrid, which is an ideal tool for showing dense grids of data, and the DataPager, which gives it the ability to split results into separately viewable groups called *pages*.
- *System.Windows.Controls.Data.Input.dll*: This assembly holds a few controls that are helpful when building data-bound forms, including a Label, DescriptionViewer, and ValidationSummary.
- *System.Windows.Controls.Input.dll*: This assembly includes the AutoCompleteBox—a text box that drops down a list of suggestions as the user types.
- *System.Windows.Controls.Navigation.dll*: This assembly contains the Frame and Page controls that are the basis of Silverlight's navigation system.

All of these assemblies add new controls to your Silverlight Toolkit. Microsoft also makes many more add-in controls available through the Silverlight Toolkit, which you can download at www.codeplex.com/Silverlight.

When you add a control from an add-on assembly onto a Silverlight page, Visual Studio automatically adds the assembly reference you need. If you select that reference and look in the Properties window, you'll see that the Copy Local property is set to True, which is different from the other assemblies that make up the core Silverlight runtime. As a result, when you compile your application, the assembly will be embedded in the final package. Visual Studio is intelligent enough to recognize assemblies that aren't part of the core Silverlight runtime—even if you add them by hand, it automatically sets Copy Local to True.

Assembly Caching

Assembly caching is a deployment technique that allows you to leave dependent assemblies out of your XAP file. Instead, you deploy dependent assemblies *alongside* your XAP file, placing them in separate ZIP files in the same folder. The goal is to reduce application startup time by letting clients keep cached copies of frequently used assemblies.

By default, the Silverlight applications you create in Visual Studio are not configured to use assembly caching. To turn this feature on, double-click the My Project node in the Solution Explorer. Then, in the project properties window shown in Figure 1-10, switch on the setting "Reduce XAP size by using application library caching." To see the results, recompile your application, click the Show All Files button at the top of the Solution Explorer, and expand the Bin\Debug folder. You'll see a ZIP file for each cacheable assembly. For example, if your application uses System.Windows.Controls.dll, you'll see a file named System.Windows.Controls.zip next to your XAP file. This file holds a compressed copy of the System.Windows.Controls.dll assembly. The XAP, which held this assembly before you enabled assembly caching, no longer has a copy of it.

If you're using an ASP.NET test website, Visual Studio copies the XAP file and all the cacheable assemblies to the ClientBin folder in the website. Figure 1-12 shows the result after compiling an application that uses the System.Windows.Controls.dll and System.Windows.Controls.Navigation.dll assemblies.

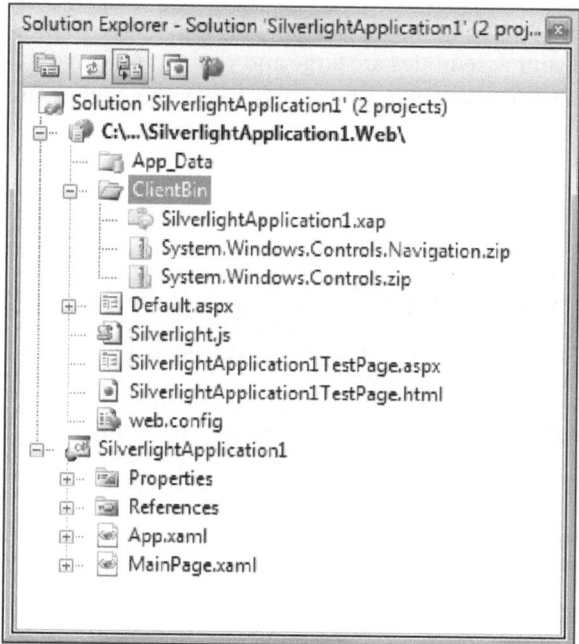

Figure 1-12. *Dependent assemblies that support assembly caching*

Assembly caching decreases the size of your XAP file. Smaller files can be downloaded more quickly, so shrinking the XAP file improves application startup time. But initially, assembly caching won't produce any performance improvement. That's because the first time clients run your Silverlight application, they'll need to download both the slimmed-down XAP and the separate ZIP files with the dependent assemblies. The total amount of downloaded data is the same.

However, the benefit appears when the user returns to run the application a second time. Once again, the browser will download the application XAP file. However, because the dependent assemblies are still in the browser cache, the client won't need to download them.

Here are a few considerations to help you get the most out of assembly caching:

- The downloaded assembly only lasts as long as the browser cache. If the user explicitly clears the cache, all the cached assemblies will be removed.

- Every time the client runs the application, the application checks for new versions of the cached assembly. If it spots a new version, it downloads it and replaces the previously cached version.

- If one application downloads an assembly and places it in the browser cache, another application that uses assembly caching can use it.

- The benefits of assembly caching are greatest for large, infrequently changed assemblies. Many assemblies aren't that big, and the cost of downloading them each time the application starts isn't significant. In this case, using assembly caching will simply complicate the deployment of your application.

- With a bit of work, you can use assembly caching with your own class library assemblies. Once again, this makes most sense if your assemblies are large and you don't change them frequently. You'll learn how to create assemblies that support assembly caching in Chapter 6.

The HTML Entry Page

The last ingredient in the deployment picture is the HTML test page. This page is the entry point into your Silverlight content—in other words, the page the user requests in the web browser. In a stand-alone Silverlight project, Visual Studio names this file TestPage.html. In an ASP.NET-hosted Silverlight project, Visual Studio names it to match your project name. Either way, you'll probably want to rename it to something more appropriate.

The HTML test page doesn't actually contain Silverlight markup or code. Instead, it simply sets up the content region for the Silverlight plug-in, using a small amount of JavaScript. (For this reason, browsers that have JavaScript disabled won't be able to see Silverlight content.) Here's a slightly shortened version of the HTML test page that preserves the key details:

```
<html xmlns="http://www.w3.org/1999/xhtml">
<!-- saved from url=(0014)about:internet -->
<head>
    <title>SilverlightApplication1</title>

    <style type="text/css">
        ...
    </style>

    <script type="text/javascript">
        ...
    </script>
</head>

<body>
    <form id="form1" runat="server" style="height:100%">

    <!-- Silverlight content will be displayed here. -->
    <div id="silverlightControlHost">
        <object data="data:application/x-silverlight-2,"
          type="application/x-silverlight-2" width="100%" height="100%">
            <param name="source" value="SilverlightApplication1.xap" />
            <param name="onError" value="onSilverlightError" />
            <param name="background" value="white" />
            <param name="minRuntimeVersion" value="4.0.50401.0" />
            <param name="autoUpgrade" value="true" />

            <a href="http://go.microsoft.com/fwlink/?LinkID=149156&v=4.0.50401.0"
              style="text-decoration:none">
                <img src="http://go.microsoft.com/fwlink/?LinkId=108181"
                  alt="Get Microsoft Silverlight" style="border-style:none"/>
            </a>
        </object>
        <iframe id="_sl_historyFrame"
          style="visibility:hidden;height:0px;width:0px;border:0px"></iframe>
```

```
    </div>
  </body>
</html>
```

The key detail is the <div> element that represents the Silverlight content region. It contains an <object> element that loads the Silverlight plug-in. The <object> element includes four key attributes. You won't change the data and type attributes—they indicate that the <object> element represents a Silverlight content region using version 2 or later. However, you may want to modify the height and width attributes, which determine the dimensions of the Silverlight content region, as described next.

■ **Note** Be cautious about changing seemingly trivial details in the HTML test page. Some minor quirks are required to ensure compatibility with certain browsers. For example, the comma at the end of the data attribute in the <object> element ensures Firefox support. The invisible <iframe> at the bottom of the Silverlight <div> allows navigation to work with Safari. As a general guideline, the only test page content you should change are the width and height settings, the list of parameters, and the alternate content.

CHANGING THE TEST PAGE

If you're using an ASP.NET website, the test page is generated once, when the ASP.NET website is first created. As a result, you can modify the HTML page without worrying that your changes will be overwritten.

If you're using a stand-alone project without an ASP.NET website, Visual Studio generates the test page each time you run the project. As a result, any changes you make to it will be discarded. If you want to customize the test page, the easiest solution is to create a new test page for your project. Here's how:

1. Run your project at least once to create the test page.

2. Click the Show All Files icon at the top of the Solution Explorer.

3. Expand the Bin\Debug folder in the Solution Explorer.

4. Find the TestPage.html file, right-click it, and choose Copy. Then right-click the Bin\Debug folder and choose Paste. This duplicate will be your custom test page. Right-click the new file and choose Rename to give it a better name.

5. To make the custom test page part of your project, right-click it and choose Include in Project.

6. To tell Visual Studio to navigate to your test page when you run the project, right-click your test page, and choose Set As Start Page.

Sizing the Silverlight Content Region

By default, the Silverlight content region is given a width and height of 100 percent, so the Silverlight content can consume all the available space in the browser window. You can constrain the size of Silverlight content region by hard-coding pixel sizes for the height and width (which is limiting and usually avoided). Or, you can place the <div> element that holds the Silverlight content region in a more restrictive place on the page—for example, in a cell in a table, in another fixed-sized element, or between other <div> elements in a multicolumn layout.

Even though the default test page sizes the Silverlight content region to fit the available space in the browser window, your XAML pages may include hard-coded dimensions. You set these by adding the Height and Width attributes to the root UserControl element and specifying a size in pixels. If the browser window is larger than the hard-coded page size, the extra space won't be used. If the browser window is smaller than the hard-coded page size, part of the page may fall outside the visible area of the window.

Hard-coded sizes make sense when you have a graphically rich layout with absolute positioning and little flexibility. If you don't, you might prefer to remove the Width and Height attributes from the <UserControl> start tag. That way, the page will be sized to match the Silverlight content region, which in turn is sized to fit the browser window, and your Silverlight content will always fit itself into the currently available space.

To get a better understanding of the actual dimensions of the Silverlight content region, you can add a border around it by adding a simple style rule to the <div>, like this:

```
<div id="silverlightControlHost" style="border: 1px red solid">
```

You'll create resizable and scalable Silverlight pages in Chapter 3, when you explore layout in more detail.

Silverlight Parameters

The <object> element contains a series of <param> elements that specify additional options to the Silverlight plug-in.

Table 1-1 lists some of basic the parameters that you can use. You'll learn about many other specialized parameters in examples throughout this book, as you delve into features such as HTML access, splash screens, transparency, and animation.

Table 1-1. Basic Parameters for the Silverlight Plug-In

Name	Value
source	A URI that points to the XAP file for your Silverlight application. This parameter is required.
onError	A JavaScript event handler that's triggered when a unhandled error occurs in the Silverlight plug-in or in your code. The onError event handler is also called if the user has Silverlight installed but doesn't meet the minRuntimeVersion parameter.
background	The color that's used to paint the background of the Silverlight content region, behind any content that you display (but in front of any HTML content that occupies the same space). If you set the Background property of a page, it's painted over this background.

Name	Value
minRuntimeVersion	This is the minimum version of Silverlight that the client must have in order to run your application. If you need the features of Silverlight 4, set this to 4.0.50401.0 (because slightly earlier versions may correspond to beta builds). If Silverlight 3 is sufficient, use 3.0.40624.0. Make sure this matches the version of Silverlight you used to compile your application.
autoUpgrade	A Boolean that specifies whether Silverlight should (if it's installed and has an insufficient version number) attempt to update itself. The default is true. You may choose to set this to false to deal with version problems on your own using the onError event, as described in the "Creating a Friendly Install Experience" section.
enableHtmlAccess	A Boolean that specifies whether the Silverlight plug-in has access to the HTML object model. Use true if you want to be able to interact with the HTML elements on the test page through your Silverlight code (as demonstrated in Chapter 14).
initParams	A string that you can use to pass custom initialization information. This technique (which is described in Chapter 6) is useful if you plan to use the same Silverlight application in different ways on different pages.
splashScreenSource	The location of a XAML splash screen to show while the XAP file is downloading. You'll learn how to use this technique in Chapter 6.
windowless	A Boolean that specifies whether the plug-in renders in windowed mode (the default) or windowless mode. If you set this true, the HTML content underneath your Silverlight content region can show through. This is ideal if you're planning to create a shaped Silverlight control that integrates with HTML content, and you'll see how to use it in Chapter 14.
onSourceDownload ProgressChanged	A JavaScript event handler that's triggered when a piece of the XAP file has been downloaded. You can use this event handler to build a startup progress bar, as in Chapter 6.
onSourceDownload Complete	A JavaScript event handler that's triggered when the entire XAP file has been downloaded.
onLoad	A JavaScript event handler that's triggered when the markup in the XAP file has been processed and your first page has been loaded.
onResize	A JavaScript event handler that's triggered when the size of a Silverlight content region has changed.

Alternative Content

The <div> element also has some HTML markup that will be shown if the <object> tag isn't understood or the plug-in isn't available. In the standard test page, this markup consists of a Get Silverlight picture, which is wrapped in a hyperlink that, when clicked, takes the user to the Silverlight download page.

```
<a href="http://go.microsoft.com/fwlink/?LinkID=149156&v=4.0.50401.0"
  style="text-decoration:none">
    <img src="http://go.microsoft.com/fwlink/?LinkId=108181"
    alt="Get Microsoft Silverlight" style="border-style:none"/>
</a>
```

Creating a Friendly Install Experience

Some of the users who reach your test page will not have Silverlight installed, or they won't have the correct version. The standard behavior is for the Silverlight test page to detect the problem and notify the user. However, this may not be enough to get the user to take the correct action.

For example, consider a user who arrives at your website for the first time and sees a small graphic asking them to install Silverlight. That user may be reluctant to install an unfamiliar program, confused about why it's needed, and intimidated by the installation terminology. Even if they do click ahead to install Silverlight, they'll face still more prompts asking them to download the Silverlight installation package and then run an executable. At any point, they might get second thoughts and surf somewhere else.

■ **Tip** Studies show that web surfers are far more likely to make it through an installation process on the Web if they're guided to do it as part of an *application*, rather than prompted to install it as a *technology*.

To give your users a friendlier install experience, begin by customizing the alternative content. As you learned in the previous section, if the user doesn't have any version of Silverlight installed, the browser shows the Silverlight badge—essentially, a small banner with a logo and a Get Silverlight button. This indicator is obvious to developers but has little meaning to end users. To make it more relevant, add a custom graphic that clearly has the name and logo of your application, include some text underneath that explaining that the Silverlight plug-in is required to power your application, and *then* include the download button.

The second area to address is versioning issues. If the user has Silverlight, but it doesn't meet the minimum version requirement, the alternative content isn't shown. Instead, the Silverlight plug-in triggers the onError event with args.ErrorCode set to 8001 (upgrade required) or 8002 (restart required) and then displays a dialog box prompting the user to get the updated version. A better, clearer approach is to handle this problem yourself.

First, disable the automatic upgrading process by setting the autoUpgrade parameter to false:

```
<param name="autoUpgrade" value="false" />
```

Then, check for the version error code in the onSilverlightError function in the test page. If you detect a version problem, you can then use JavaScript to alter the content of the <div> element that holds the Silverlight plug-in. Swap in a more meaningful graphic that clearly advertises your application, along with the download link for the correct version of Silverlight.

```
function onSilverlightError(sender, args) {
    if (args.ErrorCode == 8001)
    {
        // Find the Silverlight content region.
        var hostContainer = document.getElementById("silverlightControlHost");

        // Change the content. You can supply any HTML here.
        hostContainer.innerHTML = "...";
    }
    // (Deal with other types of errors here.)
}
```

To test your code, just set the minRuntimeVersion parameter absurdly high:

```
<param name="minRuntimeVersion" value="5" />
```

■ **Note** If you've created a stand-alone Silverlight project, and you haven't gone to the trouble of designing a custom test page, Visual Studio genereates a new

---

MATCHING THE MINIMUM RUNTIME WITH THE TARGET VERSION

As you learned earlier, you can change the version of Silverlight that your application targets at any time (just double-click the My Project node in the Solution Explorer and pick a new version in the Target Silverlight Version list). However, it's important to understand what happens to the test page when you make a change.

If you're using a stand-alone Silverlight application with an automatically generated test page, Visual Studio sets the minRuntimeVersion attribute to match the target version the next time it generates the HTML entry page. This behavior is what you want.

If you're using an ASP.NET-hosted Silverlight application or a custom HTML entry page, Visual Studio won't make any changes to the minRuntimeVersion attribute. This is a problem, because it can lead to a situation where your application is compiled with a later version of Silverlight than indicated by the minRuntimeVersion attribute. In this situation, a user might meet the minRuntimeVersion requirement without having the version of Silverlight that's actually needed to run the application. In this case, the user won't see the alternate content with the link to the install site. Instead, the user will receive a less helpful error in a message box when the browser plug-in attempts to start the Silverlight application.

Thus, if you change the target version of your Silverlight application, it's always a good idea to make sure the minRuntimeVersion attribute matches.

The Mark of the Web

One of the stranger details in the HTML test page is the following comment, which appears in the second line:

```
<!-- saved from url=(0014)about:internet -->
```

Although this comment appears to be little more than an automatically generated stamp that the browser ignores, it actually has an effect on the way you debug your application. This comment is known as the *mark of the Web*, and it's a specialized flag that forces Internet Explorer to run pages in a more restrictive security zone than it would normally use.

Ordinarily, the mark of the Web indicates the website from which a locally stored page was originally downloaded. But in this case, Visual Studio has no way of knowing where your Silverlight application will eventually be deployed. It falls back on the URL about:internet, which simply signals that the page is from some arbitrary location on the public Internet. The number (14) simply indicates the number of characters in this URL. For a more detailed description of the mark of the Web and its standard uses, see http://tinyurl.com/2ctnsj9.

All of this raises an obvious question—namely, why is Visual Studio adding a marker that's typically reserved for downloaded pages? The reason is that without the mark of the Web, Internet Explorer will load your page with the relaxed security settings of the local machine zone. This wouldn't cause a problem, except that Internet Explorer also includes a safeguard that disables scripts and ActiveX controls in this situation. As a result, if you run a test page that's stored on your local hard drive and this test page doesn't have the mark of the Web, you'll see the irritating warning message shown in Figure 1-13, and you'll need to explicitly allow the blocked content. Worst of all, you'll need to repeat this process every time you open the page.

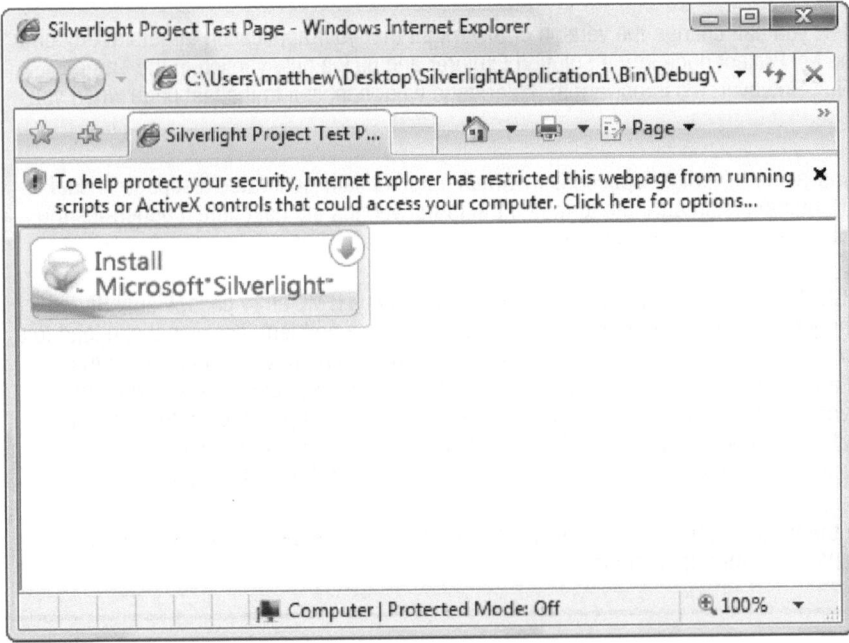

Figure 1-13. A page with disabled Silverlight content

This problem will disappear when you deploy the web page to a real website, but it's a significant inconvenience while testing. To avoid headaches such as these, make sure you add a similar mark of the Web comment if you design your own custom test pages.

The Last Word

In this chapter, you took your first look at the Silverlight application model. You saw how to create a Silverlight project in Visual Studio, add a simple event handler, and test it. You also peered behind the scenes to explore how a Silverlight application is compiled and deployed.

In the following chapters, you'll learn much more about the full capabilities of the Silverlight platform. Sometimes, you might need to remind yourself that you're coding inside a lightweight browser-hosted framework, because much of Silverlight coding feels like the full .NET platform, despite that it's built on only a few megabytes of compressed code. Out of all of Silverlight's many features, its ability to pack a miniature modern programming framework into a slim 5MB download is surely its most impressive.

CHAPTER 2

■ ■ ■

XAML

XAML (short for Extensible Application Markup Language and pronounced *zammel*) is a markup language used to instantiate .NET objects. Although XAML is a technology that can be applied to many different problem domains, it was initially designed as part of Windows Presentation Foundation (WPF), where it allows Windows developers to construct rich user interfaces. You use the same standard to build user interfaces for Silverlight applications.

Conceptually, XAML plays a role that's a lot like HTML and is even closer to its stricter cousin, XHTML. XHTML allows you to define the elements that make up an ordinary web page. Similarly, XAML allows you to define the elements that make up a XAML content region. To manipulate XHTML elements, you can use client-side JavaScript. To manipulate XAML elements, you write client-side VB code. Finally, XAML and XHTML share many of the same syntax conventions. Like XHTML, XAML is an XML-based language that consists of elements that can be nested in any arrangement you like.

In this chapter, you'll get a detailed introduction to XAML and consider a simple single-page application. Once you understand the broad rules of XAML, you'll know what is and isn't possible in a Silverlight user interface—and how to make changes by hand. By exploring the tags in a Silverlight XAML document, you'll also learn more about the object model that underpins Silverlight user interfaces and get ready for the deeper exploration to come.

Finally, at the end of this chapter, you'll consider two *markup extensions* that extend XAML with Silverlight-specific features. First, you'll see how you can streamline code and reuse markup with XAML resources and the StaticResource extension. Next, you'll learn how to link two elements together with the Binding extension. Both techniques are a core part of Silverlight development, and you'll see them at work throughout this book.

■ **What's New** The XAML standard is essentially unchanged in Silverlight 4. However, there is one minor refinement. Silverlight 4 now allows direct content (in other words, text between the opening and closing tags of an element, instead of in an attribute) in most controls. This change is intended to improve compatibility with WPF, which has allowed direct content since version 1.

XAML Basics

The XAML standard is quite straightforward once you understand a few ground rules:

- Every element in a XAML document maps to an instance of a Silverlight class. The name of the element matches the name of the class *exactly*. For example, the element <Button> instructs Silverlight to create a Button object.

- As with any XML document, you can nest one element inside another. As you'll see, XAML gives every class the flexibility to decide how it handles this situation. However, nesting is usually a way to express *containment*—in other words, if you find a Button element inside a Grid element, your user interface probably includes a Grid that contains a Button inside.

- You can set the properties of each class through attributes. However, in some situations, an attribute isn't powerful enough to handle the job. In these cases, you'll use nested tags with a special syntax.

■ **Tip** If you're completely new to XML, you'll probably find it easier to review the basics before you tackle XAML. To get up to speed quickly, try the free tutorial at www.w3schools.com/xml.

Before continuing, take a look at this bare-bones XAML document, which represents a blank page (as created by Visual Studio). The lines have been numbered for easy reference:

```
1   <UserControl x:Class="SilverlightApplication1.MainPage"
2     xmlns="http://schemas.microsoft.com/winfx/2006/xaml/presentation"
3     xmlns:x="http://schemas.microsoft.com/winfx/2006/xaml"
4     xmlns:d="http://schemas.microsoft.com/expression/blend/2008"
5     xmlns:mc="http://schemas.openxmlformats.org/markup-compatibility/2006"
6     mc:Ignorable="d" d:DesignWidth="640" d:DesignHeight="480">
7
8       <Grid x:Name="LayoutRoot">
9       </Grid>
10  </UserControl>
```

This document includes only two elements—the top-level UserControl element, which wraps all the Silverlight content on the page, and the Grid, in which you can place all your elements.

As in all XML documents, there can only be one top-level element. In the previous example, that means that as soon as you close the UserControl element with the </UserControl> tag, you end the document. No more content can follow.

XAML Namespaces

When you use an element like <UserControl> in a XAML file, the Silverlight parser recognizes that you want to create an instance of the UserControl class. However, it doesn't necessarily know *what* UserControl class to use. After all, even if the Silverlight namespaces include only a single class with that name, there's no guarantee that you won't create a similarly named class of your own. Clearly, you need a way to indicate the Silverlight namespace information in order to use an element.

In Silverlight, classes are resolved by mapping XML namespaces to Silverlight namespaces. In the sample document shown earlier, four namespaces are defined:

```
2   xmlns="http://schemas.microsoft.com/winfx/2006/xaml/presentation"
3   xmlns:x="http://schemas.microsoft.com/winfx/2006/xaml"
4   xmlns:d="http://schemas.microsoft.com/expression/blend/2008"
5   xmlns:mc="http://schemas.openxmlformats.org/markup-compatibility/2006"
```

The xmlns attribute is a specialized attribute in the world of XML, and it's reserved for declaring namespaces. This snippet of markup declares four namespaces that you'll find in every page you create with Visual Studio or Expression Blend.

■ **Note** XML namespaces are declared using attributes. These attributes can be placed inside any element start tag. However, convention dictates that all the namespaces you need to use in a document should be declared in the very first tag, as they are in this example. Once a namespace is declared, it can be used anywhere in the document.

Core Silverlight Namespaces

The first two namespaces are the most important. You'll need them to access essential parts of the Silverlight runtime:

- *http://schemas.microsoft.com/winfx/2006/xaml/presentation* is the core Silverlight namespace. It encompasses all the essential Silverlight classes, including the UserControl and Grid. Ordinarily, this namespace is declared without a namespace prefix, so it becomes the default namespace for the entire document. In other words, every element is automatically placed in this namespace unless you specify otherwise.

- *http://schemas.microsoft.com/winfx/2006/xaml* is the XAML namespace. It includes various XAML utility features that allow you to influence how your document is interpreted. This namespace is mapped to the prefix *x*. That means you can apply it by placing the namespace prefix before the name of an XML element or attribute (as in <x:ElementName> and x:Class="ClassName").

The namespace information allows the XAML parser to find the right class. For example, when it looks at the UserControl and Grid elements, it sees that they are placed in the default http://schemas.microsoft.com/winfx/2006/xaml/presentation namespace. It then searches the corresponding Silverlight namespaces, until it finds the matching classes System.Windows.UserControl and System.Windows.Controls.Grid.

XML NAMESPACES AND SILVERLIGHT NAMESPACES

The XML namespace name doesn't correspond to a single Silverlight namespace. Instead, all the Silverlight namespaces share the same XML namespace. There are a couple of reasons the creators of XAML chose this design. By convention, XML namespaces are often URIs (as they are here). These URIs look like they point to a location on the Web, but they don't. The URI format is used because it makes it unlikely that different organizations will inadvertently create different XML-based languages with the same namespace. Because the domain schemas.microsoft.com is owned by Microsoft, only Microsoft will use it in an XML namespace name.

The other reason that there isn't a one-to-one mapping between the XML namespaces used in XAML and Silverlight namespaces is because it would significantly complicate your XAML documents. If each Silverlight namespace had a different XML namespace, you'd need to specify the right namespace for each and every control you use, which would quickly get messy. Instead, the creators of Silverlight chose to map all the Silverlight namespaces that include user interface elements to a single XML namespace. This works because within the different Silverlight namespaces, no two classes share the same name.

Design Namespaces

Along with these core namespaces are two more specialized namespaces, neither of which is essential:

- *xmlns:mc="http://schemas.openxmlformats.org/markup-compatibility/2006* is the XAML compatibility namespace. You can use it to tell the XAML parser what information must to process and what information to ignore.
- *http://schemas.microsoft.com/expression/blend/2008* is a namespace reserved for design-specific XAML features that are supported in Expression Blend (and now Visual Studio 2010). It's used primarily to set the size of the design surface for a page.

Both of these namespaces are used in the single line shown here:

```
6    mc:Ignorable="d" d:DesignWidth="640" d:DesignHeight="480">
```

The DesignWidth and DesignHeight properties are part of the http://schemas.microsoft.com/expression/blend/2008 namespace. They tell the design tool to make the page 640×480 pixels large at design time. Without this detail, you would be forced to work with a squashed-up design surface that doesn't give a realistic preview of your user interface or set a hard-coded size using the Width and Height properties (which isn't ideal, because it prevents your page from resizing to fit the browser window at runtime).

The Ignorable property is part of the http://schemas.openxmlformats.org/markup-compatibility/2006 namespace. It tells the XAML design tool that it's safe to ignore the parts of the document that are prefixed with a *d* and placed in the http://schemas.microsoft.com/expression/blend/2008. In other words, if the XAML parser doesn't understand the DesignWidth and DesignHeight details, it's safe to continue because they aren't critical.

■ **Note** In the examples in this book, you'll rarely see either of these namespaces, because they aren't terribly important. They're intended for design tools and XAML readers only, not the Silverlight runtime.

Custom Namespaces

In many situations, you'll want to have access to your own namespaces in a XAML file. The most common example is if you want to use a custom Silverlight control that you (or another developer) have created. In this case, you need to define a new XML namespace prefix and map it to your assembly. Here's the syntax you need:

```
<UserControl x:Class="SilverlightApplication1.MainPage"
  xmlns:w="clr-namespace:Widgets;assembly=WidgetLibrary"
  ... >
```

The XML namespace declaration sets three pieces of information:

- *The XML namespace prefix*: You'll use the namespace prefix to refer to the namespace in your XAML page. In this example, that's *w*, although you can choose anything you want that doesn't conflict with another namespace prefix.
- *The .NET namespace*: In this case, the classes are located in the Widgets namespace. If you have classes that you want to use in multiple namespaces, you can map them to different XML namespaces or to the same XML namespace (as long as there aren't any conflicting class names).
- *The assembly*: In this case, the classes are part of the WidgetLibrary.dll assembly. (You don't include the .dll extension when naming the assembly.) Silverlight will look for that assembly in the same XAP package where your project assembly is placed.

■ **Note** Remember, Silverlight uses a lean, stripped-down version of the CLR. For that reason, a Silverlight application can't use a full .NET class library assembly. Instead, it needs to use a Silverlight class library. You can easily create a Silverlight class library in Visual Studio by choosing the Silverlight Class Library project template.

If you want to use a custom control that's located in the current application, you can omit the assembly part of the namespace mapping, as shown here:

```
xmlns:w="clr-namespace:Widgets"
```

Once you've mapped your .NET namespace to an XML namespace, you can use it anywhere in your XAML document. For example, if the Widgets namespace contains a control named HotButton, you could create an instance like this:

```
<w:HotButton Content="Click Me!" Click="DoSomething"></w:HotButton>
```

You'll use this technique throughout this book to access controls in the Silverlight add-on assemblies and the Silverlight Toolkit.

The Code-Behind Class

XAML allows you to construct a user interface, but in order to make a functioning application, you need a way to connect the event handlers that contain your application code. XAML makes this easy using the Class attribute that's shown here:

```
1   <UserControl x:Class="SilverlightApplication1.MainPage"
```

The *x* namespace prefix places the Class attribute in the XAML namespace, which means this is a more general part of the XAML language, not a specific Silverlight ingredient.

In fact, the Class attribute tells the Silverlight parser to generate a new class with the specified name. That class derives from the class that's named by the XML element. In other words, this example creates a new class named SilverlightProject1.MainPage, which derives from the UserControl class. The automatically generated portion of this class is merged with the code you've supplied in the code-behind file.

Usually, every XAML file will have a corresponding code-behind class with client-side VB code. Visual Studio creates a code-behind class for the MainPage.xaml file named MainPage.xaml.vb. Here's what you'll see in the MainPage.xaml.vb file:

```
Partial Public Class MainPage
    Inherits UserControl

    Public Sub New()
        InitializeComponent()
    End Sub

End Class
```

Currently, the MainPage class code doesn't include any real functionality. However, it does include one important detail—the default constructor, which calls InitializeComponent() when you create an instance of the class. This parses your markup, creates the corresponding objects, sets their properties, and attaches any event handlers you've defined.

■ **Note** The InitializeComponent() method plays a key role in Silverlight content. For that reason, you should never delete the InitializeComponent() call from the constructor. Similarly, if you add another constructor to your page, make sure it also calls InitializeComponent().

Naming Elements

There's one more detail to consider. In your code-behind class, you'll often want to manipulate elements programmatically. For example, you might want to read or change properties or attach and detach event handlers on the fly. To make this possible, the control must include a XAML Name attribute. In the previous example, the Grid control already includes the Name attribute, so you can manipulate it in your code-behind file.

```
6       <Grid x:Name="LayoutRoot">
7       </Grid>
```

The Name attribute tells the XAML parser to add a field like this to the automatically generated portion of the MainPage class:

```
Friend WithEvents LayoutRoot As System.Windows.Controls.Grid
```

Now you can interact with the grid in your page class code by using the name LayoutRoot.

■ **Tip** In a traditional Windows Forms application, every control has a name. In a Silverlight application, there's no such requirement. If you don't want to interact with an element in your code, you're free to remove its Name attribute from the markup. The examples in this book usually omit element names when they aren't needed, which makes the markup more concise.

Properties and Events in XAML

So far, you've considered a relatively unexciting example—a blank page that hosts an empty Grid control. Before going any further, it's worth introducing a more realistic page that includes several elements. Figure 2-1 shows an example with an automatic question answerer.

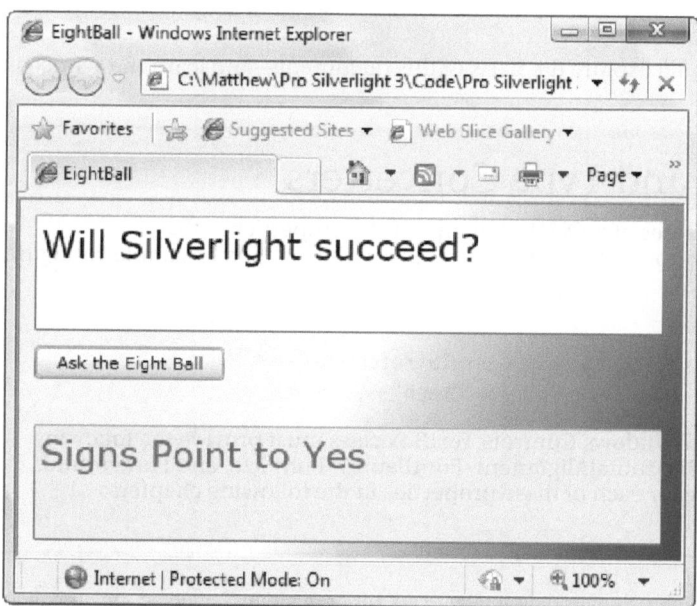

Figure 2-1. *Ask the eight ball, and all will be revealed.*

The eight ball page includes four elements: a Grid (the most common tool for arranging layout in Silverlight), two TextBox objects, and a Button. The markup that's required to arrange and configure these elements is significantly longer than the previous examples.

Here's an abbreviated listing that replaces some of the details with an ellipsis (. . .) to expose the overall structure:

```
<UserControl x:Class="EightBall.MainPage"
 xmlns="http://schemas.microsoft.com/winfx/2006/xaml/presentation"
 xmlns:x="http://schemas.microsoft.com/winfx/2006/xaml"
 Width="400" Height="300">
  <Grid x:Name="grid1">
    <Grid.Background>
      ...
    </Grid.Background>
    <Grid.RowDefinitions>
      ...
    </Grid.RowDefinitions>

    <TextBox x:Name="txtQuestion" ... >
    </TextBox>

    <Button x:Name="cmdAnswer" ... >
    </Button>

    <TextBox x:Name="txtAnswer" ... >
    </TextBox>
  </Grid>
</UserControl>
```

In the following sections, you'll explore the parts of this document—and learn the syntax of XAML along the way.

Simple Properties and Type Converters

As you've already seen, the attributes of an XML element set the properties of the corresponding Silverlight object. For example, the text boxes in the eight ball example configure the alignment, margin, and font:

```
<TextBox x:Name="txtQuestion"
 VerticalAlignment="Stretch" HorizontalAlignment="Stretch"
 FontFamily="Verdana" FontSize="24" Foreground="Green" ... >
```

For this to work, the System.Windows.Controls.TextBox class must provide the following properties: VerticalAlignment, HorizontalAlignment, FontFamily, FontSize, and Foreground. You'll learn the specific meaning for each of these properties in the following chapters.

■ **Tip** Several special characters can't be entered directly into an attribute string, including the quotation mark, the ampersand (&), and the two angle brackets. To use these values, you must replace them with the equivalent XML character entity. That's " for a quotation mark, & for the ampersand, < for the < (less than) character, and > for the > (greater than) character. Of course, this limitation is an XML detail, and it won't affect you if you set a property in code.

To make the property system work, the XAML parser needs to perform a bit more work than you might initially realize. The value in an XML attribute is always a plain-text string. However, object properties can be any .NET type. In the previous example, there are two properties that use enumerations (VerticalAlignment and HorizontalAlignment), one string (FontFamily), one integer (FontSize), and one Brush object (Foreground).

To bridge the gap between string values and nonstring properties, the XAML parser needs to perform a conversion. The conversion is performed by *type converters*, a basic piece of infrastructure that's borrowed from the full .NET Framework.

Essentially, a type converter has one role in life—it provides utility methods that can convert a specific .NET data type to and from any other .NET type, such as a string representation in this case. The XAML parser follows two steps to find a type converter:

1. It examines the property declaration, looking for a TypeConverter attribute. (If present, the TypeConverter attribute indicates what class can perform the conversion.) For example, when you use a property such as Foreground, .NET checks the declaration of the Foreground property.

2. If there's no TypeConverter attribute on the property declaration, the XAML parser checks the class declaration of the corresponding data type. For example, the Foreground property uses a Brush object. The Brush class (and its derivatives) uses the BrushConverter because the Brush class is decorated with the TypeConverter(typeof(BrushConverter)) attribute declaration.

If there's no associated type converter on the property declaration or the class declaration, the XAML parser generates an error.

This system is simple but flexible. If you set a type converter at the class level, that converter applies to every property that uses that class. On the other hand, if you want to fine-tune the way type conversion works for a particular property, you can use the TypeConverter attribute on the property declaration instead.

It's technically possible to use type converters in code, but the syntax is a bit convoluted. It's almost always better to set a property directly—not only is it faster, but it also avoids potential errors from mistyping strings, which won't be caught until runtime. This problem doesn't affect XAML, because the XAML is parsed and validated at compile time.

■ **Note** XAML, like all XML-based languages, is *case-sensitive*. That means you can't substitute <button> for <Button>. However, type converters usually aren't case-sensitive, which means both Foreground="White" and Foreground="white" have the same result.

Some classes define a content property, which allows you to provide the property value between the start and end tags. For example, the Button class designates Content as its content property, meaning this markup:

```
<Button>Click Me!</Button>
```

is equivalent to this:

```
<Button Content="Click Me!"></Button>
```

Despite this convenience, you're more likely to see the second approach in markup. Not only is it the standard Visual Studio uses when you configure elements with the Properties window, it was also the only supported option for many controls up until Silverlight 4.

Complex Properties

As handy as type converters are, they aren't practical for all scenarios. For example, some properties are full-fledged objects with their own set of properties. Although it's possible to create a string representation that the type converter could use, that syntax might be difficult to use and prone to error.

Fortunately, XAML provides another option: *property-element syntax*. With property-element syntax, you add a child element with a name in the form *Parent.PropertyName*. For example, the Grid has a Background property that allows you to supply a brush that's used to paint the area behind the elements. If you want to use a complex brush—one more advanced than a solid color fill—you'll need to add a child tag named Grid.Background, as shown here:

```
<Grid x:Name="grid1">
  <Grid.Background>
    ...
  </Grid.Background>
  ...
</Grid>
```

The key detail that makes this work is the period (.) in the element name. This distinguishes properties from other types of nested content.

This still leaves one detail—namely, once you've identified the complex property you want to configure, how do you set it? Here's the trick. Inside the nested element, you can add another tag to instantiate a specific class. In the eight ball example (shown in Figure 2-1), the background is filled with a gradient. To define the gradient you want, you need to create a LinearGradientBrush object.

Using the rules of XAML, you can create the LinearGradientBrush object using an element with the name LinearGradientBrush:

```
<Grid x:Name="grid1">
  <Grid.Background>
    <LinearGradientBrush>
    </LinearGradientBrush>
  </Grid.Background>
  ...
</Grid>
```

The LinearGradientBrush is part of the Silverlight set of namespaces, so you can keep using the default XML namespace for your tags.

However, it's not enough to simply create the LinearGradientBrush—you also need to specify the colors in that gradient. You do this by filling the LinearGradientBrush.GradientStops property with a collection of GradientStop objects. Once again, the GradientStops property is too complex to be set with an attribute value alone. Instead, you need to rely on the property-element syntax:

```
<Grid x:Name="grid1">
  <Grid.Background>
    <LinearGradientBrush>
      <LinearGradientBrush.GradientStops>
```

```
      </LinearGradientBrush.GradientStops>
    </LinearGradientBrush>
  </Grid.Background>
  ...
</Grid>
```

Finally, you can fill the GradientStops collection with a series of GradientStop objects. Each GradientStop object has an Offset and Color property. You can supply these two values using the ordinary property-attribute syntax:

```
<Grid x:Name="grid1">
  <Grid.Background>
    <LinearGradientBrush>
      <LinearGradientBrush.GradientStops>
        <GradientStop Offset="0.00" Color="Yellow" />
        <GradientStop Offset="0.50" Color="White" />
        <GradientStop Offset="1.00" Color="Purple" />
      </LinearGradientBrush.GradientStops>
    </LinearGradientBrush>
  </Grid.Background>
  ...
</Grid>
```

■ **Note** You can use property-element syntax for any property. But usually you'll use the simpler property-attribute approach if the property has a suitable type converter. Doing so results in more compact code.

Any set of XAML tags can be replaced with a set of code statements that performs the same task. The tags shown previously, which fill the background with a gradient of your choice, are equivalent to the following code:

```
Dim brush As New LinearGradientBrush()

Dim gradientStop1 As New GradientStop()
gradientStop1.Offset = 0
gradientStop1.Color = Colors.Yellow
brush.GradientStops.Add(gradientStop1)

Dim gradientStop2 As New GradientStop()
gradientStop2.Offset = 0.5
gradientStop2.Color = Colors.White
brush.GradientStops.Add(gradientStop2)

Dim gradientStop3 As New GradientStop()
gradientStop3.Offset = 1
gradientStop3.Color = Colors.Purple
brush.GradientStops.Add(gradientStop3)

grid1.Background = brush
```

Attached Properties

Along with ordinary properties, XAML also includes the concept of *attached properties*—properties that may apply to several elements but are defined in a different class. In Silverlight, attached properties are frequently used to control layout.

Here's how it works. Every control has its own set of intrinsic properties. For example, a text box has a specific font, text color, and text content as dictated by properties such as FontFamily, Foreground, and Text. When you place a control inside a container, it gains additional features, depending on the type of container. For example, if you place a text box inside a grid, you need to be able to choose the grid cell where it's positioned. These additional details are set using attached properties.

Attached properties always use a two-part name in this form: *DefiningType.PropertyName*. This two-part naming syntax allows the XAML parser to distinguish between a normal property and an attached property.

In the eight ball example, attached properties allow the individual elements to place themselves on separate rows in the (invisible) grid:

```
<TextBox ... Grid.Row="0">
</TextBox>

<Button ... Grid.Row="1">
</Button>

<TextBox ... Grid.Row="2">
</TextBox>
```

Attached properties aren't really properties at all. They're actually translated into method calls. The XAML parser calls the shared method that has this form: *DefiningType.SetPropertyName()*. For example, in the previous XAML snippet, the defining type is the Grid class, and the property is Row, so the parser calls Grid.SetRow().

When calling SetPropertyName(), the parser passes two parameters: the object that's being modified and the property value that's specified. For example, when you set the Grid.Row property on the TextBox control, the XAML parser executes this code:

```
Grid.SetRow(txtQuestion, 0)
```

This pattern (calling a shared method of the defining type) is a convenience that conceals what's really taking place. To the casual eye, this code implies that the row number is stored in the Grid object. However, the row number is actually stored in the object that it *applies to*—in this case, the TextBox object.

This sleight of hand works because the TextBox derives from the DependencyObject base class, as do all Silverlight elements. The DependencyObject is designed to store a virtually unlimited collection of dependency properties (and attached properties are one type of dependency property).

In fact, the Grid.SetRow() method is actually a shortcut that's equivalent to calling the DependencyObject.SetValue() method, as shown here:

```
txtQuestion.SetValue(Grid.RowProperty, 0)
```

Attached properties are a core ingredient of Silverlight. They act as an all-purpose extensibility system. For example, by defining the Row property as an attached property, you guarantee that it's usable with any control. The other option, making it part of a base class such as FrameworkElement, complicates life. Not only would it clutter the public interface with properties that have meaning only in certain circumstances (in this case, when an element is being used inside a Grid), it also makes it impossible to add new types of containers that require new properties.

Nesting Elements

As you've seen, XAML documents are arranged as a heavily nested tree of elements. In the current example, a UserControl element contains a Grid element, which contains TextBox and Button elements.

XAML allows each element to decide how it deals with nested elements. This interaction is mediated through one of three mechanisms that are evaluated in this order:

- If the parent implements IList<T>, the parser calls the IList<T>.Add() method and passes in the child.
- If the parent implements IDictionary<T>, the parser calls IDictionary<T>.Add() and passes in the child. When using a dictionary collection, you must also set the x:Key attribute to give a key name to each item.
- If the parent is decorated with the ContentProperty attribute, the parser uses the child to set that property.

For example, earlier in this chapter you saw how a LinearGradientBrush can hold a collection of GradientStop objects using syntax like this:

```
<LinearGradientBrush>
  <LinearGradientBrush.GradientStops>
    <GradientStop Offset="0.00" Color="Yellow" />
    <GradientStop Offset="0.50" Color="White" />
    <GradientStop Offset="1.00" Color="Purple" />
  </LinearGradientBrush.GradientStops>
</LinearGradientBrush>
```

The XAML parser recognizes the LinearGradientBrush.GradientStops element is a complex property because it includes a period. However, it needs to process the tags inside (the three GradientStop elements) a little differently. In this case, the parser recognizes that the GradientStops property returns a GradientStopCollection object, and the GradientStopCollection implements the IList interface. Thus, it assumes (quite rightly) that each GradientStop should be added to the collection using the IList.Add() method:

```
Dim gradientStop1 As New GradientStop()
gradientStop1.Offset = 0
gradientStop1.Color = Colors.Yellow
Dim list As IList = brush.GradientStops
list.Add(gradientStop1)
```

Some properties might support more than one type of collection. In this case, you need to add a tag that specifies the collection class, like this:

```
<LinearGradientBrush>
  <LinearGradientBrush.GradientStops>
    <GradientStopCollection>
      <GradientStop Offset="0.00" Color="Yellow" />
      <GradientStop Offset="0.50" Color="White" />
      <GradientStop Offset="1.00" Color="Purple" />
```

```
    </GradientStopCollection>
  </LinearGradientBrush.GradientStops>
</LinearGradientBrush>
```

■ **Note** If the collection defaults to a null reference, (Nothing), you need to include the tag that specifies the collection class, thereby creating the collection object. If there's a default instance of the collection and you simply need to fill it, you can omit that part.

Nested content doesn't always indicate a collection. For example, consider the Grid element, which contains several other elements:

```
<Grid x:Name="grid1">
  ...
  <TextBox x:Name="txtQuestion" ... >
  </TextBox>

  <Button x:Name="cmdAnswer" ... >
  </Button>

  <TextBox x:Name="txtAnswer" ... >
  </TextBox>
</Grid>
```

These nested tags don't correspond to complex properties, because they don't include the period. Furthermore, the Grid control isn't a collection and so it doesn't implement IList or IDictionary. What the Grid *does* support is the ContentProperty attribute, which indicates the property that should receive any nested content. Technically, the ContentProperty attribute is applied to the Panel class, from which the Grid derives, and looks like this:

```
<ContentPropertyAttribute("Children")> _
Public MustInherit Class Panel
    Inherits FrameworkElement
    ...
End Class
```

This indicates that any nested elements should be used to set the Children property. The XAML parser treats the content property differently depending on whether it's a collection property (in which case it implements the IList or IDictionary interface). Because the Panel.Children property returns a UIElementCollection and because UIElementCollection implements IList, the parser uses the IList.Add() method to add nested content to the grid.

In other words, when the XAML parser meets the previous markup, it creates an instance of each nested element and passes it to the Grid using the Grid.Children.Add() method:

```
txtQuestion = New TextBox()
...
grid1.Children.Add(txtQuestion)

cmdAnswer = New Button()
...
grid1.Children.Add(cmdAnswer)
```

```
txtAnswer = New TextBox()
...
grid1.Children.Add(txtAnswer)
```

What happens next depends entirely on how the control implements the content property. The Grid displays all the elements it holds in an invisible layout of rows and columns, as you'll see in Chapter 3.

BROWSING NESTED ELEMENTS WITH VISUALTREEHELPER

Silverlight provides a VisualTreeHelper class that allows you to walk through the hierarchy elements. The VisualTreeHelper class provides three shared methods for this purpose: GetParent(), which returns the element that contains a specified element; GetChildrenCount(), which indicates how many elements are nested inside the specified element; and GetChild(),which retrieves one of the nested elements, by its index number position.

The advantage of VisualTreeHelper is that it works in a generic way that supports all Silverlight elements, no matter what content model they use. For example, you may know that list controls expose items through an Items property, layout containers provide their children through a Children property, and content controls expose the nested content element through a Content property, but only the VisualTreeHelper can dig through all three with the same seamless code.

The disadvantage to using the VisualTreeHelper is that it gets every detail of an element's visual composition, including some that aren't important to its function. For example, when you use VisualTreeHelper to browse through a ListBox, you'll come across a few low-level details that probably don't interest you, such as the Border that outlines it, the ScrollViewer that makes it scrollable, and the Grid that lays out items in discrete rows. For this reason, the only practical way to use the VisualTreeHelper is with recursive code—in essence, you keep digging through the tree until you find the type of element you're interested in, and then you act on it. The following example uses this technique to clear all the text boxes in a hierarchy of elements:

```
Private Sub Clear(ByVal element As DependencyObject)
    ' If this is a text box, clear the text.
    Dim txt As TextBox = TryCast(element, TextBox)
    If Not txt Is Nothing Then
        txt.Text = ""
    End If

    ' Check for nested children.
    Dim children As Integer = VisualTreeHelper.GetChildrenCount(element)
    For i As Integer = 0 To children - 1
        Dim child As DependencyObject = VisualTreeHelper.GetChild(element, i)
        Clear(child)
    Next
End Sub
```

To set it in motion, call the Clear() method with the topmost object you want to examine. Here's how to dissect the entire current page:

```
Clear(Me)
```

Events

So far, all the attributes you've seen map to properties. However, attributes can also be used to attach event handlers. The syntax for this is *EventName="EventHandlerMethodName"*.

For example, the Button control provides a Click event. You can attach an event handler like this:

```
<Button ... Click="cmdAnswer_Click">
```

This assumes that there is a method with the name cmdAnswer_Click in the code-behind class. The event handler must have the correct signature (that is, it must match the delegate for the Click event). Here's the method that does the trick:

```
Private Sub cmdAnswer_Click(ByVal sender As Object, _
  ByVal e As RoutedEventArgs)
    Dim generator As New AnswerGenerator()
    txtAnswer.Text = generator.GetRandomAnswer(txtQuestion.Text)
End Sub
```

In many situations, you'll use attributes to set properties and attach event handlers on the same element. Silverlight always follows the same sequence: first it sets the Name property (if set), then it attaches any event handlers, and lastly it sets the properties. This means that any event handlers that respond to property changes will fire when the property is set for the first time.

There's one additional option for handling events in Silverlight. As you no doubt know, Visual Basic defines a Handles statement that allows you to hook up events declaratively in code. Silverlight supports this convention.

To use the Handles statement, your element must have a name. Provided this requirement is met, you can use the Handles statement in the familiar form, supplying *ElementName.EventName*, as shown here:

```
Private Sub cmdAnswer_Click(ByVal sender As Object, _
  ByVal e As RoutedEventArgs e) Handles cmdAnswer.Click
    ...
End Sub
```

In this case, no event attribute is required in the XAML. (In fact, you must be careful not to supply an event attribute, or you'll end up connecting the same event handler twice.)

It's up to you which event-handling approach you want to use. In this book, all the code examples use event attributes.

The Full Eight Ball Example

Now that you've considered the fundamentals of XAML, you know enough to walk through the definition for the page in Figure 2-1. Here's the complete XAML markup:

```
<UserControl x:Class="EightBall.MainPage"
 xmlns="http://schemas.microsoft.com/winfx/2006/xaml/presentation"
 xmlns:x="http://schemas.microsoft.com/winfx/2006/xaml">
  <Grid x:Name="grid1">
    <Grid.RowDefinitions>
      <RowDefinition Height="*" />
      <RowDefinition Height="Auto" />
      <RowDefinition Height="*" />
```

```
    </Grid.RowDefinitions>
    <TextBox VerticalAlignment="Stretch" HorizontalAlignment="Stretch"
     Margin="10,10,13,10" x:Name="txtQuestion"
     TextWrapping="Wrap" FontFamily="Verdana" FontSize="24"
     Grid.Row="0" Text="[Place question here.]">
    </TextBox>
    <Button VerticalAlignment="Top" HorizontalAlignment="Left"
     Margin="10,0,0,20" Width="127" Height="23" x:Name="cmdAnswer"
     Click="cmdAnswer_Click" Grid.Row="1" Content="Ask the Eight Ball">
    </Button>
    <TextBox VerticalAlignment="Stretch" HorizontalAlignment="Stretch"
     Margin="10,10,13,10" x:Name="txtAnswer" TextWrapping="Wrap"
     IsReadOnly="True" FontFamily="Verdana" FontSize="24" Foreground="Green"
     Grid.Row="2" Text="[Answer will appear here.]">
    </TextBox>

    <Grid.Background>
      <LinearGradientBrush>
        <LinearGradientBrush.GradientStops>
          <GradientStop Offset="0.00" Color="Yellow" />
          <GradientStop Offset="0.50" Color="White" />
          <GradientStop Offset="1.00" Color="Purple" />
        </LinearGradientBrush.GradientStops>
      </LinearGradientBrush>
    </Grid.Background>
  </Grid>
</Window>
```

Remember, you probably won't write the XAML for a graphically rich user interface by hand—doing so would be unbearably tedious. However, you might have good reason to edit the XAML code to make a change that would be awkward to accomplish in the designer. You might also find yourself reviewing XAML to get a better idea of how a page works.

XAML Resources

Silverlight includes a resource system that integrates closely with XAML. Using resources, you can do the following:

- *Create nonvisual objects*: This is useful if other elements use these objects. For example, you could create a data object as a resource and then use data binding to display its information in several elements.
- *Reuse objects*: Once you define a resource, several elements can draw on it. For example, you can define a single brush that's used to color in several shapes. Later in this book, you'll use resources to define styles and templates that are reused among elements.
- *Centralize details*: Sometimes, it's easier to pull frequently changed information into one place (a resources section) rather than scatter it through a complex markup file, where it's more difficult to track down and change.

The resource system shouldn't be confused with *assembly resources*, which are blocks of data that you can embed in your compiled Silverlight assembly. For example, the XAML files you add to your project are embedded as assembly resources. You'll learn more about assembly resources in Chapter 6.

The Resources Collection

Every element includes a Resources property, which stores a dictionary collection of resources. The resources collection can hold any type of object, indexed by string.

Although every element includes the Resources property, the most common way to define resources is at the page level. That's because every element has access to the resources in its own resource collection and the resources in all of its parents' resource collections. So if you define a resource in the page, all the elements on the page can use it.

For example, consider the eight ball example. Currently, the GradientBrush that paints the background of the Grid is defined inline (in other words, it's defined and set in the same place). However, you might choose to pull the brush out of the Grid markup and place it in the resources collection instead:

```
<UserControl x:Class="EightBall.MainPage"
 xmlns="http://schemas.microsoft.com/winfx/2006/xaml/presentation"
 xmlns:x="http://schemas.microsoft.com/winfx/2006/xaml">
  <UserControl.Resources>
    <LinearGradientBrush x:Key="BackgroundBrush">
      <LinearGradientBrush.GradientStops>
        <GradientStop Offset="0.00" Color="Yellow" />
        <GradientStop Offset="0.50" Color="White" />
        <GradientStop Offset="1.00" Color="Purple" />
      </LinearGradientBrush.GradientStops>
    </LinearGradientBrush>
  </UserControl.Resources>
  ...
</UserControl>
```

The only important new detail is the Key attribute that's been added to the brush (and preceded by the *x:* namespace prefix, which puts it in the XAML namespace rather than the Silverlight namespace). The Key attribute assigns the name under which the brush will be indexed in the resources collection. You can use whatever you want, so long as you use the same name when you need to retrieve the resource. It's a good idea to name resources based on their functions (which won't change) rather than the specific details of their implementations (which might). For that reason, BackgroundBrush is a better name than LinearGradientBrush or ThreeColorBrush.

■ **Note** You can instantiate any .NET class in the resources section (including your own custom classes), as long as it's XAML friendly. That means it needs to have a few basic characteristics, such as a public zero-argument constructor and writeable properties.

To use a resource in your XAML markup, you need a way to refer to it. This is accomplished using a *markup extension*—a specialized type of syntax that sets a property in a nonstandard way. Markup extensions extend the XAML language and can be recognized by their curly braces. To use a resource, you use a markup extension named StaticResource:

```
<Grid x:Name="grid1" Background="{StaticResource BackgroundBrush}">
```

This refactoring doesn't shorten the markup you need for the eight ball example. However, if you need to use the same brush in multiple elements, the resource approach is the best way to avoid duplicating the same details. And even if you don't use the brush more than once, you might still prefer this approach if your user interface includes a number of graphical details that are likely to change. For example, by placing all the brushes front and center in the resources collection, you'll have an easier time finding them and changing them. Some developers use the resources collection for virtually every complex object they create to set a property in XAML.

■ **Note** The word *static* stems from the fact that WPF has two types of resources, static and dynamic. However, Silverlight includes only static resources.

The Hierarchy of Resources

Every element has its own resource collection, and Silverlight performs a recursive search up your element tree to find the resource you want. For example, imagine you have the following markup:

```xml
<UserControl x:Class="Resources.ResourceHierarchy"
 xmlns="http://schemas.microsoft.com/winfx/2006/xaml/presentation"
 xmlns:x="http://schemas.microsoft.com/winfx/2006/xaml"
 Width="400" Height="300">
  <Grid x:Name="LayoutRoot" Background="White">
    <StackPanel>
      <StackPanel.Resources>
        <LinearGradientBrush x:Key="ButtonFace">
          <GradientStop Offset="0.00" Color="Yellow" />
          <GradientStop Offset="0.50" Color="White" />
          <GradientStop Offset="1.00" Color="Purple" />
        </LinearGradientBrush>
      </StackPanel.Resources>

      <Button Content="Click Me First" Margin="5"
       Background="{StaticResource ButtonFace}"></Button>
      <Button Content="Click Me Next" Margin="5"
       Background="{StaticResource ButtonFace}"></Button>
    </StackPanel>
  </Grid>
</UserControl>
```

Figure 2-2 shows the page this markup creates.

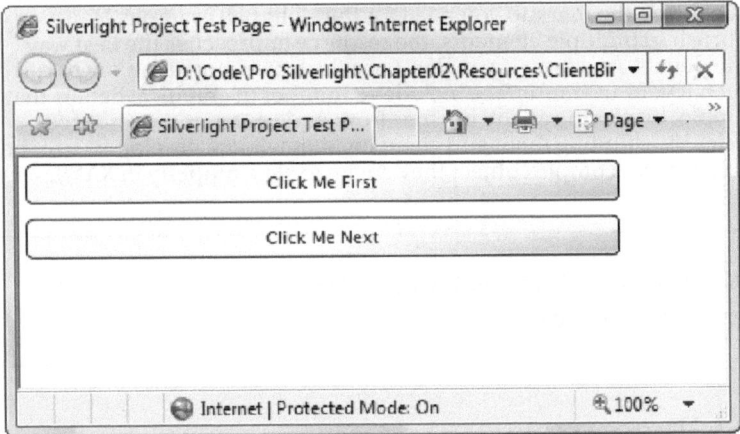

Figure 2-2. Using one brush to color two buttons

Here, both buttons set their backgrounds to the same resource. When encountering this markup, Silverlight will check the resources collection of the button itself and then the StackPanel (where it's defined). If the StackPanel didn't include the right resource, Silverlight would continue its search with the resources collection of the Grid and then the UserControl. If it still hasn't found a resource with the right name, Silverlight will end by checking the application resources that are defined in the <Application.Resources> section of the App.xaml file:

```
<Application xmlns="http://schemas.microsoft.com/client/2007"
 xmlns:x="http://schemas.microsoft.com/winfx/2006/xaml"
 x:Class="SilverlightApplication1.App">
  <Application.Resources>
    <LinearGradientBrush x:Key="ButtonFace">
      <LinearGradientBrush.GradientStops>
        <GradientStop Offset="0.00" Color="Yellow" />
        <GradientStop Offset="0.50" Color="White" />
        <GradientStop Offset="1.00" Color="Purple" />
      </LinearGradientBrush.GradientStops>
    </LinearGradientBrush>
  </Application.Resources>
</Application>
```

The advantage of placing resources in the application collection is that they're completely removed from the markup in your page, and they can be reused across an entire application. In this example, it's a good choice if you plan to use the brush in more than one page.

■ **Note** Before creating an application resource, consider the trade-off between complexity and reuse. Adding an application resource gives you better reuse, but it adds complexity because it's not immediately clear which pages use a given resource. (It's conceptually the same as an old-style C++ program with too many global variables.) A good guideline is to use application resources if your object is reused widely. If it's used in just two or three pages, consider defining the resource in each page.

Order is important when defining a resource in markup. The rule of thumb is that a resource must appear before you refer to it in your markup. That means that even though it's perfectly valid (from a markup perspective) to put the <StackPanel.Resources> section after the markup that declares the buttons, this change will break the current example. When the XAML parser encounters a reference to a resource it doesn't know, it throws an exception.

Interestingly, resource names can be reused as long as you don't use the same resource name more than once in the same collection. In this case, Silverlight uses the resource it finds first. This allows you to define a resource in your application resources collection and then selectively override it with a replacement in some pages with a replacement.

Accessing Resources in Code

Usually, you'll define and use resources in your markup. However, if the need arises, you can work with the resources collection in code. The most straightforward approach is to look up the resource you need in the appropriate collection by name. For example, if you store a LinearGradientBrush in the <UserControl.Resources> section with the key name ButtonFace, you could use code like this:

```
Dim brush As LinearGradientBrush = CType(Me.Resources("ButtonFace"), _
    LinearGradientBrush)

' Swap the color order.
Dim color As Color = brush.GradientStops(0).Color
brush.GradientStops(0).Color = brush.GradientStops(2).Color
brush.GradientStops(2).Color = color
```

When you change a resource in this way, every element that uses the resource updates itself automatically (see Figure 2-3). In other words, if you have four buttons using the ButtonFace brush, they will all get the reversed colors when this code runs.

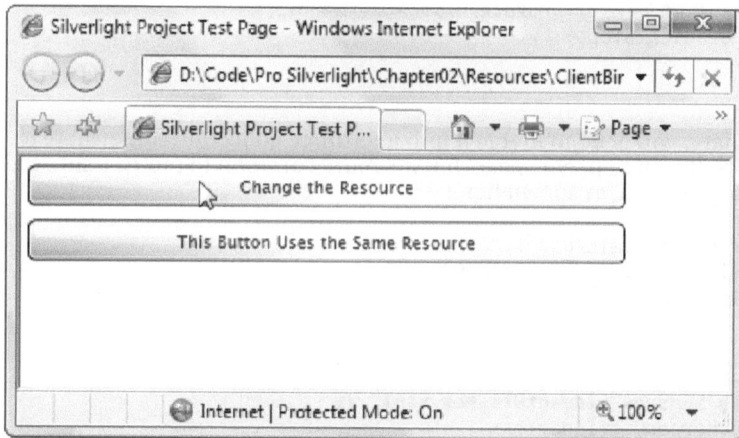

Figure 2-3. Altering a resource

However, there's one limitation. Because Silverlight doesn't support dynamic resources, you aren't allowed to change the resource reference. That means you can't replace a resource with a new object. Here's an example of code that breaks this rule and will generate a runtime error:

```
Dim brush As New SolidColorBrush(Colors.Yellow)
Me.Resources("ButtonFace") = brush
```

Rather than dig through the Resources collection to find the object you want, you can give your resource a name by adding the Name attribute. You can then access it directly by name in your code. However, you can't set both a name and a key on the same object, and the StaticResource markup extension recognizes keys only. Thus, if you create a named resource, you won't be able to use it in your markup with a StaticResource reference. For that reason, it's more common to use keys.

Organizing Resources with Resource Dictionaries

If you want to share resources between multiple projects or just improve the organization of a complex, resource-laden project, you can create a *resource dictionary*. A resource dictionary is simply a XAML document that does nothing but store a set of resources. To create a resource dictionary in Visual Studio, right-click your project in the Solution Explorer, choose Add ➤ New Item, pick the Silverlight Resource Dictionary template, supply any name you like, and click Add.

Here's an example of a resource dictionary named ElementBrushes.xaml that defines one resource:

```
<ResourceDictionary
 xmlns="http://schemas.microsoft.com/winfx/2006/xaml/presentation"
 xmlns:x="http://schemas.microsoft.com/winfx/2006/xaml">

  <LinearGradientBrush x:Key="ButtonFace">
    <LinearGradientBrush.GradientStops>
      <GradientStop Offset="0.00" Color="Yellow" />
      <GradientStop Offset="0.50" Color="White" />
      <GradientStop Offset="1.00" Color="Purple" />
    </LinearGradientBrush.GradientStops>
  </LinearGradientBrush>

</ResourceDictionary>
```

To use a resource dictionary, you need to merge it into a resource collection somewhere in your application. You could do this in a specific page, but it's more common to merge it into the resources collection for the application, as shown here:

```
<Application xmlns="http://schemas.microsoft.com/client/2007"
 xmlns:x="http://schemas.microsoft.com/winfx/2006/xaml"
 x:Class="SilverlightApplication1.App">
  <Application.Resources>
    <ResourceDictionary>
      <ResourceDictionary.MergedDictionaries>
        <ResourceDictionary Source="ElementBrushes.xaml" />
      </ResourceDictionary.MergedDictionaries>
    </ResourceDictionary>
  </Application.Resources>
</Application>
```

The MergedDictionaries collection is a collection of ResourceDictionary objects that you want to use to supplement your resource collection. In this example, there's just one, but you can combine as many as you want. And if you want to add your own resources *and* merge in resource dictionaries, you simply need to place your resources before or after the MergedProperties section, as shown here:

```
<Application.Resources>
  <ResourceDictionary>
    <ResourceDictionary.MergedDictionaries>
      <ResourceDictionary Source="BasicBrushes.xaml" />
      <ResourceDictionary Source="ButtonBrushes.xaml" />
    </ResourceDictionary.MergedDictionaries>
    <LinearGradientBrush x:Key="GraphicalBrush1" ... ></LinearGradientBrush>
    <LinearGradientBrush x:Key="GraphicalBrush2" ... ></LinearGradientBrush>
  </ResourceDictionary>
</Application.Resources>
```

■ **Note** As you learned earlier, it's perfectly reasonable to have resources with the same name stored in different but overlapping resource collections. However, it's not acceptable to merge resource dictionaries that use the same resource names. If there's a duplicate, you'll receive an exception when you compile your application.

One reason to use resource dictionaries is to define the styles for application skins that you can apply dynamically to your controls. (You'll learn how to develop this technique in Chapter 12.) Another reason is to store content that needs to be localized (such as error message strings).

Element-to-Element Binding

In the previous section, you saw how to use the StaticResource markup extension, which gives XAML additional capabilities (in this case, the ability to easily refer to a resource that's defined elsewhere in your markup). You'll see the StaticResource at work throughout the examples in this book. Another markup extension that gets heavy use is the Binding expression, which sets up a relationship that funnels information from a source object to a target control.

In Chapter 16, you'll use binding expressions to create data-bound pages that allow the user to review and edit the information in a linked data object. But in this chapter, you'll take a quick look at a more basic skill—the ability to connect two elements together with a binding expression.

One-Way Binding

To understand how you can bind an element to another element, consider the simple window shown in Figure 2-4. It contains two controls: a Slider and a TextBlock with a single line of text. If you pull the thumb in the slider to the right, the font size of the text is increased immediately. If you pull it to the left, the font size is reduced.

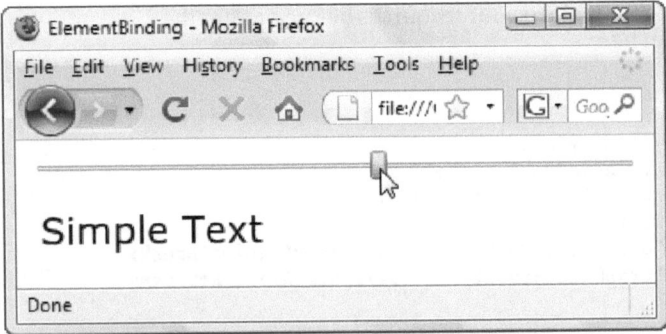

Figure 2-4. Linked controls through data binding

Clearly, it wouldn't be difficult to create this behavior using code. You would simply react to the Slider.ValueChanged event and copy the current value from the slider to the TextBlock. However, data binding makes it even easier.

When using data binding, you don't need to make any change to your source object (which is the Slider in this example). Just configure it to take the right range of values, as you would ordinarily.

```
<Slider x:Name="sliderFontSize" Margin="3"
 Minimum="1" Maximum="40" Value="10">
</Slider>
```

The binding is defined in the TextBlock element. Instead of setting the FontSize using a literal value, you use a binding expression, as shown here:

```
<TextBlock Margin="10" Text="Simple Text" x:Name="lblSampleText"
 FontSize="{Binding ElementName=sliderFontSize, Path=Value}" >
</TextBlock>
```

Data binding expressions use a XAML markup extension (and hence have curly braces). You begin with the word *Binding*, followed by any constructor arguments (there are none in this example) and then a list of the properties you want to set by name—in this case, ElementName and Path. ElementName indicates the source element. Path indicates the property in the source element. Thus, this binding expression copies the value from the Slider.Value property to the TextBlock.FontSize property.

■ **Tip** The Path can point to a property of a property (for example, FontFamily.Source) or an indexer used by a property (for example, Content.Children[0]). You can also refer to an attached property (a property that's defined in another class but applied to the bound element) by wrapping the property name in parentheses. For example, if you're binding to an element that's placed in a Grid, the path (Grid.Row) retrieves the row number where you've placed it.

One of the neat features of data binding is that your target is updated automatically, no matter how the source is modified. In this example, the source can be modified in only one way—by the user's interaction with the slider thumb. However, consider a slightly revamped version of this example that adds a few buttons, each of which applies a preset value to the slider. Click one of these buttons, and this code runs:

```
Private Sub cmd_SetLarge(ByVal sender As Object, ByVal e As RoutedEventArgs)
    sliderFontSize.Value = 30
End Sub
```

This code sets the value of the slider, which in turn forces a change to the font size of the text through data binding. It's the same as if you had moved the slider thumb yourself.

However, this code wouldn't work as well:

```
Private Sub cmd_SetLarge(ByVal sender As Object, ByVal e As RoutedEventArgs)
    lblSampleText.FontSize = 30
End Sub
```

It sets the font of the text box directly. As a result, the slider position isn't updated to match. Even worse, this has the effect of wiping out your font size binding and replacing it with a literal value. If you move the slider thumb now, the text block won't change at all.

Two-Way Binding

Interestingly, there's a way to force values to flow in both directions: from the source to the target *and* from the target to the source. The trick is to set the Mode property of the Binding. Here's a revised bidirectional binding that allows you to apply changes to either the source or the target and have the other piece of the equation update itself automatically:

```
<TextBlock Margin="10" Text="Simple Text" Name="lblSampleText"
 FontSize="{Binding ElementName=sliderFontSize, Path=Value, Mode=TwoWay}" >
</TextBlock>
```

In this example, there's no reason to use a two-way binding, because you can solve the problem by manipulating the value of the slider rather than changing the font size of the TextBlock. However, consider a variation of this example that includes a text box where the user can set the font size precisely (see Figure 2-5).

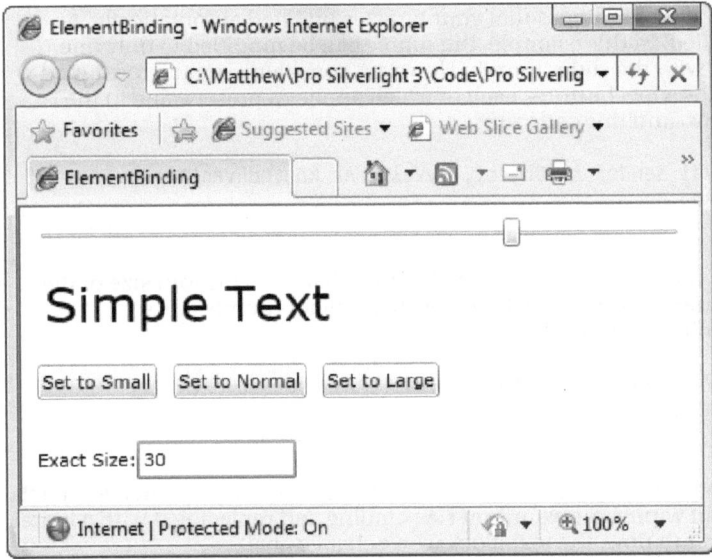

Figure 2-5. *Two-way binding with a text box*

Here, the text box needs to use a two-way binding, because it both receives the bound data value and sets it. When the user drags the slider (or clicks a button), the text box receives the new slider value. And when the user types a new value in the text box, the binding copies the value to the slider.

Here's the two-way binding expression you need:

```
<TextBox Text="{Binding ElementName=lblSampleText, Path=FontSize, Mode=TwoWay}">
</TextBox>
```

■ **Note** If you experiment with this example, you'll discover that the text box applies its value to the slider only once it loses focus. This is the default update behavior in Silverlight, but you can change it by forcing immediate updates as the user types—a trick you'll pick up in Chapter 16.

You'll learn far more about data binding in Chapter 16, when you add data objects and collections into the mix. But this example illustrates two important points—how the Binding extension enhances XAML with the ability to tie properties from different objects together and how you can create basic element synchronization effects with no code required.

The Last Word

In this chapter, you took a tour through a simple XAML file and learned the syntax rules of XAML at the same time. You also considered two markup extensions that Silverlight uses to enhance XAML: the StaticResource extension for referencing resources and the Binding extension for connecting properties in different objects.

When you're designing an application, you don't need to write all your XAML by hand. Instead, you can use a tool such as Visual Studio or Expression Blend to drag and drop your pages into existence. Based on that, you might wonder whether it's worth spending so much time studying the syntax of XAML. The answer is a resounding *yes*. Understanding XAML is critical to Silverlight application design. Understanding XAML will help you learn key Silverlight concepts and ensure that you get the markup you really want. More importantly, there is a host of tasks that are far easier to accomplish with at least some handwritten XAML. In Visual Studio, these tasks include defining resources, creating control templates, writing data binding expressions, and defining animations. Expression Blend has better design support, but on many occasions, it's still quicker to make a change by hand than wade through a sequence of windows.

■ ■ ■

CHAPTER 3

Layout

Half the battle in user interface design is organizing the content in a way that's attractive, practical, and flexible. In a browser-hosted application, this is a particularly tricky task, because your application may be used on a wide range of different computers and devices (all with different display hardware), and you have no control over the size of the browser window in which your Silverlight content is placed.

Fortunately, Silverlight inherits the most important part of WPF's extremely flexible layout model. Using the layout model, you organize your content in a set of different layout *containers*. Each container has its own layout logic—one stacks elements, another arranges them in a grid of invisible cells, and another uses a hard-coded coordinate system. If you're ambitious, you can even create your own containers with custom layout logic.

In this chapter, you'll learn how to use layout containers to create the visual skeleton for a Silverlight page. You'll spend most of your time exploring Silverlight's core layout containers, including the StackPanel, Grid, and Canvas. Once you've mastered these basics, you'll see how to extend your possibilities by creating new layout containers with custom layout logic. You'll also see how you can create an application that breaks out of the browser window and uses the full screen.

■ **What's New** Silverlight 4 doesn't change the layout system in any way. However, it adds the handy Viewbox control (which was formerly a part of the Silverlight Toolkit) to the core framework,. You'll learn more in the section "Scaling with the Viewbox."

The Layout Containers

A Silverlight window can hold only a single element. To fit in more than one element and create a more practical user interface, you need to place a container in your page and then add other elements to that container. Your layout is determined by the container that you use.

All the Silverlight layout containers are panels that derive from the MustInherit System.Windows.Controls.Panel class (see Figure 3-1).

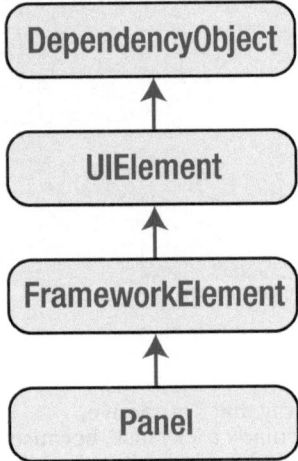

Figure 3-1. *The hierarchy of the Panel class*

The Panel class adds two public properties: Background and Children. Background is the brush that's used to paint the panel background. Children is the collection of items that's stored in the panel. (This is the first level of elements—in other words, these elements may themselves contain more elements.) The Panel class also has a bit of internal plumbing you can use to create your own layout container, as you'll learn later in this chapter.

On its own, the base Panel class is nothing but a starting point for other more specialized classes. Silverlight provides three Panel-derived classes that you can use to arrange layout, and the Silverlight Toolkit adds two more. All of them are listed in Table 3-1, in the order you'll meet them in this chapter. As with all Silverlight controls and most visual elements, these classes are found in the System.Windows.Controls namespace.

Table 3-1. *Core Layout Panels*

Name	Description
StackPanel	Places elements in a horizontal or vertical stack. This layout container is typically used for small sections of a larger, more complex page.
WrapPanel	Places elements in a series of wrapped lines. In horizontal orientation, the WrapPanel lays items out in a row from left to right and then onto subsequent lines. In vertical orientation, the WrapPanel lays out items in a top-to-bottom column and then uses additional columns to fit the remaining items. This layout container is available in the Silverlight Tookit.
DockPanel	Aligns elements against an entire edge of the container. This layout container is available in the Silverlight Tookit.
Grid	Arranges elements in rows and columns according to an invisible table. This is one of the most flexible and commonly used layout containers.
Canvas	Allows elements to be positioned absolutely using fixed coordinates. This layout container is the simplest but least flexible.

Layout containers can be nested. A typical user interface begins with the Grid, Silverlight's most capable container, and contains other layout containers that arrange smaller groups of elements, such as captioned text boxes, items in a list, icons on a toolbar, a column of buttons, and so on.

■ **Note** There's one specialized layout panel that doesn't appear in Table 3-1: the VirtualizingStackPanel. It arranges items in the same way as the StackPanel, but it uses a memory-optimization technique called *virtualization*. The VirtualizingStackPanel allows list controls like the ListBox to hold tens of thousands of items without a dramatic slowdown, because the VirtualizingStackPanel creates objects only for the currently visible items. But although you might use the VirtualizingStackPanel to build custom templates and controls (see Chapter 13), you won't use it to arrange the elements in a page, and so it isn't covered in this chapter.

The Panel Background

All Panel elements introduce the concept of a background by adding a Background property. It's natural to expect that the Background property would use some sort of color object. However, the Background property actually uses something much more versatile: a Brush object. This design gives you the flexibility to fill your background and foreground content with a solid color (by using the SolidColorBrush) or something more exotic (for example, a gradient or a bitmap, by using a LinearGradientBrush or ImageBrush). In this section, you'll consider only the simple solid-color fills provided by the SolidColorBrush, but you'll try fancier brushwork in Chapter 9.

■ **Note** All of Silverlight's Brush classes are found in the System.Windows.Media namespace.

For example, if you want to give your entire page a light blue background, you could adjust the background of the root panel. Here's the code that does the trick:

```
layoutRoot.Background = New SolidColorBrush(Colors.AliceBlue)
```

Technically, every Color object is an instance of the Color structure in the System.Windows.Media namespace. You can get a wide range of ready-made colors from the Colors class, which provides a shared property for each one. (The property names are based on the color names supported by web browsers.) The code shown here uses one of these colors to create a new SolidColorBrush. It then sets the brush as the background brush for the root panel, which causes its background to be painted a light shade of blue.

━━━

■ **Tip** Silverlight also includes a SystemColors class that provides Color objects that match the current system preferences. For example, SystemColors.ActiveColorBorder gets the color that's used to fill the border of the foreground window. In some cases, you might choose to ensure your application blends in better with the current color scheme, particularly if you're building an out-of-browser application, as described in Chapter 21.

━━━

The Colors and SystemColors classes offer handy shortcuts, but they aren't the only way to set a color. You can also create a Color object by supplying the red, green, and blue (RGB) values, along with an alpha value that indicates transparency. Each one of these values is a number from 0 to 255:

```
Dim red As Integer = 0
Dim green As Integer = 255
Dim blue As Integer = 0
layoutRoot.Background = New SolidColorBrush(Color.FromArgb(255, red, green, blue))
```

You can also make a color partly transparent by supplying an alpha value when calling the Color.FromArgb() method. An alpha value of 255 is completely opaque, while 0 is completely transparent.

Often, you'll set colors in XAML rather than in code. Here, you can use a helpful shortcut. Rather than define a Brush object, you can supply a color name or color value. The type converter for the Background property will automatically create a SolidColorBrush object using the color you specify. Here's an example that uses a color name:

```
<Grid x:Name="layoutRoot" Background="Red">
```

It's equivalent to this more verbose syntax:

```
<Grid x:Name="layoutRoot">
  <Grid.Background>
    <SolidColorBrush Color="Red"></SolidColorBrush>
  </Grid.Background>
</Grid>
```

You need to use the longer form if you want to create a different type of brush, such as a LinearGradientBrush, and use that to paint the background.

If you want to use a color code, you need to use a slightly less convenient syntax that puts the R, G, and B values in hexadecimal notation. You can use one of two formats—either #rrggbb or #aarrggbb (the difference being that the latter includes the alpha value). You need only two digits to supply the A, R, G, and B values because they're all in hexadecimal notation. Here's an example that creates the same color as in the previous code snippets using #aarrggbb notation:

```
<Grid x:Name="layoutRoot" Background="#FFFF0000">
```

Here the alpha value is FF (255), the red value is FF (255), and the green and blue values are 0.

By default, the Background of a layout panel is set to a null reference (Nothing), which is equivalent to this:

```
<Grid x:Name="layoutRoot" Background="{x:Null}">
```

When your panel has a null background, any content underneath will show through (similar to if you set a fully transparent background color). However, there's an important difference—the layout container won't be able to receive mouse events.

■ **Note** Brushes support automatic change notification. In other words, if you attach a brush to a control and change the brush, the control updates itself accordingly.

Borders

The layout containers allow you to paint a background, but not a border outline. However, there's an element that fills in the gap—the Border.

The Border class is pure simplicity. It takes a single piece of nested content (which is often a layout panel) and adds a background or border around it. To master the Border, you need nothing more than the properties listed in Table 3-2.

Table 3-2. Properties of the Border Class

Name	Description
Background	Sets a background that appears behind all the content in the border using a Brush object. You can use a solid color or something more exotic.
BorderBrush	Sets the fill of the border that appears around the edge of the Border object, using a Brush object. The most straightforward approach is to use a SolidColorBrush to create a solid border.
BorderThickness	Sets the width (in pixels) of the border on each side. The BorderThickness property holds an instance of the System.Windows.Thickness structure, with separate components for the top, bottom, left, and right edges.
CornerRadius	Rounds the corners of your border. The greater the CornerRadius, the more dramatic the rounding effect is.
Padding	Adds spacing between the border and the content inside. (By contrast, Margin adds spacing outside the border.)

Here's a straightforward, slightly rounded border around a basic button:

```
<Border Margin="25"  Background="LightYellow"
 BorderBrush="SteelBlue" BorderThickness="8" CornerRadius="15">
  <Button Margin="10 Content="Click  Me"></Button>
</Border>
```

This example adds a little bit or margin space around the border and the button, which is a feature you'll learn about in the next section. Figure 3-2 shows the result.

Figure 3-2. *A basic border*

Simple Layout with the StackPanel

The StackPanel is one of the simplest layout containers. It simply stacks its children in a single row or column. These elements are arranged based on their order.

For example, consider this page, which contains a stack with one TextBlock and four buttons:

```
<UserControl x:Class="Layout.SimpleStack"
 xmlns="http://schemas.microsoft.com/winfx/2006/xaml/presentation"
 xmlns:x="http://schemas.microsoft.com/winfx/2006/xaml">
  <StackPanel Background="White">
    <TextBlock Text="A Button Stack"></TextBlock>
    <Button Content="Button 1"></Button>
    <Button Content="Button 2"></Button>
    <Button Content="Button 3"></Button>
    <Button Content="Button 4"></Button>
  </StackPanel>
</UserControl>
```

Figure 3-3 shows the result.

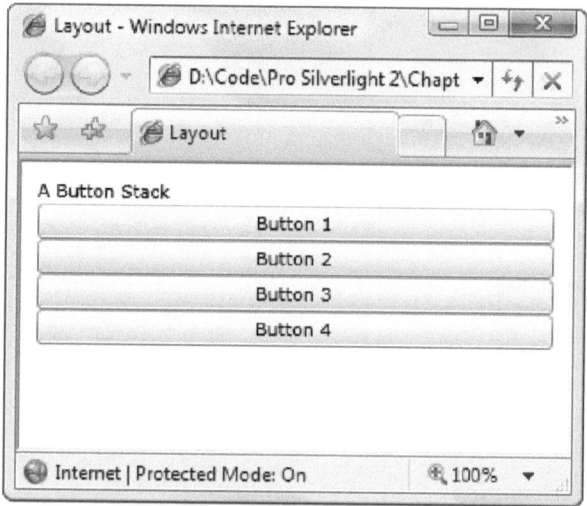

Figure 3-3. *The StackPanel in action*

By default, a StackPanel arranges elements from top to bottom, making each one as tall as is necessary to display its content. In this example, that means the TextBlock and buttons are sized just high enough to comfortably accommodate the text inside. All the elements are then stretched to the full width of the StackPanel, which is the width of your page.

In this example, the Height and Width properties of the page are not set. As a result, the page grows to fit the full Silverlight content region (in this case, the complete browser window). Most of the examples in this chapter use this approach, because it makes it easier to experiment with the different layout containers. You can then see how a layout container resizes itself to fit different page sizes simply by resizing the browser window.

■ **Note** Once you've examined all the layout containers, you'll take a closer look at the issue of page sizes, and you'll learn about your different options for dealing content that doesn't fit in the browser window.

The StackPanel can also be used to arrange elements horizontally by setting the Orientation property:

```
<StackPanel Orientation="Horizontal" Background="White">
```

Now elements are given their minimum width (wide enough to fit their text) and are stretched to the full height of the containing panel (see Figure 3-4).

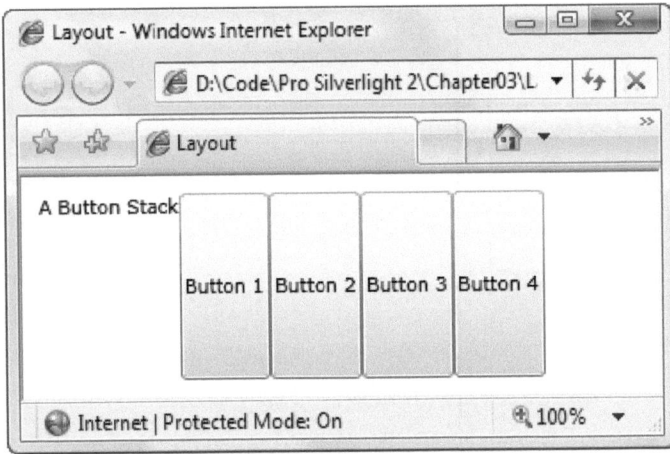

Figure 3-4. The StackPanel with horizontal orientation

Clearly, this doesn't provide the flexibility real applications need. Fortunately, you can fine-tune the way the StackPanel and other layout containers work using layout properties, as described next.

Layout Properties

Although layout is determined by the container, the child elements can still get their say. In fact, layout panels work in concert with their children by respecting a small set of layout properties, as listed in Table 3-3.

Table 3-3. Layout Properties

Name	Description
HorizontalAlignment	This property determines how a child is positioned inside a layout container when there's extra horizontal space available. You can choose Center, Left, Right, or Stretch.
VerticalAlignment	This one determines how a child is positioned inside a layout container when there's extra vertical space available. You can choose Center, Top, Bottom, or Stretch.
Margin	Use Margin to add a bit of breathing room around an element. The Margin property holds an instance of the System.Windows.Thickness structure, with separate components for the top, bottom, left, and right edges.
MinWidth and MinHeight	These properties set the minimum dimensions of an element. If an element is too large for its layout container, it will be cropped to fit.

Name	Description
MaxWidth and MaxHeight	These two properties set the maximum dimensions of an element. If the container has more room available, the element won't be enlarged beyond these bounds, even if the HorizontalAlignment and VerticalAlignment properties are set to Stretch.
Width and Height	Use these properties to explicitly set the size of an element. This setting overrides a Stretch value for the HorizontalAlignment and VerticalAlignment properties. However, this size won't be honored if it's outside of the bounds set by the MinWidth, MinHeight, MaxWidth, and MaxHeight.

All of these properties are inherited from the base FrameworkElement class and are therefore supported by all the graphical widgets you can use in a Silverlight page.

■ **Note** As you learned in Chapter 2, different layout containers can provide attached properties to their children. For example, all the children of a Grid object gain Row and Column properties that allow them to choose the cell where they're placed. Attached properties allow you to set information that's specific to a particular layout container. However, the layout properties in Table 3-3 are generic enough that they apply to many layout panels. Thus, these properties are defined as part of the base FrameworkElement class.

Alignment

To understand how these properties work, take another look at the simple StackPanel shown in Figure 3-3. In this example—a StackPanel with vertical orientation—the VerticalAlignment property has no effect because each element is given as much height as it needs and no more. However, the HorizontalAlignment *is* important. It determines where each element is placed in its row.

Ordinarily, the default HorizontalAlignment is Left for a label and Stretch for a Button. That's why every button takes the full column width. However, you can change these details:

```
<StackPanel Background="White">
  <TextBlock HorizontalAlignment="Center" Text="A Button Stack"></TextBlock>
  <Button HorizontalAlignment="Left" Content="Button 1"></Button>
  <Button HorizontalAlignment="Right" Content="Button 2"></Button>
  <Button Content="Button 3"></Button>
  <Button Content="Button 4"></Button>
</StackPanel>
```

Figure 3-5 shows the result. The first two buttons are given their minimum sizes and aligned accordingly, while the bottom two buttons are stretched over the entire StackPanel. If you resize the page, you'll see that the label remains in the middle and the first two buttons stay stuck to either side.

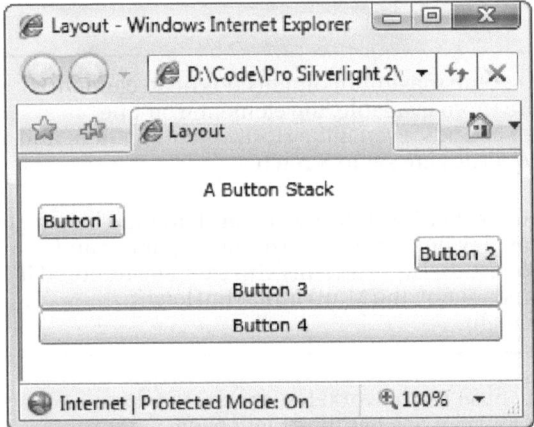

Figure 3-5. *A StackPanel with aligned buttons*

■ **Note** The StackPanel also has its own HorizontalAlignment and VerticalAlignment properties. By default, both of these are set to Stretch, and so the StackPanel fills its container completely. In this example, that means the StackPanel fills the page. If you use a different value for VerticalAlignment, the StackPanel will be made just large enough to fit the widest control.

Margins

There's an obvious problem with the StackPanel example in its current form. A well-designed page doesn't just contain elements—it also includes a bit of extra space in between the elements. To introduce this extra space and make the StackPanel example less cramped, you can set control margins.

When setting margins, you can set a single width for all sides, like this:

```
<Button Margin="5" Content="Button 3"></Button>
```

Alternatively, you can set different margins for each side of a control in the order *left, top, right, bottom*:

```
<Button Margin="5,10,5,10" Content="Button 3"></Button>
```

In code, you can set margins using the Thickness structure:

```
cmd.Margin = New Thickness(5)
```

Getting the right control margins is a bit of an art, because you need to consider how the margin settings of adjacent controls influence one another. For example, if you have two buttons stacked on top of each other and the topmost button has a bottom margin of 5 and the bottommost button has a top margin of 5, you have a total of 10 pixels of space between the two buttons.

Ideally, you'll be able to keep different margin settings as consistent as possible and avoid setting distinct values for the different margin sides. For instance, in the StackPanel example, it makes sense to use the same margins on the buttons and on the panel itself, as shown here:

```
<StackPanel Margin="3" Background="White">
  <TextBlock Margin="3" HorizontalAlignment="Center"
  Text="A Button Stack"></TextBlock>
  <Button Margin="3" HorizontalAlignment="Left" Content="Button 1"></Button>
  <Button Margin="3" HorizontalAlignment="Right" Content="Button 2"></Button>
  <Button Margin="3" Content="Button 3"></Button>
  <Button Margin="3" Content="Button 4"></Button>
</StackPanel>
```

This way, the total space between two buttons (the sum of the two button margins) is the same as the total space between the button at the edge of the page (the sum of the button margin and the StackPanel margin). Figure 3-6 shows this more respectable page, and Figure 3-7 shows how the margin settings break down.

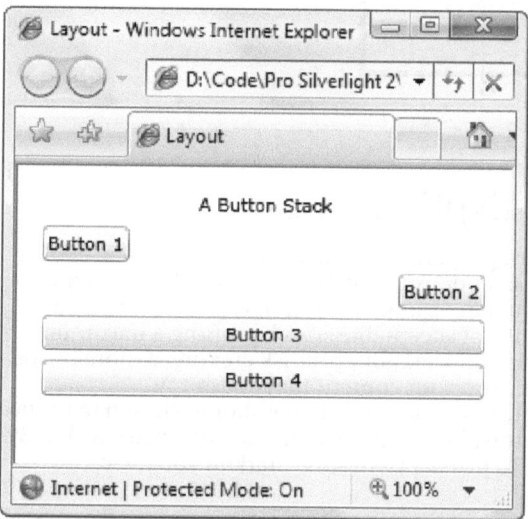

Figure 3-6. *Adding margins between elements*

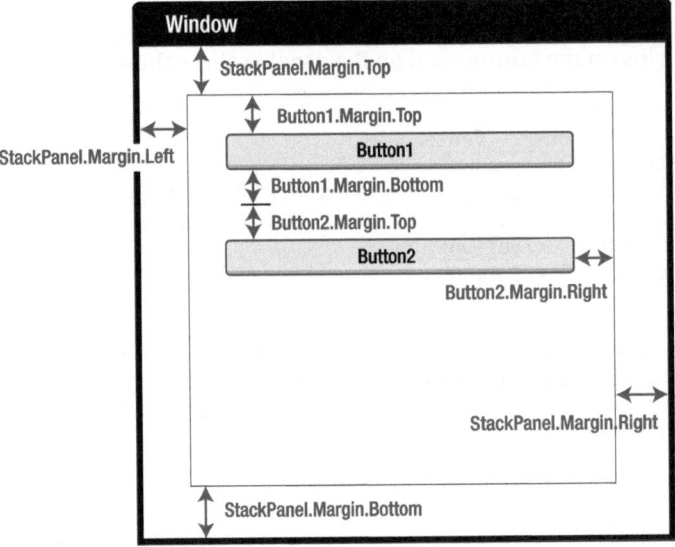

Figure 3-7. *How margins are combined*

Minimum, Maximum, and Explicit Sizes

Finally, every element includes Height and Width properties that allow you to give it an explicit size. However, just because you can set explicit sizes doesn't mean you *should*. In most cases, it's better to let elements grow to fit their content. For example, a button expands as you add more text. You can lock your elements into a range of acceptable sizes by setting a maximum and minimum size, if necessary. If you do add size information, you risk creating a more brittle layout that can't adapt to changes and (at worst) truncates content that doesn't fit.

For example, you might decide that the buttons in your StackPanel should stretch to fit the StackPanel but be made no larger than 200 pixels wide and no smaller than 100 pixels wide. (By default, buttons start with a minimum width of 75 pixels.) Here's the markup you need:

```
<StackPanel Margin="3">
  <TextBlock Margin="3" HorizontalAlignment="Center"
   Text="A Button Stack"></TextBlock>
  <Button Margin="3" MaxWidth="300" MinWidth="200" Content="Button 1"></Button>
  <Button Margin="3" MaxWidth="300" MinWidth="200" Content="Button 2"></Button>
  <Button Margin="3" MaxWidth="300" MinWidth="200" Content="Button 3"></Button>
  <Button Margin="3" MaxWidth="300" MinWidth="200" Content="Button 4"></Button>
</StackPanel>
```

■ **Tip** At this point, you might be wondering if there's an easier way to set properties that are standardized across several elements, such as the button margins in this example. The answer is *styles*—a feature that allows you to reuse property settings. You'll learn about styles in Chapter 12.

When the StackPanel sizes a button that doesn't have a hard-coded size, it considers several pieces of information:

- *The minimum size*: Each button will always be at least as large as the minimum size.
- *The maximum size*: Each button will always be smaller than the maximum size (unless you've incorrectly set the maximum size to be smaller than the minimum size).
- *The content*: If the content inside the button requires a greater width, the StackPanel will attempt to enlarge the button.
- *The size of the container*: If the minimum width is larger than the width of the StackPanel, a portion of the button will be cut off. But if the minimum width isn't set (or is less than the width of the StackPanel), the button will not be allowed to grow wider than the StackPanel, even if it can't fit all its text on the button surface.
- *The horizontal alignment*: Because the button uses a HorizontalAlignment of Stretch (the default), the StackPanel will attempt to enlarge the button to fill the full width of the StackPanel.

The trick to understanding this process is to realize that the minimum and maximum size set the absolute bounds. Within those bounds, the StackPanel tries to respect the button's desired size (to fit its content) and its alignment settings.

Figure 3-8 sheds some light on how this works with the StackPanel. On the left is the page at its minimum size. The buttons are 200 pixels each, and the page cannot be resized to be narrower. If you shrink the page from this point, the right side of each button will be clipped off. (You can deal with this situation using scrolling, as discussed later in this chapter.)

As you enlarge the page, the buttons grow with it until they reach their maximum of 300 pixels. From this point on, if you make the page any larger, the extra space is added to either side of the button (as shown on the right in Figure 3-8).

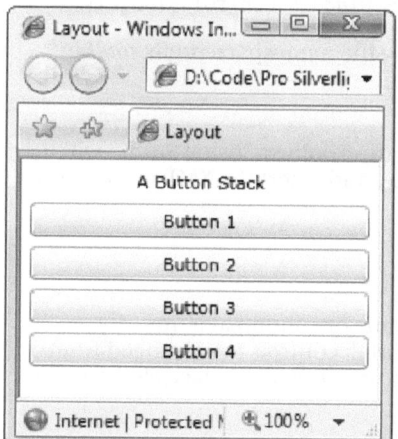 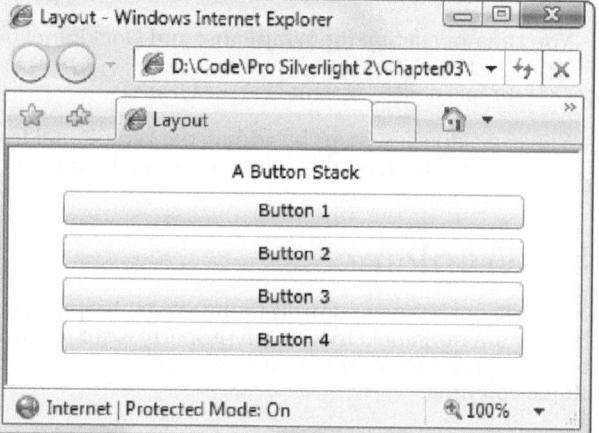

Figure 3-8. Constrained button sizing

■ **Note** In some situations, you might want to use code that checks how large an element is in a page. The Height and Width properties are no help because they indicate your desired size settings, which might not correspond to the actual rendered size. In an ideal scenario, you'll let your elements size to fit their content, and the Height and Width properties won't be set at all. However, you can find out the actual size used to render an element by reading the ActualHeight and ActualWidth properties. But remember, these values may change when the page is resized or the content inside it changes.

The WrapPanel and DockPanel

Obviously, the StackPanel alone can't help you create a realistic user interface. To complete the picture, the StackPanel needs to work with other more capable layout containers. Only then can you assemble a complete window.

The most sophisticated layout container is the Grid, which you'll consider later in this chapter. But first, it's worth looking at the WrapPanel and DockPanel, which are two simple layout containers that are available as part of the Silverlight Toolkit. Both complement the StackPanel by offering different layout behavior.

To use the WrapPanel or the DockPanel, you need to add a reference to the System.Windows.Controls.Toolkit.dll assembly where they are defined. To get this assembly, you must install the Silverlight Toolkit, which is available at `http://silverlight.codeplex.com`.

Once you've added the assembly reference, you need to map the namespace so it's available in your markup, as shown here:

```
<UserControl x:Class="Layout.WrapAndDock" ...
  xmlns:toolkit=
  "clr-namespace:System.Windows.Controls;assembly=System.Windows.Controls.Toolkit">
```

You can now define the WrapPanel and DockPanel using the namespace prefix *toolkit*:

```
<toolkit:WrapPanel ...></toolkit:WrapPanel>
```

You can skip a few steps by adding the WrapPanel from the Toolbox. Visual Studio will then add the appropriate assembly reference, map the namespace, and insert the XML markup for the control.

The WrapPanel

The WrapPanel lays out controls in the available space, one line or column at a time. By default, the WrapPanel.Orientation property is set to Horizontal; controls are arranged from left to right and then on subsequent rows. However, you can use Vertical to place elements in multiple columns.

■ **Tip** Like the StackPanel, the WrapPanel is really intended for control over small-scale details in a user interface, not complete window layouts. For example, you might use a WrapPanel to keep together the buttons in a toolbar-like control.

Here's an example that defines a series of buttons with different alignments and places them into the WrapPanel:

```
<toolkit:WrapPanel Margin="3">
  <Button VerticalAlignment="Top" Content="Top Button"></Button>
  <Button MinHeight="60" Content="Tall Button"></Button>
  <Button VerticalAlignment="Bottom" Content="Bottom Button"></Button>
  <Button Content="Stretch Button"></Button>
  <Button VerticalAlignment="Center" Content="Centered Button"></Button>
</toolkit:WrapPanel>
```

Figure 3-9 shows how the buttons are wrapped to fit the current size of the WrapPanel (which is determined by the size of the control that contains it). As this example demonstrates, a WrapPanel in horizontal mode creates a series of imaginary rows, each of which is given the height of the tallest contained element. Other controls may be stretched to fit or aligned according to the VerticalAlignment property. In the example on the left in Figure 3-9, all the buttons fit into one tall row and are stretched or aligned to fit. In the example on the right, several buttons have been bumped to the second row. Because the second row does not include an unusually tall button, the row height is kept at the minimum button height. As a result, it doesn't matter what VerticalAlignment setting the various buttons in this row use.

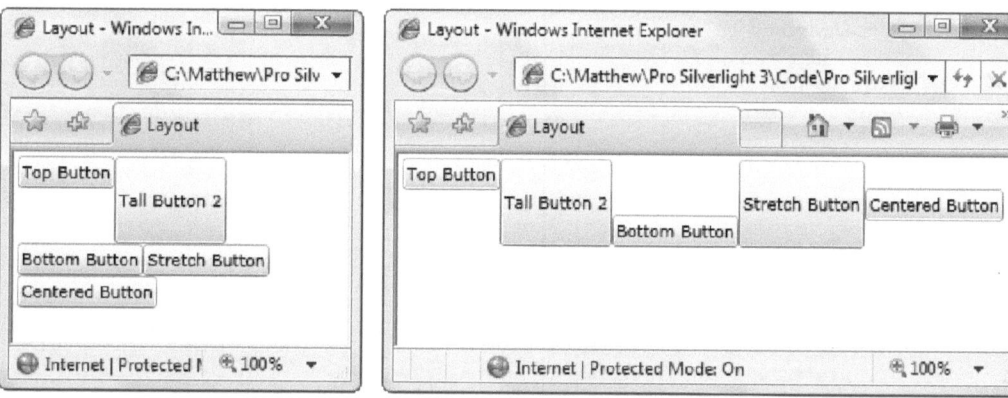

Figure 3-9. *Wrapped buttons*

■ **Note** The WrapPanel is the only one of the five Silverlight layout containers whose effects can't be duplicated with a crafty use of the Grid.

The DockPanel

The Silverlight Toolkit also includes a layout container called the DockPanel. It stretches controls against one of its outside edges. The easiest way to visualize this is to think of the toolbars that sit at the top of many Windows applications. These toolbars are docked to the top of the window. As with the StackPanel, docked elements get to choose one aspect of their layout. For example, if you dock a button to the top of a DockPanel, it's stretched across the entire width of the DockPanel but given whatever height it requires (based on the content and the MinHeight property). On the other hand, if you dock a button to the left side of a container, its height is stretched to fit the container, but its width is free to grow as needed.

The obvious question is this: How do child elements choose the side where they want to dock? The answer is through an attached property named Dock, which can be set to Left, Right, Top, or Bottom. Every element that's placed inside a DockPanel automatically acquires this property.

Here's an example that puts one button on every side of a DockPanel:

```
<toolkit:DockPanel LastChildFill="True">
  <Button toolkit:DockPanel.Dock="Top" Content="Top Button"></Button>
  <Button toolkit:DockPanel.Dock="Bottom" Content="Bottom Button"></Button>
  <Button toolkit:DockPanel.Dock="Left" Content="Left Button"></Button>
  <Button toolkit:DockPanel.Dock="Right" Content="Right Button"></Button>
  <Button Content="Remaining Space"></Button>
</toolkit:DockPanel>
```

This example also sets the LastChildFill to True, which tells the DockPanel to give the remaining space to the last element. Figure 3-10 shows the result.

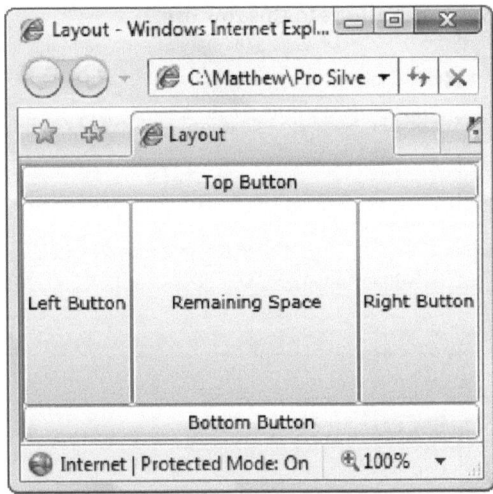

Figure 3-10. *Docking to every side*

Clearly, when docking controls, the order is important. In this example, the top and bottom buttons get the full edge of the DockPanel because they're docked first. When the left and right buttons are docked next, they fit between these two buttons. If you reversed this order, the left and right buttons would get the full sides, and the top and bottom buttons would become narrower, because they'd be docked between the two side buttons.

You can dock several elements against the same side. In this case, the elements simply stack up against the side in the order they're declared in your markup. And, if you don't like the spacing or the stretch behavior, you can tweak the Margin, HorizontalAlignment, and VerticalAlignment properties, just as you did with the StackPanel. Here's a modified version of the previous example that demonstrates:

```
<toolkit:DockPanel LastChildFill="True">
  <Button toolkit:DockPanel.Dock="Top" Content="A Stretched Top Button"></Button>
  <Button toolkit:DockPanel.Dock="Top" HorizontalAlignment="Center"
    Content="A Centered Top Button"></Button>
  <Button toolkit:DockPanel.Dock="Top" HorizontalAlignment="Left"
    Content="A Left-Aligned Top Button"></Button>
  <Button toolkit:DockPanel.Dock="Bottom" Content="Bottom Button"></Button>
  <Button toolkit:DockPanel.Dock="Left" Content="Left Button"></Button>
  <Button toolkit:DockPanel.Dock="Right" Content="Right Button"></Button>
  <Button Content="Remaining Space"></Button>
</toolkit:DockPanel>
```

The docking behavior is still the same. First, the top buttons are docked, then the bottom button is docked, and finally the remaining space is divided between the side buttons and a final button in the middle. Figure 3-11 shows the resulting window.

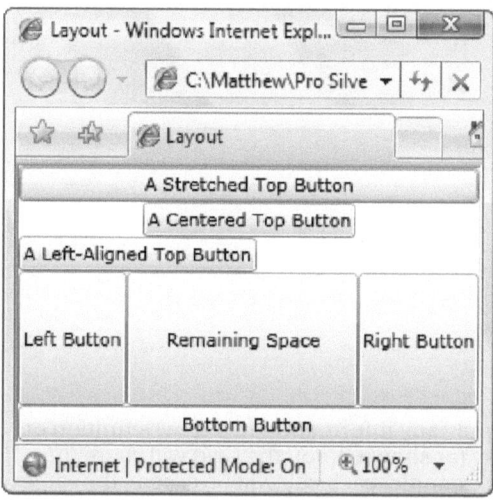

Figure 3-11. *Docking multiple elements to the top*

The Grid

The Grid is the most powerful layout container in Silverlight. In fact, the Grid is so useful that when you add a new XAML document for a page in Visual Studio, it automatically adds the Grid tags as the first-level container, nested inside the root UserControl element.

The Grid separates elements into an invisible grid of rows and columns. Although more than one element can be placed in a single cell (in which case they overlap), it generally makes sense to place just a single element per cell. Of course, that element may itself be another layout container that organizes its own group of contained controls.

■ **Tip** Although the Grid is designed to be invisible, you can set the Grid.ShowGridLines property to True to take a closer look. This feature isn't really intended for prettying up a page. Instead, it's a debugging convenience that's designed to help you understand how the Grid has subdivided itself into smaller regions. This feature is important because you have the ability to control exactly how the Grid chooses column widths and row heights.

Creating a Grid-based layout is a two-step process. First, you choose the number of columns and rows that you want. Next, you assign the appropriate row and column to each contained element, thereby placing it in just the right spot.

You create grids and rows by filling the Grid.ColumnDefinitions and Grid.RowDefinitions collections with objects. For example, if you decide you need two rows and three columns, you'd add the following tags:

```
<Grid ShowGridLines="True" Background="White">
  <Grid.RowDefinitions>
    <RowDefinition></RowDefinition>
    <RowDefinition></RowDefinition>
  </Grid.RowDefinitions>
  <Grid.ColumnDefinitions>
    <ColumnDefinition></ColumnDefinition>
    <ColumnDefinition></ColumnDefinition>
    <ColumnDefinition></ColumnDefinition>
  </Grid.ColumnDefinitions>

  ...
</Grid>
```

As this example shows, it's not necessary to supply any information in a RowDefinition or ColumnDefinition element. If you leave them empty (as shown here), the Grid will share the space evenly between all rows and columns. In this example, each cell will be exactly the same size, depending on the size of the containing page.

To place individual elements into a cell, you use the attached Row and Column properties. Both these properties take zero-based index numbers. For example, here's how you could create a partially filled grid of buttons:

```
<Grid ShowGridLines="True" Background="White">
  ...
  <Button Grid.Row="0" Grid.Column="0" Content="Top Left"></Button>
  <Button Grid.Row="0" Grid.Column="1" Content="Top Middle"></Button>
  <Button Grid.Row="1" Grid.Column="2" Content="Bottom Right"></Button>
  <Button Grid.Row="1" Grid.Column="1" Content="Bottom Middle"></Button>
</Grid>
```

Each element must be placed into its cell explicitly. This allows you to place more than one element into a cell (which rarely makes sense) or leave certain cells blank (which is often useful). It also means you can declare your elements out of order, as with the final two buttons in this example. However, it makes for clearer markup if you define your controls row by row, and from left to right in each row.

There is one exception. If you don't specify the Grid.Row property, the Grid assumes that it's 0. The same behavior applies to the Grid.Column property. Thus, you leave both attributes off an element to place it in the first cell of the Grid.

Figure 3-12 shows how this simple grid appears at two different sizes. Notice that the ShowGridLines property is set to True so that you can see the separation between each column and row.

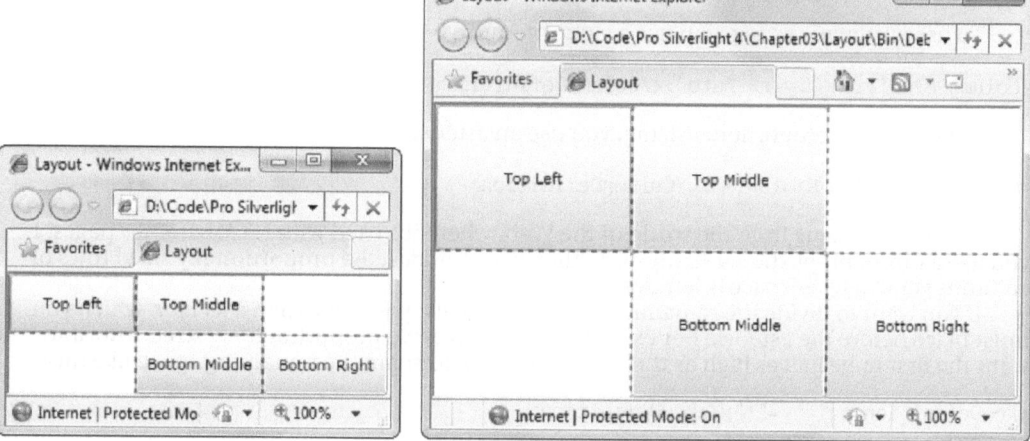

Figure 3-12. *A simple grid*

As you would expect, the Grid honors the basic set of layout properties listed in Table 3-3. That means you can add margins around the content in a cell, you can change the sizing mode so an element doesn't grow to fill the entire cell, and you can align an item along one of the edges of a cell. If you force an element to have a size that's larger than the cell can accommodate, part of the content will be chopped off.

Fine-Tuning Rows and Columns

As you've seen, the Grid gives you the ability to create a proportionately sized collection of rows and columns, which is often quite useful. However, to unlock the full potential of the Grid, you can change the way each row and column is sized.

The Grid supports three sizing strategies:

- *Absolute sizes*: You choose the exact size using pixels. This is the least useful strategy, because it's not flexible enough to deal with changing content size, changing container size, or localization.

- *Automatic sizes*: Each row or column is given exactly the amount of space it needs and no more. This is one of the most useful sizing modes.

- *Proportional sizes*: Space is divided between a group of rows or columns. This is the standard setting for all rows and columns. For example, in Figure 3-12 you can see that all cells increase in size proportionately as the Grid expands.

For maximum flexibility, you can mix and match these different sizing modes. For example, it's often useful to create several automatically sized rows and then let one or two remaining rows get the leftover space through proportional sizing.

You set the sizing mode using the Width property of the ColumnDefinition object or the Height property of the RowDefinition object to a number. For example, here's how you set an absolute width of 100 pixels:

```
<ColumnDefinition Width="100"></ColumnDefinition>
```

To use automatic sizing, you use a value of Auto:

```
<ColumnDefinition Width="Auto"></ColumnDefinition>
```

Finally, to use proportional sizing, you use an asterisk (*):

```
<ColumnDefinition Width="*"></ColumnDefinition>
```

This syntax stems from the world of the Web, where it's used with HTML frames pages. If you use a mix of proportional sizing and other sizing modes, the proportionally sized rows or columns get whatever space is left over.

If you want to divide the remaining space unequally, you can assign a *weight*, which you must place before the asterisk. For example, if you have two proportionately sized rows and you want the first to be half as high as the second, you could share the remaining space like this:

```
<RowDefinition Height="*"></RowDefinition>
<RowDefinition Height="2*"></RowDefinition>
```

This tells the Grid that the height of the second row should be twice the height of the first row. You can use whatever numbers you like to portion out the extra space.

Nesting Layout Containers

The Grid is impressive on its own, but most realistic user interfaces combine several layout containers. They may use an arrangement with more than one Grid or nay mix the Grid with other layout containers like the StackPanel.

The following markup presents a simple example of this principle. It creates a basic dialog box with OK and Cancel buttons in the bottom-right corner and a large content region that's sized to fit its content (the text in a TextBlock). The entire package is centered in the middle of the page by setting the alignment properties on the Grid.

```
<Grid ShowGridLines="True" Background="SteelBlue"
 HorizontalAlignment="Center" VerticalAlignment="Center">
  <Grid.RowDefinitions>
    <RowDefinition Height="*"></RowDefinition>
    <RowDefinition Height="Auto"></RowDefinition>
  </Grid.RowDefinitions>

  <TextBlock Margin="10" Grid.Row="0" Foreground="White"
   Text="This is simply a test of nested containers."></TextBlock>
  <StackPanel Grid.Row="1" HorizontalAlignment="Right" Orientation="Horizontal">
    <Button Margin="10,10,2,10" Padding="3" Content="OK"></Button>
    <Button Margin="2,10,10,10" Padding="3" Content="Cancel"></Button>
  </StackPanel>
</Grid>
```

You'll notice that this Grid doesn't declare any columns. This is a shortcut you can take if your grid uses just one column and that column is proportionately sized (so it fills the entire width of the Grid). Figure 3-13 shows the rather pedestrian dialog box this markup creates.

■ **Note** In this example, the Padding property adds some minimum space between the button border and the content inside (the word *OK* or *Cancel*). You'll learn more about Padding when you consider content controls in Chapter 5.

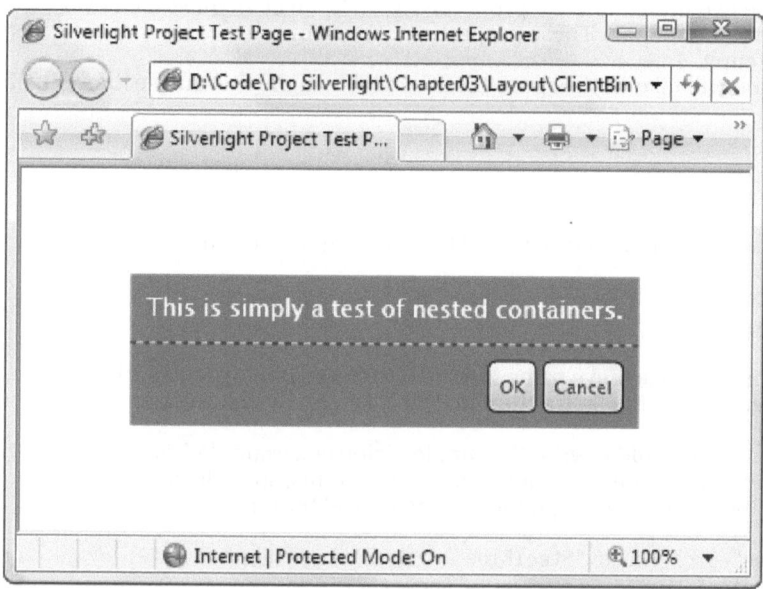

Figure 3-13. *A basic dialog box*

At first glance, nesting layout containers seems like a fair bit more work than placing controls in precise positions using coordinates. And in many cases, it is. However, the longer setup time is compensated by the ease with which you can change the user interface in the future. For example, if you decide you want the OK and Cancel buttons to be centered at the bottom of the page, you simply need to change the alignment of the StackPanel that contains them:

```
<StackPanel Grid.Row="1" HorizontalAlignment="Center" ... >
```

Similarly, if you need to change the amount of content in the first row, the entire Grid will be enlarged to fit, and the buttons will move obligingly out of the way. And if you add a dash of styles to this page (Chapter 12), you can improve it even further and remove other extraneous details (such as the margin settings) to create cleaner and more compact markup.

■ **Tip** If you have a densely nested tree of elements, it's easy to lose sight of the overall structure. Visual Studio provides a handy feature that shows you a tree representation of your elements and allows you to click your way down to the element you want to look at (or modify). This feature is the Document Outline window, and you can view it by choosing View ➤ Other Windows ➤ Document Outline from the menu.

Spanning Rows and Columns

You've already seen how to place elements in cells using the Row and Column attached properties. You can also use two more attached properties to make an element stretch over several cells: RowSpan and ColumnSpan. These properties take the number of rows or columns that the element should occupy.

For example, this button will take all the space that's available in the first and second cell of the first row:

```
<Button Grid.Row="0" Grid.Column="0" Grid.ColumnSpan="2" Content="Span Button">
</Button>
```

And this button will stretch over four cells in total by spanning two columns and two rows:

```
<Button Grid.Row="0" Grid.Column="0" Grid.RowSpan="2" Grid.ColumnSpan="2"
Content="Span Button"></Button>
```

Row and column spanning can achieve some interesting effects and is particularly handy when you need to fit elements in a tabular structure that's broken up by dividers or longer sections of content.

Using column spanning, you could rewrite the simple dialog box example from Figure 3-13 using just a single Grid. This Grid divides the page into three columns, spreads the text box over all three, and uses the last two columns to align the OK and Cancel buttons.

```
<Grid ShowGridLines="True" Background="SteelBlue"
 HorizontalAlignment="Center" VerticalAlignment="Center">
  <Grid.RowDefinitions>
    <RowDefinition Height="*"></RowDefinition>
    <RowDefinition Height="Auto"></RowDefinition>
  </Grid.RowDefinitions>
  <Grid.ColumnDefinitions>
    <ColumnDefinition Width="*"></ColumnDefinition>
    <ColumnDefinition Width="Auto"></ColumnDefinition>
    <ColumnDefinition Width="Auto"></ColumnDefinition>
  </Grid.ColumnDefinitions>
  <TextBlock Margin="10" Grid.Row="0" Grid.Column="0" Grid.ColumnSpan="3"
   Foreground="White"
   Text="This is simply a test of nested containers."></TextBlock>

  <Button Margin="10,10,2,10" Padding="3"
    Grid.Row="1" Grid.Column="1" Content="OK"></Button>
  <Button Margin="2,10,10,10" Padding="3"
    Grid.Row="1" Grid.Column="2" Content="Cancel"></Button>
</Grid>
```

Most developers will agree that this layout isn't clear or sensible. The column widths are determined by the size of the two buttons at the bottom of the page, which makes it difficult to add new content into the existing Grid structure. If you make even a minor addition to this page, you'll probably be forced to create a new set of columns.

As this shows, when you choose the layout containers for a page, you aren't simply interested in getting the correct layout behavior—you also want to build a layout structure that's easy to maintain and enhance in the future. A good rule of thumb is to use smaller layout containers such as the StackPanel for one-off layout tasks, such as arranging a group of buttons. On the other hand, if you need to apply a consistent structure to more than one area of your page, the Grid is an indispensable tool for standardizing your layout.

The GridSplitter

Every Windows user has seen *splitter bars*—draggable dividers that separate one section of a window from another. For example, when you use Windows Explorer, you're presented with a list of folders (on the left) and a list of files (on the right). You can drag the splitter bar in between to determine what proportion of the window is given to each pane.

In Silverlight, you can create a similar design and give the user the ability to resize rows or columns by adding a splitter bar to a Grid. Figure 3-14 shows a window where a GridSplitter sits between two columns. By dragging the splitter bar, the user can change the relative widths of both columns.

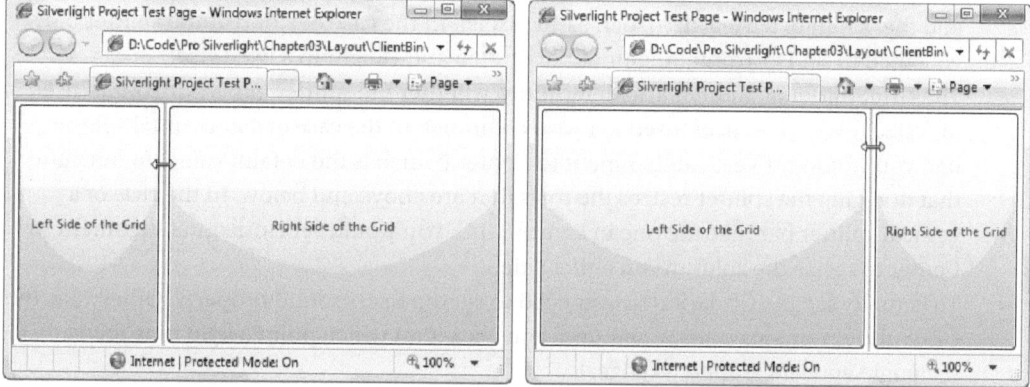

Figure 3-14. *Moving a splitter bar*

■ **Note** The GridSplitter is defined in the System.Windows.Controls.dll assembly. If you aren't already using this assembly, you'll need to add an assembly reference and an XML mapping before you get access to the GridSplitter, much as you did to use the WrapPanel and DockPanel. The easiest way to accomplish both steps is to add the GridSplitter from the Toolbox, which gets Visual Studio will do the job for you.

To use the GridSplitter effectively, you need to know a little bit more about how it works. Although the GridSplitter serves a straightforward purpose, it can be awkward at first. To get the result you want, follow these guidelines:

- The GridSplitter must be placed in a Grid cell. You can place the GridSplitter in a cell with existing content, in which case you need to adjust the margin settings so it doesn't overlap. A better approach is to reserve a dedicated column or row for the GridSplitter, with a Height or Width value of Auto.

- The GridSplitter always resizes entire rows or columns (not single cells). To make the appearance of the GridSplitter consistent with this behavior, you should stretch the GridSplitter across an entire row or column, rather than limit it to a single cell. To accomplish this, you use the RowSpan or ColumnSpan properties you considered earlier. For example, the GridSplitter in Figure 3-14 has a RowSpan of 2. As a result, it stretches over the entire column. If you didn't add this setting, it would appear only in the top row (where it's placed), *even though* dragging the splitter bar would resize the entire column.

- Initially, the GridSplitter is invisibly small. To make it usable, you need to give it a minimum size. In the case of a vertical splitter bar (like the one in Figure 3-14), you need to set the VerticalAlignment to Stretch (so it fills the whole height of the available area) and the Width to a fixed size (such as 10 pixels). In the case of a horizontal splitter bar, you need to set HorizontalAlignment to Stretch and Height to a fixed size.

- The GridSplitter alignment also determines whether the splitter bar is horizontal (used to resize rows) or vertical (used to resize columns). In the case of a horizontal splitter bar, you would set VerticalAlignment to Center (which is the default value) to indicate that dragging the splitter resizes the rows that are above and below. In the case of a vertical splitter bar (like the one in Figure 3-14), you would set HorizontalAlignment to Center to resize the columns on either side.

- To actually see the GridSplitter, you need to set the Background property. Otherwise, the GridSplitter remains transparent until you click it (at which point a light blue focus rectangle appears around its edges).

- The GridSplitter respects minimum and maximum sizes, if you've set them on your ColumnDefinition or RowDefinition objects. The user won't be allowed to enlarge or shrink a column or row outside of its allowed size range.

To reinforce these rules, it helps to take a look at the actual markup for the example shown in Figure 3-14. In the following listing, the GridSplitter details are highlighted:

```
<Grid Background="White">
  <Grid.ColumnDefinitions>
    <ColumnDefinition MinWidth="100"></ColumnDefinition>
    <ColumnDefinition Width="Auto"></ColumnDefinition>
    <ColumnDefinition MinWidth="50"></ColumnDefinition>
  </Grid.ColumnDefinitions>

  <Button Grid.Column="0" Margin="3" Content="Left Side of the Grid"></Button>
```

```
<controls:GridSplitter Grid.Column="1" Grid.RowSpan="2" Background="LightGray"
  Width="3" VerticalAlignment="Stretch" HorizontalAlignment="Center"
  ShowsPreview="False"></controls:GridSplitter>
  <Button Grid.Column="2" Margin="3" Content="Right Side of the Grid"></Button>
</Grid>
```

■ **Tip** Remember, if a Grid has just a single row or column, you can leave out the RowDefinitions section. Also, elements that don't have their row position explicitly set are assumed to have a Grid.Row value of 0 and are placed in the first row. The same holds true for elements that don't supply a Grid.Column value.

This markup includes one additional detail. When the GridSplitter is declared, the ShowsPreview property is set to False (which is the default value). As a result, when the splitter bar is dragged from one side to another, the columns are resized immediately. But if you set ShowsPreview to True, when you drag, you'll see a gray shadow follow your mouse pointer to show you where the split will be. The columns won't be resized until you release the mouse button. You can also change the fill that's used for the GridSplitter so that it isn't just a shaded gray rectangle. The trick is to set the Background property.

A Grid usually contains no more than a single GridSplitter. However, you can nest one Grid inside another, and if you do, each Grid may have its own GridSplitter. This allows you to create a page that's split into two regions (for example, left and right panes) and then further subdivide one of these regions (say, the pane on the right) into more sections (such as resizable top and bottom portions). Figure 3-15 shows an example.

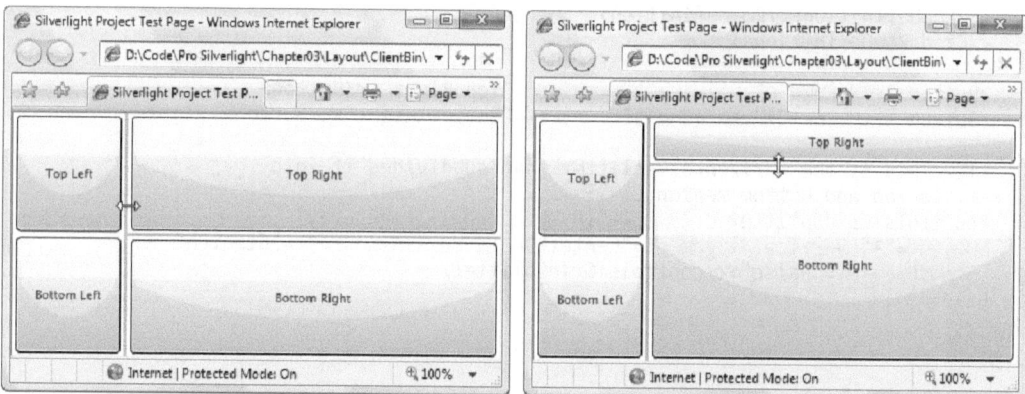

Figure 3-15. Resizing a window with two splits

Creating this page is fairly straightforward, although it's a chore to keep track of the three Grid containers that are involved: the overall Grid, the nested Grid on the left, and the nested Grid on the right. The only trick is to make sure the GridSplitter is placed in the correct cell and given the correct alignment. Here's the complete markup:

```
<!-- This is the Grid for the entire page. -->
<Grid Background="White">
  <Grid.ColumnDefinitions>
    <ColumnDefinition></ColumnDefinition>
    <ColumnDefinition Width="Auto"></ColumnDefinition>
    <ColumnDefinition></ColumnDefinition>
  </Grid.ColumnDefinitions>

  <!-- This is the nested Grid on the left.
       It isn't subdivided further with a splitter. -->
  <Grid Grid.Column="0" VerticalAlignment="Stretch">
    <Grid.RowDefinitions>
      <RowDefinition></RowDefinition>
      <RowDefinition></RowDefinition>
    </Grid.RowDefinitions>
    <Button Margin="3" Grid.Row="0" Content="Top Left"></Button>
    <Button Margin="3" Grid.Row="1" Content="Bottom Left"></Button>
  </Grid>

  <!-- This is the vertical splitter that sits between the two nested
       (left and right) grids. -->
  <controls:GridSplitter Grid.Column="1" Background="LightGray"
   Width="3" HorizontalAlignment="Center" VerticalAlignment="Stretch">
  </controls:GridSplitter>

  <!-- This is the nested Grid on the right. -->
  <Grid Grid.Column="2">
    <Grid.RowDefinitions>
      <RowDefinition></RowDefinition>
      <RowDefinition Height="Auto"></RowDefinition>
      <RowDefinition></RowDefinition>
    </Grid.RowDefinitions>

    <Button Grid.Row="0" Margin="3" Content="Top Right"></Button>
    <Button Grid.Row="2" Margin="3" Content="Bottom Right"></Button>

    <!-- This is the horizontal splitter that subdivides it into
         a top and bottom region.. -->
    <controls:GridSplitter Grid.Row="1" Background="LightGray"
     Height="3" VerticalAlignment="Center" HorizontalAlignment="Stretch"
     ShowsPreview="False"></controls:GridSplitter>
  </Grid>
</Grid>
```

Coordinate-Based Layout with the Canvas

The only layout container you haven't considered yet is the Canvas. It allows you to place elements using exact coordinates, which is a poor choice for designing rich data-driven forms and standard dialog boxes, but it's a valuable tool if you need to build something a little different (such as a drawing surface for a diagramming tool). The Canvas is also the most lightweight of the layout containers. That's because it doesn't include any complex layout logic to negotiate the sizing preferences of its children. Instead, it simply lays them all out at the position they specify, with the exact size they want.

To position an element on the Canvas, you set the attached Canvas.Left and Canvas.Top properties. Canvas.Left sets the number of pixels between the left edge of your element and the left edge of the Canvas. Canvas.Top sets the number of pixels between the top of your element and the top of the Canvas.

Optionally, you can size your element explicitly using its Width and Height properties. This is more common when using the Canvas than it is in other panels because the Canvas has no layout logic of its own. If you don't set the Width and Height properties, your element will get its desired size—in other words, it will grow just large enough to fit its content. If you change the size of the Canvas, it has no effect on the Controls inside.

Here's a simple Canvas that includes four buttons:

```
<Canvas Background="White">
  <Button Canvas.Left="10" Canvas.Top="10" Content="(10,10)"></Button>
  <Button Canvas.Left="120" Canvas.Top="30" Content="(120,30)"></Button>
  <Button Canvas.Left="60" Canvas.Top="80" Width="50" Height="50"
  Content="(60,80)"></Button>
  <Button Canvas.Left="70" Canvas.Top="120" Width="100" Height="50"
  Content="(70,120)"></Button>
</Canvas>
```

Figure 3-16 shows the result.

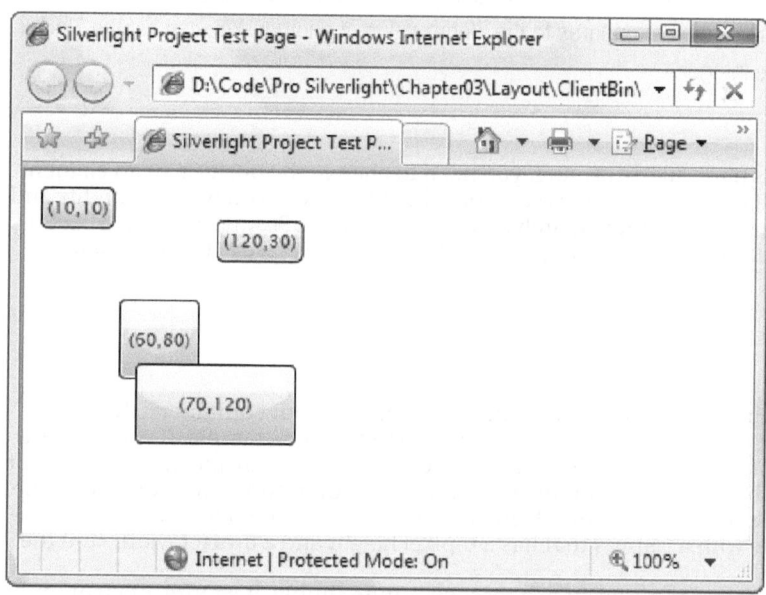

Figure 3-16. *Explicitly positioned buttons in a Canvas*

Like any other layout container, the Canvas can be nested inside a user interface. That means you can use the Canvas to draw some detailed content in a portion of your page, while using more standard Silverlight panels for the rest of your elements.

Layering with ZIndex

If you have more than one overlapping element, you can set the attached Canvas.ZIndex property to control how they are layered.

Ordinarily, all the elements you add have the same ZIndex—0. When elements have the same ZIndex, they're displayed in the same order that they exist in the Canvas.Children collection, which is based on the order that they're defined in the XAML markup. Elements declared later in the markup—such as button (70,120)—are displayed overtop of elements that are declared earlier—such as button (60,80).

However, you can promote any element to a higher level by increasing its ZIndex. That's because higher ZIndex elements *always* appear over lower ZIndex elements. Using this technique, you could reverse the layering in the previous example:

```
<Button Canvas.Left="60" Canvas.Top="80" Canvas.ZIndex="1" Width="50" Height="50"
  Content="(60,80)"></Button>
<Button Canvas.Left="70" Canvas.Top="120" Width="100" Height="50"
  Content="(70,120)"</Button>
```

■ **Note** The actual values you use for the Canvas.ZIndex property have no meaning. The important detail is how the ZIndex value of one element compares to the ZIndex value of another. You can set the ZIndex using any positive or negative integer.

The ZIndex property is particularly useful if you need to change the position of an element programmatically. Just call Canvas.SetZIndex() and pass in the element you want to modify and the new ZIndex you want to apply. Unfortunately, there is no BringToFront() or SendToBack() method—it's up to you to keep track of the highest and lowest ZIndex values if you want to implement this behavior.

Clipping

There's one aspect of the Canvas that's counterintuitive. In most layout containers, the contents are limited to the space that's available in that container. For example, if you create a StackPanel with a height of 100 pixels and place a tall column of buttons inside, those that don't fit will be chopped off the bottom. However, the Canvas doesn't follow this common-sense rule. Instead, it draws all its children, even if they fall outside its bounds. That means you could replace the earlier example with a Canvas that has a 0-pixel height and a 0-pixel width, and the result wouldn't change.

The Canvas works this way for performance reasons—quite simply, it's more efficient for the Canvas to draw all its children and then check whether each one falls insides its bounds. However, this isn't always the behavior you want. For example, Chapter 10 includes an animated game that sends bombs flying off the edge of the playing area, which is a Canvas. In this situation, the bombs must be visible only inside the Canvas—when they leave, they should disappear under the Canvas border, not drift overtop other elements.

Fortunately, the Canvas has support for *clipping*, which ensures that elements (or the portions of an element) that aren't inside a specified area are cut off, in much the same way as elements that extend beyond the edges of a StackPanel or Grid. The only inconvenience is that you need to set the shape of the clipping area manually using the Canvas.Clip property.

Technically, the Clip property takes a Geometry object, which is a useful object you'll consider in more detail when you tackle drawing in Chapter 8. Silverlight has different Geometry-derived classes for different types of shapes, including squares and rectangles (RectangleGeometry), circles and ellipses (EllipseGeometry), and more complex shapes (PathGeometry). Here's an example that sets the clipping region to a rectangular area that matches the bounds of the Canvas:

```
<Canvas x:Name="canvasBackground" Width="200" Height="500" Background="AliceBlue">
  <Canvas.Clip>
    <RectangleGeometry Rect="0,0 200,500"></RectangleGeometry>
  </Canvas.Clip>
  ...
<Canvas>
```

In this example, the clipping region can be described as a rectangle with its top-left corner at point (0, 0), a width of 200 pixels, and a height of 500 pixels. The coordinate for the top-left corner is relative to the Canvas itself, so you must always have a top-left corner of (0,0) unless you want to leave out some of the content in the upper or left region of the Canvas.

Setting the clipping region in markup isn't always the best approach. It's particularly problematic if your Canvas is sized dynamically to fit a resizable container or the browser window. In this situation, it's far more effective to set the clipping region programmatically. Fortunately, all you need is a simple event handler that changes the clipping region when the Canvas is resized by reaching the Canvas.SizeChanged event. (This event also fires when the Canvas is first created, so it also takes care of the initial clipping region setup.)

```
Private Sub canvasBackground_SizeChanged(ByVal sender As Object, _
  ByVal e As SizeChangedEventArgs)
    Dim rect As New RectangleGeometry()
    rect.Rect = New Rect(0, 0, canvasBackground.ActualWidth, _
      canvasBackground.ActualHeight)
    canvasBackground.Clip = rect
End Sub
```

You can attach that event handler like so:

```
<Canvas x:Name="canvasBackground" SizeChanged="canvasBackground_SizeChanged"
  Background="AliceBlue">
```

You'll see this technique in action with the bomb-dropping game in Chapter 10.

CHOOSING THE RIGHT LAYOUT CONTAINER

As a general rule of thumb, the Grid and StackPanel are best when dealing with business-style applications (for example, when displaying data entry forms or documents). They deal well with changing window sizes and dynamic content (for example, blocks of text that can grow or shrink depending on the information at hand). They also make it easier to modify, localize, and reskin the application, because adjacent elements will bump each other out of the way as they change size. The Grid and StackPanel are also closest to the way ordinary HTML pages work.

The Canvas is dramatically different. Because all of its children are arranged using fixed coordinates, you need to go to more work to position them (and even more work if you want to tweak the layout later on in response to new elements or new formatting.) However, the Canvas makes sense in certain types of graphically rich applications, such as games. In these applications, you need fine-grained control, text and graphics often overlap, and you often change coordinates programmatically. Here, the emphasis isn't on flexibility but on achieving a specific visual appearance, and the Canvas makes more sense.

Custom Layout Containers

Although Silverlight has a solid collection of layout containers, it can't offer everything. The developers of Silverlight left out many more specialized layout containers to keep the Silverlight download as lean as possible.

However, there's no reason you can't create some layout containers of your own. You simply need to derive a custom class from Panel and supply the appropriate layout logic. And if you're ambitious, you can combine the layout logic of a panel with other Silverlight features. For example, you can create a panel that handles mouse-over events to provide automatic drag support for the elements inside (like the dragging example shown in Chapter 4), or you can create a panel that displays its children with an animated effect.

In the following sections, you'll learn how the layout process works, and then you'll see how to build a custom layout container. The example you'll consider is the UniformGrid—a stripped-down grid control that tiles elements into a table of identically sized cells.

The Two-Step Layout Process

Every panel uses the same plumbing: a two-step process that's responsible for sizing and arranging children. The first stage is the *measure* pass, and it's at this point that the panel determines how large its children want to be. The second stage is the *layout* pass, and it's at this point that each control is assigned its bounds. Two steps are required, because the panel might need to take into account the desires of all its children before it decides how to partition the available space.

You add the logic for these two steps by overriding the oddly named MeasureOverride() and ArrangeOverride() methods, which are defined in the FrameworkElement class as part of the Silverlight layout system. The odd names represent that the MeasureOverride() and ArrangeOverride() methods replace the logic that's defined in the MeasureCore() and ArrangeCore() methods that are defined in the UIElement class. These methods are *not* overridable.

MeasureOverride()

The first step is to determine how much space each child wants using the MeasureOverride() method. However, even in the MeasureOverride() method, children aren't given unlimited room. At a bare minimum, children are confined to fit in the space that's available to the panel. Optionally, you might want to limit them more stringently. For example, a Grid with two proportionally sized rows will give children half the available height. A StackPanel will offer the first element all the space that's available and then offer the second element whatever's left, and so on.

Every MeasureOverride() implementation is responsible for looping through the collection of children and calling the Measure() method of each one. When you call the Measure() method, you supply the bounding box—a Size object that determines the maximum available space for the child control. At the end of the MeasureOverride() method, the panel returns the space it needs to display all its children and their desired sizes.

Here's the basic structure of the MeasureOverride() method, without the specific sizing details:

```
Protected Overrides Function MeasureOverride(ByVal panelSpace As Size) As Size
    ' Examine all the children.
    For Each element As UIElement In Me.Children
        ' Ask each child how much space it would like, given the
        ' availableElementSize constraint.
        Dim availableElementSize As New Size(...)
        element.Measure(availableElementSize)
        ' (You can now read element.DesiredSize to get the requested size.)
    Next

    ' Indicate how much space this panel requires.
    ' This will be used to set the DesiredSize property of the panel.
    Return New Size(...)
End Function
```

The Measure() method doesn't return a value. After you call Measure() on a child, that child's DesiredSize property provides the requested size. You can use this information in your calculations for future children (and to determine the total space required for the panel).

You *must* call Measure() on each child, even if you don't want to constrain the child's size or use the DesiredSize property. Many elements will not render themselves until you've called Measure(). If you want to give a child free rein to take all the space it wants, pass a Size object with a value of Double.PositiveInfinity for both dimensions. (The ScrollViewer is one element that uses this strategy, because it can handle any amount of content.) The child will then return the space it needs for all its content. Otherwise, the child will normally return the space it needs for its content or the space that's available—whichever is smaller.

At the end of the measuring process, the layout container must return its desired size. In a simple panel, you might calculate the panel's desired size by combining the desired size of every child.

■ **Note** You can't simply return the constraint that's passed to the MeasureOverride() method for the desired size of your panel. Although this seems like a good way to take all the available size, it runs into trouble if the container passes in a Size object with Double.PositiveInfinity for one or both dimensions (which means "take all the space you want"). Although an infinite size is allowed as a sizing constraint, it's not allowed as a sizing result, because Silverlight won't be able to figure out how large your element should be. Furthermore, you really shouldn't take more space than you need. Doing so can cause extra whitespace and force elements that occur after your layout panel to be bumped farther down the window.

If you're an attentive reader, you may have noticed that there's a close similarity between the Measure() method that's called on each child and the MeasureOverride() method that defines the first step of the panel's layout logic. In fact, the Measure() method triggers the MeasureOverride() method. Thus, if you place one layout container inside another, when you call Measure(), you'll get the total size required for the layout container and all its children.

One reason the measuring process goes through two steps (a Measure() method that triggers the MeasureOverride() method) is to deal with margins. When you call Measure(), you pass in the total available space. When Silverlight calls the MeasureOverride() method, it automatically reduces the available space to take margin space into account (unless you've passed in an infinite size).

ArrangeOverride()

Once every element has been measured, it's time to lay them out in the space that's available. The layout system calls the ArrangeOverride() method of your panel, and the panel calls the Arrange() method of each child to tell it how much space it's been allotted. (As you can probably guess, the Arrange() method triggers the ArrangeOverride() method, much as the Measure() method triggers the MeasureOverride() method.)

When measuring items with the Measure() method, you pass in a Size object that defines the bounds of the available space. When placing an item with the Arrange() method, you pass in a System.Windows.Rect object that defines the size *and* position of the item. At this point, it's as though every element is placed with Canvas-style X and Y coordinates that determine the distance between the top-left corner of your layout container and the element.

Here's the basic structure of the ArrangeOverride() method, without the specific sizing details:

```
Protected Overrides Function ArrangeOverride(ByVal panelSize As Size) As Size
    ' Examine all the children.
    For Each element As UIElement In Me.Children
        ' Assign the child its bounds.
        Dim elementBounds As New Rect(...)
        element.Arrange(elementBounds)
        ' (You can now read element.ActualHeight and element.ActualWidth
        '  to find out the size it used.)
    Next

    ' Indicate how much space this panel occupies.
```

```
' This will be used to set the ActualHeight and ActualWidth properties
' of the panel.
Return arrangeSize
End Function
```

When arranging elements, you can't pass infinite sizes. However, you can give an element its desired size by passing in the value from its DesiredSize property. You can also give an element *more* space than it requires. In fact, this happens frequently. For example, a vertical StackPanel gives a child as much height as it requests but gives it the full width of the panel itself. Similarly, a Grid might use fixed or proportionally sized rows that are larger than the desired size of the element inside. And even if you've placed an element in a size-to-content container, that element can still be enlarged if an explicit size has been set using the Height and Width properties.

When an element is made larger than its desired size, the HorizontalAlignment and VerticalAlignment properties come into play. The element content is placed somewhere inside the bounds that it has been given.

Because the ArrangeOverride() method always receives a defined size (not an infinite size), you can return the Size object that's passed in to set the final size of your panel. In fact, many layout containers take this step to occupy all the space that's been given. You aren't in danger of taking up space that could be needed for another control, because the measure step of the layout system ensures that you won't be given more space than you need unless that space is available.

The UniformGrid

Now that you've examined the layout system in a fair bit of detail, it's worth creating your own layout container that adds something you can't get with the basic set of Silverlight panels. In this section, you'll see an example straight from the WPF world: a UniformGrid that arranges its children into automatically generated, equally sized cells.

■ **Note** The UniformGrid is useful as a lightweight alternative to the regular Grid, because it doesn't require explicitly defined rows and columns, and it doesn't force you to manually place each child in the right cell. It makes particularly good sense when display a tiled set of images. In fact, WPF includes a slightly more ambitious version of this control as part of the .NET Framework.

Like all custom panels, the UniformGrid starts with a simple class declaration that inherits from the base Panel control:

```
Public Class UniformGrid
    Inherits System.Windows.Controls.Panel
    ...
End Class
```

■ **Note** You can build the UniformGrid directly inside any Silverlight application. But if you want to reuse your custom layout container in multiple applications, it's a better idea to place it in a new Silverlight class library for it. When you want to use your custom layout container in an application, simply add a reference to the compiled class library.

Conceptually, the UniformGrid is quite simple. It examines the available space, calculates how many cells are needed (and how big each cell will be), and then lays out its children one after the other. The UniformGrid allows you to customize its behavior with two properties, Rows and Columns, which can be set independently or in conjunction:

```
Private _columns As Integer
Public Property Columns() As Integer
    Get
        Return _columns
    End Get
    Set(ByVal value As Integer)
        _columns = value
    End Set
End Property

Private _rows As Integer
Public Property Rows() As Integer
    Get
        Return _rows
    End Get
    Set(ByVal value As Integer)
        _rows = value
    End Set
End Property
```

Here's how the Rows and Columns properties affect the layout logic:

- If both the Rows and Columns properties are set, the UniformGrid knows how big to make the grid. It simply needs to divide the available space proportionately to find the size of each cell. If there are more elements than cells, the extra elements aren't displayed.
- If only one of these properties is set, the UniformGrid calculates the other, assuming that you want to display all the elements inside. For example, if you set Columns to 3 and place eight elements inside, the UniformGrid will divide the available space into three rows.
- If neither of these properties is set, the UniformGrid will calculate both of them, assuming that you want to display all the elements and you want an equal number of rows and columns. (However, the UniformGrid won't create an entirely blank row or column. Instead, if it can't match the number of rows and columns exactly, the UniformGrid will add an extra column.)

To implement this system, the UniformGrid keeps track of the *real* number of columns and rows. This holds the value in the Columns and Rows properties, if they're set. If they aren't, the Grid uses a custom method called CalculateColumns() to count the child elements and determine the dimensions of the grid. This method can then be called during the first stage of layout.

```
Private realColumns As Integer
Private realRows As Integer

Private Sub CalculateColumns()
    ' Count the elements, and don't do anything
    ' if the panel is empty.
    Dim elementCount As Double = Me.Children.Count
    If elementCount = 0 Then Return

    realRows = Rows
    realColumns = Columns

    ' If the Rows and Columns properties were set, use them.
    If (realRows <> 0) AndAlso (realColumns <> 0) Then
        Return
    End If

    ' If neither property was set, start by calculating the columns.
    If (realColumns = 0) AndAlso realRows = 0 Then
        realColumns = CInt(Fix(Math.Ceiling(Math.Sqrt(elementCount))))
    End If

    ' If only Rows is set, calculate Columns.
    If realColumns = 0 Then
        realColumns = CInt(Fix(Math.Ceiling(elementCount / realRows)))
    End If

    ' If only Columns is set, calculate Rows.
    If realRows = 0 Then
        realRows = CInt(Fix(Math.Ceiling(elementCount / realColumns)))
    End If
End Sub
```

The Silverlight layout system starts the layout process by calling the MeasureOverride() method in the UniformGrid. It needs to call the column calculation method (ensuring the number of rows and columns are set) and then divide the available space into equally sized cells.

```
Protected Overrides Function MeasureOverride(ByVal constraint As Size) As Size
    CalculateColumns()

    ' Share out the available space equally.
    Dim childConstraint As New Size(constraint.Width / realColumns, _
      constraint.Height / realRows)
    ...
```

Now the elements inside the UniformGrid need to be measured. However, there's a trick—an element may return a larger value when its Measure() method is called, indicating that it's minimum size is greater than the allocated space. The UniformGrid keeps track of the largest requested width and height values. Finally, when the entire measuring process is finished, the

UniformGrid calculates the size required to make every cell big enough to accommodate the maximum width and height. It then returns that information as its requested size.

```
...
' Keep track of the largest requested dimensions for any element.
Dim largestCell As New Size()

' Examine all the elements in this panel.
For Each child As UIElement In Me.Children
    ' Get the desired size of the child.
    child.Measure(childConstraint)

    ' Record the largest requested dimensions.
    largestCell.Height = Math.Max(largestCell.Height, child.DesiredSize.Height)
    largestCell.Width = Math.Max(largestCell.Width, child.DesiredSize.Width)
Next

' Take the largest requested element width and height, and use
' those to calculate the maximum size of the grid.
Return New Size(largestCell.Width * realColumns, largestCell.Height * realRows)
End Function
```

The ArrangeOverride() code has a similar task. However, it's no longer measuring the children. Instead, it takes note of the final space measurement, calculates the cell size, and positions each child inside the appropriate bounds. If it reaches the end of the grid but there are still extra elements (which only occurs if the control consumer sets limiting values for Columns and Rows), these extra items are given a 0×0 layout box, which hides them.

```
Protected Overrides Function ArrangeOverride(ByVal arrangeSize As Size) As Size
    ' Calculate the size of each cell.
    Dim cellWidth As Double = arrangeSize.Width / realColumns
    Dim cellHeight As Double = arrangeSize.Height / realRows

    ' Determine the placement for each child.
    Dim childBounds As New Rect(0, 0, cellWidth, cellHeight)

    ' Examine all the elements in this panel.
    For Each child As UIElement In Me.Children
        ' Position the child.
        child.Arrange(childBounds)

        ' Move the bounds to the next position.
        childBounds.X += cellWidth
        If childBounds.X >= cellWidth * realColumns Then
            ' Move to the next row.
            childBounds.Y += cellHeight
            childBounds.X = 0

            ' If there are more elements than cells,
            ' hide extra elements.
            If childBounds.Y >= cellHeight * realRows Then
                childBounds = New Rect(0, 0, 0, 0)
            End If
        End If
```

```
        End If
    Next

    ' Return the size this panel actually occupies.
    Return arrangeSize
End Function
```

Using the UniformGrid is easy. You simply need to map the namespace in your XAML markup and then define the UniformGrid in the same way you define any other layout container. Here's an example that places the UniformGrid in a StackPanel with some text content. This allows you to verify that the size of the UniformGrid is correctly calculated and make sure that the content that follows it is bumped out of the way:

```
<UserControl x:Class="Layout.UniformGridTest"
 xmlns="http://schemas.microsoft.com/winfx/2006/xaml/presentation"
 xmlns:x="http://schemas.microsoft.com/winfx/2006/xaml"
 xmlns:local="clr-namespace:Layout" >
  <StackPanel Background="White">
    <TextBlock Margin="5" Text="Content above the WrapPanel."></TextBlock>

    <local:UniformGrid Margin="5" Background="LawnGreen">
      <Button Height="20" Content="Short Button"></Button>
      <Button Width="150" Content="Wide Button"></Button>
      <Button Width="80" Height="40" Content="Fixed Button"></Button>
      <TextBlock Margin="5" Text="Text in the UniformGrid cell goes here"
        TextWrapping="Wrap" Width="100"></TextBlock>
      <Button Width="80" Height="20" Content="Short Button"></Button>
      <TextBlock Margin="5" Text="More text goes in here"
        VerticalAlignment="Center"></TextBlock>
      <Button Content="Unsized Button"></Button>
      <Button Content="Unsized Button"></Button>
    </local:UniformGrid>
    <TextBlock Margin="5" Text="Content below the WrapPanel."></TextBlock>
  </StackPanel>
</UserControl>
```

Figure 3-17 shows how this markup is displayed. By examining the different sizing characteristics of the children inside the UniformGrid, you can set how its layout works in practice. For example, the first button (named Short Button) has a hard-coded Height property. As a result, its height is limited, but it automatically takes the full width of the cell. The second button (Wide Button) has a hard-coded Width property. However, it's the widest element in the UniformGrid, which means its width determines the cell width for the entire table. As a result, its dimensions match the unsized buttons exactly—both fill all the available cell space. Similarly, it's the three lines of wrapped text in the TextBlock that requires the most vertical headroom and so determines the height of all the cells in the grid.

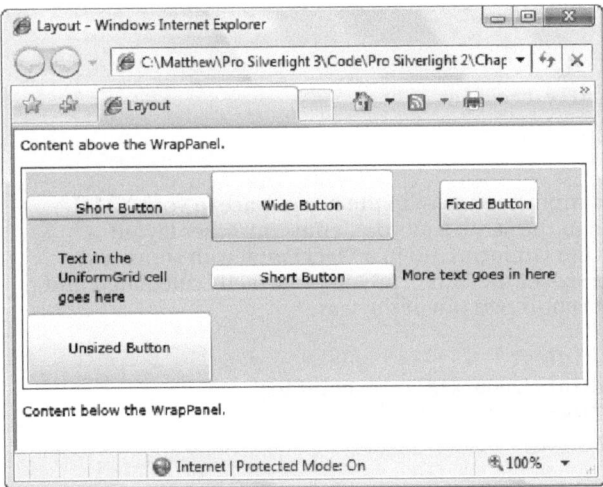

Figure 3-17. *The UniformGrid*

■ **Note** To take a look at a more ambitious (and more mathematically complex) custom layout container, check out the radial panel at http://tinyurl.com/cwk6nz, which arranges elements around the edge of an invisible circle.

Sizing Pages

So far, you've taken an extensive look at the different layout containers Silverlight offers and how you can use them to arrange groups of elements. However, there's one important part of the equation that you haven't considered yet—the top-level page that holds your entire user interface.

As you've already seen, the top-level container for each Silverlight page is a custom class that derives from UserControl. The UserControl class adds a single property, named Content, to Silverlight's basic element infrastructure. The Content property accepts a single element, which becomes the content of that user control.

User controls don't include any special functionality—they're simply a convenient way to group together a block of related elements. However, the way you size your user control can affect the appearance of your entire user interface, so it's worth taking a closer look.

You've already seen how you can use different layout containers with a variety of layout properties to control whether your elements size to fit their content, the available space, or hard-coded dimensions. Many of the same options are available when you're sizing a page, including the following:

- *Fixed size*: Set the Width and Height properties of the user control to give your page an exact size. If you have controls inside the page that exceed these dimensions, they will be truncated. When using a fixed-size window, it's common to change the HorizontalAlignment and VerticalAlignment properties of the user control to Center, so it floats in the center of the browser window rather than being locked into the top-left corner.

- *Browser size*: If you don't use the Width and Height properties of your user control, your application will take the full space allocated to it in the Silverlight content region. (And by default, the HTML entry page that Visual Studio creates sizes the Silverlight content region to take 100% of the browser window.) If you use this approach, it's still possible to create elements that stretch off the bounds of the display region, but the user can now observe the problem and resize the browser window to see the missing content. If you want to preserve some blank space between your page and the browser window when using this approach, you can set the user control's Margin property.

- *Constrained size*: Instead of using the Width and Height properties, use the MaxWidth, MaxHeight, MinWidth, and MinHeight properties. Now, the user control will resize itself to fit the browser windows within a sensible range, and it will stop resizing when the window reaches very large or very small dimensions, ensuring it's never scrambled beyond recognition.

- *Unlimited size*: In some cases, it makes sense to let your Silverlight content region take more than the full browser window. In this situation, the browser will add scroll bars, much as it does with a long HTML page. To get this effect, you need to remove the Width and Height properties and edit the entry page (TestPage.html). In the entry page, remove the width="100%" and height="100%" attributes in the <object> element. This way, the Silverlight content region will be allowed to grow to fit the size of your user control.

■ **Note** Remember, design tools such as Visual Studio and Expression Blend add the DesignWidth and DesignHeight attributes to your user control. These attributes affect the rendering of your page only at design time (where they act like the Width and Height properties). At runtime, they are ignored. Their primary purpose is to allow you to create user interfaces that follow the browser-size model, while still giving you a realistic preview of your application at design time.

All of these approaches are reasonable choices. It simply depends on the type of user interface that you're building. When you use a non-fixed-size page, your application can take advantage of the extra space in the browser window by reflowing its layout to fit. The disadvantage is that extremely large or small windows may make your content more difficult to read or use. You can design for these issues, but it takes more work. On the other hand, the disadvantage of hard-coded sizes it that your application will be forever locked in a specific window size no matter what the browser window looks like. This can lead to oceans of empty space (if you've hard-coded a size that's smaller than the browser window) or make the application unusable (if you've hard-coded a size that's bigger than the browser window).

As a general rule of thumb, resizable pages are more flexible and preferred where possible. They're usually the best choice for business applications and applications with a more traditional user interface that isn't too heavy on the graphics. On the other hand, graphically rich applications and games often need more precise control over what's taking place in the page and are more likely to use fixed page sizes.

■ **Tip** If you're testing different approaches, it helps to make the bounds of the page more obvious. One easy way to do so is to apply a nonwhite background to the top-level content element (for example, setting the Background property of a Grid to Yellow). You can't set the Background property on the user control itself, because the UserControl class doesn't provide it. Another option is to use a Border element as your top-level element, which allows you to outline the page region.

There are also a few more specialized sizing options that you'll learn about in the following sections: scrollable interfaces, scalable interfaces, and full-screen interfaces.

Scrolling with the ScrollViewer

None of the containers you've seen have provided support for *scrolling*, which is a key feature for fitting large amounts of content in a limited amount of space. In Silverlight, scrolling support is easy to get, but it requires another ingredient—the ScrollViewer content control.

To get scrolling support, you need to wrap the content you want to scroll inside a ScrollViewer. Although the ScrollViewer can hold anything, you'll typically use it to wrap a layout container. For example, here's a two-column grid of text boxes and buttons that's made scrollable. The page is sized to the full browser area, but it adds a margin to help distinguish the scroll bar from the browser window that surrounds it. The following listing shows the basic structure of this example, with the markup that creates the first row of elements:

```xml
<UserControl x:Class="Layout.Scrolling"
 xmlns="http://schemas.microsoft.com/winfx/2006/xaml/presentation"
 xmlns:x="http://schemas.microsoft.com/winfx/2006/xaml"
 Margin="20">
  <ScrollViewer Background="AliceBlue">
    <Grid Margin="3,3,10,3">
      <Grid.RowDefinitions>
        <RowDefinition Height="Auto"></RowDefinition>
        ...
      </Grid.RowDefinitions>
      <Grid.ColumnDefinitions>
        <ColumnDefinition Width="*"></ColumnDefinition>
        <ColumnDefinition Width="Auto"></ColumnDefinition>
      </Grid.ColumnDefinitions>

      <TextBox Grid.Row="0" Grid.Column="0" Margin="3"
       Height="Auto" VerticalAlignment="Center"></TextBox>
      <Button Grid.Row="0" Grid.Column="1" Margin="3" Padding="2"
       Content="Browse"></Button>
      ...
    </Grid>
  </ScrollViewer>
</UserControl>
```

Figure 3-18 shows the result.

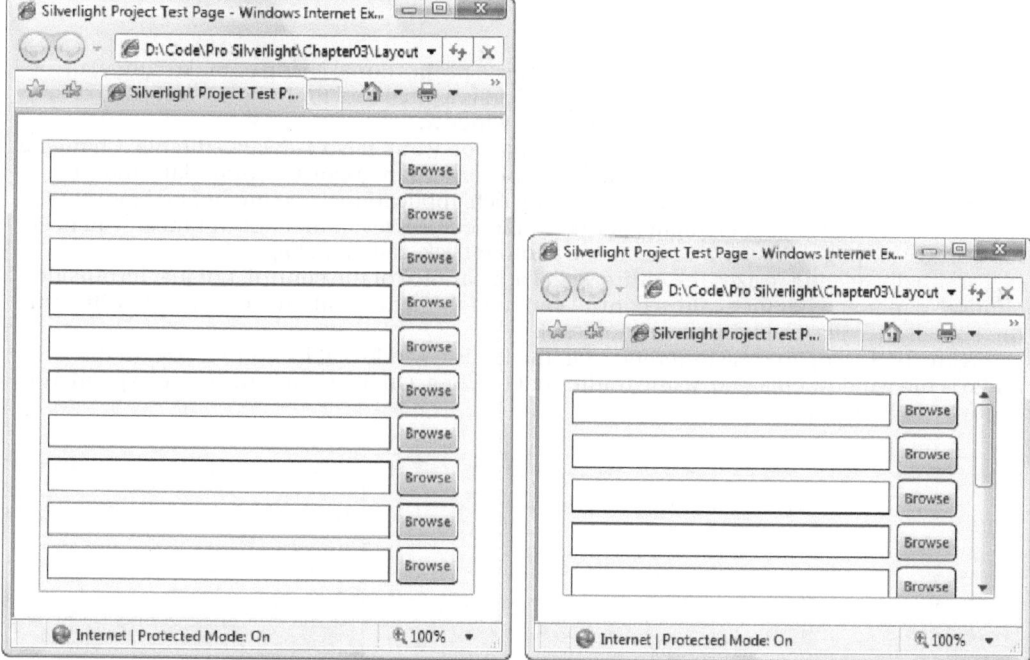

Figure 3-18. A scrollable page

If you resize the page in this example so that it's large enough to fit all its content, the scroll bar becomes disabled. However, the scroll bar will still be visible. You can control this behavior by setting the VerticalScrollBarVisibility property, which takes a value from the ScrollBarVisibility enumeration. The default value of Visible makes sure the vertical scroll bar is always present. Use Auto if you want the scroll bar to appear when it's needed and disappear when it's not. Or use Disabled if you don't want the scroll bar to appear at all.

■ **Note** You can also use Hidden, which is similar to Disabled but subtly different. First, content with a hidden scroll bar is still scrollable. (For example, you can scroll through the content using the arrow keys.) Second, the content in a ScrollViewer is laid out differently. When you use Disabled, you tell the content in the ScrollViewer that it has only as much space as the ScrollViewer itself. On the other hand, if you use Hidden, you tell the content that it has an infinite amount of space. That means it can overflow and stretch off into the scrollable region.

The ScrollViewer also supports horizontal scrolling. However, the HorizontalScrollBarVisibility property is Hidden by default. To use horizontal scrolling, you need to change this value to Visible or Auto.

Scaling with the Viewbox

Earlier in this chapter, you saw how the Grid can use proportional sizing to make sure your elements take all the available space. Thus, the Grid is a great tool for building resizable interfaces that grow and shrink to fit the browser window.

Although this resizing behavior is usually what you want, it isn't always suitable. Changing the dimensions of controls changes the amount of content they can accommodate and can have subtle layout-shifting effects. In graphically rich applications, you might need more precise control to keep your elements perfectly aligned. However, that doesn't mean you need to use fixed-size pages. Instead, you can use another trick, called *scaling*.

Essentially, scaling resizes the entire visual appearance of the control, not just its outside bounds. No matter what the scale, a control can hold the same content—it just looks different. Conceptually, it's like changing the zoom level.

Figure 3-19 compares the difference. On the left is a window at its normal size. In the middle is the window enlarged, using traditional resizing. On the right is the same expanded window using scaling.

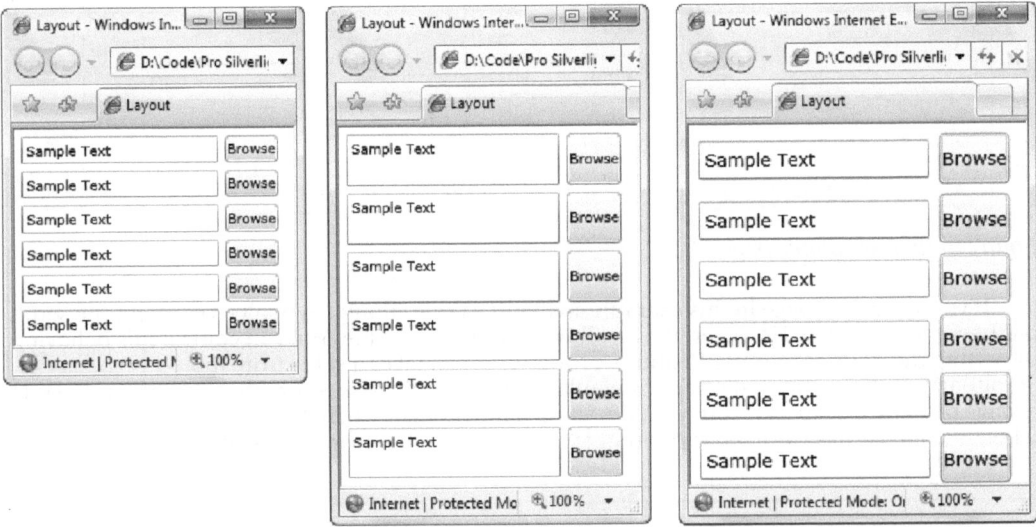

Figure 3-19. Comparing an original (left), resized (middle), and rescaled (right) page

To use scaling, you need to use a *transform*. As you'll discover in Chapter 9, transforms are a key part of Silverlight's flexible 2-D drawing framework. They allow you to rescale, skew, rotate, and otherwise change the appearance of any element. In this example, you need the help of a ScaleTransform to change the scale of your page.

You can use the ScaleTransform in two ways. The first option is a do-it-yourself approach. You respond to the UserControl.SizeChanged event, examine the current size of the page, carry out the appropriate calculations, and create the ScaleTransform by hand. Although this works, there's a far less painful alternative. You can use the Viewbox control, which performs exactly the same task but doesn't require a line of code.

Before you can write the rescaling code that you need, you need to make sure your markup is configured correctly. Here are the requirements you must meet:

- Your user control can't be explicitly sized—instead, it needs to be able to grow to fill the browser window.

- To rescale a window to the right dimensions, you need to know its ideal size, that is, the dimensions that exactly fit all of its content. Although these dimensions won't be set in your markup, they'll be used for the scaling calculations in your code.

As long as these details are in place, it's fairly easy to create a scalable page. The following markup uses a Grid that has an ideal size of 200×225 pixels and contains the stack of text boxes and buttons shown in Figure 3-19:

```xml
<UserControl x:Class="Layout.Page"
 xmlns="http://schemas.microsoft.com/winfx/2006/xaml/presentation"
 xmlns:x="http://schemas.microsoft.com/winfx/2006/xaml"

  <!-- This container is required for rescaling. -->
  <Viewbox>
    <!-- This container is the layout root of your ordinary user interface.
         Note that it uses a hard-coded size. -->
    <Grid Background="White" Width="200" Height="225" Margin="3,3,10,3">
      <Grid.RowDefinitions>
        ...
      </Grid.RowDefinitions>
      <Grid.ColumnDefinitions>
        <ColumnDefinition Width="*"></ColumnDefinition>
        <ColumnDefinition Width="Auto"></ColumnDefinition>
      </Grid.ColumnDefinitions>

      <TextBox Grid.Row="0" Grid.Column="0" Margin="3"
       Height="Auto" VerticalAlignment="Center" Text="Sample Text"></TextBox>
      <Button Grid.Row="0" Grid.Column="1" Margin="3" Padding="2"
       Content="Browse"></Button>
      ...

    </Grid>
  </Viewbox>
</UserControl>
```

In this example, the Viewbox preserves the aspect ratio of the resized content. In other words, it sizes the content to fit the smallest dimension (height or width), rather than stretching it out of proportion to fill all the available space. If you want to use a Viewbox that does stretch its contents without regard for their proportions, simply set the Stretch property to Fill. This isn't terribly useful for page scaling, but it may make sense if you're using the Viewbox for another purpose—say, to size vector graphics in a button.

Finally, it's worth noting that you can create some interesting effects by placing a Viewbox in a ScrollViewer. For example, you can manually set the size of Viewbox to be larger than the available space (using its Height and Width properties) and then scroll around inside the magnified content. You could use this technique to create a zoomable user interface increases the scale as the user drags a slider or turns the mouse wheel. You'll see an example of this technique with the mouse wheel in Chapter 4.

SILVERLIGHT SUPPORT FOR BROWSER ZOOMING

When accessed in some browsers and operating systems—currently, the most recent versions of Firefox and Internet Explorer—Silverlight applications support a feature called *autozoom*. That means the user can change the zoom percentage to shrink or enlarge a Silverlight application. (In Internet Explorer, this can be accomplished using the browser status bar of the View ➤ Zoom menu.) For example, if the user chooses a zoom percentage of 110%, the entire Silverlight application, including its text, images, and controls, will be scaled up 10 percent.

For the most part, this behavior makes sense—and it's exactly what you want. However, if you plan to create an application that provides its own zooming feature, the browser's autozoom might not be appropriate. In this situation, you can disable autozoom simply by adding the enableAutoZoom parameter to the HTML entry page and setting it to false, as shown here:

```
<div id="silverlightControlHost">
  <object data="data:application/x-silverlight-2,"
   type="application/x-silverlight-2" width="100%" height="100%">
    <param name="enableAutoZoom" value="false" />
    ...
  </object>
  <iframe style='visibility:hidden;height:0;width:0;border:0px'></iframe>
</div>
```

Full-Screen Mode

Silverlight applications also have the capability to enter a full-screen mode, which allows them to break out of the browser window altogether. In full-screen mode, the Silverlight plug-in fills the whole display area and is shown overtop of all other applications, including the browser.

Full-screen mode has some serious limitations:

- *You can only switch into full-screen mode when responding to a user input event*: In other words, you can switch into full-screen mode when the user clicks a button or presses a key. However, you can't switch into full-screen mode as soon as your application loads. (If you attempt to do so, your code will simply be ignored.) This limitation is designed to prevent a Silverlight application from fooling a user into thinking it's actually another local application or a system window.

- *While in full-screen mode, keyboard access is limited*: Your code will still respond to the following keys: Tab, Enter, Home, End, Page Up, Page Down, spacebar, and the arrow keys. All other keys are ignored. This means that you can build a simple full-screen arcade game, but you can't use text boxes or other input controls. This limitation is designed to prevent password spoofing—for example, tricking the user into entering a password by mimicking a Windows dialog box.

> ■ **Note** Full-screen mode was primarily designed for showing video content in a large window. In Silverlight 1, full-screen mode does not allow any keyboard input. In later versions, select keys are allowed—just enough to build simple graphical applications (for example, a photo browser) and games. To handle key presses outside of an input control, you simply handle the standard KeyPress event (for example, you can add a KeyPress event handler to your root layout container to capture every key press that takes place). Chapter 4 has more information about keyboard handling.

Here's an event handler that responds to a button press by switching into full-screen mode:

```
Private Sub Button_Click(ByVal sender As Object, ByVal e As RoutedEventArgs)
    Application.Current.Host.Content.IsFullScreen = True
End Sub
```

When your application enters full-screen mode, it displays a message like the one shown in Figure 3-20. This message includes the web domain where the application is situated. If you're using an ASP.NET website and the built-in Visual Studio web server, you'll see the domain http://localhost. If you're hosting your application with an HTML test page that's stored on your hard drive, you'll see the domain file://. The message also informs users that they can exit full-screen mode by pressing the Esc key. Alternatively, you can set the IsFullScreen property to False to exit full-screen mode.

Figure 3-20. The full-screen mode message

For your application to take advantage of full-screen mode, your top-level user control should not have a fixed Height or Width. That way, it can grow to fit the available space. You can also use the scaling technique described in the previous section to scale the elements in your application to larger sizes with a render transform when you enter full-screen mode.

There's one other way to get out of full-screen mode in a Silverlight application: by switching to another application. Ordinarily, this behavior makes perfect sense. But if you have a multiple monitor setup, it might not be what you want. It prevents you from having a full-screen Silverlight application running on one monitor while you work with another application on another monitor.

If you want to prevent this behavior, you can use the following line of code to "pin" a full-screen application so it stays in full-screen mode even when the application loses focus:

```
Application.Current.Host.Content.FullScreenOptions = _
    FullScreenOptions.StaysFullScreenWhenUnfocused
```

You must use this option *before* you switch into full-screen mode. Then, when you set the IsFullScreen property, the user will be prompted to give your application permission to stay pinned (Figure 3-21). The confirmation dialog also includes an option for remembering the user's choice, in which case the message won't be shown the next time the application switches into full-screen mode.

If the user accepts, the window will remain pinned in full-screen mode until the user hits the Esc key or your code sets the IsFullScreen property to False. If the user doesn't accept, the application will switch into full-screen mode with the normal behavior, which means it will lose its full-screen state when the application loses focus.

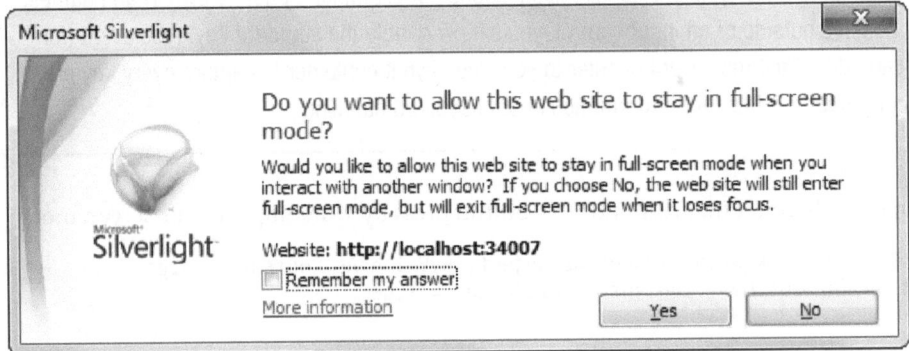

Figure 3-21. *Switching into full-screen mode with a pinned window*

The Last Word

In this chapter, you took a detailed tour of the new Silverlight layout model and learned how to place elements in stacks, grids, and other arrangements. You built more complex layouts using nested combinations of the layout containers, and you threw the GridSplitter into the mix to make resizable split pages. You even considered how to build your own layout containers to get custom effects. Finally, you saw how to take control of the top-level user control that hosts your entire layout by resizing it, rescaling it, and making it fill the entire screen.

CHAPTER 4

■■■

Dependency Properties and Routed Events

At this point, you're probably itching to dive into a realistic, practical example of Silverlight coding. But before you can get started, you need to understand a few more fundamentals. In this chapter, you'll get a whirlwind tour of two key Silverlight concepts: *dependency properties* and *routed events*.

Both of these concepts first appeared in Silverlight's big brother technology, WPF. They came as quite a surprise to most developers—after all, few expected a user interface technology to retool core parts of .NET's object abstraction. However, WPF's changes weren't designed to improve .NET but to support key WPF features. The new property model allowed WPF elements to plug into services such as data binding, animation, and styles. The new event model allowed WPF to adopt a layered content model (as described in the next chapter) without horribly complicating the task of responding to user actions such as mouse clicks and key presses.

Silverlight borrows both concepts, albeit in a streamlined form. In this chapter, you'll see how they work.

■ **What's New** Silverlight 4 dependency properties and routed events still work in the same way. However, there are some important improvements to the event model that you'll see in this chapter. First, all elements now raise a MouseRightButtonDown event, which allows you to create custom right-click context menus (see the "Right-Clicks" section). Second, many controls provide built-in scrolling support for the mouse wheel, including the TextBox, Calendar, ScrollViewer, and all list-based controls (see "The MouseWheel" section). Finally, Silverlight now includes a rudimentary command model, which gives you more flexibility about the way you structure your code (see "The Command Model" section).

Dependency Properties

Essentially, a dependency property is a property that can be set directly (for example, by your code) or by one of Silverlight's services (such as data binding, styles, or animation). The key feature of this system is the way that these different property providers are *prioritized*. For example, an animation will take precedence over all other services while it's running. These overlapping factors make for a very flexible system. They also give dependency properties their name—in essence, a dependency property *depends* on multiple property providers, each with its own level of precedence.

Most of the properties that are exposed by Silverlight elements are dependency properties. For example, the Text property of the TextBlock, the Content property of the Button, and the Background property of the Grid—all of which you saw in the simple example in Chapter 1—are all dependency properties. This hints at an important principle of Silverlight dependency properties—they're designed to be consumed in the same way as normal properties. That's because the dependency properties in the Silverlight libraries are always wrapped by ordinary property definitions.

Although dependency features can be read and set in code like normal properties, they're implemented quite differently behind the scenes. The simple reason why is performance. If the designers of Silverlight simply added extra features on top of the .NET property system, they'd need to create a complex, bulky layer for your code to travel through. Ordinary properties could not support all the features of dependency properties without this extra overhead.

■ **Tip** As a general rule, you don't need to know that a property is a dependency property in order to use it. However, some Silverlight features are limited to dependency properties. Furthermore, you'll need to understand dependency properties in order to define them in your own classes.

Defining and Registering a Dependency Property

You'll spend much more time using dependency properties than creating them. However, there are still many reasons that you'll need to create your own dependency properties. Obviously, they're a key ingredient if you're designing a custom Silverlight element. They're also required in some cases if you want to add data binding, animation, or another Silverlight feature to a portion of code that wouldn't otherwise support it.

Creating a dependency property isn't difficult, but the syntax takes a little getting used to. It's thoroughly different from creating an ordinary .NET property.

The first step is to define an object that *represents* your property. This is an instance of the DependencyProperty class (which is found in the System.Windows namespace). The information about your property needs to be available all the time. For that reason, your DependencyProperty object must be defined as a shared field in the associated class.

For example, consider the FrameworkElement class from which all Silverlight elements inherit. FrameworkElement defines a Margin dependency property that all elements share. It's defined like this:

```
Public Class FrameworkElement
    Inherits UIElement

    Public Shared ReadOnly MarginProperty As DependencyProperty

    ...
End Class
```

By convention, the field that defines a dependency property has the name of the ordinary property plus the word *Property* at the end. That way, you can separate the dependency property definition from the name of the actual property. The field is defined with the ReadOnly keyword, which means it can be set only in the shared constructor for the FrameworkElement.

■ **Note** Silverlight does not support WPF's system of property sharing—in other words, defining a dependency property in one class and reusing it in another. However, dependency properties follow the normal rules of inheritance, which means that a dependency property such as Margin that's defined in the FrameworkElement class applies to all Silverlight elements, because all Silverlight elements derive from FrameworkElement.

Defining the DependencyProperty object is just the first step. For it to become usable, you need to register your dependency property with Silverlight. This step needs to be completed before any code uses the property, so it must be performed in a shared constructor for the associated class.

Silverlight ensures that DependencyProperty objects can't be instantiated directly, because the DependencyProperty class has no public constructor. Instead, a DependencyProperty instance can be created only using the shared DependencyProperty.Register() method. Silverlight also ensures that DependencyProperty objects can't be changed after they're created, because all DependencyProperty members are read-only. Instead, their values must be supplied as arguments to the Register() method.

The following code shows an example of how a DependencyProperty can be created. Here, the FrameworkElement class uses a shared constructor to initialize the MarginProperty:

```
Shared Sub New()
    MarginProperty = DependencyProperty.Register("Margin", _
      GetType(Thickness), GetType(FrameworkElement), _
      Nothing)
    ...
End Sub
```

The DependencyProperty.Register() method accepts the following arguments:

- The property name (Margin in this example).
- The data type used by the property (the Thickness structure in this example).
- The type that owns this property (the FrameworkElement class in this example).
- A PropertyMetadata object that provides additional information. Currently, Silverlight uses the PropertyMetadata to store just optional pieces of information: a default value for the property and a callback that will be triggered when the property is changed. If you don't need to use either feature, supply a null value (Nothing), as in this example.

■ **Note** To see a dependency property that uses the PropertyMetadata object to set a default value, refer to the WrapBreakPanel example later in this chapter.

With these details in place, you're able to register a new dependency property so that it's available for use. However, whereas typical property procedures retrieve or set the value of a private field, the property procedures for a Silverlight property use the GetValue() and SetValue() methods that are defined in the base DependencyObject class. Here's an example:

```
Public Property Margin() As Thickness
    Get
        Return CType(GetValue(MarginProperty), Thickness)
    End Get
    Set(ByVal value As Thickness)
        SetValue(MarginProperty, value)
    End Set
End Property
```

When you create the property wrapper, you should include nothing more than a call to SetValue() and a call to GetValue(), as in the previous example. You should *not* add any extra code to validate values, raise events, and so on. That's because other features in Silverlight may bypass the property wrapper and call SetValue() and GetValue() directly. One example is when the Silverlight parser reads your XAML markup and uses it to initialize your user interface.

You now have a fully functioning dependency property, which you can set just like any other .NET property using the property wrapper:

```
myElement.Margin = New Thickness(5)
```

There's one extra detail. Dependency properties follow strict rules of precedence to determine their current value. Even if you don't set a dependency property directly, it may already have a value—perhaps one that's applied by a binding or a style or one that's inherited through the element tree. (You'll learn more about these rules of precedence in the next section.) However, as soon as you set the value directly, it overrides these other influences.

At some point later, you may want to remove your local value setting and let the property value be determined as though you never set it. Obviously, you can't accomplish this by setting a new value. Instead, you need to use another method that's inherited from DependencyObject: the ClearValue() method. Here's how it works:

```
myElement.ClearValue(FrameworkElement.MarginProperty)
```

This method tells Silverlight to treat the value as though you never set it, thereby returning it to its previous value. Usually, this will be the default value that's set for the property, but it could also be the value that's set through property inheritance or by a style, as described in the next section.

Dynamic Value Resolution

As you've already learned, dependency properties depend on multiple different services, called *property providers*. To determine the current value of a property, Silverlight has to decide which one takes precedence. This process is called *dynamic value resolution*.

When evaluating a property, Silverlight considers the following factors, arranged from highest to lowest precedence:

1. **Animations.** If an animation is currently running and that animation is changing the property value, Silverlight uses the animated value.

2. **Local value.** If you've explicitly set a value in XAML or in code, Silverlight uses the local value. Remember, you can set a value using the SetValue() method or the property wrapper. If you set a property using a resource (Chapter 2) or data binding (Chapter 16), it's considered to be a locally set value.

3. **Styles.** Silverlight styles (Chapter 12) allow you to configure multiple controls with one rule. If you've set a style that applies to this control, it comes into play now.

4. **Property value inheritance.** Silverlight uses *property value inheritance* with a small set of control properties, including Foreground, FontFamily, FontSize, FontStretch, FontStyle, and FontWeight. That means if you set these properties in a higher-level container (such as a Button or a ContentControl), they cascade down to the contained content elements (like the TextBlock that actually holds the text inside).

■ **Note** The limitation with property value inheritance is that the container must provide the property you want to use. For example, you might want to specify a standard font for an entire page by setting the FontFamily property on the root Grid. However, this won't work because the Grid doesn't derive from Control, and so it doesn't provide the FontFamily property. One solution is to wrap your elements in a ContentControl, which includes all the properties that use property value inheritance but has no built-in visual appearance.

---

5. **Default value.** If no other property setter is at work, the dependency property gets its default value. The default value is set with the PropertyMetadata object when the dependency property is first created, as explained in the previous section.

One of the advantages of this system is that it's very economical. For example, if the value of a property has not been set locally, Silverlight will retrieve its value from the template or a style. In this case, no additional memory is required to store the value. Another advantage is that different property providers may override one another, but they don't *overwrite* each other. For example, if you set a local value and then trigger an animation, the animation temporarily takes control. However, your local value is retained, and when the animation ends, it comes back into effect.

Attached Properties

Chapter 2 introduced a special type of dependency property called an *attached property*. An attached property is a full-fledged dependency property, and like all dependency properties, it's managed by the Silverlight property system. The difference is that an attached property applies to a class other than the one where it's defined.

The most common example of attached properties is found in the layout containers you saw in Chapter 3. For example, the Grid class defines the attached properties Row and Column, which you set on the contained elements to indicate where they should be positioned. Similarly, the Canvas defines the attached properties Left and Top that let you place elements using absolute coordinates.

To define an attached property, you use the DependencyProperty.RegisterAttached() method instead of Register(). Here's the code from the Grid class that registers the attached Grid.Row property:

```
RowProperty = DependencyProperty.RegisterAttached( _
  "Row", GetType(Integer), GetType(Grid), Nothing)
```

The parameters are exactly the same for the RegisterAttached() method as they are for the Register() method.

When creating an attached property, you don't define the .NET property wrapper. That's because attached properties can be set on *any* dependency object. For example, the Grid.Row property may be set on a Grid object (if you have one Grid nested inside another) or on some other element. In fact, the Grid.Row property can be set on an element even if that element isn't in a Grid—and even if there isn't a single Grid object in your element tree.

Instead of using a .NET property wrapper, attached properties require a pair of shared methods that can be called to set and get the property value. These methods use the familiar SetValue() and GetValue() methods (inherited from the DependencyObject class). The shared methods should be named Set*PropertyName*() and Get*PropertyName*().

The SetPropertyName() method takes two arguments: the element on which you want to set the property and the property value. Because the Grid.Row property is defined as an integer, the second parameter of the SetRow() method must be an integer:

```
Public Shared Sub SetRow(ByVal element As UIElement, ByVal value As Integer)
    element.SetValue(Grid.RowProperty, value)
End Sub
```

The GetPropertyName() method takes the element on which the property is set and returns the property value. Because the Grid.Row property is defined as an integer, the GetRow() method must return an integer:

```
Public Shared Function GetRow(ByVal element As UIElement) As Integer
    Return CInt(element.GetValue(Grid.RowProperty))
End Function
```

And here's an example that positions an element in the first row of a Grid using code:

```
Grid.SetRow(txtElement, 0)
```

This sets the Grid.Row property to 0 on the txtElement object, which is a TextBox. Because Grid.Row is an attached property, Silverlight allows you to apply it to any other element.

The WrapBreakPanel Example

Now that you understand the theory behind dependency properties, it's time to ground your knowledge in a realistic example.

In Chapter 3, you learned how to create custom panels that use different layout logic to get exactly the effect you want. For example, you took a look at a custom UniformGrid panel that organizes elements into an invisible grid of identically sized cells. The following example considers part of a different custom layout panel, which is called the WrapBreakPanel. Here is its class declaration:

```
Public Class WrapBreakPanel
    Inherits System.Windows.Controls.Panel
    ...
End Class
```

Ordinarily, the WrapBreakPanel behaves like the WrapPanel (although it doesn't inherit directly from WrapPanel, and its layout logic is written from scratch). Like the WrapPanel, the WrapBreakPanel lays out its children one after the other, moving to the next line once the width in the current line is used up. However, the WrapBreakPanel adds a new feature that the

WrapPanel doesn't offer—it allows you to force an immediate line break wherever you want, simply by using an attached property.

■ **Note** The full code for the WrapBreakPanel is available with the downloadable samples for this chapter. The only detail considered here is the properties that customize how it works.

Because the WrapBreakPanel is a Silverlight element, its properties should almost always be dependency properties so you have the flexibility to use them with other Silverlight features such as data binding and animation. For example, it makes sense to give the WrapBreakPanel an Orientation property like its relative, the basic WrapPanel. That way, you could support displays that need to flow elements into multiple columns. Here's the code you need to add to the WrapBreakPanel class to define an Orientation property that uses the data type System.Windows.Controls.Orientation:

```
Public Shared ReadOnly OrientationProperty As DependencyProperty = _
    DependencyProperty.Register("Orientation", GetType(Orientation), _
    GetType(WrapBreakPanel), New PropertyMetadata(Orientation.Horizontal))
```

This code uses one minor time-saver. Rather than define the DependencyProperty and register it with code in a shared constructor, this definition takes care of the definition and registration (and the compiled code doesn't change). It also sets the default value to Orientation.Horizontal.

Next, you need to add the property wrapper, which is perfectly straightforward:

```
Public Property Orientation() As Orientation
    Get
        Return CType(GetValue(OrientationProperty), Orientation)
    End Get
    Set(ByVal value As Orientation)
        SetValue(OrientationProperty, value)
    End Set
End Property
```

When using the WrapBreakPanel in a Silverlight page, you can set the Orientation property as you set any other property:

```
<local:WrapBreakPanel Margin="5" Orientation="Vertical">
    ...
</local:WrapBreakPanel>
```

A more interesting experiment is to create a version of the WrapBreakPanel that uses an attached property. As you've already learned, attached properties are particularly useful in layout containers, because they allow children to pass along extra layout information (such as row positioning in the Grid or coordinates and layering in the Canvas).

The WrapBreakPanel includes an attached property that allows any child element to force a line break. By using this attached property, you can ensure that a specific element begins on a new line, no matter what the current width of the WrapBreakPanel. The attached property is named LineBreakBefore, and the WrapBreakPanel defines it like this:

```
Public Shared LineBreakBeforeProperty As DependencyProperty = _
  DependencyProperty.RegisterAttached("LineBreakBefore", GetType(Boolean), _
  GetType(WrapBreakPanel), Nothing)
```

To implement the LineBreakBefore property, you need to create the shared get and set methods that call GetValue() and SetValue() on the element:

```
Public Shared Function GetLineBreakBefore(ByVal element As UIElement) As Boolean
    Return CBool(element.GetValue(LineBreakBeforeProperty))
End Function

Public Shared Sub SetLineBreakBefore(ByVal element As UIElement, _
  ByVal value As Boolean)
    element.SetValue(LineBreakBeforeProperty, value)
End Sub
```

You can then modify the MeasureOverride() and ArrangeOverride() methods to check for forced breaks, as shown here:

```
' Check if the element fits in the line, or if a line break was requested.
If (currentLineSize.Width + desiredSize.Width > constraint.Width) _
  OrElse WrapBreakPanel.GetLineBreakBefore(element) Then
  ...
End If
```

To use this functionality, you simply need to add the LineBreakBefore property to an element, as shown here:

```
<local:WrapBreakPanel Margin="5" Background="LawnGreen">
  <Button Width="50" Content="Button"></Button>
  <Button Width="150" Content="Wide Button"></Button>
  <Button Width="50" Content="Button"></Button>
  <Button Width="150" Content="Button with a Break"
    local:WrapBreakPanel.LineBreakBefore="True" FontWeight="Bold"></Button>
  <Button Width="150" Content="Wide Button"></Button>
  <Button Width="50" Content="Button"></Button>
</local:WrapBreakPanel>
```

Figure 4-1 shows the result.

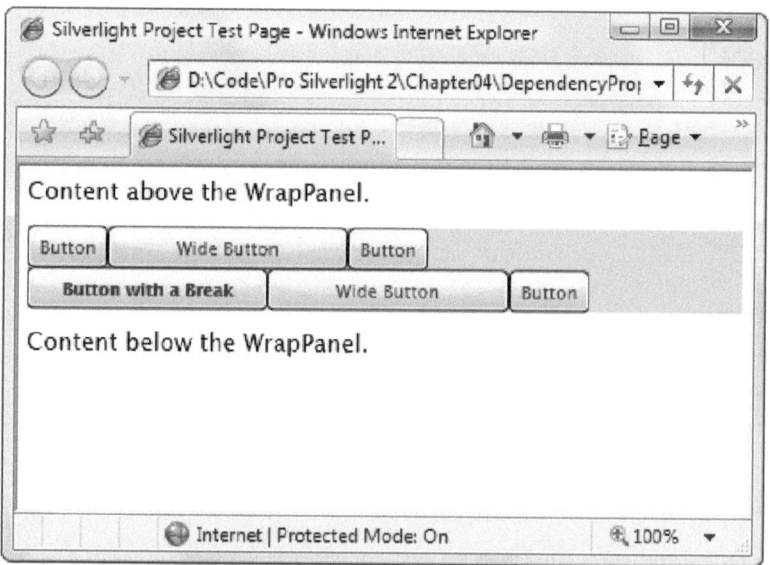

Figure 4-1. *A WrapBreakPanel that supports forced line breaks*

Routed Events

Every .NET developer is familiar with the idea of *events*—messages that are sent by an object (such as a Silverlight element) to notify your code when something significant occurs. WPF enhanced the .NET event model with a new concept of *event routing*, which allows an event to originate in one element but be raised by another one. For example, event routing allows a click that begins in a shape to rise up to that shape's container and then to the containing page before it's handled by your code.

Silverlight borrows some of WPF's routed event model but in a dramatically simplified form. While WPF supports several types of routed events, Silverlight allows only one: *bubbled events* that rise up the containment hierarchy from deeply nested elements to their containers. Furthermore, Silverlight's event bubbling is linked to a few keyboard and mouse input events (such as MouseMove and KeyDown), and it's supported by just a few low-level elements. As you'll see, Silverlight doesn't use event bubbling for higher-level control events (such as Click), and you can't use event routing with the events in your own custom controls.

The Core Element Events

Elements inherit their basic set of events from two core classes: UIElement and FrameworkElement. As Figure 4-2 shows, all Silverlight elements derive from these elements.

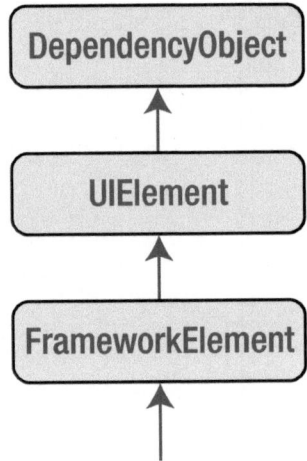

Silverlight element classes

Figure 4-2. *The hierarchy of Silverlight elements*

The UIElement class defines the most important events for handling user input and the only events that use event bubbling. Table 4-1 lists all the UIElement events. You'll see how to use these events through the rest of this chapter.

Table 4-1. *The UIElement Events*

Event	Bubbles	Description
KeyDown	Yes	Occurs when a key is pressed.
KeyUp	Yes	Occurs when a key is released.
TextInput	Yes	Occurs when the element receives a character. Usually, this is a typed-in character, and TextInput fires after KeyDown and KeyUp. However, TextInput will also fire if a character is entered through another device, like a touch pad.
GotFocus	Yes	Occurs when the focus changes to this element (when the user clicks it or tabs to it). The element that has focus is the control that will receive keyboard events first.
LostFocus	Yes	Occurs when the focus leaves this element.
MouseLeftButton Down	Yes	Occurs when the left mouse button is pressed while the mouse pointer is positioned over the element.
MouseLeftButtonUp	Yes	Occurs when the left mouse button is released.

Event	Bubbles	Description
MouseRightButton Down	Yes	Occurs when the right mouse button is pressed while the mouse pointer is positioned over the element. If you don't want to show the standard Silverlight system menu, you must set the MouseButtonEventArgs.Handled property to True in your event handler.
MouseRightButton Up	Yes	Occurs when the right mouse button is released.
MouseEnter	No	Occurs when the mouse pointer first moves onto the element. This event doesn't bubble, but if you have several nested elements, they'll all fire MouseEnter events as you move to the most deeply nested element, passing over the bounding line that delineates the others.
MouseLeave	No	Occurs when the mouse pointer moves off the element. This event doesn't bubble, but if you have several nested elements, they'll all fire MouseEnter events as you move the mouse away (in the reverse order that the MouseEnter events occurred).
MouseMove	Yes	Occurs when the mouse moves while over the element. The MouseMove event is fired frequently—for example, if the user slowly moves the mouse pointer across the face of a button, you'll quickly receive hundreds of MouseMove events. For that reason, you shouldn't perform time-consuming tasks when reacting to this event.
MouseWheel	Yes	Occurs when the user turns the mouse wheel while over the element (or while that element has focus).
DragEnter	Yes	Occurs when the user first drags a selected file (from the computer) over the element.
DragLeave	Yes	Occurs when the user drags a selected file off of the element.
DragOver	Yes	Occurs (repeatedly) as the user moves the mouse over the element, while dragging a selected file.
Drop	Yes	Occurs when the user drops a selected file onto the element. Because the DragEnter, DragLeave, DragOver, and Drop events support only dragged files (not other objects), they're discussed in Chapter 18.
LostMouseCapture	No	Occurs when an element loses its mouse capture. *Mouse capturing* is a technique that an element can use to receive mouse events even when the mouse pointer moves away, off its surface.

In some cases, higher-level events may effectively replace some of the UIElement events. For example, the Button class provides a Click event that's triggered when the user presses and releases the mouse button or when the button has focus and the user presses the space bar. Thus, when handling button clicks, you should always respond to the Click event, not MouseLeftButtonDown or MouseLeftButtonUp (which it suppresses). Similarly, the TextBox provides a TextChanged event, which fires when the text is changed by any mechanism, in addition to the basic KeyDown and KeyUp events.

The FrameworkElement class adds just a few more events to this model, as detailed in Table 4-2. None of these events uses event bubbling.

***Table 4-2.** The FrameworkElement Events*

Event	Description
Loaded	Occurs after an element has been created and added to the object tree (the hierarchy of elements in the window). After this point, you may want to perform additional customization to the element in code.
SizeChanged	Occurs after the size of an element changes. As you saw in Chapter 3, you can react to this event to implement scaling.
LayoutUpdated	Occurs after the layout inside an element changes. For example, if you create a page that uses no fixed size (and so fits the browser window) and you resize the browser window, the controls will be rearranged to fit the new dimensions, and the LayoutUpdated event will fire for your top-level layout container.
BindingValidation Error	Occurs if a bound data object throws an exception when the user attempts to change a property. You'll learn how to use the BindingValidationError event to implement validation in Chapter 16.

Event Bubbling

Bubbling events are events that travel *up* the containment hierarchy. For example, MouseLeftButtonDown is a bubbling event. It's raised first by the element that is clicked. Next, it's raised by that element's parent and then by *that* element's parent, and so on, until Silverlight reaches the top of the element tree.

Event bubbling is designed to support composition—in other words, to let you build more complex controls out of simpler ingredients. One example is Silverlight's *content controls*, which are controls that have the ability to hold a single nested element as content. These controls are usually identified by the fact that they provide a property named Content. For example, the button is a content control. Rather than displaying a line of text, you can fill it with a StackPanel that contains a whole group of elements, like this:

```
<Button BorderBrush="Black" BorderThickness="1" Click="cmd_Click">
  <StackPanel>
    <TextBlock Margin="3" Text="Image and text label"></TextBlock>
    <Image Source="happyface.jpg" Stretch="None"></Image>
    <TextBlock Margin="3" Text="Courtesy of the StackPanel"></TextBlock>
  </StackPanel>
</Button>
```

Here, the content element is a StackPanel that holds two pieces of text and an image. Figure 4-3 shows the fancy button that this markup creates.

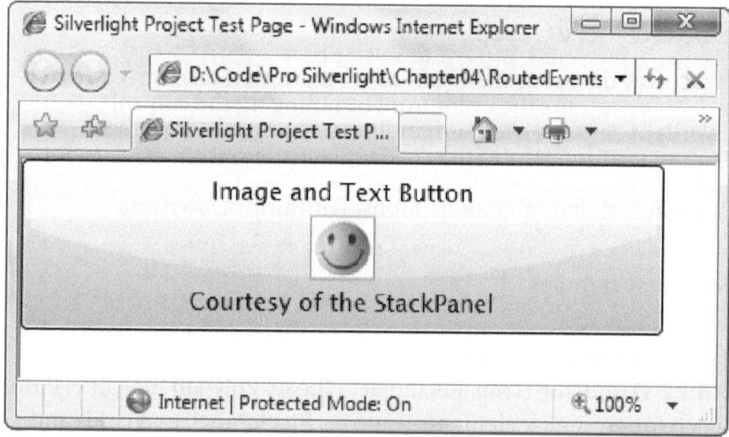

Figure 4-3. *A button with contained elements*

In this situation, it's important that the button reacts to the mouse events of its contained elements. In other words, the Button.Click event should fire when the user clicks the image, some of the text, or part of the blank space inside the button border. In every case, you'd like to respond with the same code.

Of course, you could wire up the same event handler to the MouseLeftButtonDown or MouseLeftButtonUp event of each element inside the button, but that would result in a significant amount of clutter, and it would make your markup more difficult to maintain. Event bubbling provides a better solution.

When the happy face is clicked, the MouseLeftButtonDown event fires first for the Image, then for the StackPanel, and then for the containing button. The button then reacts to the MouseLeftButtonDown by firing its own Click event, to which your code responds (with its cmd_Click event handler).

■ **Note** The Button.Click event does not use event bubbling. This is a dramatic difference from WPF. In the world of Silverlight, only a small set of basic infrastructure events support event bubbling. Higher-level control events cannot use event bubbling. However, the button *uses* the bubbling nature of the MouseLeftButtonDown event to make sure it captures clicks on any contained elements.

Handled (Suppressed) Events

When the button in Figure 4-3 receives the MouseLeftButtonDown event, it takes an extra step and marks the event as *handled*. This prevents the event from bubbling up the control hierarchy any further. Most Silverlight controls use this handling technique to suppress MouseLeftButtonDown and MouseLeftButtonUp so they can replace them with more useful, higher-level events such as Click.

However, there are a few elements that don't handle MouseLeftButtonDown and MouseLeftButtonUp:

- The Image class used to display bitmaps
- The TextBlock class used to show text
- The MediaElement class used to display video
- The shape classes used for 2-D drawing (Line, Rectangle, Ellipse, Polygon, Polyline, Path)
- The layout containers used for arranging elements (Canvas, StackPanel, and Grid) and the Border class

These exceptions allow you to use these elements in content controls such as the Button control without any limitations. For example, if you place a TextBlock in a button, when you click the TextBlock, the MouseLeftButtonUp event will bubble up to the button, which will then fire its Click event. However, if you take a control that isn't in the preceding list and place it inside the button—say, a list box, check box, or another button—you'll get different behavior. When you click that nested element, the MouseLeftButtonUp event won't bubble to the containing button, and the button won't register a click.

■ **Note** MouseLeftButtonDown and MouseLeftButtonUp are the only events that controls suppress. The bubbling key events (KeyUp, KeyDown, LostFocus, and GotFocus) aren't suppressed by any controls.

An Event Bubbling Example

To understand event bubbling and handled events, it helps to create a simple example, like the one shown in Figure 4-4. Here, as in the example you saw previously, the MouseLeftButtonDown event starts in a TextBlock or Image and travels through the element hierarchy.

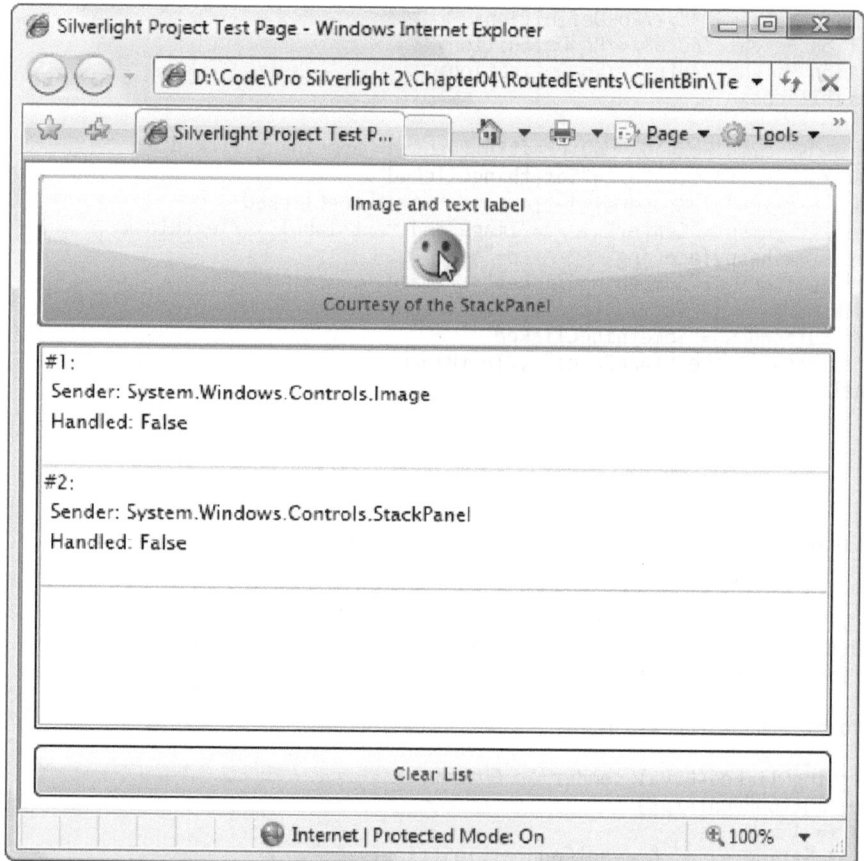

Figure 4-4. *A bubbled image click*

In this example, you can watch the MouseLeftButtonDown event bubble by attaching event handlers to multiple elements. As the event is intercepted at different levels, the event sequence is displayed in a list box. Figure 4-4 shows the display immediately after clicking the happy face image in the button. As you can see, the MouseLeftButtownDown event fires in the image and then in the containing StackPanel and is finally intercepted by the button, which handles it. The button does not fire the MouseLeftButtonDown event, and therefore the MouseLeftButtonDown event does not bubble up to the Grid that holds the button.

To create this test page, the image and every element above it in the element hierarchy are wired up to the same event handler—a method named SomethingClicked(). Here's the XAML that does it:

```
<UserControl x:Class="RoutedEvents.EventBubbling"
 xmlns="http://schemas.microsoft.com/winfx/2006/xaml/presentation"
 xmlns:x="http://schemas.microsoft.com/winfx/2006/xaml">

  <Grid Margin="3" MouseLeftButtonDown="SomethingClicked">
    <Grid.RowDefinitions>
      <RowDefinition Height="Auto"></RowDefinition>
```

```xml
      <RowDefinition Height="*"></RowDefinition>
      <RowDefinition Height="Auto"></RowDefinition>
      <RowDefinition Height="Auto"></RowDefinition>
    </Grid.RowDefinitions>

    <Button Margin="5" Grid.Row="0" MouseLeftButtonDown="SomethingClicked">
      <StackPanel MouseLeftButtonDown="SomethingClicked">
        <TextBlock Margin="3" MouseLeftButtonDown="SomethingClicked"
         HorizontalAlignment="Center" Text="Image and text label"></TextBlock>
        <Image Source="happyface.jpg" Stretch="None"
         MouseLeftButtonDown="SomethingClicked"></Image>
        <TextBlock Margin="3" HorizontalAlignment="Center"
         MouseLeftButtonDown="SomethingClicked"
         Text="Courtesy of the StackPanel"></TextBlock>
      </StackPanel>
    </Button>

    <ListBox Grid.Row="1" Margin="5" x:Name="lstMessages"></ListBox>

    <Button Grid.Row="3" Margin="5" Padding="3" x:Name="cmdClear"
     Click="cmdClear_Click" Content="Clear List"></Button>
  </Grid>
</UserControl>
```

The SomethingClicked() method simply examines the properties of the RoutedEventArgs object and adds a message to the list box:

```vb
Protected eventCounter As Integer = 0

Private Sub SomethingClicked(ByVal sender As Object, _
  ByVal e As MouseButtonEventArgs)
    eventCounter += 1
    Dim message As String = "#" & eventCounter.ToString() & ":" & _
      Environment.NewLine & " Sender: " & sender.ToString() & _
      Environment.NewLine
    lstMessages.Items.Add(message)
End Sub

Private Sub cmdClear_Click(ByVal sender As Object, ByVal e As RoutedEventArgs)
    lstMessages.Items.Clear()
End Sub
```

When dealing with a bubbled event such as MouseLeftButtonDown, the sender parameter that's passed to your event handler always provides a reference to the last link in the chain. For example, if an event bubbles up from an image to a StackPanel before you handle it, the sender parameter references the StackPanel object.

In some cases, you'll want to determine where the event originally took place. The event arguments object for a bubbled event provides a Source property that tells you the specific element that originally raised the event. In the case of a keyboard event, this is the control that had focus when the event occurred (for example, when the key was pressed). In the case of a mouse event, this is the topmost element under the mouse pointer when the event occurred (for example, when a mouse button was clicked). However, the Source property can get a bit more detailed than you want—for example, if you click the blank space that forms the background of a button, the Source property will provide a reference to the Shape or Path object that actually draws the part of background you clicked.

Along with Source, the event arguments object for a bubbled event also provides a Boolean property named Handled, which allows you to suppress the event. For example, if you handle the MouseLeftButtonDown event in the StackPanel and set Handled to True, the StackPanel will not fire the MouseLeftButtonDown event. As a result, when you click the StackPanel (or one of the elements inside), the MouseLeftButtonDown event will not reach the button, and the Click event will never fire. You can use this technique when building custom controls (for example, if you've taken care of a user action like a button click and you don't want higher-level elements to get involved).

■ **Note** WPF provides a back door that allows code to receive events that are marked handled (and would ordinarily be ignored). Silverlight does not provide this capability.

Mouse Handling

You're unlikely to see a Silverlight application that doesn't use the MouseLeftButtonDown event. However, there's still a lot more to learn about handling the mouse events. In the following sections, you'll consider how you can react to right-clicks, mouse movement, and the mouse wheel. You'll also learn how to capture the mouse—so you can continue handling its events even when it moves away—how to simulate drag and drop and how to change the mouse cursor.

Right-Clicks

By default, when you right-click anywhere in a Silverlight application, a pop-up Silverlight menu appears. This menu includes a single command named Silverlight, which opens a tabbed window where you can change Silverlight settings. Additionally, if you've created an application that supports out-of-browser installation (see Chapter 21), this menu has a second command for installing the application.

You may decide to handle right-clicks on some elements (or even the entire window) to provide more specialized functionality. For example, when the user right-clicks a specific element, you may want to show a customized context menu with commands for that element. Although Silverlight doesn't include a context menu control, you can easily get one from the Silverlight Toolkit (http://silverlight.codeplex.com). You can then use it to attach a right-click menu to any control.

However, there's a catch. Even if you show your own context menu, the MouseRightButtonDown event will bubble up to the top level of your application, causing Silverlight to show its standard system menu. To hide this menu so that yours is the only one that appears, you must handle the MouseRightButton event and set the MouseButtonEventArgs.Handled property to True. This suppresses the right-click event and the system menu.

Mouse Movements

Along with the obvious mouse clicking events (MouseLeftButtonDown, MouseLeftButtonUp, MouseRightButtonDown, and MouseRightButtonUp), Silverlight also provides mouse events that fire when the mouse pointer is moved. These events include MouseEnter (which fires when the mouse pointer moves over the element), MouseLeave (which fires when the mouse pointer moves away), and MouseMove (which fires at every point in between).

All of these events provide your code with the same information: a MouseEventArgs object. The MouseEventArgs object includes one important ingredient: a GetPosition() method that tells you the coordinates of the mouse in relation to an element of your choosing. Here's an example that displays the position of the mouse pointer:

```
Private Sub MouseMoved(ByVal sender As Object, ByVal e As MouseEventArgs)
    Dim pt As Point = e.GetPosition(Me)
    lblInfo.Text = String.Format("You are at ({0},{1}) in page coordinates", _
      pt.X, pt.Y)
End Sub
```

In this case, the coordinates are measured from the top-left corner of the page area (just below the title bar of the browser).

■ **Tip** To receive mouse events in a layout container, the Background property must be set to a non-null value—for example, a solid white fill.

The Mouse Wheel

These days, a large proportion of computer users have a mouse with a scroll wheel. You can use that fact to your advantage, by responding to with an appropriate action when the user turns the mouse wheel. The only rule of thumb is to make sure mouse wheel support is a useful extra, not an essential part of your application's behavior. After all, there are still a large proportion of users who don't have mouse wheels (for example, laptop users) or don't think to use them.

The MouseWheel event passes some basic information about the amount the wheel has turned since the last MouseWheel event, using the MouseWheelEventArgs.Delta property. Typically, each notch in the mouse wheel has a value of 120, so a single nudge of the mouse wheel will pass a Delta value of 120 to your application. The Delta value is positive if the mouse wheel was rotated away from the user and negative if it was rotated toward the user.

To get a better grip on this situation, consider the example of the interface shown in Figure 4-5. Here, the user can zoom into or out of a Grid of content just by turning the mouse wheel.

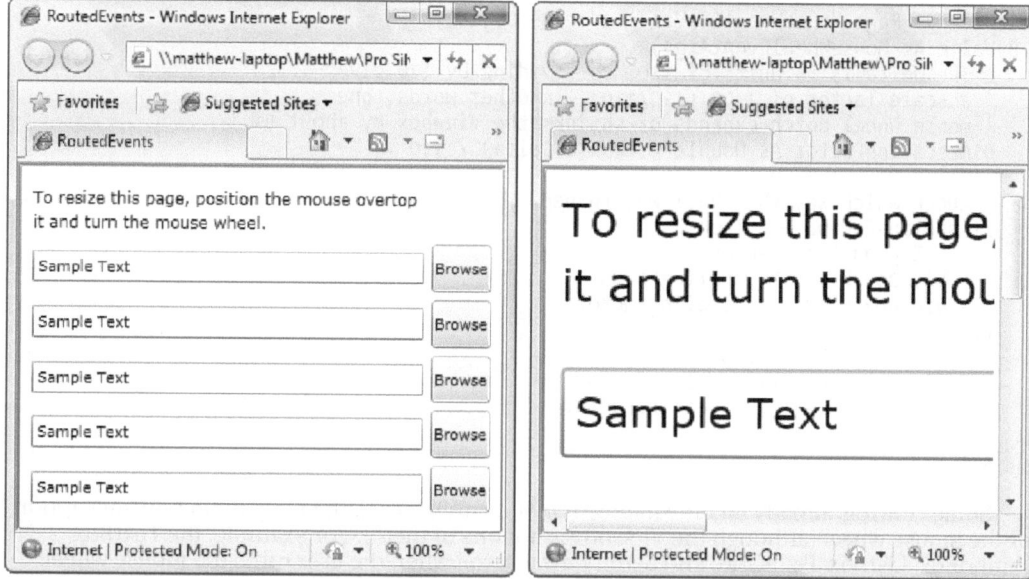

Figure 4-5. *Zooming with the mouse wheel*

To create the example, you need two controls you first considered in Chapter 3—the ScrollViewer and Viewbox. The Viewbox powers the magnification, while the ScrollViewer simply allows the user to scroll over the whole surface of the Viewbox when it's too big to fit in the browser window.

```
<UserControl x:Class="RoutedEvents.MouseWheelZoom"
 xmlns="http://schemas.microsoft.com/winfx/2006/xaml/presentation"
 xmlns:x="http://schemas.microsoft.com/winfx/2006/xaml"
MouseWheel="Page_MouseWheel">

  <ScrollViewer VerticalScrollBarVisibility="Auto"
   HorizontalScrollBarVisibility="Auto">

    <Viewbox x:Name="viewbox" Height="250" Width="350">
      <Grid Background="White" Height="250" Width="350">
        ...
      </Grid>
    </Viewbox>

  </ScrollViewer>
</UserControl>
```

Notice that initially the Viewbox is given exactly the same hard-coded size as the Grid inside. This ensures that the Viewbox doesn't need to perform any initial scaling—instead, the Grid is at its natural size when the application first starts.

When the user turns the mouse wheel, a MouseWheel event handler checks the delta and simply adjusts the Width and Height properties of the Viewbox proportionally. This expands or shrinks the Viewbox and rescales everything inside:

```
Private Sub Page_MouseWheel(ByVal sender As Object, _
   ByVal e As MouseWheelEventArgs)
      ' The Delta is in units of 120, so dividing by 120 gives
      ' a scale factor of 1.09 (120/110). In other words, one
      ' mouse wheel notch expands or shrinks the Viewbox by about 9%.
      Dim scalingFactor As Double = CDbl(e.Delta) / 110

      ' Check which way the wheel was turned.
      If scalingFactor > 0 Then
          ' Expand the viewbox.
          viewbox.Width *= scalingFactor
          viewbox.Height *= scalingFactor
      Else
          ' Shrink the viewbox.
          viewbox.Width /= -scalingFactor
          viewbox.Height /= -scalingFactor
      End If
End Sub
```

Some controls already include handle the MouseWheel event, giving them built-in support for the mouse wheel (although the Viewbox is not one of them). For example, the TextBox, ComboBox, ListBox, DataGrid, and ScrollViewer scroll when the user turns the mouse wheel. The Calendar moves from month to month.

Capturing the Mouse

Ordinarily, every time an element receives a mouse button down event, it will receive a corresponding mouse button up event shortly thereafter. However, this isn't always the case. For example, if you click an element, hold down the mouse, and then move the mouse pointer off the element, the element won't receive the mouse up event.

In some situations, you may want to have a notification of mouse up events, even if they occur after the mouse has moved off your element. To do so, you need to *capture* the mouse by calling the MouseCapture() method of the appropriate element (MouseCapture() is defined by the base UIElement class, so it's supported by all Silverlight elements). From that point on, your element will receive the MouseLeftButtonDown and MouseLeftButtonUp event until it loses the mouse capture. There are two ways to lose the mouse capture. First, you can give it up willingly by calling Mouse.Capture() again and passing in a null reference (Nothing). Second, the user can click outside of your application—on another program, on the browser menu, on HTML content on the same web page. When an element loses mouse capture, it fires the LostMouseCapture event.

While the mouse has been captured by an element, other elements won't receive mouse events. That means the user won't be able to click buttons elsewhere in the page, click inside text boxes, and so on. Mouse capturing is sometimes used to implement draggable and resizable elements.

A Mouse Event Example

You can put all these mouse input concepts together (and learn a bit about dynamic control creation) by reviewing a simple example.

Figure 4-6 shows a Silverlight application that allows you to draw small circles on a Canvas and move them around. Every time you click the Canvas, a red circle appears. To move a circle, you simply click and drag it to a new position. When you click a circle, it changes color from red to green. Finally, when you release your circle, it changes color to orange. There's no limit to how many circles you can add or how many times you can move them around your drawing surface.

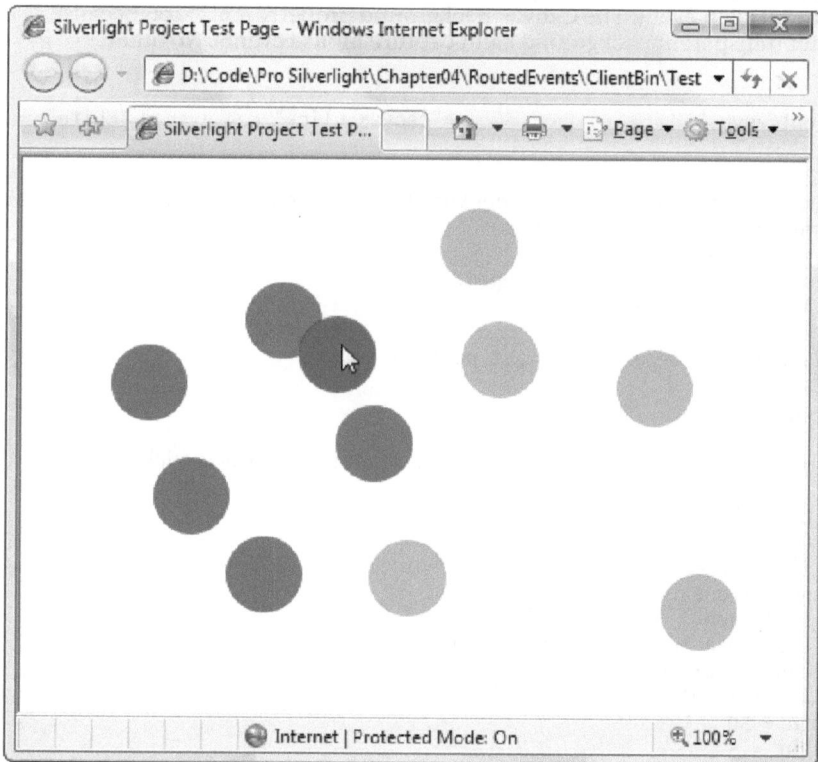

Figure 4-6. Dragging shapes

■ **Note** This example demonstrates "simulated" drag and drop, which is a drag-and-drop feature that you implement yourself, with custom code in your application. By comparison, a "true" drag-and-drop feature relies on functionality that's built into the operating system. Silverlight does include a true drag-and-drop feature, but it works in only a very limited scenario—when dragging files from the computer onto a Silverlight window. Chapter 18 demonstrates the file-based drag-and-drop feature.

Each circle is an instance of the Ellipse element, which is simply a colored shape that's a basic ingredient in 2-D drawing. Obviously, you can't define all the ellipses you need in your XAML markup. Instead, you need a way to generate the Ellipse objects dynamically each time the user clicks the Canvas.

Creating an Ellipse object isn't terribly difficult—after all, you can instantiate it like any other .NET object, set its properties, and attach event handlers. You can even use the SetValue() method to set attached properties to place it in the correct location in the Canvas. However, there's one more detail to take care of—you need a way to place the Ellipse in the Canvas. This is easy enough, because the Canvas class exposes a Children collection that holds all the child elements. Once you've added an element to this collection, it will appear in the Canvas.

The XAML page for this example uses a single event handler for the Canvas.MouseLeftButtonDown event. The Canvas.Background property is also set, because a Canvas with the default transparent background can't capture mouse events. No other elements are defined.

```
<Canvas x:Name="parentCanvas" MouseLeftButtonDown="canvas_Click" Background="White">
</Canvas>
```

In the code-behind class, you need two member variables to keep track of whether an ellipse-dragging operation is currently taking place:

```
' Keep track of when an ellipse is being dragged.
Private isDragging As Boolean = False

' When an ellipse is clicked, record the exact position
' where the click is made.
Private mouseOffset As Point
```

Here's the event-handling code that creates an ellipse when the Canvas is clicked:

```
Private Sub canvas_Click(ByVal sender As Object, _
  ByVal e As MouseButtonEventArgs)
    ' Create an ellipse (unless the user is in the process
    ' of dragging another one).
    If (Not isDragging) Then
        ' Give the ellipse a 50-pixel diameter and a red fill.
        Dim ellipse As New Ellipse()
        ellipse.Fill = New SolidColorBrush(Colors.Red)
        ellipse.Width = 50
        ellipse.Height = 50

        ' Use the current mouse position for the center of
        ' the ellipse.
        Dim point As Point = e.GetPosition(Me)
        ellipse.SetValue(Canvas.TopProperty, point.Y - ellipse.Height/2)
        ellipse.SetValue(Canvas.LeftProperty, point.X - ellipse.Width/2)

        ' Watch for left-button clicks.
        AddHandler ellipse.MouseLeftButtonDown, AddressOf ellipse_MouseDown

        ' Add the ellipse to the Canvas.
        parentCanvas.Children.Add(ellipse)
    End If
End Sub
```

Not only does this code create the ellipse, but it also connects an event handler that responds when the ellipse is clicked. This event handler changes the ellipse color and initiates the ellipse-dragging operation:

```
Private Sub ellipse_MouseDown(ByVal sender As Object, _
  ByVal e As MouseButtonEventArgs)
    ' Dragging mode begins.
```

```
        isDragging = True
        Dim ellipse As Ellipse = CType(sender, Ellipse)

        ' Get the position of the click relative to the ellipse
        ' so the top-left corner of the ellipse is (0,0).
        mouseOffset = e.GetPosition(ellipse)

        ' Change the ellipse color.
        ellipse.Fill = New SolidColorBrush(Colors.Green)

        ' Watch this ellipse for more mouse events.
        AddHandler ellipse.MouseMove, AddressOf ellipse_MouseMove
        AddHandler ellipse.MouseLeftButtonUp, AddressOf ellipse_MouseUp

        ' Capture the mouse. This way you'll keep receiveing
        ' the MouseMove event even if the user jerks the mouse
        ' off the ellipse.
        ellipse.CaptureMouse()
End Sub
```

The ellipse isn't actually moved until the MouseMove event occurs. At this point, the Canvas.Left and Canvas.Top attached properties are set on the ellipse to move it to its new position. The coordinates are set based on the current position of the mouse, taking into account the point where the user initially clicked. This ellipse then moves seamlessly with the mouse, until the left mouse button is released.

```
Private Sub ellipse_MouseMove(ByVal sender As Object, ByVal e As MouseEventArgs)
    If isDragging Then
        Dim ellipse As Ellipse = CType(sender, Ellipse)

        ' Get the position of the ellipse relative to the Canvas.
        Dim point As Point = e.GetPosition(parentCanvas)

        ' Move the ellipse.
        ellipse.SetValue(Canvas.TopProperty, point.Y - mouseOffset.Y)
        ellipse.SetValue(Canvas.LeftProperty, point.X - mouseOffset.X)
    End If
End Sub
```

When the left mouse button is released, the code changes the color of the ellipse, releases the mouse capture, and stops listening for the MouseMove and MouseUp events. The user can click the ellipse again to start the whole process over.

```
Private Sub ellipse_MouseUp(ByVal sender As Object, ByVal e As MouseButtonEventArgs)
    If isDragging Then
        Dim ellipse As Ellipse = CType(sender, Ellipse)

        ' Change the ellipse color.
        ellipse.Fill = New SolidColorBrush(Colors.Orange)

        ' Don't watch the mouse events any longer.
        RemoveHandler ellipse.MouseMove, AddressOf ellipse_MouseMove
        RemoveHandler ellipse.MouseLeftButtonUp, AddressOf ellipse_MouseUp
        ellipse.ReleaseMouseCapture()
```

```
            isDragging = False
        End If
End Sub
```

Mouse Cursors

A common task in any application is to adjust the mouse cursor to show when the application is busy or to indicate how different controls work. You can set the mouse pointer for any element using the Cursor property, which is inherited from the FrameworkElement class.

Every cursor is represented by a System.Windows.Input.Cursor object. The easiest way to get a Cursor object is to use the shared properties of the Cursors class (from the System.Windows.Input namespace). They include all the standard Windows cursors, such as the hourglass, the hand, the resizing arrows, and so on. Here's an example that sets the hourglass for the current page:

```
Me.Cursor = Cursors.Wait
```

Now when you move the mouse over the current page, the mouse pointer changes to the familiar hourglass icon (in Windows XP) or the swirl (in Windows Vista).

■ **Note** The properties of the Cursors class draw on the cursors that are defined on the computer. If the user has customized the set of standard cursors, the application you create will use those customized cursors.

If you set the cursor in XAML, you don't need to use the Cursors class directly. That's because the type converter for the Cursor property is able to recognize the property names and retrieve the corresponding Cursor object from the Cursors class. That means you can write markup like this to show the hand cursor when the mouse is positioned over a button:

```
<Button Cursor="Hand" Content="Help Me"></Button>
```

It's possible to have overlapping cursor settings. In this case, the most specific cursor wins. For example, you could set a different cursor on a button and on the page that contains the button. The button's cursor will be shown when you move the mouse over the button, and the page's cursor will be used for every other region in the page.

■ **Tip** Unlike WPF, Silverlight does not support custom mouse cursors. However, you can hide the mouse cursor (set it to Cursors.None) and then make a small image follow the mouse pointer using code like that shown in the previous section.

Keyboard Handling

Although the mouse provides the most obvious way for a user to interact with a Silverlight application, you can't ignore that other essential input device—the keyboard. Most of the time, you'll rely on controls such as the TextBox, which collects typed-in text without forcing you to pay attention to exactly *how* that text is entered. However, if you need finer-grained control— for example, if you want to perform validation or provide notifications as the user types—you'll need to pay attention to each individual key press as it happens.

In the following sections, you'll learn how to handle key presses and interpret key-event information. You'll also consider how Silverlight manages control focus, which determines the control that gets the keyboard's input.

Key Presses

As you saw in Table 4-1, Silverlight elements use KeyDown and KeyUp events to notify you when a key is pressed. These events use bubbling, so they travel up from the element that currently has focus to the containing elements.

When you react to a key press event, you receive a KeyEventArgs object that provides two additional pieces of information: Key and PlatformKeyCode. Key indicates the key that was pressed as a value from the System.Windows.Input.Key enumeration (for example, Key.S is the S key). PlatformKeyCode is an integer value that must be interpreted based on the hardware and operating system that's being used on the client computer. For example, a nonstandard key that Silverlight can't recognize will return a Key.Unknown value for the Key property but will provide a PlatformKeyCode that's up to you to interpret. An example of a platform-specific key is Scroll Lock on Microsoft Windows computers.

■ **Note** In general, it's best to avoid any platform-specific coding. But if you really do need to evaluate a nonstandard key, you can use the BrowserInformation class from the System.Windows.Browser namespace to get more information about the client computer where your application is running.

The best way to understand the key events is to use a sample program such as the one shown in Figure 4-7 a little later in this chapter. It monitors a text box for three events: KeyDown, KeyUp, and the higher-level TextChanged event (which is raised by the TextBox control), using this markup:

```
<TextBox KeyDown="txt_KeyDown" KeyUp="txt_KeyUp"
 TextChanged="txt_TextChanged"></TextBox>
```

Here, the TextBox handles the KeyDown, KeyUp, and TextChanged events explicitly. However, the KeyDown and KeyUp events bubble, which means you can handle them at a higher level. For example, you can attach KeyDown and KeyUp event handlers on the root Grid to receive key presses that are made anywhere in the page.

Here are the event handlers that react to these events:

```
Private Sub txt_KeyUp(ByVal sender As Object, ByVal e As KeyEventArgs)
    Dim message As String = "KeyUp " & " Key: " & e.Key
      lstMessages.Items.Add(message)
End Sub
```

```
Private Sub txt_KeyDown(ByVal sender As Object, ByVal e As KeyEventArgs)
    Dim message As String = "KeyDown " & " Key: " & e.Key
    lstMessages.Items.Add(message)
End Sub

Private Sub txt_TextChanged(ByVal sender As Object, _
  ByVal e As TextChangedEventArgs)
    Dim message As String = "TextChanged"
    lstMessages.Items.Add(message)
End Sub
```

Figure 4-7 shows the result of typing a lowercase *S* in the text box.

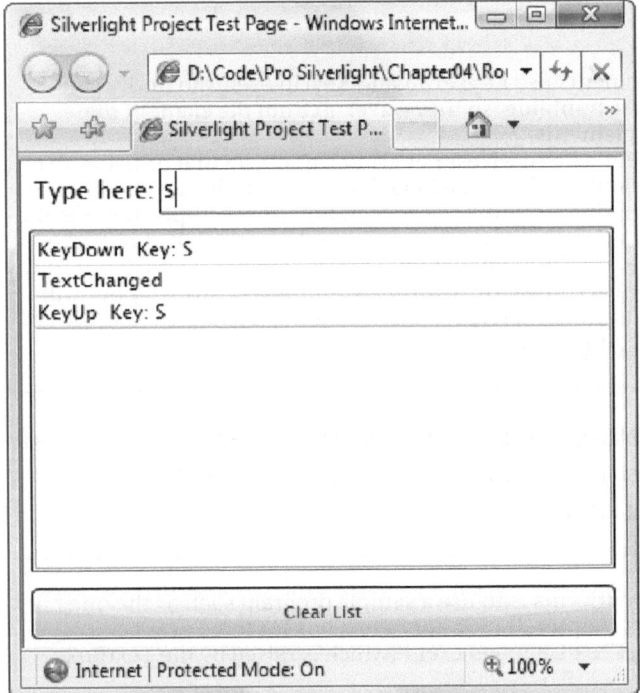

Figure 4-7. *Watching the keyboard*

Typing a single character may involve multiple key presses. For example, if you want to type a *capital* letter *S*, you must first press the Shift key and then the *S* key. On most computers, keys that are pressed for longer than a brief moment start generating repeated key presses. For that reason, if you type a capital *S*, you're likely to see a series of KeyDown events for the Shift key, as shown in Figure 4-8. However, you'll key only two KeyUp events (for the *S* and for the Shift key) and just one TextChanged event.

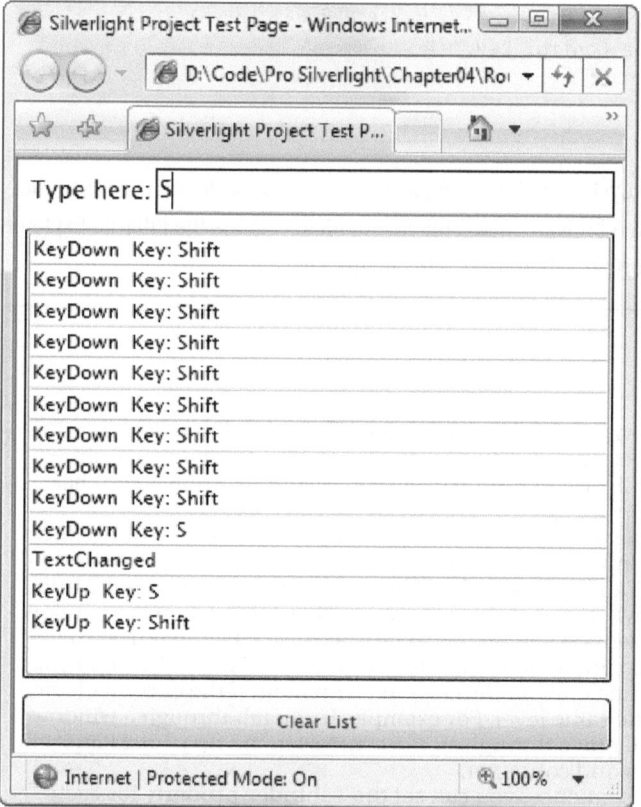

Figure 4-8. Repeated keys

■ **Note** Controls like the TextBox aren't designed for low-level keyboard handling. When dealing with a text-entry control, you should only react to its higher-level keyboard events (like TextChanged).

Key Modifiers

When a key press occurs, you often need to know more than just what key was pressed. It's also important to find out what other keys were held down at the same time. That means you might want to investigate the state of other keys, particularly modifiers such as Shift and Ctrl, both of which are supported on all platforms. Although you can handle the events for these keys separately and keep track of them in that way, it's much easier to use the shared Modifiers property of the Keyboard class.

To test for a Keyboard.Modifier, you use bitwise logic. For example, the following code checks whether the Ctrl key is currently pressed:

```
If (Keyboard.Modifiers And ModifierKeys.Control) = ModifierKeys.Control Then
    message &= "You are holding the Control key."
End If
```

■ **Note** The browser is free to intercept keystrokes. For example, in Internet Explorer you won't see the KeyDown event for the Alt key, because the browser intercepts it. The Alt key opens the Internet Explorer menu (when used alone) or triggers a shortcut (when used with another key).

Focus

In the Silverlight world, a user works with one control at a time. The control that is currently receiving the user's key presses is the control that has *focus*. Sometimes, this control is drawn slightly differently. For example, the Silverlight button uses blue shading to show that it has the focus.

To move the focus from one element to another, the user can click the mouse or use the Tab and arrow keys. In previous development frameworks, programmers have been forced to take great care to make sure that the Tab key moves focus in a logical manner (generally from left to right and then down the window) and that the right control has focus when the window first appears. In Silverlight, this extra work is seldom necessary because Silverlight uses the hierarchical layout of your elements to implement a tabbing sequence. Essentially, when you press the Tab key, you'll move to the first child in the current element or, if the current element has no children, to the next child at the same level. For example, if you tab through a window with two StackPanel containers, you'll move through all the controls in the first StackPanel and then through all the controls in the second container.

If you want to take control of tab sequence, you can set the TabIndex property for each control to place it in numerical order. The control with a TabIndex of 0 gets the focus first, followed by the next highest TabIndex value (for example, 1, then 2, then 3, and so on). If more than one element has the same TabIndex value, Silverlight uses the automatic tab sequence, which means it jumps to the nearest subsequent element.

■ **Tip** By default, the TabIndex property for all controls is set to 1. That means you can designate a specific control as the starting point for a window by setting its TabIndex to 0 but rely on automatic navigation to guide the user through the rest of the window from that starting point, according to the order that your elements are defined.

The TabIndex property is defined in the Control class, along with an IsTabStop property. You can set IsTabStop to False to prevent a control from being included in the tab sequence. A control that has IsTabStop set to False can still get the focus in another way—either programmatically (when your code calls its Focus() method) or by a mouse click.

Controls that are invisible or disabled are skipped in the tab order and are not activated regardless of the TabIndex and IsTabStop settings. To hide or disable a control, you set the Visibility and IsEnabled properties, respectively.

The Command Model

In a well-designed Silverlight application, the application logic doesn't sit in the event handlers but is coded in higher-level methods. Each one of these methods represents a single application "task." In turn, each task may rely on other libraries (such as separately compiled components that encapsulate data access, web service calls, or other types of business logic). Figure 4-9 shows this relationship.

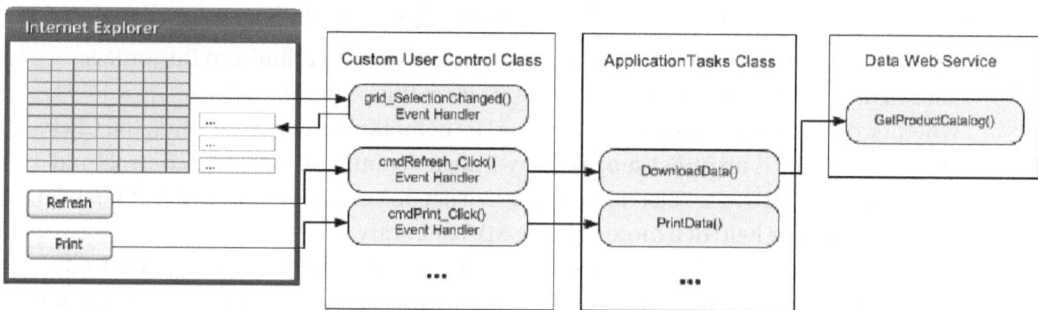

Figure 4-9. *Mapping event handlers to a task*

The most obvious way to use this design is to add event handlers wherever they're needed and use each event handler to call the appropriate application method. In essence, your Silverlight page (the custom class that derives from UserControl) becomes a stripped-down switchboard that responds to input and forwards requests to the heart of the application.

Although this design is perfectly reasonable, it still requires a fair bit of user interface code. You need to respond to a variety of events to call your application methods, and sometimes the same application task can be triggered in different ways (for example, through a button, a right-click menu command, and so on). Life gets even more complex when you need to manage the state of your user interface—for example, disabling controls that shouldn't be accessible when the tasks they trigger aren't relevant. Even if you stick to good design practices, your switchboard code can become dense and tangled.

Silverlight doesn't provide a mature solution for this problem yet. However, many Silverlight developers have started moving to a design pattern called MVVM (which stands for Model-View-ViewModel). The basic idea behind MVVM architecture is that your application is separated into distinct layers. The *model* is the content or data that your application manages. The *view* is the graphical front-end, complete with buttons, graphics, and all your Silverlight elements. In between is the *view model* that allows them to communicate. For example, if the user clicks a button in the view, it can trigger a command in the view model, which then modifies the data in the model.

True MVVM design has both fanatical adherents and more cautious critics. It requires the help of a separate library, such as MVVM Light (www.galasoft.ch/mvvm/getstarted) or Prism (http://compositewpf.codeplex.com). It's also beyond the scope of this book. However, Silverlight 4 has started to add features that will make these higher-level toolkits more practical. The most obvious feature is commands.

■ **Note** Silverlight's command feature is a modest, stripped-down extensibility point that has none of the features of WPF's rich commanding model. Instead, Silverlight commands give developers just enough functionality to begin building their own MVVM architectures.

Here are the essential details about Silverlight's command support:

- Commands are supported by only two element classes: Hyperlink and ButtonBase (although several button-like controls derive from ButtonBase).
- Whether you're using a hyperlink or button, its command is triggered when the Click event fires. There is no direct support for wiring up commands to other controls and other events (such as the selection change in a list box or key presses in a text box). For that, you need the help of a more capable MVVM library.
- Commands work through two properties that have been added to the Hyperlink and ButtonBase class: Command and CommandParameter. The Command is the action that will be triggered when the button click takes place, and the CommandParameter is a single object that will be passed to the command, with additional information.
- You must create the commands you need. Silverlight includes the ICommand interface, which all commands must implement. However, it doesn't include any command classes of its own. Although you could create dozens or hundreds of different command classes, most developers prefer to create a general command class that can be reused for different tasks.

In the following sections, you'll see a simple example of a command at work.

Building a Command

The heart of the Silverlight command model is the System.Windows.Input.ICommand interface, which defines how commands work. This interface includes two methods and an event:

```
Public Interface ICommand
    Private Event CanExecuteChanged As EventHandler
    Sub Execute(ByVal parameter As Object)

    Function CanExecute(ByVal parameter As Object) As Boolean
End Interface
```

In a simple implementation, the Execute() method would contain the application task logic (for example, printing the document). The CanExecute() method returns the state of the command: True if it's enabled and False if it's disabled. Both Execute() and CanExecute() accept an additional parameter object that you can use to pass along any extra information you need.

Finally, the CanExecuteChanged event is raised when the state changes. This is a signal to any controls using the command that they should call the CanExecute() method to check the command's state. This is part of the glue that allows command sources (such as a button) to automatically enable themselves when the command is available and to disable themselves when it's not.

For example, the following PrintTextCommand prints a single string when the command is triggered and the Execute() method is called. (In truth, the command displays the string in a message box, but you could add real printing logic using the skills you'll pick up in Chapter 8.) The CanExecute() method simply examines the string and disallows printing if that string is missing or empty. In both cases, the command receives the string as a command parameter.

Here's the complete code:

```
Public Class PrintTextCommand
  Implements ICommand

    Public Event CanExecuteChanged(ByVal sender As Object, _
      ByVal e As System.EventArgs) Implements ICommand.CanExecuteChanged

    Private _canExecute As Boolean

    Public Function CanExecute(ByVal parameter As Object) As Boolean _
      Implements ICommand.CanExecute
        ' Check if the command can execute.
        ' In order to be executable, it must have non-blank text in the
        ' command parameter.
        Dim canExecuteNow As Boolean = (parameter IsNot Nothing) _
          AndAlso (parameter.ToString() <> "")

        ' Determine if the CanExecuteChanged event should be raised.
        If _canExecute <> canExecuteNow Then
            _canExecute = canExecuteNow
            RaiseEvent CanExecuteChanged(Me, New EventArgs())
        End If

        Return _canExecute
    End Function

    Public Sub Execute(ByVal parameter As Object) _
      Implements ICommand.Execute
        MessageBox.Show("Printing: " & parameter)
    End Sub
End Class
```

Connecting a Command

To use this command, you need to set the Command and CommandParameter properties of your button. Although you could do this in code, this defeats the true purpose of the command model, which is to help you remove the event-handling code from your user control class. Instead, the ideal command implementation connects everything you need in XAML.

First, you need to map your project namespace to an XML prefix, so the custom command class you created is available in your markup. Here's an example for a page named Commands in an application named RoutedEvents:

```
<UserControl x:Class="RoutedEvents.Commands"
    xmlns:local="clr-namespace:RoutedEvents" ... >
```

Now you can add the command as a resource:

```
<UserControl.Resources>
  <local:PrintTextCommand x:Key="printCommand"></local:PrintTextCommand>
</UserControl.Resources>
```

The final step is to find your button control and set its Command property (using the resource) and its CommandParameter property (using data binding). In this example, the CommandParameter extracts the text from a nearby text box.

```
<Button Margin="5" Content="Print Command"
 Command="{StaticResource printCommand}"
 CommandParameter="{Binding ElementName=txt, Path=Text}"></Button>
<TextBox x:Name="txt" Grid.Row="1" Margin="5"></TextBox>
```

Figure 4-10 shows the two states of the button. When there's no text in the text box, the command can't execute, and the button is disabled automatically. When there is text in the text box, the button becomes enabled. If its clicked, the PrintTextCommand.Execute() method runs.

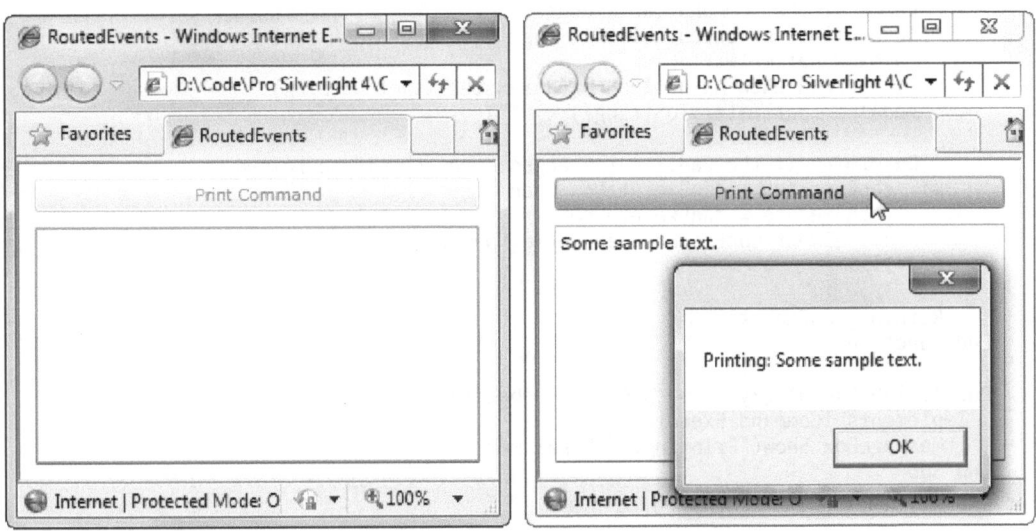

Figure 4-10. *A disabled (left) and enabled (right) command*

This automatic state management is nice, but it's limited to the enabling and disabling of controls. It's not hard to imagine situations where you'd like to manage state in other ways—for example, checking a check box or hiding a list item when other conditions become true. Unfortunately, this isn't possible in the current implementation of Silverlight commands.

The real promise of the command model is the ability to get things done without writing tedious event-handling code. In fact, this example has absolutely no code in the derived user control class. This is the right state of mind for MVVM design, but a more formalized model would introduce a distinct view model. The command would then be made part of that view model and exposed through a property, along with any other commands that you need to use for this page. For a complete example of a Silverlight application that uses this design, see http://tinyurl.com/3x9e3t8.

But don't get too excited—if you need to build anything but the simplest MVVM application, you'll need additional infrastructure. You can build it yourself or step up to one of the MVVM libraries mentioned earlier. And for more design advice about MVVM, refer to a book like *Pro Business Applications with Silverlight 4*.

The Last Word

In this chapter, you took a deep look at Silverlight dependency properties and routed events. First, you saw how dependency properties are defined and registered and how they plug into other Silverlight services. You explored event bubbling and saw how it allows an event to travel up the element hierarchy. Next, you considered the basic set of mouse and keyboard events that all elements provide. Lastly, you saw the start of a new direction in Silverlight event handling, with the command model.

■ **Tip** One of the best ways to learn more about the internals of Silverlight is to browse the code for basic Silverlight elements, such as Button, UIElement, and FrameworkElement. One of the best tools to perform this browsing is Reflector, which is available at www.red-gate.com/products/reflector. Using Reflector, you can see the definitions for dependency properties and routed events, browse through the shared constructor code that initializes them, and even explore how the properties and events are used in the class code.

■ ■ ■

CHAPTER 5

Elements

Now that you've learned the fundamentals of XAML, layout, and mouse and keyboard handling, you're ready to consider the elements that allow you to build both simple and complex user interfaces.

In this chapter, you'll get an overview of Silverlight's core elements, and you'll explore many elements that you haven't studied yet. First, you'll learn how to display wrapped, formatted text with the TextBlock and how to show images with the Image element. Next, you'll consider *content controls*, including Silverlight's many different flavors of button and the ToolTip control. Finally, you'll take a look at several more specialized elements, such as Silverlight's list, text-entry, range, and date controls. By the time you finish this chapter, you'll have a solid overview of the essential ingredients that make up Silverlight pages.

■ **What's New** Silverlight 4 gives the TextBlock a very minor text trimming feature (described in the "Trimming Text" section) and introduces programmatic support for the clipboard (see the "Text Selection" section). It also adds three new controls, which are covered in different chapters. You used the Viewbox in Chapter 3, you'll use the WebBrowser in Chapter 21, and you'll explore the RichTextBox in this chapter.

The Silverlight Elements

You've already met quite a few of Silverlight's core elements, such as the layout containers in Chapter 3. Some of the more specialized elements, such as the ones used for drawing 2-D graphics, displaying Deep Zoom images, and playing video, won't be covered until later in this book. But this chapter deals with all the basics—fundamental widgets such as buttons, text boxes, lists, and check boxes.

Table 5-1 provides an at-a-glance look at the key elements that Silverlight includes and points you to the chapters of this book where they're described. The list is ordered alphabetically to match the order of elements in the Visual Studio Toolbox. The gray shading highlights controls that are new to Silverlight 4.

Table 5-1. Silverlight Elements

Class	Description	Place in This Book	Assembly (If Not a Core Element)
AutoCompleteBox	A specialized text box that provides a list of possible matches as the user types.	This chapter	System.Windows.Controls.Input.dll
Border	A rectangular or rounded border that's drawn around a single, contained element.	Chapter 3	
Button	The familiar button, complete with a shaded gray background, which the user clicks to launch a task.	This chapter	
Calendar	A one-month-at-a-time calendar view that allows the user to select a single date.	This chapter	System.Windows.Controls.dll
Canvas	A layout container that allows you to lay out elements with precise coordinates.	Chapter 3	
CheckBox	A box that can be checked or unchecked, with optional content displayed next to it.	This chapter	
ComboBox	A drop-down list of items, out of which a single one can be selected.	This chapter	
ContentControl	The base control from which all content controls derive, such as Button, CheckBox, ToolTip, ScrollViewer, and many more. Although you can use this class directly, you're much more likely to work with its descendants.	This chapter	
DataGrid	A rich data control that shows a collection of data objects in a multicolumned grid and offers built-in features such as sorting and selection.	Chapter 17	System.Windows.Controls.Data.dll
DataPager	A data control that provides paging for other data sources and can work in conjunction with controls like the DataGrid.	Chapter 17	System.Windows.Controls.Data.dll

Class	Description	Place in This Book	Assembly (If Not a Core Element)
DatePicker	A text box for date entry, with a drop-down calendar for easy selection.	This chapter	System.Windows. Controls.dll
Ellipse	A shape drawing element that represents an ellipse.	Chapter 8	
Frame	A container that displays a separate XAML file inside an ordinary page. You can use frames in various ways to create a complex navigation system.	Chapter 7	System.Windows. Controls. Navigation.dll
Grid	A layout container that places children in an invisible grid of cells.	Chapter 3	
GridSplitter	A resizing bar that allows users to change the height or adjacent rows or width of adjacent columns in a Grid.	Chapter 3	System.Windows. Controls.dll
HyperlinkButton	A link that directs the user to another web page.	This chapter	
Image	An element that displays a supported image file.	This chapter	
Label	A text display control that's similar to the TextBlock but heavier weight. When paired up with a data-bound control, the Label can examine the bound data object to extract caption text and determine whether it should show a required field indicator or error indicator.	Chapter 16	System.Windows. Controls.dll
ListBox	A list of items, out of which a single one can be selected.	This chapter	
MediaElement	A media file, such as a video window.	Chapter 11	
MultiScaleImage	An element that supports Silverlight's Deep Zoom feature and allows the user to zoom into a precise location in a massive image.	Chapter 11	
PasswordBox	A text box that masks the text the user enters.	This chapter	

Class	Description	Place in This Book	Assembly (If Not a Core Element)
ProgressBar	A colored bar that indicates the percent completion of a given task.	This chapter	
RadioButton	A small circle that represents one choice out of a group of options, with optional content displayed next to it.	This chapter	
Rectangle	A shape drawing element that represents a rectangle.	Chapter 8	
RichTextBox	An editable text box that supports richly formatted text.	This chapter	
ScrollViewer	A container that holds any large content and makes it scrollable.	Chapter 3	
Slider	An input control that lets the user set a numeric value by dragging a thumb along a track.	This chapter	
StackPanel	A layout container that stacks items from top to bottom or left to right.	Chapter 3	
TabControl	A container that places items into separate tabs and allows the user to view just one tab at a time.	This chapter	System.Windows.Controls.dll
TextBlock	An all-purpose text display control that includes the ability to give different formatting to multiple pieces of inline text.	This chapter	
TextBox	The familiar text-entry control.	This chapter	
TreeView	A rich data control that shows the familiar tree of items, with as many hierarchical levels as you need.	Chapter 17	System.Windows.Controls.dll
Viewbox	A container that can scale its content up or down, as needed.	Chapter 3	
WebBrowser	An Internet Explorer–powered browser window that you can embed inside a Silverlight page	Chapter 21	

In Chapter 1, you learned that Silverlight includes some noncore controls that—if used—are automatically added to the compiled XAP file so they can be deployed with your application. As you can see in the last column of Table 5-1, this doesn't apply to most Silverlight controls, and even some highly specialized controls like the MultiScaleImage are part of the standard Silverlight package.

In the following sections, you'll take a closer look at many of the controls from Table 5-1, and you'll learn how to customize them in your own applications.

■ **Tip** If you're still hungering for more controls, you can find many specialized (and downright ingenious) offerings in the Silverlight Toolkit, a freely downloadable and distributable add-on that's available on Microsoft's CodePlex site at `http://silverlight.codeplex.com`. Highlights include a rich array of beautifully rendered chart controls that include nearly everything you'll find in Excel, from pie charts to scatter plots. Once you've installed the Silverlight Toolkit, you'll find the new controls packed into the Silverlight tab of the Toolbox.

Static Text

Although Silverlight includes a Label control, it's intended for data binding scenarios and discussed in Chapter 16. If you just want the best way to show blocks of formatted text, you're far better off with the lightweight, flexible TextBlock element, which you've seen at work in many of the examples over the past four chapters.

The TextBlock element is refreshingly straightforward. It provides a Text property, which accepts a string with the text you want to display.

```
<TextBlock Text="This is the content."></TextBlock>
```

Alternatively, you can supply the text as nested content:

```
<TextBlock>This is the content.</TextBlock>
```

The chief advantage of this approach is that you can add line breaks and tabs to make large sections of text more readable in your code. Silverlight follows the standard rules of XML, which means it *collapses* whitespace. Thus, a series of spaces, tabs, and hard returns is rendered using a single space character.

If you really do want to split text over lines at an explicit position, you have three options. First, you can use separate TextBlock elements. Second, you can use a LineBreak inside the TextBlock element, as shown here:

```
<TextBlock>
    This is line 1.<LineBreak/>
    This is line 2.
</TextBlock>
```

Third, you can add the `xml:space="preserve"` attribute to your TextBlock element, which tells the XAML parser to honor every space, tab, and hard return between the opening > character (which ends the start tag) and the closing < (which begins the end tag). Here's an example:

```
<TextBlock xml:space="preserve"
>This is line 1.
     This is an indented line 2.</TextBlock>
```

This approach gives you the most powerful way to micro-manage how text blocks are formatted. However, its strictness is usually too limiting for real-world applications.

■ **Note** When using inline text, you can't use the < and > characters, because these have a specific XML meaning. Instead, you need to replace the angled brackets with the character entities < (for the less than symbol) and > (for the greater than symbol), which will be rendered as < and >.

Unsurprisingly, text is colored black by default. You can change the color of your text using the Foreground property. You can set it using a color name in XAML:

```
<TextBlock x:Name="txt" Text="Hello World" Foreground="Red"></TextBlock>
```

or in code

```
txt.Foreground = New SolidColorBrush(Colors.Red)
```

Instead of using a color name, you can use RGB values. You can also use partially transparent colors that allow the background to show through. Both topics are covered in Chapter 3 when discussing how to paint the background of a panel.

■ **Tip** Ordinarily, you'll use a solid color brush to fill in text. (The default is obviously a black brush.) However, you can create more exotic effects by filling in your text with gradients and tiled patterns using the fancy brushes discussed in Chapter 9.

The TextBlock also provides a TextAlignment property (which allows you to center or right-justify text), a Padding property (which sets the space between the text and the outer edges of the TextBlock), and a few more properties for controlling fonts, inline formatting, text wrapping, and text trimming. You'll consider these properties in the following sections.

Font Properties

The TextBlock class defines font properties that determine how text appears in a control. These properties are outlined in Table 5-2.

Table 5-2. *Font-Related Properties of the Control Class*

Name	Description
FontFamily	The name of the font you want to use. Because Silverlight is a client-side technology, it's limited to just nine built-in fonts (Arial, Arial Black, Comic Sans MS, Courier New, Georgia, Lucida, Times New Roman, Trebuchet MS, and Verdana). However, you can also distribute custom fonts by going to a bit more work and packing them up with your project assembly, as you'll see shortly in the "Font Embedding" section.
FontSize	The size of the font in pixels. Ordinary Windows applications measure fonts using *points*, which are assumed to be 1/72 of an inch on a standard PC monitor, while pixels are assumed to be 1/96 of an inch. Thus, if you want to turn a Silverlight font size into a more familiar point size, you can use a handy trick—just multiply by 3/4. For example, a 20-pixel FontSize is equivalent to a traditional 15-point font size.
FontStyle	The angling of the text, as represented as a FontStyle object. You get the FontStyle preset you need from the shared properties of the FontStyles class, which includes Normal and Italic lettering. If you apply italic lettering to a font that doesn't provide an italic variant, Silverlight will simply slant the letters. However, this behavior gives only a crude approximation of a true italic typeface.
FontWeight	The heaviness of text, as represented as a FontWeight object. You get the FontWeight preset you need from the shared properties of the FontWeights class. Normal and Bold are the most obvious of these, but some typefaces provide other variations such as Bold, Light, ExtraBold, and so on. If you use Bold on a font that doesn't provide a bold variant, Silverlight will paint a thicker border around the letters, thereby simulating a bold font.
FontStretch	The amount that text is stretched or compressed, as represented by a FontStretch object. You get the FontStretch preset you need from the shared properties of the FontStretches class. For example, UltraCondensed reduces fonts to 50 percent of their normal width, while UltraExpanded expands them to 200 percent. Font stretching is an OpenType feature that is not supported by many typefaces. The built-in Silverlight fonts don't support any of these variants, so this property is relevant only if you're embedding a custom font that does.

Obviously, the most important of these properties is FontFamily. A *font family* is a collection of related typefaces—for example, Arial Regular, Arial Bold, Arial Italic, and Arial Bold Italic are all part of the Arial font family. Although the typographic rules and characters for each variation are defined separately, the operating system realizes they're related. As a result, you can configure an element to use Arial Regular, set the FontWeight property to Bold, and be confident that Silverlight will switch to the Arial Bold typeface.

When choosing a font, you must supply the full family name, as shown here:

```
<TextBlock x:Name="txt" FontFamily="Times New Roman" FontSize="18">
 Some Text</TextBlock>
```

It's much the same in code:

```
txt.FontFamily = "Times New Roman"
txt.FontSize = "18"
```

When identifying a FontFamily, a shortened string is not enough. That means you can't substitute Times or Times New instead of the full name Times New Roman.

Optionally, you can use the full name of a typeface to get italic or bold, as shown here:

```
<TextBlock FontFamily="Times New Roman Bold">Some Text</TextBlock >
```

However, it's clearer and more flexible to use just the family name and set other properties (such as FontStyle and FontWeight) to get the variant you want. For example, the following markup sets the FontFamily to Times New Roman and sets the FontWeight to FontWeights.Bold:

```
<TextBlock FontFamily="Times New Roman" FontWeight="Bold">Some Text</TextBlock >
```

Standard Fonts

Silverlight supports nine core fonts, which are guaranteed to render correctly on any browser and operating system that supports Silverlight. They're shown in Figure 5-1.

Figure 5-1. *Silverlight's built-in fonts*

In the case of Lucida, there are two variants with slightly different names. Lucida Sans Unicode is included with Windows, while Lucida Grande is an almost identical font that's included with Mac OS X. To allow this system to work, the FontFamily property supports font fallback—in other words, you can supply a comma-separated list of font names, and Silverlight will used the first supported font. The default TextBlock font is equivalent to setting the FontFamily property to the string "Lucida Sans Unicode, Lucida Grande."

You might think that you can use more specialized fonts, which may or may not be present on the client's computer. However, Silverlight doesn't allow this. If you specify a font that isn't one of the nine built-in fonts and it isn't included with your application assembly (more on that in the next section), your font setting will be ignored. This happens regardless of whether the client has an installed font with the appropriate name. This makes sense—after all, using a font that's supported on only some systems could lead to an application that's mangled or completely unreadable on others, which is an easy mistake to make.

Font Embedding

If you want to use nonstandard fonts in your application, you can embed them in your application assembly. That way, your application never has a problem finding the font you want to use.

The embedding process is simple. First, you add the font file (typically, a file with the extension .ttf) to your application and set Build Action to Resource. You can do this in Visual Studio by selecting the font file in the Solution Explorer and changing its Build Action setting in the Properties page.

Next, when you set the FontFamily property, you need to use this format:

FontFileName#FontName

For example, if you have a font file named BayernFont.ttf and it includes a font named Bayern, you would use markup like this:

```
<TextBlock FontFamily="BayernFont.ttf#Bayern">This is an embedded font</TextBlock>
```

Figure 5-2 shows the result.

Figure 5-2. *Using an embedded font*

Alternatively, you can set the font using a stream that contains the font file. In this case, you need to set the TextBlock.FontSource property with the font file stream and then set the TextBlock.FontFamily property with the font name. For example, if you added the BayernFont.ttf file as a resource to a project named FontTest, you can retrieve it programmatically using this code:

```
Dim fontUri As New Uri("FontTest;component/BayernFont.ttf", UriKind.Relative)
Dim sri As StreamResourceInfo = Application.GetResourceStream(fontUri)

lbl.FontSource = New FontSource(sri.Stream)
lbl.FontFamily = New FontFamily("Bayern")
```

To pull the resource out of the current assembly, this code uses the Application.GetResourceStream() method and a specialized URI syntax that always takes this form:

AssemblyName;component/*FontResourceName*

No matter which approach you use, the process of using a custom font is fairly easy. However, font embedding raises obvious licensing concerns. Most font vendors allow their fonts to be embedded in documents (such as PDF files) but not applications (such as Silverlight assemblies). The problem is obvious—users can download the XAP file by hand, unzip it, retrieve the font resource, and then access it on their local computers. Silverlight doesn't make any attempt to enforce font licensing, but you should make sure you're on solid legal ground before you redistribute a font.

You can check a font's embedding permissions using Microsoft's free font properties extension utility, which is available at www.microsoft.com/typography/TrueTypeProperty21.mspx. Once you install this utility, right-click any font file, and choose Properties to see more detailed information about it. In particular, check the Embedding tab for information about the allowed embedding for this font. Fonts marked with Installed Embedding Allowed are suitable for Silverlight applications, while fonts with Editable Embedding Allowed may not be. Consult with the font vendor for licensing information about a specific font.

■ **Note** If all else fails, you can get around licensing issues by changing your fonts to graphics. This works for small pieces of graphical text (for example, headings) but isn't appropriate for large blocks of text. You can save graphical text as a bitmap in your favorite drawing program, or you can convert text to a series of shapes using Silverlight's Path element (which is discussed in Chapter 8). You can convert graphical text to a path using Expression Designer or Expression Blend (simply select the TextBlock and choose Object ➤ Path ➤ Convert to Path). Interestingly, Silverlight also allows you to perform the same trick through code. Surf to http://tinyurl.com/69f74v to see an example in which a Silverlight application calls a web service that dynamically generates a path for non-Western text. The web service returns the path data to the Silverlight application, which displays it seamlessly.

Underlining

You can add underlining to any font by setting the TextDecorations property to Underline:

```
<TextBlock TextDecorations="Underline">Underlined text</TextBlock>
```

In WPF, there are several types of text decorations, including overlines and strikethrough. However, currently, Silverlight includes only underlining.

If you want to underline an individual word in a block of text, you'll need to use inline elements, as described in the next section.

Runs

In many situations, you'll want to format individual bits of text but keep them together in a single paragraph in a TextBlock. To accomplish this, you need to use a Run object inside the TextBlock element. Here's an example that formats several words differently (see Figure 5-3):

```
<TextBlock FontFamily="Georgia" FontSize="20" >
  This <Run FontStyle="Italic" Foreground="YellowGreen">is</Run> a
  <Run FontFamily="Comic Sans MS" Foreground="Red" FontSize="40">test.</Run>
</TextBlock>
```

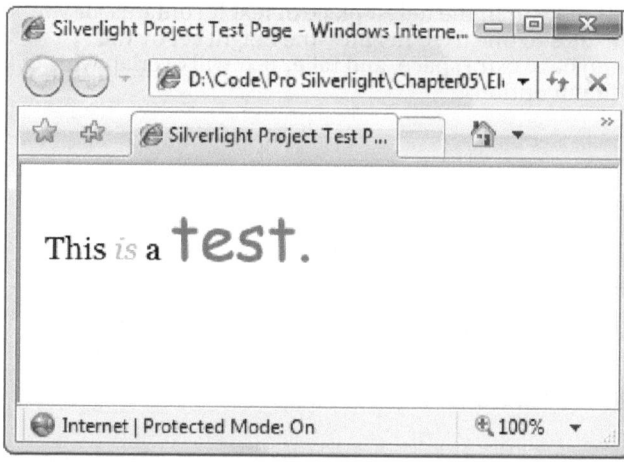

Figure 5-3. *Formatting text with runs*

A run supports the same key formatting properties as the TextBlock, including Foreground, TextDecorations, and the five font properties (FontFamily, FontSize, FontStyle, FontWeight, and FontStretch).

Technically, a Run object is not a true element. Instead, it's an *inline*. Silverlight provides two just types of inlines—the LineBreak class that you saw earlier and the Run class. You can interact with the runs in your TextBlock through the TextBlock.Inlines collection. In fact, the TextBlock actually has two overlapping content models. You can set text through the simple Text property, or you can supply it through the Inlines collection. However, the changes you make in one affect the other, so if you set the Text property, you'll wipe out the current collection of inlines.

■ **Note** The inline Run and LineBreak classes are part of the text element model, which is supported (in a more fully featured way) by the RichTextBox control discussed later in this chapter.

Wrapping Text

To wrap text over several lines, you use the TextWrapping property. Ordinarily, TextWrapping is set to TextWrapping.NoWrap, and content is truncated if it extends past the right edge of the containing element. If you use TextWrapping.Wrap, your content will be wrapped over multiple lines when the width of the TextBlock element is constrained in some way. (For example, you place it into a proportionately sized or fixed-width Grid cell.) When wrapping, the TextBlock splits lines at the nearest space. If you have a word that is longer than the available line width, the TextBlock will split that word wherever it can to make it fit.

When wrapping text, the LineHeight and LineStackingStrategy properties become important. The LineHeight property can set a fixed height (in pixels) that will be used for every line. However, the LineHeight can be used to increase the line height only—if you specify a height that's smaller than what's required to show the text, your setting will be ignored. The LineStackingStrategy determines what the TextBlock will do when dealing with multiline content that uses different fonts. You can choose to use the standard behavior, MaxHeight, which makes each line as high as it needs to be to fit the tallest piece of text it contains, or you can use BlockLineHeight, which sets the lines to one fixed height—the height set by the LineHeight property. Shorter text will then have extra space, and taller text will overlap with other lines. Figure 5-4 compares the different options.

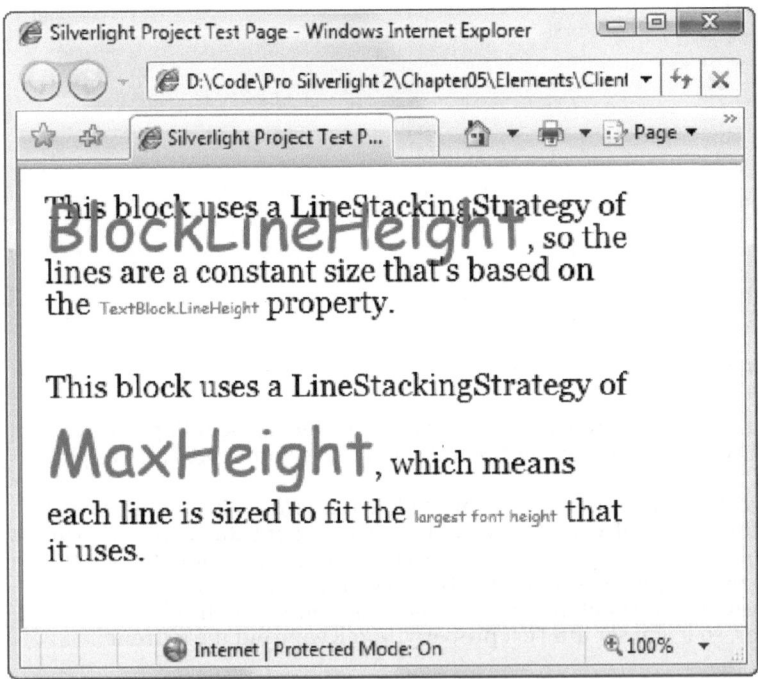

Figure 5-4. *Two different ways to calculate line height*

Trimming Text

As you learned in the previous section, text that doesn't fit the width of its container has two options: wrap or be truncated. For example, if you have a sentence like "Silverlight is a fantastic platform," it might be truncated like this:

```
Silverlight is a fant
```

The TextTrimming property gives you a slightly more graceful way to deal with this situation. If you set the TextTrimming property to WordEllipsis (the only option other than the default), Silverlight adds an ellipsis at the end of truncated text. So, the previous example might come out like this:

```
Silverlight is a ...
```

This gives the user a visual cue that the full text isn't shown.

Images

Displaying an image is one of the easier tasks in Silverlight. You simply need to add an Image element and set its Source property. However, there are some limitations that you need to understand.

The most obvious limitation is that the Image element supports just two image formats. It has full support for JPEG and fairly broad support for PNG (although it doesn't support PNG files that use 64-bit color or grayscale). The Image element does *not* support GIF files. There are two reasons for this omission—it allows the Silverlight download to remain that much slimmer, and it avoids potential confusion between the Silverlight animation model and the much more basic (and unsupported) animated GIF feature that's used on the Web.

It's also important to recognize that the Image.Source property is set with a relative or absolute URI. Usually, you'll use a relative URI to display an image that you've added to your project as a resource. For example, if you add a new image named grandpiano.jpg to your project, Visual Studio will automatically configure it to be a resource, and it will embed that resource in the compiled assembly as a block of binary data. At runtime, you can retrieve that image using its resource name (which is the file name it has in the Solution Explorer). Here's how:

```
<Image Source="grandpiano.jpg"></Image>
```

Or, assuming the image is in a project subfolder named Images, you can retrieve it like so:

```
<Image Source="Images/grandpiano.jpg"></Image>
```

Alternatively, you can construct the URI in code and set the Image.Source property programmatically:

```
img.Source = New BitmapImage(New Uri("grandpiano.jpg", UriKind.Relative))
```

You can also use image URIs to point to images that aren't embedded in your application. You can show images that are located on the same website as your Silverlight application, or images that exist on separate websites.

```
<Image Source="http://www.mysite.com/Images/grandpiano.jpg"></Image>
```

However, there's one catch. When testing a file-based website (one that doesn't use an ASP.NET website and the Visual Studio test web server), you won't be able to use absolute URLs. This limitation is a security restriction that results from the mismatch between the way you're running your application (from the file system) and the way you want to retrieve your images (from the Web, over HTTP). The same limitation comes into play if you attempt to access an image over HTTPS when your Silverlight page was accessed through HTTP (or vice versa).

For more information and to see a few examples that demonstrate your different options for using URIs and managing resources, refer to Chapter 6.

■ **Tip** Interestingly, Silverlight uses bitmap caching to reduce the number of URI requests it makes. That means you can link to an image file on a website multiple times, but your application will download it only once.

Image Sizing

Images can be resized in two ways. First, you can set an explicit size for your image using the Height and Width properties. Second, you can place your Image element in a container that uses resizing, such as a proportionately sized cell in a Grid. If neither of these factors comes into play—in other words, you don't set the Height and Width properties and you place your Image in a simple layout container like the Canvas—your image will be displayed using the native size that's defined in the image file.

To control this behavior, you can use the Stretch property. The Stretch property determines how an image is resized when the dimensions of the Image element don't match the native dimensions of the image file. Table 5-3 lists the values you can use for the Stretch property, and Figure 5-5 compares them.

Table 5-3. Values for the Stretch Enumeration

Name	Description
Fill	Your image is stretched in width and height to fit the Image element dimensions exactly.
None	The image keeps its native size.
Uniform	The image is given the largest possible size that fits in the Image element and doesn't change its aspect ratio. This is the default value.
UniformToFill	The width and height of the image are sized proportionately until the image fills all the available height and width. For example, if you place a picture with this stretch setting into an Image element that's 100×200 pixels, you'll get a 200×200 picture, and part of it will be clipped off.

Figure 5-5. *Four different ways to size an image*

Image Errors

Several factors can cause an image not to appear, such as using a URI to a nonexistent file or trying to display an image in an unsupported format. In these situations, the Image element raises the ImageFailed event. You can react to this event to determine the problem and take alternative actions. For example, if a large image is not available from the Web, you can substitute a small placeholder that's embedded in your application assembly.

Image errors are not fatal, and your application will continue running even if it can't display an image. In this situation, the Image element will remain blank. Your image will also be blank if the image data takes a significant amount of time to download. Silverlight will perform the image request asynchronously and render the rest of the layout in your page while waiting.

Content Controls

Content controls are a specialized type of controls that are designed to hold (and display) a piece of content. Technically, a content control is a control that can contain a *single* nested element. The one-child limit is what differentiates content controls from layout containers, which can hold as many nested elements as you want.

As you learned in Chapter 3, all Silverlight layout containers derive from the Panel class, which gives the support for holding multiple elements. Similarly, all content controls derive from the ContentControl class. Figure 5-6 shows the class hierarchy.

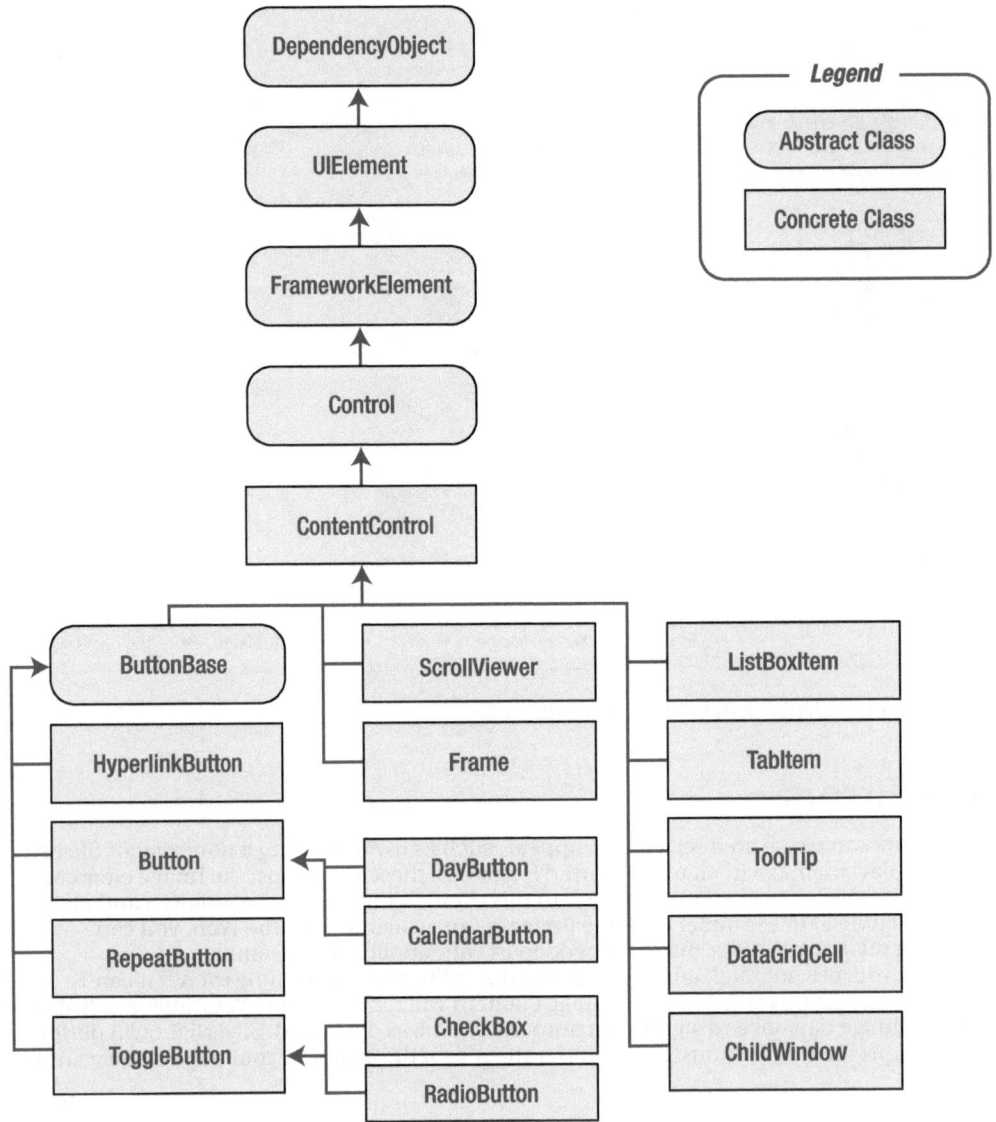

Figure 5-6. *The hierarchy of content controls*

As Figure 5-6 shows, several common controls are actually content controls, including the Label, Tooltip, Button, RadioButton, and CheckBox. There are also a few more specialized content controls, such as ScrollViewer (which you used in Chapter 3 to create a scrollable panel), and some controls that are designed for being used with another, specific control. For example, the ListBox control holds ListBoxItem content controls; the Calendar requires the DayButton and MonthButton; and the DataGrid uses the DataGridCell, DataGridRowHeader, and DataColumnHeader.

The Content Property

Whereas the Panel class adds the Children collection to hold nested elements, the ContentControl class adds a Content property, which accepts a single object. The Content property supports any type of object. It gives you three ways to show content:

- *Elements*: If you use an object that derives from UIElement for the content of a content control, that element will be rendered.
- *Other objects*: If you place a nonelement object into a content control, the control will simply call ToString() to get the text representation for that control. For some types of objects, ToString() produces a reasonable text representation. For others, it simply returns the fully qualified class name of the object, which is the default implementation.
- *Other objects, with a data template*: If you place a nonelement object into a content control and you set the ContentTemplate property with a data template, the content control will render the data template and use the expressions it contains to pull information out of the properties of your object. This approach is particularly useful when dealing with collections of data objects, and you'll see how it works in Chapter 16.

To understand how this works, consider the humble button. An ordinary button may just use a simple string object to generate its content:

```
<Button Margin="3" Content="Text content"></Button>
```

This string is set as the button content and displayed on the button surface.

■ **Tip** When filling a button with unformatted text, you may want to use the font-related properties that the Button class inherits from Control, which duplicate the TextBlock properties listed in Table 5-2.

However, you can get more ambitious by placing other elements inside the button. For example, you can place an image inside using the Image class:

```
<Button Margin="3">
  <Image Source="happyface.jpg"></Image>
</Button>
```

Or you could combine text and images by wrapping them all in a layout container like the StackPanel, as you saw in Chapter 3:

```
<Button Margin="3">
  <StackPanel>
    <TextBlock Margin="3" Text="Image and text button"></TextBlock>
    <Image Source="happyface.jpg" />
    <TextBlock Margin="3" Text="Courtesy of the StackPanel"></TextBlock>
  </StackPanel>
</Button>
```

If you want to create a truly exotic button, you could even place other content controls such as text boxes and buttons inside (and nest still elements inside these). It's doubtful that such an interface would make much sense, but it is possible.

At this point, you might be wondering if the Silverlight content model is really worth all the trouble. After all, you might choose to place an image inside a button, but you're unlikely to embed other controls and entire layout panels. However, there are a few important advantages to the content model.

For example, the previous markup placed a bitmap into a button. However, this approach isn't as flexible as creating a vector drawing out of Silverlight shapes. Using a vector drawing, you can create a button image that's scalable and can be changed programmatically (for example, with different colors, a transform, or an animation). Using a vector-based button opens you up to the possibility of creating a dynamic interface that responds to state changes and user actions.

In Chapter 8, you'll consider how you can begin building vector images in Silverlight. However, the key fact you should understand now is that the vector-drawing model integrates seamlessly with content controls because they have the ability to hold any element. For example, this markup creates a simple graphical button that contains two diamond shapes (as shown in Figure 5-7):

```
<Button Margin="3" Height="70" Width="215">
  <Grid Margin="5">
    <Polygon Points="100,25 125,0 200,25 125,50"
     Fill="LightSteelBlue" />
    <Polygon Points="100,25 75,0 0,25 75,50"
     Fill="LightGray"/>
  </Grid>
</Button>
```

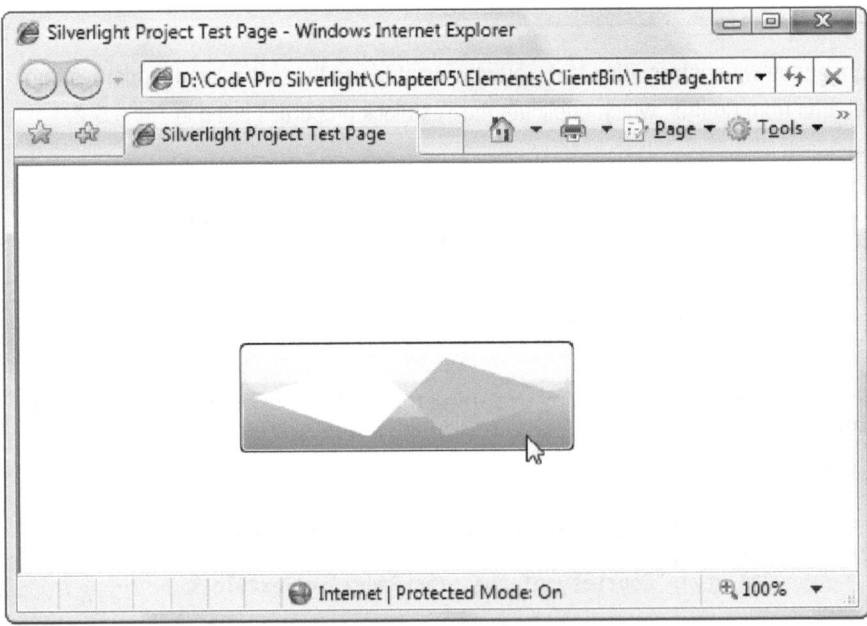

Figure 5-7. *A button with shape content*

Clearly, in this case, the nested content model is simpler than adding extra properties to the Button class to support the different types of content. Not only is the nested content model more flexible, it also allows the Button class to expose a simpler interface. And because all content controls support content nesting in the same way, there's no need to add different content properties to multiple classes.

In essence, the nested content model is a trade. It simplifies the class model for elements because there's no need to use additional layers of inheritance to add properties for different types of content. However, you need to use a slightly more complex *object* model—elements that can be built out of other nested elements.

■ **Note** You can't always get the effect you want by changing the content of a control. For example, even though you can place any content in a button, a few details never change, such as the button's shaded background, its rounded border, and the mouse-over effect that makes it glow when you move the mouse pointer over it. However, there's another way to change these built-in details—by applying a new control template. Chapter 13 shows how you can change all aspects of a control's look and feel using a control template.

Aligning Content

In Chapter 3, you learned how to align different controls in a container using the HorizontalAlignment and VerticalAlignment properties, which are defined in the base FrameworkElement class. However, once a control contains content, there's another level of organization to think about. You need to decide how the content inside your content control is aligned with its borders. This is accomplished using the HorizontalContentAlignment and VerticalContentAlignment properties.

HorizontalContentAlignment and VerticalContentAlignment support the same values as HorizontalAlignment and VerticalAlignment. That means you can line up content on the inside of any edge (Top, Bottom, Left, or Right), you can center it (Center), or you can stretch it to fill the available space (Stretch). These settings are applied directly to the nested content element, but you can use multiple levels of nesting to create a sophisticated layout. For example, if you nest a StackPanel in a Button element, the Button.HorizontalContentAlignment determines where the StackPanel is placed, but the alignment and sizing options of the StackPanel and its children will determine the rest of the layout.

In Chapter 3, you also learned about the Margin property, which allows you to add whitespace between adjacent elements. Content controls use a complementary property named Padding, which inserts space between the edges of the control and the edges of the content. To see the difference, compare the following two buttons:

```
<Button Content="Absolutely No Padding"></Button>
<Button Padding="3" Content="Well Padded"></Button>
```

The button that has no padding (the default) has its text crowded up against the button edge. The button that has a padding of 3 pixels on each side gets a more respectable amount of breathing space.

■ **Note** The HorizontalContentAlignment, VerticalContentAlignment, and Padding properties are all defined as part of the Control class, not the more specific ContentControl class. That's because there may be controls that aren't content controls but still have some sort of content. One example is the TextBox—its contained text (stored in the Text property) is adjusted using the alignment and padding settings you've applied.

Buttons

Silverlight recognizes three types of button controls: the familiar Button, the CheckBox, and the RadioButton. All of these controls are content controls that derive from ButtonBase.

The ButtonBase class includes only a few members. It defines the obviously important Click event and adds the IsFocused, IsMouseOver, and IsPressed read-only properties. Finally, the ButtonBase class adds a ClickMode property, which determines when a button fires its Click event in response to mouse actions. The default value is ClickMode.Release, which means the Click event fires when the mouse is clicked and released. However, you can also choose to fire the Click event mouse when the mouse button is first pressed (ClickMode.Press) or, oddly enough, whenever the mouse moves over the button and pauses there (ClickMode.Hover).

You've already seen how to use the ordinary button. In the following sections, you'll take a quick look at the more specialized alternatives that Silverlight provides.

The HyperlinkButton

The ordinary Button control is simple enough—you click it, and it fires a Click event that you handle in code. But what about the other variants that Silverlight offers?

One of these is the HyperlinkButton. The HyperlinkButton doesn't draw the standard button background. Instead, it simply renders the content that you supply. If you use text in the HyperlinkButton, it appears blue by default, but it's not underlined. (Use the TextDecorations property if you want that effect.) When the user moves the mouse over a HyperlinkButton, the mouse cursor changes to the pointing hand. You can override this effect by setting the Cursor property.

There are essentially three ways to use the HyperlinkButton:

- *Send the browser to an external website*: To do this, set the NavigateUri property with an absolute URL that points to the target web page. Optionally, set the TargetName property with the name of browser frame where you want to open the link. Keep in mind that if you navigate away from the current page, you'll effectively end the current Silverlight application. As a result, this technique is of relatively limited use.

- *Send a frame to another Silverlight page*: To do this, make sure you have a Frame control on your page, and set the NavigateUri with a relative URI that points to another XAML file in your project. You'll learn how to use this ability, and the rest of Silverlight's navigation features, in Chapter 7.

- *Perform some arbitrary action in code*: To do this, don't set the NavigateUri property. Instead, simply handle the Click event to carry out the appropriate action.

■ **Tip** The HTML entry page can specifically prevent navigation to external websites. To do so, simply add the enableNavigation property to the <object> element in the test page, and set it to false. You will still be allowed to use the HyperlinkButton for internal frame navigation (see Chapter 7) or to trigger an action with the Click event.

The ToggleButton and RepeatButton

Alongside Button and HyperlinkButton, two more classes derive from ButtonBase:

- *RepeatButton*: This control fires Click events continuously, as long as the button is held down. Ordinary buttons fire one Click event per user click.
- *ToggleButton*: This control represents a button that has two states (clicked or unclicked). When you click a ToggleButton, it stays in its pushed state until you click it again to release it. This is sometimes described as *sticky click* behavior.

Both RepeatButton and ToggleButton are defined in the System.Windows.Controls.Primitives namespace, which indicates they aren't often used on their own. Instead, they're used to build more complex controls by composition or extended with features through inheritance. For example, the RepeatButton is one of the ingredients used to build the higher-level ScrollBar control (which, ultimately, is part of the even higher-level ScrollViewer). The RepeatButton gives the arrow buttons at the ends of the scroll bar their trademark behavior—scrolling continues as long as you hold it down. Similarly, the ToggleButton is used to derive the more useful CheckBox and RadioButton classes described next. However, neither the RepeatButton nor the ToggleButton is a MustInherit class, so you can use both of them directly in your user interfaces or to build custom controls if the need arises.

The CheckBox

Both the CheckBox and the RadioButton are buttons of a different sort. They derive from ToggleButton, which means they can be switched on or off by the user, which is the reason for their toggle behavior. In the case of the CheckBox, switching the control on means placing a check mark in it.

The CheckBox class doesn't add any members, so the basic CheckBox interface is defined in the ToggleButton class. Most important, ToggleButton adds an IsChecked property. IsChecked is a nullable Boolean, which means it can be set to True, False, or a null value (Nothing). Obviously, True represents a checked box, while False represents an empty one. The null value is a little trickier—it represents an indeterminate state, which is displayed as a shaded box. The indeterminate state is commonly used to represent values that haven't been set or areas where some discrepancy exists. For example, if you have a check box that allows you to apply bold formatting in a text application and the current selection includes both bold and regular text, you might set the check box to Nothing to show an indeterminate state.

To assign a null value in Silverlight markup, you need to use the null markup extension, as shown here:

```
<CheckBox IsChecked="{x:Null}" Content="A check box in indeterminate state">
</CheckBox>
```

Along with the IsChecked property, the ToggleButton class adds a property named IsThreeState, which determines whether the user is able to place the check box into an indeterminate state. If IsThreeState is False (the default), clicking the check box alternates its state between checked and unchecked, and the only way to place it in an indeterminate state is through code. If IsThreeState is True, clicking the check box cycles through all three possible states.

The ToggleButton class also defines three events that fire when the check box enters specific states: Checked, Unchecked, and Indeterminate. In most cases, it's easier to consolidate this logic into one event handler by handling the Click event that's inherited from ButtonBase. The Click event fires whenever the button changes state.

The RadioButton

The RadioButton also derives from ToggleButton and uses the same IsChecked property and the same Checked, Unchecked, and Indeterminate events. Along with these, the RadioButton adds a single property named GroupName, which allows you to control how radio buttons are placed into groups.

Ordinarily, radio buttons are grouped by their container. That means if you place three RadioButton controls in a single StackPanel, they form a group from which you can select just one of the three. On the other hand, if you place a combination of radio buttons in two separate StackPanel controls, you have two independent groups on your hands.

The GroupName property allows you to override this behavior. You can use it to create more than one group in the same container or to create a single group that spans multiple containers. Either way, the trick is simple—just give all the radio buttons that belong together the same group name.

Consider this example:

```
<StackPanel>
  <Border Margin="5" Padding="5" BorderBrush="Yellow" BorderThickness="1"
  CornerRadius="5">
    <StackPanel>
      <RadioButton Content="Group 1"></RadioButton>
      <RadioButton Content="Group 1"></RadioButton>
      <RadioButton Content="Group 1"></RadioButton>
      <RadioButton GroupName="Group3" Content="Group 3"></RadioButton>
    </StackPanel>
  </Border>
  <Border Margin="5" Padding="5" BorderBrush="Yellow" BorderThickness="1"
  CornerRadius="5">
    <StackPanel>
      <RadioButton Content="Group 2"></RadioButton>
      <RadioButton Content="Group 2"></RadioButton>
      <RadioButton Content="Group 2"></RadioButton>
      <RadioButton GroupName="Group3" Content="Group 3"></RadioButton>
    </StackPanel>
  </Border>
</StackPanel>
```

Here, there are two containers holding radio buttons but three groups (see Figure 5-8). The final radio button at the bottom of each group box is part of a third group. In this example, it makes for a confusing design, but there may be some scenarios where you want to separate a specific radio button from the pack in a subtle way without causing it to lose its group membership.

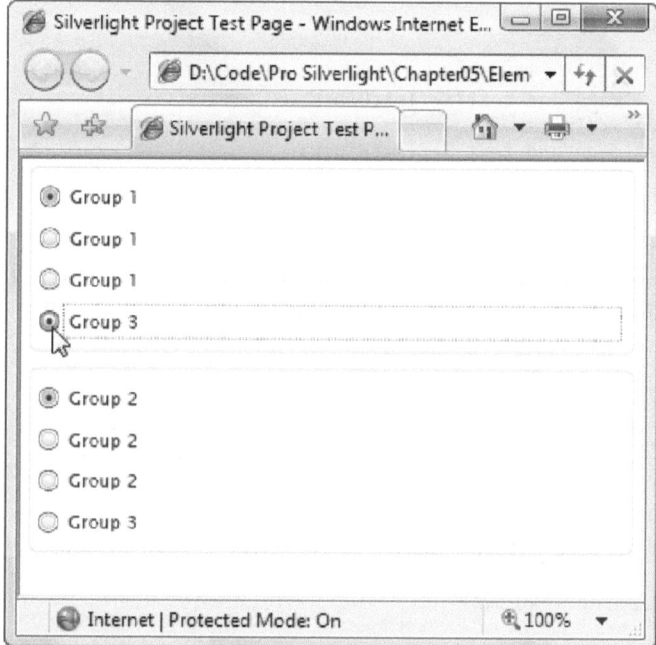

Figure 5-8. *Grouping radio buttons*

Tooltips and Pop-Ups

Silverlight has a flexible model for *tooltips* (those infamous yellow boxes that pop up when you hover over something interesting). Because tooltips in Silverlight are content controls, you can place virtually anything inside a tooltip.

Tooltips are represented by the ToolTip content control. However, you don't add the ToolTip element to your markup directly. Instead, you use the ToolTipService to configure a tooltip for an existing element, by setting attached properties. Silverlight will then create the ToolTip automatically and display it when it's needed.

The simplest example is a text-only tooltip. You can create a text-only tooltip by setting the ToolTipService.ToolTip property on another element, as shown here:

```
<Button ToolTipService.ToolTip="This is my tooltip"
 Content="I have a tooltip"></Button>
```

When you hover over this button, the text "This is my tooltip" appears in a gray pop-up box.

Customized Tooltips

If you want to supply more ambitious tooltip content, such as a combination of nested elements, you need to break the ToolTipService.ToolTip property out into a separate element. Here's an example that sets the ToolTip property of a button using more complex nested content:

```
<Button Content="I have a fancy tooltip">
  <ToolTipService.ToolTip>
    <StackPanel>
      <TextBlock Margin="3" Text="Image and text"></TextBlock>
      <Image Source="happyface.jpg"></Image>
      <TextBlock Margin="3" Text="Image and text"></TextBlock>
    </StackPanel>
  </ToolTipService.ToolTip>
</Button>
```

As in the previous example, Silverlight implicitly creates a ToolTip element. The difference is that in this case the ToolTip object contains a StackPanel rather than a simple string. Figure 5-9 shows the result.

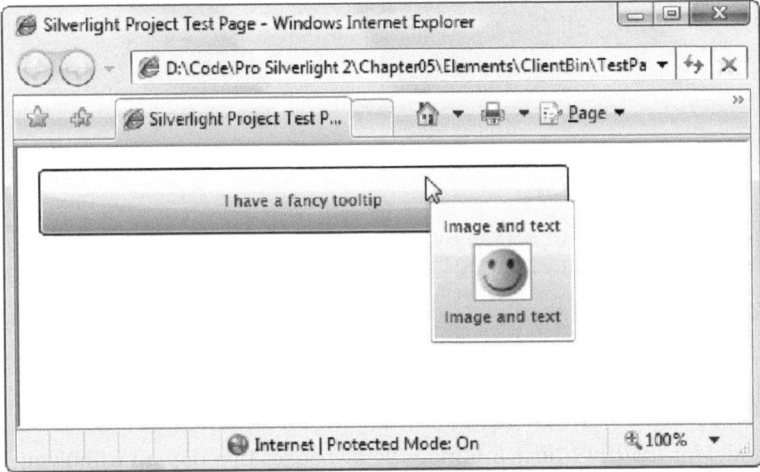

Figure 5-9. *A fancy tooltip*

■ **Note** Don't put user-interactive controls in a tooltip because the ToolTip page can't accept focus. For example, if you place a button in a ToolTip, the button will appear, but it isn't clickable. (If you attempt to click it, your mouse click will just pass through to the page underneath.) If you want a tooltip-like page that can hold other controls, consider using the Popup control instead, which is discussed shortly, in the section named "The Popup."

At this point, you might be wondering if you can customize other aspects of the tooltip's appearance, such as the standard gray background. You can get a bit more control by explicitly defining the ToolTip element when setting the ToolTipService.ToolTip property. Because the ToolTip is a content control, it provides a number of useful properties. You can adjust size and alignment properties (such as Width, Height, MaxWidth, HoriztontalContentAlignment, Padding, and so on), font (FontFamily, FontSize, FontStyle, and so on), and color (Background and Foreground). You can also use the HorizontalOffset and VerticalOffset properties to nudge the tooltip away from the mouse pointer and into the position you want, with negative or positive values.

Using the ToolTip properties, the following markup creates a tooltip that uses a red background and makes the text inside white by default:

```
<Button Content="I have a fancy tooltip">
  <ToolTipService.ToolTip>
    <ToolTip Background="DarkRed" Foreground="White">
      <StackPanel>
        <TextBlock Margin="3" Text="Image and text"></TextBlock>
        <Image Source="happyface.jpg"></Image>
        <TextBlock Margin="3" Text="Image and text"></TextBlock>
      </StackPanel>
    </ToolTip>
  </ToolTipService.ToolTip>
</Button>
```

If you assign a name to your tooltip, you can also interact with it programmatically. For example, you can use the IsEnabled property to temporarily disable a ToolTip and IsOpen to programmatically show or hide a tooltip (or just check whether the tooltip is open). You can also handle its Opened and Closed events, which is useful if you want to generate the content for a tooltip dynamically, just as it opens.

■ **Tip** If you still want more control over the appearance of a tooltip—for example, you want to remove the black border or change its shape—you simply need to substitute a new control template with the visuals you prefer. Chapter 13 has the details.

The Popup

The Popup control has a great deal in common with the ToolTip control, although neither one derives from the other.

Like the ToolTip, the Popup can hold a single piece of content, which can include any Silverlight element. (This content is stored in the Popup.Child property, rather than the ToolTip.Content property.) Also, like the ToolTip, the content in the Popup can extend beyond the bounds of the page. Lastly, the Popup can be placed using the same placement properties and shown or hidden using the same IsOpen property.

The differences between the Popup and ToolTip are more important. They include the following:

- *The Popup is never shown automatically*: You must set the IsOpen property for it to appear. The Popup does not disappear until you explicitly set its IsOpen property to False.

- *The Popup can accept focus*: Thus, you can place user-interactive controls in it, such as a Button. This functionality is one of the key reasons to use the Popup instead of the ToolTip.

Because the Popup must be shown manually, you may choose to create it entirely in code. However, you can define it just as easily in XAML markup—just make sure to include the Name property, so you can manipulate it in code. The placement of the Popup in your markup isn't important, because its top-left corner will always be aligned with the top-left corner of the Silverlight content region.

```
<StackPanel Margin="20">
  <TextBlock TextWrapping="Wrap" MouseLeftButtonDown="txt_MouseLeftButtonDown"
   Text="Click here to open the PopUp."></TextBlock>

  <Popup x:Name="popUp" MaxWidth="200">
    <Border Background="Lime" MouseLeftButtonDown="popUp_MouseLeftButtonDown">
      <TextBlock Margin="10" Text="This is the PopUp."></TextBlock>
    </Border>
  </Popup>
</StackPanel>
```

The only remaining details are the relatively trivial code that shows the Popup when the user clicks it and the code that hides the Popup when it's clicked:

```
Private Sub txt_MouseLeftButtonDown(ByVal sender As Object, _
  ByVal e As MouseButtonEventArgs)
    popUp.IsOpen = True
End Sub

Private Sub popUp_MouseLeftButtonDown(ByVal sender As Object, _
  ByVal e As MouseButtonEventArgs)
    popUp.IsOpen = False
End Sub
```

Figure 5-10 shows the Popup in action.

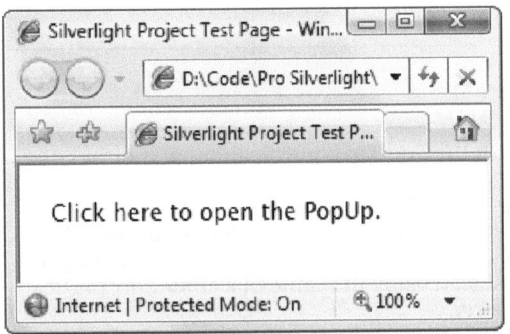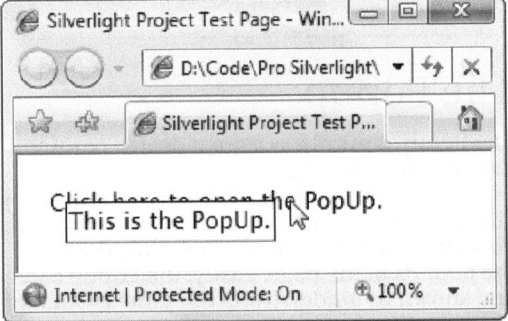

Figure 5-10. *A tooltip-like effect with the Popup*

■ **Tip** If you plan to create an extravagantly detailed Popup, you may want to consider creating a custom user control for the Popup content. You can then place an instance of that custom user control inside your pop-up. The end result is the same, but this technique simplifies your markup dramatically. And if you want your Popup to take on the characteristics of a self-contained dialog box, you should consider the ChildWindow control instead, which is described in Chapter 7.

Items Controls

Controls that wrap collections of items generally derive from the ItemsControl class. Silverlight provides four list-based controls. You'll take a look at the ListBox, the ComboBox, and the TabControl in this section. You'll explore the TreeView in Chapter 17.

The ItemsControl class fills in the basic plumbing that's used by all list-based controls. Notably, it gives you two ways to fill the list of items. The most straightforward approach is to add them directly to the Items collection, using code or XAML. This is the approach you'll see in this chapter. However, if you need to display a dynamic list, it's more common to use data binding. In this case, you set the ItemsSource property to the object that has the collection of data items you want to display. This process is covered in Chapter 16.

The ListBox

To add items to the ListBox, you can nest ListBoxItem elements inside the ListBox element. For example, here's a ListBox that contains a list of colors:

```
<ListBox>
  <ListBoxItem Content="Green"></ListBoxItem>
  <ListBoxItem Content="Blue"></ListBoxItem>
  <ListBoxItem Content="Yellow"></ListBoxItem>
  <ListBoxItem Content="Red"></ListBoxItem>
</ListBox>
```

As you'll recall from Chapter 2, different controls treat their nested content in different ways. The ListBox stores each nested object in its Items collection.

■ **Note** The ListBox class also allows multiple selection if you set the SelectionMode property to Multiple or Extended. In Multiple mode, you can select or deselect any item by clicking it. In Extended mode, you need to hold down the Ctrl key to select additional items or the Shift key to select a range of items. In either type of multiple-selection list, you use the SelectedItems collection instead of the SelectedItem property to get all the selected items.

The ListBox is a remarkably flexible control. Rather than being limited to ListBoxItem objects, it can hold any arbitrary element. This works because the ListBoxItem class derives from ContentControl, which gives it the ability to hold a single piece of nested content. If that piece of content is a UIElement-derived class, it will be rendered in the ListBox. If it's some other type of object, the ListBoxItem will call ToString() and display the resulting text.

For example, if you decided you want to create a list with images, you could create markup like this:

```
<ListBox>
  <ListBoxItem>
    <Image Source="happyface.jpg"></Image>
  </ListBoxItem>
```

```
<ListBoxItem>
  <Image Source="happyface.jpg"></Image>
</ListBoxItem>
</ListBox>
```

The ListBox is actually intelligent enough to create the ListBoxItem objects it needs implicitly. That means you can place your objects directly inside the ListBox element. Here's a more ambitious example that uses nested StackPanel objects to combine text and image content:

```
<ListBox>
  <StackPanel Orientation="Horizontal">
    <Image Source="happyface.jpg" Width="30" Height="30"></Image>
    <TextBlock VerticalAlignment="Center" Text="A happy face"></TextBlock>
  </StackPanel>
  <StackPanel Orientation="Horizontal">
    <Image Source="redx.jpg" Width="30" Height="30"></Image>
    <TextBlock VerticalAlignment="Center" Text="A warning sign"></TextBlock>
  </StackPanel>
  <StackPanel Orientation="Horizontal">
    <Image Source="happyface.jpg" Width="30" Height="30"></Image>
    <TextBlock VerticalAlignment="Center" Text="A happy face"></TextBlock>
  </StackPanel>
</ListBox>
```

In this example, the StackPanel becomes the item that's wrapped by the ListBoxItem. This markup creates the list shown in Figure 5-11.

Figure 5-11. *A list of images*

This ability to nest arbitrary elements inside list box items allows you to create a variety of list-based controls without needing to use specialized classes. For example, you can display a check box next to every item by nesting the CheckBox element inside the ListBox.

There's one caveat to be aware of when you use a list with different elements inside. When you read the SelectedItem value (and the SelectedItems and Items collections), you won't see ListBoxItem objects—instead, you'll see whatever objects you placed in the list. In the previous example, that means SelectedItem provides a StackPanel object.

When manually placing items in a list, it's up to you whether you want to place the items in directly or explicitly wrap each one in a ListBoxItem object. The second approach is often cleaner, albeit more tedious. The most important consideration is to be consistent. For example, if you place StackPanel objects in your list, the ListBox.SelectedItem object will be a StackPanel. If you place StackPanel objects wrapped by ListBoxItem objects, the ListBox.SelectedItem object will be a ListBoxItem, so code accordingly. And there's a third option—you can place data objects inside your ListBox and use a data template to display the properties you want. Chapter 16 has more about this technique.

The ListBoxItem offers a little bit of extra functionality from what you get with directly nested objects. Namely, it defines an IsSelected property that you can read (or set) and a Selected and Unselected event that tells you when that item is highlighted. However, you can get similar functionality using the members of the ListBox class, such as the SelectedItem and SelectedIndex properties and the SelectionChanged event.

■ **Note** The ListBox has support for *virtualization*, thanks to the way it uses VirtualizingStackPanel to lay out items. This means that the ListBox creates ListBoxItem objects only for the items that are currently in view, which allows it to display massive lists with tens of thousands of items without consuming ridiculous amounts of memory or slowing its performance down to a crawl. As the user scrolls, the existing set of ListBoxItem objects is reused with different data to show the appropriate items. List controls that don't support virtualization (which includes every control other than the ListBox and the DataGrid) load and scroll *much* more slowly when they're packed full of items.

The ComboBox

The ComboBox is similar to the ListBox control. It holds a collection of ComboBoxItem objects, which are created either implicitly or explicitly. As with the ListBoxItem, the ComboBoxItem is a content control that can contain any nested element. Unlike combo boxes in the Windows world, you can't type in the Silverlight ComboBox control to select an item or edit the selected value. Instead, you must use the arrow keys or the mouse to pick from the list.

The key difference between the ComboBox and ListBox classes is the way they render themselves in a window. The ComboBox control uses a drop-down list, which means only one item can be selected at a time.

One ComboBox quirk is the way it sizes itself when you use automatic sizing. The ComboBox widens itself to fit its content, which means that it changes size as you move from one item to the next. Unfortunately, there's no easy way to tell the ComboBox to take the size of its largest contained item. Instead, you may need to supply a hard-coded value for the Width property, which isn't ideal.

The TabControl

You're no doubt familiar with the TabControl, a handy container that condenses a large amount of user interface into a set of tabbed pages. In Silverlight, the TabControl is an items control that holds one or more TabItem elements.

Like several of Silverlight's more specialized controls, the TabControl is defined in a separate assembly. When you add it to a page, Visual Studio will add a reference to the System.Windows.Controls.dll assembly and map a new XML namespace, like this one:

```
<UserControl xmlns:controls=
 "clr-namespace:System.Windows.Controls;assembly=System.Windows.Controls"
 ... >
```

To use the TabControl, you must fill it with one or more TabItem elements. Each TabItem represents a separate page. Because the TabItem is a content control, it can hold another Silverlight element (like a layout container).

Here's an example of a TabControl that includes two tabs. The first tab holds a StackPanel with three check boxes:

```
<controls:TabControl>
  <controls:TabItem Header="Tab One">
    <StackPanel Margin="3">
      <CheckBox Margin="3" Content="Setting 1"></CheckBox>
      <CheckBox Margin="3" Content="Setting 2"></CheckBox>
      <CheckBox Margin="3" Content="Setting 3"></CheckBox>
    </StackPanel>
  </controls:TabItem>
  <controls:TabItem Header="Tab Two">
    ...
  </controls:TabItem>
</controls:TabControl>
```

The TabItem holds its content (in this example, a StackPanel) in the TabItem.Content property. Interestingly, the TabItem also has another property that can hold arbitrary content—the Header. In the previous example, the Header holds a simple text string. However, you just as readily fill it with graphical content or a layout container that holds a whole host of elements, as shown here:

```
<controls:TabControl>
  <controls:TabItem>
    <controls:TabItem.Header>
      <StackPanel>
        <TextBlock Margin="3">Image and Text Tab Title</TextBlock>
        <Image Source="happyface.jpg" Stretch="None" />
      </StackPanel>
    </controls:TabItem.Header>

    <StackPanel Margin="3">
      <CheckBox Margin="3" Content="Setting 1"></CheckBox>
      <CheckBox Margin="3" Content="Setting 2"></CheckBox>
      <CheckBox Margin="3" Content="Setting 3"></CheckBox>
    </StackPanel>
  </controls:TabItem>
  <controls:TabItem Header="Tab Two">
```

```
    ...
    </controls:TabItem>
</controls:TabControl>
```

Figure 5-12 shows the somewhat garish result.

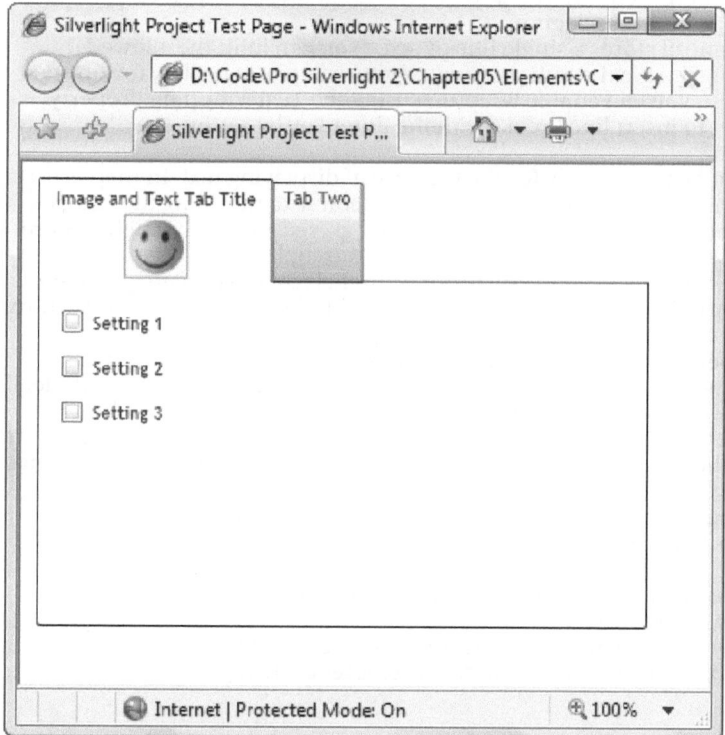

Figure 5-12. *An exotic tab title*

Like the ListBox, the TabControl includes a SelectionChanged event that fires when the visible tab changes. It also has a SelectedIndex property and a SelectedItem property, which allow you to determine or set the current tab. The TabControl adds a TabStripPlacement property, which allows you to make the tabs appear on the side or bottom of the tab control, rather than their normal location at the top.

Text Controls

Silverlight includes a standard TextBox control and several more specialized controls that derive from TextBox, including a PasswordBox (for entering text that should be concealed), an AutoCompleteBox (which shows a drop-down list of suggestions as the user types), and a RichTextBox (which allows richly formatted text, links, and pictures). You'll learn about all these variants in the following sections.

The TextBox

The basic TextBox stores a string, which is provided by the Text property. You can change the alignment of that text using the TextAlignment property, and you can use all the properties listed in Table 5-2 to control the font of the text inside the text box. The TextBox also supports many of the features of its counterpart in the Windows world, including scrolling, text wrapping, clipboard cut-and-paste, and selection.

Ordinarily, the TextBox control stores a single line of text. (You can limit the allowed number of characters by setting the MaxLength property.) However, you can allow text to span multiple lines in two ways. First, you can enable wrapping using the TextWrapping property. Second, you can allow the user to insert line breaks with the Enter key by setting the AcceptsReturn property to True.

Sometimes, you'll create a text box purely for the purpose of displaying text. In this case, set the IsReadOnly property to True to prevent editing. This is preferable to disabling the text box by setting IsEnabled to False because a disabled text box shows grayed-out text (which is more difficult to read) and does not support selection (or copying to the clipboard).

As you already know, you can select text in any text box by clicking and dragging with the mouse or holding down Shift while you move through the text with the arrow keys. The TextBox class also gives you the ability to determine or change the currently selected text programmatically, using the SelectionStart, SelectionLength, and SelectedText properties.

SelectionStart identifies the zero-based position where the selection begins. For example, if you set this property to 10, the first selected character is the 11th character in the text box. Selection Length indicates the total number of selected characters. (A value of 0 indicates no selected characters.) Finally, the SelectedText property allows you to quickly examine or change the selected text in the text box.

You can react to the selection being changed by handling the SelectionChanged event. Here's an example that reacts to this event and displays the current selection information:

```
Private Sub txt_SelectionChanged(ByVal sender As Object, ByVal e As RoutedEventArgs)
    If txtSelection IsNot Nothing Then
        txtSelection.Text = String.Format("Selection from {0} to {1} is ""{2}""", _
            txt.SelectionStart, txt.SelectionLength, txt.SelectedText)
    End If
End Sub
```

Figure 5-13 shows the result.

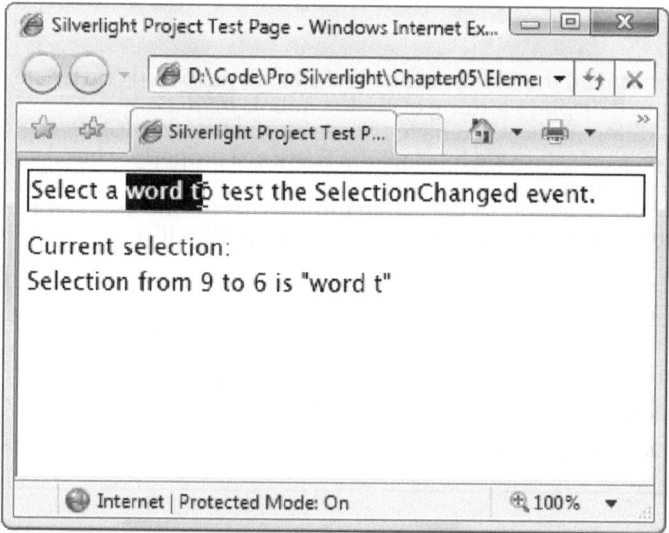

Figure 5-13. Selecting text

PROGRAMMATICALLY USING THE CLIPBOARD

Silverlight includes a Clipboard class in the System.Windows namespace. It provides three shared methods that you can call in code to work with Windows clipboard:

- *GetText()*: This method retrieves any Unicode text that's currently on the clipboard (as a string). Other types of data that could be on the clipboard, such as images and files, are not available to Silverlight applications.

- *SetText()*: This method places the text you specify on the clipboard.

- *ContainsText()*: This method returns True if the clipboard contains Unicode text content.

You can only access the clipboard in the event handler for a user-initiated action (like a mouse click or a key press). The first time your code attempts to use the clipboard with the GetText() or SetText() method, a dialog box will appear asking for clipboard access. If the user clicks Yes, this message won't appear for the rest of the session (but it will reappear the next time you run this or another Silverlight application and attempt to use clipboard again). If the user clicks No, the GetText() or SetText() method will throw a SecurityException, which you must catch in your code.

The PasswordBox

Silverlight includes a separate control called the PasswordBox to deal with password entry. The PasswordBox looks like a TextBox, but it displays a string of circle symbols to mask the characters inside. You can choose a different mask character by setting the PasswordChar property, and you can set (or retrieve) the text inside through the Password property. The PasswordBox does not provide a Text property.

Additionally, the PasswordBox does not support the clipboard. This means the user can't copy the text it contains using shortcut keys, and your code can't use properties like SelectedText.

■ **Note** The WPF PasswordBox uses in-memory encryption to ensure that passwords can't be retrieved in certain types of exploits (like memory dumps). The Silverlight Password box doesn't include this feature. It stores its contents in the same way as the ordinary TextBox.

The AutoCompleteBox

The AutoCompleteBox fuses a text entry with a drop-down list of suggestions. This feature is a common sight on the Web, powering everything from the search box on the Google homepage to the Internet Explorer address bar.

The Silverlight implementation is a surprisingly powerful control that gives you several ways to decide what items should appear in the drop-down list. The simplest approach is to start with an ordinary AutoCompleteBox:

```
<input:AutoCompleteBox x:Name="txtMonth"></input:AutoCompleteBox>
```

When you add an AutoCompleteBox from the toolbox, Visual Studio creates an XML alias named input:

```
<UserControl xmlns:input=
"clr-namespace:System.Windows.Controls;assembly=System.Windows.Controls.Input" ... >
```

Once you've added an AutoCompleteBox, create an array or list that holds the collection of possible suggestions (in no particular order), and apply this collection to the AutoCompleteBox.ItemsSource property. Typically, you'd perform this step when the page first loads by adding your code to the page constructor or handling the UserControl.Loaded event.

Here's an example that uses the set of 12 calendar months:

```
Dim monthList As String() = {"January", "February", "March", "April", _
                             "May", "June", "July", "August", "September", _
                             "October", "November", "December"}
txtMonth.ItemsSource = monthList
```

That's enough to get the default behavior. When the user types a letter in the box at runtime, a drop-down list of potential matches will appear, in alphabetical order (Figure 5-14). To select an item (and avoid typing the whole text in by hand), you can click it with the mouse or cursor down to it with the arrow keys.

■ **Note** The AutoCompleteBox offers suggestions, but it doesn't impose rules. There is no easy way to constrain users so that they can't deviate from the list of suggestions.

There's one other way for the AutoCompleteBox to behave. If you set IsTextCompletionEnabled to True, the AutoCompleteBox automatically fills in the text box as the user types. For example, if the user types J in the month example, the AutoCompleteBox finds the first matching month and fills in *anuary*. The new filled-in text is highlighted, which means that it will be overwritten if the user continues to type (or deleted if the user presses the Delete or Backspace key). Figure 5-14 compares the difference.

■ **Note** When you read the AutoCompleteBox.Text property, you get exactly the text that's currently displayed in the AutoCompleteBox. If you've set IsTextCompletionEnabled to True, you also get any text that's automatically inserted as part of a match.

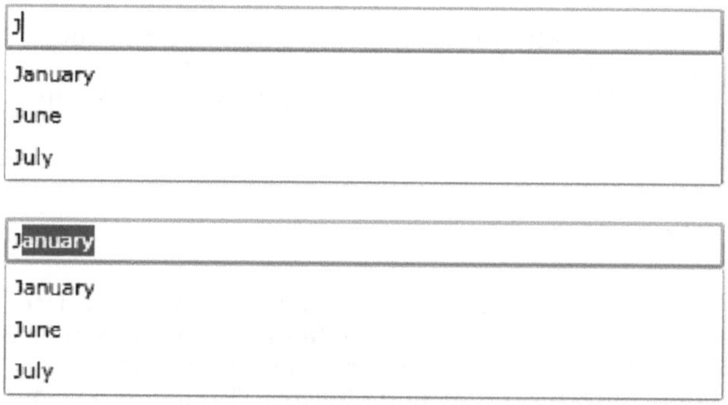

Figure 5-14. Months that start with J

Filter Mode

Ordinarily, the AutoCompleteBox filters out the list of bound items by comparing the start of each one with the text that's been typed in so far. However, you can change this behavior by setting the FilterMode property. It takes one of the values from the AutoCompleteFilterMode enumeration. The most useful ones are described in Table 5-4.

Table 5-4. Values for the AutoCompleteFilterMode Enumeration

Name	Description
None	No filtering will be performed, and all the items will appear in the list of suggestions. This is also the option you'll use if you need to fetch the collection of items yourself—for example, if you need to query them from a database or request them from a web service.
StartsWith	All the items that start with the typed-in text will appear. This is the default.
StartsWithCaseSensitive	All the items that start with the typed-in text will appear provided the capitalization also matches.
Contains	All the items that contain the typed-in text will appear. For example, typing **ember** would match *September, November*, and *December*.
ContainsCaseSensitive	All the items that contain the typed-in text will appear provided the capitalization also matches.
Custom	You must perform the filtering by applying a delegate that does the work to the TextFilter or ItemFilter property. In fact, if you set TextFilter or ItemFilter the FilterMode property is automatically switched to Custom.

Custom Filtering

To perform any sort of custom filtering, you must set the TextFilter or ItemFilter property. Use TextFilter if your ItemsSource is a collection or strings, and use ItemFilter if your ItemsSource is a collection with some other sort of object. Either way, the TextFilter or ItemFilter property takes a delegate that points to a method that performs the custom filtering. This method takes two arguments: the text that the user has entered so far and the item that you're currently testing for a match.

```
Public Function ItemFilter(ByVal text As String, ByVal item As Object) _
  As Boolean
    ...
End Function
```

The code in the filtering method should perform whatever comparison logic you need and return True if the item should be included as a drop-down suggestion based on the current text or False if it should be omitted.

Custom filtering is particularly useful if you're comparing text against a list of complex objects. That's because it allows you to incorporate the information that's stored in different properties.

For example, imagine you have this simple Product class:

```vbnet
Public Class Product
    Private _productName As String
    Public Property ProductName() As String
        Get
            Return _productName
        End Get
        Set(ByVal value As String)
            _productName = value
        End Set
    End Property

    Private _productCode As String
    Public Property ProductCode() As String
        Get
            Return _productCode
        End Get
        Set(ByVal value As String)
            _productCode = value
        End Set
    End Property

    Public Sub New(ByVal productName As String, ByVal productCode As String)
        Me.ProductName = productName
        Me.ProductCode = productCode
    End Sub

    Public Overrides Function ToString() As String
        Return ProductName
    End Function
End Class
```

You then decide to build an AutoCompleteBox that attempts to match the user's text with a Product object. In preparation for this step, you fill the AutoComplexBox.ItemsSource collection with product objects:

```vbnet
Dim products As Product() = { _
  New Product("Peanut Butter Applicator", "C_PBA-01"), _
  New Product("Pelvic Strengthener", "C_PVS-309"), ...}

acbProduct.ItemsSource = products
```

If you take no further steps, the AutoCompleteBox will use its standard behavior. As the user types, it will call ToString() on each Product object. It will then use that text to perform its suggestion filtering. Because the Product class overrides the ToString() method to return the product name, the AutoCompleteBox will attempt to match the user's text with a product name, which is perfectly reasonable.

However, if you perform custom filtering, you can get a bit more sophisticated. For example, you can check whether the user's text matches the ProductName property *or* the ProductCode property and deem the Product object as a match either way. Here's an example of the custom filtering logic that does the trick:

177

```
Public Function ProductItemFilter(ByVal text As String, ByVal item As Object) _
  As Boolean
    Dim product As Product = CType(item, Product)

    ' Call it a match if the typed-in text appears in the product code
    ' or at the beginning of the product name.
    Return product.ProductName.StartsWith(text) _
      OrElse product.ProductCode.Contains(text)
End Function
```

You simply need to connect this method to your AutoComplexBox when it's first initialized:

```
acbProduct.ItemFilter = ProductItemFilter
```

Now if the user types the text **PBA**, it matches the product code C_PBA-01 and shows the matching item Peanut Butter Applicator in the list of suggestions, as shown in Figure 5-15.

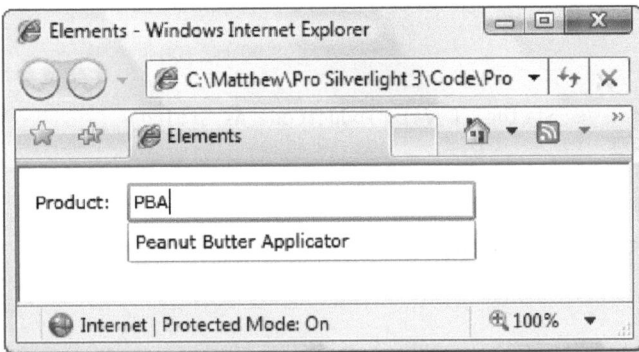

Figure 5-15. A custom search that matches product codes

Dynamic Item Lists

So far, you've used the ItemsSource property to fill the AutoCompleteBox with a collection of suggestions. For this to work, you must have the complete list, and it must be a manageable size. If you need to pull the information from somewhere else or the list is large enough that it isn't practical to load the whole thing at once, you'll need to take a different approach to filling the AutoCompleteBox. Instead of setting the ItemsSource property when the page is first created, you'll need to set it in real time, as the user types.

To do so, set the FilterMode property to None, and handle the Populating event. The Populating event fires whenever the AutoCompleteBox is ready to search for results. By default, this happens every time the user presses a key and changes the current text entry. You can make the AutoCompleteBox somewhat more relaxed using the MinimumPrefixLength and MinimumPopupDelay properties that are discussed at the end of this section.

```
<input:AutoCompleteBox x:Name="acbProducts" FilterMode="None"
  Populating="acbProducts_Populating" ></input:AutoCompleteBox>
```

When the Populating event fires, you have two choices: set the ItemsSource property immediately or launch an asynchronous process to do it. Setting the ItemsSource property immediately makes sense if you have the list of suggestions on hand or you can generate them quickly. The list of suggestions will then appear in the drop-down list right away.

But in many situations, you'll need a potentially time-consuming step to get the list of suggestions, such as performing a series of calculations or querying a web service. In this situation, you need to launch an asynchronous process. Although you can accomplish this with the multithreading support that's described in Chapter 19, you won't necessarily need to. Some Silverlight features have built-in asynchronous support. This is the case with Silverlight's implementation of web services, which is hardwired to use asynchronous calls exclusively.

When using an asynchronous operation, you need to explicitly cancel the normal processing in the Populating event handler, by setting PopulatingEventArgs.Cancel to True. You can then launch the asynchronous operation. The following example gets the suggestion list asynchronously from a web service. (You'll learn much more about coding and consuming web services in Chapter 15. For now, you can review the example code and the downloadable project with this chapter.)

```vb
Private Sub acbProducts_Populating(ByVal sender As Object, _
  ByVal e As PopulatingEventArgs)

    ' Signal that the task is being performed asynchronously.
    e.Cancel = True

    ' Create the web service object.
    Dim service As New ProductAutoCompleteClient()

    ' Attach an event handler to the completion event.
    AddHandler service.GetProductMatchesCompleted, _
      AddressOf GetProductMatchesCompleted

    ' Call the web service (asynchronously).
    service.GetProductMatchesAsync(e.Parameter)
End Sub
```

On the web server, the code in a GetProductMathes() web method runs and retrieves the matches:

```vb
Public Function GetProductMatches(ByVal inputText As String) As String()
    ' Get the products (for example, from a server-side database).
    Dim products As Product() = GetProducts()

    ' Create a collection of matches.
    Dim productMatches As New List(Of String)()
    For Each product As Product In products
        ' See if this is a match.
        If product.ProductName.StartsWith(inputText) OrElse _
          product.ProductCode.Contains(inputText) Then
            productMatches.Add(product.ProductName)
        End If
    Next

    ' Return the list of matches.
    Return productMatches.ToArray()
End Function
```

When the asynchronous operation finishes and you receive the result in your Silverlight application, you fill the ItemsSource property with the list of suggestions. Then, you must call the PopulateComplete() method to notify the AutoCompleteBox that the new data has arrived. Here's the callback handler that does the job in the current example:

```
Private Sub GetProductMatchesCompleted(ByVal sender As Object, _
  ByVal e As GetProductMatchesCompletedEventArgs)
    ' Check for a web service error.
    If e.Error IsNot Nothing Then
        lblStatus.Text = e.Error.Message
        Return
    End If

    ' Set the suggestions.
    acbProducts.ItemsSource = e.Result

    ' Notify the control that the data has arrived.
    acbProducts.PopulateComplete()
End Sub
```

When filling the AutoCompleteBox with a time-consuming or asynchronous step, there are two properties you may want to adjust: MinimumPrefixLength and MinimumPopupDelay. MinimumPrefixLength determines how much text must be typed in before the AutoCompleteBox gives its suggestions. Ordinarily, the AutoCompleteBox offers suggestions after the first letter is entered. If you want it to wait for three letters (the standard used by many of the Ajax-powered autocompletion text boxes that you'll find on the Web), set MinimumPrefixLength to 3. Similarly, you can force the AutoCompleteBox to hold off until a certain interval of time has passed since the user's last keystroke using the MinimumPopulateDelay property. This way, you won't waste time with a flurry of overlapping calls to a slow web service. Of course, this doesn't necessarily determine how long it takes for the suggestions to *appear*—that depends on the wait before initiating the query and then the time needed to contact the web server and receive a response.

The RichTextBox

If you've programmed with WPF before, you are probably familiar with its flow document model—a flexible system for displaying richly formatted, read-only content. Compared to ordinary text display (say, with the TextBlock element), flow documents support advanced features such as balanced columns, text wrapping around floating figures, and intelligent algorithms for letter spacing, word wrapping, and hyphenation. Silverlight doesn't include the flow document feature, but it does borrow part of the same model to support one of its most impressive controls: the RichTextBox.

The RichTextBox is an editable text box control that supports rich formatting. Unlike the ordinary TextBox control, the RichTextBox allows individual sections (say, a single word or an entire paragraph) to be formatted with a different font and color. The RichTextBox also supports images, links, and inline elements (like drop-down lists and buttons). Best of all, the RichTextBox is easy to use in any application.

Text Elements

Before you can understand the RichTextBox, you need to know a bit about the model it uses. While an ordinary text box shows holds a string of text, the RichTextBox holds an entire document, which is represented by a collection of *text elements* (sometimes called *flow elements*, because they are used to create *flow documents*).

Text elements have an important difference from the elements you've seen so far. They don't inherit from the familiar UIElement and FrameworkElement classes. Instead, they form an entirely separate branch of classes that derive first from DependencyObject and then from TextElement. As you might expect, they are also far simpler, supporting no events and providing a small set of mostly formatting-related properties. Figure 5-16 shows the inheritance hierarchy for text elements.

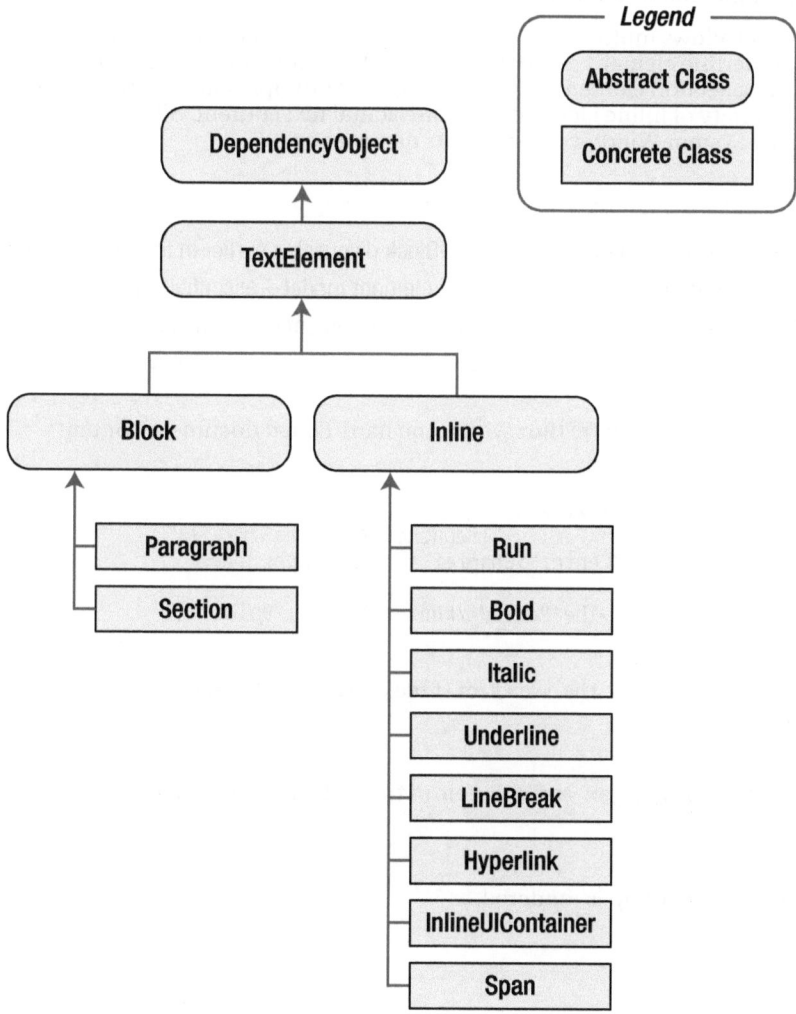

Figure 5-16. Text elements

There are two key branches of text elements:

- *Block elements*: There are two block element classes: Paragraph and Section. Paragraphs can hold text and a combination on inline elements. Sections can hold a group of paragraphs (or a group of sections), but they must be created programmatically, as they aren't usable in XAML.
- *Inline elements*: These elements are nested inside a block element (or another inline element). They include elements for formatting text (Bold, Italic, Underline, Run), making hard line breaks (LineBreak), adding hyperlinks (Hyperlink), and embedding other controls (InlineUIContainer). Finally, the Span container gives you the ability to group together multiple inline elements in one container.

This text element model allows multiple layers of nesting. For example, you can place a Bold element inside an Underline element to create text that's both bold and underlined. Similarly, you might create a Section element that wraps together multiple Paragraph elements, each of which contains a variety of inline elements with the actual text content. All of these elements are defined in the System.Windows.Documents namespace.

■ **Note** You may remember the inline elements from the TextBlock discussion earlier in this chapter. In fact, the TextBlock shares a stripped-down version of the text element model—essentially, it's a container that holds a small, read-only scrap of a document that consists entirely of inline objects.

The following example shows a RichTextBox with some hard-coded document content already filled in:

```
<RichTextBox Margin="5" x:Name="richText">
  <Paragraph Foreground="DarkRed" FontFamily="Trebuchet MS" FontSize="22"
   FontWeight="Bold" TextAlignment="Center">Chapter I</Paragraph>
  <Paragraph>
    <Bold><Italic><Run FontSize="12">The Period</Run></Italic></Bold>
  </Paragraph>
  <Paragraph>
    It was the best of times, it was the worst of times, it was the age of ...
    <LineBreak></LineBreak>
  </Paragraph>
  <Paragraph>
    There were a king with a large jaw and a queen with a plain face, on the ...
  </Paragraph>
</RichTextBox>
```

Figure 5-17 shows how this markup is rendered.

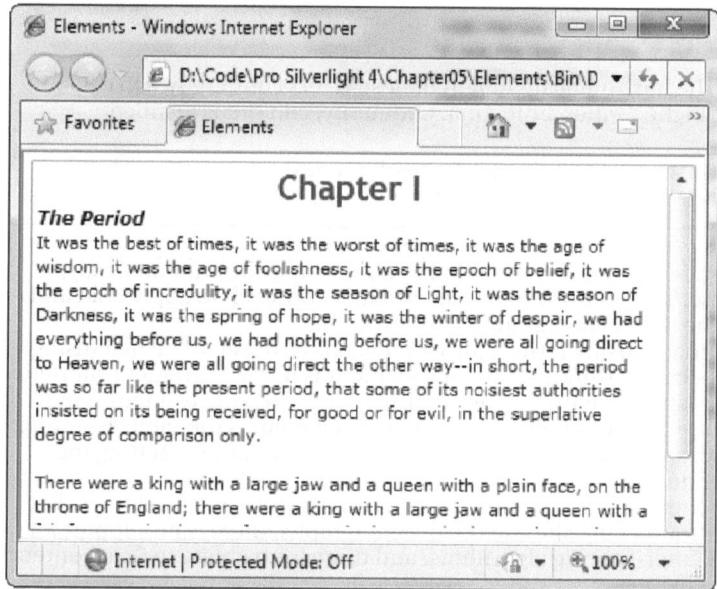

Figure 5-17. *A simple document*

This example is enough to illustrate the key details of the text element model. First, every RichTextBox holds a collection of block elements. In this case, the RichTextBox holds four Paragraph objects, which hold still more objects nested inside, such as Bold, Italic, and Run to apply formatting, and LineBreak to add some empty space between paragraph. (Ordinarily, there is none.)

Formatting Text Elements

You can use two broad approaches to applying formatting. The first is to use the properties of the appropriate text element class, most of which are inherited from the base class TextElement. The previous example used this approach to set the formatting for the chapter title in the first paragraph and to apply formatting to a Run in the second paragraph. Table 5-5 lists all the properties you can set.

Table 5-5. *Properties for Content Elements*

Name	Description
Foreground	Accept brushes that will be used to paint the foreground text. You can't set a background for individual text elements, although you can set a background for the entire document using the RichTextBox.Background property.
FontFamily, FontSize, FontStretch, FontStyle, and FontWeight	Allow you to configure the font that's used to display text.

Name	Description
TextAlignment	Sets the horizontal alignment of nested text content (which can be Left, Right, Center, or Justify). Ordinarily, content is justified.
TextDecorations	Allows you to add a line underneath the text. This property is provided only for inline objects.

The second approach is to use a nested element that applies formatting automatically, such as Bold, Italic, or Underline. The previous example used this approach to apply bold and italic formatting to the title in the second paragraph.

Truthfully, the second approach (using formatting elements) is a subset of the first (using formatting properties). That's because using a formatting element is just a convenient shortcut to using a Run or Span with the same formatting. For example, using the Bold element is the same as setting the FontWeight property to Bold, using the Italic element is the same as setting the FontStyles property to Italic, and using the Underline element is the same as setting the TextDecorations property to Underline.

Compared to WPF flow documents, the Silverlight text elements lack several features. For example, there is no way add extra spacing above or below paragraphs, there is no support for advanced text justification and letter-spacing algorithms, and there is no ability to float content or wrap text around the sides of figures.

Manipulating Text Elements in Code

Documents can also be explored and even created programmatically. The entry point to the content in the RichTextBox is the Blocks property, which holds a collection of block elements. It's analogous to the Inlines property of the TextBlock, which holds a collection of inline elements.

Creating a document programmatically is fairly tedious because of a number of disparate elements that need to be created. As with all XAML elements, you must create each element and then set all its properties, because there are no constructors to help you. You also need to create a Run object to wrap every piece of text and then add your Run objects to a suitable container (like a Paragraph object). This step isn't required when you create a document with markup, because Silverlight automatically creates a Run object to wrap the text you place inside each Paragraph object.

Here's a snippet of code that creates a document with a single paragraph and some bolded text:

```
' Create the first part of the sentence.
Dim runFirst As New Run()
runFirst.Text = "Hello world of "

' Create bolded text.
Dim bold As New Bold()
Dim runBold As New Run()
runBold.Text = "dynamically generated"
bold.Inlines.Add(runBold)

' Create last part of sentence.
Dim runLast As New Run()
runLast.Text = " documents"
```

```
' Add three parts of sentence to a paragraph, in order.
Dim paragraph As New Paragraph()
paragraph.Inlines.Add(runFirst)
paragraph.Inlines.Add(bold)
paragraph.Inlines.Add(runLast)

' Add this paragraph to the RichTextBox.
richText.Blocks.Clear()
richText.Blocks.Add(paragraph)
```

The result is the sentence "Hello world of **dynamically generated** documents."

Most of the time, you won't create flow documents programmatically. However, you might want to create an application that browses through portions of a flow document and modifies them dynamically. You can do this in the same way that you interact with any other WPF elements: by responding to element events and by attaching a name to the elements that you want to change. However, because flow documents use deeply nested content with a free-flowing structure, you may need to dig through several layers to find the actual content you want to modify. (Remember, this content is always stored in a Run object, even if the run isn't declared explicitly.)

■ **Tip** The previous example demonstrates the most straightforward way to generate a document with code. However, more complex scenarios are possible—for example, if you want to move through user-supplied content, find individual words, and tweak their formatting. This sort of task works better using the lower-level TextPointer class, which lets you find logical insertion points in the RichTextBox content. For a demonstration of this more advanced but less common technique, see http://tinyurl.com/26352z2.

Creating a Text Editor

Although you can create documents with handwritten XAML or build and manipulate them in code, the real goal of the RichTextBox is to provide a place where users can edit rich content.

On its own, the RichTextBox works just like the TextBox control. The user can type text into it, edit that text, cut and paste a selection, and so on. However, there's no built-in way for the user to apply formatting. To allow this, it's up for you to add the appropriate controls, like a button for toggling bold formatting or a list for changing the font. Figure 5-18 shows an example.

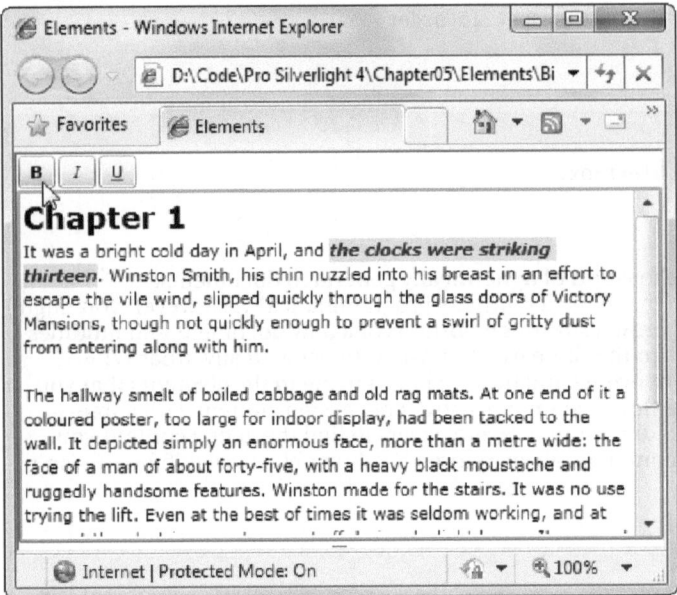

Figure 5-18. *A RichTextBox with editing controls*

In this example, the user selects some portion of text (which may span multiple inline elements and multiple paragraphs) and then uses the buttons to toggle bold, italic, and underline formatting. The code is actually quite simple. The first step is to get a reference to the TextSelection object that represents the selection from the RichTextBox.Selection property:

```
Dim selection As TextSelection = richTextBox.Selection
```

The TextSelection object provides information about the start point and end point of the selection (through the Start and End properties) and the selected content (through the Text and Xaml properties). If nothing is selected, the RichTextBox.Selection property will still return a valid TextSelection object, but its Text will be an empty string.

The TextSelection object also provides the two methods you need to manage formatting: GetPropertyValue(), which tests the selection for a given formatting characteristic; and ApplyPropertyValue(), which sets the new formatting you want. For example, if you want to make the selected text bold, you simply need to manipulate the FontWeightProperty using ApplyPropertyValue():

```
selection.ApplyPropertyValue(Run.FontWeightProperty, FontWeights.Bold)
```

Similarly, you can use GetPropertyValue() to check whether the currently selected text is bold:

```
Dim currentBoldState As FontWeight
currentBoldState = _
  CType(selection.GetPropertyValue(Run.FontWeightProperty), FontWeight)
```

There's one catch with this approach. If the selected text has mixed bold and normal text, the GetPropertyValue() method returns DependencyProperty.UnsetValue instead of a

FontWeight. It's up to you how you handle this case—typically, you'll either bold the entire selection (as this example does) or do nothing at all.

The following code shows the complete code that underpins the Bold button (represented by the button with the bold letter **B** in Figure 5-18). Depending on the selection, this code either applies or removes bold formatting.

```
Private Sub cmdBold_Click(ByVal sender As Object, ByVal e As RoutedEventArgs)
    Dim selection As TextSelection = richTextBox.Selection

    ' If no text is selected, treat it as a selection of normal text.
    Dim currentState As FontWeight = FontWeights.Normal

    ' Try to get the bold state of the selected text.
    If selection.GetPropertyValue(Run.FontWeightProperty) IsNot _
      DependencyProperty.UnsetValue Then
        currentState = _
          CType(selection.GetPropertyValue(Run.FontWeightProperty), FontWeight)
    End If

    If currentState Is FontWeights.Normal Then
        selection.ApplyPropertyValue(Run.FontWeightProperty, FontWeights.Bold)
    Else
        selection.ApplyPropertyValue(Run.FontWeightProperty, FontWeights.Normal)
    End If

    ' A nice detail is to bring the focus back to the text box, so the user
    ' can resume typing.
    richTextBox.Focus()
End Sub
```

■ **Note** You'll notice that this code doesn't check for an empty selection. If no text is currently selected, clicking the Bold button applies bold formatting to the insertion point, so that when the user starts typing, the new text will be bold. Another possible approach is to enable the Bold button only when text is selected. You can manage the state of the Bold button by responding to the RichTextBox.SelectionChanged event.

The code for the Italic and Underline buttons is virtually identical. The only difference is the property it uses—instead of checking and setting FontWeight, these code routines use the FontStyle and TextDecorations properties, respectively.

Saving and Opening Rich Text Files

If your application gives users the ability to edit rich text content, it's highly likely that you need a way to store the final result. Fortunately, the RichTextBox makes this easy. Although Silverlight doesn't have any built-in support for common rich text formats such as .rtf, .doc, or .docx, it does allow you retrieve its fully formatted contents as a XAML document using the RichTextBox.Xaml property.

Here's an example that copies the XAML markup to an ordinary text box so you can study it:

```
txtFlowDocumentMarkup.Text = richTextBox.Xaml
```

Figure 5-19 shows the result.

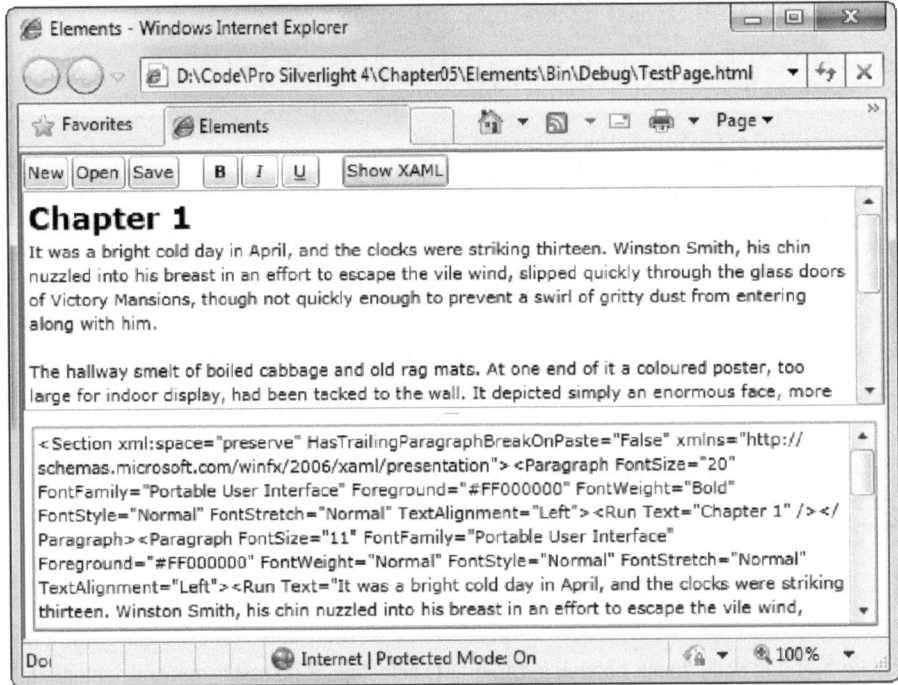

Figure 5-19. The XAML markup for a RichTextBox

The XAML document that's exposed by the RichTextBox.Xaml property consists of the text elements you learned about earlier (such as Section, Paragraph, and Run) and has the full formatting information specified as attributes. Because you can store this XAML content in a simple string, it's easy to use any other technique you need to store it, such as submitting it to a web service that wraps a back-end database (Chapter 15) or saving it in a simple text file (Chapter 18). But if you can't wait to see the details, check out the downloadable examples with this chapter, which allow you to save the content from a RichTextBox to a file on your hard drive and open it later. (The trick is the straightforward SaveFileDialog and OpenFileDialog classes, which you'll consider in Chapter 18.)

■ **Tip** To clear the content in a RichTextBox, call RichTextBox.Blocks.Clear().

Using Hyperlinks and Elements in a RichTextBox

The RichTextBox stretches its capabilities beyond the display and editing of rich text content. It also supports interactive elements, such as the Hyperlink.

The Hyperlink is an inline element that's rendered as blue underlined text. It works in the same way as the HyperlinkButton control discussed earlier in this chapter. Like the HyperlinkButton, the Hyperlink can launch a web browser to show an external page (if you set the NavigateUrl property to an absolute URL) or send a frame in your application to a new XAML page (if you set NavigateUrl to a relative URL). Or, if you attach an event handler to the Click event, it can trigger a code routine that performs any action you want. Here's an example:

```
<RichTextBox Margin="5" x:Name="richText" IsReadOnly="True">
  <Paragraph>
    <Hyperlink Click="cmdDoSomething_Click">Click this link.</Hyperlink>
  </Paragraph>
</RichTextBox>
```

There's one catch. The hyperlinks in your document are active only if you set the RichTextBox.IsReadOnly property to True. Otherwise, the hyperlink looks the same, but it can't be clicked (although its text can be edited).

■ **Tip** Hyperlinks and other types of embedded controls don't make much sense in in an editable document. But they make perfect sense if you're showing some type of read-only content in the RichTextBox, such as product documentation or company information.

The Hyperlink takes you to the very edge of what's possible with the RichTextBox model of text elements. However, the RichTextBox has a convenient back door that lets you embed any Silverlight element alongside your rich text content. That back door is the InlineUIContainer, which hosts a full-fledged Silverlight element. For example, you can use the InlineUIContainer to add a Button, CheckBox, Image, or even a DataGrid to your RichTextBox. Here's an example:

```
<RichTextBox Margin="5" x:Name="richText" IsReadOnly="True">
  <Paragraph Foreground="DarkRed" FontFamily="Trebuchet MS" FontSize="22"
    FontWeight="Bold">
    <InlineUIContainer>
      <Image Source="bookcover.jpg" Stretch="None"></Image>
    </InlineUIContainer>
    Chapter I
  </Paragraph>
  ...
</RichTextBox>
```

Figure 5-20 shows the result. Note that inline elements are placed inline with the current text line, as though they are ordinary characters. In other words, it's not possible to have several lines of text wrap around a large element like an Image.

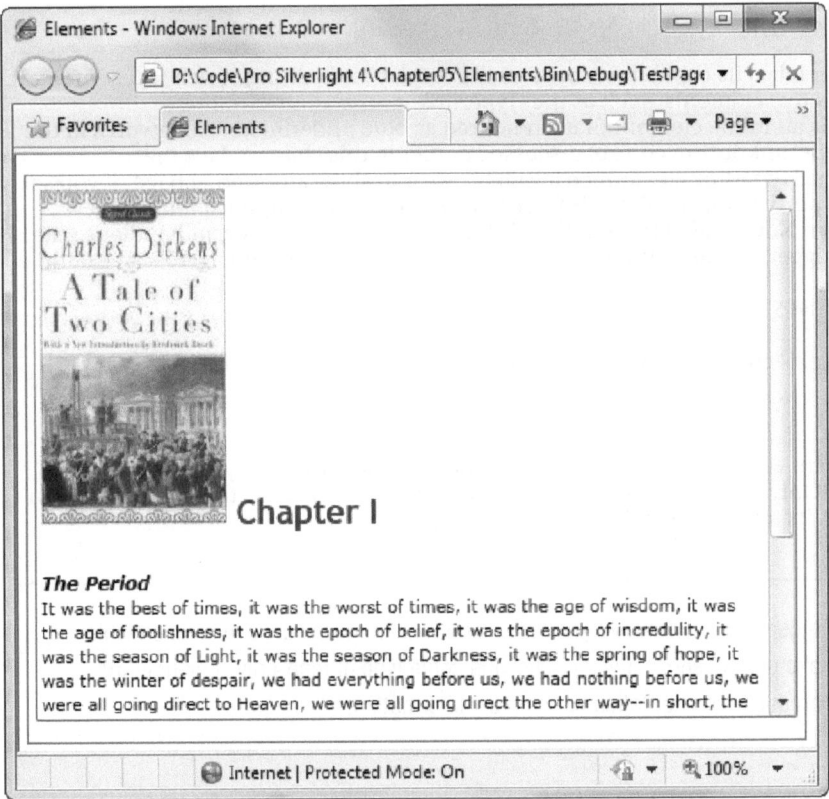

Figure 5-20. *Embedding an Image element in a RichTextBox*

The InlineUIContainer has two important limitations. First, if you want the embedded element to be interactive so that it can accept focus and receive input events, you must set the RichTextBox.IsReadOnly property to True. If you don't, the user will still see the element and will still be able to edit the RichTextBox content—for example, the user will be able to delete the InlineUIContainer element or add text around it. However, the user won't be able to click it, type into it, or otherwise interact with it. This is perfectly fine for an element like the Image but obviously less suitable for an element like the Button.

The second limitation is that the InlineUIContainer isn't represented in the document that's returned by RichTextBox.Xaml. This is a significant shortcoming, because it means you won't be able to load a ready-made XAML document that includes InlineUIContainer objects. Instead, you'll be forced to add them programmatically.

Range-Based Controls

Silverlight includes three controls that use the concept of a *range*. These controls take a numeric value that falls in between a specific minimum and maximum value. These controls—ScrollBar, Slider, and ProgressBar—derive from the RangeBase class (which itself derives from the Control class). The RangeBase class adds a ValueChanged event, a Tooltip property, and the range properties shown in Table 5-6.

Table 5-6. *Properties of the RangeBase Class*

Name	Description
Value	This is the current value of the control (which must fall between the minimum and maximum). By default, it starts at 0. Contrary to what you might expect, Value isn't an integer—it's a double, so it accepts fractional values. You can react to the ValueChanged event if you want to be notified when the value is changed.
Maximum	This is the upper limit (the largest allowed value). The default value is 1.
Minimum	This is the lower limit (the smallest allowed value). The default value is 0.
SmallChange	This is the amount the Value property is adjusted up or down for a "small change." The meaning of a small change depends on the control (and may not be used at all). For the ScrollBar and Slider, this is the amount the value changes when you use the arrow keys. For the ScrollBar, you can also use the arrow buttons at either end of the bar. The default SmallChange is 0.1.
LargeChange	This is the amount the Value property is adjusted up or down for a "large change." The meaning of a large change depends on the control (and may not be used at all). For the ScrollBar and Slider, this is the amount the value changes when you use the Page Up and Page Down keys or when you click the bar on either side of the thumb (which indicates the current position). The default LargeChange is 1.

Ordinarily, there's no need to use the ScrollBar control directly. The higher-level ScrollViewer control, which wraps two ScrollBar controls, is typically much more useful. (The ScrollViewer was covered in Chapter 3.) However, the Slider and ProgressBar are more useful on their own.

The Slider

The Slider is a specialized control that's occasionally useful. You might use it to set numeric values in situations where the number itself isn't particularly significant. For example, it makes sense to set the volume in a media player by dragging the thumb in a slider bar from side to side. The general position of the thumb indicates the relative loudness (normal, quiet, loud), but the underlying number has no meaning to the user.

Here's an example that creates the horizontal slider shown in Figure 5-21:

```
<Slider Orientation="Horizontal" Minimum="0" Maximum="10" Width="100" />
```

Figure 5-21. *A basic slider*

Unlike WPF, the Silverlight slider doesn't provide any properties for adding tick marks. However, as with any control, you can change its appearance while leaving its functionality intact using the control template feature described in Chapter 13.

The ProgressBar

The ProgressBar indicates the progress of a long-running task. Unlike the slider, the ProgressBar isn't user interactive. Instead, it's up to your code to periodically increment the Value property. By default, the Minimum value of a ProgressBar is 0, and the Maximum value is 100, so the Value corresponds to the percentage of work done. You'll see an example with the ProgressBar in Chapter 6, with a page that downloads a file from the Web and shows its progress on the way.

One neat trick that you can perform with the ProgressBar is using it to show a long-running status indicator, even if you don't know how long the task will take. You do this by setting the IsIndeterminate property to True:

```
<ProgressBar Height="18" Width="200" IsIndeterminate="True"></ProgressBar>
```

When setting IsIndeterminate, you no longer use the Minimum, Maximum, and Value properties. No matter what values these properties have, the ProgressBar will show a hatched pattern that travels continuously from left to right. This pattern indicates that there's work in progress, but it doesn't provide any information about how much progress has been made so far.

Date Controls

Silverlight adds two date controls, neither of which exists in the WPF control library. Both are designed to allow the user to choose a single date.

The Calendar control displays a calendar that's similar to what you see in the Windows operating system (for example, when you configure the system date). It shows a single month at a time and allows you to step through from month to month (by clicking the arrow buttons) or jump to a specific month (by clicking the month header to view an entire year and then clicking the month).

The DatePicker requires less space. It's modeled after a simple text box, which holds a date string in long or short date format. However, the DatePicker provides a drop-down arrow that, when clicked, pops open a full calendar view that's identical to that shown by the Calendar control. This pop-up is displayed overtop of any other content, just like a drop-down combo box.

Figure 5-22 shows the two display modes that the Calendar supports and the two date formats that the DatePicker allows.

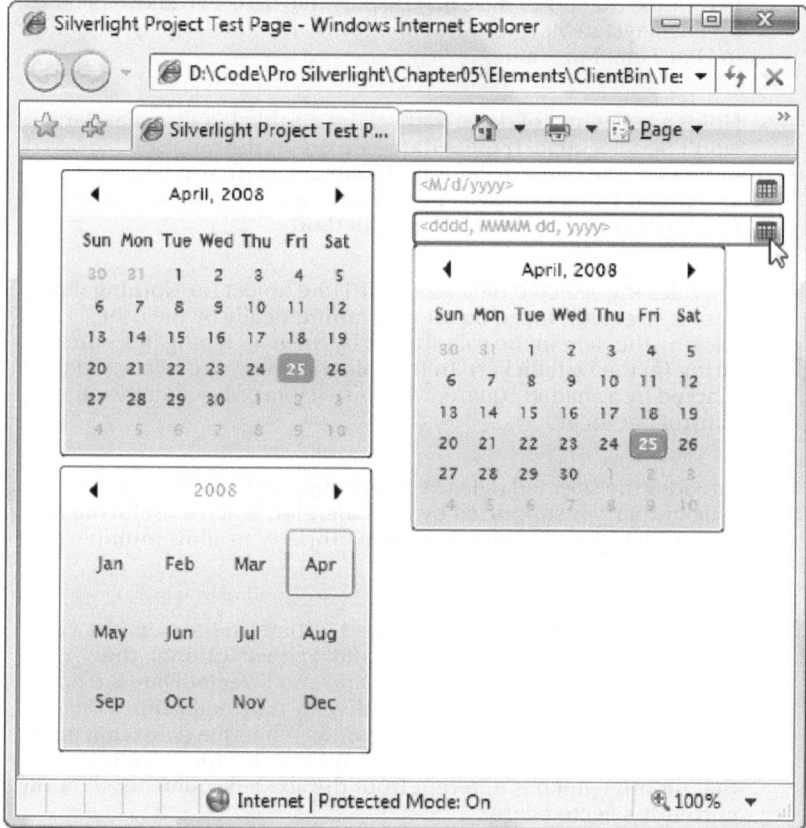

Figure 5-22. *The Calendar and DatePicker*

The Calendar and DatePicker include properties that allow you to determine which dates are shown and which dates are selectable (provided they fall in a contiguous range). Table 5-7 lists the properties you can use.

Table 5-7. Properties of the Calendar and DatePicker Classes

Property	Description
DisplayDateStart and DisplayDateEnd	Sets the range of dates that are displayed in the calendar view, from the first, earliest date (DisplayDateStart) to the last, most recent date (DisplayDateEnd). The user won't be able to navigate to months that don't have any displayable dates. To show all dates, set DisplayDateStart to DateTime.MinValue and DisplayDateEnd to DateTime.MaxValue.
BlackoutDates	Holds a collection of dates that will be disabled in the calendar and won't be selectable. If these dates are not in the range of displayed dates or if one of these dates is already selected, you'll receive an exception. To prevent selection of any date in the past, call the BlackoutDates.AddDatesInPast() method.
SelectedDate	Provides the selected date as a DateTime object (or Nothing if no date is selected). It can be set programmatically by the user clicking the date in the calendar or by the user typing in a date string (in the DatePicker). In the calendar view, the selected date is marked by a shaded square, which is visible only when the date control has focus.
SelectedDates	Provides the selected dates as a collection of DateTime objects. This property is supported by the Calendar, and it's useful only if you've changed the SelectionMode property to allow multiple date selection.
DisplayDate	Determines the date that's displayed initially in the calendar view (using a DateTime object). If it's a null value (Nothing), the SelectedDate is shown. If DisplayDate and SelectedDate are both null, the current date is used. The display date determines the initial month page of the calendar view. When the date control has focus, a square outline is displayed around the appropriate day in that month (which is different from the shaded square used for the currently selected date).
FirstDayOfWeek	Determines the day of the week that will be displayed at the start of each calendar row, in the leftmost position.
IsTodayHighlighted	Determines whether the calendar view uses highlighting to point out the current date.
DisplayMode (Calendar only)	Determines the initial display month of the calendar. If set to Month, the Calendar shows the standard single-month view. If set to Year, the Calendar shows the months in the current year (similar to when the user clicks the month header). Once the user clicks a month, the Calendar shows the full calendar view for that month.

Property	Description
SelectionMode (Calendar only)	Determines what type of date selections are allowed. The default is SingleDate, which allows a single date to be selected. Other options include None (selection is disabled entirely), SingleRange (a contiguous group of dates can be selected), and MultipleRange (any combination of dates can be selected). In SingleRange or MultipleRange modes, the user can drag to select multiple dates or click while holding down the Ctrl key. You can use the SelectedDates property to get a collection with all the selected dates.
IsDropDownOpen (DatePicker only)	Determines whether the calendar view drop-down is open in the DatePicker. You can set this property programmatically to show or hide the calendar.
SelectedDateFormat (DatePicker only)	Determines how the selected date will be displayed in the text part of the DatePicker. You can choose Short or Long. The actual display format is based on the client computer's regional settings. For example, if you use Short, the date might be rendered in the yyyy/mm/dd format or dd/mm/yyyy. The long format generally includes the month and day names.

The date controls also provide a few different events. Most useful is SelectedDateChanged (in the DatePicker) or the very similar SelectedDatesChanged (in the Calendar), which adds support for multiple date selection. You can react to these events to reject specific date selections, such as dates that fall on a weekend:

```
Private Sub Calendar_SelectedDatesChanged(ByVal sender As Object, _
  ByVal e As CalendarDateChangedEventArgs)
    ' Check all the newly added items.
    For Each selectedDate As DateTime In e.AddedItems
        If (selectedDate.DayOfWeek = DayOfWeek.Saturday) OrElse _
          (selectedDate.DayOfWeek = DayOfWeek.Sunday) Then
            lblError.Text = "Weekends are not allowed"

            ' Remove the selected date.
            CType(sender, Calendar).SelectedDates.Remove(selectedDate)
        End If
    Next
End Sub
```

You can try this with a Calendar that supports single or multiple selection. If it supports multiple selection, try dragging the mouse over an entire week of dates. All the dates will remain highlighted except for the disallowed weekend dates, which will be unselected automatically.

The Calendar also adds a DisplayDateChanged event (when the user browses to a new month). The DatePicker adds CalendarOpened and CalendarClosed events (which fire when the calendar drop-down is displayed and closed) and a DateValidationError event (which fires when the user types a value in the text entry portion that can't be interpreted as a valid date). Ordinarily, invalid values are discarded when the user opens the calendar view, but here's an option that fills in some text to alert the user of the problem:

```
Private Sub DatePicker_DateValidationError(ByVal sender As Object, _
  ByVal e As DatePickerDateValidationErrorEventArgs)
    lblError.Text = "'" & e.Text & "' is not a valid value because " & _
      e.Exception.Message
End Sub
```

The Last Word

In this chapter, you saw all the fundamental Silverlight elements. You considered several categories:

- The TextBlock, which allows you to display richly formatted text using built-in and custom fonts
- The Image, which allows you to show JPEG and PNG images
- Content controls that can contain nested elements, including various types of buttons and the ToolTip
- List controls that contain a collection of items, such as the ListBox, ComboBox, and TabControl
- Text controls, including the standard TextBox, the PasswordBox, and the AutoCompleteBox
- Range-based controls that take a numeric value from a range, such as the Slider
- The date controls, which allow the user to select one or more dates from a calendar display

Although you haven't had an exhaustive look at every detail of XAML markup, you've learned enough to reap all its benefits. Now, your attention can shift to the Silverlight technology itself, which holds some of the most interesting surprises. In the next chapter, you'll start out by considering the core of the Silverlight application model: the Application class.

CHAPTER 6

■■■

The Application Model

Over the past five chapters, you've taken a detailed look at the different visual ingredients you can put inside a Silverlight page. You've learned how to use layout containers and common controls and how to respond to mouse and keyboard events. Now, it's time to take a closer look at the Silverlight *application model*—the scaffolding that shapes how Silverlight applications are deployed, downloaded, and hosted.

You'll begin by considering the life cycle of a Silverlight application. You'll examine the events that fire when your application is created, is unloaded, or runs into trouble with an unhandled exception. Next, you'll pick up a few practical techniques that help you extend your application beyond Silverlight's basic behavior. You'll see how to pass in initialization parameters, show a custom splash screen, and break free from the confines of the browser to run your Silverlight application in a stand-alone window—even when the client computer can't get a network connection.

Finally, you'll explore the many options Silverlight provides for efficiently retrieving the large files called *binary resources*, whether they're images, video, or other assemblies that your application requires. You'll learn two strategies for dealing with resources: including them in your application package for easy deployment and downloading them on demand to streamline performance.

The Application Class

In Chapter 1, you took your first look at the App.xaml file. Much as every XAML page is a template for a custom class that derives from System.Windows.UserControl, the App.xaml file is a template for a custom class (named App by default) that derives from System.Windows.Application. You'll find the class definition in the App.xaml.vb file:

```
Partial Public Class App
    Inherits Application
    ...
End Class
```

When the Silverlight plug-in loads your application, it begins by creating an instance of the App class. From that point on, the application object serves as your entry point for a variety of application-specific features, including application events, resources, and services.

Accessing the Current Application

You can retrieve a reference to the application object at any time, at any point in your code, using the shared Application.Current property. However, this property is typed as a System.Windows.Application object. To use any custom properties or methods that you've added to the derived application class, you must cast the reference to the App type. For example, if you've added a method named DoSomething() to the App.xaml.vb file, you can invoke it with code like this:

```
CType(Application.Current, App).DoSomething()
```

This technique allows you to use your custom application class as a sort of switchboard for global tasks that affect your entire application. For example, you can add methods to your application class that control navigation or registration, and you can add properties that store global data. You'll see the App class used this way in examples throughout this book.

Application Properties

Along with the shared Current property, the Application class also provides several more members, as described in Table 6-1.

Table 6-1. Members of the Application Class

Member	Description
Host	This property lets you interact with the browser and, through it, the rest of the HTML content on the web page. It's discussed in Chapter 14.
Resources	This property provides access to the collection of XAML resources that are declared in App.xaml, as described in Chapter 2.
RootVisual	This property provides access to the root visual for your application—typically, the user control that's created when your application first starts. Once set, the root visual can't be changed, although you can manipulate the content in the root visual to change what's displayed in the page. For example, if it's the Grid control, you can remove one or more of its current children and insert new controls in their place. Chapter 7 demonstrates this technique.
IsRunningOutOfBrowser and InstallState	These properties let you recognize and monitor out-of-browser applications. IsRunningOutOfBrowser indicates whether the application is currently running out of the browser (True) or in the browser window (False). InstallState provides a value from the InstallState enumeration that indicates whether the current application is installed as an out-of-process application on the current computer (Installed), not installed (NotInstalled or InstallFailed), or in the process of being installed (Installing). You'll learn more about both properties when you consider out-of-browser applications in Chapter 21.

Member	Description
ApplicationLifetimeObjects	This property holds a collection of *application extension services*. These are objects that provide additional respond to application events, in much the same way as your event-handling code in the Application class. The difference is that the code for an application extension service is separated into its own class, which makes it easier to reuse this code in more than one Silverlight application.
Install() and CheckAndDownloadUpdate Async()	These methods provide support for out-of-browser applications. The Install() method installs the current Silverlight application on the client's computer. The CheckAndDownloadUpdateAsync() method launches an asynchronous process that checks the web server for updates. If an updated version is found, it's downloaded and used the next time the user runs the application.
GetResourceStream()	This shared method is used to retrieve resources in code. You'll see how to use it later in this chapter in the "Resources" section.
LoadComponent()	This shared method accepts a XAML file and instantiates the corresponding elements (much as Silverlight does automatically when you create a page class and the constructor calls the InitializeComponent() method).

Along with these properties and methods, the Application object also raises events at various points in the life cycle of your application. You'll explore these next.

Application Events

In Chapter 1, you took your first look at the life cycle of a Silverlight application. Here's a quick review:

1. The user requests the HTML entry page in the browser.
2. The browser loads the Silverlight plug-in. It then downloads the XAP file that contains your application.
3. The Silverlight plug-in reads the AppManifest.xml file from the XAP to find out what assemblies your application uses. It creates the Silverlight runtime environment and then loads your application assembly (along with any dependent assemblies).
4. The Silverlight plug-in creates an instance of your custom application class (which is defined in the App.xaml and App.xaml.vb files).
5. The default constructor of the application class raises the Startup event.
6. Your application handles the Startup event and creates the root visual object for your application.

From this point on, your page code takes over, until it encounters an unhandled error (UnhandledException) or finally ends (Exit). These events—Startup, UnhandledException, and

Exit—are the core events that the Application class provides. Along with these standards, the Application class includes two events—InstallStateChanged and CheckAndDownloadUpdateCompleted—that are designed for use with the out-of-browser applications you'll explore in Chapter 21.

As with the page and element events you've considered in earlier chapters, there are two ways to attach application event handlers. One option is to use the Handles keyword when declaring your event-handling method in the App.xaml.vb file. Here's an example that creates an event handler for the Application.Startup event:

```
Private Sub Application_Startup(ByVal o As Object, ByVal e As StartupEventArgs) _
    Handles Me.Startup
    ...
End Sub
```

Alternatively, you can add event attributes to the XAML markup, as shown here:

```
<Application ... x:Class="SilverlightApplication1.App"
 Startup="Application_Startup">
```

There's no reason to prefer one approach to the other. By default, Visual Studio uses the code approach shown first.

In the following sections, you'll see how you can write code that plugs into the application events.

Application Startup

By default, the Application_Startup method creates the first page and assigns it to the Application.RootVisual property, ensuring that it becomes the top-level application element—the visual core of your application:

```
Private Sub Application_Startup(ByVal o As Object, ByVal e As StartupEventArgs) _
    Handles Me.Startup

    Me.RootVisual = New MainPage()
End Sub
```

Although you can change the root visual by adding or removing elements, you can't reassign the RootVisual property at a later time. After the application starts, this property is essentially read-only.

Initialization Parameters

The Startup event passes in a StartupEventArgs object, which includes one additional detail: initialization parameters. This mechanism allows the page that hosts the Silverlight control to pass in custom information. This is particularly useful if you host the same Silverlight application on different pages or you want the Silverlight application to vary based on user-specific or session-specific information. For example, you can customize the application's view depending on whether users are entering from the customer page or the employee page. Or, you may choose to load different information based on the product the user is currently viewing. Just remember that the initialization parameters come from the tags of the HTML entry page, and a malicious user can alter them.

■ **Note** For more detailed interactions between the HTML and your Silverlight application—for example, to pass information back and forth while your Silverlight application is running—see Chapter 14.

For example, imagine you want to pass a ViewMode parameter that has two possible values, Customer or Employee, as represented by this enumeration:

```
Public Enum ViewMode
    Customer
    Employee
End Enum
```

You need to change a variety of details based on this information, so it makes sense to store it somewhere that's accessible throughout your application. The logical choice is to add a property to your custom application class, like this:

```
Private _viewMode As ViewMode = ViewMode.Customer
Public ReadOnly Property ViewMode() As ViewMode
    Get
        Return _viewMode
    End Get
End Property
```

This property defaults to customer view, so it needs to be changed only if the web page specifically requests the employee view.

To pass the parameter into your Silverlight application, you need to add a <param> element to the markup in the Silverlight content region. This parameter must have the name initParams. Its value is a comma-separated list of name-value pairs that set your custom parameters. For example, to add a parameter named viewMode, you add the following line (shown in bold) to your markup:

```
<div id="silverlightControlHost">
  <object data="data:application/x-silverlight,"
   type="application/x-silverlight-2" width="100%" height="100%">
    <param name="source" value="TransparentSilverlight.xap"/>
    <param name="onerror" value="onSilverlightError" />
    <param name="background" value="white" />
    <param name="initParams" value="viewMode=Customer" />
    ...
  </object>
  <iframe style="visibility:hidden;height:0;width:0;border:0px"></iframe>
</div>
```

Then, you can retrieve this from the StartupEventArgs.InitParams collection. However, you must check first that it exists:

```
Private Sub Application_Startup(ByVal o As Object, ByVal e As StartupEventArgs) _
   Handles Me.Startup
      ' Take the view mode setting, and store in an application property.
      If e.InitParams.ContainsKey("viewMode") Then
         Dim view As String = e.InitParams("viewMode")
         If view = "Employee" Then
            _viewMode = ViewMode.Employee
         End If
      End If

      ' Create the root page.
      Me.RootVisual = New MainPage()
End Sub
```

If you have many possible values, you can use the following leaner code to convert the string to the corresponding enumeration value, assuming the text matches exactly:

```
Dim view As String = e.InitParams("viewMode")
Try
    Me.viewMode = CType(System.Enum.Parse(GetType(ViewMode), view, True), ViewMode)
Catch
End Try
```

Now, different pages are free to pass in a different parameter and launch your application with different view settings. Because the view information is stored as a property in the custom application class (named App), you can retrieve it anywhere in your application:

```
lblViewMode.Text = "Current view mode: " & _
   (CType(Application.Current, App)).ViewMode.ToString()
```

Figure 6-1 shows what you'll see when you run the test page that uses the Customer view mode.

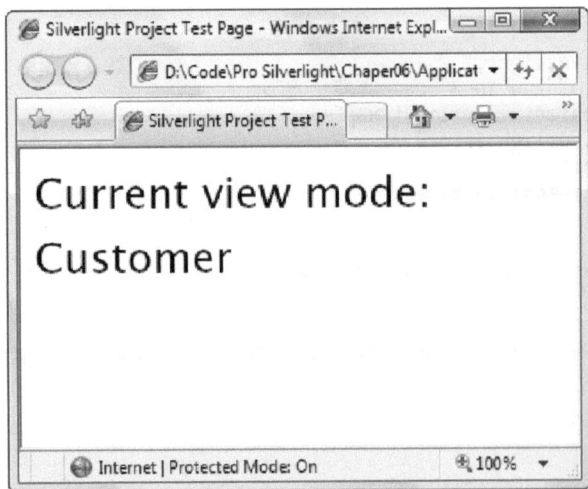

Figure 6-1. *Displaying an initialization parameter*

If you have more than one initialization parameter, pass them all in one comma-delimited string. Initialization values should be made up of alphanumeric characters. There's currently no support for escaping special characters such as commas in parameter values:

```
<param name="initParams" value="startPage=Page1,viewMode=Customer" />
```

Now, the event handler for the Startup event can retrieve the StartPage value and use it to choose the application's root page. You can load the correct page using a block of conditional logic that distinguishes between the available choices, or you can write a more general solution that uses reflection to attempt to create the class with the requested name, as shown here:

```
Dim startPage As UserControl = Nothing
If e.InitParams.ContainsKey("startPage") Then
    Dim StartPageName As String = e.InitParams("startPage")

    Try
        ' Create an instance of the page.
        Dim type As Type = Me.GetType()
        Dim currentAssembly As System.Reflection.Assembly = type.Assembly
        startPage = CType( _
          currentAssembly.CreateInstance(type.Namespace & "." & startPageName), _
          UserControl)
    Catch
        startPage = Nothing
    End Try
End If

' If no parameter was supplied or the class couldn't be created, use a default.
If startPage Is Nothing Then
    startPage = New MenuPage()
End If

Me.RootVisual = startPage
```

Application Shutdown

At some point, your Silverlight application ends. Most commonly, this occurs when the user surfs to another page in the web browser or closes the browser window. It also occurs if the users refreshes the page (effectively abandoning the current instance of the application and launching a new one), if the page runs JavaScript code that removes the Silverlight content region or changes its source, or if an unhandled exception derails your code.

Just before the application is released from memory, Silverlight gives you the chance to run some code by responding to the Application.Exit event. This event is commonly used to store user-specific information locally in isolated storage (see Chapter 18) so it's available the next time the user runs your application.

The Exit event doesn't provide any additional information in its event arguments.

Unhandled Exceptions

Although you should use disciplined exception-handling code in situations where errors are possible (for example, when reading a file, downloading web content, or accessing a web service), it's not always possible to anticipate all sources of error. If your application encounters an error that isn't handled, it will end, and the Silverlight content region will revert to a blank space. If you've included JavaScript code that reacts to potential errors from the Silverlight plug-in, that code will run. Otherwise, you won't receive any indication about the error that's just occurred.

The Application.UnhandledException event gives you a last-ditch chance to respond to an exception before it reaches the Silverlight plug-in and terminates your application. This code is notably different from the JavaScript error-handling code that you may add to the page, because it has the ability to mark an exception as handled. Doing so effectively neutralizes the exception, preventing it from rising to the plug-in and ending your application.

Here's an example that checks the exception type and decides whether to allow the application to continue:

```
Private Sub Application_UnhandledException(ByVal sender As object, _
  ByVal e As ApplicationUnhandledExceptionEventArgs) Handles Me.UnhandledException

    If TypeOf e.ExceptionObject Is FileNotFoundException Then
        ' Suppress the exception and allow the application to continue.
        e.Handled = True
    End If
End Sub
```

Ideally, an exception like this should be handled closer to where it occurs—for example, in your page code, when you're performing a task that may result in a FileNotFoundException. Application-level error handling isn't ideal, because it's difficult to identify the original process that caused the problem and it's awkward to notify the user about what went wrong. But application-level error handling does occasionally offer a simpler and more streamlined way to handle certain scenarios—for example, when a particular type of exception crops up in numerous places.

After you've neutralized the error, it makes sense to notify the user. One option is to call a custom method in your root visual. For example, this code calls a custom ReportError() method in the MainPage class, which is the root visual for this application:

```
Dim rootPage As MainPage = CType(Me.RootVisual, MainPage)
rootPage.ReportError(e.ExceptionObject)
```

Now the MainPage.ReportError() method can examine the exception object and display the appropriate message in an element on the page.

In an effort to make your applications a little more resilient, Visual Studio adds a bit of boilerplate error-handling code to every new Silverlight application. This code checks whether a debugger is currently attached (which indicates that the application is running in the Visual Studio debug environment). If there's no debugger, the code handles the error (rendering it harmless) and uses the HTML interoperability features you'll learn about in Chapter 14 to raise a JavaScript error in its place. Here's the slightly simplified code that shows how the process works:

```
Public Sub Application_UnhandledException(ByVal sender As Object, _
  ByVal e As ApplicationUnhandledExceptionEventArgs) Handles Me.UnhandledException

    If Not System.Diagnostics.Debugger.IsAttached Then
        ' Suppress the exception and allow the application to continue.
        e.Handled = True

        Try
            ' Build an error message.
            Dim errorMsg As String = e.ExceptionObject.Message & _
```

```
                e.ExceptionObject.StackTrace
                errorMsg = errorMsg.Replace("""""c, "\"c).Replace("\r\n", "\n")

                ' Use the Window.Eval() method to run a line of JavaScript code that
                ' will raise an error with the error message.
                System.Windows.Browser.HtmlPage.Window.Eval( _
                    "throw new Error(""Unhandled Error in Silverlight 2 Application " & _
                    errorMsg & """);")
            Catch
            End Try
        End If
    End Sub
```

Essentially, this code converts a fatal Silverlight error to a relatively harmless JavaScript error. The way the JavaScript error is dealt with depends on the browser. In Internet Explorer, a yellow alert icon appears in the status bar. (Double-click the alert icon to get the full error details, as shown in Figure 6-2.) In Firefox, a script error message appears. Either way, the error won't stop your application from continuing.

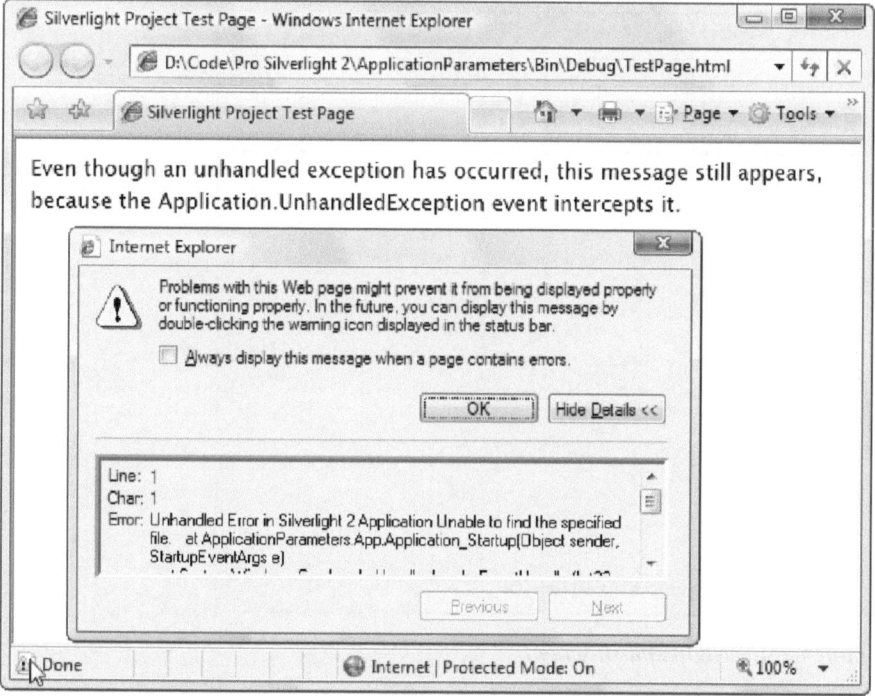

Figure 6-2. *A JavaScript error that represents an unhandled Silverlight exception*

When you finish developing your application, you need to tweak the automatically generated error-handling code. That's because it isn't acceptable to indiscriminately ignore all errors—doing so allows bugs to flourish and cause other usability problems or data errors further down the road. Instead, consider selectively ignoring errors that correspond to known error conditions and signaling the problem to the user.

■ **Caution** It's easy to forget that you need to tweak the Application.UnhandledException event handler, because it springs into action only when you run your Silverlight application without a debugger. When you're testing your application in Visual Studio, you don't see this behavior—instead, any unhandled exception ends the application immediately.

Custom Splash Screens

If a Silverlight application is small, it downloads quickly and appears in the browser. If a Silverlight application is large, it may take a few seconds to download. As long as your application takes longer than 500 milliseconds to download, Silverlight shows an animated splash screen.

The built-in splash screen isn't too exciting—it displays a ring of blinking circles and the percentage of the application that's been downloaded so far (see Figure 6-3).

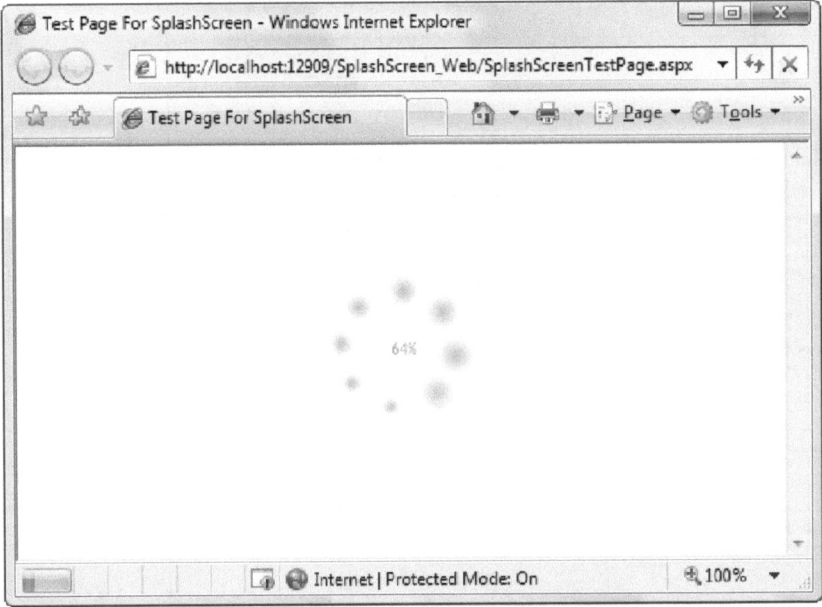

Figure 6-3. The built-in Silverlight splash screen

If you don't like the stock splash screen, you can easily create your own (see Figure 6-4). Essentially, a custom splash screen is a XAML file with the graphical content you want to display and a dash of JavaScript code that updates the splash screen as the application is downloaded. You can't use VB code at this point, because the Silverlight programming environment hasn't been initialized yet. However, this isn't a major setback, because the code you need is relatively straightforward. It lives in one or two event-handling functions that are triggered as content is being downloaded and after it's finished, respectively.

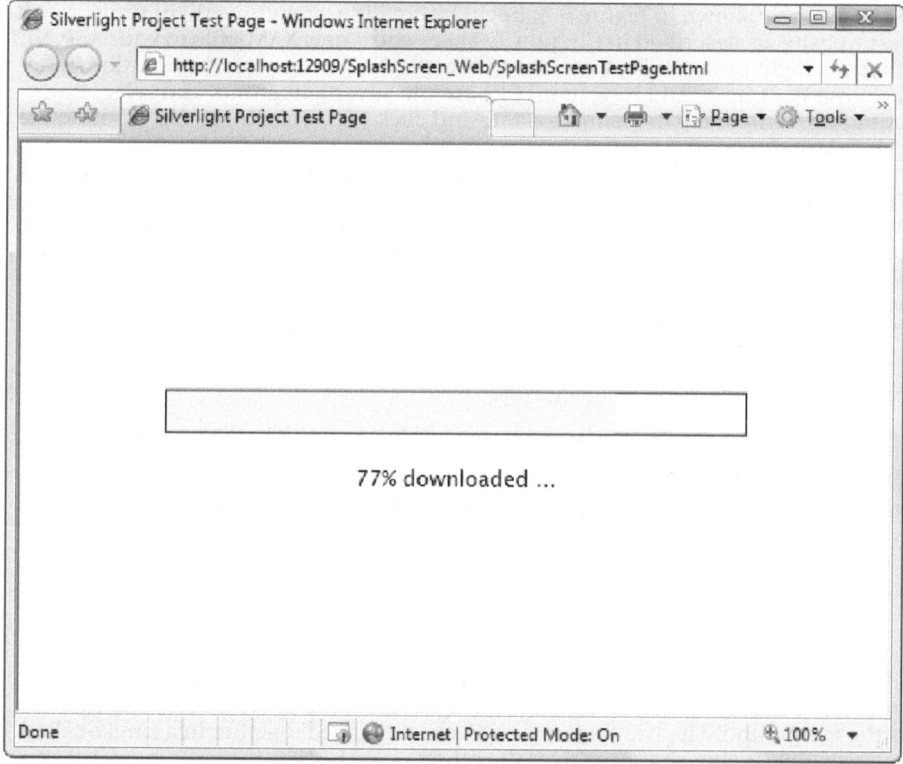

Figure 6-4. A custom splash screen

The XAML file for your splash screen can't be part of your Silverlight XAP file. That's because the splash screen needs to be shown while the XAP file is still in the process of being downloaded. For that reason, the splash screen XAML must be a separate file that's placed alongside your XAP file at the same web location.

■ **Note** Testing a custom splash screen requires some work. Ordinarily, you don't see the splash screen during testing because the application is sent to the browser too quickly. To slow down your application enough to see the splash screen, you need to first ensure that you're using an ASP.NET test website, which ensures that your Silverlight application is hosted by Visual Studio test web server (as described in Chapter 1). Then, you need to add multiple large resource files to your Silverlight project—say, a handful of MP3 files—and set the build action of each one to Resource so it's added to the XAP file. Another trick is to temporarily remove the line of code in the Application_Startup() method that sets the root visual for your application. This way, after your application has been completely downloaded, it won't display anything. Instead, the splash screen will remain visible, displaying a progress percentage of 100%.

To create the example shown in Figure 6-4, begin by creating a new Silverlight project with an ASP.NET test website, as described in Chapter 1. Then, add a new XAML file to your ASP.NET website (not the Silverlight project). To do so, select the ASP.NET website in the Solution Explorer, and choose Website ➤ Add New Item. Choose the Silverlight group, and select the Silverlight JScript page template. Then enter a name and click Add. This XAML file will hold the markup for your splash screen.

When you add a new XAML file, Visual Studio creates a basic XAML skeleton that defines a Canvas. That's because Visual Studio assumes you're building a Silverlight 1.0 application, which supports a much smaller set of elements and doesn't include any of the more advanced layout containers. But you can use any of the core Silverlight elements—that is, elements that are in the built-in assemblies and don't require a separate download. You can't use elements that are defined in the add-on System.Windows.Controls.dll assembly or those in any other assembly that needs to be packaged in the XAP and downloaded by the client.

■ **Tip** The easiest way to build a simple splash screen is to create it in your Silverlight project and then copy the markup into the splash screen file on your website. This way, you can take advantage of the Visual Studio design surface and XAML IntelliSense, which won't be available if you write the markup directly in your ASP.NET website.

Here's the XAML for the splash screen shown in Figure 6-4. It includes a Grid with a TextBlock and two Rectangle elements. (Rectangle is a shape-drawing element you'll learn about in Chapter 8.) The first rectangle paints the background of the progress bar, and the second paints the foreground. The two Rectangle objects are placed together in a single-celled grid so that one rectangle is superimposed over the other:

```
<Grid xmlns="http://schemas.microsoft.com/winfx/2006/xaml/presentation"
  xmlns:x="http://schemas.microsoft.com/winfx/2006/xaml">
  <StackPanel VerticalAlignment="Center">
    <Grid>
      <Rectangle x:Name="progressBarBackground" Fill="White" Stroke="Black"
        StrokeThickness="1" Height="30" Width="200"></Rectangle>
      <Rectangle x:Name="progressBar" Fill="Yellow" Height="28" Width="0">
      </Rectangle>
    </Grid>
    <TextBlock x:Name="progressText" HorizontalAlignment="Center"
      Text="0% downloaded ..."></TextBlock>
  </StackPanel>
</Grid>
```

Next, you need to add a JavaScript function to your HTML entry page or ASP.NET test page. (If you plan to use both, place the JavaScript function in a separate file and then link to it in both files, using the source attribute of the script block.) The JavaScript code can look up named elements on the page using the sender.findName() method and manipulate their properties. It can also determine the current progress using the eventArgs.progress property. In this example, the event-handling code updates the text and widens the progress bar based on the current progress percentage:

```
<script type="text/javascript">
  function onSourceDownloadProgressChanged(sender, eventArgs)
  {
      sender.findName("progressText").Text =
        Math.round((eventArgs.progress * 100)) + "% downloaded ...";
      sender.findName("progressBar").Width =
        eventArgs.progress * sender.findName("progressBarBackground").Width;
  }
</script>
```

■ **Note** The splash-screen example that's included with the downloadable code uses a slightly more advanced technique that draws on a transform, a concept you'll explore in Chapter 9. This approach allows you to create a progress-bar effect without hard-coding the maximum width, so the progress bar is sized to fit the current browser window.

To use this splash screen, you need to add the splashscreensource parameter to identify your XAML splash screen and the onsourcedownloadprogresschanged parameter to hook up your JavaScript event handler. If you want to react when the download is finished, you can hook up a different JavaScript event handler using the onsourcedownloadcomplete parameter:

```
<object data="data:application/x-silverlight," type="application/x-silverlight-2"
 width="100%" height="100%">
  <param name="source" value="ClientBin/SplashScreen.xap"/>
  <param name="onerror" value="onSilverlightError" />
  <param name="background" value="white" />
  <param name="splashscreensource" value="SplashScreen.xaml" />
  <param name="onsourcedownloadprogresschanged"
   value="onSourceDownloadProgressChanged" />
  ...
</object>
```

Expert designers can craft elaborate splash screens. This tradition is well-established with Flash applications. To see a taste of what's possible, visit www.smashingmagazine.com/2008/03/13/showcase-of-creative-flash-preloaders. You can duplicate many of these effects with an ordinary Silverlight splash screen, like the one described here. However, some are extremely difficult. Most would be far easier to achieve *after* you've downloaded your application, such as code-heavy animations.

If you want more flexibility to create an eye-catching splash screen, you need to use a completely different technique. Make your application as small as possible. Move its functionality to class-library assemblies, and place large resources (such as graphics and videos) in separate files or in separate class-library assemblies. Now that your application is stripped down to a hollow shell, it can be downloaded quickly. After it's downloaded, your application can show its fancy preloader and start the real work—programmatically downloading the resources and assemblies it needs to function.

Designing an application this way takes more work, but you'll get all the information you need to perform dynamic downloads in the following sections. Pay particular attention to the "Downloading Assemblies on Demand" section later in this chapter.

Binary Resources

As you learned in Chapter 1, a Silverlight application is actually a package of files that's archived using ZIP compression and stored as a single file, with the extension .xap. In a simple application, the XAP file has little more than a manifest (which list the files your project uses) and your application assembly. But you can place something else in the XAP file: *resources*.

A XAP resource is a distinct file that you want to make available to your compiled application. Common examples include graphical assets—images, sounds, and video files that you want to display in your user interface.

Using resources can be unnecessarily complicated because of the wealth of different options Silverlight provides for storing them. Here's a quick roundup of your options:

- *In the application assembly*: The resource file is embedded in the compiled DLL file for your project, such as SilverlightApplication1.dll. This is the default approach.

- *In the application package*: The resource file is placed in the XAP file alongside your application assembly. It's just as easy to deploy, but now it's easier to manage because you replace or modify your assets by editing the XAP file, without compiling your application.

- *On the site of origin*: The resource file is placed on the website alongside your XAP file. Now you have more deployment headaches, because you need to make sure you deploy both the XAP file and the resource file. However, you gain the ability to use your resource in other ways—for example, you can use images in ordinary HTML web pages or make videos available for easy downloading. You can reduce the size of the initial XAP download, which is important if the resources are large.

These aren't all your options. As you'll see later in this chapter in the "Class Library Assemblies" section, you can also place resources in other assemblies that your application uses. (This approach gives you more advanced options for controlling the way you share content between different Silverlight applications.) But before tackling that topic, it's worth taking a closer look at the more common options outlined previously. In the following sections, you'll explore each approach.

■ **Note** Binary resources shouldn't be confused with the XAML resources you explored in Chapter 2. XAML resources are objects that are declared in your markup. Binary resources are non-executable files that are inserted into your assembly or XAP file when your project is compiled.

Placing Resources in the Application Assembly

This is the standard approach, and it's similar to the approach used in other types of .NET applications (such as WPF applications). For example, if you want to show an image in Silverlight's Image element, begin by adding the image file to your project. By default, Visual Studio gives image files the Resource build action, as shown in Figure 6-5. (To change the build action of an existing file, select it in the Solution Explorer, and make a new selection in the Build Action box in the Properties pane.)

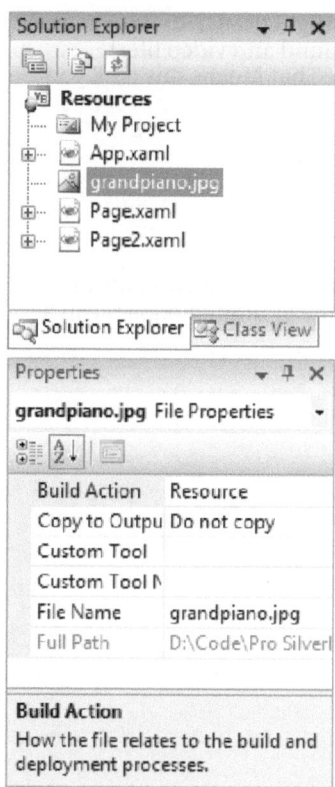

Figure 6-5. An application resource

■ **Note** Don't confuse the build action of Resource with Embedded Resource. Although both do the same thing (embed a resource in the assembly as a block of binary data), Silverlight doesn't support the Embedded Resource approach, and you can't reference files that are stored in this way using URIs.

Now, when you compile your application, the resource will be embedded in the project assembly, and the project assembly will be placed in the XAP file.

■ **Note** Although the resource option makes it most difficult for a user to extract a resource file from your application, it's still possible. To retrieve a resource, the user needs to download the XAP file, unzip it, and decompile the DLL file. Tools like Reflector (http://reflector.red-gate.com) provide plug-ins that can extract and save embedded resources from an assembly.

Using an embedded resource is easy because of the way Silverlight uses URIs. If you use a relative URI with the Image (for graphics) or MediaElement (for sound and video files), Silverlight checks the assembly for a resource with the right name. That means this is all you need to use the resource shown in Figure 6-5:

```
<Image Source="grandpiano.jpg"></Image>
```

Using Subfolders

It's possible to use the folders to group resource files in your project. This changes how the resource is named. For example, consider Figure 6-6, which puts the grandpiano.jpg file in a subfolder named Images.

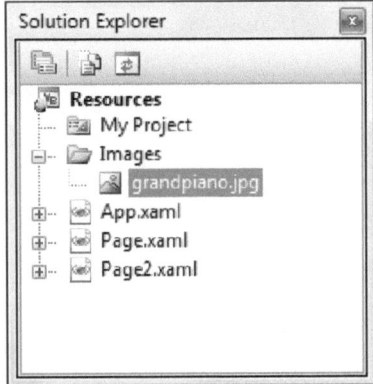

Figure 6-6. A resource in a subfolder

Now, you need to use this URI:

```
<Image Source="Images/grandpiano.jpg"></Image>
```

Programmatically Retrieving a Resource

Using resources is easy when you have an element that supports Silverlight's URI standard, such as Image or MediaElement. However, in some situations, you need to manipulate your resource in code before handing it off to an element, or you may not want to use an element at all. For example, you may have some static data in a text or binary file that's stored as a resource. In your code, you want to retrieve this file and process its data.

To perform this task, you need the help of the Application.GetResourceStream() method. It allows you to retrieve the data for a specific resource, which you indicate by supplying the correct URI. The trick is that you need to use the following URI format:

AssemblyName;component/*ResourceFileName*

For example, if you have a resource named ProductList.bin in a project named SilverlightApplication1, you use this line of code:

```
Dim sri As StreamResourceInfo = Application.GetResourceStream( _
  New Uri("SilverlightApplication1;component/ProductList.bin", UriKind.Relative))
```

The GetResourceStream() method doesn't retrieve a stream. Instead, it gets a System.Windows.Resources.StreamResourceInfo object, which wraps a Stream property (with the underlying stream) and a ContentType property (with the MIME type). Here's the code that creates a BinaryReader object for the stream:

```
Dim reader As New BinaryReader(sri.Stream)
```

You can now use the methods of the binary reader to pull each piece of data out of the file. The same approach works with StreamReader (for text-based data) and XmlReader (for XML data). But you have a slightly easier option when XML data is involved, because the XmlReader.Create() method accepts either a stream or a URI string that points to a resource. So, if you have a resource named ProductList.xml, this code works:

```
Dim sri As StreamResourceInfo = Application.GetResourceStream( _
  New Uri("SilverlightApplication1;component/ProductList.xml", UriKind.Relative))
Dim reader As XmlReader = XmlReader.Create(sri.Stream, New XmlReaderSettings())
```

So does this more streamlined approach:

```
Dim reader As XmlReader = XmlReader.Create("ProductList.xml")
```

Placing Resources in the Application Package

Your second option for resource storage is to place it in the XAP file where your application assembly is stored. To do this, you need to add the appropriate file to your project and change the build action to Content. Best of all, you can use almost the same URLs. Just precede them with a forward slash, as shown here:

```
<Image Source="/grandpiano.jpg"></Image>
```

Similarly, here's a resource in a subfolder in the XAP:

```
<Image Source="/Images/grandpiano.jpg"></Image>
```

The leading slash represents the root of the XAP file.

If you add the extension .zip to your XAP file, you can open it and verify that the resource file is stored inside, as shown in Figure 6-7.

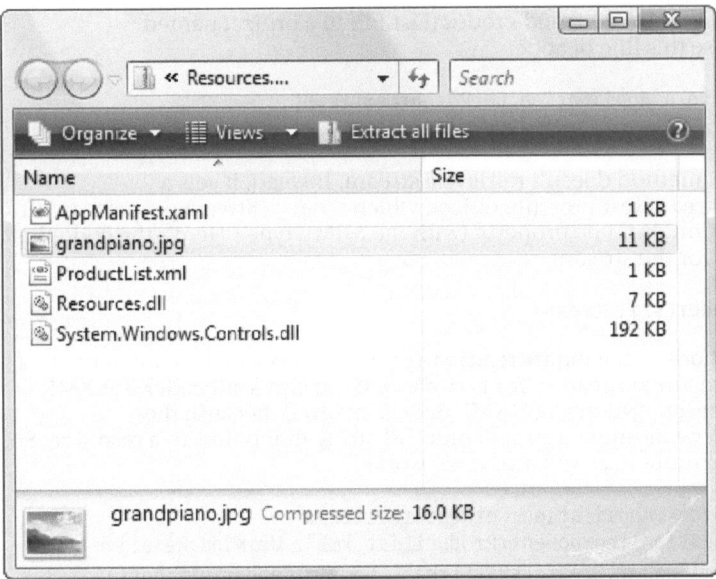

Figure 6-7. *A resource in a XAP file*

Placing resources in the XAP file gives you the same easy deployment as embedding them in the assembly. However, it adds a bit of flexibility. If you're willing to do a little more work, you can manipulate the files in the XAP file (for example, updating a graphic) without recompiling the application. Furthermore, if you have several class library assemblies in the same XAP file, they can all use the same resource files in the XAP. (This is an unlikely arrangement but a possible one.) Overall, placing resources in the application package is a similar approach to embedding them in the assembly.

Placing Resources on the Web

Your third option is to remove resource files from your application but make them available on the Web. That way, your application can download them when needed. Thanks to Silverlight's URI support, you can usually use this scenario without writing any extra code to deal with the download process.

The simplest option when deploying resources on the Web is to place them in the same web location as your Silverlight assembly. If you're using an ASP.NET test website, you can easily add a resource file to the test website—just place it in the ClientBin folder where the XAP file is located. If you're using an HTML test page, the easiest option is to tell Visual Studio to copy your resource file to the build location. To do so, begin by adding the resource file to your Silverlight project. Then, select the resource file and choose None for the build action, so it won't be compiled into the XAP. Finally, set the Copy to Output Directory setting to Copy Always.

When using web resources, you use the same URIs as when placing resources in the application package. These are relative URIs prefaced with a forward slash. Here's an example:

```
<Image Source="/grandpiano.jpg"></Image>
```

Silverlight checks the XAP file first and then checks the folder where the XAP file is located. Thus, you can freely switch between the XAP file approach and the website approach after you've compiled an application—you just need to add or remove the resource files in the XAP file.

Web-deployed resources don't need to be located at the same site as your XAP file, although that's the most common approach. If you use an absolute URL, you can show an image from any location:

```
<Image Source="http://www.mysite.com/Images/grandpiano.jpg"></Image>
```

■ **Note** When you're testing an application that uses images with absolute URLs, a small glitch can creep in. The problem is that the Image element can't perform cross-scheme access, which means that if you're running Silverlight directly from your hard drive using a simple HTML test page, you can't retrieve an image from the Web. To resolve this problem, add an ASP.NET test website to your project, as described in Chapter 1.

Web-deployed resources are treated in a significantly different way in your application. Because they aren't in the XAP file (either directly or indirectly, as part of the assembly), they aren't compressed. If you have a large, easily compressed file (say, XML data), this means the web-deployed option results in longer download times, at least for some users. More significant is the fact the web-deployed resources are downloaded on demand, when they're referenced in your application. Thus, if you have a significant number of large resources, web deployment is often much better—it trades a long delay on startup for many smaller delays when individual resources are accessed.

■ **Note** The obvious disadvantage with all of these resource-storing approaches is that they require fixed, unchanging data. In other words, there's no way for your application to modify the resource file and then save the modified version in the assembly, XAP file, or website. (In theory, the last option—website uploading—could be made possible, but it would create an obvious security hole.) The best solution when you need to change data is to use isolated storage (if storing the changed data locally is a good enough solution) or a web service (if you need a way to submit changes to the server). These approaches are discussed in Chapter 18 and Chapter 15, respectively.

Failing to Download Resources

When you use web-deployed resources, you introduce the possibility that your resources won't be where you expect them to be and that you won't be able to download them successfully. Elements that use the URI system often provide events to notify when a download can't be completed, such as ImageFailed for the Image and MediaFailed for the MediaElement.

Failing to download a resource isn't considered a critical error. For example, if the Image element fails to find the right picture, it simply remains blank. But you can react to the corresponding failure event to update your user interface.

Downloading Resources with WebClient

You can't access web-deployed resources using the handy Application.GetResourceStream() method. As a result, if you want to use the data from a web-deployed resource and you don't have an element that uses Silverlight URIs, you'll need to do more work.

In this situation, you need to use the System.Net.WebClient class to download the resource. The WebClient class provides three key methods. OpenReadAsync() is the most useful—it downloads a file as blob of binary data, which is then exposed as a stream. By comparison, DownloadStringAsync() downloads the contents into a single string. Finally, CancelAsync() halts any download that's currently underway.

WebClient does its work asynchronously. You can respond to the DownloadProgressChanged event while the download is underway to find out how many bytes have been retrieved so far. When the download is complete, you can respond to the OpenReadCompleted or DownloadStringCompleted event, depending on which operation you're using, and then retrieve your content.

WebClient has the following important limitations:

- *It doesn't support downloading from the file system*: To use the WebClient class, you must be running your application through a web server. The easiest way to do this in Visual Studio is to let Visual Studio create an ASP.NET website, which is then hosted by the integrated web server (as described in Chapter 1). If you open your Silverlight page directly from the file system, you'll get an exception when you attempt to use the downloading methods in the WebClient.

- *It doesn't support relative URIs*: To get the correct URI, you can determine the URI of the current page and then add the relative URI that points to your resource.

- *It allows only one download at a time*: If you attempt to start a second request while the first is underway, you'll receive a NotSupportedException.

■ **Note** There's one other issue: Silverlight's security model. If you plan to use WebClient to download a file from another web server (not the web server where your application is hosted), make sure that web server explicitly allows cross-domain calls. Chapter 15 discusses this issue in detail.

Here's an example that puts the pieces together. It reads binary data from the ProductList.bin file, as you saw earlier. However, in this example, ProductList.bin is hosted on the website and isn't part of the XAP file or project assembly. (When you test this example using an ASP.NET website, you need to add the ProductList.bin file to the ASP.NET website, not the Silverlight project. To see the correct setup, refer to the downloadable examples for this chapter.)

When a button is clicked, the downloading process starts. Notice that string processing is at work with the URI. To get the right path, you need to create a fully qualified URI using the current address of the entry page, which you can retrieve from the Host property of the current Application object:

```
Private Sub cmdRetrieveResource_Click(ByVal sender As Object, _
  ByVal e As RoutedEventArgs)
    ' Construct the fully qualified URI.
    ' Assume the file is in the website root, one level above the ClientBin
    ' folder. (In other words, the file has been added to the root level
    ' of the ASP.NET website.)
    Dim uri As String = Application.Current.Host.Source.AbsoluteUri
    Dim index As Integer = uri.IndexOf("/ClientBin")
    uri = uri.Substring(0, index) & "/ProductList.bin"

    ' Begin the download.
    Dim webClient As New WebClient()
    AddHandler webClient.OpenReadCompleted, AddressOf webClient_OpenReadCompleted
    webClient.OpenReadAsync(New Uri(uri))
End Sub
```

Now, you can respond when the file has been completed and manipulate the downloaded data as a stream:

```
Private Sub webClient_OpenReadCompleted(ByVal sender As Object, _
  ByVal e As OpenReadCompletedEventArgs)
    If e.Error IsNot Nothing Then
        ' (Add code to display error or downgrade gracefully.)
    Else
        Dim stream As Stream = e.Result
        Dim reader As New BinaryReader(stream)
        ' (Now process the contents of the resource.)
        reader.Close()
    End If
End Sub
```

For simplicity's sake, this code retrieves the resource every time you click the button. But a more efficient approach is to store the retrieved data in memory so it doesn't need to be downloaded more than once.

The OpenReadCompletedEventArgs provides several pieces of information along with the Result property. To determine whether the operation was cancelled using the CancelAsync() method, you can check the Cancelled property, and if an error occurred, you can get the exception object from the Error property. (In this situation, attempting to read the other properties of the OpenReadCompletedEventArgs object will result in a TargetInvocationException.) You can also use an overloaded version of the OpenReadAsync() method that accepts a custom object, which you can then retrieve from the UserState property. However, this is of limited use, because WebClient allows only one download at a time.

When you're downloading a large file, it's often worth showing a progress indicator to inform the user about what's taking place. To do so, attach an event handler to the DownloadProgressChanged event:

```
AddHandler webClient.DownloadProgressChanged, _
  AddressOf webClient_DownloadProgressChanged
```

Here's the code that calculates the percentage that's been downloaded and uses it to set the value of a progress bar and a text label:

```
Private Sub webClient_DownloadProgressChanged(ByVal sender As Object, _
   ByVal e As DownloadProgressChangedEventArgs)
      lblProgress.Text = e.ProgressPercentage.ToString() & " % downloaded."
      progressBar.Value = e.ProgressPercentage
End Sub
```

Class Library Assemblies

So far, the examples you've seen in this book have placed all their code into a single assembly. For a small or modest-sized Silverlight application, this straightforward design makes good sense. But it's not hard to imagine that you might want to factor out certain functionality and place it in a separate class library assembly. Usually, you'll take this step because you want to reuse that functionality with more than one Silverlight application. Alternatively, you may want to break it out it so it can be coded, compiled, debugged, and revised separately, which is particularly important if that code is being created by a different development team.

Creating a Silverlight class library is easy. It's essentially the same process you follow to create and use class library assemblies in ordinary .NET applications. First, create a new project in Visual Studio using the Silverlight Class Library project template. Then, add a reference in your Silverlight application that points to that project or assembly. The dependent assembly will be copied into the XAP package when you build your application.

Using Resources in an Assembly

Class libraries give you a handy way to share resources between applications. You can embed a resource in a class library and then retrieve it in your application. This technique is easy—the only trick is constructing the right URIs. To pull a resource out of a library, you need to use a URI that includes the application in this format:

/ClassLibraryName;component/*ResourceFileName*

This is the same format you learned about earlier, in the section "Programmatically Retrieving a Resource," but with one addition: now, the URI begins with a leading slash, which represents the root of the XAP file. This URI points to the dependent assembly in that file and then indicates a resource in that assembly.

For example, consider the ResourceClassLibrary assembly in Figure 6-8. It includes a resource named happyface.jpg, and that file has a build action of Resource.

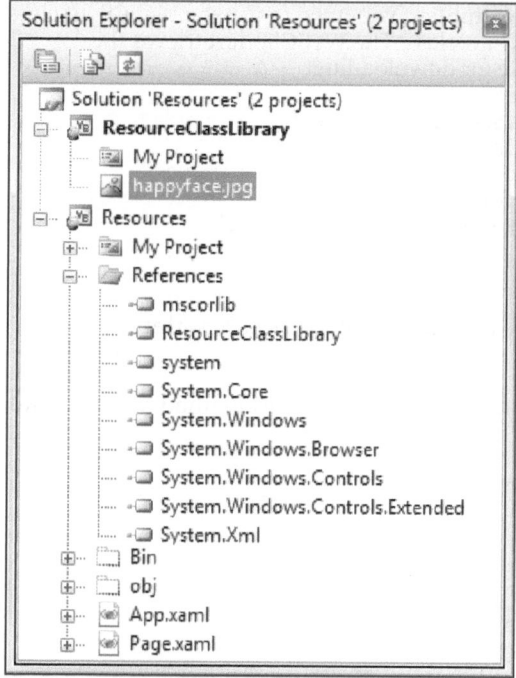

Figure 6-8. *A resource in a class library*

Here's an image file that uses the resource from the class library:

```
<Image Source="/ResourceClassLibrary;component/happyface.jpg"></Image>
```

Downloading Assemblies on Demand

In some situations, the code in a class library is used infrequently, or perhaps not at all for certain users. If the class library contains a significant amount of code or (more likely) has large embedded resources such as graphics, including it with your application will increase the size of your XAP file and lengthen download times needlessly. In this case, you may want to create a separate component assembly—one that isn't downloaded until you need it. This scenario is similar to on-demand resource downloading. You place the separate resource in a separate file outside of the XAP file but on the same website.

Before you use assembly downloading, you need to make sure the dependent assembly isn't placed in the XAP file. To do so, first click the Show All Files button at the top of the Solution Explorer, expand the References group, and select the project reference that points to the assembly. In the Properties window, set Copy Local to False. Next, make sure the assembly is copied to the same location as your website. If you're using an ASP.NET test website, that means you must add the assembly to the ClientBin folder in the test website. (You can't try this example with a simple HTML test page, because WebClient doesn't work when you run a Silverlight application from the file system.)

CHAPTER 6 ■ THE APPLICATION MODEL

To implement on-demand downloading of assemblies, you need to use the WebClient class you saw earlier, in conjunction with the AssemblyPart class. The WebClient retrieves the assembly, and the AssemblyPart makes it available for downloading:

```
Dim uri As String = Application.Current.Host.Source.AbsoluteUri
Dim index As Integer = uri.IndexOf("/ClientBin")
' In this example, the URI includes the /ClientBin portion, because we've
' decided to place the DLL in the ClientBin folder.
uri = uri.Substring(0, index) & "/ClientBin/ResourceClassLibrary.dll"

' Begin the download.
Dim webClient As New WebClient()
AddHandler webClient.OpenReadCompleted, AddressOf webClient_OpenReadCompleted
webClient.OpenReadAsync(New Uri(uri))
```

When the assembly is downloaded, you use the AssemblyPart.Load() method to load it into the current application domain:

```
Private Sub webClient_OpenReadCompleted(ByVal sender As Object, _
    ByVal e As OpenReadCompletedEventArgs)
    If e.Error IsNot Nothing Then
        ' (Add code to display error or degrade gracefully.)
    Else
        Dim assemblypart As New AssemblyPart()
        assemblypart.Load(e.Result)
    End If
End Sub
```

After you've performed this step, you can retrieve resources from your assembly and instantiate types from it. It's as though your assembly was part of the XAP file from the start. You can try a demonstration of this technique with the sample code for this chapter.

Once again, it's important to keep track of whether you've downloaded an assembly so you don't attempt to download it more than once. Some applications daisy-chain assemblies: one application downloads other dependent assemblies on demand, and these assemblies download additional assemblies when *they* need them.

■ **Tip** If you attempt to use an assembly that hasn't been downloaded, you'll receive an exception. But the exception won't be raised to the code that is attempting to use the assembly. Instead, that code will be aborted, and the exception will pass to the event handler for the Application.UnhandledException event. The exception is a FileNotFoundException object, and the message includes the name of the missing assembly.

Supporting Assembly Caching

As you learned in Chapter 1, assembly caching is a feature that allows Silverlight to download class library assemblies and store them in the browser cache. This way, these assemblies don't need to be downloaded every time the application is launched.

■ **Note** A common misconception is that assembly caching replaces the on-demand assembly loading technique that's described in the previous section. However, both approaches have different effects. Assembly caching reduces the startup time on repeat visits to the same application (or when running applications that share some of the same functionality). On-demand assembly loading reduces the startup time on every visit, regardless of what's in the browser cache and whether the application has been used before. Assembly caching is particularly useful with large, frequently used assemblies that your application is sure to use. On-demand assembly loading is particularly useful for large, infrequently used assembly that your application may not need to download *ever*.

By default, the assemblies you build won't support assembly caching. However, you can add this support by satisfying two requirements. First, your assembly must have a strong name. Second, your assembly needs a special type of XML file that describes its contents, called an *.extmap.xml* file. The following sections walk you through both requirements, and you can refer to the downloadable code for this chapter to assembly caching in action with a custom assembly.

The Strong Key Name

To support assembly caching, your class library assembly needs a strong name, which will uniquely identify it in the browser cache and prevent naming conflicts. To create a strong key for your assembly, follow these steps:

1. Double-click the My Project item in the Solution Explorer.
2. Click the Signing tab.
3. Select the "Sign the assembly" option.
4. In the "Choose a strong key name" list, choose <New...> to show the Create Strong Name Key dialog box.
5. To finish creating your key, you'll need to supply a file name (like MyKey.snk) and, optionally, a password.
6. Click OK. Visual Studio will create the new key file and add it to your class library project.

This creates a strong key file and uses it for your assembly. From this point on, every time you compile your project, Visual Studio uses the strong key to sign the final assembly.

Before you can continue to the next step, you need to know public key token of the key pair that's used to sign your assembly. Unfortunately, Visual Studio doesn't provide an easy way to get this information (at least not without a plug-in of some sort). Instead, you need to resort to the sn.exe command-line tool. First, choose Microsoft Visual Studio 2010 ➤ Visual Studio Tools ➤ Visual Studio Command Prompt. Once you've loaded the Visual Studio command prompt, change to the directory that holds your key file. Then, run the following two commands (replacing MyKey.snk with the name of your key):

```
sn -p MyKey.snk MyKey.bin
sn -t MyKey.bin
```

When you complete the second command, you'll see a message like this:

```
Microsoft (R) .NET Framework Strong Name Utility  Version 3.5.30729.1
Copyright (c) Microsoft Corporation.  All rights reserved.

Public key token is e6a351dca87c1032
```

The bold part is the piece of information you need for the next step: creating a *.extmap.xml* file for your assembly.

The .extmap.xml File

The .extmap.xml file is an ordinary text file that holds XML content. It's named to match your assembly. For example, if you have a class library assembly named CacheableAssembly.dll, you'll need to create a file named CacheableAssembly.extmap.xml. The presence of this file tells Silverlight that your assembly supports assembly caching.

To make life easy, you can add the .extmap.xml file to your class library project. Select it in the Solution Explorer, and set Build Action to None and the Copy to Output Directory setting to "Copy always." This ensures that the file will be placed in the same directory as your assembly file when you compile it. Figure 6-9 shows a class library with the appropriate .extmap.xml file.

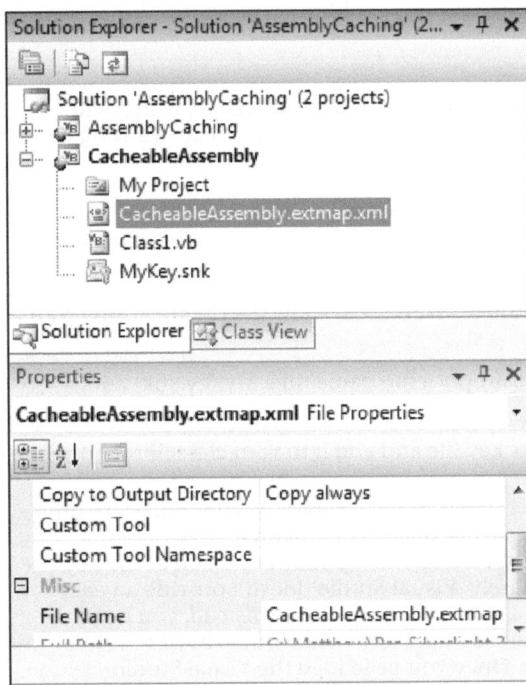

Figure 6-9. The .extmap.xml file for CacheableAssembly.dll

The easiest way to create an .extmap.xml file is to take a sample (like the one shown next), and modify it for your assembly. In the following listing, the details you need to change are in bold:

```xml
<?xml version="1.0"?>
<manifest xmlns:xsi="http://www.w3.org/2001/XMLSchema-instance"
          xmlns:xsd="http://www.w3.org/2001/XMLSchema">
  <assembly>
    <name>CacheableAssembly</name>
    <version>1.0.0.0</version>
    <publickeytoken>e6a351dca87c1032</publickeytoken>
    <relpath>CacheableAssembly.dll</relpath>
    <extension downloadUri="CacheableAssembly.zip" />
  </assembly>
</manifest>
```

The name and version details are obvious, and they should match your assembly. The public key token is the identifying fingerprint of the strong key that was used to sign your assembly, and you collected it with the sn.exe tool in the previous section. The relative path (relpath) is the exact file name of the assembly. Finally, the downloadUri attribute provides the most important piece of information—it tells the application where to find the packaged, downloadable assembly.

You have two options for setting downloadUri. The easiest approach is the one that's used in the previous example—simply supply a file name. When you switch on assembly caching in your application, Visual Studio will take your class library assembly (in this case, CacheableAssembly.dll), compress it, and place the compressed file (in this case, CacheableAssembly.zip) alongside the compiled XAP file. As you saw in Chapter 1, this is the approach that Silverlight's add-on assemblies use.

■ **Note** Although Visual Studio compresses your assembly using ZIP compression, it's not necessary to use a file name that ends with the extension .zip. If your web server requires a different extension, feel free to use that for the downloadUri. And if you have use the same downloadUri file name for more than one assembly, Visual Studio compresses all the assemblies into a single ZIP file.

Your other option is to use an absolute URI for the downloadUri:

```xml
<extension
 downloadUri="http://www.mysite.com/assemblies/v1.0/CacheableAssembly.zip" />
```

In this case, Visual Studio won't package up the assembly when you compile the application. Instead, it expects you to have already placed the assembly at the web location you've specified. This gives you a powerful way to share libraries between multiple applications. However, the download location must be on the same domain as the Silverlight application, or it must explicitly allow cross-domain access, as described in Chapter 15.

With the .extmap.xml file shown earlier, you're ready to use assembly caching. To try it, create an application that uses your class library. Turn on assembly caching for your application by opening the project properties and selecting the "Reduce XAP size by using application library caching" option. Finally, build your application. If you check out the contents of your Debug folder, you'll find the packaged up ZIP file for your assembly (as shown in Figure 6-10).

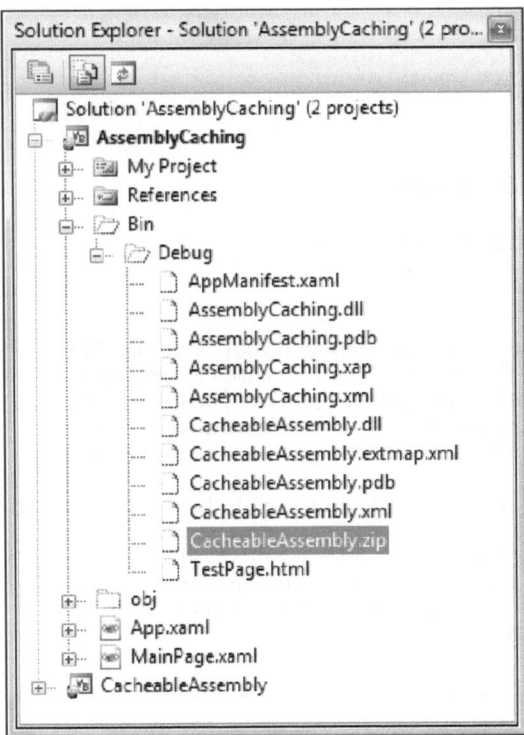

Figure 6-10. *The compressed assembly, ready for caching*

The Last Word

In this chapter, you explored the Silverlight application model in detail. You reexamined the application object and the events it fires. You learned how to pass initialization parameters from different web pages and how to display a custom splash screen while your application is being downloaded. Finally, you explored the resource system that Silverlight uses and considered the many options for deploying resources and class libraries, from placing them alongside your assembly to downloading them only when needed.

CHAPTER 7

Navigation

With the know-how you've picked up so far, you're ready to create applications that use a variety of different controls and layouts. However, there's still something missing: the ability to transition from one page to another. After all, traditional rich-client applications are usually built around different windows that encapsulate distinct tasks. To create this sort of application in Silverlight, you need a way to move beyond the single-page displays you've seen so far.

You can use two basic strategies to perform page changes in a Silverlight application, and each one has its proper place. The first option is to do it yourself by directly manipulating the user interface. For example, you can use code to access the root visual, remove the user control that represents the first page, and add another user control that represents a different page. This technique is straightforward, simple, and requires relatively little code. It also gives you the ability to micromanage details such as state management and to apply animated transition effects.

The second option is to use Silverlight's navigation system, which revolves around two new controls: Frame and Page. The basic idea is that a single *frame* container can switch between multiple *pages*. Although this approach to navigation is really no easier than managing the user interface manually, it provides a number of value-added features that would be extremely tedious to implement on your own. These include meaningful URIs, page tracking, and integration with the browser's history list.

In this chapter, you'll start by learning the basic do-it-yourself method of navigation. Next, you'll take a quick detour to consider the ChildWindow class, which gives you a straightforward way to simulate a *modal dialog box* (a window that temporary blocks the current page but doesn't replace it). Finally, you'll step up to the Frame and Page controls and see how they plug into Silverlight's built-in navigation system.

■ **What's New** Silverlight 4 adds a feature called *custom content loaders*, which allows you to extend the navigation system with your own code. Although custom content loaders can be quite complex, they give you (and third-party component developers) a way to implement specialized navigation behaviors, such as authentication and download-on-demand pages. For the full details, see the "Custom Content Loaders" section later in this chapter.

Loading User Controls

The basic idea of do-it-yourself navigation is to programmatically change the content that's shown in the Silverlight page, usually by manipulating layout containers or content controls. Of course, you don't want to be forced to create and configure huge batches of controls in code—that task is easier to complete using XAML. Instead, you need a way to create and load distinct user controls, each of which represents a page, and each of which is prepared at design time as a separate XAML file.

In the following sections, you'll see two related variations of this technique. First, you'll see an example that loads user controls into an existing page. This approach is best suited to user interfaces that need to keep some common elements (for example, a toolbar at the top or information panel at the side) as they load new content. Next, you'll see how to swap out the entire content of the current page.

Embedding User Controls in a Page

Many Silverlight applications are based around a single central page that acts as the main window for the entire application. You can change part of this page to load new content and simulate navigation.

One example of this design is the menu page that's used for most of the sample projects that accompany this book. This page uses the Grid control to divide itself into two main sections (separated by a horizontal GridSplitter). At the top is a list of all the pages you can visit. When you select one of the items from this list, it's loaded into the larger content region underneath, as shown in Figure 7-1.

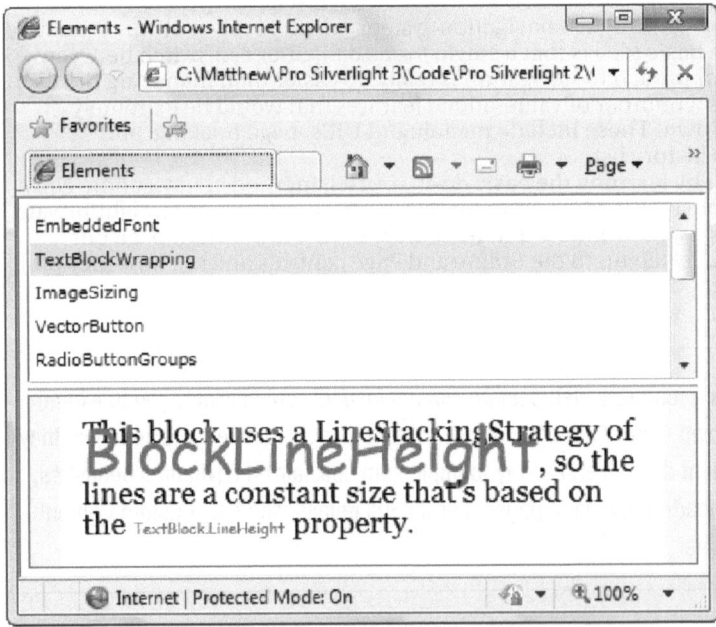

Figure 7-1. A window that loads user controls dynamically

Dynamically loading a user control is easy—you simply need to create an instance of the appropriate class and then add it to a suitable container. Good choices include the Border, ScrollViewer, StackPanel, or Grid control. The example shown previously uses the Border element, which is a content control that adds the ability to paint a border around its edges using the BorderBrush and BorderThickness properties.

Here's the markup (without the list of items in the list box):

```
<UserControl x:Class="Navigation.MenuPage"
 xmlns="http://schemas.microsoft.com/winfx/2006/xaml/presentation"
 xmlns:x="http://schemas.microsoft.com/winfx/2006/xaml"
 xmlns:basics=
 "clr-namespace:System.Windows.Controls;assembly=System.Windows.Controls">
  <Grid x:Name="LayoutRoot" Background="White" Margin="5">
    <Grid.RowDefinitions>
      <RowDefinition Height="*"></RowDefinition>
      <RowDefinition Height="Auto"></RowDefinition>
      <RowDefinition Height="3*"></RowDefinition>
    </Grid.RowDefinitions>

    <ListBox Grid.Row="0" SelectionChanged="lstPages_SelectionChanged">
      ...
    </ListBox>

    <basics:GridSplitter Grid.Row="1" Margin="0 3" HorizontalAlignment="Stretch"
     Height="2"></basics:GridSplitter>

    <Border Grid.Row="2" BorderBrush="SlateGray" BorderThickness="1"
     x:Name="borderPlaceholder" Background="AliceBlue"></Border>
  </Grid>
</UserControl>
```

In this example, the Border is named borderPlaceholder. Here's how you might display a new custom user control named Page2 in the borderPlaceholder region:

```
Dim newPage As New Page2()
borderPlaceholder.Child = newPage
```

If you're using a different container, you may need to set a different property instead. For example, Silverlight's layout panels can hold multiple controls and so provide a Children collection instead of a Child property. You need to clear this collection and then add the new control to it. Here's an example that duplicates the previous code, assuming you've replaced the Border with a single-celled Grid:

```
Dim newPage As New Page2()
gridPlaceholder.Children.Clear()
gridPlaceholder.Children.Add(newPage)
```

If you create a Grid without declaring any rows or columns, the Grid has a single proportionately sized cell that fits all the available space. Thus, adding a control to that Grid produces the same result as adding it to a Border.

The actual code that's used in the examples is a bit different because it needs to work for different types of controls. To determine which type of user control to create, the code examines the ListBoxItem object that was just clicked. It then uses reflection to create the corresponding user-control object:

```
Private Sub lstPages_SelectionChanged(ByVal sender As Object, _
  ByVal e As SelectionChangedEventArgs)
    ' Get the selected item.
    Dim newPageName As String = (CType(e.AddedItems(0), _
      ListBoxItem)).Content.ToString()

    ' Create an instance of the page named
    ' by the current button.
    Dim type As Type = Me.GetType()
    Dim assm As System.Reflection.Assembly = type.Assembly
    Dim newPage As UserControl = CType( _
      assm.CreateInstance(type.Namespace & "." & newPageName), _
      UserControl)

    ' Show the page.
    borderPlaceholder.Child = newPage
End Sub
```

Despite the reflection code, the process of *showing* the newly created user control—that is, setting the Border.Child property—is exactly the same.

Hiding Elements

If you decide to create a dynamic page like the one shown in the previous example, you aren't limited to adding and removing content. You can also temporarily *hide* it. The trick is to set the Visibility property, which is defined in the base UIElement class and inherited by all elements:

```
panel.Visibility = Visibility.Collapsed
```

The Visibility property uses an enumeration that provides just two values: Visible and Collapsed. (WPF included a third value, Hidden, which hides an element but keeps a blank space where it should be. However, this value isn't supported in Silverlight.) Although you can set the Visibility property of individual elements, usually you'll show and hide entire containers (for example, Border, StackPanel, or Grid objects) at once.

When an element is hidden, it takes no space in the page and doesn't receive any input events. The rest of your interface resizes itself to fill the available space, unless you've positioned your other elements with fixed coordinates using a layout container such as the Canvas.

■ **Tip** Many applications use panels that collapse or slide out of the way. To create this effect, you can combine this code with a dash of Silverlight animation. The animation changes the element you want to hide—for example, shrinking, compressing, or moving it. When the animation ends, you can set the Visibility property to hide the element permanently. You'll see how to use this technique in Chapter 10.

Managing the Root Visual

The page-changing technique shown in the previous example is common, but it's not suited for all scenarios. Its key drawback is that it slots new content into an existing layout. In the previous example, that means the list box remains fixed at the top of the page. This is handy if you need to make sure a toolbar or panel always remains accessible, but it isn't as convenient if you want to switch to a completely new display for a different task.

An alternative approach is to change the entire page from one control to another. The basic technique is to use a simple layout container as your application's root visual. You can then load user controls into the root visual when required and unload them afterward. (The root visual itself can never be replaced after the application has started.)

As you learned in Chapter 6, the startup logic for a Silverlight application usually creates an instance of a user control, as shown here:

```
Private Sub Application_Startup(ByVal o As Object, ByVal e As StartupEventArgs) _
    Handles Me.Startup

    Me.RootVisual = New MainPage()
End Sub
```

The trick is to use something more flexible—a simple container like the Border or a layout panel like the Grid. Here's an example of the latter approach:

```
' This Grid will host your pages.
Private rootGrid As New Grid()


Private Sub Application_Startup(ByVal o As Object, ByVal e As StartupEventArgs) _
    Handles Me.Startup

    ' Load the first page.
    Me.RootVisual = rootGrid
    rootGrid.Children.Add(New MainPage())
End Sub
```

Now, you can switch to another page by removing the first page from the Grid and adding a different one. To make this process relatively straightforward, you can add a shared method like this to the App class:

```
Public Shared Sub Navigate(ByVal newPage As UserControl)
    ' Get the current application object and cast it to
    ' an instance of the custom (derived) App class.
    Dim currentApp As App = CType(Application.Current, App)

    ' Change the currently displayed page.
    currentApp.rootGrid.Children.Clear()
    currentApp.rootGrid.Children.Add(newPage)
End Sub
```

You can navigate at any point using code like this:

```
App.Navigate(New Page2())
```

■ **Tip** You can add a dash of Silverlight animation and graphics to create a more pleasing transition between pages, such as a gentle fade or wipe. You'll learn how to use this technique in Chapter 10.

Retaining Page State

If you plan to allow the user to navigate frequently between complex pages, it makes sense to create each page once and keep the page instance in memory until later. This approach also has the sometimes-important side effect of maintaining that page's current state, including all the values in any input controls.

To implement this pattern, you first need a system to identify pages. You could fall back on string names, but an enumeration gives you better error prevention. Here's an enumeration that distinguished between three pages:

```
Public Enum Pages
    MainWindow
    ReviewPage
    AboutPage
End Enum
```

You can then store the pages of your application in private fields in your custom application class. Here's a simple dictionary that does the trick:

```
Private Shared pageCache As New Dictionary(Of Pages, UserControl)
```

In your Navigate() method, create the page only if it needs to be created—in other words, the corresponding object doesn't exist in the collection of cached pages:

```
Public Shared Sub Navigate(ByVal newPage As Pages)
    ' Get the current application object and cast it to
    ' an instance of the custom (derived) App class.
    Dim currentApp As App = CType(Application.Current, App)

    ' Check if the page has been created before.
    If Not pageCache.ContainsKey(newPage) Then
        ' Create the first instance of the page,
        ' and cache it for future use.
        Dim type As Type = currentApp.GetType()
        Dim currentAssembly As System.Reflection.Assembly = type.Assembly
        pageCache(newPage) = CType(currentAssembly.CreateInstance( _
          type.Namespace & "." & newPage.ToString()), UserControl)
    End If

    ' Change the currently displayed page.
    currentApp.rootGrid.Children.Clear()
    currentApp.rootGrid.Children.Add(pageCache(newPage))
End Sub
```

Now, you can navigate by indicating the page you want with the Pages enumeration:

```
App.Navigate(Pages.MainWindow)
```

Because only one version of the page is ever created and it's kept in memory over the lifetime of the application, all of the page's state remains intact when you navigate away and back again (see Figure 7-2).

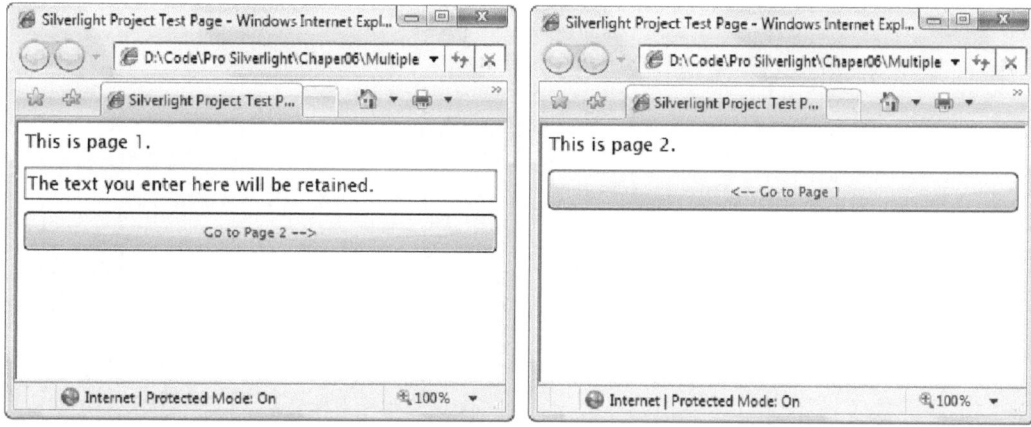

Figure 7-2. *Moving from one page to another*

Browser History

The only limitation with the navigation methods described in this section is that the browser has no idea you've changed from one page to another. If you want to let the user go back, it's up to you to add the controls that do it. The browser's Back button will only send you to the previous HTML page (thereby exiting your Silverlight application).

If you want to create an application that integrates more effectively with the browser and supports the Back button, you'll need to use the Frame and Page classes discussed later in this chapter.

Child Windows

In many situations, you don't need a way to *change* the page—you just need to temporarily show some sort of content before allowing the user to return to the main application page. The obvious example is a confirmation dialog box, but Windows and web applications use pop-up windows to collect information, show basic program information, and provide access to configuration settings.

In Silverlight, you can create this sort of design using a handy content control called ChildWindow. Essentially, ChildWindow mimics the modal dialog boxes you've seen on the Windows platform. When you show a child window, the rest of the application information is disabled (and a gray shaded overlay is displayed over of it as a user cue). Then, the child window appears centered on top of the page. After the user completes a task in the child window, your code closes it, and the rest of the application becomes responsive again.

Figure 7-3 shows an example. Here, the page includes a single button that, when clicked, pops open a child window requesting more information. When the user clicks a button (or clicks the *X* in the top-right corner), the window vanishes.

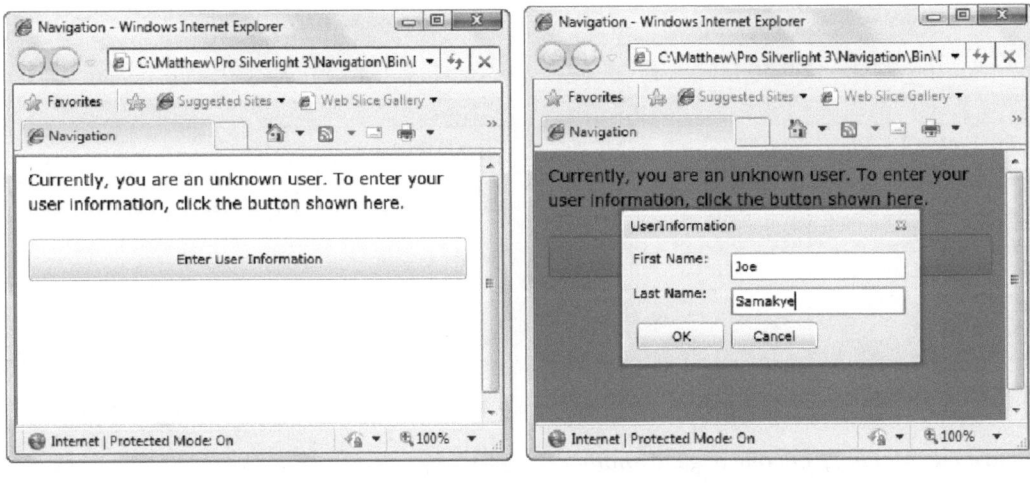

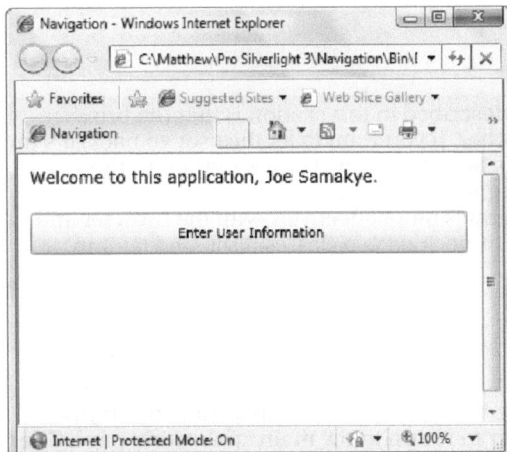

Figure 7-3. *Showing a child window*

The child window pops into view with a subtle but attractive expansion effect. It also behaves like a real window, allowing you to click its title bar and drag it around the page (but not out of the browser display area).

Although the ChildWindow control provides the illusion of a separate pop-up window that appears on top of your application, it's actually just another element that's added to your existing page. However, the ChildWindow control is clever enough to disable the rest of the content in the root visual of your application and position itself appropriately, making it look and behave like a traditional pop-up window. Finally, it's worth noting that when you show a child window, the user interface underneath remains active, even though the user can't interact with it. For example, if you have an animation running or a video playing, it continues in the background while the child window is visible (unless you explicitly stop it).

■ **Note** The ChildWindow control always blocks the main user interface. However, you can download a free FloatableWindow control that doesn't share this characteristic at `http://floatablewindow.codeplex.com`. You can use the FloatableWindow to display one or more pop-up windows over your main Silverlight page and keep them there while the user interacts with the rest of the application. You can use this design to implement a notification window, separate task area, or floating tool panel, but tread with caution. If not handled carefully, floating windows can be confusing for the end user.

Designing a ChildWindow

Before you can show a child window, you need to create one with a XAML template, in the same way you design a user control. To add a bare-bones starter in Visual Studio, right-click the project name in the Solution Explorer, and choose Add ➤ New Item. Then, pick the Silverlight Child Window template, enter a name, and click Add. Visual Studio creates the new XAML template and a code-behind file, and it adds a reference to the System.Windows.Controls.dll assembly where the ChildWindow control is defined.

■ **Note** ChildWindow is a control that derives from ContentControl. It adds two new properties (Title and DialogResult), two methods (Show and Close), and two events (Closing and Closed).

After you've added a child window, you can design it in exactly the same way you design an ordinary user control. To make your life easier, Visual Studio automatically creates a two-row Grid in the new child window template and places OK and Cancel buttons in the bottom row, complete with event handlers that close the window. (Of course, you can remove or reconfigure these buttons to suit your application design.)

Here's the markup for the child window shown in Figure 7-3. It provides two text boxes for user information and adds the standard OK and Cancel buttons underneath:

```
<controls:ChildWindow x:Class="Navigation.UserInformation"
  xmlns="http://schemas.microsoft.com/winfx/2006/xaml/presentation"
  xmlns:x="http://schemas.microsoft.com/winfx/2006/xaml"
  xmlns:controls=
"clr-namespace:System.Windows.Controls;assembly=System.Windows.Controls"
  Title="UserInformation">
  <Grid x:Name="LayoutRoot" Margin="2">
    <Grid.RowDefinitions>
      <RowDefinition Height="Auto"></RowDefinition>
      <RowDefinition Height="Auto"></RowDefinition>
      <RowDefinition Height="Auto"></RowDefinition>
    </Grid.RowDefinitions>
    <Grid.ColumnDefinitions>
      <ColumnDefinition></ColumnDefinition>
      <ColumnDefinition></ColumnDefinition>
```

```
        </Grid.ColumnDefinitions>

        <TextBlock>First Name:</TextBlock>
        <TextBox x:Name="txtFirstName" Grid.Column="1" Margin="3" Width="150"></TextBox>
        <TextBlock Grid.Row="1">Last Name:</TextBlock>
        <TextBox x:Name="txtLastName" Grid.Row="1" Grid.Column="1" Margin="3"></TextBox>

        <Button Grid.Row="2" Margin="3" Width="75" Height="23"
         HorizontalAlignment="Right" Content="OK" Click="cmdOK_Click"></Button>
        <Button Grid.Row="2" Grid.Column="1" Margin="3" Width="75" Height="23"
         HorizontalAlignment="Left" Content="Cancel" Click="cmdCancel_Click"></Button>
    </Grid>
</controls:ChildWindow>
```

The event handlers for the two buttons set the ChildWindow.DialogResult property. This property is a nullable Boolean value that indicates whether the user accepted the action represented by this window (True), cancelled it (False), or did neither (Nothing).

```
Private Sub cmdOK_Click(ByVal sender As Object, ByVal e As RoutedEventArgs)
    Me.DialogResult = True
End Sub

Private Sub cmdCancel_Click(ByVal sender As Object, ByVal e As RoutedEventArgs)
    Me.DialogResult = False
End Sub
```

Setting the DialogResult property also closes the window, returning control to the root visual. In some cases, the DialogResult property may not be relevant to your application (for example, if you're showing an About window that includes a single Close button). In this case, you can close the window by using the ChildWindow.Close() method rather than setting the DialogResult property.

Showing a ChildWindow

Showing a child window is easy. You need to create an instance of your custom ChildWindow class and call the Show() method:

```
Dim childWindow As New UserInformation()
childWindow.Show()
```

It's important to realize that although the child window blocks the main user interface, the Show() method doesn't block the execution of your code. Thus, if you put code after the call to the Show() method, that code runs immediately.

This presents a problem if you need to react when the user closes the child window, which is usually the case. In the example shown in Figure 7-3, the application needs to gather the entered user name and use it to update the display in the main page. To perform this task, your code must respond to the ChildWindow.Closed event. (The ChildWindow class also provides a Closing event that fires when the window begins to close, but this is intended for scenarios when you need to cancel the close operation—for example, if necessary information hasn't been entered.)

Remember to attach an event handler to the Closed event before you show the child window:

```
Dim childWindow As New UserInformation()
AddHandler childWindow.Closed, AddressOf childWindow_Closed
childWindow.Show()
```

There's still more to think about. If your child window is anything more than a simple confirmation box, you'll probably need to return additional information to the rest of your application. In the current example, that information consists of the user's first and last names. In theory, your application code could grab the ChildWindow object and directly extract this information from the appropriate controls. However, this sort of interaction is fragile. It creates tight dependencies between the main page and the child window, and these dependencies aren't always obvious. If you change the design of your application—for example, swapping the first name and last name text boxes for different controls—the code breaks. A far better approach is to create an extra layer of public properties and methods in your child window. Your main application page can call on these members to get the information it needs. Because these methods are stored in the custom ChildWindow class, you'll know to tweak them so they continue to work if you revamp the child window's user interface.

For example, in the current example, you can add this property to the UserInformation class to expose the full name information:

```
Public ReadOnly Property UserName() As String
    Get
        Return txtFirstName.Text & " " & txtLastName.Text
    End Get
End Property
```

Now, you can access this detail when you respond to the Closed event:

```
Private Sub childWindow_Closed(ByVal sender As Object, ByVal e As EventArgs)
    Dim childWindow As UserInformation = CType(sender, UserInformation)
    If childWindow.DialogResult = True Then
        lblInfo.Text = "Welcome to this application, " & childWindow.UserName & "."
    End If
End Sub
```

One final improvement is worth making. Currently, the child window is created each time the user clicks the Enter User Information button. As a result, the first name and last name text boxes always remain empty, even if the user has entered name information previously. To correct this, you can add a property setter for the UserName property or, even better, you can keep the lightweight UserInformation object in memory. In this example, the ChildWindow object is created it once, as a member variable of the main page:

```
Private childWindow As New UserInformation()
```

You must now attach the ChildWindow.Closed event handler in the page constructor:

```
Public Sub New()
    InitializeComponent()
    AddHandler childWindow.Closed, AddressOf childWindow_Closed
End Sub
```

The UserInformation object will keep its state, meaning that every time you show it, the previously entered name information will remain in place.

■ **Tip** Here's a neat way to use ChildWindow: to pop up rich error messages that have more polish than the basic MessageBox. In fact, if you create a new project in Visual Studio using the Silverlight Navigation Application template, you'll find that it uses this design. The navigation template includes markup and code for a custom ChildWindow called ErrorWindow. When the Application.UnhandledException event occurs, the application uses the ErrorWindow to politely explain the problem.

The Frame and Page

Changing the user interface by hand is a good approach if your application has very few pages (like an animated game that revolves around a main screen and a configuration window). It also makes sense if you need complete control over the navigation process (perhaps so you can implement page-transition effects, like the ones you'll see in Chapter 10). But if you're building a more traditional application and you expect the user to travel back and forth through a long sequence of pages, Silverlight's navigation system can save you some significant work.

The navigation system is built into two controls: Frame and Page. Of the two, the Frame control is the more essential, because it's responsible for creating the container in which navigation takes place. The Page control is an optional sidekick—it gives you a convenient way to show different units of content in a frame. Both classes have members that expose the navigation features to your code.

Frames

The Frame is a content control—a control that derives from ContentControl and contains a single child element. This child is exposed through the Content property.

Other content controls include Button, ListBoxItem, ToolTip, and ScrollViewer. However, the Frame control has a notable difference: if you're using it right, you'll almost never touch the Content property directly. Instead, you'll change the content using the higher-level Navigate() method. The Navigate() method changes the Content property, but it also triggers the navigation services that are responsible for tracking the user's page history and updating the browser's URI.

For example, consider the following page markup. It defines a Grid that has two rows. In the top row is a Border that holds a Frame. (Although the Frame class has the BorderBrush and BorderThickness properties, it lacks the CornerRadius property, so you need to use a Border element if you want a rounded border around your content.) In the bottom row is a button that triggers navigation. Figure 7-4 shows the page.

```
<UserControl x:Class="Navigation.MainPage"
    xmlns="http://schemas.microsoft.com/winfx/2006/xaml/presentation"
    xmlns:x="http://schemas.microsoft.com/winfx/2006/xaml" xmlns:navigation=
"clr-namespace:System.Windows.Controls;assembly=System.Windows.Controls.Navigation">
    <Grid>
        <Grid.RowDefinitions>
            <RowDefinition></RowDefinition>
            <RowDefinition Height="Auto"></RowDefinition>
        </Grid.RowDefinitions>
```

```
<Border Margin="10" Padding="10" BorderBrush="DarkOrange" BorderThickness="2"
   CornerRadius="4">
  <navigation:Frame x:Name="mainFrame"></navigation:Frame>
</Border>
<Button Grid.Row="1" Margin="5" Padding="5" HorizontalAlignment="Center"
   Content="Navigate to a New Page" Click="cmdNavigate_Click"></Button>
  </Grid>
</UserControl>
```

To use the Frame class, you must map the System.Windows.Controls namespace from the System.Windows.Controls.Navigation.dll assembly to an XML namespace prefix. This example uses the prefix *navigation*.

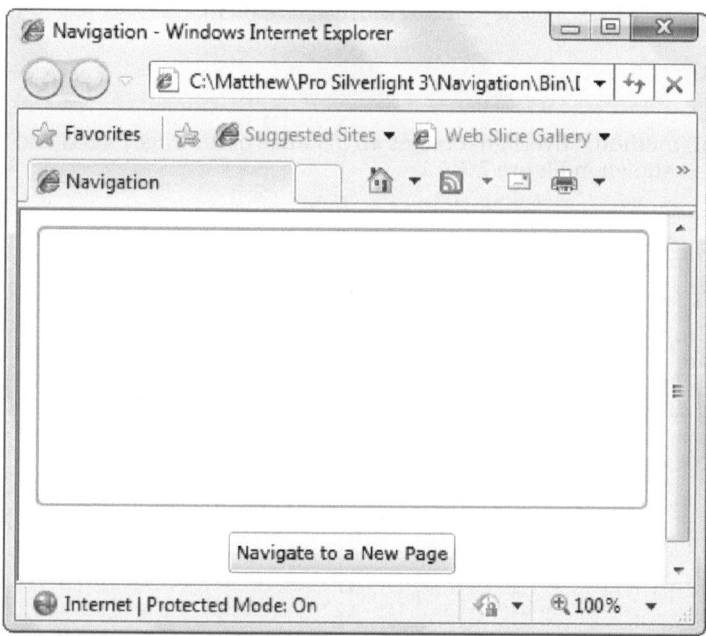

Figure 7-4. An empty frame

Currently, the frame is empty. But if the user clicks the button, an event handler runs and calls the Navigate() method. The Navigate() method takes a single argument—a URI pointing to a compiled XAML file in your application:

```
Private Sub cmdNavigate_Click(ByVal sender As Object, ByVal e As RoutedEventArgs)
    mainFrame.Navigate(New Uri("/Page1.xaml", UriKind.Relative))
End Sub
```

This code works because the application includes a user control named Page1.xaml. Note that the URI always begins with a forward slash, which represents the application root.

■ **Note** You cannot use the Navigate() method with URIs that point to other types of content or to pages outside your application (for example, external websites).

Here's the markup for the Page1.xaml user control:

```
<UserControl x:Class="Navigation.Page1"
 xmlns="http://schemas.microsoft.com/winfx/2006/xaml/presentation"
 xmlns:x="http://schemas.microsoft.com/winfx/2006/xaml">
  <Grid x:Name="LayoutRoot" Background="White">
    <TextBlock TextWrapping="Wrap">This is the unremarkable content in
Page1.xaml.</TextBlock>
  </Grid>
</UserControl>
```

When you call the Navigate() method, Silverlight creates an instance of the Page1 class and uses it to set the frame content, as shown in Figure 7-5.

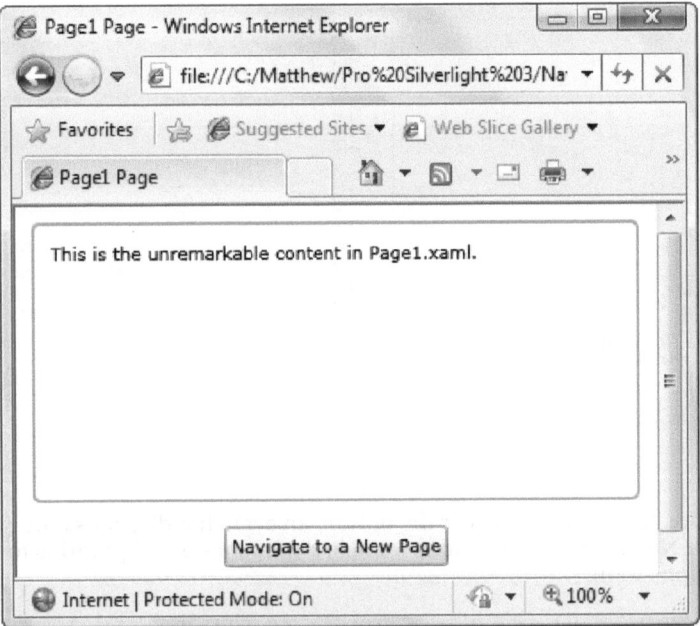

Figure 7-5. Filling a frame with content through navigation

If you were performing navigation by hand, you could replace the call to Navigate() with this code:

```
' Create the user control.
Dim newPage As New Page1()

' Show the user control, replacing whatever content is currently visible.
mainFrame.Content = newPage
```

However, this code changes *only* the content, whereas the Navigate() method treats the action as a higher-level navigation event that hooks into some additional features. When you call Navigate(), you'll notice two significant differences—browser URI integration and history support—which are described in the following sections.

■ **Tip** You can get the URI of the current page at any time using the Frame.Source property. You can also set the Source property as an alternative to calling Navigate().

Browser URI Integration

When you change the content of a Frame control through the Navigate() method, the name of the XAML resource is appended to the current URI, after the fragment marker (#). So if your application lives at this URI:

```
localhost://Navigation/TestPage.html
```

and you perform navigation with code like this:

```
mainFrame.Navigate(New Uri("/Page1.xaml", UriKind.Relative))
```

you'll now see this URI in your browser:

```
localhost://Navigation/TestPage.html#/Page1.xaml
```

This system has many implications—some good, some potentially bad (or at least complicating). Essentially, when you use Silverlight's frame-based navigation system, each page you load into the frame has a distinct URI, which also means it's a separate history item and a new entry point into your application.

For example, if you close the browser and reopen it later, you can type in the newly constructed navigation URI with *#/Page1.xaml* at the end to request TestPage.html, load the Silverlight application, and insert the content from Page1.xaml into the frame, all in one step. Similarly, users can create a bookmark with this URI that lets them return to the application with the correct page loaded in the frame. This feature is sometimes called *deep linking*, because it allows you to use links that link not just to the entry point of an application but also to some record or state *inside* that application.

■ **Tip** With a little more effort, you can use deep linking as a starting point for *search engine optimization* (SEO). The basic idea is to create multiple HTML or ASP.NET pages that lead to different parts of your Silverlight application. Each page will point to the same XAP file, but the URI will link to a different page *inside* that application. Web search engines can then add multiple index entries for your application, one for each HTML or ASP.NET page that leads into it.

URI integration is obviously a convenient feature, but it also raises a few questions, which are outlined in the following sections.

What Happens If the Page Has More Than One Frame?

The URI fragment indicates the page that should appear in the frame, but it doesn't include the frame name. It turns out that this system really works only for Silverlight applications that have a single frame. (Applications that contain two or more frames are considered to be a relatively rare occurrence.)

If you have more than one frame, they will all share the same navigation path. As a result, when your code calls Navigate() in one frame or when the user enters a URI that includes a page name as a fragment, the same content will be loaded into every frame. To avoid this problem, you must pick a single frame that represents the main application content. This frame will control the URI and the browser history list. Every other frame will be responsible for tracking its navigation privately, with no browser interaction. To implement this design, set the JournalOwnership property of each additional frame to OwnJournal. From that point on, the only way to perform navigation in these frames is with code that calls the Navigate() method.

What Happens If the Startup Page Doesn't Include a Frame Control?

Pages with multiple frames aren't the only potential problem with the navigation system's use of URIs. Another issue occurs if the application can't load the requested content because there's no frame in the application's root visual. This situation can occur if you're using one of the dynamic user interface tricks described earlier—for example, using code to create the Frame object or swap in another page that contains a frame. In this situation, the application starts normally; but because no frame is available, the fragment part of the URI is ignored.

To remedy this problem, you need to either simplify your application so the frame is available in the root visual at startup or add code that responds to the Application.Startup event (see Chapter 6) and checks the document fragment portion of the URI, using code like this:

```
Dim fragment As String
fragment = System.Windows.Browser.HtmlPage.Document.DocumentUri.Fragment
```

If you find that the URI contains fragment information, you can then add code by hand to restore the application to its previous state. Although this is a relatively rare design, take the time to make sure it works properly. After all, when a fragment URI appears in the browser's address bar, the user naturally assumes it's a suitable bookmark point. And if you don't want to provide this service, consider disabling the URI system altogether by setting the JournalOwnership property to OwnJournal.

What About Security?

In a very real sense, the URI system is like a giant back door into your application. For example, a user can enter a URI that points to a page you don't want that user to access—even one that you never load with the Navigate() method. Silverlight doesn't attempt to impose any measure of security to restrict this scenario. In other words, adding a Frame control to your application provides a potential path of access to any other page in your application.

Fortunately, you can use several techniques to clamp down on this ability. First, you can detach the frame from the URI system by setting the JournalOwnership property to OwnJournal, as described earlier. However, this gives up the ability to use descriptive URIs for any of the pages in your application, and it also removes the integration with the browser history list that's described in the next section. A better approach is to impose selective restriction by handling the Frame.Navigating event. At this point, you can examine the URI (through the NavigatingCancelEventArgs object) and, optionally, cancel navigation:

```
Private Sub mainFrame_Navigating(ByVal sender As Object, _
  ByVal e As NavigatingCancelEventArgs)

    If e.Uri.ToString().ToLower().Contains("RestrictedPage.xaml") Then
        e.Cancel = True
    End If
End Sub
```

You'll notice that this code doesn't match the entire URI but simply checks for the presence of a restricted page name. This is to avoid potential *canonicalization problems*—in other words, allowing access to restricted pages by failing to account for the many different ways the same URI can be written. Here's an example of functionally equivalent but differently written URIs:

```
localhost://Navigation/TestPage.html#/Page1.xaml
localhost://Navigation/TestPage.html#/FakeFolder/.../Page1.xaml
```

This example assumes that you never want to perform navigation to RestrictedPage.xaml. The Navigating event does not distinguish whether the user has edited the URI or whether the navigation attempt is the result of the user clicking the link or your code calling the Navigate() method. Presumably, the application will use RestrictedPage.xaml in some other way—for example, with code that manually instantiates the user control and loads it into another container.

Finally, it's worth noting that there's one other way to deal with restricted pages. You could create a custom content loader (as described in the "Custom Content Loaders" section at the end of this chapter), which would check the requested URI against a collection of allowed pages. This approach is similar to handling the Frame.Navigating event, but it's a bit more general (and a bit more complex). If you get your design right, you can use it in any application that needs it—all you need to do is add the custom content loader to the appropriate frame and configure the collection of allowed pages. The disadvantage is that it will take you significantly more coding to arrive at a workable solution.

History Support

The navigation features of the Frame control also integrate with the browser. Each time you call the Navigate() method, Silverlight adds a new entry in the history list (see Figure 7-6). The first page of your application appears in the history list first, with the title of the HTML entry page. Each subsequent page appears under that in the history list, using the user-control file name for the display text (such as Page1.xaml). In the "Pages" section later in this chapter, you'll learn how you can supply your own, more descriptive title text using a custom page.

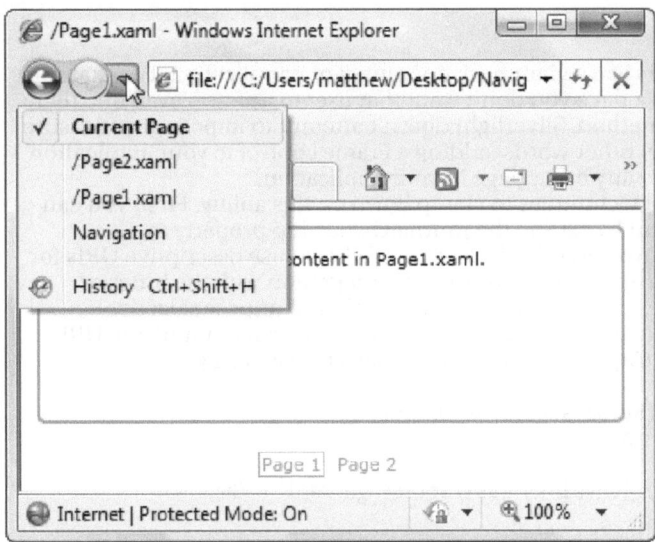

Figure 7-6. *The navigation history of the frame*

The browser's history list works exactly the way you'd expect. The user can click the Back or Forward button or pick an entry in the history list to load a previous page into the frame. Best of all, this doesn't cause your application to restart. As long as the rest of the URI stays the same (everything except the fragment), Silverlight simply loads the appropriate page into the frame. On the other hand, if the user travels to another website and then uses the Back button to return, the Silverlight application is reloaded, the Application.Startup event fires, and *then* Silverlight attempts to load the requested page into the frame.

Incidentally, you can call the Frame.Navigate() method multiple times in succession with different pages. The user ends up on the last page, but all the others are added to the history list in between. Finally, the Navigate() method does nothing if the page is already loaded—it doesn't add a duplicate entry to the history list.

Navigation Failure

Silverlight has a quirk that affects how it deals with the Back button when using navigation. The problem occurs if you use the browser history list to return to a page that doesn't include the fragment portion of the URI. For example, you will encounter this problem in the previous example if you launch the application, surf to a new page, and then press the browser's Back button to return to the initial page. When you do, you'll receive an ArgumentException that explains "Content for the URI cannot be loaded." In other words, the URI did not specify the content for the frame, and Silverlight didn't know what to put there.

There are two easy ways to deal with this problem. The first approach is to handle the Frame.NavigationFailed event. You can then examine the exception object (provided through the NavigationFailedEventArgs.Exception property) and set the NavigationFailedEventArgs.Handled property to True to gracefully ignore this nonissue and carry on.

Your other option is to use the UriMapper to set the initial content of the frame. With this technique, you map the ordinary, empty URI to a valid page, which will then be shown in the frame (although that page can be blank). You'll see how to use this technique in the next section.

URI Mapping

As you've seen, the fragment URI system puts the page name in the URI. In some situations, you'd prefer not to make this detail as glaring. Perhaps you don't want to expose the real page name, you don't want to tack on the potentially confusing .xaml extension, or you want to use a URI that's easier to remember and type in by hand. In all these situations, you can use URI mapping to define different, simpler URIs that map to the standard versions you've seen so far.

To use URI mapping, you first need to add a UriMapper object as a XAML resource. Typically, you'll define the UriMapper in the resources collection of the main page or the App.xaml file, as shown here:

```
<Application xmlns="http://schemas.microsoft.com/winfx/2006/xaml/presentation"
 xmlns:x="http://schemas.microsoft.com/winfx/2006/xaml"
 x:Class="Navigation.App" xmlns:navigation=
"clr-namespace:System.Windows.Navigation;assembly=System.Windows.Controls.Navigation">
  <Application.Resources>
    <navigation:UriMapper x:Key="PageMapper">
    </navigation:UriMapper>
  </Application.Resources>
</Application>
```

You then need to link your UriMapper to your frame by setting the Frame.UriMapper property:

```
<navigation:Frame x:Name="mainFrame" UriMapper="{StaticResource PageMapper}">
</navigation:Frame>
```

Now, you can add your URI mappings inside the UriMapper. Here's an example:

```
<navigation:UriMapper x:Key="PageMapper">
  <navigation:UriMapping Uri="Home" MappedUri="/Views/HomePage.xaml" />
</navigation:UriMapper>
```

If your application is located here

```
localhost://Navigation/TestPage.html
```

you can use this simplified URI

```
localhost://Navigation/TestPage.html#Home
```

which is mapped to this URI:

```
localhost://Navigation/TestPage.html#/Views/HomePage.xaml
```

The only catch is that it's up to you to use the simplified URI when you call the Navigate() method, as shown here:

```
mainFrame.Navigate(New Uri("Home", UriKind.Relative))
```

Note that you don't need to include a forward slash at the beginning of a mapped URI.

After mapping, both the original and the new URI will work, allowing you to reach the same page. If you use the original URI format when calling the Navigate() method (or in a link or in a bookmark), that's what the user sees in the browser's address bar.

You can also use the UriMapper to set the initial content in a frame. The trick is to map a Uri that's just an empty string, as shown here:

```
<navigation:UriMapper x:Key="PageMapper">
  <navigation:UriMapping Uri="" MappedUri="/InitialPage.xaml" />
  <navigation:UriMapping Uri="Home" MappedUri="/Views/HomePage.xaml" />
</navigation:UriMapper>
```

Now, when the page first appears, the frame will show the content from InitialPage.xaml.

■ **Note** Remember, if you don't use the UriMapper to set the initial page, you must handle the Frame.NavigationFailed event. Otherwise, users may receive an error when they step back to the first page using the browser's Back button.

The UriMapper object also supports URIs that take query-string arguments. For example, consider the following mapping:

```
<navigation:UriMapping Uri="Products/{id}"
 MappedUri="/Views/ProductPage.xaml?id={id}"></navigation:UriMapping>
```

In this example, the {id} portion in curly brackets is a placeholder. You can use any URI that has this basic form but supplies an arbitrary value for the id. For example, this URI

```
localhost://Navigation/TestPage.html#Products/324
```

will be mapped to this:

```
localhost://Navigation/TestPage.html#/Views/ProductPage.xaml?id=324
```

The easiest way to retrieve the id query-string argument in the ProductPage.xaml code is to use the NavigationContext object described later in the "Pages" section.

Forward and Backward Navigation

As you've learned, you can set the Frame.JournalOwnership property to determine whether the frame uses the browser's history-tracking system (the default) or is responsible for keeping the record of visited pages on its own (which is called the *journal*). If you opt for the latter by setting the JournalOwnership property to OwnJournal, your frame won't integrate with the browser history or use the URI system described earlier. You'll need to provide a way for the user to navigate through the page history. The most common way to add this sort of support is to create your own Forward and Backward buttons.

Custom Forward and Backward buttons are also necessary if you're building an out-of-browser application, like the sort described in Chapter 21. That's because an application running in a stand-alone window doesn't have access to any browser features and doesn't include any browser user interface (including the Back and Forward buttons). In this situation, you're forced to supply your own navigation buttons for programmatic navigation, even if you haven't changed the JournalOwnership property.

If you're not sure whether your application is running in a browser or in a stand-alone window, check the Application.IsRunningOutOfBrowser property. For example, the following code shows a panel with navigation buttons when the application is hosted in a stand-alone window. You can use this in the Loaded event handler for your root visual.

```
If App.Current.IsRunningOutOfBrowser Then
  pnlNavigationButtons.Visibility = Visibility.Visible
End If
```

Designing Forward and Backward buttons is easy. You can use any element you like—the trick is simply to step forward or backward through the page history by calling the GoBack() and GoForward() methods of the Frame class. You can also check the CanGoBack property (which is True if there are pages in the backward history) and the CanGoForward property (which is True if there are pages in the forward history) and use that information to selectively enable and disable your custom navigation buttons. Typically, you'll do this when responding to the Frame.Navigated event:

```
Private Sub mainFrame_Navigated(ByVal sender As Object, _
  ByVal e As NavigationEventArgs)

    If mainFrame.CanGoBack Then
        cmdBack.Visibility = Visibility.Visible
    Else
        cmdBack.Visibility = Visibility.Collapsed
    End If

    If mainFrame.CanGoForward Then
        cmdForward.Visibility = Visibility.Visible
    Else
        cmdForwawrd.Visibility = Visibility.Collapsed
    End If
End Sub
```

Rather than hide the buttons (as done here), you may choose to disable them and change their visual appearance (for example, changing the color, opacity, or picture, or adding an animated effect). Unfortunately, there's no way to get a list of page names from the journal, which means you can't display a history list like the one shown in the browser.

Hyperlinks

In the previous example, navigation was performed through an ordinary button. However, it's a common Silverlight design to use a set of HyperlinkButton elements for navigation. Thanks to the URI system, it's even easier to use the HyperlinkButton than an ordinary button. You simply need to set the NavigateUri property to the appropriate URI. You can use URIs that point directly to XAML pages, or you can use mapped URIs that go through the UriMapper.

Here's a StackPanel that creates a strip of three navigation links:

```
<StackPanel Margin="5" HorizontalAlignment="Center" Orientation="Horizontal">
  <HyperlinkButton NavigateUri="/Page1.xaml" Content="Page 1" Margin="3" />
  <HyperlinkButton NavigateUri="/Page2.xaml" Content="Page 2" Margin="3" />
  <HyperlinkButton NavigateUri="Home" Content="Home" Margin="3" />
</StackPanel>
```

Although the concept hasn't changed, this approach allows you to keep the URIs in the XAML markup and leave your code simple and uncluttered by extraneous details.

Pages

The previous examples all used navigation to load user controls into a frame. Although this design works, it's far more common to use a custom class that derives from Page instead of a user control, because the Page class provides convenient hooks into the navigation system and (optionally) automatic state management.

To add a page to a Visual Studio project, right-click the project name in the Solution Explorer, and choose Add ➤ New Item. Then, select the Silverlight Page template, enter a page name, and click Add. Aside from the root element, the markup you place in a page is the same as the markup you put in a user control. Here's a reworked example that changes Page1.xaml from a user control into a page by modifying the root element and setting the Title property:

```
<navigation:Page x:Class="Navigation.Page1"
 xmlns="http://schemas.microsoft.com/winfx/2006/xaml/presentation"
 xmlns:x="http://schemas.microsoft.com/winfx/2006/xaml"
 xmlns:navigation=
"clr-namespace:System.Windows.Controls;assembly=System.Windows.Controls.Navigation"
 Title="Sample Page">
  <Grid x:Name="LayoutRoot" Background="White">
    <TextBlock TextWrapping="Wrap">This is the unremarkable content in
Page1.xaml.</TextBlock>
    </Grid>
</navigation:Page>
```

■ **Tip** It's a common design convention to place pages in a separate project folder from your user controls. For example, you can place all your pages in a folder named Views and use navigation URIs like /Views/Page1.xaml.

Technically, Page is a class that derives from UserControl and adds a small set of members. These include a set of methods you can override to react to navigation actions and four properties: Title, NavigationService, NavigationContext, and NavigationCacheMode. The Title property is the simplest. It sets the text that's used for the browser history list, as shown in the previous example. The other members are described in the following sections.

Navigation Properties

Every page provides a NavigationService property that offers an entry point into Silverlight's navigation system. The NavigationService property provides a NavigationService object, which supplies the same navigational methods as the Frame class, including Navigate(), GoBack(), and GoForward(), and properties such as CanGoBack, CanGoForward, and CurrentSource. That means you can trigger page navigation from inside a page by adding code like this:

```
Me.NavigationService.Navigate(New Uri("/Page2.xaml", UriKind.Relative))
```

The Page class also includes a NavigationContext property that provides a NavigationContext object. This object exposes two properties: Uri gets the current URI, which was used to reach the current page, and QueryString gets a collection that contains any query-string arguments that were tacked on to the end of the URI. This way, the code that triggers the navigation can pass information to the destination page. For example, consider the following code, which embeds two numbers into a URI as query-string arguments:

```
Dim uriText As String
uriText = String.Format("/Product.xaml?productID={0}&type={1}", _
  productID, productType)
mainFrame.Navigate(New Uri(uriText), UriKind.Relative)
```

A typical completed URI might look something like this:

```
/Product.xaml?productID=402&type=12
```

You can retrieve the product ID information in the destination page with code like this:

```
Dim productID, type As Integer

If Me.NavigationContext.QueryString.ContainsKey("productID") Then
    productID = Int32.Parse(Me.NavigationContext.QueryString("productID"))
End If
If Me.NavigationContext.QueryString.ContainsKey("type") Then
    type = Int32.Parse(Me.NavigationContext.QueryString("type"))
End If
```

Of course, there are other ways to share information between pages, such as storing it in the application object. The difference is that query-string arguments are preserved in the URI, so users who bookmark the link can reuse it later to return to an exact point in the application (for example, the query string allows you to create links that point to particular items in a catalog of data). On the downside, query-string arguments are visible to any user who takes the time to look at the URI, and they can be tampered with.

State Storage

Ordinarily, when the user travels to a page using the Forward and Backward buttons or the history list, the page is re-created from scratch. When the user leaves the page, the page object is discarded from memory. One consequence of this design is that if a page has user input controls (for example, a text box), they're reset to their default values on a return visit. Similarly, any member variables in the page class are reset to their initial values.

The do-it-yourself state-management approach described earlier lets you avoid this issue by caching the entire page object in memory. Silverlight allows a similar trick with its own navigation system using the Page.NavigationCacheMode property.

The default value of NavigationCacheMode is Disabled, which means no caching is performed. Switch this to Required, and the Frame will keep the page object in memory after the user navigates away. If the user returns, the already instantiated object is used instead of a newly created instance. The page constructor will not run, but the Loaded event will still fire.

There's one other option for NavigationCacheMode. Set it to Enabled, and pages will be cached—up to a point. The key detail is the Frame.CacheSize property, which sets the maximum number of optional pages that can be cached. For example, when this value is 10 (the default), the Frame will cache the ten most recent pages that have a NavigationCacheMode of Enabled. When an eleventh page is added to the cache, the first (oldest) page object will be discarded from the cache. Pages with NavigationCacheMode set to Required don't count against the CacheSize total.

Typically, you'll set NavigationCacheMode to Required when you want to cache a page to preserve its current state. You'll set NavigationCacheMode to Enabled if you want the option of caching a page to save time and improve performance—for example, if your page includes time-consuming initialization logic that performs detailed calculations or calls a web service. In this case, make sure you place this logic in the constructor, not in an event handler for the Loaded event (which still fires when a page is served from the cache).

Navigation Methods

The Page class also includes a small set of methods that are triggered during different navigation actions. They include the following:

- *OnNavigatedTo()*: This method is called when the frame navigates to the page (either for the first time or on a return trip through the history).

- *OnNavigatingFrom()*: This method is called when the user is about to leave the page; it allows you to cancel the navigation action.

- *OnNavigatedFrom()*: This method is called when the user has left the page, just before the next page appears.

You could use these methods to perform various actions when a page is being left or visited, such as tracking and initialization. For example, you could use them to implement a more selective form of state management that stores just a few details from the current page in memory, rather than caching the entire page object. Simply store page state when OnNavigatedFrom() is called and retrieve it when OnNavigatedTo() is called. Where you store the state is up to you—you can store it in the App class, or you can use shared members in your custom page class, as done here with a single string:

```
Public Partial Class CustomCachedPage
    Inherits Page

    ...

    Private Shared _textBoxState As String
    Public Shared Property TextBoxState() As String
        Get
            Return _textBoxState
        End Get
        Set(ByVal value As String)
            _textBoxState = value
        End Set
    End Property

End Class
```

Here's the page code that uses this property to store the data from a single text box and retrieve it when the user returns to the page later:

```
Protected Overrides Sub OnNavigatedFrom(ByVal e As NavigationEventArgs)
    ' Store the text box data.
    CustomCachedPage.TextBoxState = txtCached.Text
    MyBase.OnNavigatedFrom(e)
End Sub

Protected Overrides Sub OnNavigatedTo(ByVal e As NavigationEventArgs)
    ' Retrieve the text box data.
    If CustomCachedPage.TextBoxState IsNot Nothing Then
        txtCached.Text = CustomCachedPage.TextBoxState
    End If
    MyBase.OnNavigatedTo(e)
End Sub
```

Navigation Templates

You now know everything you need to use Silverlight's Frame and Page classes to create a navigable application. However, there's no small gap between simply using these features and actually making them look good, with a slick, visually consistent display. There are two ways to bridge this gap. One option is to gradually build your design skills, review other people's example, experiment, and eventually end up with the perfectly customized user interface you want. The other option is to use a ready-made *navigation template* as a starting point. These are starter project templates that you can use in Visual Studio, and they give you basic project structure and a few finer points of style.

Figure 7-7 shows what you start with if you create a new project using the Silverlight Navigation Application project template instead of the general-purpose Silverlight Application template.

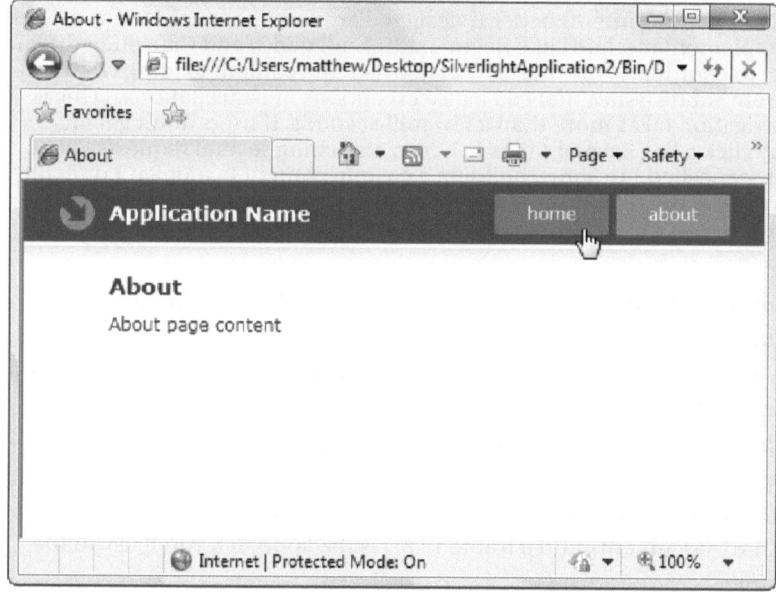

Figure 7-7. An application created with a navigation template

The basic structure of this application is simple enough—there's a page header at the top of a page with a group of link buttons on the left for navigation. Underneath is the Frame that performs the navigation. Pages are mapped through the UriMapper and placed in a project subfolder named Views.

Silverlight ships with just one navigation template, which is shown in Figure 7-7. However, the Silverlight team posted several alternative templates when Silverlight 3 was released, and you can get them at http://tinyurl.com/ktv4vu. These templates use the same basic structure, but they tweak the visual styles and the placement of the link buttons.

Custom Content Loaders

As you've seen, Silverlight's navigation system is powerful but not overly flexible. It makes certain assumptions about the way URIs will be handled. (Essentially, every URI provides the name of a XAML page that will be loaded in a frame.) This system works well for applications that use straightforward navigation, but it doesn't give you the opening to explore more interesting extensions, like URIs that prompt your application to contact remote services or start a download. It also doesn't let you plug additional features into the navigation system, like user authentication and authorization.

Some of these features may appear in future versions of Silverlight. Many of them will be provided in separate libraries (some free, some not) by third-party developers. But now, you can begin to extend the navigation on your own, using a feature called custom content loaders.

The idea of custom content loaders is simple. You create a class that implements the INavigationContentLoader interface. This class receives a URI and then handles it appropriately (for example, by providing the page content or starting an entirely different task). You can then attach your custom content loader to a frame in any application. You can even chain content loaders together so that one content loader can pass a requested URI to the next content loader in a sequence.

Although the idea of content loaders is fairly simple, programming them is not. Here, there are two issues. First, the INavigationContentLoader is designed around an asynchronous model. That means you need to implement methods like BeginLoad(),EndLoad(), and CancelLoad() and use an IAsyncResult object to coordinate the interaction between the content loading code and your frame. Second, there are subtle issues that can cause exceptions or cancellations. (This is particularly true if your navigation takes more than a few milliseconds. If users don't get an instant response, they may click a link several times in a row, launching several requests, all of which will be subsequently cancelled.) In short, building a custom content loader isn't difficult. But building one that's as robust and reliable as it should be often is.

So with these complications in mind, how should you approach custom content loaders? You have three options:

- *Leave infrastructure programming to the experts*: You can still benefit from the custom content loader feature, but not by building your own implementations. Instead, you'll download and use the custom content loaders that are created by other developers—those who live to write low-level plumbing code.

- *Create simple implementations*: You can handle several simple scenarios by building a custom content loader that offloads most of the real work to the PageResourceContentLoader class. (PageResourceContentLoader is the built-in content loader that loads XAML documents into a frame.) This is the approach you'll use in the example in this chapter.

- *Use someone else's higher-level class*: The race is on to create a simple content loading model overtop of the low-level INavigationContentLoader interface. Ideally, you'll simply need to derive from a straightforward synchronous base class to create your custom content loader. David Hill has a solid first step at http://tinyurl.com/37rqc7x.

In the following sections, you'll see how to build a simple content loader that implements user authentication.

Authentication and Navigation

If you've ever programmed an ASP.NET application, you're already familiar with the way authentication, authorization, and navigation fit together.

In a traditional web application, some pages are accessible to everybody, while others are protected. The *authorization* rules describe these differences, marking certain pages as public and others as restricted (which means they are available only to logged-in users or only to specific users or groups). If an anonymous user attempts to request a restricted page, that user is redirected to a login page for *authentication*. Usually, this step involves entering a valid user name and password, after which the user is sent back to the originally requested page.

Using a custom content loader, you can create exactly this sort of system. Your custom content loader will intercept each page request, check whether the user is authenticated, and then take appropriate action. To do this, the application needs to distinguish between public pages and those that require authentication. In this example, the pages that require authentication are placed in a project subfolder named SecurePages (Figure 7-8).

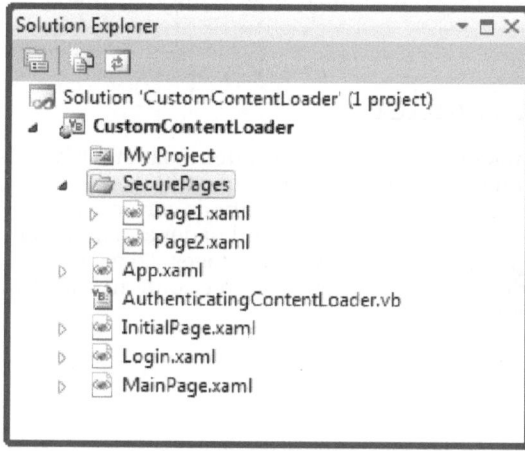

Figure 7-8. Pages that need authentication

Figure 7-9 shows a simple application that tests the finished authentication system. The first page shown in the frame is InitialPage.xaml (Figure 7-9, left). The user can travel to another page by clicking a button (which triggers the Navigate() method) or a link. However, if the application attempts to navigate to a page in the SecurePages folder, the custom content loader redirects the user to the login page (Figure 7-9, right).

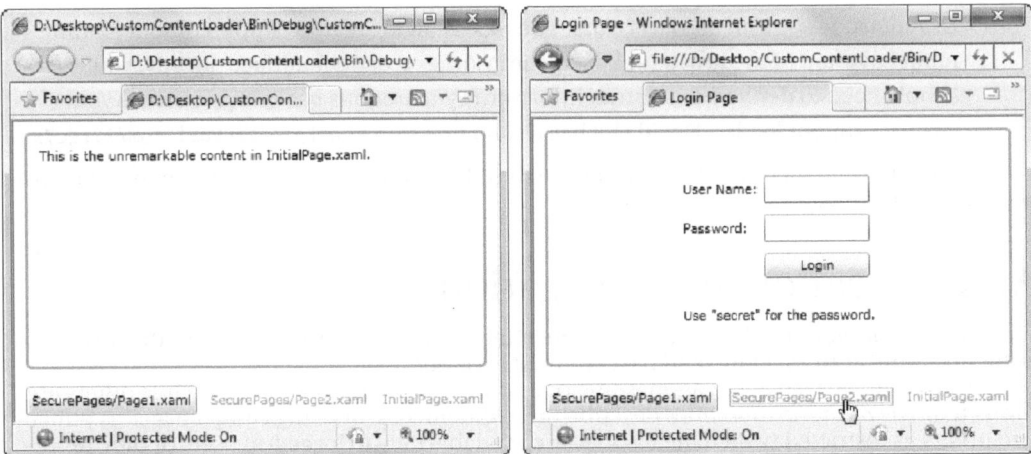

Figure 7-9. Requesting a secured page

While at the login page, the user has several choices. The user can enter a valid user name and password combination and click Login. The login page will then redirect the user to the originally requested page. Or, the user can navigate to a page that isn't secured (which works as it normally does) or to a page that is secured (in which case the user returns immediately to the login page).

Now that you understand how this application works, it's time to look at the code that underpins it.

Creating a Custom Content Loader

A custom content loader is simply a class that implements INavigationContentLoader. This interface requires that you supply the following methods: BeginLoad(), CanLoad(), CancelLoad(), and EndLoad(). Although it's tricky to implement these methods correctly, you can use a shortcut. Simply define an instance of the PageResourceContentLoader as a member field, and call its methods as the content loader passes through its various stages. To get the idea, consider the following custom content loader, which uses PageResourceContentLoader to duplicate the standard navigation behavior *exactly*:

```
Public Class CustomContentLoader
    Implements INavigationContentLoader

    Private _pageResourceContentLoader As New PageResourceContentLoader()

    Public Function BeginLoad(ByVal targetUri As Uri, ByVal currentUri As Uri, _
      ByVal userCallback As AsyncCallback, ByVal asyncState As Object) _
      As IAsyncResult _
      Implements INavigationContentLoader.BeginLoad

        Return _pageResourceContentLoader.BeginLoad(targetUri, currentUri, _
          userCallback, asyncState)
    End Function
```

```
    Public Sub CancelLoad(ByVal asyncResult As IAsyncResult) _
        Implements INavigationContentLoader.CancelLoad

        _pageResourceContentLoader.CancelLoad(asyncResult)
    End Sub

    Public Function CanLoad(ByVal targetUri As Uri, _
      ByVal currentUri As System.Uri) As Boolean _
        Implements INavigationContentLoader.CanLoad

        Return _pageResourceContentLoader.CanLoad(targetUri, currentUri)
    End Function

    Public Function EndLoad(ByVal asyncResult As IAsyncResult) As LoadResult _
        Implements INavigationContentLoader.EndLoad

        Return _pageResourceContentLoader.EndLoad(asyncResult)
    End Function
End Class
```

As simple as this seems, it provides the basic structure for the
AuthenticatingContentLoader that's used in Figure 7-9. But now you need a few more details,
starting with two new properties that allow the user to designate a login page and the folder that
holds secured pages:

```
Public Class AuthenticatingContentLoader
    Implements INavigationContentLoader

    Private _pageResourceContentLoader As New PageResourceContentLoader()

    Public Property LoginPage As String
    Public Property SecuredFolder As String

    ...
End Class
```

The CanLoad(), CancelLoad(), and EndLoad() methods are unchanged. But the
BeginLoad() needs a bit of fine-tuning. It needs to determine whether the user is currently
logged in. If the user isn't logged in and the request is for a page in the secured folder, the user is
redirected to the login page.

```
Public Function BeginLoad(ByVal targetUri As Uri, ByVal currentUri As Uri, _
  ByVal userCallback As AsyncCallback, ByVal asyncState As Object) As IAsyncResult _
    Implements INavigationContentLoader.BeginLoad

    If Not App.UserIsAuthenticated Then
        If (System.IO.Path.GetDirectoryName( _
          targetUri.ToString()).Trim("\") = SecuredFolder) _
          And (targetUri.ToString() <> LoginPage) Then
            ' Redirect the request to the login page.
            targetUri = New Uri(LoginPage, UriKind.Relative)
        End If
    End If
```

```
            Return _pageResourceContentLoader.BeginLoad(targetUri, currentUri, _
                userCallback, asyncState)
End Function
```

To track whether a user is logged in, you need to add a Boolean flag to the App class, like this:

```
Partial Public Class App
    Inherits Application

    Public Shared Property UserIsAuthenticated As Boolean
End Class
```

Now that you have the content loader finished, it's time to put it to work in the application.

Using the Custom Content Loader

Using a custom content loader is likely using any other custom ingredient in your XAML. The first step is to map an XML namespace for your project namespace so the content loader is available. In this example, the project is named CustomContentLoader:

```
<UserControl x:Class="CustomContentLoader.MainPage"
  xmlns:local="clr-namespace:CustomContentLoader" ... >
```

Next, you need to set the Frame.ContentLoader property with an instance of your custom content loader. You also set the properties of the custom content loader at the same time:

```
<navigation:Frame x:Name="mainFrame" UriMapper="{StaticResource UriMapper}">
  <navigation:Frame.ContentLoader>
    <local:AuthenticatingContentLoader LoginPage="/Login.xaml"
      SecuredFolder="SecurePages"></local:AuthenticatingContentLoader>
  </navigation:Frame.ContentLoader>
</navigation:Frame>
```

The only remaining step is to write the code for the login page. In this example, the login page simply checks for a hard-coded password, but a more realistic approach is to call some sort of remote service (as described in Chapter 15) that authenticates the user on the web server. The important detail is that once the user is authenticated, the UserIsAuthenticated flag must be set to True, and the application must call the NavigationService.Refresh() method, which repeats the entire navigation sequence. This time, because the user is authenticated, no redirection will be performed, so the user will end up at the originally requested page.

```
Private Sub cmdLogin_Click(ByVal sender As Object, ByVal e As RoutedEventArgs)
    ' Use a hard-coded password. A more realistic application would call a remote
    ' authentication service that runs on an ASP.NET website.
    If txtPassword.Text = "secret" Then
        App.UserIsAuthenticated = True
        navigationService.Refresh()
    End If
End Sub
```

Now the content loader works seamlessly. Of course, there's much more you could add to this basic example. For example, you could configure the AuthenticatingContentLoader to accept a collection of authorization rules, which it could then evaluate to determine whether a specific user can access a specific page. To delve into more complex authentication like this, and to see examples of more sophisticated custom content loaders, be sure to check out David Hill's blog (www.davidpoll.com). As a Silverlight program manager specializing in navigation, he provides helpful insight and several advanced custom content loader examples.

The Last Word

In this chapter, you considered how to step up from single-page displays to true applications using a range of techniques. First, you considered simple content hiding and swapping techniques, which give you unlimited flexibility and allow you to simulate navigation. Next, you considered the ChildWindow, which allows you to create a pop-up window that appears over the rest of your application. After that, you took a detailed look at the Frame and Page classes and Silverlight's built-in Silverlight navigation system, which enables features such as history tracking and deep linking. And finally, you saw how you can extend the navigation system using custom content loaders.

CHAPTER 8

■ ■ ■

Shapes and Geometries

Silverlight's 2-D drawing support is the basic foundation for many of its more sophisticated features, such as custom-drawn controls, interactive graphics, and animation. Even if you don't plan to create customized art for your application, you need to have a solid understanding of Silverlight's drawing fundamentals. You'll use it to add professional yet straightforward touches, such as reflection effects. You'll also need it to add interactivity to your graphics—for example, to make shapes move or change in response to user actions.

Silverlight supports a surprisingly large subset of the drawing features from WPF. In this chapter, you'll explore the shape model, which allows you to construct graphics out of rectangles, ellipses, lines, and curves. You'll also see how you can convert existing vector art to the XAML format you need, which lets you reuse existing graphics rather than build them from scratch.

Basic Shapes

The simplest way to draw 2-D graphical content in a Silverlight user interface is to use *shapes*: dedicated classes that represent simple lines, ellipses, rectangles, and polygons. Technically, shapes are known as drawing *primitives*. You can combine these basic ingredients to create more complex graphics.

The most important detail about shapes in Silverlight is that they all derive from FrameworkElement. As a result, shapes *are* elements. This has a number of important consequences:

- *Shapes draw themselves*: You don't need to manage the invalidation and painting process. For example, you don't need to manually repaint a shape when content moves, the page is resized, or the shape's properties change.

- *Shapes are organized in the same way as other elements*: In other words, you can place a shape in any of the layout containers you learned about in Chapter 3. (The Canvas is obviously the most useful container because it lets you place shapes at specific coordinates, which is important when you're building a complex drawing out of multiple pieces.)

- *Shapes support the same events as other elements*: That means you don't need to do any extra work to deal with key presses, mouse movements, and mouse clicks. You can use the same set of events you'd use with any element.

Silverlight uses a number of optimizations to make 2-D drawing as fast as possible. For example, because shapes often overlap in complex drawings, Silverlight uses sophisticated algorithms to determine when part of a shape won't be visible and thereby avoid the overhead of rendering it and then overwriting it with another shape.

The Shape Classes

Every shape derives from the MustInherit System.Windows.Shapes.Shape class. Figure 8-1 shows the inheritance hierarchy for shapes.

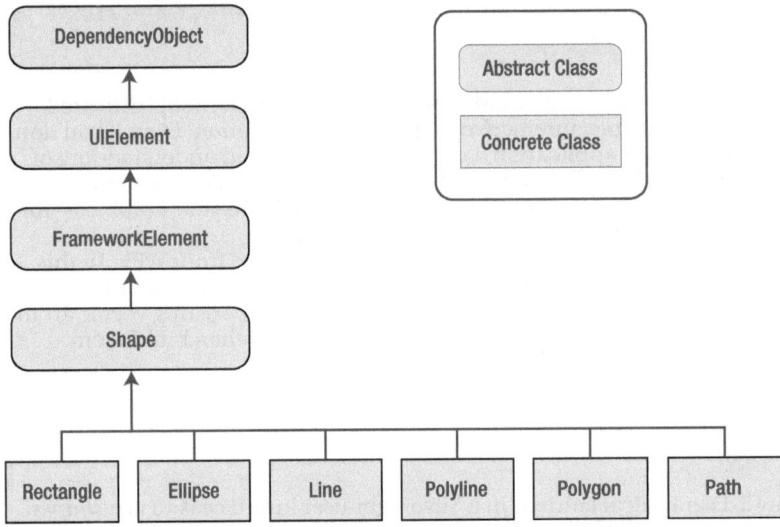

Figure 8-1. *The Silverlight shape classes*

As you can see, a relatively small set of classes derives from the Shape class. Line, Ellipse, and Rectangle are all straightforward; Polyline is a connected series of straight lines; and Polygon is a closed shape made up of a connected series of straight lines. Finally, the Path class is an all-in-one superpower that can combine basic shapes in a single element.

Although the Shape class can't do anything on its own, it defines a small set of important properties, which are listed in Table 8-1.

Table 8-1. *Shape Properties*

Name	Description
Fill	Sets the brush object that paints the surface of the shape (everything inside its borders).
Stroke	Sets the brush object that paints the edge of the shape (its border).
StrokeThickness	Sets the thickness of the border, in pixels.

Name	Description
StrokeStartLineCap and StrokeEndLineCap	Determine the contour of the edge of the beginning and end of the line. These properties have an effect only for the Line, Polyline, and (sometimes) Path shapes. All other shapes are closed and so have no starting and ending point.
StrokeDashArray, StrokeDashOffset, and StrokeDashCap	Allow you to create a dashed border around a shape. You can control the size and frequency of the dashes and how the edge where each dash line begins and ends is contoured.
StrokeLineJoin and StrokeMiterLimit	Determine the contour of the corners of a shape. Technically, these properties affect the *vertices* where different lines meet, such as the corners of a Rectangle element. These properties have no effect for shapes without corners, such as the Line and Ellipse elements.
Stretch	Determines how a shape fills its available space. You can use this property to create a shape that expands to fit its container. However, you'll rarely set the Stretch property, because each shape uses the default value that makes most sense for it.
GeometryTransform	Allows you to apply a transform object that changes the coordinate system used to draw a shape. This lets you skew, rotate, or displace a shape. Transforms are particularly useful when you're animating graphics. You'll learn about transforms in Chapter 9.

Rectangle and Ellipse

Rectangle and Ellipse are the two simplest shapes. To create either one, set the familiar Height and Width properties (inherited from FrameworkElement) to define the size of your shape, and then set the Fill or Stroke property (or both) to make the shape visible. You're also free to use properties such as MinHeight, MinWidth, HorizontalAlignment, VerticalAlignment, and Margin.

■ **Note** If you fail to supply a brush for the Stroke or Fill property, your shape won't appear.

Here's a simple example that stacks an ellipse on a rectangle (see Figure 8-2) using a StackPanel:

```
<StackPanel>
  <Ellipse Fill="Yellow" Stroke="Blue"
   Height="50" Width="100" Margin="5" HorizontalAlignment="Left"></Ellipse>
  <Rectangle Fill="Yellow" Stroke="Blue"
   Height="50" Width="100" Margin="5" HorizontalAlignment="Left"></Rectangle>
</StackPanel>
```

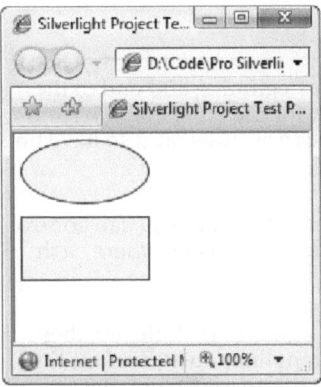

Figure 8-2. *Two simple shapes*

The Ellipse class doesn't add any properties. The Rectangle class adds two: RadiusX and RadiusY. When set to nonzero values, these properties allow you to create nicely rounded corners.

You can think of RadiusX and RadiusY as describing an ellipse that's used to fill in the corners of the rectangle. For example, if you set both properties to 10, Silverlight draws your corners using the edge of a circle that's 10 pixels wide. As you make your radius larger, more of your rectangle is rounded off. If you increase RadiusY more than RadiusX, your corners round off more gradually along the left and right sides and more sharply along the top and bottom edges. If you increase the RadiusX property to match your rectangle's width and increase RadiusY to match its height, you end up converting your rectangle into an ordinary ellipse.

Figure 8-3 shows a few rectangles with rounded corners.

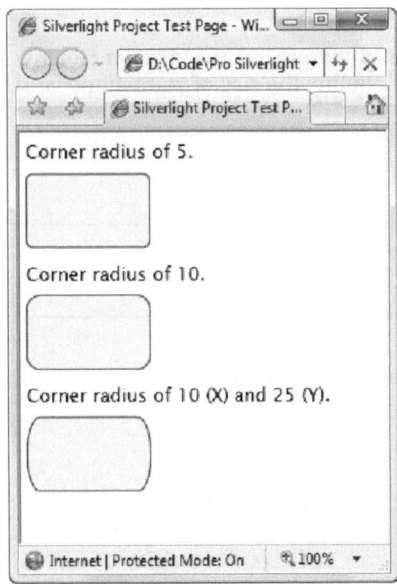

Figure 8-3. *Rounded corners*

Sizing and Placing Shapes

As you already know, hard-coded sizes usually aren't the ideal approach to creating user interfaces. They limit your ability to handle dynamic content, and they make it more difficult to localize your application into other languages.

When you're drawing shapes, these concerns don't always apply. Often, you need tighter control over shape placement. However, in some cases, you can make your design a little more flexible with proportional sizing. Both the Ellipse and Rectangle elements have the ability to size themselves to fill the available space.

If you don't supply the Height and Width properties, the shape is sized based on its container. For example, you can use this stripped-down markup to create an ellipse that fills a page:

```
<Grid>
  <Ellipse Fill="Yellow" Stroke="Blue"></Ellipse>
</Grid>
```

Here, the Grid contains a single proportionately sized row. The ellipse fills the entire row, the row fills the Grid, and the Grid fills the page.

This sizing behavior depends on the value of the Stretch property (which is defined in the Shape class). By default, it's set to Fill, which stretches a shape to fill its container if an explicit size isn't indicated. Table 8-2 lists all your possibilities.

Table 8-2. *Values for the Stretch Enumeration*

Name	Description
Fill	Your shape is stretched in width and height to fit its container exactly. (If you set an explicit height and width, this setting has no effect.)
None	The shape isn't stretched. Unless you set a nonzero width and height (using the Height and Width or MinHeight and MinWidth properties), your shape doesn't appear.
Uniform	The width and height are increased proportionately until the shape reaches the edge of the container. If you use this with an ellipse, you end up with the biggest circle that fits in the container. If you use it with a rectangle, you get the biggest possible square. (If you set an explicit height and width, your shape is sized within those bounds. For example, if you set a Width of 10 and a Height of 100 for a rectangle, you get only a 10×10 square.)
UniformToFill	The width and height are sized proportionately until the shape fills all the available height and width. For example, if you place a rectangle with this Stretch setting into a page that's 100×200 pixels, you get a 200×200 rectangle, and part of it is clipped off. (If you set an explicit height and width, your shape is sized within those bounds. For example, if you set a Width of 10 and a Height of 100 for a rectangle, you get a 100×100 rectangle that's clipped to fit a 10×100 box.)

Figure 8-4 shows the difference between Fill, Uniform, and UniformToFill.

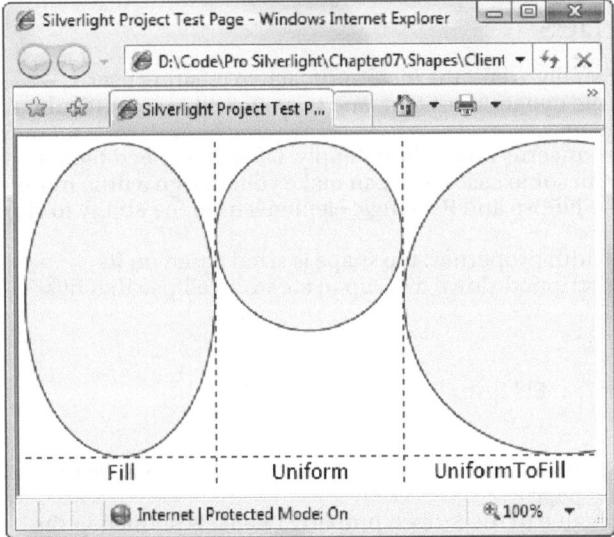

Figure 8-4. *Filling three cells in a Grid*

Usually, a Stretch value of Fill is the same as setting both HorizontalAlignment and VerticalAlignment to Stretch. The difference occurs if you choose to set a fixed Width or Height value on your shape. In this case, the HorizontalAlignment and VerticalAlignment values are ignored. But the Stretch setting still has an effect: it determines how your shape content is sized within the bounds you've given it.

■ **Tip** In most cases, you'll size a shape explicitly or allow it to stretch to fit. You won't combine both approaches.

So far, you've seen how to size a rectangle and an ellipse; but what about placing them where you want them? Silverlight shapes use the same layout system as any other element. However, some layout containers aren't as appropriate. For example, StackPanel, DockPanel, and WrapPanel often aren't what you want because they're designed to separate elements. Grid is more flexible because it allows you to place as many elements as you want in the same cell (although it doesn't let you position them in different parts of that cell). The ideal container is the Canvas, which forces you to specify the coordinates of each shape using the attached Left, Top, Right, and Bottom properties. This gives you complete control over how shapes overlap:

```
<Canvas>
  <Ellipse Fill="Yellow" Stroke="Blue" Canvas.Left="100" Canvas.Top="50"
    Width="100" Height="50"></Ellipse>
  <Rectangle Fill="Yellow" Stroke="Blue" Canvas.Left="30" Canvas.Top="40"
    Width="100" Height="50"></Rectangle>
</Canvas>
```

With the Canvas, the order of your tags is important. In the previous example, the rectangle is superimposed on the ellipse because the ellipse appears first in the list and so is drawn first (see Figure 8-5). If this isn't what you want, you can rearrange the markup or use the Canvas.ZIndex attached property to move an element to a specific layer.

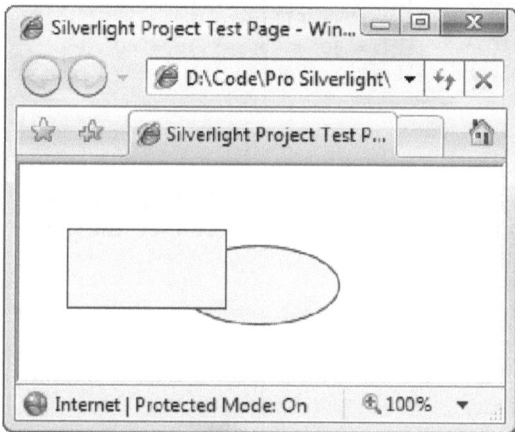

Figure 8-5. *Overlapping shapes in a Canvas*

Remember, a Canvas doesn't need to occupy an entire page. For example, there's no reason why you can't create a Grid that uses a Canvas in one of its cells. This gives you the perfect way to lock down fixed bits of drawing logic in a dynamic, free-flowing user interface.

Sizing Shapes Proportionately with a Viewbox

The only limitation to using the Canvas is that you won't be able to resize your graphics to fit larger or smaller windows. This makes perfect sense for some content (for example, buttons don't usually change size when the window is expanded) but not necessarily for others. For example, you might create a complex graphic that you want to be resizable so it can take advantage of the available space.

In situations like these, Silverlight has an easy solution. If you want to combine the precise control of the Canvas with easy resizability, you can use the Viewbox. The Viewbox is a simple class that stretches a single element (provided in the Child property) according to the stretching behavior you set (using the Stretch and StretchDirection properties). You first saw it in Chapter 3, where it was used to create a rescalable page.

Although you can place a single shape in the Viewbox, that doesn't provide any advantage over the behavior you get naturally. Instead, the Viewbox shines when you need to wrap a group of shapes that make up a drawing. Then, you place the layout container for your drawing (typically, the Canvas) inside the Viewbox.

The following example puts a Viewbox in the second row of a Grid. The Viewbox takes the full height and width of the row. The row takes whatever space is left over after the first autosized row is rendered. Here's the markup:

```
<Grid Margin="5">
  <Grid.RowDefinitions>
    <RowDefinition Height="Auto"></RowDefinition>
    <RowDefinition Height="*"></RowDefinition>
  </Grid.RowDefinitions>
</Grid.RowDefinitions>
```

```xml
<TextBlock>The first row of a Grid.</TextBlock>

<Viewbox Grid.Row="1" HorizontalAlignment="Left" >
  <Canvas Width="200" Height="150">
    <Ellipse Fill="Yellow" Stroke="Blue" Canvas.Left="10"  Canvas.Top="50"
      Width="100" Height="50" HorizontalAlignment="Left"></Ellipse>
    <Rectangle Fill="Yellow" Stroke="Blue" Canvas.Left="30"  Canvas.Top="40"
      Width="100" Height="50" HorizontalAlignment="Left"></Rectangle>
  </Canvas>
</Viewbox>
</Grid>
```

Figure 8-6 shows how the Viewbox adjusts itself as the window is resized. The first row is unchanged. However, the second row expands to fill the extra space. As you can see, the shape in the Viewbox changes proportionately as the page grows.

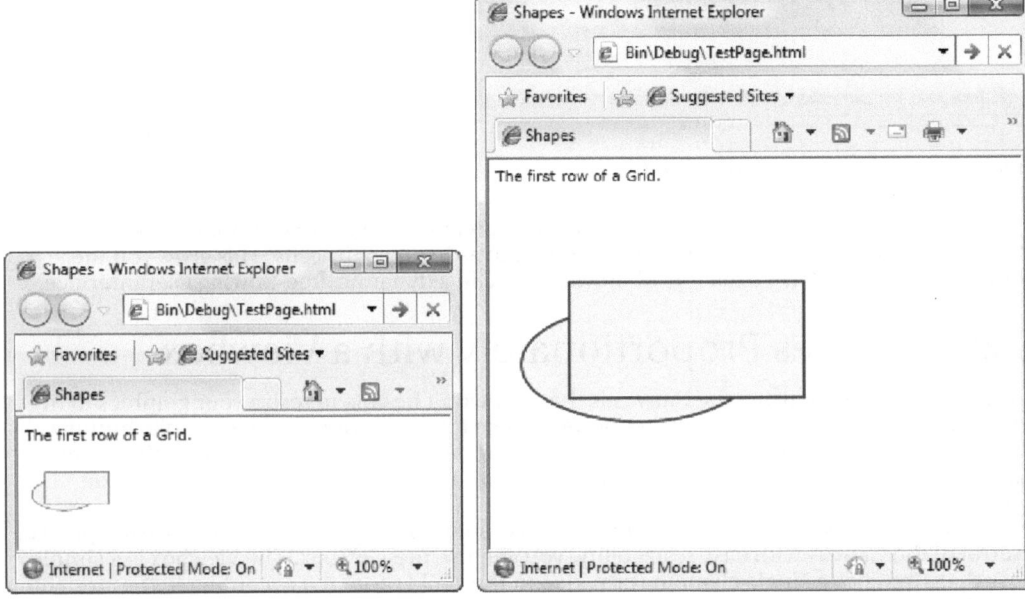

Figure 8-6. *Resizing with a viewbox*

Like all shapes, Viewbox has a Stretch property, which takes a default value of Uniform. However, you can use any of the other values from Table 8-2. You can also get slightly more control by using the StretchDirection property. By default, this property takes the value Both, but you can use UpOnly to create content that can grow but won't shrink beyond its original size, and you can use DownOnly to create content that can shrink but not grow.

■ **Note** When a shape is resized, Silverlight resizes its inside area and its border proportionately. That means the larger your shape grows, the thicker its border is.

For a Viewbox to perform its magic, it needs to be able to determine two pieces of information: the ordinary size that your content would have (if it weren't in a Viewbox) and the new size you want it to have. The second detail—the new size—is simple enough. The Viewbox gives the inner content all the space that's available, based on its Stretch property. That means the bigger the Viewbox, the bigger your content.

The first detail—the ordinary, non-Viewbox size is implicit in the way you define the nested content. In the previous example, the Canvas is given an explicit size of 200 by 150 units. Thus, the Viewbox scales the image from that starting point. For example, the ellipse is initially 100 units wide, which means it takes up half the allotted Canvas drawing space. As the Canvas grows larger, the Viewbox respects these proportions, and the ellipse continues to take half the available space.

However, consider what happens if you remove the Width and Height properties from the Canvas. Now, the Canvas is given a size of 0 by 0 units, so the Viewbox can't resize it and your nested content doesn't appear. (This is different from the behavior you get if you have a Canvas on its own. That's because even though the Canvas is still given a size of 0 by 0, your shapes are allowed to draw outside the Canvas area. The Viewbox isn't as tolerant of this error.)

Now, consider what happens if you wrap the Canvas inside a proportionately sized cell in a Grid, and you don't specify the size of the Canvas. If you aren't using a Viewbox, this approach works perfectly well—the Canvas is stretched to fill the cell, and the content inside is visible. But if you place all this content in a Viewbox, this strategy fails. The Viewbox can't determine the initial size, so it can't resize the Canvas appropriately.

You can get around this problem by placing certain shapes (such as rectangles and ellipses) directly in an autosized container (such as the Grid). The Viewbox can then evaluate the minimum size the Grid needs to fit its content and then scale it up to fit what's available. However, the easiest way to get the size you really want in a Viewbox is to wrap your content in an element that has a fixed size, whether it's a Canvas, a Button, or something else. This fixed size then becomes the initial size that the Viewbox uses for its calculations. Hard-coding a size this way doesn't limit the flexibility of your layout because the Viewbox is sized proportionately based on the available space and its layout container.

Line

The Line shape represents a straight line that connects one point to another. The starting and ending points are set by four properties: X1 and Y1 (for the first point) and X2 and Y2 (for the second). For example, here's a line that stretches from (0, 0) to (10, 100):

```
<Line Stroke="Blue" X1="0" Y1="0" X2="10" Y2="100"></Line>
```

The Fill property has no effect for a line. You must set the Stroke property.

The coordinates you use in a line are relative to the upper-left corner where the line is placed. For example, if you place the previous line in a StackPanel, the coordinate (0, 0) points to wherever that item in the StackPanel is placed. It may be the upper-left corner of the page, but it probably isn't. If the StackPanel uses a nonzero margin or if the line is preceded by other elements, the line begins at a point (0, 0) some distance down from the top of the page.

It's perfectly reasonable to use negative coordinates for a line. You can use coordinates that take your line out of its allocated space and draw over of any other part of the page. This isn't possible with the Rectangle and Ellipse elements that you've seen so far. However, this behavior also has a drawback: lines can't use the flow content model. That means there's no point setting properties such as Margin, HorizontalAlignment, and VerticalAlignment on a line, because they won't have any effect. The same limitation applies to the Polyline and Polygon shapes.

If you place a Line element in a Canvas, the attached position properties (such as Top and Left) still apply. They determine the starting position of the line. In other words, the two line coordinates are offset by that amount. Consider this line:

```
<Line Stroke="Blue" X1="0" Y1="0" X2="10" Y2="100"
 Canvas.Left="5" Canvas.Top="100"></Line>
```

It stretches from (0, 0) to (10, 100), using a coordinate system that treats the point (5, 100) on the Canvas as (0, 0). That makes it equivalent to this line that doesn't use the Top and Left properties:

```
<Line Stroke="Blue" X1="5" Y1="100" X2="15" Y2="200"></Line>
```

It's up to you whether you use the position properties when you draw a line on a Canvas. Often, you can simplify your line drawing by picking a good starting point. You also make it easier to move parts of your drawing. For example, if you draw several lines and other shapes at a specific position in a Canvas, it's a good idea to draw them relative to a nearby point (by using the same Top and Left coordinates). That way, you can shift that entire part of your drawing to a new position as needed.

■ **Note** There's no way to create a curved line with Line or Polyline shapes. Instead, you need the more advanced Path class described later in this chapter.

Polyline

The Polyline class lets you draw a sequence of connected straight lines. You supply a list of X and Y coordinates using the Points property. Technically, the Points property requires a PointCollection object, but you fill this collection in XAML using a lean string-based syntax. You need to supply a list of points and add a space or a comma between each coordinate.

A polyline can have as few as two points. For example, here's a polyline that duplicates the first line you saw in this section, which stretches from (5, 100) to (15, 200):

```
<Polyline Stroke="Blue" Points="5 100 15 200"></Polyline>
```

For better readability, use commas between each X and Y coordinate:

```
<Polyline Stroke="Blue" Points="5,100 15,200"></Polyline>
```

And here's a more complex polyline that begins at (10, 150). The points move steadily to the right, oscillating between higher Y values such as (50, 160) and lower ones such as (70, 130):

```
<Canvas>
  <Polyline Stroke="Blue" StrokeThickness="5" Points="10,150 30,140 50,160 70,130
90,170 110,120 130,180 150,110 170,190 190,100 210,240">
  </Polyline>
</Canvas>
```

Figure 8-7 shows the final line.

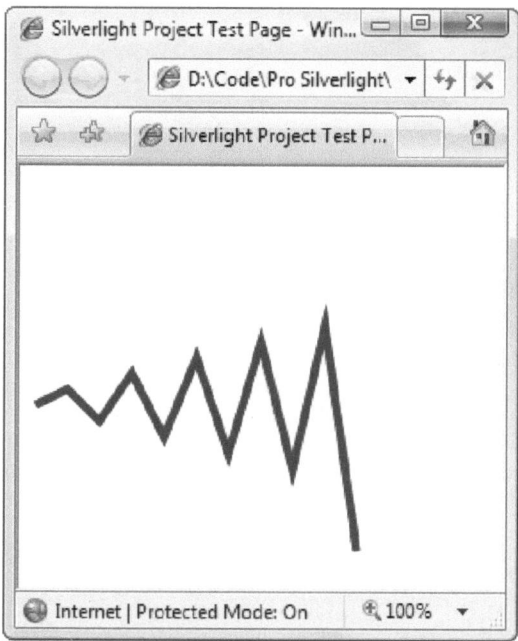

Figure 8-7. *A line with several segments*

At this point, it may occur to you that it would be easier to fill the Points collection programmatically, using some sort of loop that automatically increments X and Y values accordingly. This is true if you need to create highly dynamic graphics—for example, a chart that varies its appearance based on a set of data you extract from a database. But if you want to build a fixed piece of graphical content, you don't want to worry about the specific coordinates of your shapes. Instead, you (or a designer) will use another tool, such as Expression Design, to draw the appropriate graphics and then export them to XAML.

Polygon

Polygon is virtually the same as Polyline. Like the Polyline class, the Polygon class has a Points collection that takes a list of coordinates. The only difference is that the Polygon adds a final line segment that connects the final point to the starting point. You can fill the interior of this shape using the Fill property. Figure 8-8 shows the previous polyline as a polygon with a yellow fill.

```
<Polygon Stroke="Blue" StrokeThickness="5" Points="10,150 30,140 50,160 70,130
90,170 110,120 130,180 150,110 170,190 190,100 210,240" Fill="Yellow">
</Polygon>
```

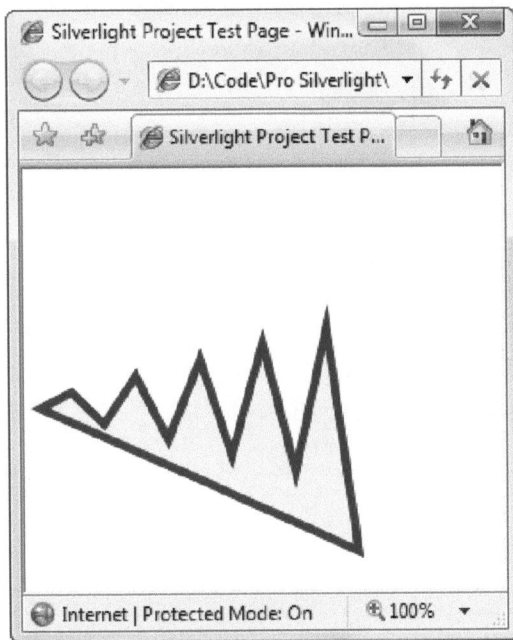

Figure 8-8. A filled polygon

■ **Note** Technically, you can set the Fill property of a Polyline class as well. In this situation, the polyline fills itself as though it were a polygon—in other words, as though it has an invisible line segment connecting the last point to the first point. This effect is of limited use.

In a simple shape where the lines never cross, it's easy to fill the interior. However, sometimes you have a more complex polygon where it's not necessarily obvious what portions are "inside" the shape (and should be filled) and what portions are outside.

For example, consider Figure 8-9, which features a line that crosses more than one other line, leaving an irregular region at the center that you may or may not want to fill. Obviously, you can control exactly what gets filled by breaking this drawing down into smaller shapes. But you may not need to do so.

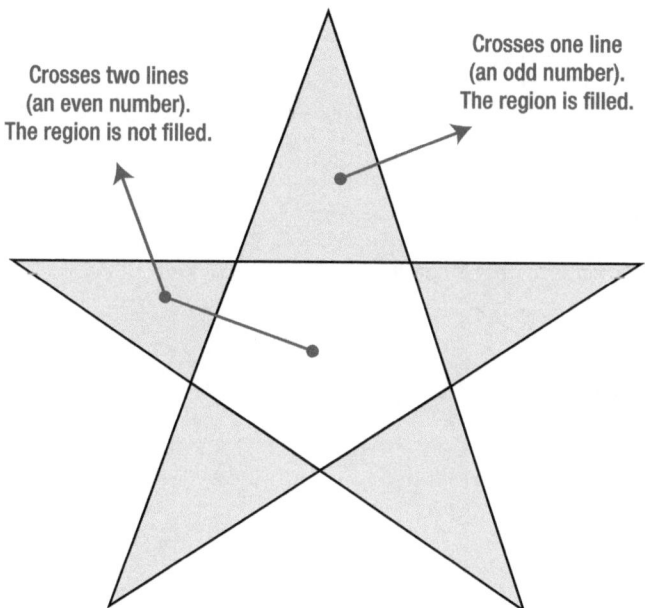

Crosses two lines
(an even number).
The region is not filled.

Crosses one line
(an odd number).
The region is filled.

Figure 8-9. Determining fill areas when FillRule is EvenOdd

Every polygon and polyline includes a FillRule property that lets you choose between two different approaches for filling in regions using the FillRule enumeration. By default, FillRule is set to EvenOdd. To decide whether to fill a region, Silverlight counts the number of lines that must be crossed to reach the outside of the shape. If this number is odd, the region is filled in; if it's even, the region isn't filled. In the center area of Figure 8-9, you must cross two lines to get out of the shape, so it's not filled.

Silverlight also supports the Nonzero fill rule, which is a little trickier. Essentially, with Nonzero, Silverlight follows the same line-counting process as EvenOdd, but it takes into account the direction that each line flows. If the number of lines going in one direction (say, left to right) is equal to the number going in the opposite direction (right to left), the region isn't filled. If the difference between these two counts isn't zero, the region is filled. In the shape from the previous example, the interior region is filled if you set the FillRule to Nonzero. Figure 8-10 shows why. (In this example, the points are numbered in the order they're drawn, and arrows show the direction in which each line is drawn.)

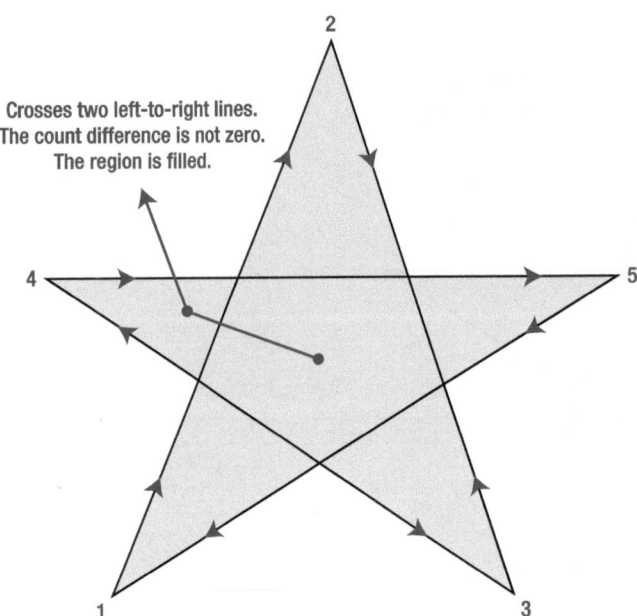

Figure 8-10. Determining fill areas when FillRule is Nonzero

■ **Note** If there is an odd number of lines, the difference between the two counts can't be zero. Thus, the Nonzero fill rule always fills at least as much as the EvenOdd rule, plus possibly a bit more.

The tricky part about Nonzero is that its fill settings depend on *how* you draw the shape, not what the shape looks like. For example, you can draw the same shape in such a way that the center isn't filled (although doing so is much more awkward—you'd begin by drawing the inner region and then draw the outside spikes in the reverse direction).

Here's the markup that draws the star shown in Figure 8-10:

```
<Polygon Stroke="Blue" StrokeThickness="1" Fill="Yellow"
 Canvas.Left="10" Canvas.Top="175" FillRule="Nonzero"
 Points="15,200 68,70 110,200 0,125 135,125">
</Polygon>
```

Line Caps and Line Joins

When you're drawing with the Line and Polyline elements, you can choose how the starting and ending edges of the line are drawn using the StrokeStartLineCap and StrokeEndLineCap properties. (These properties have no effect on other shapes because the shapes are closed.)

Ordinarily, both StartLineCap and EndLineCap are set to Flat, which means the line ends immediately at its final coordinate. Your other choices are Round (which rounds off the corner gently), Triangle (which draws the two sides of the line together in a point), and Square (which ends the line with a sharp edge). All of these values add length to the line—they take it beyond the position where it would otherwise end. The extra distance is half the thickness of the line.

■ **Note** The only difference between Flat and Square is the fact that a square-edged line extends this extra distance. In all other respects, the edge looks the same.

Figure 8-11 shows different line caps at the end of a line.

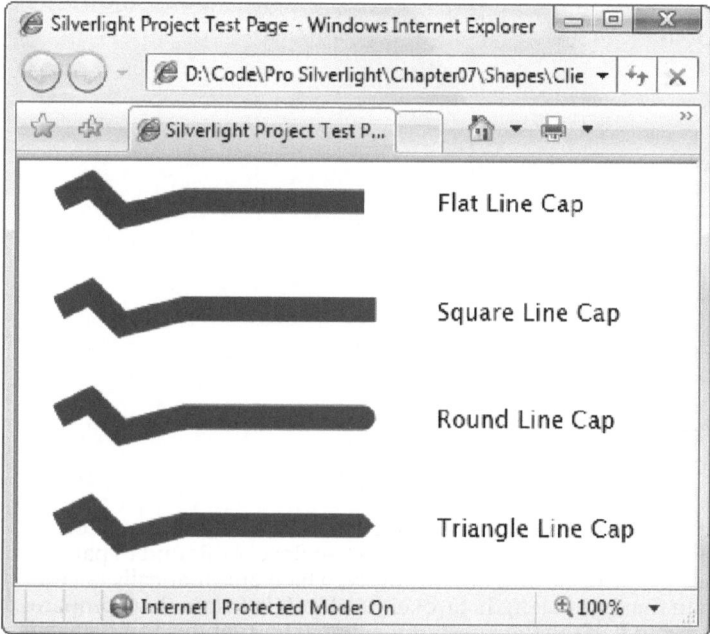

Figure 8-11. Line caps

All shape classes except Line allow you to tweak how their corners are shaped using the StrokeLineJoin property, which takes a value from the PenLineJoin enumeration. You have three choices: the default value, Miter, uses sharp edges; Bevel cuts off the point edge; and Round rounds it out gently. Figure 8-12 shows the difference.

Figure 8-12. Line joins

When you're using mitered edges with thick lines and very small angles, the sharp corner can extend an impractically long distance. In this case, you can use Bevel or Round to pare down the corner. Or, you can use the StrokeMiterLimit property, which automatically bevels the edge when it reaches a certain maximum length. StrokeMiterLimit is a ratio that compares the length used to miter the corner to half the thickness of the line. If you set this to 1 (which is the default value), you let the corner extend half the thickness of the line. If you set it to 3, you let the corner extend to 1.5 times the thickness of the line. The last line in Figure 8-12 uses a higher miter limit with a narrow corner.

Dashes

Instead of drawing boring solid lines for the borders of your shape, you can draw *dashed lines*—lines that are broken with spaces according to a pattern you specify.

When you create a dashed line in Silverlight, you aren't limited to specific presets. Instead, you choose the length of the solid segment of the line and the length of the broken (blank) segment by setting the StrokeDashArray property. For example, consider this line:

```
<Polyline Stroke="Blue" StrokeThickness="14" StrokeDashArray="1 2"
  Points="10,30 60,0 90,40 120,10 350,10">
</Polyline>
```

It has a line value of 1 and a gap value of 2. These values are interpreted relative to the thickness of the line. So, if the line is 14 pixels thick (as in this example), the solid portion is 14 pixels followed by a blank portion of 28 pixels. The line repeats this pattern for its entire length.

On the other hand, if you swap these values around like so

```
StrokeDashArray="2 1"
```

you get a line that has 28-pixel solid portions broken by 14-pixel spaces. Figure 8-13 shows both lines. As you'll notice, when a very thick line segment falls on a corner, it may be broken unevenly.

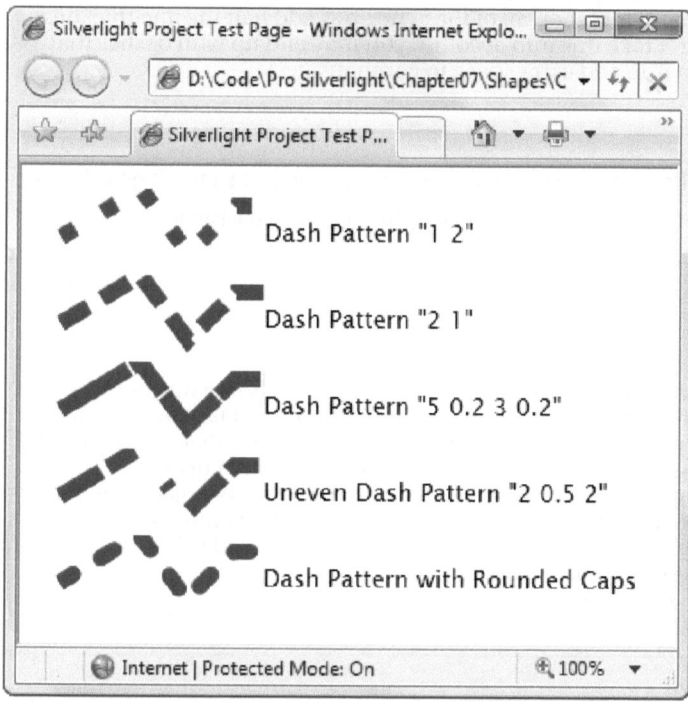

Figure 8-13. *Dashed lines*

There's no reason that you need to stick with whole number values. For example, this StrokeDashArray is perfectly reasonable:

```
StrokeDashArray="5 0.2 3 0.2"
```

It supplies a more complex sequence: a dashed line that's 5×14 long, then a break that's 0.2×15 long, followed by a 3×14 line and another 0.2×14 break. At the end of this sequence, the line repeats the pattern from the beginning.

An interesting thing happens if you supply an odd number of values for the StrokeDashArray. Take this one, for example:

```
StrokeDashArray="3 0.5 2"
```

When drawing this line, Silverlight begins with a 3-times-thickness line, followed by a 0.5-times-thickness space, followed by a 2-times-thickness-line. But when it repeats the pattern, it starts with a gap, meaning you get a 3-times-thickness *space*, followed by a 0.5-times-thickness line, and so on. Essentially, the dashed line alternates its pattern between line segments and spaces.

If you want to start midway into your pattern, you can use the StrokeDashOffset property, which is a 0-based index number that points to one of the values in your StrokeDashArray. For example, if you set StrokeDashOffset to 1 in the previous example, the line begins with the 0.5-thickness space. Set it to 2, and the line begins with the 2-thickness segment.

Finally, you can control how the broken edges of your line are capped. Ordinarily, the edge is straight, but you can set StrokeDashCap to the Bevel, Square, and Triangle values you considered in the previous section. Remember, all of these settings add half the line thickness to the end of your dash. If you don't take this into account, you may end up with dashes that overlap one another. The solution is to add extra space to compensate.

■ **Tip** When you're using the StrokeDashCap property with a line (not a shape), it's often a good idea to set StartLineCap and EndLineCap to the same value. This makes the line look consistent.

Paths and Geometries

So far, you've looked at a number of classes that derive from Shape, including Rectangle, Ellipse, Line, Polygon, and Polyline. However, there's one Shape-derived class that you haven't considered yet, and it's the most powerful by far. The Path class has the ability to encompass any simple shape, groups of shapes, and more complex ingredients such as curves.

The Path class includes a single property, Data, that accepts a Geometry object that defines the shape (or shapes) the path includes. You can't create a Geometry object directly because it's a MustInherit class. Instead, you need to use one of the derived classes listed in Table 8-3. All of these classes are found in the System.Windows.Media namespace.

Table 8-3. Geometry Classes

Name	Description
LineGeometry	Represents a straight line. The geometry equivalent of the Line shape.
RectangleGeometry	Represents a rectangle (optionally with rounded corners). The geometry equivalent of the Rectangle shape.
EllipseGeometry	Represents an ellipse. The geometry equivalent of the Ellipse shape.
GeometryGroup	Adds any number of Geometry objects to a single path, using the EvenOdd or Nonzero fill rule to determine what regions to fill.
PathGeometry	Represents a more complex figure that's composed of arcs, curves, and lines, and can be open or closed.

■ **Note** Silverlight doesn't include all the geometry classes that WPF supports. Notably absent is the CombinedGeometry class, which allows two geometries to be fused together (although the effect can be duplicated with the more powerful PathGeometry class). Also missing is StreamGeometry, which provides a lightweight read-only equivalent to PathGeometry.

You may wonder what the difference is between a path and a geometry. The geometry *defines* a shape. A path allows you to *draw* the shape. Thus, the Geometry object defines details such as the coordinates and size of your shape, whereas the Path object supplies the Stroke and Fill brushes you use to paint it. The Path class also includes the features it inherits from the UIElement infrastructure, such as mouse and keyboard handling.

In the following sections, you'll explore all the classes that derive from Geometry.

Line, Rectangle, and Ellipse Geometries

The LineGeometry, RectangleGeometry, and EllipseGeometry classes map directly to the Line, Rectangle, and Ellipse shapes you learned about in the first half of this chapter. For example, you can convert this markup that uses the Rectangle element

```
<Rectangle Fill="Yellow" Stroke="Blue"
  Width="100" Height="50"></Rectangle>
```

to this markup that uses the Path element:

```
<Path Fill="Yellow" Stroke="Blue">
  <Path.Data>
    <RectangleGeometry Rect="0,0 100,50"></RectangleGeometry>
  </Path.Data>
</Path>
```

The only real difference is that the Rectangle shape takes Height and Width values, whereas RectangleGeometry takes four numbers that describe the size *and* location of the rectangle. The first two numbers describe the X and Y coordinates point where the upper-left corner is placed, and the last two numbers set the width and height of the rectangle. You can start the rectangle at (0, 0) to get the same effect as an ordinary Rectangle element, or you can offset the rectangle using different values. The RectangleGeometry class also includes RadiusX and RadiusY properties that let you round the corners (as described earlier).

Similarly, you can convert the following line

```
<Line Stroke="Blue" X1="0" Y1="0" X2="10" Y2="100"></Line>
```

to this LineGeometry:

```
<Path Stroke="Blue">
  <Path.Data>
    <LineGeometry StartPoint="0,0" EndPoint="10,100"></LineGeometry>
  </Path.Data>
</Path>
```

And you can convert this ellipse

```
<Ellipse Fill="Yellow" Stroke="Blue" Width="100" Height="50"></Ellipse>
```

to this EllipseGeometry:

```
<Path Fill="Yellow" Stroke="Blue">
  <Path.Data>
    <EllipseGeometry RadiusX="50" RadiusY="25" Center="50,25"></EllipseGeometry>
  </Path.Data>
</Path>
```

Notice that the two radius values are half the width and height values. You can also use the Center property to offset the location of the ellipse. In this example, the center is placed in the exact middle of the ellipse bounding box so that it's drawn exactly the same way as the Ellipse shape.

Overall, these simple geometries work the same way as the corresponding shapes. You get the added ability to offset rectangles and ellipses; but that's not necessary if you're placing your shapes on a Canvas, which already gives you the ability to position your shapes at a specific position. If this were all you could do with geometries, you probably wouldn't bother to use the Path element. The difference appears when you decide to group more than one geometry in the same path and when you step up to more complex curves, as described in the following sections.

Combining Shapes with GeometryGroup

The simplest way to combine geometries is to use GeometryGroup and nest the other Geometry-derived objects inside. Here's an example that places an ellipse next to a square:

```
<Path Fill="Yellow" Stroke="Blue" Margin="5" Canvas.Top="10" Canvas.Left="10">
  <Path.Data>
    <GeometryGroup>
      <RectangleGeometry Rect="0,0 100,100"></RectangleGeometry>
      <EllipseGeometry Center="150,50" RadiusX="35" RadiusY="25"></EllipseGeometry>
    </GeometryGroup>
  </Path.Data>
</Path>
```

The effect of this markup is the same as if you supplied two Path elements, one with RectangleGeometry and one with EllipseGeometry (and that's the same as if you used Rectangle and Ellipse shapes instead). However, this approach offers one advantage: you've replaced two elements (Rectangle and Ellipse) with one (Path), which means you've reduced the overhead of your user interface. In general, a page that uses a smaller number of elements with more complex geometries performs faster than a page that has a large number of elements with simpler geometries. This effect isn't apparent in a page that has just a few dozen shapes, but it may become significant in one that requires hundreds or thousands.

Of course, combining geometries in a single Path element also has a drawback: you can't perform event handling of the different shapes separately. Instead, the Path element fires all mouse events. And although Silverlight provides a way to perform hit testing to find out whether a point is on an element (through the HitTest() method that's built into all elements), it doesn't include a way to hit-test geometries.

But even when you combine geometries, you still have the ability to manipulate the nested RectangleGeometry and EllipseGeometry objects independently. For example, each geometry provides a Transform property that you can set to stretch, skew, or rotate that part of the path.

■ **Note** Unlike WPF, Silverlight doesn't allow you to reuse a single geometry object with more than one path. If two objects share the same geometry, you must create a distinct copy for each one.

GeometryGroup becomes more interesting when your shapes intersect. Rather than treat your drawing as a combination of solid shapes, GeometryGroup uses its FillRule property (which can be EvenOdd or Nonzero, as described earlier) to decide what shapes to fill. Consider what happens if you alter the markup shown earlier like this, placing the ellipse over the square:

```
<Path Fill="Yellow" Stroke="Blue" Margin="5" Canvas.Top="10" Canvas.Left="10">
  <Path.Data>
    <GeometryGroup>
      <RectangleGeometry Rect="0,0 100,100"></RectangleGeometry>
      <EllipseGeometry Center="50,50" RadiusX="35" RadiusY="25"></EllipseGeometry>
    </GeometryGroup>
  </Path.Data>
</Path>
```

This markup creates a square with an ellipse-shaped hole in it. If you change FillRule to Nonzero, you get a solid ellipse over a solid rectangle, both with the same yellow fill.

You can create the square-with-a-hole effect by superimposing a white-filled ellipse over your square. However, the GeometryGroup class becomes more useful if you have content underneath, which is typical in a complex drawing. Because the ellipse is treated as a hole in your shape (not another shape with a different fill), any content underneath shows through. For example, consider what happens if you add this line of text behind the square-with-a-hole shape, by placing it before the Path in your markup:

```
<TextBlock Canvas.Top="50" Canvas.Left="20" FontSize="25" FontWeight="Bold">
  Hello There</TextBlock>
```

Now, you get the result shown in Figure 8-14.

Figure 8-14. Text under a Path

Curves and Lines with PathGeometry

PathGeometry is the superpower of geometries. It can draw anything the other geometries can, and much more. The only drawback is a lengthier (and somewhat more complex) syntax.

Every PathGeometry object is built out of one or more PathFigure objects (which are stored in the PathGeometry.Figures collection). A PathFigure is a continuous set of connected lines and curves that can be closed or open. You supply as many line segments as you need, and each one continues from the end of the previous line segment. The figure is closed if the end of the last line in the figure connects to the beginning of the first line.

The PathFigure class has four key properties, as described in Table 8-4.

Table 8-4. *PathFigure Properties*

Name	Description
StartPoint	This is a point that indicates where the line for the figure begins.
Segments	This is a collection of PathSegment objects that are used to draw the figure.
IsClosed	If True, Silverlight adds a straight line to connect the starting and ending points (if they aren't the same).
IsFilled	If True, the area inside the figure is filled in using the Path.Fill brush.

So far, this sounds straightforward. The PathFigure is a shape that's drawn using an unbroken line that consists of a number of segments. However, the trick is that there are several type of segments, all of which derive from the PathSegment class. Some are simple, like LineSegment, which draws a straight line. Others, like BezierSegment, draw curves and are correspondingly more complex.

You can mix and match different segments freely to build your figure. Table 8-5 lists the segment classes you can use.

Table 8-5. *PathSegment Classes*

Name	Description
LineSegment	Creates a straight line between two points.
ArcSegment	Creates an elliptical arc between two points.
BezierSegment	Creates a Bézier curve between two points.
QuadraticBezierSegment	Creates a simpler form of Bézier curve that has one control point instead of two and is faster to calculate.
PolyLineSegment	Creates a series of straight lines. You can get the same effect using multiple LineSegment objects, but a single PolyLineSegment object is more concise.
PolyBezierSegment	Creates a series of Bézier curves.
PolyQuadraticBezierSegment	Creates a series of simpler quadratic Bézier curves.

Straight Lines

It's easy to create simple lines using the LineSegment and PathGeometry classes. You set the StartPoint and add one LineSegment for each section of the line. The LineSegment.Point property identifies the end point of each segment.

For example, the following markup begins at (10, 100), draws a straight line to (100, 100), and then draws a line from that point to (100, 50). Because the PathFigure.IsClosed property is set to True, a final line segment adds the connection from (100, 50) to (10, 100). The final result is a right-angled triangle:

```
<Path Stroke="Blue">
  <Path.Data>
    <PathGeometry>
      <PathFigure IsClosed="True" StartPoint="10,100">
        <LineSegment Point="100,100" />
        <LineSegment Point="100,50" />
      </PathFigure>
    </PathGeometry>
  </Path.Data>
</Path>
```

Silverlight lets you manipulate figures in your code. For example, you can add or remove path segments, or you can dynamically warp a shape by modifying existing line segments or changing the shape's start point. You can even use animation to modify the points in your path smoothly and incrementally, as described in Chapter 10.

■ **Note** Remember, each PathGeometry object can contain an unlimited number of PathFigure objects. That means you can create several separate open or closed figures that are all considered part of the same path.

Arcs

Arcs are a little more interesting than straight lines. You identify the end point of the line using the ArcSegment.Point property, just as you would with a line segment. However, the PathFigure draws a curved line from the starting point (or the end point of the previous segment) to the end point of your arc. This curved connecting line is actually a portion of the edge of an ellipse.

Obviously, the end point isn't enough information to draw the arc, because many curves (some gentle, some more extreme) could connect two points. You also need to indicate the size of the imaginary ellipse that's being used to draw the arc. You do this using the ArcSegment.Size property, which supplies the ellipse's X radius and Y radius. The larger the imaginary ellipse, the more gradually its edge curves.

■ **Note** For any two points, the ellipse has a practical maximum and minimum size. The maximum occurs when you create an ellipse so large the line segment you're drawing appears straight. Increasing the size beyond this point has no effect. The minimum occurs when the ellipse is small enough that a full semicircle connects the two points. Shrinking the size beyond this point also has no effect.

Here's an example that creates the gentle arc shown in Figure 8-15:

```
<Path Stroke="Blue" StrokeThickness="3">
  <Path.Data>
    <PathGeometry>
      <PathFigure IsClosed="False" StartPoint="10,100">
        <ArcSegment Point="250,150" Size="200,300" />
      </PathFigure>
    </PathGeometry>
  </Path.Data>
</Path>
```

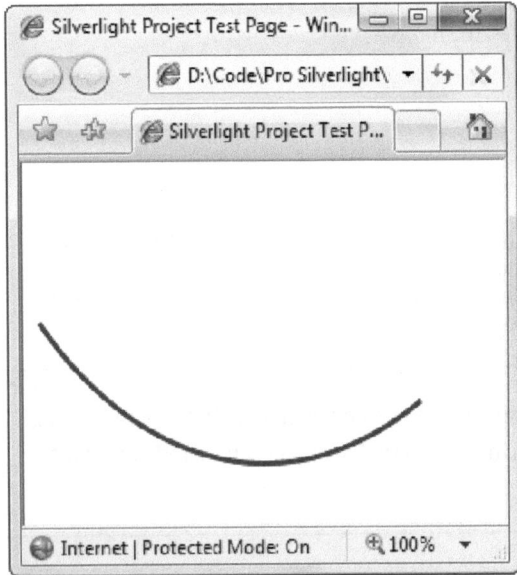

Figure 8-15. *A simple arc*

So far, arcs sound straightforward. But it turns out that even with the start and end points and the size of the ellipse, you still don't have all the information you need to draw an arc unambiguously. In the previous example, you rely on two default values that may not be set to your liking.

To understand the problem, you need to consider the other ways an arc can connect the same two points. If you picture two points on an ellipse, it's clear that you can connect them in two ways: by going around the short side or by going around the long side. Figure 8-16 illustrates.

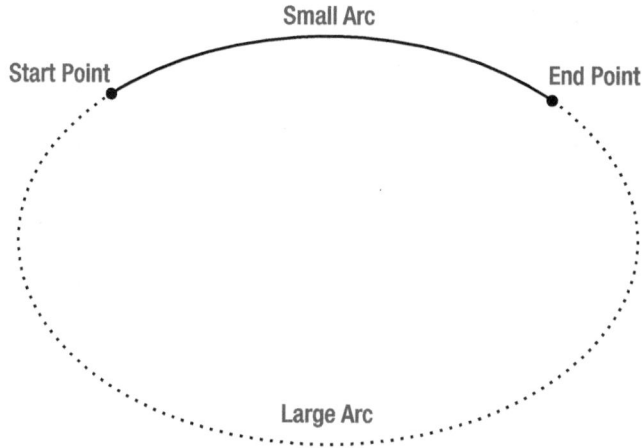

Figure 8-16. *Two ways to trace a curve along an ellipse*

You set the direction using the ArcSegment.IsLargeArc property, which can be True or False. The default value is False, which means you get the shorter of the two arcs.

Even after you've set the direction, one point of ambiguity exists: where the ellipse is placed. For example, imagine you draw an arc that connects a point on the left with a point on the right, using the shortest possible arc. The curve that connects these two points could be stretched down and then up (as it does in Figure 8-15), or it could be flipped so that it curves up and then down. The arc you get depends on the order in which you define the two points in the arc and the ArcSegment.SweepDirection property, which can be Counterclockwise (the default) or Clockwise. Figure 8-17 shows the difference.

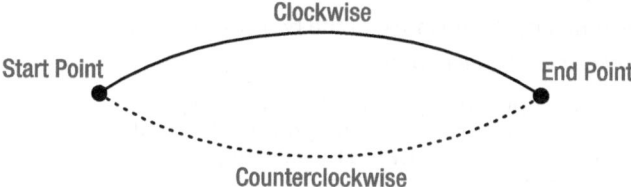

Figure 8-17. *Two ways to flip a curve*

Bézier Curves

Bézier curves connect two line segments using a complex mathematical formula that incorporates two *control points* that determine how the curve is shaped. Bézier curves are an ingredient in virtually every vector drawing application ever created because they're remarkably flexible. Using nothing more than a start point, an end point, and two control points, you can create a surprisingly wide variety of smooth curves (including loops). Figure 8-18 shows a classic Bézier curve. Two small circles indicate the control points, and a dashed line connects each control point to the end of the line it affects the most.

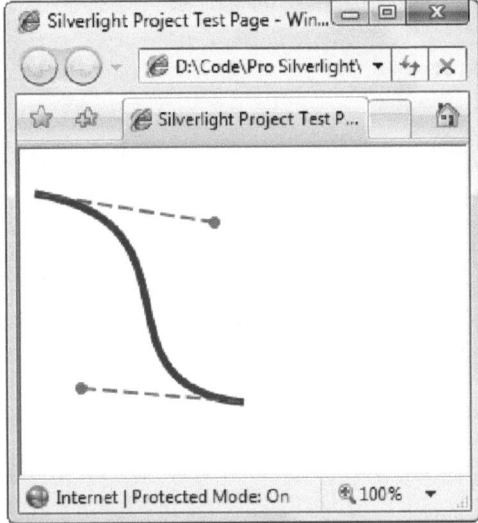

Figure 8-18. A Bézier curve

Even without understanding the math underpinnings, it's fairly easy to get a feel for how Bézier curves work. Essentially, the two control points do all the magic. They influence the curve in two ways:

- At the starting point, a Bézier curve runs parallel with the line that connects it to the first control point. At the ending point, the curve runs parallel with the line that connects it to the end point. (In between, it curves.)
- The degree of curvature is determined by the distance to the two control points. If one control point is farther away, it exerts a stronger "pull."

To define a Bézier curve in markup, you supply three points. The first two points (BezierSegment.Point1 and BezierSegment.Point2) are the control points. The third point (BezierSegment.Point3) is the end point of the curve. As always, the starting point is that starting point of the path or wherever the previous segment leaves off.

The example shown in Figure 8-18 includes three separate components, each of which uses a different stroke and thus requires a separate Path element. The first path creates the curve, the second adds the dashed lines, and the third applies the circles that indicate the control points. Here's the complete markup:

```
<Canvas>
  <Path Stroke="Blue" StrokeThickness="5" Canvas.Top="20">
    <Path.Data>
      <PathGeometry>
        <PathFigure StartPoint="10,10">
          <BezierSegment Point1="130,30" Point2="40,140"
          Point3="150,150"></BezierSegment>
        </PathFigure>
      </PathGeometry>
    </Path.Data>
  </Path>
```

```
<Path Stroke="Green" StrokeThickness="2" StrokeDashArray="5 2" Canvas.Top="20">
  <Path.Data>
    <GeometryGroup>
      <LineGeometry StartPoint="10,10" EndPoint="130,30"></LineGeometry>
      <LineGeometry StartPoint="40,140" EndPoint="150,150"></LineGeometry>
    </GeometryGroup>
  </Path.Data>
</Path>
<Path Fill="Red" Stroke="Red" StrokeThickness="8"  Canvas.Top="20">
  <Path.Data>
    <GeometryGroup>
      <EllipseGeometry Center="130,30"></EllipseGeometry>
      <EllipseGeometry Center="40,140"></EllipseGeometry>
    </GeometryGroup>
  </Path.Data>
</Path>
</Canvas>
```

Trying to code Bézier paths is a recipe for many thankless hours of trial-and-error computer coding. You're much more likely to draw your curves (and many other graphical elements) in a dedicated drawing program that has an export-to-XAML feature or Microsoft Expression Blend.

■ **Tip** To learn more about the algorithm that underlies the Bézier curve, you can read an informative Wikipedia article on the subject at `http://en.wikipedia.org/wiki/Bezier_curve`.

The Geometry Mini-Language

The geometries you've seen so far have been relatively concise, with only a few points. More complex geometries are conceptually the same but can easily require hundreds of segments. Defining each line, arc, and curve in a complex path is extremely verbose and unnecessary— after all, it's likely that complex paths will be generated by a design tool rather than written by hand, so the clarity of the markup isn't all that important. With this in mind, the creators of Silverlight added a more concise alternative syntax for defining geometries that lets you represent detailed figures with much less markup. This syntax is often described as the *geometry mini-language* (and sometimes the *path mini-language* because of its application with the Path element).

To understand the mini-language, you need to realize that it's essentially a long string holding a series of commands. These commands are read by a type converter that then creates the corresponding geometry. Each command is a single letter and is optionally followed by a few bits of numeric information (such as X and Y coordinates) separated by spaces. Each command is also separated from the previous command with a space.

For example, earlier you created a basic triangle using a closed path with two line segments. Here's the markup that did the trick:

```
<Path Stroke="Blue">
  <Path.Data>
    <PathGeometry>
      <PathFigure IsClosed="True" StartPoint="10,100">
        <LineSegment Point="100,100" />
        <LineSegment Point="100,50" />
      </PathFigure>
    </PathGeometry>
  </Path.Data>
</Path>
```

To duplicate this figure using the mini-language, you write this:

```
<Path Stroke="Blue" Data="M 10,100 L 100,100 L 100,50 Z"/>
```

This path uses a sequence of four commands. The first command (M) creates the PathFigure element and sets the starting point to (10, 100). The following two commands (L) create line segments. The final command (Z) ends the PathFigure and sets the IsClosed property to True. The commas in this string are optional, as are the spaces between the command and its parameters, but you must leave at least one space between adjacent parameters and commands. That means you can reduce the syntax even further to this less-readable form:

```
<Path Stroke="Blue" Data="M10 100 L100 100 L100 50 Z"/>
```

The geometry mini-language is easy to grasp. It uses a fairly small set of commands, which are detailed in Table 8-6. Parameters are shown in italics.

Table 8-6. Commands for the Geometry Mini-Language

Command	Description
F *value*	Sets the Geometry.FillRule property. Use 0 for EvenOdd or 1 for Nonzero. This command must appear at the beginning of the string (if you decide to use it).
M *x,y*	Creates a new PathFigure element for the geometry and sets its start point. This command must be used before any other commands except F. You can also use it during your drawing sequence to move the origin of your coordinate system. (The M stands for *move*.)
L *x,y*	Creates a LineSegment to the specified point.
H *x*	Creates a horizontal LineSegment using the specified X value and keeping the Y value constant.
V *y*	Creates a vertical LineSegment using the specified Y value and keeping the X value constant.
A *radiusX, radiusY degrees isLargeArc, isClockwise x,y*	Creates an ArcSegment to the indicated point. You specify the radii of the ellipse that describes the arc, the number of degrees the arc is rotated, and Boolean flags that set the IsLargeArc and SweepDirection properties described earlier.

Command	Description
C *x1,y1 x2,y2 x,y*	Creates a BezierSegment to the indicated point, using control points at (*x1*, *y1*) and *(x2, y2)*.
Q *x1, y1 x,y*	Creates a QuadraticBezierSegment to the indicated point, with one control point at *(x1, y1)*.
S *x2,y2 x,y*	Creates a smooth BezierSegment by using the second control point from the previous BezierSegment as the first control point in the new BezierSegment.
T *x2,y2 x,y*	Creates a smooth QuadraticBezierSegment by using the second control point from the previous QuadraticBezierSegment as the first control point in the new QuadraticBezierSegment.
Z	Ends the current PathFigure element and sets IsClosed to True. You don't need to use this command if you don't want to set IsClosed to True—instead, use M if you want to start a new PathFigure or end the string.

■ **Tip** The geometry mini-language offers one more trick. You can use a command in lowercase if you want its parameters to be evaluated relative to the previous point rather than using absolute coordinates.

Clipping with Geometry

As you've seen, geometries are the most powerful way to create a shape. However, geometries aren't limited to the Path element. They're also used anywhere you need to supply the abstract definition of a shape (rather than draw a real, concrete shape in a page).

Geometries are also used to set the Clip property, which is provided by all elements. The Clip property lets you constrain the outer bounds of an element to fit a specific geometry. You can use the Clip property to create a number of exotic effects. Although it's commonly used to trim down image content in an Image element, you can use the Clip property with any element. The only limitation is that you need a closed geometry if you want to see anything—individual curves and line segments aren't much use.

The following example uses the same geometry to clip two elements: an Image element that contains a bitmap, and a standard Button element. Figure 8-19 shows the results.

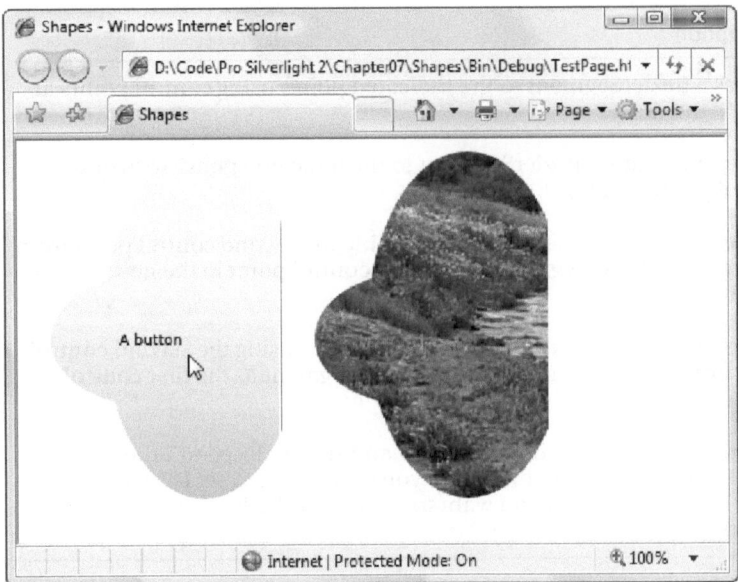

Figure 8-19. *Clipping two elements*

Here's the markup for this example:

```
<Grid>
  <Grid.ColumnDefinitions>
    <ColumnDefinition></ColumnDefinition>
    <ColumnDefinition></ColumnDefinition>
  </Grid.ColumnDefinitions>

  <Button Content="A button">
    <Button.Clip>
      <GeometryGroup FillRule="Nonzero">
        <EllipseGeometry RadiusX="75" RadiusY="50" Center="100,150" />
        <EllipseGeometry RadiusX="100" RadiusY="25" Center="200,150" />
        <EllipseGeometry RadiusX="75" RadiusY="130" Center="140,140" />
      </GeometryGroup>
    </Button.Clip>
  </Button>
  <Image Grid.Column="1" Stretch="None" Source="creek.jpg">
    <Image.Clip>
      <GeometryGroup FillRule="Nonzero">
        <EllipseGeometry RadiusX="75" RadiusY="50" Center="100,150" />
        <EllipseGeometry RadiusX="100" RadiusY="25" Center="200,150" />
        <EllipseGeometry RadiusX="75" RadiusY="130" Center="140,140" />
      </GeometryGroup>
    </Image.Clip>
  </Image>
</Grid>
```

The clipping you set doesn't take the size of the element into account. In this example, that means that if the button is enlarged, the clipped region will remain at the same position and show a different portion of the button.

Exporting Clip Art

In most cases, you won't create Silverlight art by hand. Instead, you (or a graphic designer) will use a design tool to create vector art, and then export it to XAML. The exported XAML document you end up with is essentially a Canvas that contains a combination of shape elements. You can place that Canvas inside an existing Canvas to show your artwork.

Although many drawing programs don't have built-in support for XAML export, you still have many options for getting the graphics you need. The following sections outline the options you can use to get vector art out of virtually any application.

Expression Design

Expression Design, Microsoft's illustration and graphic design program, has built-in XAML export. It can import a variety of vector-art file formats, including Adobe Illustrator (.ai) files, and it can export to XAML.

When exporting to XAML, follow these steps:

1. Choose File ➤ Export from the menu.

2. In the Export dialog box, in the Save As Type list, choose XAML. Then, enter a file name, and click Save. The Export XAML window appears (see Figure 8-20), which shows you the image you're exporting and a preview of the XAML content it will create (click the XAML tab).

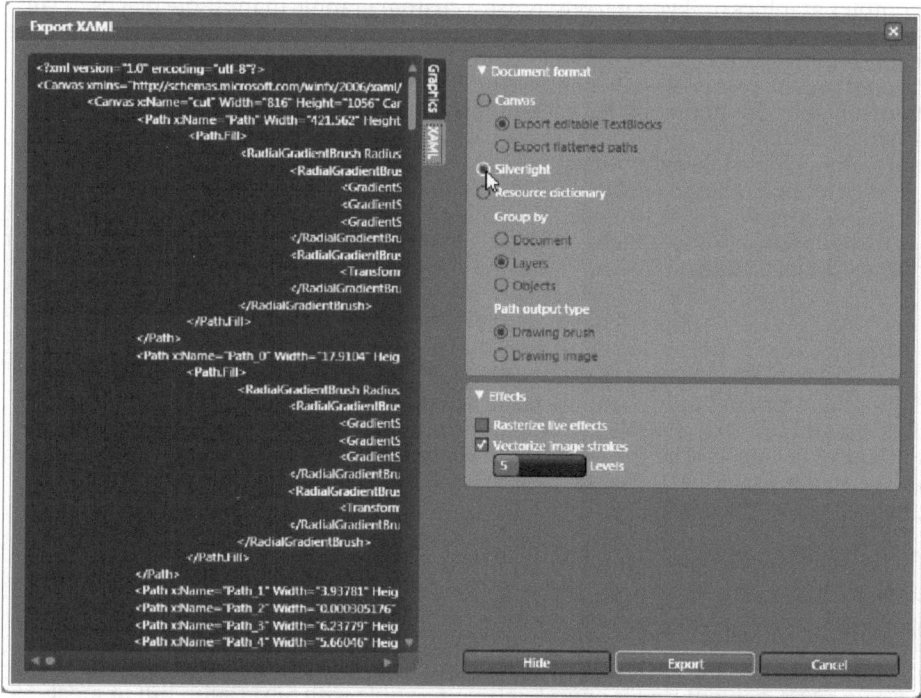

Figure 8-20. Creating a Silverlight-compatible XAML file

3. In the Document Format group of settings, click Silverlight to be sure you're creating a Silverlight-compatible XAML file. This ensures that XAML features that are supported in WPF but not in Silverlight aren't used.

■ **Note** Usually, the standard XAML export option (Canvas) works with Silverlight applications with minimal changes, such as manually removing a few unsupported attributes. However, the Resource Dictionary export option creates XAML files that don't work with Silverlight. That's because this option stores the graphic in a collection of DrawingBrush resources instead of a Canvas. This makes it easier to efficiently reuse the drawing in WPF, but it's useless in Silverlight, because Silverlight doesn't include the Drawing or DrawingBrush class.

---

4. Click Export to save the file.

The generated XAML file includes a root-level Canvas element. Inside the Canvas, you'll find dozens of Path elements, each positioned at a specific place in the Canvas and with its own data and brushes.

You can cut and paste this entire block of markup into any Silverlight page to reproduce the graphic. However, this approach is inconvenient if you have a page that includes a large number of complex graphics or if you need to reuse a custom graphic in multiple places. If you use the cut-and-paste approach here, you'll clutter your markup beyond recognition and create duplicate sections that are much more difficult to debug or modify later.

Ideally, you'd use the resources collection in the App.xaml file to share frequently used graphics. Unfortunately, this approach isn't possible, because Silverlight doesn't allow you to store and reuse entire elements (such as a Canvas with graphical content), and it doesn't provide a way to define drawings without using elements. The most common workaround is to create a separate user control for each important graphic. You can then insert these user controls into other pages, wherever you need them. You'll see this technique in action in Chapter 10, which presents a simple bomb-dropping game that uses dedicated user controls for its bomb graphic and its title logo.

Conversion

Microsoft Expression Design is one example of a design tool that supports XAML natively. However, plug-ins and conversion tools are available for many other popular formats. Mike Swanson, a Microsoft evangelist, has an old but still useful page at http://blogs.msdn.com/mswanson/articles/WPFToolsAndControls.aspx, with links to many free converters, including the following:

- An Adobe Illustrator (.ai) to XAML converter
- A Flash (.swf) to XAML converter
- A Visio plug-in for exporting XAML

You can also find more nonfree XAML conversion tools on the Web. These tools won't necessarily create XAML content that is completely compatible with Silverlight. But in most cases, you'll need to make only minor edits to fix markup errors.

Save or Print to XPS

The XML Paper Specification (XPS) is a Microsoft standard for creating fixed, print-ready documents. It's similar to the Adobe PDF standard, and support is included in modern versions of Office and Windows. The XPS standard is based on XAML, which makes it possible to transfer content from an XPS document to a Silverlight page.

For example, Figure 8-21 shows a document in Word after performing a clip-art search and dragging a vector image (a stack of money) onto the page. You have two ways to convert this to an XPS document. The easiest way is to save this graphic in the XPS format using the File ➤ Save As command. (XPS support is built into Office 2010, but Office 2007 users need to install the free Save As PDF or XPS add-in that Microsoft provides at http://tinyurl.com/y69y7g.)

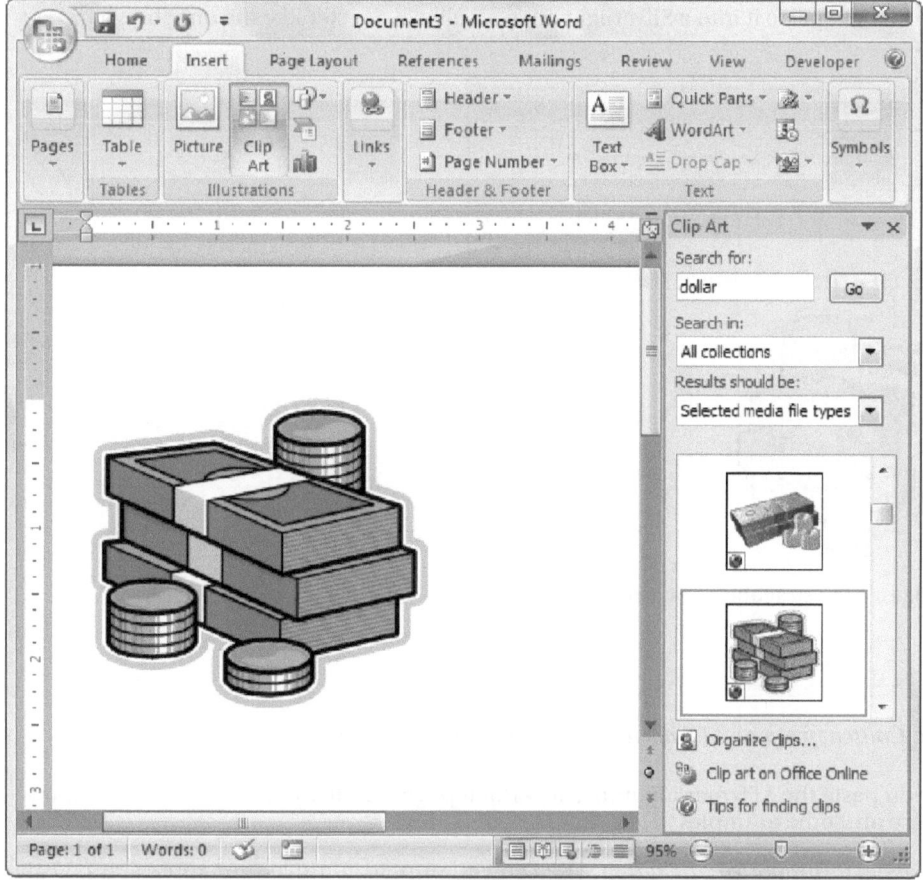

Figure 8-21. *Exporting pictures to XAML through XPS*

If you're using Windows Vista or Windows 7, you have another option—you can choose to print your document to the Microsoft XPS Document Writer print device. This approach isn't limited to Office. Instead, it gives you a convenient back door to get XAML output from virtually any Windows application.

Whether you save or print to XPS, you'll end up with a file that has the extension .xps. This file is actually a ZIP archive (somewhat like the XAP files that Silverlight uses). To extract the XAML inside, you need to begin by changing the extension to .zip and opening the archive to view the files inside. Bitmaps are included as separate files in the Resources folder. Vector art, like the money stack shown in Figure 8-21, is defined in XAML inside a page in the Documents\1\Pages folder. There, you'll find a file for each page in your document, with file names in the format [*PageNumber*].fpage. For example, in the XPS file that's generated for the previous example, you'll find a single file named 1.fpage that defines the page with the money graphic.

If you extract that file and open it in a text editor, you'll see that it's legitimate XAML. The root element is named FixedPage, which isn't recognized in Silverlight; but inside that is an ordinary Canvas that you can cut and paste into a Silverlight window. For the example shown in Figure 8-21, the Canvas holds a series of Path elements that define the different parts of the shape. After you paste it into a Silverlight page, you'll get a result like the one shown in Figure 8-22.

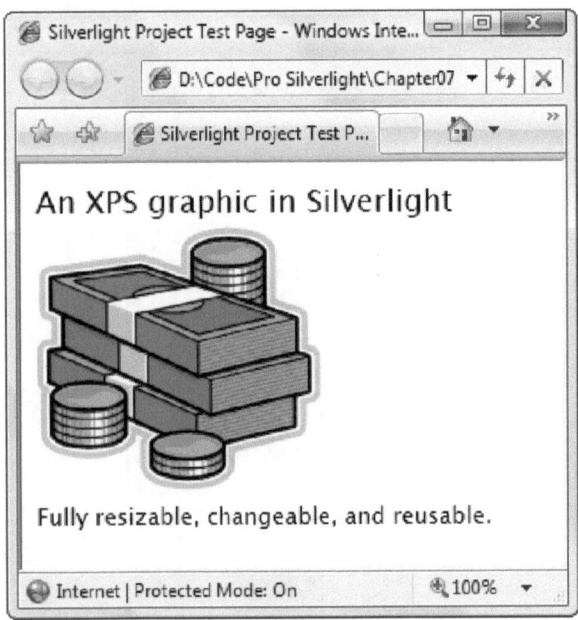

Figure 8-22. Content from an XPS document in Silverlight

When you paste the XPS markup into a Silverlight page, you'll often need to make minor changes. Here are some examples:

- *Removing unsupported attributes*: When you attempt to compile your application, Visual Studio points out any problems in your markup and flags them as compile errors.

- *Replacing the Glyphs element with a TextBlock*: The Glyphs element is a low-level way to show text. Unlike with a TextBlock, when you use the Glyphs element, you need to supply several details (including a font file) or your text won't appear. When you create an XPS document that includes text, it uses the Glyphs element. But to make your text appear, you must find the font file in your XPS archive, extract it, add it to your project, and change the Glyphs.FontUri property to match. An easier approach is to replace the Glyphs element with the higher-level TextBlock element and use the Glyphs.UnicodeString property to set the TextBlock.Text property.

- *Changing the transforms*: Sometimes, the exported art uses transforms to resize and position the graphic. (This is most common when you use the Save As XPS feature in Word rather than the XPS print driver in Windows Vista.) By removing or modifying these transforms, you can free the image from the printed layout so it can fit your Silverlight page perfectly. You'll learn all about transforms in Chapter 9.

The Last Word

In this chapter, you took a detailed look at Silverlight's support for basic 2-D drawing. You began by considering the simple shape classes and continued to Path, the most sophisticated of the shape classes, which lets you add arcs and curves.

But your journey isn't complete. In the next chapter, you'll consider how you can create better drawings by using the right brushes, controlling opacity, and applying with transforms.

■ ■ ■

CHAPTER 9

Brushes, Transforms, and Bitmaps

In the previous chapter, you started your exploration into Silverlight's 2-D drawing model. You considered how you can use Shape-derived classes such as Rectangle, Ellipse, Polygon, Polyline, and Path to create a variety of different drawings. However, shapes alone fall short of what you need to create detailed 2-D vector art for a graphically rich application. In this chapter, you'll pick up the missing pieces.

First, you'll learn about the Silverlight brushes that allow you to create gradients, tiled patterns, and bitmap fills in any shape. Next, you'll see how you can use Silverlight's effortless support for transparency to blend multiple images and elements together. Next, you'll consider transforms and projections—specialized objects that can change the visual appearance of any element by scaling, rotating, or skewing it. You'll also explore pixel shaders, which apply complex visual effects (such as blurs and color tuning) to any element. As you'll see, when you combine these features—for example, tossing together a dash of blur with the warping effect of a transform—you can create popular effects such as reflections, glows, shadows, and even simulated 3-D. And if these prebuilt tools don't give you all the control you want, you can take the next step to fully customized rendering with the WriteableBitmap class, which allows you to modify the individual pixels of a bitmap image, even while it's displayed in a Silverlight page.

■ **What's New** At the end of this chapter, you'll pick up an important graphics-powered feature that's new to Silverlight 4: printing. Although it might seem strange to cover printing in a chapter about graphics, Silverlight uses the same plumbing to render graphics as it does to generate printouts.

Brushes

As you know, brushes fill an area, whether it's the background, foreground, or border of an element, or the fill or stroke of a shape. For elements, you use brushes with the Foreground, Background, and BorderBrush properties. For shapes, you use the Fill and Stroke properties.

You've used brushes throughout this book, but so far you've done most of your work with the straightforward SolidColorBrush. Although SolidColorBrush is indisputably useful, several other classes inherit from System.Windows.Media.Brush and give you more exotic effects. Table 9-1 lists them all.

Table 9-1. *Brush Classes*

Name	Description
SolidColorBrush	Paints an area using a solid single-color fill.
LinearGradientBrush	Paints an area using a gradient fill: a gradually shaded fill that changes from one color to another (and, optionally, to another and then another, and so on).
RadialGradientBrush	Paints an area using a radial gradient fill, which is similar to a linear gradient but radiates out in a circular pattern starting from a center point.
ImageBrush	Paints an area using an image that can be stretched, scaled, or tiled.
VideoBrush	Paints an area using a MediaElement (which gets its content from a video file). This lets you play video in any shape or element.
WebBrowserBrush	Paints an area using the HTML content from a WebBrowser control. This feature is only available in out-of-browser applications, and it's described in Chapter 21.

In this chapter, you'll learn how to use LinearGradientBrush, RadialGradientBrush, and ImageBrush. VideoBrush is discussed in Chapter 11, when you explore Silverlight's media support.

The LinearGradientBrush Class

The LinearGradientBrush allows you to create a blended fill that changes from one color to another.

Here's the simplest possible gradient. It shades a rectangle diagonally from blue (in the upper-left corner) to white (in the lower-right corner):

```
<Rectangle Width="150" Height="100">
  <Rectangle.Fill>
    <LinearGradientBrush>
      <GradientStop Color="Blue" Offset="0"/>
      <GradientStop Color="White" Offset="1" />
    </LinearGradientBrush>
  </Rectangle.Fill>
</Rectangle>
```

The top gradient in Figure 9-1 shows the result.

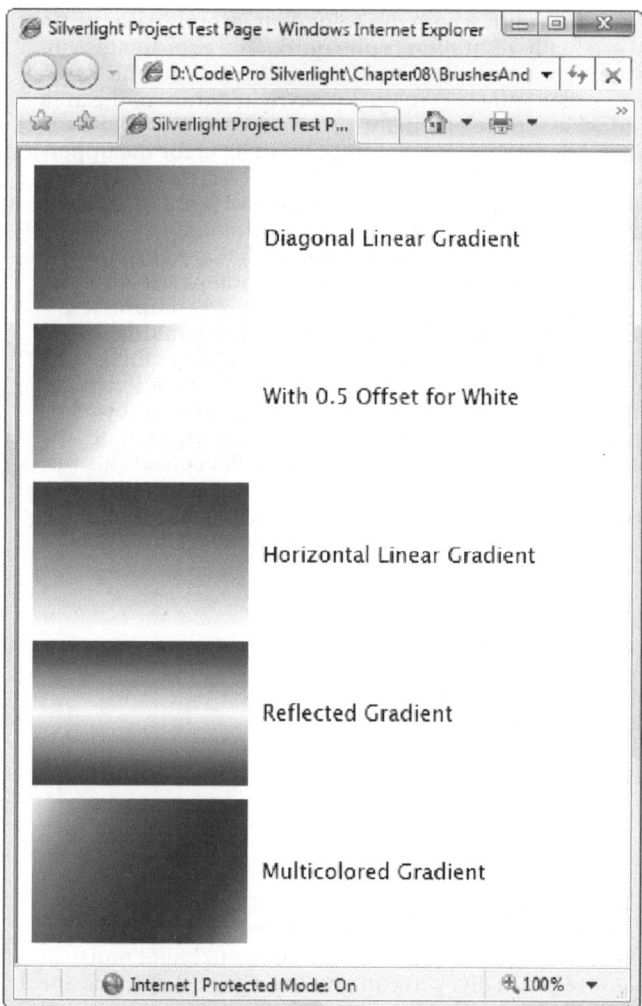

Figure 9-1. *A rectangle with different linear gradients*

To create the first gradient, you need to add one GradientStop for each color. You also need to place each color in your gradient using an Offset value from 0 to 1. In this example, the GradientStop for the blue color has an offset of 0, which means it's placed at the very beginning of the gradient. The GradientStop for the white color has an offset of 1, which places it at the end. If you change these values, you can adjust how quickly the gradient switches from one color to the other. For example, if you set the GradientStop for the white color to 0.5, the gradient will blend from blue (in the upper-left corner) to white in the middle (the point between the two corners). The right side of the rectangle will be completely white. (The second gradient in Figure 9-1 shows this example.)

The previous markup creates a gradient with a diagonal fill that stretches from one corner to another. However, you may want to create a gradient that blends from top to bottom or side to side or that uses a different diagonal angle. You control these details using the StartPoint and EndPoint properties of LinearGradientBrush. These properties allow you to choose the point where the first color begins to change and the point where the color change ends with the final color. (The area in between is blended gradually.) But there's one quirk. The coordinates you use for the starting and ending points aren't real coordinates. Instead, LinearGradientBrush assigns the point (0, 0) to the upper-left corner and (1, 1) to the lower-right corner of the area you want to fill, no matter how high and wide the area actually is.

To create a top-to-bottom horizontal fill, you can use a start point of (0, 0) for the upper-left corner and an end point of (0, 1), which represents the lower-left corner. To create a side-to-side vertical fill (with no slant), you can use a start point of (0, 0) and an end point of (1, 0) for the upper-right corner. Figure 9-1 shows a horizontal gradient (it's the third one).

You can get a little craftier by supplying start points and end points that aren't quite aligned with the corners of your gradient. For example, you can have a gradient stretch from (0, 0) to (0, 0.5), which is a point on the left edge, halfway down. This creates a compressed linear gradient—one color starts at the top, blending to the second color in the middle. The bottom half of the shape is filled with the second color. But wait—you can change this behavior using the LinearGradientBrush.SpreadMethod property. It's Pad by default (which means areas outside the gradient are given a solid fill with the appropriate color), but you can also use Reflect (to reverse the gradient, going from the second color back to the first) or Repeat (to duplicate the same color progression). Figure 9-1 shows the Reflect effect (it's the fourth gradient).

A LinearGradientBrush also lets you create gradients with more than two colors by adding more than two GradientStop objects. For example, here's a gradient that moves through a rainbow of colors:

```
<Rectangle Width="150" Height="100">
  <Rectangle.Fill>
    <LinearGradientBrush StartPoint="0,0" EndPoint="1,1">
      <GradientStop Color="Yellow" Offset="0.0" />
      <GradientStop Color="Red" Offset="0.25" />
      <GradientStop Color="Blue" Offset="0.75" />
      <GradientStop Color="LimeGreen" Offset="1.0" />
    </LinearGradientBrush>
  </Rectangle.Fill>
</Rectangle>
```

The only trick is to set the appropriate offset for each gradient stop. For example, if you want to transition through five colors, you can give your first color an offset of 0, the second 0.25, the third 0.5, the fourth 0.75, and the fifth 1. Or, if you want the colors to blend more quickly at the beginning and then end more gradually, you can give the offsets 0, 0.1, 0.5, and 1.

Remember, brushes aren't limited to shape drawing. You can substitute a LinearGradientBrush anytime you would use a SolidColorBrush—for example, when filling the background surface of an element (using the Background property), the foreground color of its text (using the Foreground property), or the fill of a border (using the BorderBrush property). Figure 9-2 shows an example of a gradient-filled TextBlock.

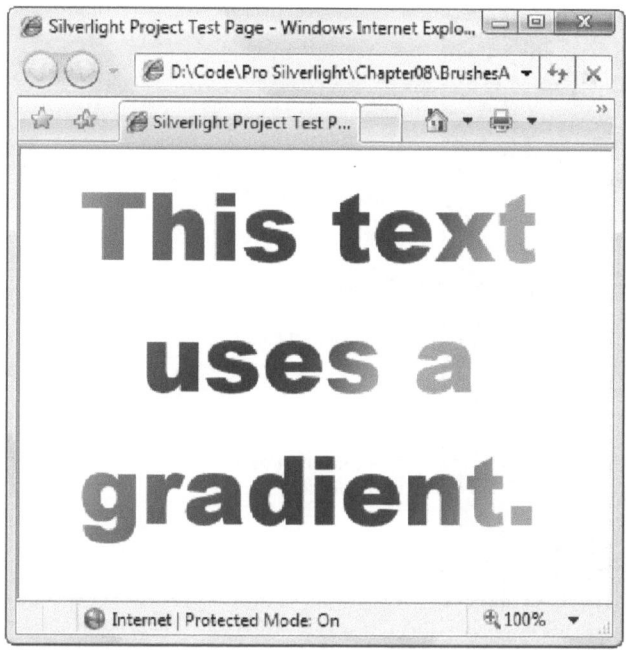

Figure 9-2. Using a linear gradient brush to set the TextBlock.Foreground property

The RadialGradientBrush Class

RadialGradientBrush works similarly to the LinearGradientBrush. It also takes a sequence of colors with different offsets. As with the LinearGradientBrush, you can use as many colors as you want. The difference is how you place the gradient.

To identify the point where the first color in the gradient starts, you use the GradientOrigin property. By default, it's (0.5, 0.5), which represents the middle of the fill region.

■ **Note** As with LinearGradientBrush, RadialGradientBrush uses a proportional coordinate system that acts as though the upper-left corner of your rectangular fill area is (0, 0) and the lower-right corner is (1, 1). That means you can pick any coordinate from (0, 0) to (1, 1) to place the starting point of the gradient. You can even go beyond these limits if you want to locate the starting point outside the fill region.

The gradient radiates out from the starting point in a circular fashion. Eventually, your gradient reaches the edge of an inner gradient circle, where it ends. The center of this circle may or may not line up with the gradient origin, depending on the effect you want. The area beyond the edge of the inner gradient circle and the outermost edge of the fill region is given a solid fill using the last color that's defined in RadialGradientBrush.GradientStops collection, as Figure 9-3 illustrates.

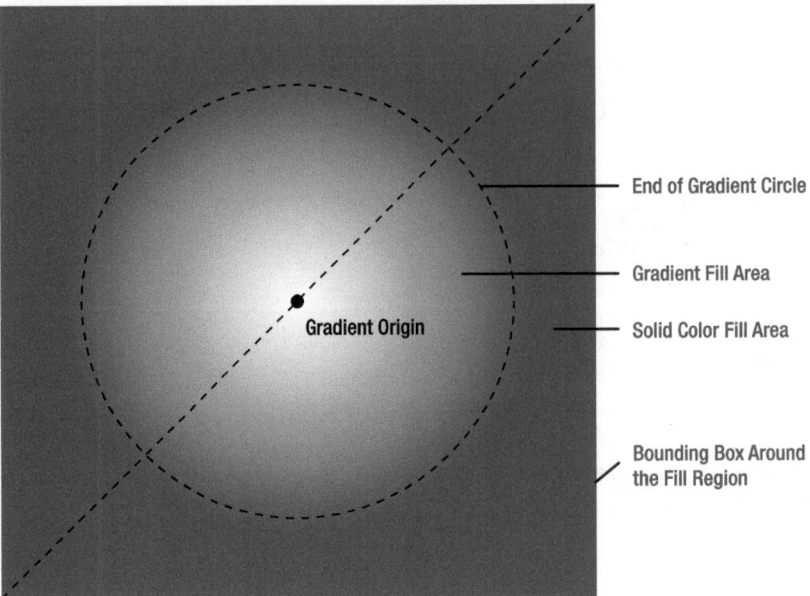

Figure 9-3. *How a radial gradient is filled*

You set the edge of the inner gradient circle using three properties: Center, RadiusX, and RadiusY. By default, the Center property is (0.5, 0.5), which places the center of the limiting circle in the middle of your fill region and in the same position as the gradient origin.

The RadiusX and RadiusY values determine the size of the limiting circle, and by default they're both set to 0.5. These values can be a bit unintuitive—they're measured in relation to the *diagonal* span of your fill area (the length of an imaginary line stretching from the upper-left corner to the lower-right corner of your fill area). That means a radius of 0.5 defines a circle that has a radius that's half the length of this diagonal. If you have a square fill region, you can use a dash of Pythagoras to calculate that this is about 0.7 times the width (or height) of your region. Thus, if you're filling a square region with the default settings, the gradient begins in the center and stretches to its outermost edge at about 0.7 times the width of the square.

■ **Note** If you trace the largest possible ellipse that fits in your fill area, that's the place where the gradient ends with your second color.

The RadialGradientBrush is a particularly good choice for filling rounded shapes and creating lighting effects. (Master artists use a combination of gradients to create buttons with a glow effect.) A common trick is to offset the GradientOrigin point slightly to create an illusion of depth in your shape. Here's an example:

```
<Ellipse Margin="5" Stroke="Black" StrokeThickness="1" Width="200" Height="200">
  <Ellipse.Fill>
    <RadialGradientBrush RadiusX="1" RadiusY="1" GradientOrigin="0.7,0.3">
      <GradientStop Color="White" Offset="0" />
```

```
        <GradientStop Color="Blue" Offset="1" />
      </RadialGradientBrush>
    </Ellipse.Fill>
</Ellipse>
```

Figure 9-4 shows this gradient, along with an ordinary radial gradient that has the standard GradientOrigin value (0.5, 0.5).

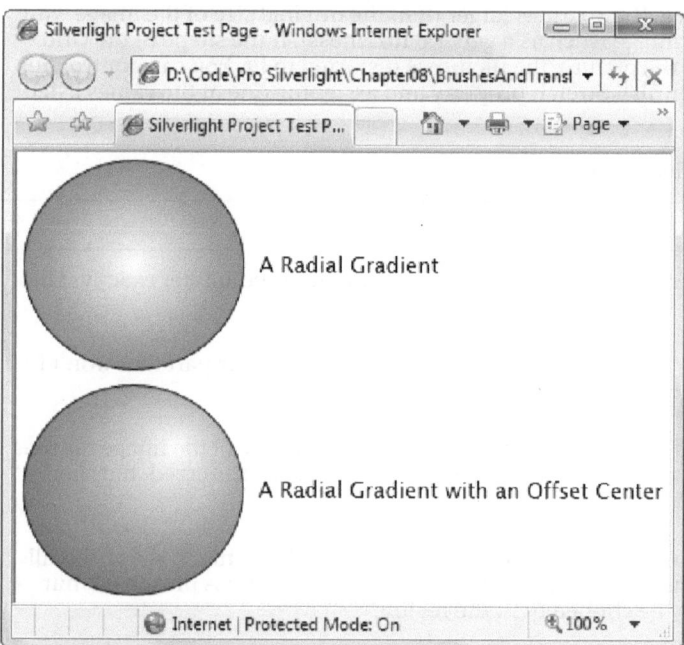

Figure 9-4. *Radial gradients*

The ImageBrush

The ImageBrush allows you to fill an area with a bitmap image using any file type that Silverlight supports (BMP, PNG, and JPEG files). You identify the image you want to use by setting the ImageSource property. For example, this brush paints the background of a Grid using an image named logo.jpg that's included in your project as a resource (and therefore embedded in your application's XAP file):

```
<Grid>
  <Grid.Background>
    <ImageBrush ImageSource="logo.jpg"></ImageBrush>
  </Grid.Background>
</Grid>
```

The ImageSource property of the ImageBrush works the same way as the Source property of the Image element, which means you can also set it using a URI that points to an embedded file in your project or a web location.

299

■ **Note** Silverlight respects any transparency information that it finds in an image. For example, Silverlight supports transparent areas in a PNG file.

In this example, the ImageBrush is used to paint the background of a cell. As a result, the image is stretched to fit the fill area. If the Grid is larger than the original size of the image, you may see resizing artifacts in your image (such as a general fuzziness). If the shape of the Grid doesn't match the aspect ratio of the picture, the picture is distorted to fit. You can control this behavior by modifying the ImageBrush.Stretch property and assigning one of the values listed in Table 9-2.

Table 9-2. *Values for the Stretch Enumeration*

Name	Description
Fill	Your image is stretched in width and height to fit its container exactly. This is the default.
None	The image isn't stretched. Its native size is used (and any part that won't fit is clipped).
Uniform	The width and height are increased proportionately until the image reaches the edge of the container. The image's aspect ratio is preserved, but there may be extra blank space.
UniformToFill	The width and height are increased proportionately until the shape fills all the available height and width. The image's aspect ratio is preserved, but the image may be clipped to fit the region.

If the image is painted smaller than the fill region, the image is aligned according to the AlignmentX and AlignmentY properties. The unfilled area is left transparent. This occurs if you're using Uniform scaling and the region you're filling has a different shape (in which case you'll get blank bars on the top or the sides). It also occurs if you're using None and the fill region is larger than the image.

Transparency

In the examples you've considered so far, the shapes have been completely opaque. However, Silverlight supports true transparency. That means if you layer several elements on top of one another and give them all varying layers of transparency, you'll see exactly what you expect. At its simplest, this feature gives you the ability to create graphical backgrounds that "show through" the elements you place on top. At its most complex, this feature lets you create multilayered animations and other effects.

There are several ways to make an element partly transparent:

- *Set the Opacity property of the element*: Opacity is a fractional value from 0 to 1, where 1 is completely solid (the default) and 0 is completely transparent. The Opacity property is defined in the UIElement class, so it applies to all elements.

- *Set the Opacity property of the brush*: Like elements, the various brush classes include an Opacity property that allows you to make their fill partially transparent. You can then use these brushes to paint part of with an element.

- *Use a semitransparent color*: Any color that has an alpha value less than 255 is semitransparent. You can use a semitransparent color when setting the foreground, background, or border of an element.

- *Set the OpacityMask property*: This lets you make specific regions of an element transparent or partially transparent. For example, you can use it to fade a shape gradually into transparency.

Figure 9-5 shows an example that uses the first two approaches to create transparent elements.

Figure 9-5. A page with several semitransparent elements

In this example, the top-level layout container is a Grid that uses an ImageBrush that sets a picture for the background. The opacity of the Grid is reduced to 70%, allowing the solid color underneath to show through. (In this case, it's a white background, which lightens the image.)

```
<Grid Margin="5" Opacity="0.7">
  <Grid.Background>
```

```
    <ImageBrush ImageSource="celestial.jpg" />
  </Grid.Background>
  ...
</Grid>
```

The first element inside the Grid is a button, which uses a partially transparent red background color (set through the Background property). The image shows through in the button background, but the text is opaque. (Had the Opacity property been set, both the foreground and background would have become semitransparent.)

```
<Button Foreground="Green" Background="#60AA4030" FontSize="16" Margin="10"
  Padding="20" Content="A Semi-Transparent Button"></Button>
```

■ **Note** Silverlight supports the ARGB color standard, which uses four values to describe every color. These four values (each of which ranges from 0 to 255) record the alpha, red, green, and blue components, respectively. The *alpha* component is a measure of how transparent the color is—0 is fully transparent, and 255 is fully opaque.

The next element is a TextBlock. By default, all TextBlock elements have a completely transparent background color, so the content underneath can show through. This example doesn't change that detail, but it does use the Opacity property to make the text partially transparent. You could accomplish the same effect by setting a white color with a nonzero alpha value for the Foreground property.

```
<TextBlock Grid.Row="1" Margin="10" TextWrapping="Wrap"
  Foreground="White" Opacity="0.3" FontSize="38" FontFamily="Arial Black"
  Text="SEMI-TRANSPARENT TEXT"></TextBlock>
```

Last is a nested Grid that places two elements in the same cell, one over the other. (You could also use a Canvas to overlap two elements and control their positions more precisely.) On the bottom is a partially transparent Image element that shows a happy face. It also uses the Opacity property to allow the other image to show through underneath. Over that is a TextBlock element with partially transparent text. If you look carefully, you can see both backgrounds show through under some letters:

```
<Image Grid.Row="2" Margin="10" Source="happyface.jpg" Opacity="0.5"></Image>
```

You can extend the layering and tile multiple images or elements on top of each other, making each one partially transparent. Of course, if you add enough transparent layers, performance will suffer, particularly if your application uses dynamic effects such as animation. Furthermore, you're unlikely to perceive the difference with more than two or three layers of transparency. However, Silverlight imposes no limits on how you use transparency.

Opacity Masks

You can use the OpacityMask property to make specific regions of an element transparent or partially transparent. The OpacityMask property allows you to achieve a variety of common and exotic effects. For example, you can use it to fade a shape gradually into transparency.

The OpacityMask property accepts any brush. The alpha channel of the brush determines where the transparency occurs. For example, if you use a SolidColorBrush that's set to a transparent color for your opacity mask (a color that has an alpha value of 0), your entire element disappears. If you use a SolidColorBrush that's set to use a nontransparent color, your element remains completely visible. If you use a SolidColorBrush that uses a semitransparent color (for example, an alpha value of 100), the element will be partially visible. The other details of the color (the red, green, and blue components) aren't important and are ignored when you set the OpacityMask property.

Using an opacity mask with a SolidColorBrush doesn't make much sense because you can accomplish the same effect more easily with the Opacity property. But an opacity mask becomes more useful when you use more exotic types of brushes, such as LinearGradientBrush or RadialGradientBrush. Using a gradient that moves from a solid to a transparent color, you can create a transparency effect that fades in over the surface of your element, like the one used by this button:

```xml
<Button FontSize="14" FontWeight="Bold" Content="A Partially Transparent Button">
  <Button.OpacityMask>
    <LinearGradientBrush StartPoint="0,0" EndPoint="1,0">
      <GradientStop Offset="0" Color="Transparent"></GradientStop>
      <GradientStop Offset="0.8" Color="Black"></GradientStop>
    </LinearGradientBrush>
  </Button.OpacityMask>
</Button>
```

Figure 9-6 shows this button over a page that displays a picture of a grand piano.

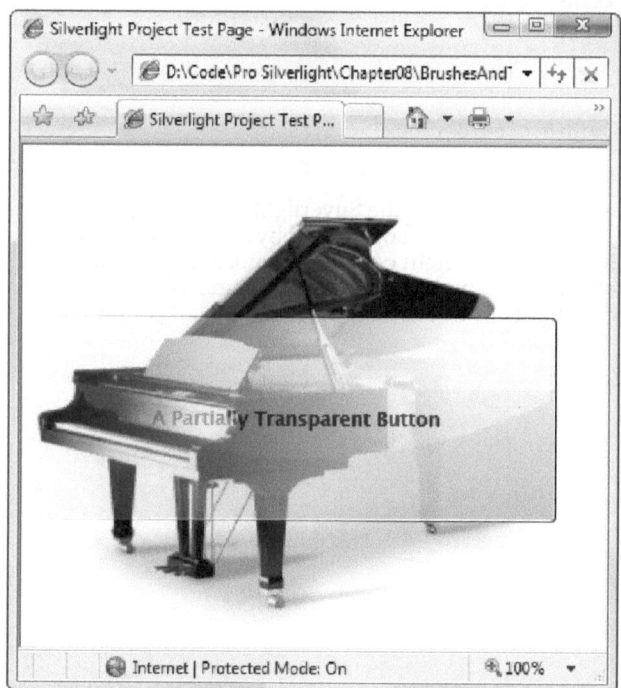

Figure 9-6. *A button that fades from transparent (left) to solid (right)*

Making the Silverlight Control Transparent

So far, you've seen how to make different elements in a Silverlight region transparent. But you can use one more transparency trick: making the Silverlight content region *windowless*, so its background allows HTML content to show through.

To configure Silverlight to use windowless rendering, you need to follow several steps. First, you must edit your XAML to make sure your markup doesn't set an opaque background. Ordinarily, when you create a new page with Visual Studio, it adds a single Grid container that fills the entire page. This Grid is the layout root for the page, and Visual Studio explicitly gives it a white background, as shown here:

```
<Grid x:Name="LayoutRoot" Background="White">
```

To make the page transparent, you need to remove the Background property setting so the Grid can revert to its default transparent background.

Next, you need to edit your HTML entry page. Find the <div> element that holds the Silverlight content region. Now, you need to make two alterations: change the background parameter from white to transparent, and add a windowless parameter with a value of true. Here's the modified HTML markup:

```
<div id="silverlightControlHost">
  <object data="data:application/x-silverlight,"
    type="application/x-silverlight-2" width="100%" height="100%">
    <param name="source" value="TransparentSilverlight.xap"/>
    <param name="onerror" value="onSilverlightError" />
    <param name="background" value="transparent" />
    <param name="windowless" value="true" />
    ...
  </object>
  <iframe id="_sl_historyFrame"
    style="visibility:hidden;height:0;width:0;border:0px"></iframe>
</div>
```

Figure 9-7 and Figure 9-8 show an example that places the Silverlight content region in the left column of a multicolumned page. Each column is represented by a <div> element with different style settings. Figure 9-7 shows the Silverlight control as it normally appears, with an opaque background. Figure 9-8 shows the same example with a windowless Silverlight content region. Because the Silverlight control is transparent, the tiled column background can show through.

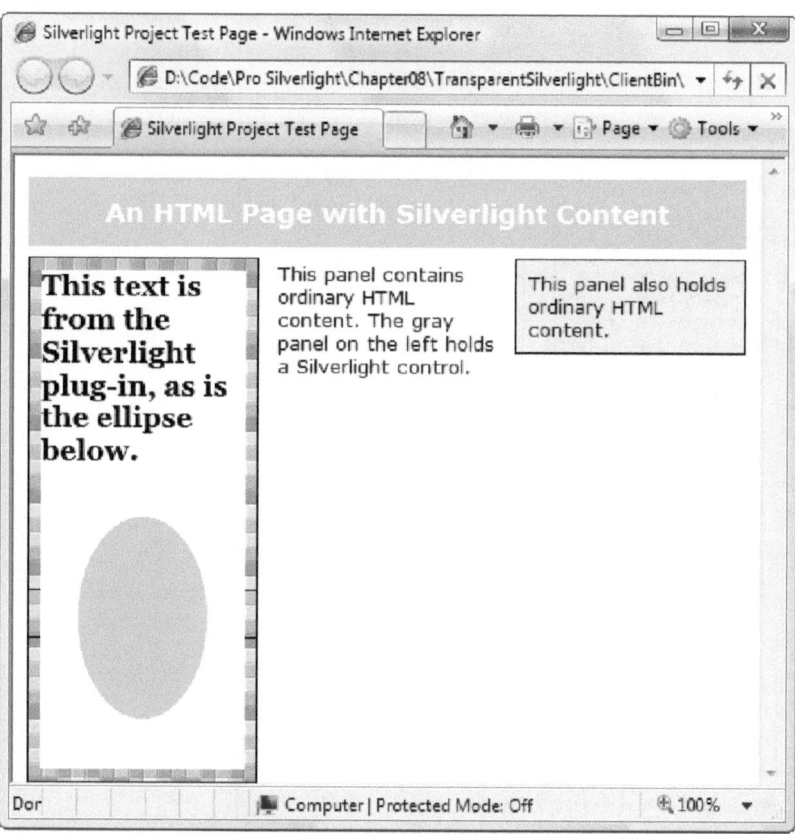

Figure 9-7. *Normal Silverlight content*

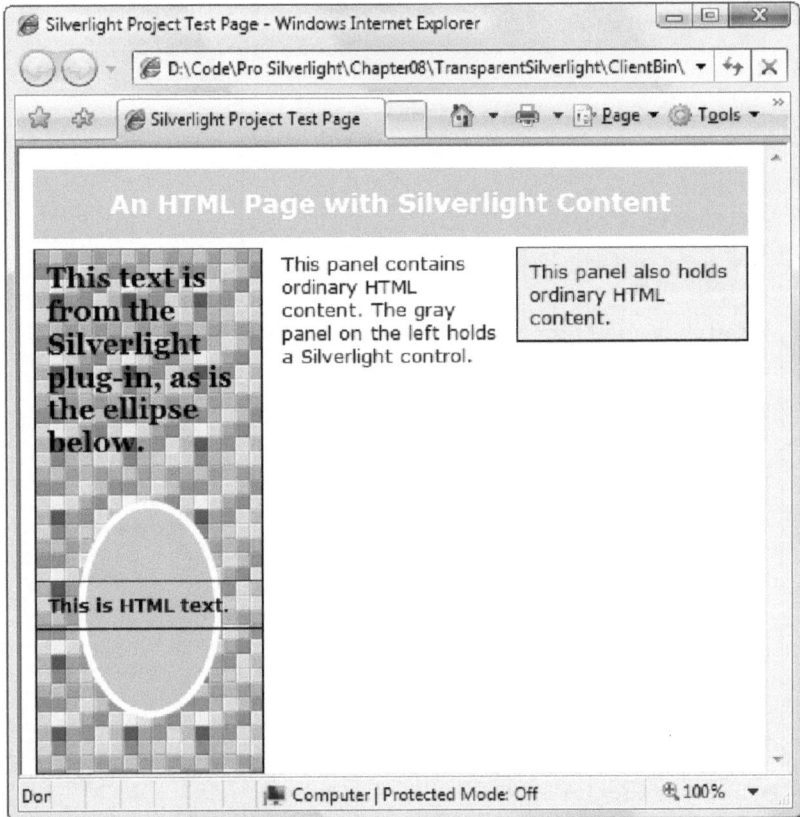

Figure 9-8. *A windowless Silverlight content region*

A windowless Silverlight content region has two important differences. Not only does it allow HTML content underneath to show through, but it also allows HTML content above to overlap. Figure 9-8 demonstrates this fact with a small snippet of floating HTML that appears over the Silverlight content region and displays the message "This is HTML text."

To create this effect, you position two <div> elements using absolute coordinates on the left side of the page, using these two style classes:

```
.SilverlightLeftPanel
{
  background-image: url('tiles5x5.png');
  background-repeat:repeat;
  position: absolute;
  top: 70px;
  left: 10px;
  width: 142px;
  height: 400px;
  border-width: 1px;
  border-style: solid;
  border-color: black;
```

```
    padding: 8px;
}

.HtmlLeftPanel
{
    background-color: Transparent;
    position: absolute;
    top: 300px;
    left: 10px;
    width: 142px;
    font-weight: bold;
    border-width: 1px;
    border-style: solid;
    border-color: black;
    padding: 8px;
}
```

The first <div> element holds the Silverlight content region, and the second <div> holds the overlapping HTML content, as shown here:

```
<div class="SilverlightLeftPanel">
  <div id="silverlightControlHost">...</div>
</div>

<div class="HtmlLeftPanel" >
  <p>This is HTML text.</p>
</div>
```

To see the complete HTML for this page, refer to the downloadable code for this chapter.

■ **Tip** The most common reason to use a windowless Silverlight control is because you want nonrectangular Silverlight content to blend in seamlessly with the web-page background underneath. However, you can also use a windowless Silverlight control to put HTML elements and Silverlight elements side by side. This is particularly useful if these elements interact (as described in Chapter 14). For example, to create a Silverlight media player with HTML playback buttons, you'll probably use a windowless Silverlight control.

Only use a windowless Silverlight content region if you need it. It requires extra overhead, which can reduce performance in applications that require frequent redrawing or use a large number of animations. When you aren't using a windowless content region, don't assume your Silverlight control will automatically get a solid white background. When running on Mac computers, Silverlight always uses windowless mode, regardless of the parameters you pass. That's why the default entry page explicitly sets the "background" parameter to white.

Transforms

Many drawing tasks can be simplified by using a *transform*—an object that alters the way a shape or element is drawn by secretly shifting the coordinate system it uses. In Silverlight, transforms are represented by classes that derive from the MustInherit System.Windows. Media.Transform class, as listed in Table 9-3.

Table 9-3. *Transform Classes*

Name	Description	Important Properties
Translate-Transform	Displaces your coordinate system by some amount. This transform is useful if you want to draw the same shape in different places.	X, Y
Rotate-Transform	Rotates your coordinate system. The shapes you draw normally are turned around a center point you choose.	Angle, CenterX, CenterY
Scale-Transform	Scales your coordinate system up or down so that your shapes are drawn smaller or larger. You can apply different degrees of scaling in the X and Y dimensions, thereby stretching or compressing your shape. When a shape is resized, Silverlight resizes its inside area and its border proportionately. That means the larger your shape grows, the thicker its border is. (Incidentally, ScaleTransform powers the Viewbox you learned about in Chapter 8 and so has some obvious similarities.)	ScaleX, ScaleY, CenterX, CenterY
Skew-Transform	Warps your coordinate system by slanting it a number of degrees. For example, if you draw a square, it becomes a parallelogram.	AngleX, AngleY, CenterX, CenterX
Composite-Transform	Fuses together a ScaleTransform, SkewTransform, RotateTransform, and TransaleTransform (which are applied in that order). This transform is really just a XAML convenience—if you need to use a combination of these transforms on a single element, the CompositeTransform can save some markup.	CenterX, CenterY, Rotation, ScaleX, ScaleY, SkewX, SkewY, TranslateX, TranslateY
Matrix-Transform	Modifies your coordinate system using matrix multiplication with the matrix you supply. This is the most complex option—it requires some mathematical skill.	Matrix

Name	Description	Important Properties
Transform-Group	Combines multiple transforms so they can all be applied at once. The order in which you apply transformations is important—it affects the final result. For example, rotating a shape (with RotateTransform) and then moving it (with TranslateTransform) sends the shape off in a different direction than if you move it and *then* rotate it. Note that although you can always replicate a CompositeTransform with a TransformGroup, the reverse is not true. For example, a TransformGroup can apply transforms in any order, it can use the same type of transform multiple times, and it can include a MatrixTransform, all unlike the CompositeTransform.	N/A

Technically, all transforms use matrix math to alter the coordinates of your shape. But using the prebuilt transforms classes—TranslateTransform, RotateTransform, ScaleTransform, SkewTransform—is far simpler than using MatrixTransform and trying to work out the right matrix for the operation you want to perform. When you perform a series of transforms with a CompositeTransform or TransformGroup, Silverlight fuses your transforms together into a single MatrixTransform, ensuring optimal performance.

■ **Note** All transforms have automatic change-notification support. If you change a transform that's being used in a shape, the shape redraws itself immediately.

Transforms are one of those quirky concepts that turns out to be extremely useful in a variety of different contexts. Some examples include the following:

- *Angling a shape*: Using RotateTransform, you can turn your coordinate system to create certain shapes more easily.
- *Repeating a shape*: Many drawings are built using a similar shape in several different places. Using a transform, you can take a shape and then move it, rotate it, resize it, and so on.

■ **Tip** To use the same shape in multiple places, you need to duplicate the shape in your markup (which isn't ideal), use code (to create the shape programmatically), or use the Path shape described in Chapter 8. The Path shape accepts Geometry objects, and you can store a geometry object as a resource so it can be reused throughout your markup.

- *Dynamic effects and animation*: You can create a number of sophisticated effects with the help of a transform, such as rotating a shape, moving it from one place to another, and warping it dynamically.

In Chapter 10, you'll use transforms to build powerful animations. But for now, you'll take a quick look at how transforms work by considering how you can apply a basic transform to an ordinary shape.

Transforming Shapes

To transform a shape, you assign the RenderTransform property to the transform object you want to use. Depending on the transform object you're using, you'll need to fill in different properties to configure it, as detailed in Table 9-3.

For example, if you're rotating a shape, you need to use the rotate transform and supply the angle in degrees. Here's an example that rotates a rectangle by 25 degrees:

```
<Rectangle Width="80" Height="10" Stroke="Blue" Fill="Yellow"
  Canvas.Left="100" Canvas.Top="100">
  <Rectangle.RenderTransform>
    <RotateTransform Angle="25" />
  </Rectangle.RenderTransform>
</Rectangle>
```

When you rotate a shape this way, you rotate it about the shape's origin (the upper-left corner). Figure 9-9 illustrates by rotating the same square 25, 50, 75, and then 100 degrees.

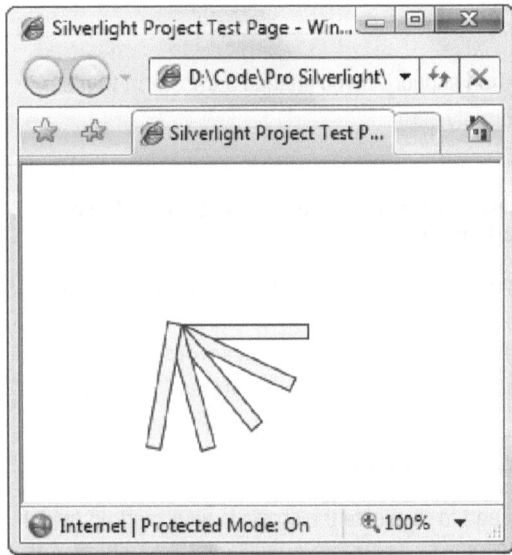

Figure 9-9. *Rotating a rectangle four times*

Sometimes you'll want to rotate a shape around a different point. RotateTransform, like many other transform classes, provides a CenterX property and a CenterY property. You can use these properties to indicate the center point around which the rotation should be performed. Here's a rectangle that uses this approach to rotate itself 25 degrees around its center point:

```
<Rectangle Width="80" Height="10" Stroke="Blue" Fill="Yellow"
  Canvas.Left="100" Canvas.Top="100">
  <Rectangle.RenderTransform>
```

```
    <RotateTransform Angle="25" CenterX="45" CenterY="5" />
  </Rectangle.RenderTransform>
</Rectangle>
```

Figure 9-10 shows the result of performing the same sequence of rotations featured in Figure 9-9, but around the designated center point.

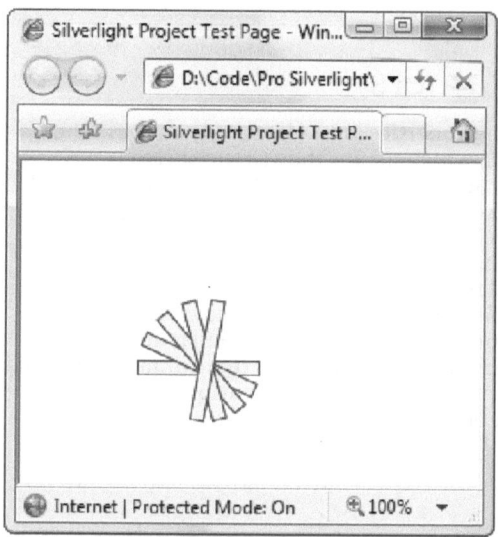

***Figure 9-10.** Rotating a rectangle around its middle*

There's a clear limitation to using the CenterX and CenterY properties of RotateTransform. These properties are defined using absolute coordinates, which means you need to know the exact center point of your content. If you're displaying dynamic content (for example, pictures of varying dimensions or elements that can be resized), this introduces a problem. Fortunately, Silverlight has a solution with the handy RenderTransformOrigin property, which is supported by all shapes. This property sets the center point using a proportional coordinate system that stretches from 0 to 1 in both dimensions. In other words, the point (0, 0) is designated as the upper-left corner and (1, 1) is the lower-right corner. (If the shape region isn't square, the coordinate system is stretched accordingly.)

With the help of the RenderTransformOrigin property, you can rotate any shape around its center point using markup like this:

```
<Rectangle Width="80" Height="10" Stroke="Blue" Fill="Yellow"
  Canvas.Left="100" Canvas.Top="100" RenderTransformOrigin="0.5,0.5">
  <Rectangle.RenderTransform>
    <RotateTransform Angle="25" />
  </Rectangle.RenderTransform>
</Rectangle>
```

This works because the point (0.5, 0.5) designates the center of the shape, regardless of its size. In practice, RenderTransformOrigin is generally more useful than the CenterX and CenterY properties, although you can use either one (or both) depending on your needs.

> ■ **Tip** You can use values greater than 1 or less than 0 when setting the RenderTransformOrigin
> property to designate a point that appears outside the bounding box of your shape. For example, you can
> use this technique with a RotateTransform to rotate a shape in a large arc around a very distant point,
> such as (5, 5).

Transforms and Layout Containers

The RenderTransform and RenderTransformOrigin properties aren't limited to shapes. The
Shape class inherits them from the UIElement class, which means they're supported by all
Silverlight elements, including buttons, text boxes, the TextBlock; entire layout containers full
of content; and so on. Amazingly, you can rotate, skew, and scale any piece of Silverlight user
interface (although in most cases you shouldn't).

It's important to note that when you apply transforms to the elements in a layout
container, the transforming is performed after the layout. For the simple Canvas, which uses
coordinate-based layout, this distinction has no effect. But for other layout containers, which
position elements *relatively* based on the placement and size of other elements, the effect is
important. For instance, consider Figure 9-11, which shows a StackPanel that contains a rotated
button. Here, the StackPanel lays out the two buttons as though the first button is positioned
normally, and the rotation happens just before the button is rendered. As a result, the rotated
button overlaps the one underneath.

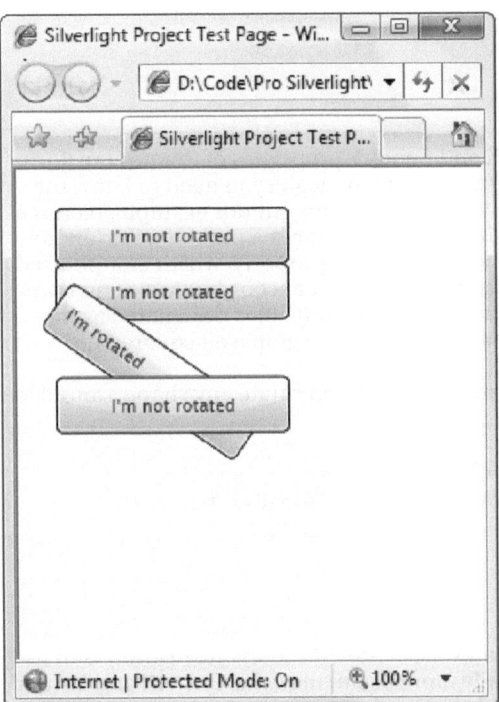

Figure 9-11. Rotating buttons

WPF also has the ability to use layout transforms, which are applied before the layout pass. This means the layout container uses the transformed dimensions of an element when positioning other elements. However, Silverlight doesn't provide this ability.

■ **Tip** You can also use transforms to change a wide range of Silverlight ingredients, such as brushes, geometries, and clipping regions.

A Reflection Effect

Transforms are important for applying many types of effects. One example is a reflection effect, such as the one demonstrated in Figure 9-12.

Figure 9-12. A reflection effect

To create a reflection effect in Silverlight, you first explicitly duplicate the content that will use the effect. For example, to create the reflection shown in Figure 9-11, you need to begin with two identical Image elements—one of which shows the original image and the other of which shows the reflected copy:

```
<Grid x:Name="LayoutRoot" Background="White">
  <Grid.RowDefinitions>
    <RowDefinition></RowDefinition>
    <RowDefinition></RowDefinition>
  </Grid.RowDefinitions>
  <Image Grid.Row="0" Source="harpsichord.jpg"></Image>
  <Image Grid.Row="1" Source="harpsichord.jpg"></Image>
</Grid>
```

Because this technique forces you to duplicate your content, it generally isn't practical to add a reflection effect to controls. But it's possible to create a reflection of a live video playback with the help of the VideoBrush class, which is described in Chapter 11.

The second step is to modify the copy of your content to make it look more like a reflection. To accomplish this, you use a combination of two ingredients—a transform, which flips the image into place, and an opacity mask, which fades it gently out of sight.

```
<Image Grid.Row="1" Source="harpsichord.jpg" RenderTransformOrigin="0,0.4">
  <Image.RenderTransform>
    <ScaleTransform ScaleY="-0.8"></ScaleTransform>
  </Image.RenderTransform>
  <Image.OpacityMask>
    <LinearGradientBrush StartPoint="0,0" EndPoint="0,1">
      <GradientStop Offset="0" Color="Transparent"></GradientStop>
      <GradientStop Offset="1" Color="#44000000"></GradientStop>
    </LinearGradientBrush>
  </Image.OpacityMask>
</Image>
```

Here, a ScaleTransform flips the image over by using a negative value for ScaleY. To flip an image horizontally, you use –1. Using a fractional value (in this case, –0.8) simultaneously flips the image over and compresses it, so it's shorter than the original image. To make sure the flipped copy appears in the right place, you must position it exactly (using a layout container like the Canvas) or use the RenderTransformOrigin property, as in this example. Here, the image is flipped around the point (0, 0.4). In other words, it keeps the same left alignment (x = 0) but is moved down (y = 0.4). Essentially, it's flipped around an imaginary horizontal line that's a bit higher than the midpoint of the image.

This example uses a LinearGradientBrush that fades between a completely transparent color and a partially transparent color to make the reflected content more faded. Because the image is upside down, you must define the gradient stops in reverse order.

Perspective Transforms

Silverlight doesn't include a true toolkit for 3-D drawing. However, it does have a feature called *perspective transforms* that lets you simulate a 3-D surface. Much like a normal transform, a perspective transform takes an existing element and manipulates its visual appearance. But with a perspective transform, the element is made to look as though it's on a 3-D surface.

Perspective transforms can come in handy, but they're a long way from real 3-D. First, and most obvious, they give you only a single shape to work with—essentially, a flat, rectangular plane, like a sheet of paper, on which you can place your elements and then tilt them away from the viewer. By comparison, a true 3-D framework allows you to fuse tiny triangles together to build more complex surfaces, ranging from cubes and polyhedrons to spheres and entire topographic maps. True 3-D frameworks also use complex math to calculate proper lighting and shading, determine what shapes are obscuring other shapes, and so on. (For an example, consider Silverlight's Windows-only big brother, WPF, which has rich 3-D support.)

■ **Note** The bottom line is this—if you're looking for a few tricks to create some 3-D eye candy without the hard work, you'll like Silverlight's perspective-transform feature. (Perspective transforms are particularly useful when combined with animation, as you'll see in the next chapter.) But if you're hoping for a comprehensive framework to model a 3-D world, you'll be sorely disappointed.

Much as Silverlight includes a MustInherit Transform class from which all transforms derive, it uses a MustInherit System.Windows.Media.Projection class from which all projections derive. Currently, Silverlight includes just two projections: the practical PlaneProjection that you'll use in this chapter and the far more complex Matrix3DProjection, which suits those who are comfortable using heavy-duty math to construct and manipulate 3D matrices. Matrix3DProjection is beyond the scope of this book. However, if you'd like to experiment with it and explore the underlying math, Charles Petzold provides a good two-part introduction with sample code at http://tinyurl.com/m29v3q and http://tinyurl.com/laalp6.

The PlaneProjection Class

PlaneProjection gives you two complementary abilities. First, you can rotate the 3-D plane around the x-axis (side-to-side), the y-axis (up-and-down), or the z-axis (which looks like a normal rotational transform). Figure 9-13 illustrates the difference, with 45-degree rotations around the three different axes.

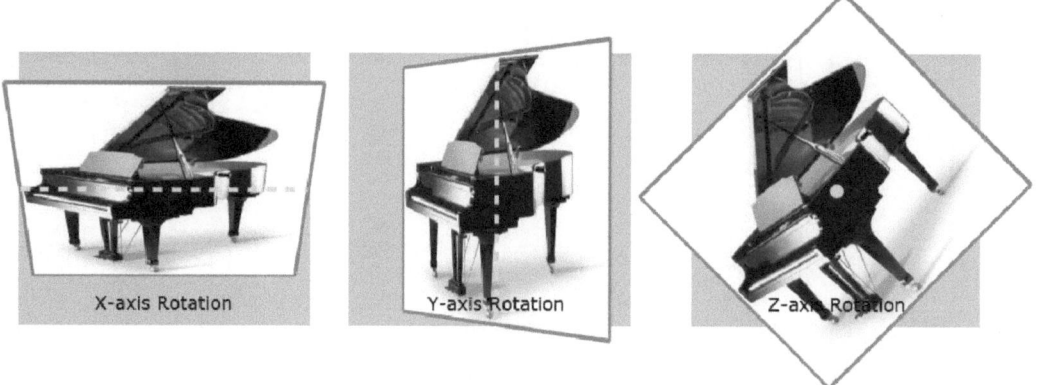

Figure 9-13. Rotations with the PlaneProjection class

In Figure 9-13, the picture is rotated around its center point. But you can explicitly choose to rotate the element around a different point by setting the right property. Here's how:

- For an x-axis rotation, use RotationX to control the amount of rotation (as an angle from 0 to 360 degrees). Use CenterOfRotationX to set the x coordinate of the center

315

point in relative terms, where 0 is the far left, 1 is the far right, and 0.5 is the middle point (and default).

- For a y-axis rotation, use RotationY to set the angle of rotation. Use CenterOfRotationY to set the y coordinate of the center point, where 0 is the top, 1 is the bottom, and 0.5 is the middle (and default).
- For a z-axis rotation, use RotationZ to set the angle of rotation. Use CenterOfRotationZ to set the z coordinate of the center point, where 0 is the middle (and default), positive numbers are in front of the element, and negative numbers are behind it.

In many cases, the rotation properties will be the only parts of the PlaneProjection that you'll want to use. However, you can also shift the element in any direction. There are two ways to move it:

- Use the GlobalOffsetX, GlobalOffsetY, and GlobalOffsetZ properties to move the element using screen coordinates, *before* the projection is applied.
- Use the LocalOffsetX, LocalOffsetY, and LocalOffsetZ properties to move the element using its transformed properties, *after* the projection is applied.

For example, consider the case where you haven't rotated the element. In this case, the global and the local properties will have the same effect. Increasing GlobalOffsetX or LocalOffsetX shifts the element to the right. Now, consider the case where the element has been rotated around the y-axis using the RotationY property (shown in Figure 9-14). In this situation, increasing GlobalOffsetX shifts the rendered content to the right, exactly the same way it does when the element hasn't been rotated. But increasing LocalOffsetX moves the content along the x-axis, which now points in a virtual 3-D direction. As a result, the content appears to move to the right and backward.

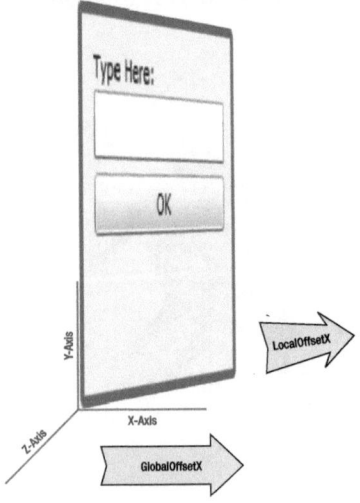

Figure 9-14. Translation with the PlaneProjection

These two details—rotation and translation—encompass everything the PlaneProjection does.

Applying a Projection

Projections works on virtually any Silverlight element, because every class that derives from UIElement includes the required Projection property. To add a perspective effect to an element, you create a PlaneProjection and use it to set the Projection property, either in code or in XAML markup.

For example, here's the PlaneProjection that rotates the first figure in Figure 9-13:

```
<Border BorderBrush="SlateGray" CornerRadius="2" BorderThickness="4">
  <Border.Projection>
    <PlaneProjection RotationY="45"></PlaneProjection>
  </Border.Projection>
  <Image Source="grandpiano.jpg"></Image>
</Border>
```

As with ordinary transforms, perspective transforms are performed after layout. Figure 9-13 illustrates this fact by using a shaded border that occupies the original position of the transformed element. Even though the element now sticks out in new places, the bounds of the shaded background are used for layout calculations. As with all elements, if more than one element overlaps, the one declared last in the markup is placed on top. (Some layout controls offer more sophisticated layering, as the Canvas does with the ZIndex property discussed in the previous chapter.)

To get a feeling for how the different PlaneProjection properties interact, it helps to play with a simple test application, like the one shown in Figure 9-15. Here, the user can rotate an element around its x-axis, y-axis, or z-axis (or any combination). In addition, the element can be displaced locally or globally along the x-axis using the LocalOffsetX and GlobalOffsetX properties described earlier.

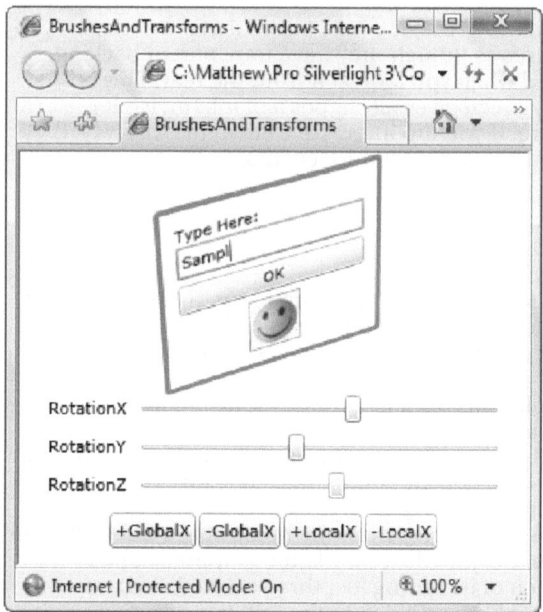

Figure 9-15. Rotating ordinary elements in 3-D

Although you can use a projection on any element, it's often useful to apply it to some sort of container, such as a layout panel or the Border element, as in this example. That way, you can place more elements inside. This example is particularly interesting because among the projected elements are interactive controls such as a button and text box. These controls continue to work in their standard ways, responding to mouse clicks, allowing focus and typing, and so on, even as you rotate the containing Border element.

```
<Border BorderBrush="SlateGray" CornerRadius="2" BorderThickness="4" Padding="10">
  <Border.Projection>
    <PlaneProjection x:Name="projection"></PlaneProjection>
  </Border.Projection>
  <StackPanel>
    <TextBlock>Type Here:</TextBlock>
    <TextBox></TextBox>
    <Button Margin="0,5" Content="OK"></Button>
    <Image Source="happyface.jpg" Stretch="None"></Image>
  </StackPanel>
</Border>
```

Although you can adjust the PlaneProjection object using code, this example uses the data-binding feature you learned about in Chapter 2. However, because the PlaneProjection isn't an element, it can't use binding expressions. Instead, you need to place the binding in the linked Slider controls and use a two-way binding to ensure that the new angles are passed backed to the projection as the user drags the tab. Here's an example with the x-axis slider:

```
<TextBlock Margin="5">RotationX</TextBlock>
<Slider Grid.Column="1" Minimum="-180" Maximum="180"
 Value="{Binding RotationX, Mode=TwoWay, ElementName=projection}"></Slider>
```

If you rotate an element far enough around the x-axis or y-axis (more than 90 degrees), you begin to see its back. Silverlight treats all elements as though they have transparent backing, which means your element's content is reversed when you look at it from the rear. This is notably different from the 3-D support in WPF, which gives all shapes a blank (invisible) backing unless you explicitly place content there. If you flip interactive elements this way, they keep working, and they continue capturing all the standard mouse events.

Pixel Shaders

One of the most impressive and most understated features in Silverlight is its support for *pixel shaders*—objects that transform the appearance of any element by manipulating its pixels just before they're displayed in the Silverlight content region. (Pixel shaders kick in after the transforms and projections you've just learned about.)

A crafty pixel shader is as powerful as the plug-ins used in graphics software like Adobe Photoshop. It can do anything from adding a basic drop shadow to imposing more ambitious effects such as blurs, glows, watery ripples, embossing, sharpening, and so on. Pixel shaders can also create eye-popping effects when they're combined with animation that alters their parameters in real time, as you'll see in Chapter 10.

Every pixel shader is represented by a class that derives from the MustInherit Effect class in the System.Windows.Media.Effects namespace. Despite the remarkable potential of pixel shaders, Silverlight takes the restrained approach of including just three derived classes in the core runtime: BlurEffect, DropShadowEffect, and ShaderEffect. In the following sections, you'll look at each one and learn how you can incorporate more dazzling effects from a free library.

BlurEffect

Silverlight's simplest effect is the BlurEffect class. It blurs the content of an element, as though you're looking at it through an out-of-focus lens. You increase the level of blur by increasing the value of the Radius property. (The default value is 5.)

To use any pixel-shader effect, you create the appropriate effect object and set the Effect property of the corresponding element:

```
<Button Content="Blurred (Radius=2)" Padding="5" Margin="3">
  <Button.Effect>
    <BlurEffect Radius="2"></BlurEffect>
  </Button.Effect>
</Button>
```

Figure 9-16 shows three different blurs (where Radius is 2, 5, and 20) applied to a stack of buttons.

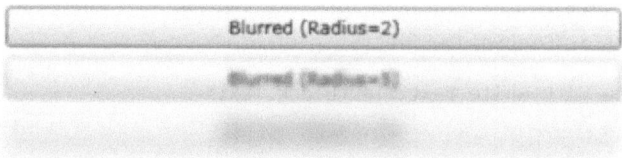

Figure 9-16. *Blurred buttons*

DropShadowEffect

DropShadowEffect adds a slightly offset shadow behind an element. You have several properties to play with, as listed in Table 9-4.

Table 9-4. *DropShadowEffect Properties*

Name	Description
Color	Sets the color of the drop shadow (the default is Black).
ShadowDepth	Determines how far the shadow is from the content, in pixels (the default is 5).
BlurRadius	Blurs the drop shadow, much like the Radius property of BlurEffect (the default is 5).
Opacity	Makes the drop shadow partially transparent, using a fractional value between 1 (fully opaque, the default) and 0 (fully transparent).
Direction	Specifies where the drop shadow should be positioned relative to the content, as an angle from 0 to 360. Use 0 to place the shadow on the right side, and increase the value to move the shadow counterclockwise. The default is 315, which places it to the lower right of the element.

319

Figure 9-17 shows several different drop-shadow effects on a TextBlock. Here's the markup for all of them:

```
<TextBlock FontSize="20" Margin="3">
  <TextBlock.Effect>
    <DropShadowEffect></DropShadowEffect>
  </TextBlock.Effect>
  <TextBlock.Text>Basic dropshadow</TextBlock.Text>
</TextBlock>

<TextBlock FontSize="20" Margin="3">
  <TextBlock.Effect>
    <DropShadowEffect Color="SlateBlue"></DropShadowEffect>
  </TextBlock.Effect>
  <TextBlock.Text>Light blue dropshadow</TextBlock.Text>
</TextBlock>

<TextBlock FontSize="20" Foreground="White" Margin="3">
  <TextBlock.Effect>
    <DropShadowEffect BlurRadius="15"></DropShadowEffect>
  </TextBlock.Effect>
  <TextBlock.Text>Blurred dropshadow with white text</TextBlock.Text>
</TextBlock>

<TextBlock FontSize="20" Foreground="Magenta" Margin="3">
  <TextBlock.Effect>
    <DropShadowEffect ShadowDepth="0"></DropShadowEffect>
  </TextBlock.Effect>
  <TextBlock.Text>Close dropshadow</TextBlock.Text>
</TextBlock>

<TextBlock FontSize="20" Foreground="LimeGreen" Margin="3">
  <TextBlock.Effect>
    <DropShadowEffect ShadowDepth="25"></DropShadowEffect>
  </TextBlock.Effect>
  <TextBlock.Text>Distant dropshadow</TextBlock.Text>
</TextBlock>
```

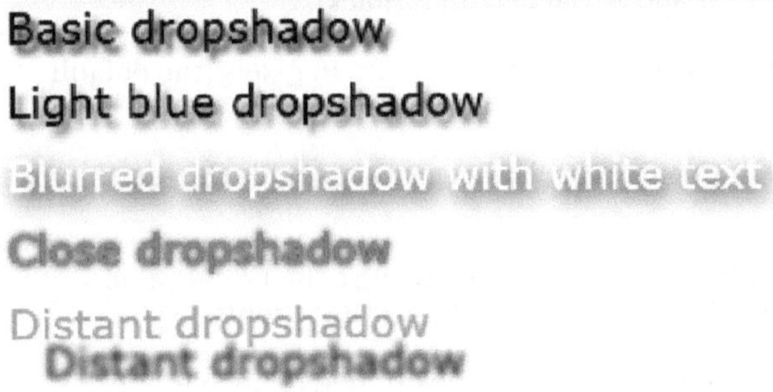

Figure 9-17. Different drop shadows

There is no class for grouping effects, which means you can apply only a single effect to an element at a time. However, you can sometimes simulate multiple effects by adding them to higher-level containers (for example, using the drop-shadow effect for a TextBlock and then placing it in a Stack Panel that uses the blur effect). In most cases, you should avoid this workaround, because it multiplies the rendering work and reduces performance. Instead, look for a single effect that can does everything you need.

ShaderEffect

The ShaderEffect class doesn't represent a ready-to-use effect. Instead, it's a MustInherit class from which you derive to create your own custom pixel shaders. By using ShaderEffect (or third-party custom effects that derive from it), you gain the ability to go far beyond mere blurs and drop shadows.

Contrary to what you may expect, the logic that implements a pixel shader isn't written in VB code directly in the effect class. Instead, pixel shaders are written using High Level Shader Language (HLSL), which was created as part of DirectX. (The benefit is obvious—because DirectX and HLSL have been around for many years, graphics developers have already created scores of pixel-shader routines that you can use in your own code.)

To create a pixel shader, you need to create the right HLSL code. The first step is to install the DirectX SDK (go to http://msdn.microsoft.com/directx). This gives you enough to create and compile HLSL code to a .ps file (using the fxc.exe command-line tool), which is what you need to use a custom ShaderEffect class. But a more convenient option is to use the free Shazzam tool (http://shazzam-tool.com). Shazzam provides an editor for HLSL files, which includes the ability to try them on sample images. It also includes several sample pixel shaders that you can use as the basis for custom effects.

Although authoring your own HLSL files is beyond the scope of this book, using an existing HLSL file isn't. Once you've compiled your HLSL file to a .ps file, you can use it in a project. Simply add the file to an existing Silverlight project, select it in the Solution Explorer, and set its Build Action to Resource. Finally, you must create a custom class that derives from ShaderEffect and uses this resource.

For example, if you're using a custom pixel shader that's compiled in a file named Effect.ps, you can use the following code:

```
Public Class CustomEffect
    Inherits ShaderEffect

    Public Sub New()
        ' Use the URI syntax described in Chapter 6 to refer to your resource.
        ' AssemblyName;component/ResourceFileName
        Dim pixelShaderUri As New Uri( _
          "CustomEffectTest;component/Effect.ps", UriKind.Relative)

        ' Load the information from the .ps file.
        PixelShader = New PixelShader()
        PixelShader.UriSource = pixelShaderUri
    End Sub
End Class
```

You can now use the custom pixel shader in any page. First, make the namespace available by adding a mapping like this:

```
<UserControl xmlns:local="clr-namespace:CustomEffectTest" ...>
```

Now, create an instance of the custom effect class, and use it to set the Effect property of an element:

```
<Image>
  <Image.Effect>
    <local:CustomEffect></local:CustomEffect>
  </Image.Effect>
</Image>
```

You can get a bit more complicated than this if you use a pixel shader that takes certain input arguments. In this case, you need to create the corresponding dependency properties by calling the shared RegisterPixelShaderSamplerProperty() method.

■ **Tip** Unless you're a hard-core graphics programmer, the best way to get more advanced pixel shaders isn't to write the HLSL yourself. Instead, look for existing HLSL examples or—even better—third-party Silverlight components that provide custom effect classes. The gold standard is the free Windows Presentation Foundation Pixel Shader Effects Library (which also works with Silverlight) at `http://wpffx.codeplex.com`. It includes a long list of dazzling effects such as swirls, color inversion, and pixelation. Even more useful, it includes transition effects that combine pixel shaders with the animation capabilities described in Chapter 10.

The WriteableBitmap Class

In Chapter 5, you learned to show bitmaps with the Image element. However, displaying a picture this way is a strictly one-way affair. Your application takes a ready-made bitmap, reads it, and displays it in the page. On its own, the Image element doesn't give you a way to create or edit bitmap information.

This is where WriteableBitmap fits in. It derives from BitmapSource, which is the class you use when setting the Image.Source property (either directly, when you set the image in code, or implicitly, when you set it in XAML). But whereas BitmapSource is a read-only reflection of bitmap data, WriteableBitmap is a modifiable array of pixels that opens up many interesting possibilities.

Generating a Bitmap

The most direct way to use WriteableBitmap is to create an entire bitmap by hand. This process may seem labor intensive, but it's an invaluable tool if you want to create fractals or create a visualization for music or scientific data. In these scenarios, you need to use a code routine to dynamically draw some sort of data, whether it's a collection of 2-D shapes (using the shape elements introduced in Chapter 8) or a raw bitmap (using WriteableBitmap).

To generate a bitmap with WriteableBitmap, you follow a fairly straightforward set of steps. First, you create the in-memory bitmap. At this time, you supply its width and height in pixels. Here's an example that creates an image as big as the current page:

```
Dim wb As New WriteableBitmap(CInt(Me.ActualWidth), CInt(Me.ActualHeight))
```

Next, you need to fill the pixels. To do so, you use the Pixels property, which provides a one-dimensional array of pixels. The pixels in this array stretch from left to right to fill each row, from top to bottom. To find a specific pixel, you need to use the following formula, which steps down the number of rows and then moves to the appropriate position in that row:

$$y * wb.PixelWidth + x$$

For example, to set the pixel (40, 100), you use this code:

```
wb.Pixels(100 * wb.PixelWidth + 40) = ...
```

The color of each pixel is represented by a single unsigned integer. However, to construct this integer, you need to pack together several pieces of information: the alpha, red, green, and blue values of the color, each of which is a single byte from 0 to 255. The easiest way to calculate the right pixel value is to use this bit-shifting code:

```
Dim alpha As Integer = 255
Dim red As Integer = 100
Dim green As Integer = 200
Dim blue As Integer = 75

Dim pixelColorValue As Integer
pixelColorValue = (alpha << 24) Or (red << 16) Or (green << 8) Or (blue << 0)

wb.Pixels(pixelPosition) = pixelColorValue
```

Here's a complete routine that steps through the entire set of available pixels, filling them with a mostly random pattern interspersed with regular gridlines (shown in Figure 9-18):

```
Dim rand As New Random()
For y As Integer = 0 To wb.PixelHeight - 1
    Dim red As Integer = 0
    Dim green As Integer = 0
    Dim blue As Integer = 0

    ' Differentiate the color to create a vertical gridline every 5 pixels
    ' and a horizontal gridline every 7 pixels.
    If (x Mod 5 = 0) OrElse (y Mod 7 = 0) Then
        ' The color is randomly chosen, but influenced by the x and y position,
        ' which creates a gradient-like effect.
        red = CInt(y / wb.PixelHeight * 255)
        green = rand.Next(100, 255)
        blue = CInt(x / wb.PixelWidth * 255)
    Else
        ' A slightly different color calculation is used for non-gridline pixels.
        red = CInt(x / wb.PixelWidth * 255)
        green = rand.Next(100, 255)
        blue = CInt(y / wb.PixelHeight * 255)
    End If

    ' Set the pixel value.
    Dim pixelColorValue As Integer
    pixelColorValue = (alpha << 24) Or (red << 16) Or (green << 8) Or (blue << 0)
```

```
        wb.Pixels(y * wb.PixelWidth + x) = pixelColorValue
Next
```

Figure 9-18. *A dynamically generated bitmap*

After that process is finished, you need to display the final bitmap. Typically, you'll use an Image element to do the job:

```
img.Source = wb
```

Even after writing and displaying a bitmap, you're still free to read from the Pixels array and modify pixel values. This gives you the ability to build more specialized routines for bitmap editing and bitmap hit testing. For an example of the bitmap hit testing in a game, check out http://tinyurl.com/mroklb.

■ **Note** In some situations, you might want to transmit the content of a WriteableBitmap to a web service or store it in a file. Although this is technically possible using the array exposed by the WriteableBitmap.Pixels property, it's not practical, because the bitmap data for a typical image is enormous. Instead, you'll almost certainly want to change it into a more bandwidth-friendly format, such as JPEG. Unfortunately, Silverlight doesn't have programmatic support for manipulating JPEG data, but you can use a third-party library like fjcore (http://code.google.com/p/fjcore) or the .NET Image Tools (http://imagetools.codeplex.com).

Capturing Content from Other Elements

In the previous example, the image was generated pixel by pixel using code. However, WriteableBitmap gives you another option: you can steal its content from an existing element.

Before you use this trick, you must begin by creating a WriteableBitmap in the familiar way—by instantiating it and declaring its size. Then, you copy the element content you want into the WriteableBitmap using the Render() method. This method takes two parameters: the element with the content you want to capture and a transform (or group of transforms) that you want to use to alter it. If you don't want to transform the content, you simply need to supply a Nothing value.

Here's an example that grabs the content from the entire page, simulating a screen-capture feature. (You can't actually capture the entire screen, because accessing content outside of the Silverlight region would constitute a security risk.)

```
' Find the top-level page.
Dim mainPage As UserControl = CType(Application.Current.RootVisual, UserControl)

' Create the bitmap.
Dim wb As New WriteableBitmap( _
  CInt(mainPage.ActualWidth), CInt(mainPage.ActualHeight))

' Copy the content into the bitmap.
wb.Render(mainPage, Nothing)
wb.Invalidate()

' Show the bitmap.
img.Source = wb
```

You'll notice that you need to call the Invalidate() method after you call Render(). The Invalidate() method tells the bitmap to actually generate its content, which allows you to hold off on this more time-consuming step until it's necessary.

Once you've filled your bitmap, you can display it the same way as before, using an existing Image element.

■ **Note** The WriteableBitmap.Render() method is particularly useful with MediaElement, where it lets you capture a frame from a currently running video. You'll learn more about video in Chapter 11.

Printing

Although graphics and printing seem to be a world apart, the behind-the-scenes bits that make it work have a lot in common. In fact, when Silverlight sends data to the printer, it sends it as a rendered graphic using the WriteableBitmap class you just met. This has certain consequences—for example, it means Silverlight printing is slower and uses more memory than printing text or vector content. However, neither one of these issues is likely to be much of a problem on a modern computer.

■ **Note** Because Silverlight printing is bitmap-based, it isn't a good tool for creating printouts that will be turned into PDFs. Not only will the generated PDF files be bigger than necessary, but any text inside won't be selectable or editable. If your goal is to create PDFs, you'll have better luck with a dedicated PDF-generating component, such as the free silverPDF library at `http://silverpdf.codeplex.com`.

Silverlight's printing support is straightforward and somewhat limited. For example, it's up to you to decide how to break content over multiple pages and how to keep track of your position in a set of data. You also need to calculate the appropriate coordinates for each line, measure the size of the content you render, and adjust for headers and footers all on your own. However, Silverlight provides you with a useful shortcut by allowing you to print the same elements that you use to create user interfaces. In fact, Silverlight printing is based around the concept of elements, but rather than laying them out in a page, you render them onto a virtual print document.

In the following sections, you'll take a look at Silverlight's printing feature. You'll use it to print a single-page image and a multiple-page list, with the test application shown in Figure 9-19.

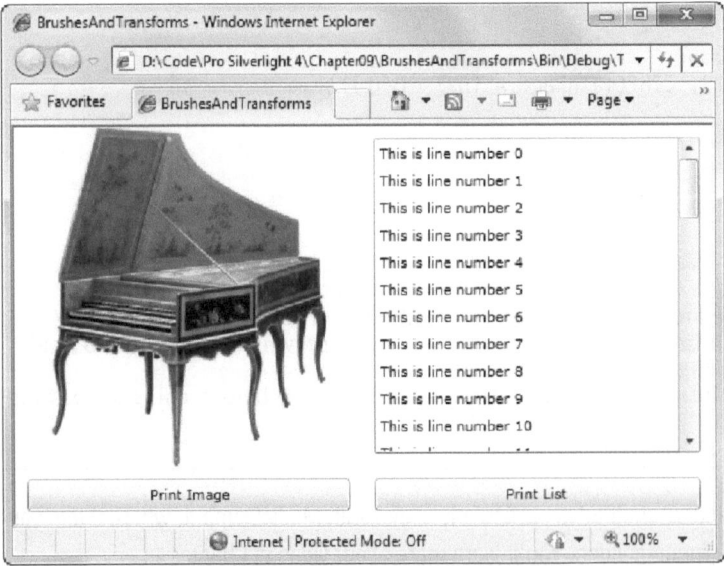

Figure 9-19. *Testing different print routines*

Printing a Single Element

All printing begins with the PrintDocument class in the System.Windows.Printing namespace. To begin a printout, you create an instance of this class, attach an event handler to its PrintPage event, and call the Print() method to get things started. Here's an example:

```
Dim document As New PrintDocument()
AddHandler document.PrintPage, AddressOf documentImage_PrintPage
document.Print("Image Document")
```

The name you supply in the Print() method is the document name. The user won't see this name in your application, but it will appear in the Windows print queue (where the user can pause or cancel it).

When you call Print(), your application shows the standard Print dialog box, where the user can pick a printer and (optionally) change printer settings. If the user clicks the Print button in the Print dialog box, your application continues running, and the PrintDocument.PrintPage event fires on a background thread to take care of the actual printing work.

In the PrintPage event handler, you do all your work using the properties of the PrintPageEventArgs object, as listed in Table 9-5. Understanding these properties is the key to mastering Silverlight's simple printing model.

Table 9-5. *PrintPageEventArgs Properties*

Property	Description
PrintableArea	Returns a Size object that describes the Height and Width of the available printable area where you can render content. If your content extends outside this area, the printed content is clipped.
PageMargins	Returns a Thickness object that details the Top, Bottom, Left, and Right margins around the current page.
PageVisual	Sets the visual element you want to print on this page. This could be a simple element (such as a TextBlock or Image) or a more complex container (such as a Grid or Canvas). For best results, this visual should exist in memory only, not in an on-screen window.
HasMorePages	Sets whether the printing should stop after this page (False) or continue to another new page (True).

During the printing process, the PrintDocument fires a series of events. It begins with BeginPrint (where you can perform preliminary calculations or initialization), then PrintPage (where the actual printing takes place), and then EndPrint (where you can clean up). Often, you'll choose to handle only the PrintPage event.

The PrintPage event fires once for each page. In the current example, just one page is printed, so the PrintPage event fires just once. When it fires, the event handler creates an Image object, configures its content, and sizes it to take up as much of the page as possible. This image is then set as the print content for the page.

```
Private Sub documentImage_PrintPage(ByVal sender As Object, _
  ByVal e As PrintPageEventArgs)
    ' Stretch the image to the size of the printed page.
    Dim imgToPrint As New Image()
    imgToPrint.Source = imgInWindow.Source
    imgToPrint.Width = e.PrintableArea.Width
    imgToPrint.Height = e.PrintableArea.Height

    ' Choose to print the image.
    e.PageVisual = imgToPrint

    ' Do not fire this event again.
    e.HasMorePages = False
End Sub
```

■ **Note** Silverlight automatically provides the PrintableArea as a number of "screen pixels." Thanks to this automatic conversion, you can use the same rendering and layout logic that you use on-screen with a printout, even though most printers have a much higher resolution than your monitor.

Keen eyes will notice that the content of the printed image (imgToPrint) is drawn from another image object (imgInWindow), which is displayed in the current window. However, it's not a good idea to print imgInWindow directly. If you do, you'll need to make do with the size that it currently has in the window, which may not suit your printout. Although you can try to adjust the size of this image, this will change the layout of your window, and it probably won't work anyway because of the way Silverlight uses clipping. For example, if you enlarge imgInWindow to the printable page size, it won't fit in the current window, and some of its content will be clipped. This clipped content will also be missing in the printout, even though the printout has more space.

Figure 9-20 shows the printout this code generates. Because the image is set to use proportional sizing, it fits the page as near as possible, but it won't be stretched to the exact dimensions of the page. (If you want, you can change this by modifying the Image.Stretch property.)

Figure 9-20. Printing an image

Printing Over Multiple Pages

The previous example did all its printing in a single page. But in most scenarios, you'll have content that will occupy an unknown number of pages. Often, this content is dynamic (for example, text the user has entered, data you've generated on the fly, records you've fetched from a changing database), but even if it isn't, you may need to use dynamic printing to deal with the possibility of different page sizes and orientations.

In the following example, the code prints the content from a list. To do this, the code must loop through the collection, keeping track of its position and requesting a new page pages whenever the current page fills up.

The code for starting the printout is virtually identical. The only difference is that you need a member field to keep track of the position in the list as the printout moves from one page to the next:

```
' Keep track of the position in the list.
Private listPrintIndex As Integer

Private Sub cmdPrintList_Click(ByVal sender As Object, ByVal e As RoutedEventArgs)
    ' Reset the position, in case a previous printout has changed it.
```

```
        listPrintIndex = 0

    Dim document As New PrintDocument()
    AddHandler document.PrintPage, AddressOf documentList_PrintPage
    document.Print("List Document")
End Sub
```

The printing code uses a Canvas for the print visual. This is a common design choice, because the Canvas can hold any combination of elements. It allows you to place elements using fixed coordinates, and it doesn't introduce any additional layout considerations.

```
' Add some extra margin space.
Private extraMargin As Integer = 50

Private Sub documentList_PrintPage(ByVal sender As Object, _
  ByVal e As PrintPageEventArgs)
    ' Use a Canvas for the printing surface.
    Dim printSurface As New Canvas()
    e.PageVisual = printSurface
    ...
```

Note that it's not necessary to explicitly size the Canvas to the print area, because the Canvas doesn't clip its content unless you set the Clip property. (In other words, content is always rendered, even when it's positioned outside the bounds of its Canvas container.)

Once the Canvas has been created, the code begins to loop through the collection of data. It creates a single TextBlock for each printed line:

```
    ...
    ' Find the starting coordinate.
    Dim topPosition As Double = e.PageMargins.Top + extraMargin

    ' Begin looping through the list.
    Do While listPrintIndex < lst.Items.Count
        ' Create a TextBlock for each line, with a 30-pixel font.
        Dim txt As New TextBlock()
        txt.FontSize = 30
        txt.Text = lst.Items(listPrintIndex).ToString()
        ...
```

Before the TextBlock can be placed in the Canvas, you need to be sure that it fits on the page. The code calculates this detail by taking into account the total available height, the current position, the rendered size of the TextBlock, and the page margins. If the TextBlock doesn't fit, the code stops printing the page and sets PrintPageEventArgs.HasMorePages to True, which requests a new page.

```
        ...
        ' If the new line doesn't fit, stop printing this page,
        ' but request another page.
        Dim measuredHeight As Double = txt.ActualHeight
        If measuredHeight > _
          (e.PrintableArea.Height - topPosition - extraMargin) Then
            e.HasMorePages = True
            Return
        End If
        ...
```

When HasMorePages is set to True and the method ends, PrintDocument object will fire its PrintPage event again, immediately. Now the code will create a new Canvas for the new page and begin placing elements on it.

However, if the current TextBlock does fit, the code simply adds it to the Canvas, moves to the next line in the list, and shifts the coordinates down to the next row on the printed page:

```
...
' Place the TextBlock on the Canvas.
txt.SetValue(Canvas.TopProperty, topPosition)
txt.SetValue(Canvas.LeftProperty, e.PageMargins.Left + extraMargin)
printSurface.Children.Add(txt)

' Move to the next line.
listPrintIndex += 1
topPosition = topPosition + measuredHeight
...
```

When the code reaches the last item of the list, the loop ends. Now the complete, multiple-page printout has finished.

```
    ...
Loop

' The printing code has reached the end of the list.
' No more pages are needed.
e.HasMorePages = False
End Sub
```

Figure 9-21 shows the complete printout spaced out over three pages.

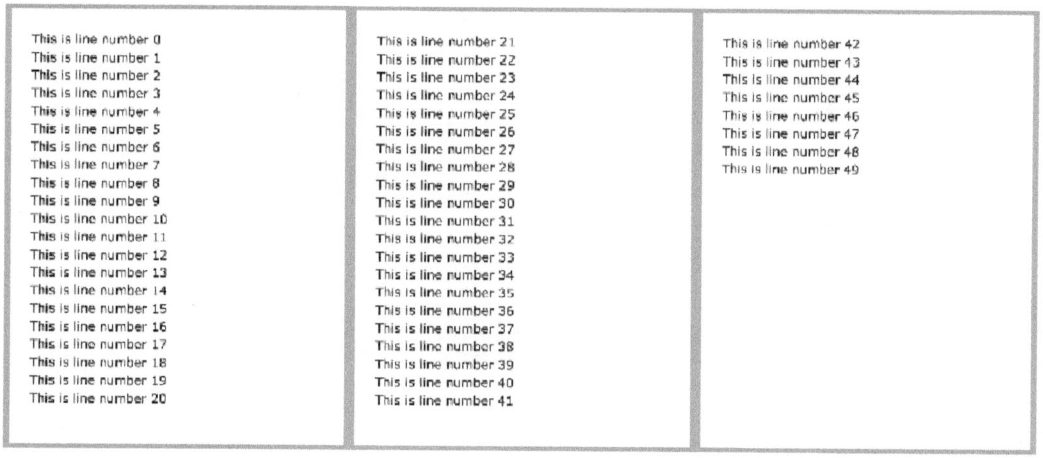

Figure 9-21. *Printing a list*

Although this example is fairly easy, you can see how the printing logic will grow more intricate if you need to print mixed content (for example, pictures and text) or multiple lines of wrapped text.

■ **Note** If you set the TextBlock.TextWrapping property to Wrap and you give it a size that's large enough to fit multiple lines of text, the TextBlock will automatically wrap your printed text. However, life isn't as easy if you need to wrap text over multiple pages, because you have no way of knowing where the TextBlock runs out of space on any given page. The only alternative is to print your documents one *word* at a time so that you can determine the exact amount of text that fits on a page. Sadly, this technique is slow and tedious. It shows the limits of Silverlight's simplified printing model.

Creating a Print Preview

Although Silverlight doesn't have a built-in print preview, you can build one of your own fairly easily. The basic technique is to write your root element to a WriteableBitmap (as demonstrated earlier this chapter). You can then show that WriteableBitmap on-screen—for example, in a ScrollViewer inside a pop-up ChildWindow. Figure 9-22 shows an example.

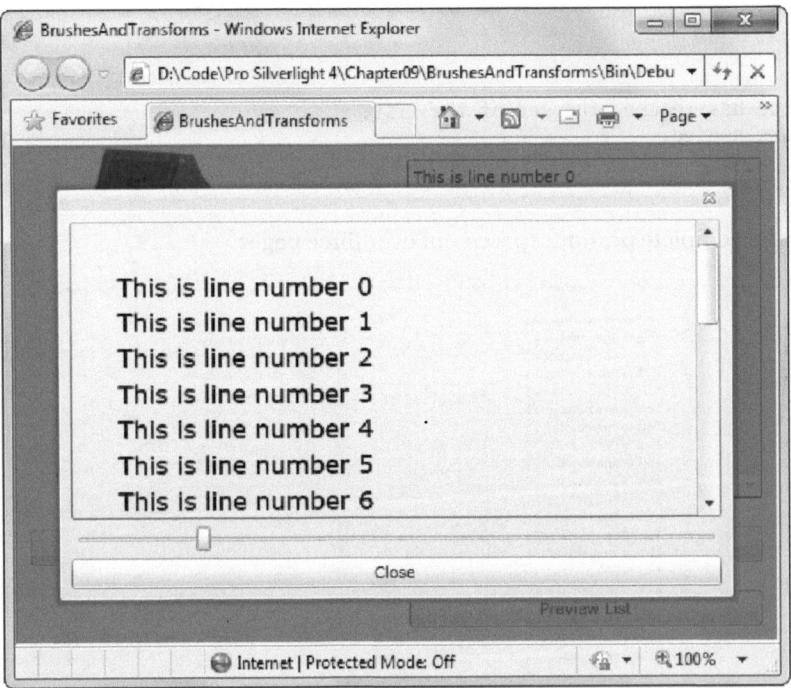

***Figure 9-22.** A simple print preview*

The first step to create an example like this is to refactor your printing code so that you can call it as part of a print operation *or* to create a visual to use in a preview.

Consider the previous example, which did the bulk of its work in an event handler named documentList_PrintPage(). To modify this example so it can more easily support a print preview feature, you first separate the code that generates the visuals into a separate method:

```
Private Function GeneratePage(ByVal printSurface As Canvas, _
  ByVal pageMargins As Thickness, ByVal pageSize As Size) As Boolean
    ...
End Function
```

Here, the GeneratePage() method accepts the three pieces of information it needs as arguments: the Canvas that acts as the root visual for the page, a Thickness object that represents the page margins, and a Size object that represents the page size. The code inside doesn't need to directly interface with the printing system (through the PrintPageEventArgs object). Instead, it simply adds elements to the Canvas until the page is full. It then returns True if a new page is needed or False if the printout is complete. (To see the full code, refer to the downloadable examples for this chapter. You'll find that the code in the GeneratePage() method is virtually identical to the original example—it's simply been transplanted from the documentList_PrintPage() event handler.)

And here's the revised documentList_PrintPage() event handler that uses the GeneratePage() method:

```
Private Sub documentList_PrintPage(ByVal sender As Object, _
  ByVal e As PrintPageEventArgs)
    ' Create the printing surface.
    Dim printSurface As New Canvas()
    e.PageVisual = printSurface

    ' File the page and determine if a new page is needed.
    e.HasMorePages = GeneratePage(printSurface, e.PageMargins, e.PrintableArea)
End Sub
```

With this rearrangement, it's easy enough to add a print preview feature. Here's one that creates a new Canvas to represent the printed page and then uses that Canvas to create a WriteableBitmap. Finally, the WriteableBitmap is passed to a custom ChildWindow class (named PrintPreview), like the ones you learned to create in Chapter 7.

```
Private Sub cmdPreviewPrintList_Click(ByVal sender As Object, _
  ByVal e As RoutedEventArgs)
    listPrintIndex = 0
    Dim printSurface As New Canvas()

    ' The page information isn't available without starting a real printout,
    ' so we hard-code a typical page size.
    Dim width As Integer = 816
    Dim height As Integer = 1056
    printSurface.Width = width
    printSurface.Height = height
    GeneratePage(printSurface, New Thickness(0), New Size(width, height))

    ' Wrap the Canvas in a WriteableBitmap.
    Dim printPreviewBitmap As New WriteableBitmap(printSurface, Nothing)

    ' Pass the bitmap to the PrintPreview window.
    Dim preview As New PrintPreview(printPreviewBitmap)
    preview.Show()
End Sub
```

The PrintPreview window contains an Image wrapped in a ScrollViewer:

```
<ScrollViewer x:Name="scrollContainer">
  <Image x:Name="imgPreview" VerticalAlignment="Top"
  HorizontalAlignment="Left"></Image>
</ScrollViewer>
```

When it first loads, it puts the WriteableBitmap into the Image element. It also stores the original dimensions of the bitmap to facilitate zooming with the sider.

```
Private originalSize As New Point()

Public Sub New(ByVal printPreviewBitmap As WriteableBitmap)
    InitializeComponent()

    imgPreview.Source = printPreviewBitmap
    imgPreview.Height = printPreviewBitmap.PixelHeight
    imgPreview.Width = printPreviewBitmap.PixelWidth

    originalSize.X = imgPreview.Width
    originalSize.Y = imgPreview.Height
End Sub
```

When the slider is changed, the size of the Image is adjusted accordingly:

```
Private Sub sliderZoom_ValueChanged(ByVal sender As Object, _
  ByVal e As RoutedPropertyChangedEventArgs(Of Double))
    imgPreview.Height = originalSize.Y * sliderZoom.Value
    imgPreview.Width = originalSize.X * sliderZoom.Value
End Sub
```

As written, this code generates only the first page of the printout. However, you could develop a more advanced example that allows the user to step from one page to another in the preview.

The Last Word

In this chapter, you delved deeper into Silverlight's 2-D drawing model. This is important, because understanding the plumbing behind 2-D graphics makes it far easier for you to manipulate them.

For example, you can alter a standard 2-D graphic by modifying the brushes used to paint various shapes, applying transforms and 3-D projections, altering the opacity, and using pixel shader effects. For still more impressive results, you can combine these techniques with Silverlight's animation features. For example, it's easy to rotate a Geometry object by modifying the Angle property of a RotateTransform object, fade a layer of shapes into existence using DrawingGroup.Opacity, or create a swirling dissolve effect by animating a custom pixel shader. You'll see examples of techniques like these in the next chapter.

At the same time, you also learned about lower-level approaches to generating and modifying graphics. You saw how you could change images on the pixel level with pixel shaders and the WriteableBitmap. Finally, you considered how Silverlight repurposes its bitmap rendering engine to provide a modest printing feature.

CHAPTER 10

■ ■ ■

Animation

Animation allows you to create truly *dynamic* user interfaces. It's often used to apply effects—for example, icons that grow when you move over them, logos that spin, text that scrolls into view, and so on. Sometimes, these effects seem like excessive glitz. But used properly, animations can enhance an application in a number of ways. They can make an application seem more responsive, natural, and intuitive. (For example, a button that slides in when you click it feels like a real, physical button—not just another gray rectangle.) Animations can also draw attention to important elements and guide the user through transitions to new content. (For example, an application could advertise new content with a twinkling, blinking, or pulsing icon.)

Animations are a core part of the Silverlight model. That means you don't need to use timers and event-handling code to put them into action. Instead, you can create and configure them declaratively, using XAML markup. Animations also integrate themselves seamlessly into ordinary Silverlight pages. For example, if you animate a button so it drifts around the page, the button still behaves like a button. It can be styled, it can receive focus, and it can be clicked to fire off the typical event-handling code.

In this chapter, you'll consider the set of animation classes that Silverlight provides. You'll see how to construct them with XAML and (more commonly) how to control them with code. Along the way, you'll see a wide range of animation examples, including page transitions and a simple catch-the-bombs game.

Understanding Silverlight Animation

Often, an animation is thought of as a series of frames. To perform the animation, these frames are shown one after the other, like a stop-motion video.

Silverlight animations use a dramatically different model. Essentially, a Silverlight animation is a way to modify the value of a dependency property over an interval of time. For example, to make a button that grows and shrinks, you can modify its Width property in an animation. To make it shimmer, you can change the properties of the LinearGradientBrush that it uses for its background. The secret to creating the right animation is determining what properties you need to modify.

If you want to make other changes that can't be made by modifying a property, you're out of luck. For example, you can't add or remove elements as part of an animation. Similarly, you can't ask Silverlight to perform a transition between a starting scene and an ending scene (although some crafty workarounds can simulate this effect). And finally, you can use animation only with a dependency property, because only dependency properties use the dynamic value-resolution system (described in Chapter 4) that takes animations into account.

■ **Note** Silverlight animation is a scaled-down version of the WPF animation system. It keeps the same conceptual framework, the same model for defining animations with animation classes, and the same storyboard system. However, WPF developers will find some key differences, particularly in the way animations are created and started in code. (For example, Silverlight elements lack the built-in BeginAnimation() method that they have in WPF.)

GOING BEYOND SILVERLIGHT ANIMATION

At first glance, the property-focused nature of Silverlight animations seems terribly limiting. But as you work with Silverlight, you'll find that it's surprisingly capable. You can create a wide range of animated effects using common properties that every element supports. In this chapter, you'll even see how you can use it to build a simple game.

That said, in some cases the property-based animation system won't suit. As a rule of thumb, the property-based animation is a great way to add dynamic effects to an otherwise ordinary application (such as buttons that glow, pictures that expand when you move over them, and so on). However, if you need to use animations as part of the core purpose of your application and you want them to continue running over the lifetime of your application, you may need something more flexible and more powerful. For example, if you're creating a complex arcade game or using physics calculations to model collisions, you'll need greater control over the animation.

Later in this chapter, you'll learn how to take a completely different approach with frame-based animations. In a frame-based animation, your code runs several times a second, and each time it runs, you have a chance to modify the content of your window. For more information, see the section "Frame-Based Animation."

The Rules of Animation

To understand Silverlight animation, you need to be aware of the following key rules:

- *Silverlight animations are time-based*: You set the initial state, the final state, and the duration of your animation. Silverlight calculates the frame rate.
- *Animations act on properties*: A Silverlight animation can do only one thing: modify the value of a property over an interval of time. This sounds like a significant limitation (and it many ways, it is), but you can create a surprisingly large range of effects by modifying properties.
- *Every data type requires a different animation class*: For example, the Button.Width property uses the double data type. To animate it, you use the DoubleAnimation class. If you want to modify the color that's used to paint the background of a Canvas, you need to use the ColorAnimation class.

Silverlight has relatively few animation classes, so you're limited in the data types you can use. Currently, you can use animations to modify properties with the following data types: Double, Object, Color, and Point. However, you can also craft your own animation classes that work for different data types—all you need to do is derive from System.Windows.Media.Animation and indicate how the value should change as time passes.

Many data types don't have a corresponding animation class because it wouldn't be practical. A prime example is enumerations. For example, you can control how an element is placed in a layout panel using the HorizontalAlignment property, which takes a value from the HorizontalAlignment enumeration. But the HorizontalAlignment enumeration allows you to choose among only four values (Left, Right, Center, and Stretch), which greatly limits its use in an animation. Although you can swap between one orientation and another, you can't smoothly transition an element from one alignment to another. For that reason, there's no animation class for the HorizontalAlignment data type. You can build one yourself, but you're still constrained by the four values of the enumeration.

Reference types aren't usually animated. However, their subproperties are. For example, all content controls sport a Background property that lets you set a Brush object that's used to paint the background. It's rarely efficient to use animation to switch from one brush to another, but you can use animation to vary the properties of a brush. For example, you can vary the Color property of a SolidColorBrush (using the ColorAnimation class) or the Offset property of a GradientStop in a LinearGradientBrush (using the DoubleAnimation class). Doing so extends the reach of Silverlight animation, allowing you to animate specific aspects of an element's appearance.

■ **Tip** As you'll see, DoubleAnimation is by far the most useful of Silverlight's animation classes. Most of the properties you'll want to change are doubles, including the position of an element on a Canvas, its size, its opacity, and the properties of the transforms it uses.

Creating Simple Animations

Creating an animation is a multistep process. You need to create three separate ingredients: an animation object to perform your animation, a storyboard to manage your animation, and an event handler (an event trigger) to start your storyboard. In the following sections, you'll tackle each of these steps.

The Animation Class

Silverlight includes two types of animation classes. Each type of animation uses a different strategy for varying a property value:

- *Linear interpolation*: The property value varies smoothly and continuously over the duration of the animation. (You can use animation easing to create more complex patterns of movement that incorporate acceleration and deceleration, as described later in this chapter.) Silverlight includes three such classes: DoubleAnimation, PointAnimation, and ColorAnimation.

- *Key-frame animation*: Values can jump abruptly from one value to another, or they can combine jumps and periods of linear interpolation (with or without animation easing).

Silverlight includes four such classes: ColorAnimationUsingKeyFrames, DoubleAnimationUsingKeyFrames, PointAnimationUsingKeyFrames, and ObjectAnimationUsingKeyFrames.

In this chapter, you'll begin by focusing on the indispensable DoubleAnimation class, which uses linear interpolation to change a double from a starting value to its ending value.

Animations are defined using XAML markup. Although the animation classes aren't elements, they can be created with the same XAML syntax. For example, here's the markup required to create a DoubleAnimation:

```
<DoubleAnimation From="160" To="300" Duration="0:0:5"></DoubleAnimation>
```

This animation lasts five seconds (as indicated by the Duration property, which takes a time value in the format *Hours:Minutes:Seconds.FractionalSeconds*). While the animation is running, it changes the target value from 160 to 300. Because the DoubleAnimation uses linear interpolation, this change takes place smoothly and continuously.

There's one important detail that's missing from this markup. The animation indicates how the property will be changed, but it doesn't indicate *what* property to use. This detail is supplied by another ingredient, which is represented by the Storyboard class.

The Storyboard Class

The storyboard manages the timeline of your animation. You can use a storyboard to group multiple animations, and it also has the ability to control the playback of animation—pausing it, stopping it, and changing its position. But the most basic feature provided by the Storyboard class is its ability to point to a specific property and specific element using the TargetProperty and TargetName properties. In other words, the storyboard bridges the gap between your animation and the property you want to animate.

Here's how you can define a storyboard that applies a DoubleAnimation to the Width property of a button named cmdGrow:

```
<Storyboard x:Name="storyboard"
  Storyboard.TargetName="cmdGrow" Storyboard.TargetProperty="Width">
  <DoubleAnimation From="160" To="300" Duration="0:0:5"></DoubleAnimation>
</Storyboard>
```

The Storyboard.TargetProperty property identifies the property you want to change. (In this example, it's Width.) If you don't supply a class name, the storyboard uses the parent element. If you want to set an attached property (for example, Canvas.Left or Canvas.Top), you need to wrap the entire property in brackets, like this:

```
<Storyboard x:Name="storyboard"
  Storyboard.TargetName="cmdGrow" Storyboard.TargetProperty="(Canvas.Left)">
  ...
</Storyboard>
```

Both TargetName and TargetProperty are attached properties. That means you can apply them directly to the animation, as shown here:

```
<Storyboard x:Name="storyboard">
  <DoubleAnimation
  Storyboard.TargetName="cmdGrow" Storyboard.TargetProperty="Width"
  From="160" To="300" Duration="0:0:5"></DoubleAnimation>
</Storyboard>
```

This syntax is more common, because it allows you to put several animations in the same storyboard but set each animation to act on a different element and property. Although you can't animate the same property at the same time with multiple animations, you can (and often will) animate different properties of the same element at once.

Starting an Animation with an Event Trigger

Defining a storyboard and an animation are the first steps to creating an animation. To actually put this storyboard into action, you need an event trigger. An *event trigger* responds to an event by performing a storyboard action. The only storyboard action that Silverlight currently supports is BeginStoryboard, which starts a storyboard (and hence all the animations it contains).

The following example uses the Triggers collection of a page to attach an animation to the Loaded event. When the Silverlight content is first rendered in the browser and the page element is loaded, the button begins to grow. Five seconds later, its width has stretched from 160 pixels to 300.

```
<UserControl ... >
  <UserControl.Triggers>
    <EventTrigger>
      <EventTrigger.Actions>
        <BeginStoryboard>
          <Storyboard>
            <DoubleAnimation Storyboard.TargetName="cmdGrow"
              Storyboard.TargetProperty="Width"
              From="160" To="300" Duration="0:0:5"></DoubleAnimation>
          </Storyboard>
        </BeginStoryboard>
      </EventTrigger.Actions>
    </EventTrigger>
  </UserControl.Triggers>

  <Grid x:Name="LayoutRoot" Background="White">
    <Button x:Name="cmdGrow" Width="160" Height="30"
     Content="This button grows"></Button>
  </Grid>
</UserControl>
```

Unfortunately, Silverlight event triggers are dramatically limited—much more so than their WPF counterparts. Currently, Silverlight only allows event triggers to respond to the Loaded event when your page is first created. They can't react to other events, like clicks, keypresses, and mouse movements. For those, you need the code described in the next section.

Starting an Animation with Code

You can start a Silverlight animation in response to any event using code that interacts with the storyboard. The first step is to move your storyboard out of the Triggers collection and place it in another collection of the same element: the Resources collection.

As you learned in Chapter 1, Silverlight elements provide a Resources property, which holds a collection where you can store miscellaneous objects. The primary purpose of the Resources collection is to let you define objects in XAML that aren't elements and so can't be placed into the visual layout of your content region. For example, you may want to declare a Brush object as a resource so it can be used by more than one element. You can retrieve resources in your code or use them elsewhere in your markup.

Here's an example that defines the button-growing animation as a resource:

```
<UserControl ... >
  <UserControl.Resources>
    <Storyboard x:Name="storyboard">
      <DoubleAnimation
        Storyboard.TargetName="cmdGrow" Storyboard.TargetProperty="Width"
        From="160" To="300" Duration="0:0:5"></DoubleAnimation>
    </Storyboard>
  </UserControl.Resources>

  <Grid x:Name="LayoutRoot" Background="White">
    <Button x:Name="cmdGrow" Width="160" Height="30" Click="cmdGrow_Click"
      Content="This button grows"></Button>
  </Grid>
</UserControl>
```

Notice that the storyboard is now given a name, so you can manipulate it in your code. (You can also add a name to the DoubleAnimation if you want to tweak its properties programmatically before launching the animation.)

Now, you need to call the methods of the Storyboard object in an event handler in your Silverlight code-behind file. The methods you can use include Begin(), Stop(), Pause(), Resume(), and Seek(), all of which are fairly self-explanatory.

```
Private Sub cmdGrow_Click(ByVal sender As Object, ByVal e As RoutedEventArgs)
    storyboard.Begin()
End Sub
```

Clicking the button launches the animation, and the button stretches from 160 to 300 pixels, as shown in Figure 10-1.

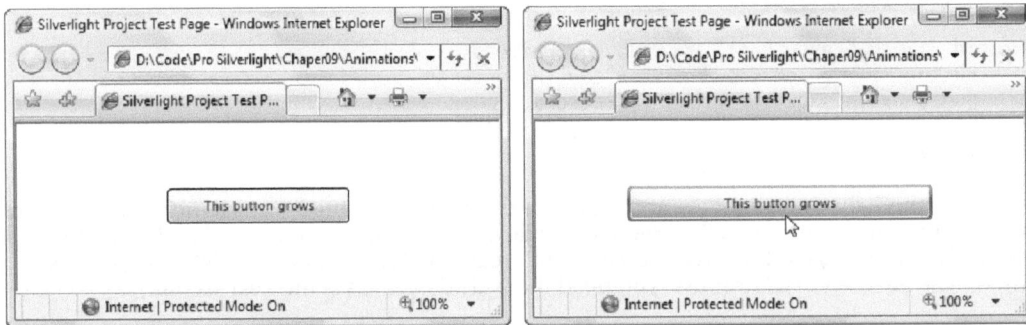

Figure 10-1. *Animating a button's width*

Configuring Animation Properties

To get the most out of your animations, you need to take a closer look at the seemingly simple animation class properties that were set in the previous example, including From, To, and Duration. As you'll see, there's a bit more subtlety—and a few more possibilities—than you may initially expect.

From

The From value is the starting value. In the previous example, the animation starts at 160 pixels. Thus, each time you click the button and start the animation, the Width property is reset to 160, and the animation runs again. This is true even if you click the button while an animation is underway.

■ **Note** This example exposes another detail about Silverlight animations: every dependency property can be acted on by only one animation at a time. If you start a second animation, the first one is discarded.

In many situations, you don't want an animation to begin at the original From value. There are two common reasons:

- *You have an animation that can be triggered multiple times in a row for a cumulative effect*: For example, you may want to create a button that grows a bit more each time it's clicked.

- *You have animations that can overlap*: For example, you may use the MouseEnter event to trigger an animation that expands a button and the MouseLeave event to trigger a complementary animation that shrinks it back. (This is often known as a *fish-eye* effect.) If you move the mouse over and off this sort of button several times in quick succession, each new animation interrupts the previous one, causing the button to jump back to the size that's set by the From property.

If you leave out the From value in the button-growing example, you can click the button multiple times without resetting its progress. Each time, a new animation starts, but it continues from the current width. When the button reaches its maximum width, further clicks have no effect, unless you add another animation to shrink it back.

```
<DoubleAnimation Storyboard.TargetName="cmdGrow"
 Storyboard.TargetProperty="Width" To="300" Duration="0:0:5"></DoubleAnimation>
```

There's one catch. For this technique to work, the property you're animating must have a previously set value. In this example, that means the button must have a hard-coded width (whether it's defined directly in the button tag or applied through a style setter). The problem is that in many layout containers, it's common not to specify a width and to allow the container to control the width based on the element's alignment properties. In this case, the default width applies, which is the special value Double.NaN (where NaN stands for "not a number"). You can't use linear interpolation to animate a property that has this value.

What's the solution? In many cases, the answer is to hard-code the button's width. As you'll see, animations often require more fine-grained control of element sizing and positioning than you'd otherwise use. The most common layout container for animatable content is the Canvas, because it makes it easy to move content around (with possible overlap) and resize it. The Canvas is also the most lightweight layout container, because no extra layout work is needed when you change a property like Width.

In the current example, you have another option. You can retrieve the current value of the button using its ActualWidth property, which indicates the current rendered width. You can't animate ActualWidth (it's read-only), but you can use it to set the From property of your animation programmatically, before you start the animation.

You need to be aware of another issue when you use the current value as a starting point for an animation: doing so may change the speed of your animation. That's because the duration isn't adjusted to take into account the smaller spread between the initial value and the final value. For example, imagine you create a button that doesn't use the From value and instead animates from its current position. If you click the button when it has almost reached its maximum width, a new animation begins. This animation is configured to take five seconds (through the Duration property), even though there are only a few more pixels to go. As a result, the growth of the button seems to slow down.

This effect appears only when you restart an animation that's almost complete. Although it's a bit odd, most developers don't bother trying to code around it. Instead, it's considered an acceptable quirk.

To

Just as you can omit the From property, you can omit the To property. You can leave out both the From and To properties to create an animation like this:

```
<DoubleAnimation Storyboard.TargetName="cmdGrow"
 Storyboard.TargetProperty="Width" Duration="0:0:5"></DoubleAnimation>
```

At first glance, this animation seems like a long-winded way to do nothing at all. It's logical to assume that because both the To and From properties are omitted, they both use the same value. But there's a subtle and important difference.

When you leave out From, the animation uses the current value and takes animation into account. For example, if the button is midway through a grow operation, the From value uses the expanded width. However, when you omit To, the animation uses the current value *without taking animation into account*. Essentially, that means the To value becomes the *original* value—whatever you last set in code, on the element tag, or through a style. (This works thanks to Silverlight's property-resolution system, which is able to calculate a value for a property based on several overlapping property providers without discarding any information. Chapter 4 describes this system in more detail.)

In the button example, if you start a grow animation and then interrupt it with the animation shown previously (perhaps by clicking another button), the button shrinks from its partially expanded size until it reaches the original width set in the XAML markup. On the other hand, if you run this code while no other animation is underway, nothing happens. That's because the From value (the animated width) and the To value (the original width) are the same.

By

Instead of using To, you can use the By property. The By property is used to create an animation that changes a value *by* a set amount, rather than *to* a specific target. For example, you can create an animation that enlarges a button by 10 pixels more than its current size, as shown here:

```
<DoubleAnimation Storyboard.TargetName="cmdGrow" By="10"
 Storyboard.TargetProperty="Width" Duration="0:0:5"></DoubleAnimation>
```

Clicking this button always enlarges the button, no matter how many times you've run the animation and how large the button has already grown.

The By property isn't offered with all animation classes. For example, it doesn't make sense with non-numeric data types, such as a Color structure (as used by ColorAnimation).

Duration

The Duration property is straightforward—it takes the time interval (in milliseconds, minutes, hours, or whatever else you'd like to use) between the time the animation starts and the time it ends. The Duration property requires a Duration object, which is similar to a TimeSpan. In fact, the Duration structure defines an implicit cast that can convert a System.TimeSpan object to System.Windows.Duration object as needed. That's why code like this is reasonable:

```
widthAnimation.Duration = TimeSpan.FromSeconds(5)
```

So, why did Microsoft bother to introduce a whole new Duration type, rather than just use the standard TimeSpan? The Duration type can be set to two special values that can't be represented by a TimeSpan object: Duration.Automatic and Duration.Forever. Neither of these values is useful in the current example. Automatic sets the animation to a 1-second duration, and Forever makes the animation infinite in length, which prevents it from having any effect.

But Duration.Forever becomes useful if you're creating a reversible animation. To do so, set the AutoReverse property to True. Now, the animation will play out in reverse once it's complete, reverting to the original value (and doubling the time the animation takes). Because a reversible animation returns to its initial state, Duration.Forever makes sense—it forces the animation to repeat endlessly.

Animation Lifetime

Technically, Silverlight animations are *temporary*, which means they don't change the value of the underlying property. While an animation is active, it overrides the property value. This is because of the way that dependency properties work (as described in Chapter 4), and it's an often-overlooked detail that can cause significant confusion.

A one-way animation (like the button-growing animation) remains active after it finishes running. That's because the animation needs to hold the button's width at the new size. This can lead to an unusual problem: if you try to modify the value of the property using code after the animation has completed, your code will appear to have no effect. Your code assigns a new local value to the property, but the animated value still takes precedence.

You can solve this problem in several ways, depending on what you're trying to accomplish:

- *Create an animation that resets your element to its original state*: You do this by not setting the To property. For example, the button-shrinking animation reduces the width of the button to its last set size, after which you can change it in your code.

- *Create a reversible animation*: You do this by setting the AutoReverse property to True. For example, when the button-growing animation finishes widening the button, it will play out the animation in reverse, returning it to its original width. The total duration of your animation is doubled.

- *Change the FillBehavior property*: Ordinarily, FillBehavior is set to HoldEnd, which means that when an animation ends, it continues to apply its final value to the target property. If you change FillBehavior to Stop, then as soon as the animation ends, the property reverts to its original value.

- *Remove the animation object when the animation ends*: To do so, handle the Completed event of the animation object or the containing storyboard.

The first three options change the behavior of your animation. One way or another, they return the animated property to its original value. If this isn't what you want, you need to use the last option.

First, before you launch the animation, attach an event handler that reacts when the animation finishes. You can do this when the page first loads:

```
AddHandler storyboard.Completed, AddressOf storyboard_Completed
```

When the Completed event fires, you can retrieve the storyboard that controls the animation and stop it:

```
Private Sub storyboard_Completed(ByVal sender As Object, ByVal e As EventArgs)
    Dim storyboard As Storyboard = CType(sender, Storyboard)
    storyboard.Stop()
End Sub
```

When you call Storyboard.Stop(), the property returns to the value it had before the animation started. If this isn't what you want, you can take note of the current value that's being applied by the animation, remove the animation, and then manually set the new property:

```
Dim currentWidth As Double = cmdGrow.Width
storyboard.Stop()
cmdGrow.Width = currentWidth
```

Keep in mind that this changes the local value of the property. That may affect how other animations work. For example, if you animate this button with an animation that doesn't specify the From property, it uses this newly applied value as a starting point. In most cases, this is the behavior you want.

RepeatBehavior

The RepeatBehavior property allows you to control how an animation is repeated. If you want to repeat it a fixed number of times, indicate the number of times to repeat, followed by an *x*. For example, this animation repeats twice:

```
<DoubleAnimation Storyboard.TargetName="cmdGrow" RepeatBehavior="2x"
 Storyboard.TargetProperty="Width" To="300" Duration="0:0:5"></DoubleAnimation>
```

Or in code, pass the number of times to the RepeatBehavior constructor:

```
widthAnimation.RepeatBehavior = New RepeatBehavior(2)
```

When you run this animation, the button increases in size (over five seconds), jumps back to its original value, and then increases in size again (over five seconds), ending at the full width of the page. If you've set AutoReverse to True, the behavior is slightly different: the entire animation is completed forward and backward (meaning the button expands and then shrinks), and *then* it's repeated again.

Rather than using RepeatBehavior to set a repeat count, you can use it to set a repeat *interval*. To do so, set the RepeatBehavior property with a time value instead of a single number. For example, the following animation repeats itself for 13 seconds:

```
<DoubleAnimation Storyboard.TargetName="cmdGrow" RepeatBehavior="0:0:13"
 Storyboard.TargetProperty="Width" To="300" Duration="0:0:5"></DoubleAnimation>
```

And here's the same change made in code:

```
widthAnimation.RepeatBehavior = New RepeatBehavior(TimeSpan.FromSeconds(13))
```

In this example, the Duration property specifies that the entire animation takes 5 seconds. As a result, the RepeatBehavior of 13 seconds triggers two repeats and then leaves the button halfway through a third repeat (at the 3-second mark).

■ **Tip** You can use RepeatBehavior to perform just part of an animation. To do so, use a fractional number of repetitions, or use a TimeSpan that's less than the duration.

Finally, you can cause an animation to repeat itself endlessly with the RepeatBehavior.Forever value:

```
<DoubleAnimation Storyboard.TargetName="cmdGrow" RepeatBehavior="Forever"
  Storyboard.TargetProperty="Width" To="300" Duration="0:0:5"></DoubleAnimation>
```

Simultaneous Animations

The Storyboard class has the ability to hold more than one animation. Best of all, these animations are managed as one group—meaning they're started at the same time.

To see an example, consider the following storyboard. It wraps two animations, one that acts on a button's Width property and another that acts on the Height property. Because the animations are grouped into one storyboard, they increment the button's dimensions in unison:

```
<Storyboard x:Name="storyboard" Storyboard.TargetName="cmdGrow">
  <DoubleAnimation Storyboard.TargetProperty="Width"
    To="300" Duration="0:0:5"></DoubleAnimation>
  <DoubleAnimation Storyboard.TargetProperty="Height"
    To="300" Duration="0:0:5"></DoubleAnimation>
</Storyboard>
```

This example moves Storyboard.TargetName property from the DoubleAnimation to the Storyboard. This is an optional change, but it saves you from setting the property twice, once on each animation object. (Obviously, if your animation objects need to act on different elements, you couldn't use this shortcut.)

In this example, both animations have the same duration, but this isn't a requirement. The only consideration with animations that end at different times is their FillBehavior. If an animation's FillBehavior property is set to HoldEnd (the default), it holds the value until all the animations in the storyboard are completed. At this point, the storyboard's FillBehavior comes into effect, either continuing to hold the values from both animations (HoldEnd) or reverting them to their initial values (Stop). On the other hand, if you have multiple animations and one of them has a FillBehavior of Stop, this animated property will revert to its initial value when the animation is complete, even if other animations in the storyboard are still running.

When you're dealing with more than one simultaneous animation, two more animation class properties become useful: BeginTime and SpeedRatio. BeginTime sets a delay that is added before the animation starts (as a TimeSpan). This delay is added to the total time, so a five-second animation with a five-second delay takes ten seconds. BeginTime is useful when you're synchronizing different animations that start at the same time but should apply their effects in sequence. SpeedRatio increases or decreases the speed of the animation. Ordinarily, SpeedRatio is 1. If you increase it, the animation completes more quickly (for example, a SpeedRatio of 5 completes five times faster). If you decrease it, the animation is slowed down (for example, a SpeedRatio of 0.5 takes twice as long). Although the overall effect is the same as changing the Duration property of your animation, setting the SpeedRatio makes it easier to control how simultaneous animations overlap.

Controlling Playback

You've already seen how to start an animation using the Storyboard.Begin() method. The Storyboard class also provides a few more methods that allow you to stop or pause an animation. You'll see them in action in the following example, shown in Figure 10-2. This page superimposes two Image elements in exactly the same position, using a grid. Initially, only the topmost image—which shows a day scene of a Toronto city landmark—is visible. But as the animation runs, it reduces the opacity from 1 to 0, eventually allowing the night scene to show through completely. The effect makes it seem that the image is changing from day to night, like a sequence of time-lapse photography.

Figure 10-2. A controllable animation

Here's the markup that defines the Grid with its two images:

```
<Grid>
  <Image Source="night.jpg"></Image>
  <Image Source="day.jpg" x:Name="imgDay"></Image>
</Grid>
```

And here's the storyboard that fades from one to the other, which is placed in the page's Resources collection:

```
<Storyboard x:Name="fadeStoryboard">
  <DoubleAnimation x:Name="fadeAnimation"
    Storyboard.TargetName="imgDay" Storyboard.TargetProperty="Opacity"
    From="1" To="0" Duration="0:0:10">
  </DoubleAnimation>
</Storyboard>
```

To make this example more interesting, it includes several buttons at the bottom that let you control the playback of this animation. Using these buttons, you can perform the typical media player actions, such as starting, pausing, resuming, and stopping, and seeking. The event-handling code uses the appropriate methods of the Storyboard object, as shown here:

```
Private Sub cmdStart_Click(ByVal sender As Object, ByVal e As RoutedEventArgs)
    fadeStoryboard.Begin()
End Sub

Private Sub cmdPause_Click(ByVal sender As Object, ByVal e As RoutedEventArgs)
    fadeStoryboard.Pause()
End Sub

Private Sub cmdResume_Click(ByVal sender As Object, ByVal e As RoutedEventArgs)
    fadeStoryboard.Resume()
End Sub

Private Sub cmdStop_Click(ByVal sender As Object, ByVal e As RoutedEventArgs)
    fadeStoryboard.Stop()
End Sub

Private Sub cmdMiddle_Click(ByVal sender As Object, ByVal e As RoutedEventArgs)
    ' Start the animation, in case it's not currently underway.
    fadeStoryboard.Begin()

    ' Move to the time position that represents the middle of the animation.
    fadeStoryboard.Seek( _
      TimeSpan.FromSeconds(fadeAnimation.Duration.TimeSpan.TotalSeconds/2))
End Sub
```

■ **Note** Remember, stopping an animation isn't equivalent to completing the animation (unless FillBehavior is set to Stop). That's because even when an animation reaches the end of its timeline, it continues to apply its final value. Similarly, when an animation is paused, it continues to apply the most recent intermediary value. However, when an animation is stopped, it no longer applies any value, and the property reverts to its preanimation value.

If you drag the thumb on the slider, the Slider.ValueChanged event fires and triggers another event handler. This event handler then takes the current value of the slider (which ranges from 0 to 3) and uses it to apply a new speed ratio:

```
Private Sub sldSpeed_ValueChanged(ByVal sender As Object, _
  ByVal e As RoutedEventArgs)
    ' To nothing if the page is still being initialized.
    If sldSpeed Is Nothing Then Return

    ' This also restarts the animation if it's currently underway.
    fadeStoryboard.SpeedRatio = sldSpeed.Value
    lblSpeed.Text = sldSpeed.Value.ToString("0.0")
End Sub
```

Unlike in WPF, the Storyboard class in Silverlight doesn't provide events that allow you to monitor the progress of an event. For example, there's no CurrentTimeInvalidated event to tell you the animation is ticking forward.

Animation Easing

One of the shortcomings of linear animation is that it often feels mechanical and unnatural. By comparison, sophisticated user interfaces have animated effects that model real-world systems. For example, they may use tactile push-buttons that jump back quickly when clicked but slow down as they come to rest, creating the illusion of true movement. Or, they may use maximize and minimize effects like Windows Vista, where the speed at which the window grows or shrinks accelerates as the window nears its final size. These details are subtle, and you're not likely to notice them when they're implemented well. However, you'll almost certainly notice the clumsy feeling of less refined animations that lack these finer points.

The secret to improving your animations and creating more natural animations is to vary the rate of change. Instead of creating animations that change properties at a fixed, unchanging rate, you need to design animations that speed up or slow down along the way. Silverlight gives you several good options.

For the most control, you can create a frame-based animation (as discussed later in the "Frame-Based Animation" section). This approach is useful if you must have absolute control over every detail, which is the case if your animation needs to run in a specific way (for example, an action game or a simulation that follows the rules of physics). The drawback is that frame-based animations take a lot of work, because the Silverlight animation model does very little to help you.

If your animations aren't quite as serious and you just want a way to make them look more professional, you can use a simpler approach. One option is a key-frame animation, which divides the animation into multiple segments and (optionally) uses key splines to add acceleration or deceleration to different segments. This approach works well (and you'll learn about it later in the "Key-Frame Animation" section). But it's tedious to implement and often requires a significant amount of XAML markup. It makes the most sense when you're using some sort of design tool that helps you create the key frames and key splines—for example, by drawing on a graph, as you can in Expression Blend.

If you don't have a design tool like Expression Blend or you don't want to go the trouble of clicking your way to a complex key-frame animation, you have one more choice: you can use a prebuilt animation-easing function. In this case, you can still define your animation normally by specifying the starting and ending property values. But in addition to these details, you add a ready-made mathematical function that alters the progression of your animation, causing it to accelerate or decelerate at different points. This is the technique you'll study in the following sections.

Using an Easing Function

The best part about animation easing is that it requires much less work than other approaches such as frame-based animation and key frames. To use animation easing, you set the EasingFunction property of an animation object with an instance of an easing function class (a class that derives from EasingFunctionBase). You'll usually need to set a few properties on the easing function, and you may be forced to play around with different settings to get the effect you want, but you'll need no code and very little additional XAML.

For example, consider the two animations shown here, which act on a button. When the user moves the mouse over the button, a small snippet of code calls the growStoryboard animation into action, stretching the button to 400 pixels. When the user moves the mouse off the button, the buttons shrinks back to its normal size.

```
<Storyboard x:Name="growStoryboard">
  <DoubleAnimation
    Storyboard.TargetName="cmdGrow" Storyboard.TargetProperty="Width"
    To="400" Duration="0:0:1.5"></DoubleAnimation>
</Storyboard>

<Storyboard x:Name="revertStoryboard">
  <DoubleAnimation
    Storyboard.TargetName="cmdGrow" Storyboard.TargetProperty="Width"
    Duration="0:0:3"></DoubleAnimation>
</Storyboard>
```

Right now, the animations use linear interpolation, which means the growing and shrinking happen in a steady, mechanical way. For a more natural effect, you can add an easing function. The following example adds an easing function named ElasticEase. The end result is that the button springs beyond its full size, snaps back to a value that's somewhat less, swings back over its full size again (but a little less than before), snaps back a bit less, and so on, repeating its bouncing pattern as the movement diminishes. It gradually comes to rest ten oscillations later. The Oscillations property controls the number of bounces at the end. The ElasticEase class provides one other property that's not used in this example: Springiness. This higher this value, the more each subsequent oscillation dies down (the default value is 3).

```
<Storyboard x:Name="growStoryboard">
  <DoubleAnimation
    Storyboard.TargetName="cmdGrow" Storyboard.TargetProperty="Width"
    To="400" Duration="0:0:1.5">
    <DoubleAnimation.EasingFunction>
      <ElasticEase EasingMode="EaseOut" Oscillations="10"></ElasticEase>
    </DoubleAnimation.EasingFunction>
  </DoubleAnimation>
</Storyboard>
```

To really appreciate the difference between this markup and the earlier example that didn't use an easing function, you need to try this animation (or run the companion examples for this chapter). It's a remarkable change. With one line of XAML, a simple animation changes from amateurish to a slick effect that would feel at home in a professional application.

■ **Note** Because the EasingFunction property accepts a single easing function object, you can't combine different easing functions for the same animation.

Easing In and Easing Out

Before you consider the different easing functions, it's important to understand *when* an easing function is applied. Every easing function class derives from EasingFunctionBase and inherits a single property named EasingMode. This property has three possible values: EaseIn (which means the effect is applied to the beginning of the animation), EaseOut (which means it's applied to the end), and EaseInOut (which means it's applied at both the beginning and end—the easing in takes place in the first half of the animation, and the easing out takes place in the second half).

In the previous example, the animation in the growStoryboard animation uses EaseOut mode. Thus, the sequence of gradually diminishing bounces takes place at the end of the animation. If you were to graph the changing button width as the animation progresses, you'd see something like the graph shown in Figure 10-3.

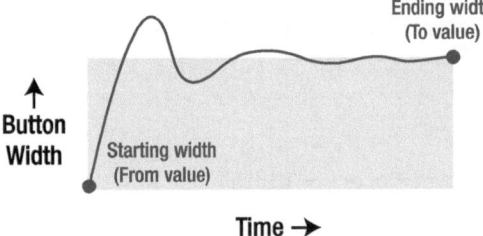

Figure 10-3. Oscillating to a stop using EaseOut with ElasticEase

■ **Note** The duration of an animation doesn't change when you apply an easing function. In the case of the growStoryboard animation, the ElasticEase function doesn't just change the way the animation ends—it also makes the initial portion of the animation (when the button expands normally) run more quickly so that there's more time left for the oscillations at the end.

If you switch the ElasticEase function to use EaseIn mode, the bounces happen at the beginning of the animation. The button shrinks below its starting value a bit, expands a bit over, shrinks back a little more, and continues this pattern of gradually increasing oscillations until it finally breaks free and expands the rest of the way. (You use the ElasticEase.Oscillations property to control the number of bounces.) Figure 10-4 shows this very different pattern of movement.

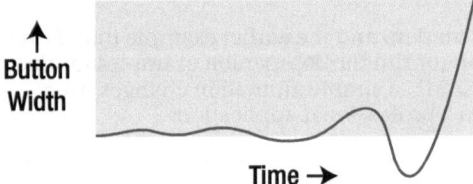

Figure 10-4. Oscillating to a start using EaseIn with ElasticEase

Finally, EaseInOut creates a stranger effect, with oscillations that start the animation in its first half followed by oscillations that stop it in the second half. Figure 10-5 illustrates.

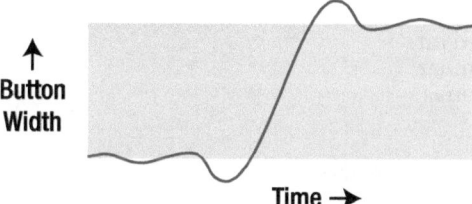

Figure 10-5. *Oscillating to a start and to a stop using EaseInOut with ElasticEase*

Easing Function Classes

Silverlight has 11 easing functions, all of which are found in the familiar System.Windows.Media.Animation namespace. Table 10-1 describes them all and lists their important properties. Remember, every animation also provides the EasingMode property, which allows you to control whether it affects that animation as it starts (EaseIn), ends (EaseOut), or both (EaseInOut).

Table 10-1. *Easing Functions*

Name	Description	Properties
BackEase	When applied with EaseIn, pulls the animation back before starting it. When applied with EaseOut, this function allows the animation to overshoot slightly and then pulls it back.	Amplitude determines the amount of pullback or overshoot. The default value is 1, and you can decrease it (to any value greater than 0) to reduce the effect or increase it to amplify the effect.
ElasticEase	When applied with EaseOut, makes the animation overshoot its maximum and swing back and forth, gradually slowing. When applied with EaseIn, the animation swings back and forth around its starting value, gradually increasing.	Oscillations controls the number of times the animation swings back and forth (the default is 3), and Springiness controls how quickly which the oscillations increase or diminish (the default is 3).
BounceEase	Performs an effect similar to ElasticEase, except the bounces never overshoot the initial or final values.	Bounces controls the number of times the animation bounces back (the default is 2), and Bounciness determines how quickly the bounces increase or diminish (the default is 2).
CircleEase	Accelerates (with EaseIn) or decelerates (with EaseOut) the animation using a circular function.	None.

Name	Description	Properties
CubicEase	Accelerates (with EaseIn) or decelerates (with EaseOut) the animation using a function based on the cube of time. The effect is similar to CircleEase, but the acceleration is more gradual.	None.
QuadraticEase	Accelerates (with EaseIn) or decelerates (with EaseOut) the animation using a function based on the square of time. The effect is similar to CubicEase, but even more gradual.	None.
QuarticEase	Accelerates (with EaseIn) or decelerates (with EaseOut) the animation using a function based on time to the power of 4. The effect is similar to CubicEase and QuadraticEase, but the acceleration is more pronounced.	None.
QuinticEase	Accelerates (with EaseIn) or decelerates (with EaseOut) the animation using a function based on time to the power of 5. The effect is similar to CubicEase, QuadraticEase, and QuinticEase, but the acceleration is more pronounced.	None.
SineEase	Accelerates (with EaseIn) or decelerates (with EaseOut) the animation using a function that includes a sine calculation. The acceleration is very gradual and closer to linear interpolation than any of the other easing functions.	None.
PowerEase	Accelerates (with EaseIn) or decelerates (with EaseOut) the animation using the power function $f(t) = t^p$. Depending on the value you use for the exponent p, you can duplicate the effect of the Cubic, QuadraticEase, QuarticEase, and QuinticEase functions.	Power, which sets the value of the exponent in the formula. Use 2 to duplicate QuadraticEase ($f(t) = t^2$), 3 for CubicEase ($f(t) = t^3$), 4 for QuarticEase ($f(t) = t^4$), and 5 for QuinticEase ($f(t) = t^5$), or choose something different. The default is 2.
ExponentialEase	Accelerates (with EaseIn) or decelerates (with EaseOut) the animation using the exponential function $f(t)=(e(at) - 1)/(e(a) - 1)$.	Exponent allows you to set the value of the exponent (2 is the default).

Many of the easing functions provide similar but subtly different results. To use animation easing successfully, you need to decide which easing function to use and how to configure it. Often, this process requires a bit of trial-and-error experimentation. Two good resources can help you out.

First, the Silverlight documentation charts example behavior for each easing function, showing how the animated value changes as time progresses. Reviewing these charts is a good way to develop a sense of what the easing function does. Figure 10-6 shows the charts for the most popular easing functions.

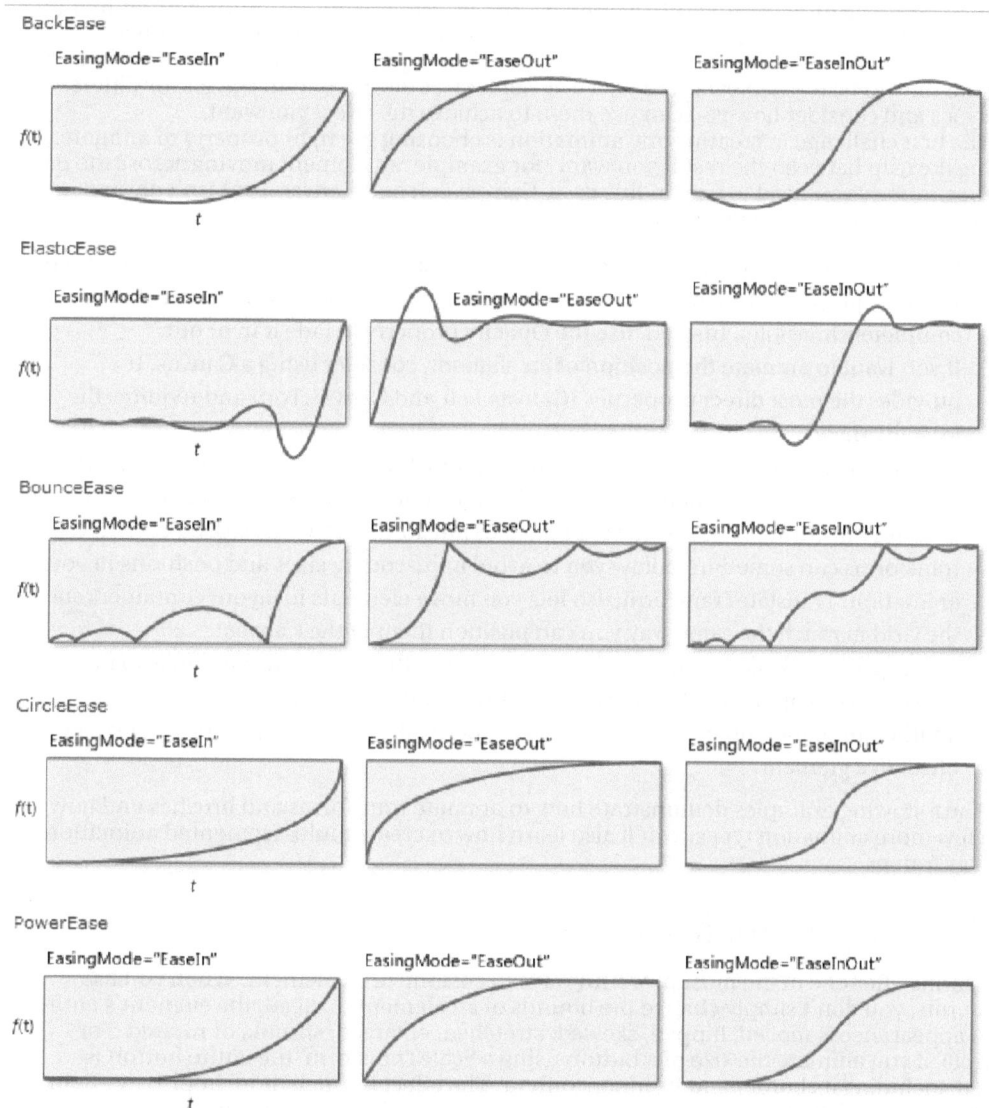

Figure 10-6. The effect of different easing functions

Second, Microsoft provides several sample applications that you can use to play with the different easing functions and try different property values. The most useful of these lets you observe the effect of any easing function on a falling square, complete with the automatically generated XAML markup needed to duplicate the effect. You can try it online at http://tinyurl.com/animationeasing.

Animation Types Revisited

You now know the fundamentals of Silverlight's property animation system—how animations are defined, how they're connected to elements, how you can control playback with a storyboard, and how you can incorporate animation easing to create more realistic effects. Now is a good time to take a step back and take a closer look at the animation classes for different data types and consider how you can use them to achieve the effect you want.

The first challenge in creating any animation is choosing the right property to animate. Making the leap between the result you want (for example, an element moving across the page) and the property you need to use (in this case, Canvas.Left and Canvas.Top) isn't always intuitive. Here are a few guidelines:

- If you want to use an animation to make an element appear or disappear, don't use the Visibility property (which allows you to switch only between completely visible or completely invisible). Instead, use the Opacity property to fade it in or out.

- If you want to animate the position of an element, consider using a Canvas. It provides the most direct properties (Canvas.Left and Canvas.Top) and requires the least overhead.

- The most common properties to animate are transforms, which you first explored in Chapter 9. You can use them to move or flip an element (TranslateTransform), rotate it (RotateTransform), resize or stretch it (ScaleTransform), and more. Used carefully, transforms can sometimes allow you to avoid hard-coding sizes and positions in your animation. TranslateTransform also lets you move elements in layout containers such as the Grid in much the same way you can position them in the Canvas.

- One good way to change the surface of an element through an animation is to modify the properties of the brush. You can use a ColorAnimation to change the color or another animation object to transform a property of a more complex brush, like the offset in a gradient.

The following examples demonstrate how to animate transforms and brushes and how to use a few more animation types. You'll also learn how to create multi-segmented animations with key frames.

Animating Transforms

Transforms offer one of the most powerful ways to customize an element. When you use transforms, you don't simply change the bounds of an element. Instead, the element's entire visual appearance is moved, flipped, skewed, stretched, enlarged, shrunk, or rotated. For example, if you animate the size of a button using a ScaleTransform, the entire button is resized, including its border and its inner content. The effect is much more impressive than if you animate its Width and Height or the FontSize property that affects its text.

To use a transform in animation, the first step is to define the transform. (An animation can change an existing transform but not create a new one.) For example, imagine you want to allow a button to rotate. This requires the RotateTransform:

```
<Button Content="A Button">
  <Button.RenderTransform>
    <RotateTransform x:Name="rotateTransform"></RotateTransform>
  </Button.RenderTransform>
</Button>
```

■ **Tip** You can use transforms in combination. It's easy—use a TransformGroup object to set the RenderTransform property. You can nest as many transforms as you need inside the transform group. You'll see an example in the bomb game that's shown later in this chapter.

Here's an animation that makes a button rotate when the mouse moves over it. It acts on the Button.RotateTransform object and uses the target property Angle. The fact that the RenderTransform property can hold a variety of different transform objects, each with different properties, doesn't cause a problem. As long as you're using a transform that has an Angle property, this animation will work.

```
<Storyboard x:Name="rotateStoryboard">
  <DoubleAnimation Storyboard.TargetName="rotateTransform"
   Storyboard.TargetProperty="Angle"
   To="360" Duration="0:0:0.8" RepeatBehavior="Forever"></DoubleAnimation>
</Storyboard>
```

If you place this animation in the page's Resources collection, you can trigger it when the user moves the mouse over the button:

```
Private Sub cmd_MouseEnter(ByVal sender As Object, ByVal e As MouseEventArgs)
    rotateStoryboard.Begin()
End Sub
```

The button rotates one revolution every 0.8 seconds and continues rotating perpetually. While the button rotates, it's completely usable—for example, you can click it and handle the Click event.

To make sure the button rotates around its center point (not the upper-left corner), you need to set the RenderTransformOrigin property as shown here:

```
<Button Content="One" Margin="5" RenderTransformOrigin="0.5,0.5"
 MouseEnter="cmd_MouseEnter">
  <Button.RenderTransform>
    <RotateTransform x:Name="rotateTransform"></RotateTransform>
  </Button.RenderTransform>
</Button>
```

Remember, the RenderTransformOrigin property uses relative units from 0 to 1, so 0.5 represents a midpoint.

To stop the rotation, you can react to the MouseLeave event. You could stop the storyboard that performs the rotation, but doing so would cause the button to jump back to its original orientation in one step. A better approach is to start a second animation that replaces the first. This animation leaves out the From property, which allows it to seamlessly rotate the button from its current angle to its original orientation in a snappy 0.2 seconds:

```
<Storyboard x:Name="unrotateStoryboard">
  <DoubleAnimation Storyboard.TargetName="rotateTransform"
    Storyboard.TargetProperty="Angle" To="0" Duration="0:0:0.2"></DoubleAnimation>
</Storyboard>
```

Here's the event handler:

```
Private Sub cmd_MouseLeave(ByVal sender As Object, ByVal e As MouseEventArgs)
    unrotateStoryboard.Begin()
End Sub
```

With a little more work, you can make these two animations and the two event handlers work for a whole stack of rotatable buttons, as shown in Figure 10-7. The trick is to handle the events of all the buttons with the same code, and dynamically assign the target of the storyboard to the current button using the Storyboard.SetTarget() method:

```
Private Sub cmd_MouseEnter(ByVal sender As Object, ByVal e As MouseEventArgs)
    rotateStoryboard.Stop()
    Storyboard.SetTarget(rotateStoryboard, (CType(sender, Button)).RenderTransform)
    rotateStoryboard.Begin()
End Sub

Private Sub cmd_MouseLeave(ByVal sender As Object, ByVal e As MouseEventArgs)
    unrotateStoryboard.Stop()
    Storyboard.SetTarget(unrotateStoryboard, _
      (CType(sender, Button)).RenderTransform)
    unrotateStoryboard.Begin()
End Sub
```

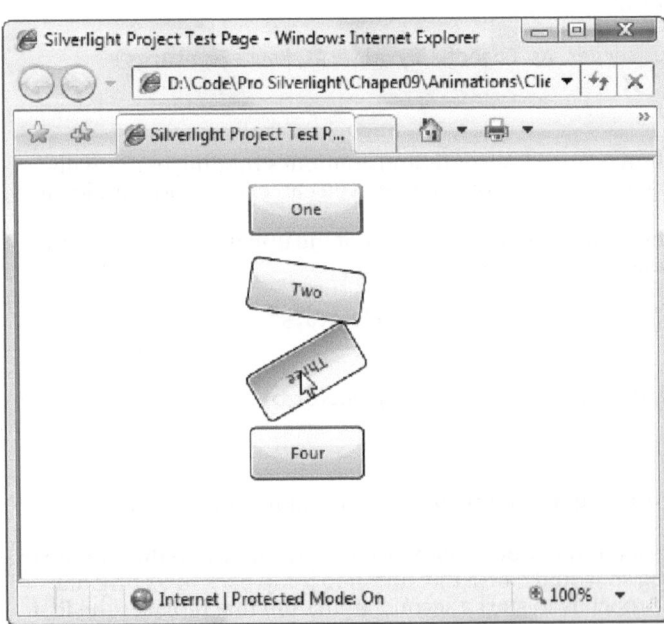

Figure 10-7. Animating a render transform

This approach has two limitations. First, because the code reuses the same storyboards for all the buttons, there's no way to have two buttons rotating at once. For example, if you quickly slide the mouse over several buttons, the buttons you leave first may not rotate all the way back to their initial position, because the storyboard is commandeered by another button. If this behavior is a problem, you can code around it by creating the storyboards you need dynamically in code. You'll see how to implement this technique later in this chapter, when you consider the bomb game.

The other shortcoming in this example is the fact that you need a fair bit of markup to define the margins, event handlers, and transforms for all the buttons. You can streamline this markup by using styles to apply the same settings to various buttons (see Chapter 12) or by configuring the buttons programmatically.

Animation Perspective Projections

Just as you can animate transforms, you can also animate perspective projections—namely, the PlaneProjection class you studied in Chapter 9, which allows you to simulate a flat, tilted 3-D surface. For example, imagine you have a group of elements wrapped in a Border control and that border uses a PlaneProjection, as shown here:

```
<Border CornerRadius="2" Padding="10" Height="140" Width="170"
 BorderBrush="SlateGray" BorderThickness="4">
  <Border.Projection>
    <PlaneProjection x:Name="projection"></PlaneProjection>
  </Border.Projection>
  ...
</Border>
```

Currently, the PlaneProjection in this example doesn't do anything. To change the way the elements are rendered, you need to modify the RotateX, RotateY, and RotateZ properties of the PlaneProjection object, which turns the 2-D surface of the border around the appropriate axis. You saw how to pull this off in Chapter 9, but now you'll use an animation to change these properties gradually and continuously.

Here's an animation that modifies all three rotation properties at different speeds, which gives the dizzying impression that the border is tumbling through 3-D space:

```
<Storyboard x:Name="spinStoryboard">
  <DoubleAnimation Storyboard.TargetName="projection" RepeatBehavior="Forever"
   Storyboard.TargetProperty="RotationY" From="0" To="360" Duration="0:0:3">
  </DoubleAnimation>
  <DoubleAnimation Storyboard.TargetName="projection" RepeatBehavior="Forever"
   Storyboard.TargetProperty="RotationZ" From="0" To="360" Duration="0:0:30">
  </DoubleAnimation>
  <DoubleAnimation Storyboard.TargetName="projection" RepeatBehavior="Forever"
   Storyboard.TargetProperty="RotationX" From="0" To="360" Duration="0:0:40">
  </DoubleAnimation>
</Storyboard>
```

Figure 10-8 shows the rotating border captured at two different points in its animation.

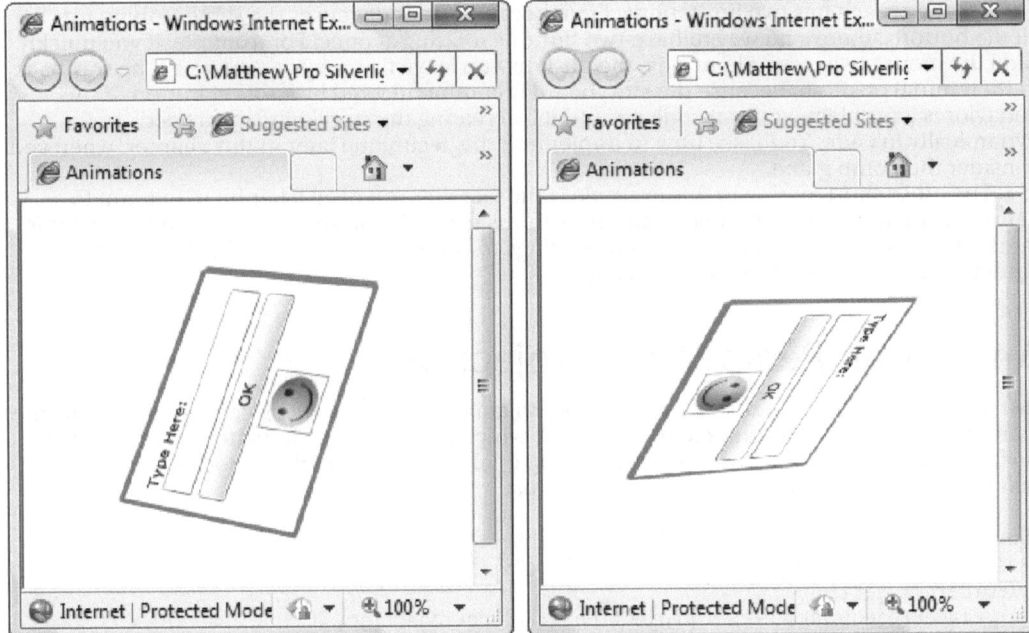

Figure 10-8. Spinning an element in 3-D

Although this technique may seem like gratuitous eye candy, a little 3-D rotation can go a long way. It's particularly useful when you're implementing transitions between different content. For example, you can create a panel that flips over and reveals different content on its back side. To do so, you take one panel and rotate it around the x- or y-axis from 0 to 90 degrees (at which point it appears to disappear because it's edge-on). You then continue with a second animation that rotates a different panel from -90 degrees to 0 degrees, exposing the new content.

Animating Brushes

Animating brushes is another common technique in Silverlight animations, and it's just as easy as animating transforms. Again, the technique is to dig into the particular subproperty you want to change, using the appropriate animation type.

Figure 10-9 shows an example that tweaks a RadialGradientBrush you studied in Chapter 8. As the animation runs, the center point of the radial gradient drifts along the ellipse, giving it a three-dimensional effect. At the same time, the outer color of the gradient changes from blue to black.

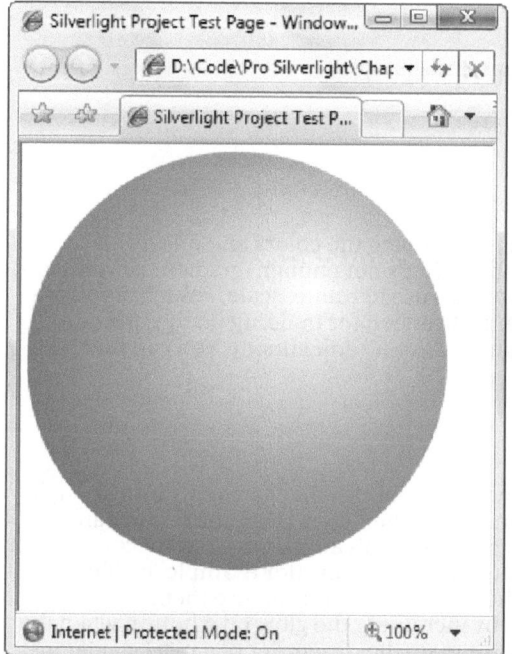 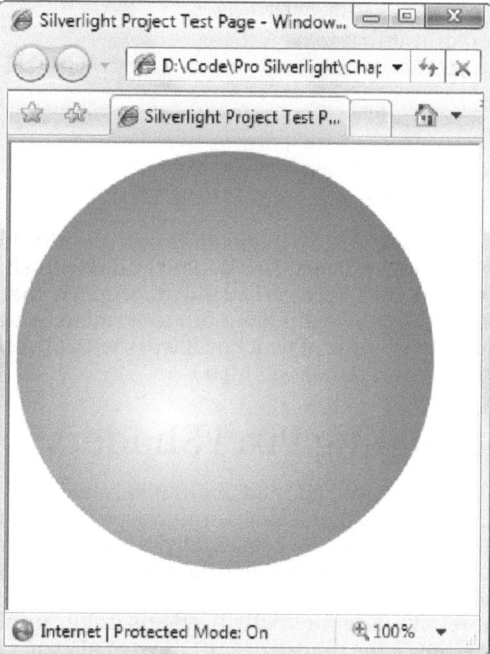

Figure 10-9. Altering a radial gradient

To perform this animation, you need to use two animation types that you haven't considered yet. ColorAnimation blends gradually between two colors, creating a subtle color-shift effect. PointAnimation allows you to move a point from one location to another. (It's essentially the same as if you modified both the x coordinate and the y coordinate using a separate DoubleAnimation, with linear interpolation.) You can use a PointAnimation to deform a figure that you've constructed out of points or to change the location of the radial gradient's center point, as in this example.

Here's the markup that defines the ellipse and its brush:

```
<Ellipse x:Name="ellipse" Margin="5" Grid.Row="1" Stretch="Uniform">
  <Ellipse.Fill>
    <RadialGradientBrush x:Name="ellipseBrush"
     RadiusX="1" RadiusY="1" GradientOrigin="0.7,0.3">
      <GradientStop x:Name="ellipseBrushStop" Color="White"
       Offset="0"></GradientStop>
      <GradientStop Color="Blue" Offset="1"></GradientStop>
    </RadialGradientBrush>
  </Ellipse.Fill>
</Ellipse>
```

And here are the two animations that move the center point and change the second color in the gradient:

```
<Storyboard x:Name="ellipseStoryboard">
  <PointAnimation Storyboard.TargetName="ellipseBrush"
   Storyboard.TargetProperty="GradientOrigin"
   From="0.7,0.3" To="0.3,0.7" Duration="0:0:10" AutoReverse="True"
```

```
      RepeatBehavior="Forever">
    </PointAnimation>
    <ColorAnimation Storyboard.TargetName="ellipseBrushStop"
      Storyboard.TargetProperty="Color"
      To="Black" Duration="0:0:10" AutoReverse="True"
      RepeatBehavior="Forever">
    </ColorAnimation>
  </Storyboard>
```

You can create a huge range of hypnotic effects by varying the colors and offsets in LinearGradientBrush and RadialGradientBrush. And if that's not enough, gradient brushes also have their own RelativeTransform property that you can use to rotate, scale, stretch, and skew them. (The WPF team has a fun tool called Gradient Obsession for building gradient-based animations, most of which will work with Silverlight with some adjustment. You can find it at http://tinyurl.com/yc5fjpm.)

Animating Pixel Shaders

In Chapter 9, you learned about pixel shaders—low-level routines that can apply bitmap-style effects such as blurs, glows, and warps to any element. On their own, pixel shaders are an interesting but only occasionally useful tool. But combined with animation, they become much more versatile. You can use them to design eye-catching transitions (for example, by blurring one control out, hiding it, and then blurring another one in). Or, you can use them to create impressive user-interactivity effects (for example, by increasing the glow on a button when the user moves the mouse over it). Best of all, you can animate the properties of a pixel shader just as easily as you animate anything else.

Figure 10-10 shows a page that's based on the rotating button example shown earlier. It contains a sequence of buttons, and when the user moves the mouse over one of the buttons, an animation is attached and started. The difference is that the animation in this example doesn't rotate the button—instead, it reduces the blur radius to 0. The result is that as you move the mouse, the nearest control slides sharply and briskly into focus.

Figure 10-10. Animating a pixel shader

The code is the same as in the rotating button example. You need to give each button a BlurEffect instead of a RotateTransform:

```
<Button Content="One" Margin="10"
 MouseEnter="cmd_MouseEnter" MouseLeave="cmd_MouseLeave">
  <Button.Effect>
    <BlurEffect Radius="10"></BlurEffect>
  </Button.Effect>
</Button>
```

You also need to change the animation accordingly:

```
<Storyboard x:Name="blurStoryboard">
  <DoubleAnimation Storyboard.TargetProperty="Radius"
   To="0" Duration="0:0:0.4"></DoubleAnimation>
</Storyboard>
<Storyboard x:Name="unblurStoryboard">
  <DoubleAnimation Storyboard.TargetProperty="Radius" To="10"
   Duration="0:0:0.2"></DoubleAnimation>
</Storyboard>
```

In this example, the Storyboard.TargetElement property isn't set in XAML because it's set in code, when the MouseEnter or MouseLeave event fires. This is exactly the same technique you saw in the rotating button example.

You could use the same approach in reverse to highlight a button. For example, you could use a pixel shader that applies a glow effect to highlight the moused-over button. And if you're interested in using pixel shaders to animate page transitions, check out the WPF Shader Effects Library (which also works with Silverlight) at http://wpffx.codeplex.com. It includes a range of eye-popping pixel shaders beyond Silverlight's standard plain BlurEffect and DropShadowEffect, as well as a set of helper classes for performing transitions with them.

Key-Frame Animation

All the animations you've seen so far have used interpolation to move from a starting point to an ending point. But what if you need to create an animation that has multiple segments and moves less regularly? For example, you may want to create an animation that quickly slides an element partially into view, then moves it more slowly, and then speeds up again to move it the rest of the way into place. Animation easing won't help—it's intended to create pleasing, natural motion, not provide a specifically tuned animation that changes speed multiple times or at precise points. You could achieve this effect by creating a sequence of two animations and using the BeginTime property to start the second animation after the first one. However, there's an easier approach: you can use a key-frame animation.

A *key-frame animation* is an animation that's made up of many short segments. Each segment represents an initial, final, or intermediary value in the animation. When you run the animation, it moves smoothly from one value to another.

For example, consider the Point animation that allowed you to move the center point of a RadialGradientBrush from one spot to another:

```
<PointAnimation Storyboard.TargetName="ellipseBrush"
 Storyboard.TargetProperty="GradientOrigin"
 From="0.7,0.3" To="0.3,0.7" Duration="0:0:10" AutoReverse="True"
 RepeatBehavior="Forever">
</PointAnimation>
```

You can replace this PointAnimation object with an equivalent PointAnimationUsingKeyFrames object, as shown here:

```
<PointAnimationUsingKeyFrames Storyboard.TargetName="ellipseBrush"
 Storyboard.TargetProperty="GradientOrigin"
 AutoReverse="True" RepeatBehavior="Forever" >
  <LinearPointKeyFrame Value="0.7,0.3" KeyTime="0:0:0"></LinearPointKeyFrame>
  <LinearPointKeyFrame Value="0.3,0.7" KeyTime="0:0:10"></LinearPointKeyFrame>
</PointAnimationUsingKeyFrames>
```

This animation includes two key frames. The first sets the Point value when the animation starts. (If you want to use the current value that's set in the RadialGradientBrush, you can leave out this key frame.) The second key frame defines the end value, which is reached after 10 seconds. The PointAnimationUsingKeyFrames object performs linear interpolation to move smoothly from the first key-frame value to the second, just as the PointAnimation does with the From and To values.

■ **Note** Every key-frame animation uses its own key-frame animation object (like LinearPointKeyFrame). For the most part, these classes are the same—they include a Value property that stores the target value and a KeyTime property that indicates when the frame reaches the target value. The only difference is the data type of the Value property. In a LinearPointKeyFrame it's a Point, in a DoubleKeyFrame it's a double, and so on.

You can create a more interesting example using a series of key frames. The following animation walks the center point through a series of positions that are reached at different times. The speed at which the center point moves changes depending on the duration between key frames and how much distance needs to be covered:

```
<PointAnimationUsingKeyFrames Storyboard.TargetName="ellipseBrush"
 Storyboard.TargetProperty="GradientOrigin"
 RepeatBehavior="Forever" >
  <LinearPointKeyFrame Value="0.7,0.3" KeyTime="0:0:0"></LinearPointKeyFrame>
  <LinearPointKeyFrame Value="0.3,0.7" KeyTime="0:0:5"></LinearPointKeyFrame>
  <LinearPointKeyFrame Value="0.5,0.9" KeyTime="0:0:8"></LinearPointKeyFrame>
  <LinearPointKeyFrame Value="0.9,0.6" KeyTime="0:0:10"></LinearPointKeyFrame>
  <LinearPointKeyFrame Value="0.8,0.2" KeyTime="0:0:12"></LinearPointKeyFrame>
  <LinearPointKeyFrame Value="0.7,0.3" KeyTime="0:0:14"></LinearPointKeyFrame>
</PointAnimationUsingKeyFrames>
```

This animation isn't reversible, but it does repeat. To make sure there's no jump between the final value of one iteration and the starting value of the next iteration, the animation ends at the same center point where it began.

Discrete Key Frames

The key frame animation you saw in the previous example uses *linear* key frames. As a result, it transitions smoothly between the key-frame values. Another option is to use *discrete* key frames. In this case, no interpolation is performed. When the key time is reached, the property changes abruptly to the new value.

Linear key-frame classes are named in the form Linear*DataType*KeyFrame. Discrete key-frame classes are named in the form Discrete*DataType*KeyFrame. Here's a revised version of the RadialGradientBrush example that uses discrete key frames:

```
<PointAnimationUsingKeyFrames Storyboard.TargetName="ellipseBrush"
 Storyboard.TargetProperty="GradientOrigin"
 RepeatBehavior="Forever" >
  <DiscretePointKeyFrame Value="0.7,0.3" KeyTime="0:0:0"></DiscretePointKeyFrame>
  <DiscretePointKeyFrame Value="0.3,0.7" KeyTime="0:0:5"></DiscretePointKeyFrame>
  <DiscretePointKeyFrame Value="0.5,0.9" KeyTime="0:0:8"></DiscretePointKeyFrame>
  <DiscretePointKeyFrame Value="0.9,0.6" KeyTime="0:0:10"></DiscretePointKeyFrame>
  <DiscretePointKeyFrame Value="0.8,0.2" KeyTime="0:0:12"></DiscretePointKeyFrame>
  <DiscretePointKeyFrame Value="0.7,0.3" KeyTime="0:0:14"></DiscretePointKeyFrame>
</PointAnimationUsingKeyFrames>
```

When you run this animation, the center point jumps from one position to the next at the appropriate time. It's a dramatic (but jerky) effect.

All key-frame animation classes support discrete key frames, but only some support linear key frames. It all depends on the data type. The data types that support linear key frames are the same ones that support linear interpolation and provide a *DataType*Animation class. These are Point, Color, and Double. The only other animatable data type is object, which doesn't support linear interpolation. (Essentially, "animating" an object means replacing it with completely new values at specific times in a discrete key-frame animation.)

■ **Tip** You can combine both types of key frame—linear and discrete—in the same key-frame animation, as long as they're both supported for that data type.

Easing Key Frames

Earlier in this chapter, you saw how easing functions can improve ordinary animations. Even though key-frame animations are split into multiple segments, each of these segments uses ordinary, boring linear interpolation.

If this isn't what you want, you can use animation easing to add acceleration or deceleration to individual key frames. However, the ordinary linear key frame and discrete key-frame classes don't support this feature. Instead, you need to use an *easing* key frame, such as EasingDoubleKeyFrame, EasingColorKeyFrame, or EasingPointKeyFrame. Each one works the same way as its linear counterpart but exposes an additional EasingFunction property.

Here's an example that uses animation easing to apply an accelerating effect to the first five seconds of the key-frame animation:

```
<PointAnimationUsingKeyFrames Storyboard.TargetName="ellipseBrush"
 Storyboard.TargetProperty="GradientOrigin"
 RepeatBehavior="Forever" >
  <LinearPointKeyFrame Value="0.7,0.3" KeyTime="0:0:0"></LinearPointKeyFrame>
  <EasingPointKeyFrame Value="0.3,0.7" KeyTime="0:0:5">
    <EasingPointKeyFrame.EasingFunction>
      <CircleEase></CircleEase>
    </EasingPointKeyFrame.EasingFunction>
  </EasingPointKeyFrame>
  <LinearPointKeyFrame Value="0.5,0.9" KeyTime="0:0:8"></LinearPointKeyFrame>
```

```
  <LinearPointKeyFrame Value="0.9,0.6" KeyTime="0:0:10"></LinearPointKeyFrame>
  <LinearPointKeyFrame Value="0.8,0.2" KeyTime="0:0:12"></LinearPointKeyFrame>
  <LinearPointKeyFrame Value="0.7,0.3" KeyTime="0:0:14"></LinearPointKeyFrame>
</PointAnimationUsingKeyFrames>
```

The combination of key frames and animation easing is a convenient way to model complex animations, but it still may not give you the control you need. Instead of using animation easing, you can create a mathematical formula that dictates the progression of your animation. This is the technique you'll learn in the next section.

Spline Key Frames

There's one more type of key frame: a *spline* key frame. Every class that supports linear key frames also supports spline key frames, and they're named in the form Spline*DataType*KeyFrame.

Like linear key frames, spline key frames use interpolation to move smoothly from one key value to another. The difference is that every spline key frame sports a KeySpline property. Using the KeySpline property, you define a cubic Bézier curve that influences the way interpolation is performed. Although it's tricky to get the effect you want (at least without an advanced design tool to help you), this technique gives you the ability to create more seamless acceleration and deceleration and more lifelike motion.

As you may remember from Chapter 8, a Bézier curve is defined by a start point, an end point, and two control points. In the case of a key spline, the start point is always (0,0), and the end point is always (1,1). You supply the two control points. The curve that you create describes the relationship between time (in the x-axis) and the animated value (in the y-axis).

Here's an example that demonstrates a key-spline animation by comparing the motion of two ellipses across a Canvas. The first ellipse uses a DoubleAnimation to move slowly and evenly across the page. The second ellipse uses a DoubleAnimationUsingKeyFrames with two SplineDoubleKeyFrame objects. It reaches the destination at the same times (after 10 seconds), but it accelerates and decelerates during its travel, pulling ahead and dropping behind the other ellipse:

```
<DoubleAnimation Storyboard.TargetName="ellipse1"
 Storyboard.TargetProperty="(Canvas.Left)"
 To="500" Duration="0:0:10">
</DoubleAnimation>

<DoubleAnimationUsingKeyFrames Storyboard.TargetName="ellipse2"
 Storyboard.TargetProperty="(Canvas.Left)" >
  <SplineDoubleKeyFrame KeyTime="0:0:5" Value="250"
   KeySpline="0.25,0 0.5,0.7"></SplineDoubleKeyFrame>
  <SplineDoubleKeyFrame KeyTime="0:0:10" Value="500"
   KeySpline="0.25,0.8 0.2,0.4"></SplineDoubleKeyFrame>
</DoubleAnimationUsingKeyFrames>
```

The fastest acceleration occurs shortly after the five-second mark, when the second SplineDoubleKeyFrame kicks in. Its first control point matches a relatively large y-axis value, which represents the animation progress (0.8) against a correspondingly smaller x-axis value, which represents the time. As a result, the ellipse increases its speed over a small distance before slowing down again.

Figure 10-11 shows a graphical depiction of the two curves that control the movement of the ellipse. To interpret these curves, remember that they chart the progress of the animation from top to bottom. Looking at the first curve, you can see that it follows a fairly even progress downward, with a short pause at the beginning and a gradual leveling off at the end. However, the second curve plummets downward quickly, achieving the bulk of its progress, and then levels off for the remainder of the animation.

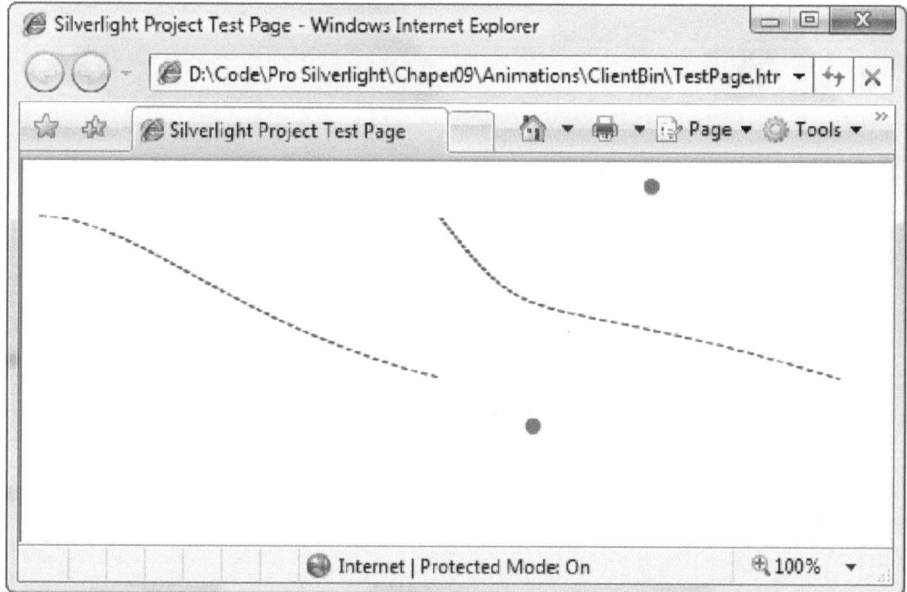

Figure 10-11. Charting the progress of a key-spline animation

Animations in Code

Sometimes, you'll need to create every detail of an animation programmatically in code. In fact, this scenario is fairly common. It occurs any time you have multiple animations to deal with, and you don't know in advance how many animations there will be or how they should be configured. (This is the case with the simple bomb-dropping game you'll see in this section.) It also occurs if you want to use the same animation in different pages or you simply want the flexibility to separate all the animation-related details from your markup for easier reuse. (This is the case with animation that's used for page transitions later in this chapter.)

It isn't difficult to create, configure, and launch an animation programmatically. You just need to create the animation and storyboard objects, add the animations to the storyboard, and start the storyboard. You can perform any cleanup work after your animation ends by reacting to the Storyboard.Completed event.

In the following example, you'll see how to create the game shown in Figure 10-12. Here, a series of bombs are dropped at ever-increasing speeds. The player must click each bomb to defuse it. When a set limit is reached—by default, five dropped bombs—the game ends.

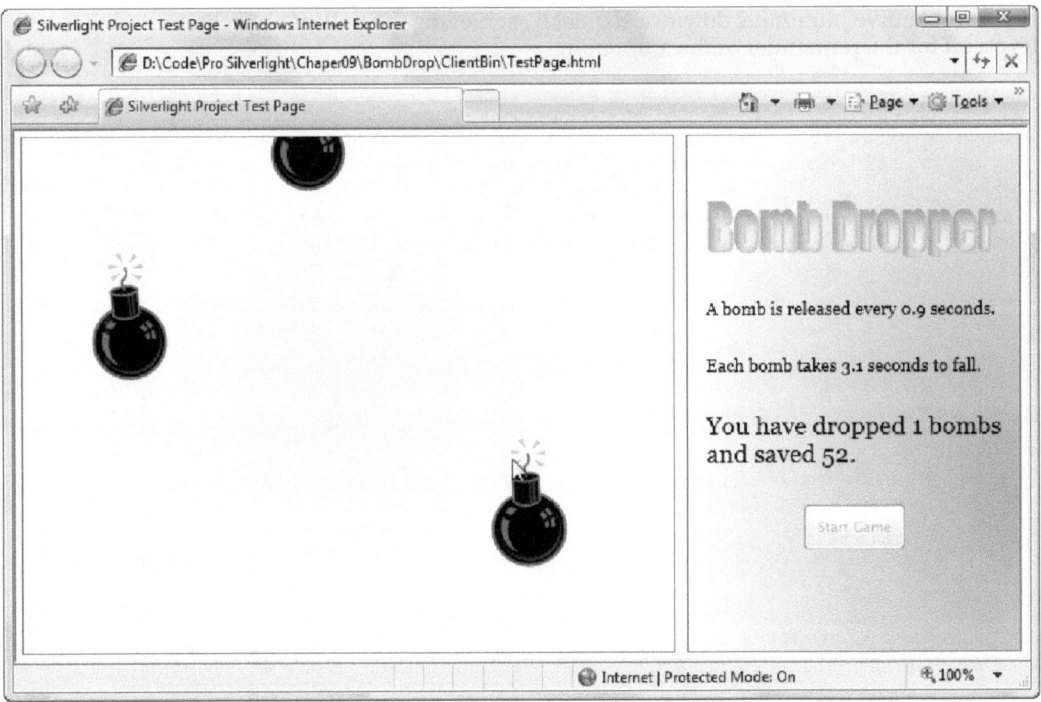

Figure 10-12. Catching bombs

In this example, every dropped bomb has its own storyboard with two animations. The first animation drops the bomb (by animating the Canvas.Top property), and the second animation rotates the bomb slightly back and forth, giving it a realistic wiggle effect. If the user clicks a bomb, these animations are halted, and two more take place to send the bomb careening harmlessly off the side of the Canvas. Finally, every time an animation ends, the application checks to see whether it represents a bomb that fell down or one that was saved and updates the count accordingly.

In the following sections, you'll see how to create each part of this example.

The Main Page

The main page in the BombDropper example is straightforward. It contains a two-column Grid. On the left side is a Border element, which contains the Canvas that represents the game surface:

```
<Border Grid.Column="0" BorderBrush="SteelBlue" BorderThickness="1" Margin="5">
  <Grid>
    <Canvas x:Name="canvasBackground" SizeChanged="canvasBackground_SizeChanged"
      MinWidth="50">
      <Canvas.Background>
        <RadialGradientBrush>
          <GradientStop Color="AliceBlue" Offset="0"></GradientStop>
          <GradientStop Color="White" Offset="0.7"></GradientStop>
        </RadialGradientBrush>
      </Canvas.Background>
```

```
      </Canvas.Background>
    </Canvas>
  </Grid>
</Border>
```

When the Canvas is sized for the first time or resized (when the user changes the size of the browser window), the following code runs and sets the clipping region:

```
Private Sub canvasBackground_SizeChanged(ByVal sender As Object, _
  ByVal e As SizeChangedEventArgs)
    ' Set the clipping region to match the current display region of the Canvas.
    Dim rect As New RectangleGeometry()
    rect.Rect = New Rect(0, 0, _
      canvasBackground.ActualWidth, canvasBackground.ActualHeight)
    canvasBackground.Clip = rect
End Sub
```

This is required because otherwise the Canvas draws its children even if they lie outside its display area. In the bomb-dropping game, this would cause the bombs to fly out of the box that delineates the Canvas.

■ **Note** Because the user control is defined without explicit sizes, it's free to resize itself to match the browser window. The game logic uses the current window dimensions without attempting to compensate for them in any way. Thus, if you have a very wide window, bombs are spread across a wide area, making the game more difficult. Similarly, if you have a very tall window, bombs fall faster so they can complete their trajectory in the same interval of time. You could get around this issue by using a fixed-size region, which you could then center in the middle of your user control. However, a resizable window makes the example more adaptable and more interesting.

On the right side of the main window is a panel that shows the game statistics, the current bomb-dropped and bomb-saved count, and a button for starting the game:

```
<Border Grid.Column="1" BorderBrush="SteelBlue" BorderThickness="1" Margin="5">
  <Border.Background>
    <RadialGradientBrush GradientOrigin="1,0.7" Center="1,0.7"
      RadiusX="1" RadiusY="1">
      <GradientStop Color="Orange"  Offset="0"></GradientStop>
      <GradientStop Color="White" Offset="1"></GradientStop>
    </RadialGradientBrush>
  </Border.Background>

  <StackPanel Margin="15" VerticalAlignment="Center" HorizontalAlignment="Center">
    <bomb:Title></bomb:Title>
    <TextBlock x:Name="lblRate" Margin="0,30,0,0" TextWrapping="Wrap"
      FontFamily="Georgia" FontSize="14"></TextBlock>
    <TextBlock x:Name="lblSpeed" Margin="0,30" TextWrapping="Wrap"
      FontFamily="Georgia" FontSize="14"></TextBlock>
    <TextBlock x:Name="lblStatus" TextWrapping="Wrap"
```

```
        FontFamily="Georgia" FontSize="20">No bombs have dropped.</TextBlock>
      <Button x:Name="cmdStart" Padding="5" Margin="0,30" Width="80"
       Content="Start Game" Click="cmdStart_Click"></Button>
    </StackPanel>
  </Border>
```

You'll notice that the right-side column contains one unusual ingredient: an element named Title. This is a custom user control that shows the BombDropper title in fiery orange letters. Technically, it's a piece of vector art. It was created in Microsoft Word using the WordArt feature, saved as an XPS file, and then exported to XAML using the technique described in Chapter 8. Although you could insert the markup for the BombDropper title directly into the main page, defining it as a separate user control allows you to separate the title markup from the rest of the user interface.

To use the Title user control, you need to map your project namespace to an XML namespace, thereby making it available in your page (as described in Chapter 2). Assuming the project is named BombDropper, here's what you need to add:

```
<UserControl x:Class="BombDropper.Page"
    xmlns="http://schemas.microsoft.com/winfx/2006/xaml/presentation"
    xmlns:x="http://schemas.microsoft.com/winfx/2006/xaml"
    xmlns:bomb="clr-namespace:BombDropper;assembly=BombDropper">
```

Now you can use the XML prefix "bomb" to insert any custom controls from your project, including user controls:

```
<bomb:Title></bomb:Title>
```

The Bomb User Control

The next step is to create the graphical image of the bomb. Although you can use a static image (as long as it has a transparent background), it's always better to deal with more flexible Silverlight shapes. By using shapes, you gain the ability to resize the bomb without introducing distortion, and you can animate or alter individual parts of the drawing. The bomb shown in this example is drawn straight from Microsoft Word's online clip-art collection. The bomb was converted to XAML by inserting it into a Word document and then saving that document as an XPS file, a process described in Chapter 8. The full XAML, which uses a combination of Path elements, isn't shown here. But you can see it by downloading the BombDropper game along with the samples for this chapter.

The XAML for the Bomb class was then simplified slightly (by removing the unnecessary extra Canvas elements around it and the transforms for scaling it). The XAML was then inserted into a new user control named Bomb. This way, the main page can show a bomb by creating the Bomb user control and adding it to a layout container (like a Canvas).

Placing the graphic in a separate user control makes it easy to instantiate multiple copies of that graphic in your user interface. It also lets you encapsulate related functionality by adding to the user control's code. In the bomb-dropping example, only one detail is added to the code—a Boolean property that tracks whether the bomb is currently falling:

```
Public Partial Class Bomb
    Inherits UserControl

    Public Sub New()
        InitializeComponent()
    End Sub
```

```
    Private _isFalling As Boolean
    Public Property IsFalling() As Boolean
        Get
            Return _isFalling
        End Get
        Set(ByVal value As Boolean)
            _isFalling = value
        End Set
    End Property
End Class
```

The markup for the bomb includes a RotateTransform, which the animation code can use to give the bomb a wiggling effect as it falls. Although you could create and add this RotateTransform programmatically, it makes more sense to define it in the XAML file for the bomb:

```
<UserControl x:Class="BombDrop.Bomb"
    xmlns="http://schemas.microsoft.com/winfx/2006/xaml/presentation"
    xmlns:x="http://schemas.microsoft.com/winfx/2006/xaml"
    >
  <UserControl.RenderTransform>
    <TransformGroup>
      <RotateTransform Angle="20" CenterX="50" CenterY="50"></RotateTransform>
      <ScaleTransform ScaleX="0.5" ScaleY="0.5"></ScaleTransform>
    </TransformGroup>
  </UserControl.RenderTransform>

  <Canvas>
    <!-- The Path elements that draw the bomb graphic are defined here. -->
  </Canvas>
</UserControl>
```

With this code in place, you could insert a bomb into your window using a <bomb:Bomb> element, much as the main window inserts the Title user control (as described in the previous section). However, in this case it makes far more sense to create the bombs programmatically.

Dropping the Bombs

To drop the bombs, the application uses DispatcherTimer, a timer that plays nicely with Silverlight user interface because it triggers events on the user-interface thread (saving you the effort of *marshalling* or *locking*, two multithreaded programming techniques that are described in Chapter 19). You choose a time interval, and then the DispatcherTimer fires a periodic Tick event at that interval.

```
Private bombTimer As New DispatcherTimer()

Public Sub New()
    InitializeComponent()
    AddHandler bombTimer.Tick, AddressOf bombTimer_Tick
End Sub
```

In the BombDropper game, the timer initially fires every 1.3 seconds. When the user clicks the button to start the game, the timer is started:

```
' Keep track of how many bombs are dropped and stopped.
Private droppedCount As Integer = 0
Private savedCount As Integer = 0

' Initially, bombs fall every 1.3 seconds, and hit the ground after 3.5 seconds.
Private initialSecondsBetweenBombs As Double = 1.3
Private initialSecondsToFall As Double = 3.5
Private secondsBetweenBombs As Double
Private secondsToFall As Double

Private Sub cmdStart_Click(ByVal sender As Object, ByVal e As RoutedEventArgs)
    cmdStart.IsEnabled = False

    ' Reset the game.
    droppedCount = 0
    savedCount = 0
    secondsBetweenBombs = initialSecondsBetweenBombs
    secondsToFall = initialSecondsToFall

    ' Start the bomb-dropping timer.
    bombTimer.Interval = TimeSpan.FromSeconds(secondsBetweenBombs)
    bombTimer.Start()
End Sub
```

Every time the timer fires, the code creates a new Bomb object and sets its position on the Canvas. The bomb is placed just above the top edge of the Canvas, so it can fall seamlessly into view. It's given a random horizontal position that falls somewhere between the left and right sides:

```
Private Sub bombTimer_Tick(ByVal sender As Object, ByVal e As EventArgs)
    ' Create the bomb.
    Dim bomb As New Bomb()
    bomb.IsFalling = True

    ' Position the bomb.
    Dim random As New Random()
    bomb.SetValue(Canvas.LeftProperty, _
      CDbl(random.Next(0, canvasBackground.ActualWidth - 50)))
    bomb.SetValue(Canvas.TopProperty, -100.0)

    ' Add the bomb to the Canvas.
    canvasBackground.Children.Add(bomb)
    ...
```

The code then dynamically creates a storyboard to animate the bomb. Two animations are used: one that drops the bomb by changing the attached Canvas.Top property and one that wiggles the bomb by changing the angle of its rotate transform. Because Storyboard.TargetElement and Storyboard.TargetProperty are attached properties, you must set them using the Storyboard.SetTargetElement() and Storyboard.SetTargetProperty() methods:

```
    ...
    ' Attach mouse click event (for defusing the bomb).
    AddHandler bomb.MouseLeftButtonDown, AddressOf bomb_MouseLeftButtonDown

    ' Create the animation for the falling bomb.
```

```
Dim storyboard As New Storyboard()
Dim fallAnimation As New DoubleAnimation()
fallAnimation.To = canvasBackground.ActualHeight
fallAnimation.Duration = TimeSpan.FromSeconds(secondsToFall)

Storyboard.SetTarget(fallAnimation, bomb)
Storyboard.SetTargetProperty(fallAnimation, New PropertyPath("(Canvas.Top)"))
storyboard.Children.Add(fallAnimation)

' Create the animation for the bomb "wiggle."
Dim wiggleAnimation As New DoubleAnimation()
wiggleAnimation.To = 30
wiggleAnimation.Duration = TimeSpan.FromSeconds(0.2)
wiggleAnimation.RepeatBehavior = RepeatBehavior.Forever
wiggleAnimation.AutoReverse = True

Storyboard.SetTarget(wiggleAnimation, _
  (CType(bomb.RenderTransform, TransformGroup)).Children(0))
Storyboard.SetTargetProperty(wiggleAnimation, New PropertyPath("Angle"))
storyboard.Children.Add(wiggleAnimation)
...
```

Both of these animations could use animation easing for more realistic behavior, but this example keeps the code simple by using basic linear animations.

The newly created bomb and storyboard are stored in two dictionary collections so they can be retrieved easily in other event handlers. The collections are stored as fields in the main page class and are defined like this:

```
' Make it possible to look up a bomb based on a storyboard, and vice versa.
Private storyboards As New Dictionary(Of Bomb, Storyboard) ()
Private bombs As New Dictionary(Of Storyboard, Bomb)()
```

Here's the code that adds the bomb and storyboard to these two collections:

```
...
bombs.Add(storyboard, bomb)
storyboards.Add(bomb, storyboard)
...
```

Next, you attach an event handler that reacts when the storyboard finishes the fallAnimation, which occurs when the bomb hits the ground. Finally, the storyboard is started, and the animations are put in motion:

```
...
storyboard.Duration = fallAnimation.Duration
AddHandler storyboard.Completed, AddressOf storyboard_Completed
storyboard.Begin()
...
```

The bomb-dropping code needs one last detail. As the game progresses, it becomes more difficult. The timer begins to fire more frequently, the bombs begin to appear more closely together, and the fall time is reduced. To implement these changes, the timer code makes adjustments whenever a set interval of time has passed. By default, BombDropper makes an adjustment every 15 seconds. Here are the fields that control the adjustments:

```
' Perform an adjustment every 15 seconds.
Private secondsBetweenAdjustments As Double = 15
Private lastAdjustmentTime As DateTime = DateTime.MinValue

' After every adjustment, shave 0.1 seconds off both.
Private secondsBetweenBombsReduction As Double = 0.1
Private secondsToFallReduction As Double = 0.1
```

And here's the code at the end of the DispatcherTimer.Tick event handler, which checks whether an adjustment is needed and makes the appropriate changes:

```
...
' Perform an "adjustment" when needed.
If (DateTime.Now.Subtract(lastAdjustmentTime).TotalSeconds > _
  secondsBetweenAdjustments) Then
    lastAdjustmentTime = DateTime.Now

    secondsBetweenBombs -= secondsBetweenBombsReduction
    secondsToFall -= secondsToFallReduction

    ' (Technically, you should check for 0 or negative values.
    ' However, in practice these won't occur because the game will
    ' always end first.)

    ' Set the timer to drop the next bomb at the appropriate time.
    bombTimer.Interval = TimeSpan.FromSeconds(secondsBetweenBombs)

    ' Update the status message.
    lblRate.Text = String.Format("A bomb is released every {0} seconds.", _
      secondsBetweenBombs)
    lblSpeed.Text = String.Format("Each bomb takes {0} seconds to fall.", _
      secondsToFall)
End If
End Sub
```

With this code in place, there's enough functionality to drop bombs at an ever-increasing rate. However, the game still lacks the code that responds to dropped and saved bombs.

Intercepting a Bomb

The user saves a bomb by clicking it before it reaches the bottom of the Canvas and explodes. Because each bomb is a separate instance of the Bomb user control, intercepting mouse clicks is easy—all you need to do is handle the MouseLeftButtonDown event, which fires when any part of the bomb is clicked (but doesn't fire if you click somewhere in the background, such as around the edges of the bomb circle).

When a bomb is clicked, the first step is to get appropriate bomb object and set its IsFalling property to indicate that it's no longer falling. (The IsFalling property is used by the event handler that deals with completed animations.)

```
Private Sub bomb_MouseLeftButtonDown(ByVal sender As Object, _
  ByVal e As MouseButtonEventArgs)
    ' Get the bomb.
    Dim bomb As Bomb = CType(sender, Bomb)
```

```
bomb.IsFalling = False

' Record the bomb's current (animated) position.
Dim currentTop As Double = Canvas.GetTop(bomb)
...
```

The next step is to find the storyboard that controls the animation for this bomb so it can be stopped. To find the storyboard, you need to look it up in the collection that this game uses for tracking. Currently, Silverlight doesn't include any standardized way to find the animations that are acting on a given element.

```
...
' Stop the bomb from falling.
Dim storyboard As Storyboard = storyboards(bomb)
storyboard.Stop()
...
```

After a button is clicked, another set of animations moves the bomb off the screen, throwing it up and left or right (depending on which side is closest). Although you could create an entirely new storyboard to implement this effect, the BombDropper game clears the current storyboard that's being used for the bomb and adds new animations to it. When this process is completed, the new storyboard is started:

```
...
' Reuse the existing storyboard, but with new animations.
' Send the bomb on a new trajectory by animating Canvas.Top
' and Canvas.Left.
storyboard.Children.Clear()

Dim riseAnimation As New DoubleAnimation()
riseAnimation.From = currentTop
riseAnimation.To = 0
riseAnimation.Duration = TimeSpan.FromSeconds(2)

Storyboard.SetTarget(riseAnimation, bomb)
Storyboard.SetTargetProperty(riseAnimation, New PropertyPath("(Canvas.Top)"))
storyboard.Children.Add(riseAnimation)

Dim slideAnimation As New DoubleAnimation()
Dim currentLeft As Double = Canvas.GetLeft(bomb)

' Throw the bomb off the closest side.
If currentLeft < canvasBackground.ActualWidth / 2 Then
    slideAnimation.To = -100
Else
    slideAnimation.To = canvasBackground.ActualWidth + 100
End If
slideAnimation.Duration = TimeSpan.FromSeconds(1)
Storyboard.SetTarget(slideAnimation, bomb)
Storyboard.SetTargetProperty(slideAnimation, New PropertyPath("(Canvas.Left)"))
storyboard.Children.Add(slideAnimation)

' Start the new animation.
storyboard.Duration = slideAnimation.Duration
storyboard.Begin()
End Sub
```

Now the game has enough code to drop bombs and bounce them off the screen when the user saves them. However, to keep track of what bombs are saved and which ones are dropped, you need to react to the Storyboard.Completed event that fires at the end of an animation.

Counting Bombs and Cleaning Up

As you've seen, the BombDropper uses storyboards in two ways: to animate a falling bomb and to animate a defused bomb. You could handle the completion of these storyboards with different event handlers, but to keep things simple the BombDropper uses just one. It tells the difference between an exploded bomb and a rescued bomb by examining the Bomb.IsFalling property.

```
' End the game when 5 bombs have fallen.
Private maxDropped As Integer = 5

Private Sub storyboard_Completed(ByVal sender As Object, ByVal e As EventArgs)
    Dim completedStoryboard As Storyboard = CType(sender, Storyboard)
    Dim completedBomb As Bomb = bombs(completedStoryboard)

    ' Determine if a bomb fell or flew off the Canvas after being clicked.
    If completedBomb.IsFalling Then
        droppedCount += 1
    Else
        savedCount += 1
    End If
    ...
```

Either way, the code then updates the display test to indicate how many bombs have been dropped and saved. It then performs some cleanup, removing the bomb from the Canvas, and removing both the bomb and the storyboard from the collections that are used for tracking.

```
    ...
    ' Update the display.
    lblStatus.Text = String.Format("You have dropped {0} bombs and saved {1}.", _
        droppedCount, savedCount)

    ' Clean up.
    completedStoryboard.Stop()
    canvasBackground.Children.Remove(completedBomb)

    ' Update the tracking collections.
    storyboards.Remove(completedBomb)
    bombs.Remove(completedStoryboard)
    ...
```

At this point, the code checks to see whether the maximum number of dropped bombs has been reached. If it has, the game ends, the timer is stopped, and all the bombs and storyboards are removed:

```
    ...
    ' Check if it's game over.
    If droppedCount >= maxDropped Then
        bombTimer.Stop()
        lblStatus.Text &= Environment.NewLine & Environment.NewLine & "Game over."
```

```
        ' Find all the storyboards that are underway.
        For Each item As KeyValuePair(Of Bomb, Storyboard) In storyboards
            Dim storyboard As Storyboard = item.Value
            Dim bomb As Bomb = item.Key

            storyboard.Stop()
            canvasBackground.Children.Remove(bomb)
        Next

        ' Empty the tracking collections.
        storyboards.Clear()
        bombs.Clear()

        ' Allow the user to start a new game.
        cmdStart.IsEnabled = True
    End If
End Sub
```

This completes the code for BombDropper game. However, you can make plenty of refinements. Some examples include the following:

- *Animate a bomb explosion effect*: This effect can make the flames around the bomb twinkle or send small pieces of shrapnel flying across the Canvas.

- *Animate the background*: This change is easy, and it adds pizzazz. For example, you can create a linear gradient that shifts up, creating an impression of movement, or one that transitions between two colors.

- *Add depth*: It's easier than you think. The basic technique is to give the bombs different sizes. Bombs that are bigger should have a higher ZIndex, ensuring that they overlap smaller bombs, and should be given a shorter animation time, ensuring that they fall faster. You can also make the bombs partially transparent, so as one falls the others behind it are visible.

- *Add sound effects*: In Chapter 11, you'll learn to use sound and other media in Silverlight. You can use well-timed sound effects to punctuate bomb explosions or rescued bombs.

- *Use animation easing*: If you want bombs to accelerate as they fall, bounce off the screen, or wiggle more naturally, you can add easing functions to the animations used here. And, as you'd expect, easing functions can be constructed in code just as easily as in XAML.

- *Fine-tune the parameters*: You can provide more dials to tweak behavior (for example, variables that set how the bomb times, trajectories, and frequencies are altered as the game processes). You can also inject more randomness (for example, allowing saved bombs to bounce off the Canvas in slightly different ways).

You can find countless examples of Silverlight game programming on the Web. Microsoft's Silverlight community site includes game samples with full source code at www.silverlight.net/showcase. You can also check out Andy Beaulieu's website, which provides Silverlight games and an impressive physics simulator, at www.andybeaulieu.com.

Encapsulating Animations

When you create animations dynamically in code, a fair bit of boilerplate code is required to create the animations, set the storyboard properties, and handle the Completed event to clean up. For this reason, Silverlight developers often wrap animations in higher-level classes that take care of the low-level details.

For example, you can create an animation class named FadeElementEffect and fade an element out of view using code like this:

```
Dim fade As New FadeElementEffect()
fade.Animate(canvas)
```

Creating classes like this is fairly straightforward, although the exact design depends on the needs of your application. In the rest of this section, you'll consider one possible way to create animation helper classes that provide transitional animations when the user navigates between pages.

Page Transitions

In Chapter 7, you saw different ways to support page navigation in a Silverlight application. One technique is to use some sort of layout container as your application's root element. You can then add user controls to this container and remove them when needed. Navigating from one page to another consists of removing the user control for the current page and adding the user control for the next page.

One advantage of this technique is that it allows you to use an animated effect to switch between the two pages. For example, you can create an animation that fades in or slides in the new page. To make this work, you add both pages to the root visual at once, one over the other. (The easiest way to do this is to place both user controls in the same cell of a Grid, but a Canvas works equally well.) Then, you animate the properties of the topmost page. For example, you can change the Opacity property to fade the page in, alter the properties of a TranslateTransform to move it, and so on. You can even apply multiple effects at once—for example, to create a "blow up" effect that expands a page from the corner to fill the entire display area.

In the rest of this chapter, you'll learn how to use a simple wipe effect that unveils the new page on top of the current one. Figure 10-13 shows the wipe in action.

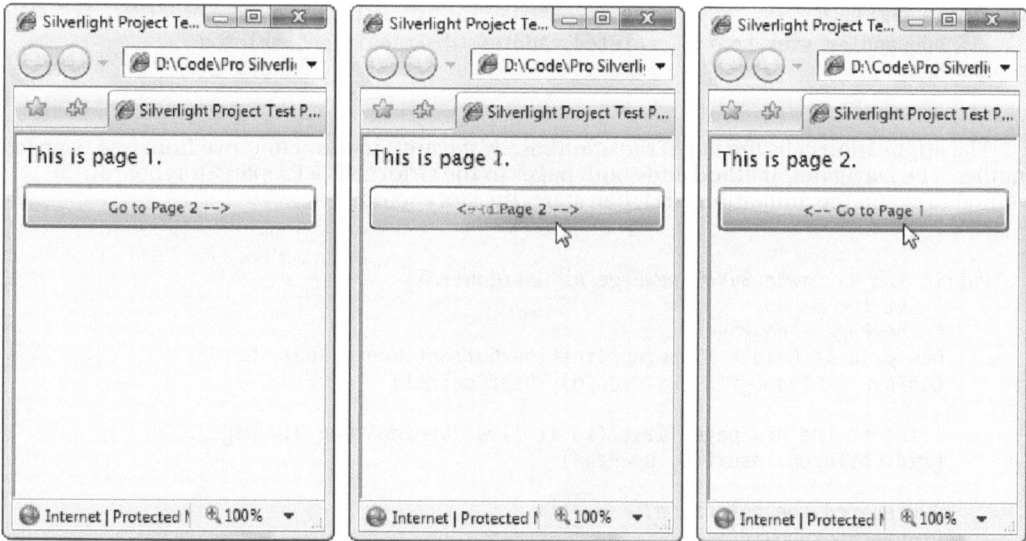

Figure 10-13. *Transitioning between pages with a wipe*

This example assumes the root element in your application is a Grid. In other words, your Application class requires code like this:

```
' This Grid will host your pages.
Private rootGrid As New Grid()

Private Sub Application_Startup(ByVal sender As Object, ByVal e As StartupEventArgs)
    ' Load the first page.
    Me.RootVisual = rootGrid
    rootGrid.Children.Add(New Page())
End Sub
```

This technique is discussed in Chapter 7.

The Base Class

The most straightforward way to animate a transition between pages is to code it directly in the App class, using a custom Navigate() method. However, it's far more flexible (and just a bit more effort) to place the animation code in a separate class. And if you standardize your animations with a MustInherit class or an interface, you'll gain far more flexibility to swap in the new effects.

In this example, all transitions inherit from a MustInherit class named PageTransitionBase. This class stores the storyboard, the previous page, and the new page as fields:

```
Public MustInherit Class PageTransitionBase

    Protected storyboard As New Storyboard()
    Protected oldPage As UserControl
    Protected newPage As UserControl
```

```
    Public Sub New()
        AddHandler storyboard.Completed, AddressOf TransitionCompleted
    End Sub
    ...
```

The application calls the PageTransitionBase.Navigate() method to move from one page to another. The Navigate() method adds both pages to the Grid, calls a PrepareStoryboard() method to set up the animation, and then starts the storyboard:

```
    ...
    Public Sub Navigate(ByVal newPage As UserControl)
        ' Set the pages.
        Me.newPage = newPage
        Dim grid As Grid = CType(Application.Current.RootVisual, Grid)
        oldPage = CType(grid.Children(0), UserControl)

        ' Insert the new page first (so it lies "behind" the old page).
        grid.Children.Insert(0, newPage)

        ' Prepared the animation.
        PrepareStoryboard()

        ' Perform the animation.
        storyboard.Begin()
    End Sub
    ...
```

The PrepareStoryboard() method is marked with the MustOverride keyword. It must be overridden in derived classes, which creates the specific animation objects they want.

```
    ...
    Protected MustOverride Sub PrepareStoryboard()
    ...
```

The TransitionCompleted() event handler responds when the animation is complete. It removes the old page:

```
    ...
    Private Sub TransitionCompleted(ByVal sender As Object, ByVal e As EventArgs)
        ' Remove the old page, which is not needed any longer.
        Dim grid As Grid = CType(Application.Current.RootVisual, Grid)
        grid.Children.Remove(oldPage)
    End Sub
End Class
```

You can also use this method to perform cleanup. However, in this example, the animation acts on the old page, which is discarded after the navigation. No extra cleanup is needed.

The Wipe Transition

To use a page transition, you need at least one derived class that creates animations. In this section, you'll consider one example: a WipeTransition class that wipes away the old page, revealing the new one underneath.

The trick to creating a wipe effect is animating a brush that uses an opacity mask. (As you learned in Chapter 9, an opacity mask determines what portions of an image or element should be visible and which ones should be transparent.) To use an animation as a page transition, you need to use a LinearGradientBrush for the opacity mask. As the animation progresses, you move the offsets in the opacity mask, gradually making more of the topmost element transparent and revealing more of the content underneath. In a page transition, the topmost element is the old page, and underneath is the new page. Wipes commonly work from left to right or top to bottom, but more creative effects are possible if you use different opacity masks.

To perform its work, the WipeTransition class overrides the PrepareStoryboard() method. Its first task is to create the opacity mask and add it to the old page (which is topmost in the grid). This opacity mask uses a gradient that defines two gradient stops: Black (the image is completely visible) and Transparent (the image is completely transparent). Initially, both stops are positioned at the left edge of the image. Because the visible stop is declared last, it takes precedence, and the image is completely opaque.

```
Public Class WipeTransition
    Inherits PageTransitionBase

    Protected Overrides Sub PrepareStoryboard()
        ' Create the opacity mask.
        Dim mask As New LinearGradientBrush()
        mask.StartPoint = New Point(0,0)
        mask.EndPoint = New Point(1,0)

        Dim transparentStop As New GradientStop()
        transparentStop.Color = Colors.Transparent
        transparentStop.Offset = 0
        mask.GradientStops.Add(transparentStop)
        Dim visibleStop As New GradientStop()
        visibleStop.Color = Colors.Black
        visibleStop.Offset = 0
        mask.GradientStops.Add(visibleStop)

        oldPage.OpacityMask = mask
        ...
```

Next, you need to perform your animation on the offsets of the LinearGradientBrush. In this example, both offsets are moved from the left side to the right side, allowing the image underneath to appear. To make this example a bit fancier, the offsets don't occupy the same position while they move. Instead, the visible offset leads the way, followed by the transparent offset after a short delay of 0.2 seconds. This creates a blended fringe at the edge of the wipe while the animation is underway.

```
        ...
        ' Create the animations for the opacity mask.
        Dim visibleStopAnimation As New DoubleAnimation()
        Storyboard.SetTarget(visibleStopAnimation, visibleStop)
        Storyboard.SetTargetProperty(visibleStopAnimation, _
          New PropertyPath("Offset"))
        visibleStopAnimation.Duration = TimeSpan.FromSeconds(1.2)
        visibleStopAnimation.From = 0
        visibleStopAnimation.To = 1.2

        Dim transparentStopAnimation As New DoubleAnimation()
        Storyboard.SetTarget(transparentStopAnimation, transparentStop)
```

```
    Storyboard.SetTargetProperty(transparentStopAnimation, _
      New PropertyPath("Offset"))
    transparentStopAnimation.BeginTime = TimeSpan.FromSeconds(0.2)
    transparentStopAnimation.From = 0
    transparentStopAnimation.To = 1
    transparentStopAnimation.Duration = TimeSpan.FromSeconds(1)
    ...
```

There's one odd detail here. The visible stop moves to 1.2 rather than 1, which denotes the right edge of the image. This ensures that both offsets move at the same speed, because the total distance that each one must cover is proportional to the duration of its animation.

The final step is to add the animations to the storyboard, which is defined in the PageTransitionBase class. You don't need to start the storyboard, because the PageTransitionBase class performs this step as soon as the PrepareStoryboard() method returns.

```
    ...
    ' Add the animations to the storyboard.
    storyboard.Children.Add(transparentStopAnimation)
    storyboard.Children.Add(visibleStopAnimation)
End Sub

End Class
```

Now, you can use code like this to navigate between pages:

```
Dim transition As New WipeTransition()
transition.Navigate(New Page2())
```

As with the BombDropper, there are plenty of imaginative ways to extend this example:

- *Add transition properties*: You could enhance the WipeTransition class with more possibilities, allowing a configurable wipe direction, a configurable wipe time, and so on.
- *Create more transitions*: Creating a new animated page transition is as simple as deriving a class from PageTransitionBase and overriding PrepareStoryboard().
- *Refactor the PageTransitionBase code*: The current example is designed to be as simple as possible. However, a more elaborate design would pull out the code that adds and removes pages and place it in the custom application class. This opens up new possibilities. It allows you to use different layouts. (For example, you can use a transition animation in one panel rather than for the entire window.) It also lets the application class add application services. (For example, you can keep pages alive in a cache after you navigate away from them, as described in Chapter 7. This lets you retain the current state of all your elements.)

For fancier effects, check out the collection of custom pixel shaders and transitions in the free WPF Shader Effects Library at http://codeplex.com/wpffx.

Frame-Based Animation

Along with the property-based animation system, Silverlight provides a way to create frame-based animation using nothing but code. All you need to do is respond to the shared CompositionTarget.Rendering event, which is fired to get the content for each frame. This is a far lower-level approach, which you shouldn't tackle unless you're sure the standard property-based animation model won't work for your scenario (for example, if you're building a simple side-scrolling game, creating physics-based animations, or modeling particle effects such as fire, snow, and bubbles).

The basic technique for building a frame-based animation is easy. You attach an event handler to the shared CompositionTarget.Rendering event. After you do, Silverlight begins calling this event handler continuously. (As long as your rendering code executes quickly enough, Silverlight will call it 60 times each second.) In the rendering event handler, it's up to you to create or adjust the elements in the window accordingly. In other words, you need to manage all the work yourself. When the animation has ended, detach the event handler.

Figure 10-14 shows a straightforward example. Here, a random number of circles fall from the top of a Canvas to the bottom. They fall at different speeds (based on a random starting velocity), but they accelerate downward at the same rate. The animation ends when all the circles reach the bottom.

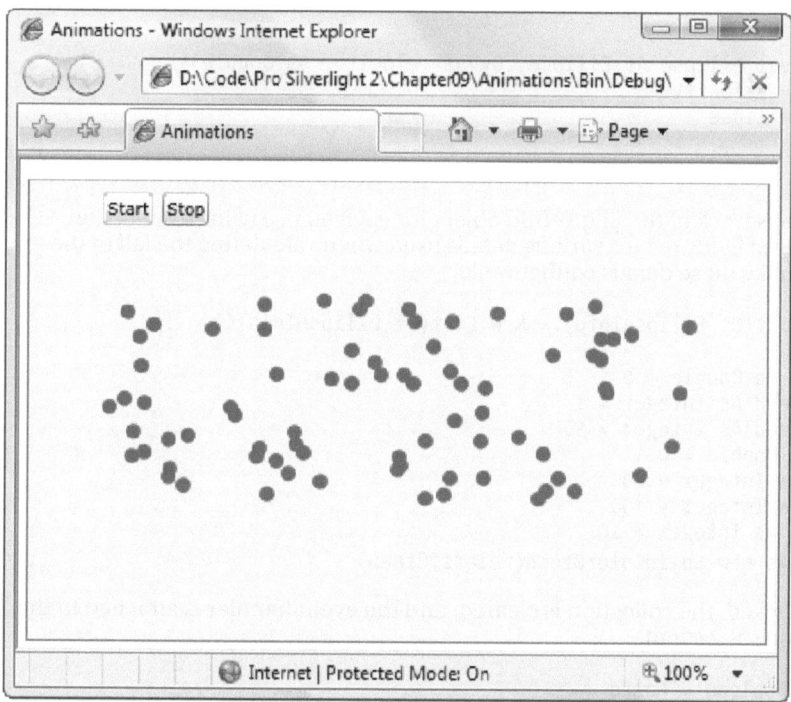

Figure 10-14. A frame-based animation of falling circles

In this example, each falling circle is represented by an Ellipse element. A custom class named EllipseInfo keeps a reference to the ellipse and tracks the details that are important for the physics model. In this case, there's only one piece of information: the velocity at which the ellipse is moving along the y-axis. (You could easily extend this class to include a velocity along the x-axis, additional acceleration information, and so on.)

```
Public Class EllipseInfo
    Private _ellipse As Ellipse
    Public Property Ellipse() As Ellipse
        Get
```

```
                    Return _ellipse
                End Get
                Set(ByVal value As Ellipse)
                    _ellipse = value
                End Set
        End Property

        Private _velocityY As Double
        Public Property VelocityY() As Double
            Get
                    Return _velocityY
            End Get
            Set(ByVal value As Double)
                    _velocityY = value
            End Set
        End Property

        Public Sub New(ByVal ellipse As Ellipse, ByVal velocityY As Double)
            Me.VelocityY = velocityY
            Me.Ellipse = ellipse
        End Sub
    End Class
End Class
```

The application keeps track of the EllipseInfo object for each ellipse using a collection. Several more window-level fields record various details used when calculating the fall of the ellipse. You can easily make these details configurable.

```
Private ellipses As List(Of EllipseInfo) = New List(Of EllipseInfo)()

Private accelerationY As Double = 0.1
Private minStartingSpeed As Integer = 1
Private maxStartingSpeed As Integer = 50
Private speedRatio As Double = 0.1
Private minEllipses As Integer = 20
Private maxEllipses As Integer = 100
Private ellipseRadius As Integer = 10
Private ellipseBrush As New SolidColorBrush(Colors.Green)
```

When a button is clicked, the collection is cleared, and the event handler is attached to the CompositionTarget.Rendering event:

```
Private rendering As Boolean = False

Private Sub cmdStart_Clicked(ByVal sender As Object, ByVal e As RoutedEventArgs)
    If Not rendering Then
        ellipses.Clear()
        canvas.Children.Clear()

        AddHandler CompositionTarget.Rendering, AddressOf RenderFrame
        rendering = True
    End If
End Sub
```

If the ellipses don't exist, the rendering code creates them automatically. It creates a random number of ellipses (currently, between 20 and 100) and gives each of them the same size and color. The ellipses are placed at the top of the Canvas, but they're offset randomly along the x-axis, and each one is given a random starting speed:

```
Private Sub RenderFrame(ByVal sender As Object, ByVal e As EventArgs)
    If ellipses.Count = 0 Then
        ' Animation just started. Create the ellipses.
        Dim halfCanvasWidth As Integer = CInt(Fix(canvas.ActualWidth)) / 2

        Dim rand As New Random()
        Dim ellipseCount As Integer = rand.Next(minEllipses, maxEllipses + 1)
        For i As Integer = 0 To ellipseCount - 1
            ' Create the ellipse.
            Dim ellipse As New Ellipse()
            ellipse.Fill = ellipseBrush
            ellipse.Width = ellipseRadius
            ellipse.Height = ellipseRadius

            ' Place the ellipse.
            canvas.SetLeft(ellipse, _
              halfCanvasWidth + rand.Next(-halfCanvasWidth, halfCanvasWidth))
            canvas.SetTop(ellipse, 0)
            canvas.Children.Add(ellipse)

            ' Track the ellipse.
            Dim info As New EllipseInfo(ellipse, _
              speedRatio * rand.Next(minStartingSpeed, maxStartingSpeed))
            ellipses.Add(info)
        Next
        ...
```

If the ellipses already exist, the code tackles the more interesting job of animating them. Each ellipse is moved slightly using the Canvas.SetTop() method. The amount of movement depends on the assigned velocity.

```
        ...
    Else
        For i As Integer = ellipses.Count - 1 To 0 Step -1
            Dim info As EllipseInfo = ellipses(i)
            Dim top As Double = canvas.GetTop(info.Ellipse)
            canvas.SetTop(info.Ellipse, top + 1 * info.VelocityY)
            ...
```

To improve performance, the ellipses are removed from the tracking collection as soon as they've reached the bottom of the Canvas. That way, you don't need to process them again. To allow this to work without causing you to lose your place while stepping through the collection, you need to iterate backward, from the end of the collection to the beginning.

If the ellipse hasn't yet reached the bottom of the Canvas, the code increases the velocity. (Alternatively, you could set the velocity based on how close the ellipse is to the bottom of the Canvas for a magnet-like effect.)

```
            ...
            If top >= (canvas.ActualHeight - ellipseRadius * 2 - 10) Then
                ' This circle has reached the bottom.
                ' Stop animating it.
                ellipses.Remove(info)
```

```
            Else
                ' Increase the velocity.
                info.VelocityY += accelerationY
            End If
            ...
```

Finally, if all the ellipses have been removed from the collection, the event handler is removed, allowing the animation to end:

```
            ...
            If ellipses.Count = 0 Then
                ' End the animation.
                ' There's no reason to keep calling this method
                ' if it has no work to do.
                RemoveHandler CompositionTarget.Rendering, AddressOf RenderFrame
                rendering = False
            End If
        Next
    End If
End Sub
```

Obviously, you can extend this animation to make the circles bounce, scatter, and so on. The technique is the same—you need to use more complex formulas to arrive at the velocity.

There's one caveat to consider when building frame-based animations: they aren't time-dependent. In other words, your animation may run faster on fast computers, because the frame rate will increase, and your CompositionTarget.Rendering event will be called more frequently. To compensate for this effect, you need to write code that takes the current time into account.

Animation Performance

Often, an animated user interface requires little more than creating and configuring the right animation and storyboard objects. But in other scenarios, particularly ones in which you have multiple animations taking place at the same time, you may need to pay more attention to performance. Certain effects are more likely to cause these issues—for example, those that involve video, large bitmaps, and multiple levels of transparency typically demand more from the computer's CPU. If they're not implemented carefully, they may run with notable jerkiness, or they may steal CPU time away from other applications that are running at the same time.

Fortunately, Silverlight has a few tricks that can help you out. In the following sections, you'll learn to slow down the maximum frame rate and use cache bitmaps on the computer's video card, two techniques that can lessen the load on the CPU. You'll also learn about a few diagnostic tricks that can help you determine whether your animation is running at its best or facing potential problems.

Desired Frame Rate

As you've already learned, much of Silverlight animation uses interpolation, which modifies a property smoothly from its starting point to its end point. For example, if you set a starting value of 1 and an ending value of 10, your property may be rapidly changed from 1 to 1.1, 1.2, 1.3, and so on, until the value reaches 10.

You may wonder how Silverlight determines the increments it uses when performing interpolation. Happily, this detail is taken care of automatically. Silverlight uses whatever increment it needs to ensure a smooth animation at the currently configured frame rate. The standard frame rate Silverlight uses is 60 frames per second. In other words, every 1/60th of a second, Silverlight calculates all animated values and updates the corresponding properties. A rate of 60 frames per second ensures smooth, fluid animations from start to finish. (Of course, Silverlight may not be able to deliver on its intentions, depending on its performance and the client's hardware.)

Silverlight makes it possible for you to decrease the frame rate. You may choose to do this if you know your animation looks good at a lower frame rate, so you don't want to waste the extra CPU cycles. Or, you may find that your animation performs better on lesser-powered computers when it runs at a slower frame rate. On the Web, many animations run at a more modest 15 frames per second.

To adjust the frame rate, you need to add the maxFramerate parameter to the entry page for your application, as shown here:

```
<div id="silverlightControlHost">
  <object data="data:application/x-silverlight-2,"
   type="application/x-silverlight-2" width="100%" height="100%">
    <param name="maxFramerate" value="15" />
    ...
  </object>
  <iframe style="visibility:hidden;height:0;width:0;border:0px"></iframe>
</div>
```

■ **Tip** For the best animation performance, use transparency sparingly, avoid animating text size (because font smoothing and hinting slow down performance), and don't use the windowless setting discussed in Chapter 9 (which lets HTML elements show through the Silverlight content region).

Hardware Acceleration

The holy grail of graphics programming is to offload most of the work to the graphics processing unit (GPU) on the computer's video card. After all, video cards are specially designed to be able to handle certain types of graphical tasks (for example, bitmap scaling) quickly and efficiently. But when running a typical web application, your video card is hardly working at all. Surely, it makes sense to enlist their help and free up the much more valuable CPU.

■ **Note** Technically, offloading work to the GPU is called *hardware acceleration*, because this technique speeds up complex video tasks such as 3-D rendering in cutting-edge computer games. In a Silverlight application, hardware acceleration can reduce the load on the CPU and it may improve the frame rate of your animations (allowing them to run more smoothly).

Unfortunately, implementing hardware acceleration is not as easy as it seems. The first problem is that hardware acceleration requires an extra layer of video card support on the platform that's running the application. For a Silverlight application running on a Windows computer, that means you'll need a DirectX 9–compatible video card and drivers. On Mac OS X, you'll need an OpenGL2-compatible video card with drivers. Furthermore, hardware acceleration works on a Mac hardware only when your application is running in full-screen mode (as described in Chapter 3). The Windows implementation of Silverlight doesn't have the same limitation.

The second problem is that video cards are designed to accelerate certain specific graphic operations (for example, shading in the tiny triangles that make up 3D scenes). Many of these optimizations aren't suited to Silverlight applications. In fact, Silverlight applications use just one type of optimization: the ability of a video card to cache some visual element as a bitmap and (optionally) scale it, clip it, rotate it, or make it partially transparent. Other types of hardware acceleration might be possible, but they aren't currently implemented in Silverlight.

Enabling Hardware Acceleration

Before you can even consider using hardware acceleration in a portion of your application, you need to configure the test page to support it. You do this by adding the enableGPUAcceleration parameter and setting it true, as shown here:

```
<div id="silverlightControlHost">
  <object data="data:application/x-silverlight-2,"
    type="application/x-silverlight-2" width="100%" height="100%">
    <param name="enableGPUAcceleration" value="true" />
    <param name="enableCacheVisualization" value="true" />
    <param name="enableFrameRateCounter" value="true" />
    ...
  </object>
  <iframe style="visibility:hidden;height:0;width:0;border:0px"></iframe>
</div>
```

You'll notice this example also adds two optional parameters that work in conjunction with hardware acceleration. The enableCacheVisualization parameter uses tinting to highlight areas of your application that aren't taking advantage of bitmap caching on the video card. The enableFrameRateCounter parameter displays a frame rate counter that updates itself continuously as your animations run. Both of these parameters give you helpful diagnostic tools that allow you to evaluate performance during testing. You'll remove them in the final version of your application.

Setting the enableGPUAcceleration property has no immediate effect. It gives you the ability to switch on bitmap caching for individual elements. But until you take this step, you won't notice any change in your application's performance.

Bitmap Caching

Bitmap caching tells Silverlight to take a bitmap image of your content as it currently is and copy that to the memory on your video card. From this point on, the video card can take charge of manipulating the bitmap and refreshing the display. This process is far faster than getting the Silverlight runtime to do all the work and communicate continuously with the video card.

However, there's a catch. The video card is limited in what it can do with the bitmap. It supports the following operations:

- Scaling the bitmap (with a RenderTransform)
- Rotating the bitmap (with a RenderTransform)
- Changing the opacity of the bitmap (using the Opacity property)
- Clipping the bitmap with a rectangular clipping region (using the Clip property)

Thus, if you have animations that perform scaling, rotation, or fading on an element, you'll get a benefit from hardware acceleration. However, if you have animations that do anything else to change the way an element looks—for example, skewing an element, changing its color, rotating it in 3D space with a perspective transform, applying a pixel shader, and so on, you should definitely *not* use bitmap caching. In this sort of situation, Silverlight will be forced to keep passing an updated copy of the bitmap back to the video card, updating its cache several times a second. This process will actually *decrease* performance.

To switch on bitmap caching, you set the CacheMode property of the corresponding element to BitmapCache. Every element provides this property, which means you have a fine-grained ability to choose exactly which elements use this feature.

■ **Note** If you cache an element that contains other elements, like a layout container, all the elements will be cached in a single bitmap. Thus, you need to be extremely careful about adding caching to something like a Canvas—only do it if all the children are limited to the allowed transformations in the previous list.

To get a better understanding, it helps to play with a simple example. Figure 10-15 shows a project that's included with the downloadable samples for this chapter. Here, two animations are at work. The first rotates an Image element that contains the picture of a phone booth. The second one changes the size of a button using a ScaleTransform, endlessly expanding and shrinking it. Both animations are clear candidates for bitmap caching.

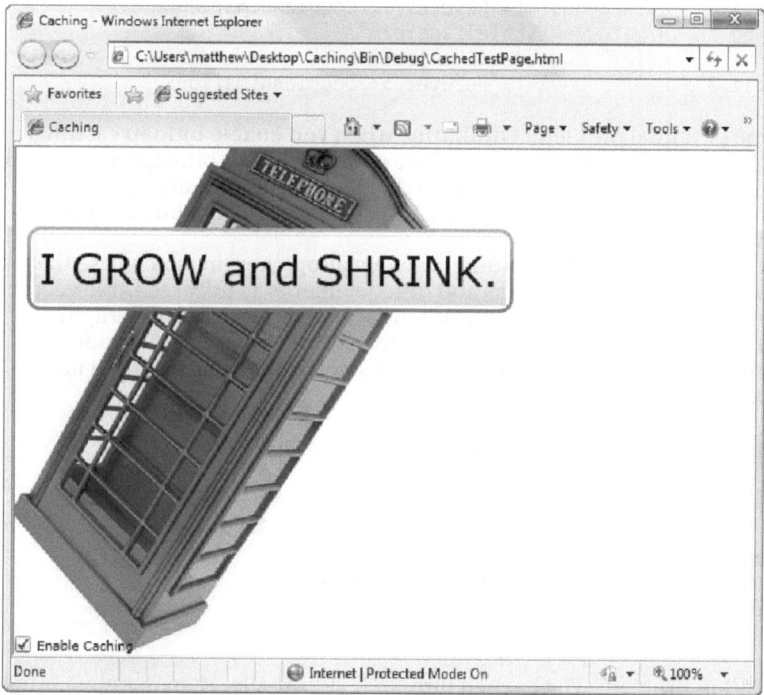

Figure 10-15. A test page with two animated elements

Here's the markup that switches on bitmap caching for both:

```
<Canvas>
  <Image x:Name="img" Source="phone_booth.jpg" Stretch="None"
   CacheMode="BitmapCache">
    <Image.RenderTransform>
      <RotateTransform x:Name="rotateTransform"></RotateTransform>
    </Image.RenderTransform>
  </Image>

  <Button x:Name="cmd" Content="I GROW and SHRINK." Canvas.Top="70" Canvas.Left="10"
   CacheMode="BitmapCache">
    <Button.RenderTransform>
      <ScaleTransform x:Name="scaleTransform"></ScaleTransform>
    </Button.RenderTransform>
  </Button>
</Canvas>
```

And here's the markup that declares the animations:

```
<Storyboard x:Name="storyboard">
  <DoubleAnimation Storyboard.TargetName="rotateTransform"
   Storyboard.TargetProperty="Angle" To="360" Duration="0:0:2"
   RepeatBehavior="Forever"></DoubleAnimation>
  <DoubleAnimation Storyboard.TargetName="scaleTransform"
   Storyboard.TargetProperty="ScaleX" AutoReverse="True"
   To="20" Duration="0:0:1.8" RepeatBehavior="Forever"></DoubleAnimation>
  <DoubleAnimation Storyboard.TargetName="scaleTransform"
   Storyboard.TargetProperty="ScaleY" AutoReverse="True"
   To="20" Duration="0:0:1.8" RepeatBehavior="Forever"></DoubleAnimation>
</Storyboard>
```

Bitmap caching has one potential problem. Ordinarily, when you enable bitmap caching, Silverlight takes a snapshot of the element at its current size and copies that bitmap to the video card. If you then use a ScaleTransform to make the bitmap bigger, you'll be enlarging the cached bitmap, not the actual element. In the current example, that means the button will grow fuzzy and pixelated as it grows.

To solve this problem, you could switch on bitmap caching altogether (in which case the effect disappears, because Silverlight treats buttons and other elements as fully resizable vector images). However, another option is to explicitly indicate the size of bitmap that Silverlight should cache on the video card, using the BitmapCache.RenderAtScale property. Ordinarily, this property is set to 1, and the element is taken at its current size. But the markup here takes a snapshot of the button at five times its current size:

```
<Button x:Name="cmd" Content="I GROW and SHRINK." Canvas.Top="70" Canvas.Left="10">
  <Button.CacheMode>
    <BitmapCache RenderAtScale="5"></BitmapCache>
  </Button.CacheMode>
  <Button.RenderTransform>
    <ScaleTransform x:Name="scaleTransform"></ScaleTransform>
  </Button.RenderTransform>
</Button>
```

This resolves the pixelation problem. The cached bitmap is still smaller than the maximum animated size of the button (which reaches ten times its original size), but the video card is able to double the size of the bitmap from five to ten times the size without any obvious scaling artifacts. There are only two potential disadvantages to increasing the RenderAtScale property. First, you're forcing Silverlight to transfer more data to the video card (which slows the initial rendering step). Second, you're asking the video card to use more of its onboard video memory. Different video cards have different amounts of memory, and when the available memory is used up, the video card won't be able to cache any more bitmaps, and Silverlight will fall back on software rendering.

Evaluating Hardware Acceleration

The easiest way to evaluate the success of your bitmap caching is to run your application both with and without hardware acceleration. In most cases, the difference won't be obvious until you check the CPU usage of your computer or the frame rate of your animation. To check the CPU usage, load Task Manager and watch the Performance tab. In an informal test with the previous example, CPU usage on a single-processor computer dropped from about 50% to about 20% when caching was switched on. The downloadable samples for this chapter include an example that allows you to switch caching on and off using a check box. The change is performed programmatically using code like this:

```
img.CacheMode = New BitmapCache()
```

Another useful tool is Silverlight's built-in diagnostic support. Earlier, you learned about the enableCacheVisualization and enableFrameRateCounter parameters, which you can add to your test page to capture some extra diagnostic information. Figure 10-16 shows an example where both parameters are switched on and caching is turned off.

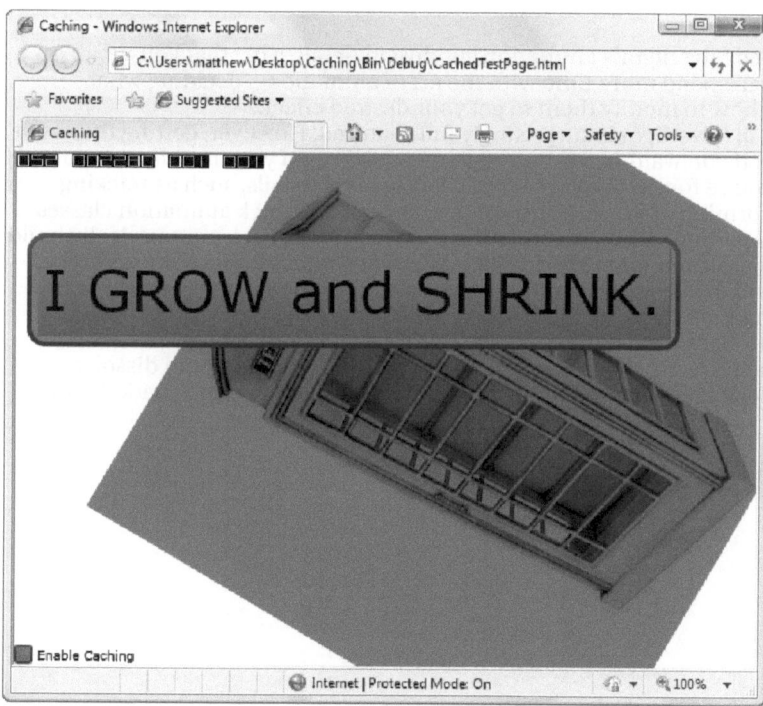

Figure 10-16. Using cache visualization and the frame rate counter

Here, the Image and Button elements are tinted red to indicate that they aren't being cached (thanks to enableCacheVisualization). The set of numbers in the top-left corner provides frame rate information (thanks to enableFrameRateCounter), as follows:

- The first number shows the animation frame rate. In this example, switching off caching drops it from 55 to 35. (Remember, the default maximum frame rate is 60.)
- The second number shows how many kilobytes of video card memory are used. This increases when caching is turned on.
- The third number shows the total number of hardware-accelerated surfaces. Remember, switching bitmap caching on for one element will usually affect several surfaces—even in the case of the button, there is a TextBlock with content inside.
- The fourth number shows the number of implicit hardware-accelerated surfaces. In some situations, switching caching on for one element may necessitate turning it on for another (for example, if the second element overlaps the first one). In this case, Silverlight will automatically perform caching for the additional element, which is known as an implicit surface.

The bottom line is that you can quickly size up an example like this and determine that bitmap caching makes sense. In this scenario, it both reduces the CPU load and improves the frame rate.

The Last Word

In this chapter, you explored Silverlight's animation support in detail. Now that you've mastered the basics, you can spend more time with the art of animation—deciding what properties to animate and how to modify them to get your desired effect.

The animation model in Silverlight is surprisingly full-featured. However, getting the result you want isn't always easy. If you want to animate separate portions of your interface as part of a single animated scene, you're forced to take care of a few tedious details, such as tracking animated objects and performing cleanup. Furthermore, none of the stock animation classes accepts arguments in their parameters. As a result, the code required to programmatically build a new animation is often simple but long. The future of Silverlight animation promises higher-level classes that are built on the basic plumbing you've learned about in this chapter. Ideally, you'll be able to plug animations into your application by using prebuilt animation classes, wrapping your elements in specialized containers, and setting a few attached properties. The actual implementation that generates the effect you want—whether it's a smooth dissolve between two images or a series of animated fly-ins that builds a page—will be provided for you.

■■■

CHAPTER 11

Sound, Video, and Deep Zoom

In this chapter, you'll tackle one of Silverlight's most mature features: audio and video support.

Since version 1.0, Silverlight has distinguished itself as a technology that brings high-end multimedia support to the limited world of the browser. And though Silverlight can't support the full range of media codecs (because that would multiply the size of the Silverlight download and increase its licensing costs), Silverlight still gives you everything you need to incorporate high-quality audio and video in your applications. Even more remarkable is the way that Silverlight allows you to *use* multimedia, particularly video. For example, you can use video to fill thousands of elements at once and combine it with other effects, such as animation, transforms, and transparency.

In this chapter, you'll learn how to incorporate ordinary audio and video into your applications, and you'll consider the best way to encode and host video files for Silverlight. Next, you'll see how Silverlight's VideoBrush class allows you to create impressive effects such as video-filled text and video reflections. Finally, you'll look at Deep Zoom—a different interactive multimedia technology that lets users zoom into massive images in real time.

■ **What's New** Silverlight adds support for webcam and microphone input. However, unless you're willing to embark of a serious coding challenge (for example, handling chunks of video and encoding them in real time), you won't find this support terribly practical. For the full details, see the "Webcam and Microphone Input" section in this chapter.

Supported File Types

Because Silverlight needs to ensure compatibility on a number of different operating systems and browsers, it can't support the full range of media files that you'll find in a desktop application such as Windows Media Player. Before you get started with Silverlight audio and video, you need to know exactly what media types it supports.

For audio, Silverlight supports the following:

- Windows Media Audio (WMA) versions 7, 8, and 9
- MP3 with fixed or variable bit rates from 8 to 320 Kbps

When it comes to video, Silverlight supports the follow standards:

- Windows Media Video 7 (WMV1)
- Windows Media Video 8 (WMV2)
- Windows Media Video 9 (WMV3)
- Windows Media Video Advanced Profile, non-VC-1 (WMVA)
- Windows Media Video Advanced Profile, VC-1 (WMVC1)
- H.264 video and AAC audio (also known as MPEG-4 Part 10 or MPEG-4 AVC)

Often, you can recognize Windows Media Video by the file extension .wmv. Other video formats—for example, MPEG and QuickTime—need not apply.

The last two formats in this list—VC-1 and H.264—are widely supported industry standards. Notable places where they're used include Blu-ray, HD DVD, and the Xbox 360. They're also the most common choice for Silverlight applications. (Of course, these standards support different bit rates and resolutions, so your Silverlight application isn't forced to include DVD-quality video just because it uses VC-1 or H.264.)

Silverlight doesn't support other Windows Media formats (such as Windows Media Screen, Windows Media Audio Professional, and Windows Media Voice), nor does it support the combination of Windows Media Video with MP3 audio. Finally, it doesn't support video files that use frames with odd-number dimensions (dimensions that aren't divisible by 2), such as 127×135.

■ **Note** Adding audio to a Silverlight application is fairly easy, because you can throw in just about any MP3 file. Using a video file is more work. Not only must you make sure you're using one of the supported WMV formats, but you also need to carefully consider the quality you need and the bandwidth your visitors can support. Later in this chapter, you'll consider how to encode video for a Silverlight application. But first, you'll consider how to add basic audio.

The MediaElement

In Silverlight, all the audio and video functionality is built into a single class: MediaElement.

Like all elements, a media element is placed directly in your user interface. If you're using the MediaElement to play audio, this fact isn't important, because the MediaElement remains invisible. If you're using the MediaElement for video, you place it where the video window should appear.

A simple MediaElement tag is all you need to play a sound. For example, add this markup to your user interface:

```
<MediaElement Source="test.mp3"></MediaElement>
```

Now, once the page is loaded, it will download the test.mp3 file and begin playing it automatically.

Of course, for this to work, your Silverlight application needs to be able to find the test.mp3 file. The MediaElement class uses the same URL system as the Image class. That means you can embed a media file in your XAP package or deploy it to the same website alongside the XAP file. Generally, it's best to keep media files separate, unless they're extremely small. Otherwise, you'll bloat the size of your application and lengthen the initial download time.

■ **Note** When you first add a media file like test.mp3 to a project, Visual Studio sets its Build Action setting to None and its Copy To Output Directory setting to "Do not copy." To deploy your media file alongside your XAP file, you must change the Copy To Output Directory setting to "Copy always." To deploy your media file inside the XAP package, change Build Action to Resource. The downloadable code for this chapter uses the first of these two approaches.

Controlling Playback

The previous example starts playing an audio file immediately when the page with the MediaElement is loaded. Playback continues until the audio file is complete.

Although this example is straightforward, it's also a bit limiting. Usually, you'll want the ability to control playback more precisely. For example, you may want it to be triggered at a specific time, repeated indefinitely, and so on. One way to achieve this result is to use the methods of the MediaElement class at the appropriate time.

The startup behavior of the MediaElement is determined by its AutoPlay property. If this property is set to False, the audio file is loaded, but your code takes responsibility for starting the playback at the right time:

```
<MediaElement x:Name="media" Source="test.mp3" AutoPlay="False"></MediaElement>
```

When using this approach, you must make sure to give the MediaElement a name so that you can interact with it in code. Generally, interaction consists of calling the Play(), Pause(), and Stop() methods. You can also use the SetSource() method to load new media content from a stream (which is useful if you're downloading media files asynchronously using the WebClient class, as described in Chapter 6), and you can change the Position property to move through the audio.

Here's a simple event handler that seeks to the beginning of the current audio file and then starts playback:

```
Private Sub cmdPlay_Click(ByVal sender As Object, ByVal e As RoutedEventArgs)
    media.Position = TimeSpan.Zero
    media.Play()
End Sub
```

If this code runs while playback is already under way, the first line resets the position to the beginning, and playback continues from that point. In this case, the second line has no effect because the media file is already being played.

■ **Note** Depending on the types of media files you support, you may want to check the CanPause and CanSeek properties before you attempt to pause playback or jump to a new position. Some types of streamed media files don't support pausing and seeking.

Handling Errors

MediaElement doesn't throw an exception if it can't find or load a file. Instead, it's up to you to handle the MediaFailed event. Fortunately, this task is easy. First, tweak your MediaElement tag as shown here:

```
<MediaElement ... MediaFailed="media_MediaFailed"></MediaElement>
```

Then, in the event handler, you can use the ExceptionRoutedEventArgs.ErrorException property to get an exception object that describes the problem. Here's an example that displays the appropriate error message:

```
Private Sub media_MediaFailed(ByVal sender As Object, _
  ByVal e As ExceptionRoutedEventArgs)
    lblErrorText.Text = e.ErrorException.Message
End Sub
```

Playing Multiple Sounds

The MediaElement is limited to playing a single media file. If you change the Source property (or call the SetSource() method), any playback that's currently taking place stops immediately. However, this limitation doesn't apply to Silverlight as a whole. Silverlight can quite easily play multiple media files at once, as long as each one has its own MediaElement.

You can use two approaches to create an application with multiple sounds. Your first option is to create all the MediaElement objects you need at design time. This approach is useful if you plan to reuse the same two or three MediaElement objects. For example, you can define two MediaElement objects and flip between them each time you play a new sound. (You can keep track of which object you used last using a Boolean variable in your page class.) To make this technique really effortless, you can store the audio file names in the Tag property of the appropriate element, so all your event-handling code needs to do is read the file name from the Tag property, find the right MediaElement to use, set its Source property, and then call its Play() method. Because this example uses two MediaElement objects, you're limited to two simultaneous sounds, which is a reasonable compromise if you don't think the user will be able to pick out a third sound out over the din anyway.

Your other option is to create every MediaElement object you need dynamically. This approach requires more overhead, but the difference is minimal (unless you go overboard and play dozens of simultaneous media files). When you create a MediaElement in code, you need to remember to add it to a container in your application. Assuming you haven't changed the AutoPlay property, the MediaElement will begin playing as soon as you add it. If you set AutoPlay to False, you'll need to use the Play() method. Finally, it's also a good idea to handle the MediaEnded event to remove the MediaElement after playback is finished.

Here's some code for a button that starts a new playback of the same sound file each time it's clicked:

```
Private Sub cmdPlay_Click(ByVal sender As Object, ByVal e As RoutedEventArgs)
    Dim media As New MediaElement()
    media.Source = New Uri("test.mp3", UriKind.Relative)
    AddHandler media.MediaEnded, AddressOf media_MediaEnded
    LayoutRoot.Children.Add(media)
End Sub

Private Sub media_MediaEnded(ByVal sender As Object, ByVal e As RoutedEventArgs)
    LayoutRoot.Children.Remove(CType(sender, MediaElement))
End Sub
```

To make it easier to keep track of a batch of dynamically generated MediaElement objects, you can add them all to a designated container (for example, an invisible stack panel). This allows you to quickly examine all the currently playing media files and stop them all. Figure 11-1 shows an example that uses this approach and displays the element count of the invisible StackPanel every time a MediaElement is inserted or removed.

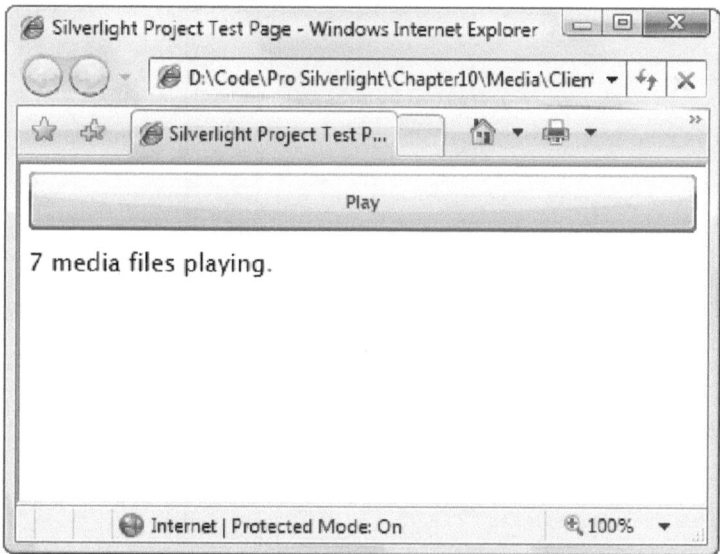

Figure 11-1. Playing media files simultaneously

Changing Volume, Balance, and Position

The MediaElement exposes a number of properties that allow you to control your playback. The most fundamental are as follows:

- *Volume:* Sets the volume as a number from 0 (completely muted) to 1 (full volume). The default value is 0.5. To temporarily mute playback without pausing it or changing the volume setting, set IsMuted to True.

- *Balance:* Sets the balance between the left and right speaker as a number from -1 (left speaker only) to 1 (right speaker only). The default is 0, which splits the sound evenly.

- *CurrentState:* Indicates whether the player is currently Playing, Paused, Stopped, downloading a media file (Opening), buffering it (Buffering), or acquiring a license for DRM content (AcquiringLicense). If no media file was supplied, CurrentState is Closed.

- *Position:* Provides a TimeSpan object that indicates the current location in the media file. You can set this property to skip to a specific time position.

Figure 11-2 shows a simple page that allows the user to control playback.

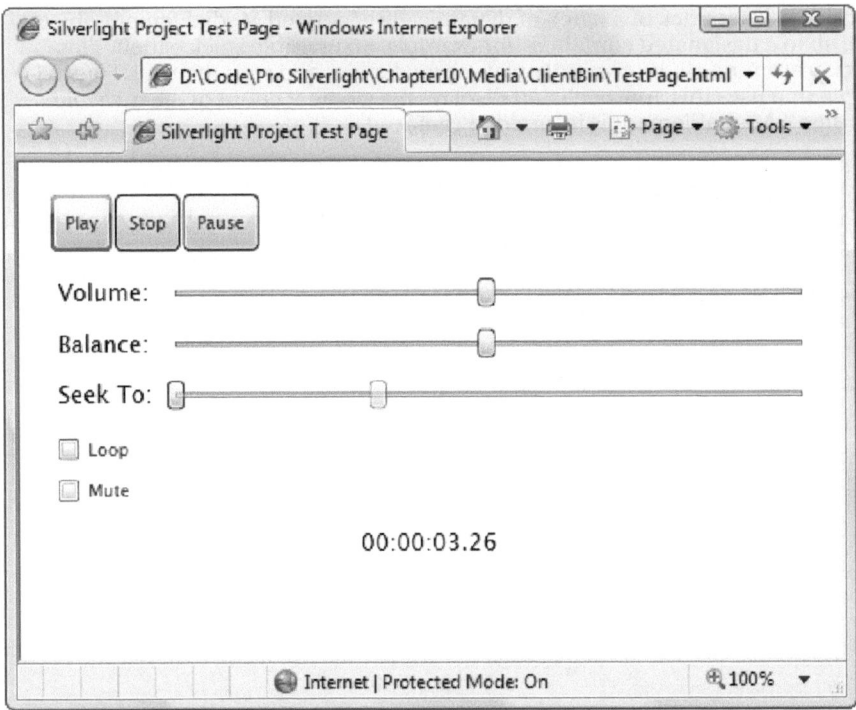

Figure 11-2. *Controlling more playback details*

At the top of the window are three buttons for controlling playback. They use rather unremarkable code—they call the Start(), Stop(), and Play() methods of the MediaElement when clicked.

Underneath are two sliders for adjusting volume and balance. These sliders are set to the appropriate ranges (0 to 1 and -1 to 1):

```
<Slider Grid.Column="1" x:Name="sliderVolume" Minimum="0" Maximum="1" Value="0.5"
  ValueChanged="sliderVolume_ValueChanged" ></Slider>

<Slider Grid.Row="1" Grid.Column="1" x:Name="sliderBalance" Minimum="-1" Maximum="1"
  ValueChanged="sliderBalance_ValueChanged"></Slider>
```

When the user drags the thumb in the slider, the change is applied to the MediaElement:

```
Private Sub sliderVolume_ValueChanged(ByVal sender As Object, _
  ByVal e As RoutedPropertyChangedEventArgs(Of Double))
    media.Volume = sliderVolume.Value
End Sub

Private Sub sliderBalance_ValueChanged(ByVal sender As Object, _
  ByVal e As RoutedPropertyChangedEventArgs(Of Double))
    media.Balance = sliderBalance.Value
End Sub
```

The third slider lets the user jump to a new position. It actually consists of two sliders that are superimposed on top of one another. The slider in the background (the one defined first) is the position slider that the user drags to jump to a new part of the audio file:

```
<Slider Minimum="0" Grid.Column="1" Grid.Row="2" x:Name="sliderPosition"
 ValueChanged="sliderPosition_ValueChanged"></Slider>
```

In front is a slider that ignores mouse activity (because its IsHitTestVisible property is set to False) and is partially transparent (because its Opacity property is set to 0.5). As a result, the slider appears to be a faint image behind the position slider:

```
<Slider Minimum="0" Grid.Column="1" Grid.Row="2" x:Name="sliderPositionBackground"
 IsHitTestVisible="False" Opacity="0.5"></Slider>
```

This slider (sliderPositionBackground) represents the current position of the audio file. As the audio advances, the code moves the thumb in sliderPositionBackground along the track to give the user a visual indication of how far playback has progressed. You could do much the same trick by moving the sliderPosition slider, but this could become problematic because your code would need to distinguish between user-initiated changes (when the user drags the slider, at which point your code should change the current position of the MediaElement) and playback synchronization (at which point your code should do nothing).

The code sets up the position sliders by reading the full running time from the NaturalDuration property after the media file has been opened:

```
Private Sub media_MediaOpened(ByVal sender As Object, ByVal e As RoutedEventArgs)
    sliderPosition.Maximum = media.NaturalDuration.TimeSpan.TotalSeconds
    sliderPositionBackground.Maximum = media.NaturalDuration.TimeSpan.TotalSeconds
End Sub
```

You can then jump to a specific position when the topmost slider tab is moved:

```
Private Sub sliderPosition_ValueChanged(ByVal sender As Object, _
  ByVal e As RoutedEventArgs)
    ' Pausing the player before moving it reduces audio "glitches"
    ' when the value changes several times in quick succession.
    media.Pause()
    media.Position = TimeSpan.FromSeconds(sliderPosition.Value)
    media.Play()
End Sub
```

Incidentally, the MediaElement doesn't fire any sort of event to notify you that playback is underway. Thus, if you want to move the thumb for sliderPositionBackground along the track or you want to update the TextBlock with the current time offset at the bottom of the page, you need to use a timer.

The DispatcherTimer is a perfect solution. You can create one when the page loads, use a short 0.1-second interval, and start and stop it along with your playback:

```
Private timer As New DispatcherTimer()

Public Sub New()
    InitializeComponent()
    timer.Interval = TimeSpan.FromSeconds(0.1)
    AddHandler timer.Tick, AddressOf timer_Tick
End Sub

Private Sub cmdPlay_Click(ByVal sender As Object, ByVal e As RoutedEventArgs)
    media.Play()
    timer.Start()
End Sub
```

When the DispatcherTimer.Tick event fires, you can update your user interface by displaying the current time position in a TextBlock and moving the position indicator (the semi-transparent noninteractive thumb of the background slider):

```
Private Sub timer_Tick(ByVal sender As Object, ByVal e As EventArgs)
    lblStatus.Text = media.Position.ToString().TrimEnd(New Char(){"0"c})
    sliderPositionBackground.Value = media.Position.TotalSeconds
End Sub
```

The two check boxes on the page are the last ingredient in this media player and one of the simplest details. The Mute check box sets the corresponding IsMuted property of the MediaElement:

```
Private Sub chkMute_Click(ByVal sender As Object, ByVal e As RoutedEventArgs)
    media.IsMuted = CBool(chkMute.IsChecked)
End Sub
```

The MediaElement has no built-in support for looping playback. If the Loop check box is set, the code in the page restarts playback when the MediaEnded event fires:

```
Private Sub media_MediaEnded(ByVal sender As Object, ByVal e As RoutedEventArgs)
    If CBool(chkLoop.IsChecked) Then
        media.Position = TimeSpan.Zero
        media.Play()
    Else
        timer.Stop()
    End If
End Sub
```

Although relatively simple, this example could be the springboard for a more advanced player—all you need is a heavy dose of animation, transparency, and eye candy. You'll see some examples of more stylized media players that have mostly the same functionality when you consider Expression Encoder later in this chapter.

THE RAW AUDIO/VIDEO PIPELINE

One of the best kept secrets in Silverlight is its support for raw audio and video. This support allows a Silverlight application to decode chunks of audio and stream them to a MediaElement for playback. Needless to say, the process is tedious, quite complex, and sometimes hampered by latency issues. It's also far beyond the scope of this chapter.

Although most developers are unlikely to ever deal directly with the raw audio and video pipeline, you may well use other components that are based on this support. For example, third-party developers can use the raw audio and video pipeline to create libraries for playing back new media formats, implementing cutting-edge applications such as a virtual synthesizer, or supporting practical features such as seamless audio looping. For an example, check out the free MediaStreamSource that allows Silverlight to play PCM-encoded WAV audio at http://code.msdn.microsoft.com/wavmss.

Playing Video

Everything you've learned about using the MediaElement class applies equally well when you use a video file instead of an audio file.

The key difference with video files is that the visual and layout-related properties of the MediaElement are suddenly important. The original size of the video is provided through the NaturalVideoHeight and NaturalVideoWidth properties of the MediaElement. You can also scale or stretch a video to fit different page sizes using the Stretch property. Use None to keep the native size (which is recommended for optimum performance), Uniform to stretch the video to fit its container without changing its aspect ratio (which is the default), Fill to stretch it to fit its container in both dimensions (even if that means stretching the picture), and UniformToFill to resize the picture to fit the largest dimension of its container while preserving its aspect ratio (which guarantees that part of the video page will be clipped out if the container doesn't have the same aspect ratio as the video).

■ **Tip** The MediaElement's preferred size is based on the native video dimensions. For example, if you create a MediaElement with a Stretch value of Uniform (the default) and place it inside a Grid row with a Height value of Auto, the row will be sized just large enough to keep the video at its standard size, so no scaling is required.

Client-Side Playlists

Silverlight also supports Windows Media *metafiles*, which are essentially playlists that point to one or more other media files. Windows Media metafiles typically have the file extension .wax, .wvx, .wmx, .wpl, or .asx. Certain features of these files, such as script commands, aren't supported and cause errors if used. For the full list of unsupported features, refer to the Silverlight documentation.

Here's a basic playlist that refers to two video files:

```
<asx version="3.0">
  <title>Two Video Playlist</title>
    <entry>
    <title>Video 1</title>
      <ref href="Video1.wmv" />
  </entry>
  <entry>
    <title>Video 2</title>
    <ref href="Video2.wmv" />
  </entry>
</asx>
```

If you point the Source property of the MediaElement to this file, it will begin playing Video1.wmv (assuming it exists) and then play Video2.wmv immediately after. In this case, both files are in the same location on the server (and in the same folder as the playlist), but you can adjust the href attribute to point to files in other folders or servers.

Typically, .asx files are used with .asf streaming files. In this case, the .asx file includes a link to the .asf streaming file.

Server-Side Playlists

If you're streaming video using Windows Media Services, you can also create a server-side playlist. Server-side playlists are processed on the server. They let you combine more than one video into a single stream without revealing the source of each video to the user. Server-side playlists offer one technique for integrating advertisements into your video stream: create a server-side playlist that places an ad before the requested video.

Server-side playlists often have the file extension .wsx. As with client-side playlists, they contain XML markup:

```
<?wsx version="1.0"?>
<smil>
  <seq id="sq1">
    <media id="video2" src="Video1.wmv" />
    <media id="video1" src="Advertisement.wmv" />
    <media id="video2" src="Video2.wmv" />
  <seq>
</smil>
```

The root element is <smil>. Here, the <smil> element contains an ordered sequence of video files represented by the <seq> element, with each video represented by the <media> element. More sophisticated server-side playlists can repeat videos, play clips of longer videos, and specify videos that will be played in the event of an error. For more information about the standard for .wsx files (and the elements that are supported and unsupported in Silverlight), see http://msdn.microsoft.com/library/cc645037.aspx.

Progressive Downloading and Streaming

Ordinarily, if you take no special steps, Silverlight plays media files using *progressive downloading*. This means that the client downloads media files one chunk at a time, using the standard HTTP protocol. When the client has accumulated enough of a buffer to provide for a few seconds of playback, it begins playing the media file and continues downloading the rest of the file in the background.

Thanks to progressive downloading, the client can begin playing a media file almost immediately. In fact, the total length of the file has no effect on the initial playback delay. The only factor is the *bit rate*—how many bytes of data it takes to play five seconds of media. Progressive downloading also has a second, not-so-trivial advantage: it doesn't require any special server software, because the client handles all the buffering. Thus, you can use progressive downloading with any web server.

The same isn't true of *streaming*, a technology that uses a specialized stateful protocol to send data from the web server to the client. Streaming has the instant-playback ability of progressive downloading, but it's more efficient. There are numerous factors at work, but switching from progressive downloading to streaming can net your web server a two- or three-times improvement in scalability—in other words, it may be able to serve the same video content to three times as many simultaneous users. This is the reason streaming is usually adopted.

However, streaming also has one significant disadvantage: it needs dedicated server-side software. (With Silverlight, this software is Windows Media Services, which is available as a free download for Windows Server 2008.) Unfortunately, it's considerably more complex to configure and maintain a media streaming server than it is to host an application that uses progressive downloading.

■ **Note** If you use a MediaElement with a URL that starts with http:// or https://, Silverlight begins a progressive download. If you use a MediaElement with a URL that starts with mms://, Silverlight attempts to stream it and falls back on a progressive download if streaming fails.

It's worth noting that the word *streaming* isn't always used in the technical sense described here. For example, Microsoft provides a fantastic free Silverlight hosting service called Silverlight Streaming. It provides 10 GB of hosting space for Silverlight applications and media files. But despite its name, Silverlight Streaming doesn't use streaming—instead, it simply serves video files and allows the client to perform progressive downloading.

IMPROVING PROGRESSIVE DOWNLOADING

If you don't want the complexity of configuring and maintaining a server with Windows Media Services or you use a web host that doesn't provide this service, your applications will use progressive downloading. You'll get the most out of progressive downloading if you follow these best practices:

- *Consider providing multiple versions of the same media file*: If you have huge media files and you need to support users with a wide range of connection speeds, consider including an option in your application that lets users specify their bandwidth. If a user specifies a low-speed bandwidth, you can seamlessly load smaller media files into the MediaElement. (The only problem is that average users don't always know their bandwidth, and the amount of video data a computer can handle can be influenced by other factors, such as the current CPU load or the quality of a wireless connection.)

- *Adjust the BufferingTime property on the MediaElement*: You can control how much content Silverlight buffers in a progressive download by setting the BufferingTime property of the MediaElement. The default is five seconds of playback, but higher-quality videos that will be played over lower-bandwidth connections will need different rates. A longer BufferingTime value won't allow a slow connection to play a high–bit rate video file (unless you buffer virtually the entire file), but it will smooth over unreliable connections and give a bit more breathing room.

- *Keep the user informed about the download*: It's often useful to show the client how much of a particular media file has been downloaded. For example, websites such as YouTube and players such as Media Player use a progress bar that has a shaded background, indicating how much of the file is available. To create a similar effect in a Silverlight application, you can use the DownloadProgressChanged event. It fires each time Silverlight crosses a 5% download threshold (for example, when it downloads the first 5%, when it reaches 10%, when it reaches 15%, and so on). It fires again when the file is completely downloaded. When the DownloadProgressChanged event fires, you can read the DownloadProgress property to determine how much of the file is currently available (as a value from 0 to 1). Use this information to set the width of a rectangle, and you're well on the way to creating a download progress bar.

- *Consider informing the user about the buffer*: You can react as the buffer is filled using the BufferingProgressChanged event and read the BufferingProgress property to find out how much content is in the buffer (as a value from 0 to 1). For example, with a BufferingTime value of 5 seconds, a BufferingProgress value of 1 means the client has its full 5 seconds of media, whereas a BufferingProgress value of 0.5 means the buffer is half full, with just 2.5 seconds available. This may be too much information to display, or it may be a useful way to show the user why a media file can't be buffered successfully over the current connection.

- *Use bit-rate throttling and IIS smooth streaming*: Bit-rate throttling can improve the scalability of your web server and smooth streaming can improve the performance of your video—sometimes dramatically. Both features are described in the "Adaptive Streaming" section that follows.

Adaptive Streaming

In recent years, the tide has shifted from true streaming to *adaptive streaming*, which is really a way to mimic the benefits of streaming while still using progressive downloading and ordinary HTTP behind the scenes. Currently, about 65% of all web content is delivered by progressive download, with YouTube leading the way as the single most popular deliverer of video content. IIS now supports two features that make adaptive streaming work more efficiently and help to close the performance gap with traditional streaming:

- *Bit-rate throttling*: Bit-rate throttling prevents people with good connections from downloading a video file really quickly, which can swamp the server if a large number of people request the file simultaneously. Typically, when using bit-rate throttling, you configure IIS to begin by sending a burst of content when a video file is requested. This ensures that the user can start playback as quickly as possible. However, after this burst—for example, after the user has downloaded 10 seconds of video—the rest of the video data is sent much more slowly. Limiting the transfer rate has no real effect on the client's ability to play the media, as long as the client can download the content faster than the application can play it. (In other words, a 700 Kbps transfer limit would be a disaster if you had a high-quality video with a bit rate greater than 700 Kbps.)

> ■ **Note** Bit-rate throttling also saves bandwidth overall. That's because most web surfers won't watch a video form start to finish. It's estimated that 80% of users navigate to a new page before finishing a video, effectively throwing away any extra unwatched video data they've downloaded in advance.

- *IIS smooth streaming*: With smooth streaming, the web server customizes the bit rate of the media file to suit the client. If the situation changes—for example, the network starts to slow down—the server deals with the issue seamlessly, automatically adjusting the bit rate down, and bringing it back up again when the connection improves. The player won't have to stop and refill its buffer. Similarly, clients with more CPU resources are given chunks higher-bit-rate video, while more limited clients are given reduced-bit-rate video.

To use either of these features, you need to download the IIS Media Services, which Microsoft provides as a free download at www.iis.net/media. To create video files that support smooth streaming, you'll also need Expression Encoder Pro (rather than the free version). To learn more about bit-rate throttling and how to configure it, read the walk-through at http://tinyurl.com/r7h6hp.To learn more about smooth streaming and its architecture, see http://tinyurl.com/cszay7.

Advanced Video Playback

You now know enough to play audio and video in a Silverlight application. However, a few finer details can help you get the result you want when dealing with video. First, you need to start with the right type of video—that means a file in the right format and with the right dimensions and bit rate (the number of bytes of data required per second). You may also want to consider a streamed video file for optimum network efficiency. Next, you may be interested in additional features such as markers. And finally, some of the most dazzling Silverlight effects depend on an artful use of the VideoBrush, which allows you to paint an ordinary Silverlight element with live video. You'll explore all these topics in the following sections.

Video Encoding

To get the best results, you should prepare your files with Silverlight in mind. For example, you should use video files that won't overwhelm the bandwidth of your visitors. This is particularly true if you plan to use large media files (for example, to display a 30-minute lecture).

Typically, the WMV files you use in your Silverlight application will be a final product based on larger, higher-quality original video files. Often, the original files will be in a non-WMV format. However, this detail isn't terribly important, because you'll need to reencode them anyway to reduce their size and quality to web-friendly proportions.

To get the right results when preparing video for the Web, you need the right tool. Microsoft provides two options:

- *Windows Movie Maker:* Included with some versions of Windows (such as Windows Vista) and aimed squarely at the home user, Windows Movie Maker is too limiting for professional use. Although it can work in a pinch, its lack of control and its basic features makes it more suitable for authoring home movies than preparing web video content.

- *Expression Encoder:* Available as a premium part of Microsoft's Expression Suite, Expression Encoder boasts some heavyweight features. Best of all, it's designed for Silverlight, which means it provides valuable features such as the automatic generation of custom-skinned Silverlight video pages. Best of all, Expression Encoder is available in a free version that you can download at www.microsoft.com/expression/products/Encoder4_Overview.aspx.

■ **Note** The premium version of Expression Encoder (called Expression Encoder Pro) adds support for H.264 encoding, unlimited screen-capture recording (the free version is capped at ten minutes), and IIS Smooth Streaming (a feature that lets your web server adjust the quality of streamed video based on changing network conditions and the client's CPU resources). If you don't need these features, the free version of Expression Encoder is a remarkably polished and powerful tool.

To learn more about video encoding, you can browse the product documentation, website articles, or a dedicated book. The following sections outline the basics to get you started with Expression Encoder.

Encoding in Expression Encoder

Expression Encoder gives you basic encoding ability, with a few nifty extra features:

- *Simple video editing:* You can cut out sections of video, insert a lead-in, and perform other minor edits.
- *Overlays:* You can watermark videos with a still or animated logo that stays superimposed over the video for as long as you want.
- *A/B compare:* To test the effect of a change or a new encoding, you can play the original and preview the converted video at the same time. Expression Encoder keeps both videos synchronized, so you can get a quick sense of quality differences.
- *Silverlight-ready:* Expression Encoder ships with suitable profiles for a Silverlight application. Additionally, Expression Encoder allows you to create a fully skinned Silverlight video player, complete with nifty features like image thumbnails.

To encode a video file in Expression Encoder, follow these steps:

1. When the program starts, choose Silverlight Project, and click OK.
2. To specify the source file, choose File ➤ Import. Browse to the appropriate media file, select it, and click Open. There will be a short delay while Expression Encoder analyzes the file before it appears in the list in the Sources tab at the bottom of the window. At this point, you can perform any other edits you want, such as trimming out unwanted video, inserting a lead-in, or adding an overlay. (Many of these changes are made through the Enhance tab, which you can show by choosing Window ➤ Enhance.)

3. To specify the destination file, look for the Output tab on the right side of the window. In the Job Output section, you can specify the directory where the new file will be placed and its name.

4. To choose the bit rate, look in the Presets tab (in the top-right corner of the window), and expand the Encoding for Silverlight section. Then, expand the VC-1 section inside. If you're using progressive downloads, you need to select a format from the Variable bit-rate group. If you're using streaming with Windows Media Services, choose a format from the Constant bit-rate group instead. Different formats result in different bitrates, video quality, and video size—to get more details, hover over a format in the list (as shown in Figure 11-3). When you've picked the format you want (or if you just want to preview the effect it will have on your video), click the Apply button at the bottom of the Presets tab.

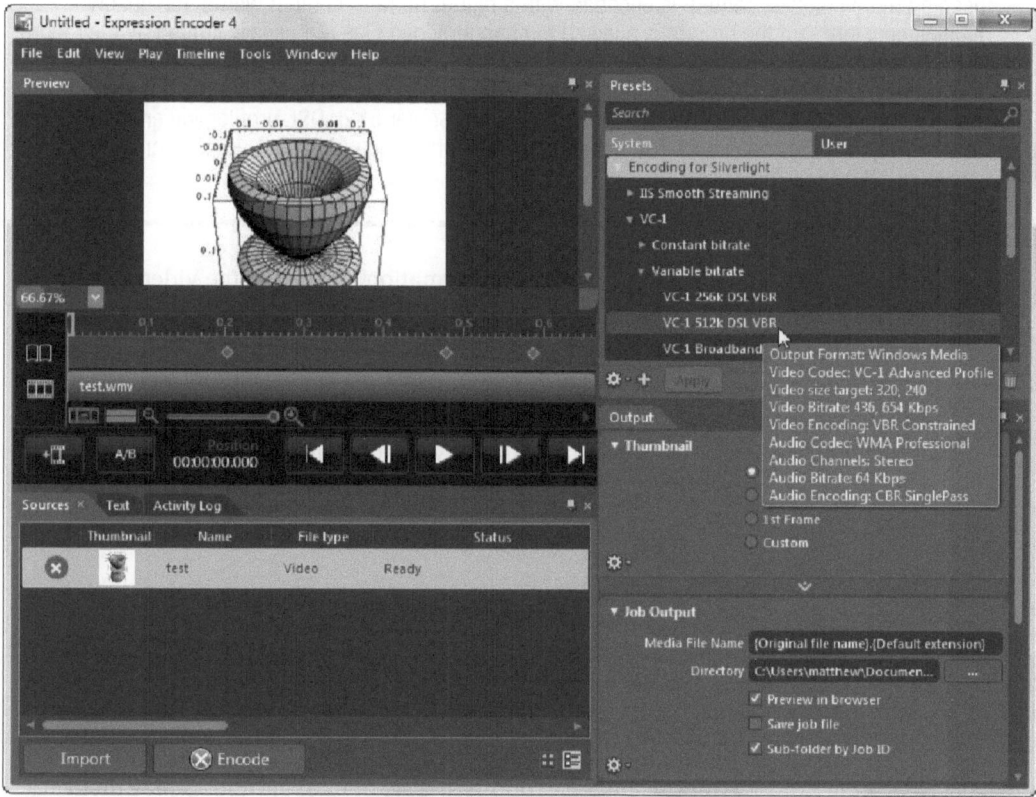

Figure 11-3. *Choosing the type of encoding*

SILVERLIGHT COMPRESSION: CBR AND VBR

Depending on whether you're planning to use streaming or simple progressive downloads, Silverlight chooses between two compression modes:

- *Constant Bit-Rate Encoding (CBR)*: This is the best choice if you plan to allow video streaming. With CBR encoding, the average bit rate and the peak bit rate are the same, which means the data flow remains relatively constant at all times. Another way of looking at this is that the quality of the encoding may vary in order to preserve a constant bit rate, ensuring that the user gets smooth playback. (This isn't necessary if your application is using progressive downloading, because then it will cache as much of the media file as it can.)

- *Variable Bit-Rate Encoding (VBR)*: This is the best choice if you plan to use progressive downloading. With VBR encoding, the bit rate varies throughout the file depending on the complexity of the video, meaning more complex content is encoded with a higher bit rate. In other words, the quality remains constant, but the bit rate is allowed to change. Video files are usually limited by their worst parts, so a VBR-encoded file generally requires a smaller total file size to achieve the same quality as a CBR-encoded file. When you use VBR encoding with Silverlight, the maximum bit rate is still constrained. For example, if you choose the VC-1 Web Server 512k DSL profile, you create encoded video with an average bit rate of 350 Kbps (well within the range of the 512 Kbps connection) and a maximum bit rate of 750 Kbps.

---

5. After you choose an encoding, the relevant information appears in the Video section of the Encode tab. Before you perform the encoding, you can tweak these details. For example, you can adjust the dimensions of the video output using the Size box. You can also preview what the file will look like by playing it in the video window on the left.

6. To encode your video, click the Encode button at the bottom of the window, in the Media Content panel. If you want, you can save your job when the encoding is finished so you can reuse its settings later (perhaps to encode an updated version of the same file).

Markers

Markers are text annotations that are embedded in a media file and linked to a particular time. Technically, the WMV format supports text markers and script commands (used to do things like launch web pages while playback is underway), but Silverlight treats both of these the same: as timed bookmarks with a bit of text.

Markers provide some interesting possibilities for creating smarter Silverlight-based media players. For example, you can embed captions as a set of markers and display them at the appropriate times. (You could even use this technique to build a poor man's subtitling system.) Or, you can embed other types of instructions, which your application can then read and act on.

Although it's up to you to write the code that reacts to markers, Silverlight gives you two tools: a MarkerReached event and the Markers collection in the MediaElement. But before you can investigate these details, you first need to consider how to add markers to your media file in the first place.

Adding Markers with Expression Encoder

Expression Encoder has a built-in feature for adding markers. Here's how to use it:

1. After you've imported a media file, choose the Window ➤ Metadata to show the Metadata tab at the bottom right of the window.
2. Drag the playback bar under the video file to the position where you want to place the marker.
3. In the Metadata tab, find the Markers box. At the bottom of the Markers box, click the Add button to create a new marker, which is added to the list (see Figure 11-4).

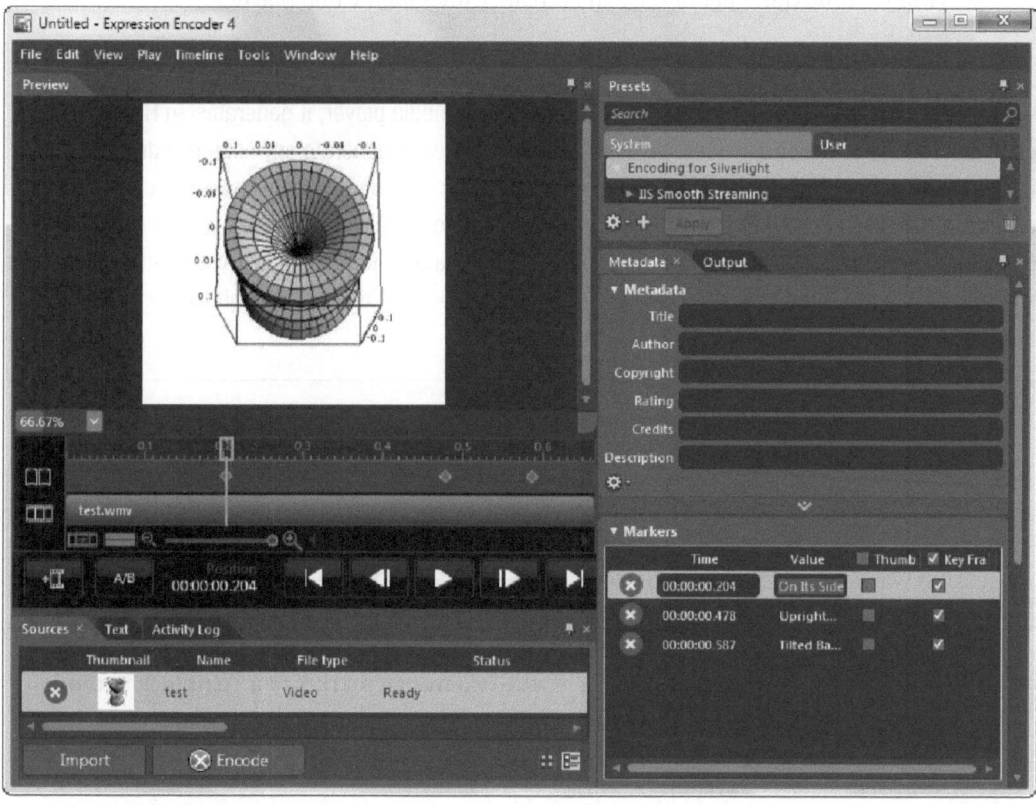

Figure 11-4. Adding a new marker in Expression Encoder

4. Adjust the time if necessary, and supply the marker text in the Value column.
5. If you want to use a marker for indexed navigation, you may want to select the Key Frame and Thumbnail check boxes next to your new marker. If you create a key frame at this location, playback can resume at precisely this location with minimal delay. If you create a thumbnail, you can show that thumbnail to the user. The user can click that thumbnail to tell your application to seek to the corresponding marker location. Both of

these features apply only if you use Expression Encoder to generate a Silverlight video page (see step 7), although you can build similar features on your own.

6. Return to step 2, and repeat to add more markers. You can also edit existing markers and click Remove to delete the currently selected marker.

7. Expression Encoder can build a complete Silverlight 4–based media player to go along with your encoded video. To use this feature, choose Window ➤ Templates to show the Templates tab at the right. Then, choose an item from the Template list. The template determines the Silverlight version and the visual skin that the Silverlight player page uses—you see a thumbnail preview when you make your selection. If you choose (None), Expression Encoder doesn't create a Silverlight video player.

■ **Tip** When Expression Encoder creates a template-based media player, it generates an HTML entry page and several compiled XAP files. Expression Encoder doesn't generate any source code or project files, so you won't be able to fine-tune the media player after you create it. However, you can take an existing template and use it as the basis for your own custom template, which you can then use in conjunction with any Expression Encoder project. To do so, first pick the template you want to use as the starting point in the Templates tab. Then, click the tiny gear icon in the bottom-right corner of the tab, and choose Edit Copy of Template In ➤ Visual Studio.

---

8. When you're finished, click Encode to start encoding your video.

Using Markers in a Silverlight Application

The easiest way to show marker information is to handle the MarkerReached event of the MediaElement. The TimelineMarkerRoutedEventArgs.Marker property provides a TimelineMarker object. The TimelineMarker object includes the text of the marker (through the Text property) and the exact time where it's placed (through the Time property).

Here's a simple event handler that copies the text from a marker to a TextBlock in the Silverlight page, as shown in Figure 11-5:

```
Private Sub media_MarkerReached(ByVal sender As Object, _
  ByVal e As TimelineMarkerRoutedEventArgs)
    lblMarker.Text = e.Marker.Text & " at " & _
      e.Marker.Time.TotalSeconds & " seconds"
End Sub
```

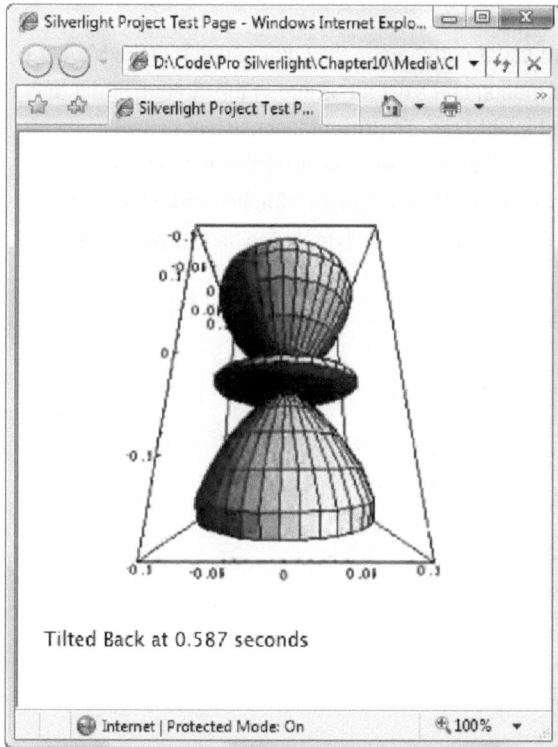

Figure 11-5. Displaying a marker

Rather than setting text, you can examine it and then determine the appropriate action to perform.

Instead of waiting for the MarkerReached event, you can examine the Markers collection of the MediaElement. This technique is particularly useful if you want to use markers for navigation. For example, you can react to the MediaOpened event (at which point the Markers collection has been populated) and then display the marker information in a list:

```
Private Sub media_MediaOpened(ByVal sender As Object, ByVal e As RoutedEventArgs)
    For Each marker As TimelineMarker In media.Markers
        lstMarkers.Items.Add(marker.Text & " (" & marker.Time.Minutes & ":" & _
          marker.Time.Seconds & ":" & marker.Time.Milliseconds & ")")
    Next
End Sub
```

■ **Note** If your media file includes separate-stream script commands, they don't appear in the Markers collection. That's because this type of marker information can exist anywhere in the stream, and it may not have been downloaded when the MediaOpened event fires. To prevent inconsistent behavior, these types of markers are never added to the Markers collection. However, the MediaElement still detects them and fires the MarkerReached event at the appropriate time. If this isn't the behavior you want, use the more common header-embedded script commands, which place them in the header (which *is* read before MediaOpened fires).

You can also use the TimelineMarker.Time property to perform navigation:

```
media.Position = selectedMarker.Time
media.Play()
```

Figure 11-6 shows the result.

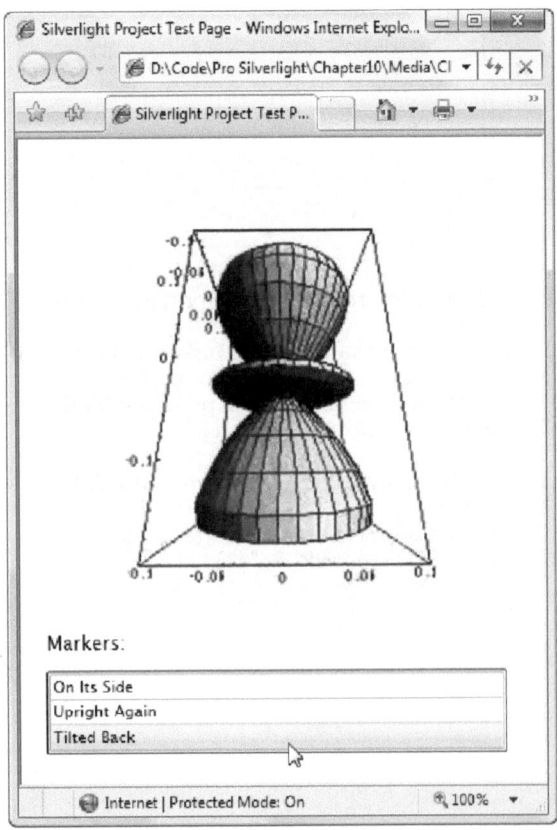

Figure 11-6. Navigating with a marker list

In this example, the code reads the markers from the media file. You can also create TimelineMarker objects programmatically and add them to the Markers collection after the media file has been loaded and the MediaOpened event has fired. In this case, the marker acts as a normal marker in all respects—for example, the MediaElement fires the MarkerReached event when it's reached. However, the marker isn't persisted in the video file when you close and reload it. This behavior gives you the ability to load marker information from another source, like a text file.

■ **Note** Expression Encoder includes a feature that lets you create image thumbnails for your markers. These images are embedded in your video file or linked to it in any way. If you use this feature, it's up to you to show the images in your page and use code to navigate to the right position. If you look at the code for the video player application that Expression Encoder can create, you'll find that it hard-codes the image file names and the marker positions, which is a suitable approach for automatically generated code but not as good an idea in application code that you need to maintain.

VideoBrush

VideoBrush is a Silverlight brush that paints an area with the video content that's currently playing in a specified MediaElement. Like other Silverlight brushes, you can use VideoBrush to fill anything from a basic shape to a complex path or element.

The basic approach to using a VideoBrush is straightforward. First, create a MediaElement for the file you want to play:

```
<MediaElement x:Name="fireMovie" Source="fire.wmv"
 Height="0" Width="0"></MediaElement>
```

Notice that this example sets the Height and Width of the MediaElement to 0. This way, the original video window doesn't appear, and it won't take up any space in your layout. The only video that will appear is the video that's being painted by the VideoBrush. You can't get the same result by setting the Visibility property—if you hide the MediaElement by setting its Visibility to Collapsed, you also end up hiding the content that the VideoBrush is painting.

■ **Tip** In some situations, you may want to display the original video window (which is shown in the MediaElement) *and* the video content that's painted by the VideoBrush. For example, you'll want the original video window to remain visible if you're using the VideoBrush to create a reflection effect.

The next step is to choose the element you want to paint with the VideoBrush. You can use the VideoBrush anywhere an element expects a brush. If you're dealing with the shape elements, you'll set properties like Fill and Stroke. If you're dealing with other elements, you'll look for properties like Foreground and Background. The following example uses the VideoBrush to fill the text in a large TextBlock:

```
<TextBlock Text="Fiery Letters" FontFamily="Arial Black" FontSize="80">
  <TextBlock.Foreground>
    <VideoBrush SourceName="fireMovie"></VideoBrush>
  </TextBlock.Foreground>
</TextBlock>
```

The SourceName property links the VideoBrush to the corresponding MediaElement. Figure 11-7 shows the result—text that's filled with roaring flames.

Figure 11-7. *Using video to fill text*

When you use the VideoBrush, playback is still controlled through the MediaElement. In the current example, the video file begins to play automatically, because AutoPlay is True by default. Alternatively, you can set AutoPlay to False and control playback using the familiar Play(), Stop(), and Pause() methods of the MediaElement.

It's also worth noting that you can set certain details in the MediaElement without affecting the VideoBrush. Properties that affect the visual appearance of the MediaElement, such as Height, Width, Opacity, Stretch, RenderTransform, and Clip, have no effect on the VideoBrush. (The obvious exception is Visibility.) Instead, if you want to alter the video output, you can modify similar properties of the VideoBrush or the element you're painting with the VideoBrush.

Video Effects

Because the MediaElement works like any other Silverlight element and the VideoBrush works like any other Silverlight brush, you have the ability to manipulate video in some surprising ways. Here are some examples:

- You can use a MediaElement as the content inside a content control, such as a button.
- You can set the content for thousands of content controls at once with multiple MediaElement objects—although the client's CPU may not bear up very well under the strain.
- You can combine video with transformations through the RenderTransform property. This lets you move your video page, stretch it, skew it, or rotate it.

- You can set the Clipping property of the MediaElement to cut down the video page to a specific shape or path and show only a portion of the full frame.
- You can set the Opacity property to allow other content to show through behind your video. You can even stack multiple semitransparent video pages on top of each other.
- You can use an animation to change a property of the MediaElement (or one of its transforms) dynamically.
- You can copy the current content of the video page to another place in your user interface using a VideoBrush, which allows you to create specific effects like reflection.
- You can use the same VideoBrush to paint multiple elements (or create multiple VideoBrush objects that use the same MediaElement). Both of these techniques let you fill multiple objects with the same video or transformed versions of the same video.

For example, Figure 11-8 shows a video with a reflection effect underneath. It does so by creating a Grid with two rows. The top row holds a MediaElement that plays a video file. The bottom row holds a rectangle that's painted with a VideoBrush. The video content is flipped over by using the RelativeTransform property and then faded out gradually toward the bottom using an OpacityMask gradient:

```
<Grid Margin="15" HorizontalAlignment="Center">
  <Grid.RowDefinitions>
    <RowDefinition></RowDefinition>
    <RowDefinition></RowDefinition>
  </Grid.RowDefinitions>

  <MediaElement Grid.Row="0" x:Name="media" Source="test.wmv"
   Stretch="Uniform"></MediaElement>

  <Rectangle Grid.Row="1" Stretch="Uniform">
    <Rectangle.Fill>
      <VideoBrush SourceName="media">
        <VideoBrush.RelativeTransform>
          <ScaleTransform ScaleY="-1" CenterY="0.5"></ScaleTransform>
        </VideoBrush.RelativeTransform>
      </VideoBrush>
    </Rectangle.Fill>

    <Rectangle.OpacityMask>
      <LinearGradientBrush StartPoint="0,0" EndPoint="0,1">
        <GradientStop Color="Black" Offset="0"></GradientStop>
        <GradientStop Color="Transparent" Offset="0.6"></GradientStop>
      </LinearGradientBrush>
    </Rectangle.OpacityMask>
  </Rectangle>
</Grid>
```

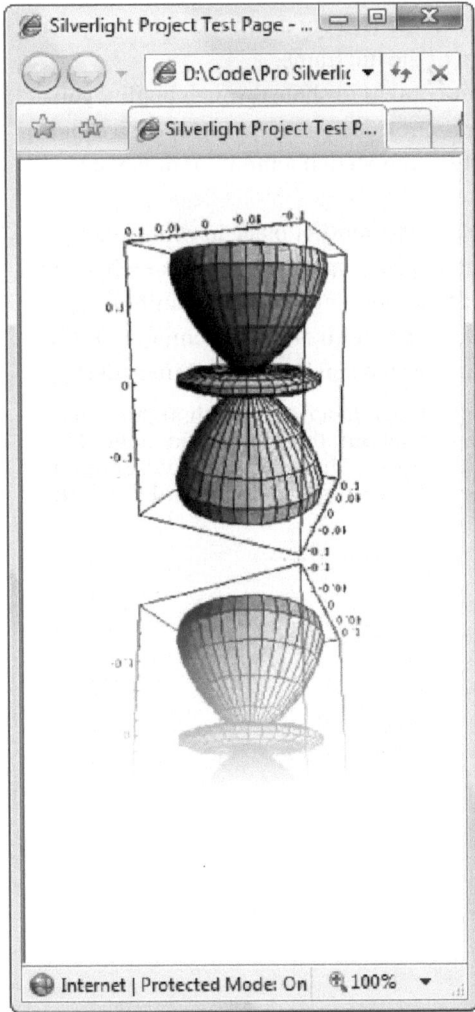

Figure 11-8. *Reflected video*

This example performs fairly well. Each frame must be copied to the lower rectangle, and each frame needs to be flipped and faded to create the reflection effect. (Silverlight uses an intermediary rendering surface to perform these transformations.) But the work required to download and decode the frame of video is performed just once, and on a modern computer, the extra overhead is barely noticeable.

One of the most impressive effects in the early days of Silverlight development was a video puzzle. It took a high-resolution video file and split it into a grid of interlocking puzzle pieces, which the user could then drag apart. The effect—separate puzzle pieces, each playing a completely synchronized portion of a single video—was stunning.

With the help of the VideoBrush, creating an effect like this is almost trivial. The following example shows a slightly simplified version of the original puzzle demonstration. It starts with a single window of puzzle pieces that's divided into a configurable number of squares. When the user clicks a square in the video window, an animation moves it to a random position (as shown in Figure 11-9). Several clicks later, the video image is completely scrambled, but all the pieces are still playing the synchronized video.

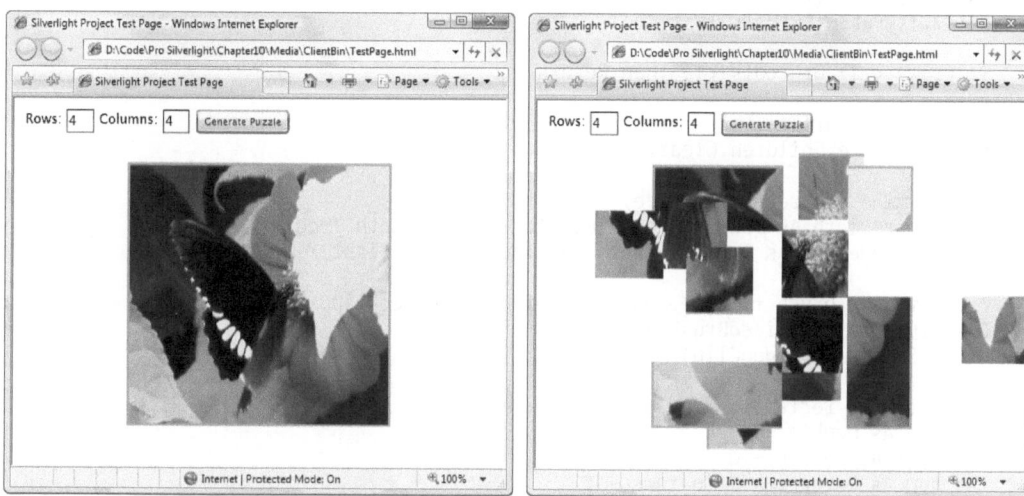

Figure 11-9. *Scrambling a video while it's playing*

To create this example, you first need the MediaElement that plays the video. Because all the puzzle pieces are showing portions of the same video and you want the playback synchronized, you need just one MediaElement. It's given a Height and Width of 0 to make it invisible, so it appears only when used through the VideoBrush:

```
<MediaElement x:Name="videoClip" Source="Butterfly.wmv" Height="0" Width="0"
MediaEnded="videoClip_MediaEnded"></MediaElement>
```

When the media ends, it's started again, providing a looping playback:

```
Private Sub videoClip_MediaEnded(ByVal sender As Object, ByVal e As RoutedEventArgs)
    videoClip.Stop()
    videoClip.Play()
End Sub
```

Next, you need a layout container that will hold the puzzle pieces. In this case, a Canvas makes the most sense because the animation needs to move the pieces around the page when they're clicked:

```
<Canvas Margin="20" x:Name="puzzleSurface" Width="300" Height="300"
 Background="White" HorizontalAlignment="Center" VerticalAlignment="Center">
</Canvas>
```

The most interesting code happens when a user clicks the Generate Puzzle button. This code calculates the size of rectangle needed to make a puzzle piece and then dynamically creates each piece as a simple Rectangle element. Here's the code that starts it off:

```
Private Sub cmdGeneratePuzzle_Click(ByVal sender As Object, _
  ByVal e As RoutedEventArgs)
    ' Get the requested dimensions.
```

415

```
Dim rows As Integer
Dim cols As Integer
Int32.TryParse(txtRows.Text, rows)
Int32.TryParse(txtCols.Text, cols)

If (rows < 1) Or (cols <1) Then Return

' Clear the surface.
puzzleSurface.Children.Clear()

' Determine the rectangle size.
Dim squareWidth As Double = puzzleSurface.ActualWidth / cols
Dim squareHeight As Double = puzzleSurface.ActualHeight / rows

' Create the brush for the MediaElement named videoClip.
Dim brush As New VideoBrush()
brush.SetSource(videoClip)

' Create the rectangles.
Dim top As Double = 0
Dim left As Double = 0
For row As Integer = 0 To rows - 1
    For col As Integer = 0 To cols - 1
        ...
```

The next step is to make sure each Rectangle element shows only the region that's assigned to it. You could accomplish this by applying a transform to the VideoBrush, but then you'd need to use a different VideoBrush object for each square. An alternate approach is to tweak the clipping region of rectangle. In this case, each rectangle gets the size of the full video window, but it's clipped to show just the appropriate region. Here's the code that creates the rectangles and sets the clipping:

```
...
' Create the rectangle. Every rectangle is sized to match the Canvas.
Dim rect As New Rectangle()
rect.Width = puzzleSurface.ActualWidth
rect.Height = puzzleSurface.ActualHeight

rect.Fill = brush
Dim rectBrush As New SolidColorBrush(Colors.Blue)
rect.StrokeThickness = 3
rect.Stroke = rectBrush

' Clip the rectangle to fit its portion of the puzzle.
Dim clip As New RectangleGeometry()

' A 1-pixel correction factor ensures there are never lines in between.
clip.Rect = New Rect(left, top, squareWidth+1, squareHeight+1)
rect.Clip = clip

' Handle rectangle clicks.
AddHandler rect.MouseLeftButtonDown, AddressOf rect_MouseLeftButtonDown

puzzleSurface.Children.Add(rect)

' Go to the next column.
```

```
            left += squareWidth
        Next
        ' Go to the next row.
        left = 0
        top += squareHeight
    Next
    ' (If the video is not already playing, you can start it now.)
End Sub
```

When a rectangle is clicked, the code responds by starting two animations that move it to a new, random position. Although you could create these animations manually, it's easier to define them in the resources collection. That's because the application requires just two animations and can reuse them for whatever square is clicked.

Here are the two animations. The animation that shifts the rectangle sideways takes 0.25 seconds, and the animation that moves it up or down takes 0.15 seconds:

```
<UserControl.Resources>
  <Storyboard x:Name="squareMoveStoryboard">
    <DoubleAnimation x:Name="leftAnimation" Duration="0:0:0.25"
     Storyboard.TargetProperty="(Canvas.Left)"></DoubleAnimation>
    <DoubleAnimation x:Name="topAnimation" Duration="0:0:0.15"
     Storyboard.TargetProperty="(Canvas.Top)"></DoubleAnimation>
  </Storyboard>
</UserControl.Resources>
```

You'll notice that this code uses a single storyboard for all its animations. You must take extra care when reusing this storyboard. Before you can start a new animation, you must manually place the current square in its new position and then stop the storyboard. The alternative is to dynamically create a new storyboard every time a square is clicked. (You saw this technique in action in Chapter 10, with the bomb-dropping game.)

Here's the code that manages the storyboard and moves the square when it's clicked, sending it drifting to a new, random location:

```
Private previousRectangle As Rectangle

Private Sub rect_MouseLeftButtonDown(ByVal sender As Object, _
  ByVal e As MouseButtonEventArgs)
    ' Get the square.
    Dim rectangle As Rectangle = CType(sender, Rectangle)

    ' Stop the current animation.
    If previousRectangle IsNot Nothing Then
        Dim left As Double = Canvas.GetLeft(rectangle)
        Dim top As Double = Canvas.GetTop(rectangle)
        squareMoveStoryboard.Stop()
        Canvas.SetLeft(rectangle, left)
        Canvas.SetTop(rectangle, top)
    End If

    ' Attach the animation.
    squareMoveStoryboard.Stop()
    Storyboard.SetTarget(squareMoveStoryboard, rectangle)
```

```
    ' Choose a random direction and movement amount.
    Dim rand As New Random()
    Dim sign As Integer = 1
    If rand.Next(0, 2) = 0 Then
        sign = -1
    End If
    leftAnimation.To = Canvas.GetLeft(rectangle) + rand.Next(60,150) * sign
    topAnimation.To = Canvas.GetTop(rectangle) + rand.Next(60, 150) * sign

    ' Store a reference to the square that's being animated.
    previousRectangle = rectangle

    ' Start the animation.
    squareMoveStoryboard.Begin()
End Sub
```

This is all the code you need to complete the example, combining video, interactivity, and a rather dramatic effect that's leagues beyond other browser-based application platforms.

Webcam and Microphone Input

One of the more hyped additions to Silverlight 4 is support for video and audio input, through a user's webcam and microphone. This opens the door to new types of rich Internet applications, from video conferencing to real-time chat in a multiplayer game. However, although this door is open, the opening is a very narrow one. To create a realistic application that incorporates user-recorded audio or video, you'll be forced to write pages of custom code, perform complex encoding and decoding, and rely on external libraries to get support for the formats you need.

In the following sections, you'll see both sides of Silverlight's webcam and microphone support. You'll learn how you can use the built-in support to accomplish a few relatively trivial tasks (such as showing live video in a portion of your page), and you'll consider the next steps you must take to build more practical applications that use interactive video and audio.

Accessing a Capture Device

Fortunately, you don't need to worry about hardware configuration or device specifics to use Silverlight's support for webcams and microphones. Instead, Silverlight automatically recognizes video and audio capture devices and makes them available to your application. To see what Silverlight's discovered on your computer, right-click any currently running Silverlight application to show the Silverlight menu, and then click Silverlight. Then, choose the Webcam/Mic tab of the Silverlight Configuration window, which is shown in Figure 11-10.

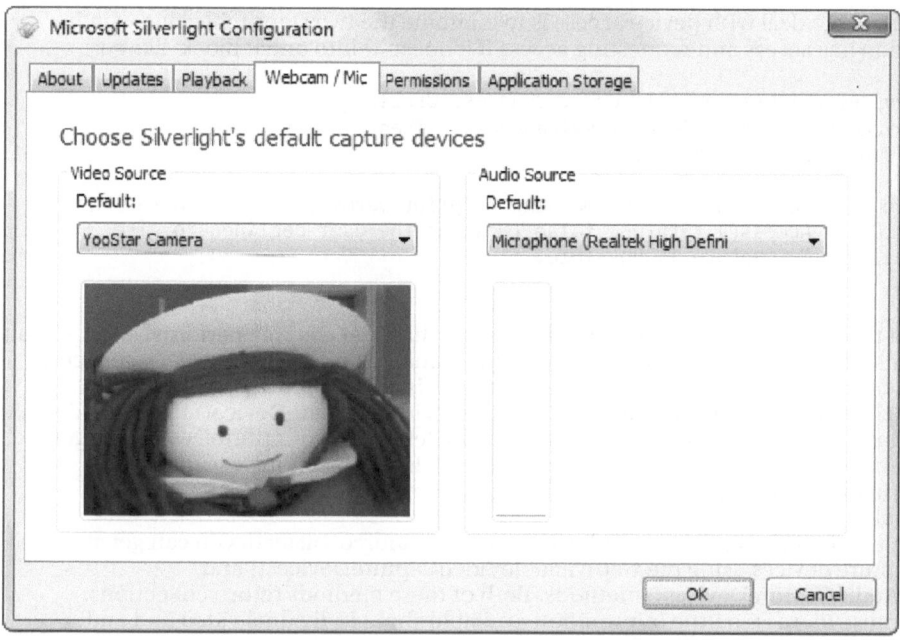

Figure 11-10. Recognized webcam and microphone devices

If you have more than one suitable device, you can choose what Silverlight designates as the default in an application. (Usually, Silverlight just follows the lead of the operating system, which has its own way to designate the default video and audio capture devices.)

All capture operations begin with the CaptureDeviceConfiguration class from the System.Windows.Media namespace. It provides a handful of shared members that you can use to get started. First up is the Boolean property AllowedDeviceAccess, which returns True if the user has given your application permission to use Silverlight's audio and video capture features. Initially, this property will be False, and your application must call the RequestDeviceAccess() method to prompt the user for permission (Figure 11-11). However, if the user checks the "Remember my answer" check box and clicks Yes, your application will automatically get permission next time and won't need to call RequestDeviceAccess().

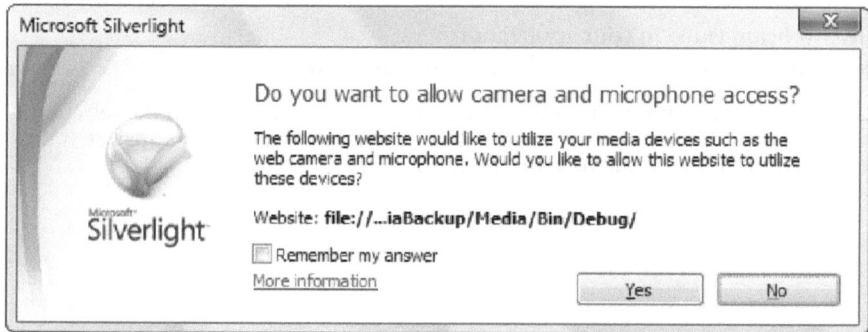

Figure 11-11. Requesting access to webcam and microphone devices

The usual way to deal with device access is to combine the two steps (checking whether your application has access and requesting access if it doesn't) into one If block, like so:

```
If CaptureDeviceConfiguration.AllowedDeviceAccess OrElse _
    CaptureDeviceConfiguration.RequestDeviceAccess() Then
    ' Permission has been granted.
Else
    ' AllowedDeviceAccess was false (meaning no prior permission was granted) and
    ' RequestDeviceAccess() returned false (meaning the user declined to give
    ' permission.) No capture is possible.
End If
```

Note that this code uses the OrElse operator rather than just Or. This performs short-circuit evaluation, ensuring that the second condition isn't evaluated and the RequestDeviceAccess() method isn't called if the AllowedDeviceAccess property is already True.

Once you have permission, you can call GetDefaultVideoCaptureDevice() to get the default video device (as a VideoCaptureDevice object) and GetDefaultAudioCaptureDevice() to get the default audio device (as an AudioCaptureDevice object). Both objects expose just a few properties: FriendlyName, SupportedFormats, AudioFrameSize (for audio), and DesiredFormat.

You aren't limited to using the default video or audio source. Instead, you can get all the supported capture devices using the GetAvailableVideoCaptureDevices() and GetAvailableAudioCaptureDevices() methods. Both of these methods return collections. However, because there is so little information available about individual video and audio devices, and because it's impossible for your code to know much about the user's hardware configuration in advance, it's very rare for an application to choose a device from these collections using any sort of programmatic logic. Instead, these methods are useful if you want to give the user the choice. For example, you can bind the results returned from GetAvailableVideoCaptureDevices() to some sort of list control (displaying the FriendlyName of each one) and then use whichever device the user selects.

You've now seen all the methods of the CaptureDeviceConfiguration class. It gives you the tools you need to request device access and retrieve limited information about every available video and audio device. However, to actually *use* these devices, you need the help of the CaptureSource class described in the next section.

Basic Webcam Support

Silverlight includes built-in support for two tasks with the webcam:

- Showing live webcam video in your application
- Taking a snapshot (as a bitmap) from the video stream of a webcam

Figure 11-12 shows a sample application that uses both of these features. On the left, it shows a live display of what the webcam is capturing. When the user clicks the Snapshot button, it copies the current frame to the box on the right.

Figure 11-12. A basic webcam application

The first step to use either of these features is to create a CaptureSource object. The CaptureSource provides the Start() and Stop() methods that allow you to control when your application is capturing audio and video. It also includes a State property that indicates what the CaptureSource is currently doing, as a value for the CaptureStates enumeration (Started, Stopped, or Failed).

Usually, you'll define the CaptureSource object as a member field, so it's available in all your event handlers:

```
Private capture As New CaptureSource()
```

Before you start a capture with CaptureSource, you must assign the VideoCaptureDevice property (if you're capturing video), the AudioCaptureDevice (if you're capturing audio), or both. Here's an example that starts capturing video with the default video capture device:

```
capture.VideoCaptureDevice = _
  CaptureDeviceConfiguration.GetDefaultVideoCaptureDevice()
capture.Start()
```

This starts the capture process and ensures that the CaptureSource object gets the stream of video or audio input. But how do you actually use that video input? If you want to display it in your page, you need the help of the VideoBrush, which takes the CaptureSource stream and lets you use it to paint another element. (In Figure 11-12, that element is a basic Rectangle object.)

You now know enough to create the code that initiates video capture in the current example (Figure 11-12). To start it off, the user clicks the Start button. At this point, the application checks for device access (and requests it if necessary), selects the default video capture device, starts the capture, and creates the VideoBrush that copies the video output to the screen. Here's the complete code:

```
Private Sub cmdStartCapture_Click(ByVal sender As Object, _
  ByVal e As RoutedEventArgs)
    If CaptureDeviceConfiguration.AllowedDeviceAccess OrElse _
      CaptureDeviceConfiguration.RequestDeviceAccess() Then
        ' It's always safe to call Stop(), even if no capture is running.
        ' However, attempting to call Start() while a capture is already running,
        ' without calling Stop() first, causes an exception.
        capture.Stop()

        ' Get the default webcam.
        capture.VideoCaptureDevice = _
          CaptureDeviceConfiguration.GetDefaultVideoCaptureDevice()

        If capture.VideoCaptureDevice Is Nothing Then
            MessageBox.Show("Your computer does not have a video capture device.")
        Else
            ' Start a new capture.
            capture.Start()

            ' Map the live video to a VideoBrush.
            Dim videoBrush As New VideoBrush()
            videoBrush.Stretch = Stretch.Uniform
            videoBrush.SetSource(capture)

            ' Use the VideoBrush to paint the fill of a Rectangle.
            rectWebcamDisplay.Fill = videoBrush
        End If
    End If
End Sub
```

Stopping the capture is easy:

```
Private Sub cmdStopCapture_Click(ByVal sender As Object, ByVal e As RoutedEventArgs)
    capture.Stop()
End Sub
```

The current example also uses the other feature of the CaptureSource class—its ability to take a snapshot of the current frame and provide it as a WriteableBitmap object. To use this feature, you call the CaptureImageAsync() method, which starts the frame grab, and handle the CaptureImageCompleted event, which fires when the bitmap is ready. Here's the code you need:

```
Private Sub cmdSnapshot_Click(ByVal sender As Object, _
  ByVal e As RoutedEventArgs)
    If capture.State = CaptureState.Started Then
        AddHandler capture.CaptureImageCompleted, _
          AddressOf capture_CaptureImageCompleted
        capture.CaptureImageAsync()
    End If
End Sub

Private Sub capture_CaptureImageCompleted(ByVal sender As Object, _
```

```
    ByVal e As CaptureImageCompletedEventArgs)
      ' Show the frame in an Image control.
      imgSnapshot.Source = e.Result
End Sub
```

This hints at additional possibilities. For example, once you have a bitmap of a single frame, you have the ability to store it (perhaps using the isolated storage features described in Chapter 18), alter it with pixel shader effects (Chapter 9), upload it to a web service (Chapter 15), or even send it directly to a networked server (Chapter 20). However, before you start storing or sending captured frames, you need to think seriously about size, because bitmaps are horrendously, wastefully huge. A realistic application would certainly convert the WriteableBitmap content to another format, such as PNG or JPEG, before storing or sending it. Unfortunately, this functionality isn't built into Silverlight, but it's fairly easy to get using separately available libraries like fjcore (http://code.google.com/p/fjcore) or the .NET Image Tools (http://imagetools.codeplex.com).

Recording Audio Snippets

The CaptureSource is a simple wrapper that provides a small amount of basic functionality. To do anything more ambitious, you need to dig deeper into the processing pipeline and write a custom video sink or audio sink.

Video sinks and audio sinks are custom classes that have a chance to process the raw data as it's being captured from a video or audio device, one chunk at a time. If you plan to create a custom video sink, you must create a custom class that derives from VideoSink. To create an audio sink, you create a class that derives from AudioSink.

For example, imagine you want to create an application that allows the user to record a short snippet of audio or video and store it in memory. You can then play that audio pack, save it, or pass it to the server. Figure 11-13 shows just this short of application, which stores a short clip of audio and plays it back at the click of a button.

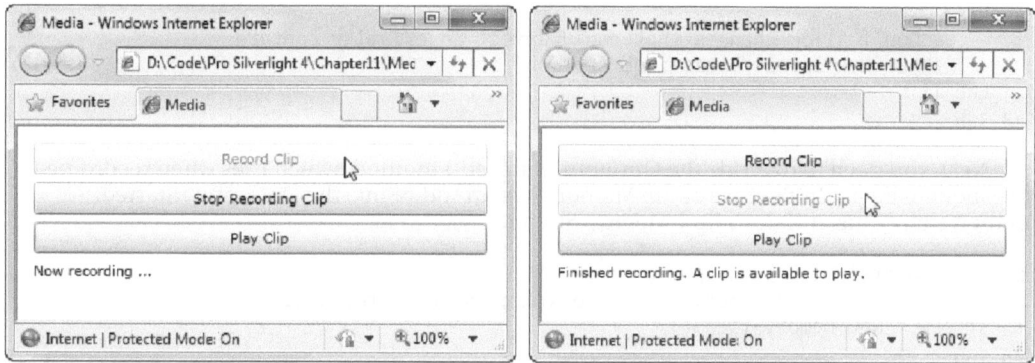

Figure 11-13. Recording and playing an audio clip

Although you might expect the CaptureSource to handle this task, it's limited to playing live video or capturing static images. As awkward as it seems (and is), you need to handle the raw data yourself to implement this sort of solution.

The following code listing shows the simplest possible custom audio sink—one that merely copies each sample of data to an in-memory stream. Here's how the class is defined:

```
Public Class MemoryStreamAudioSink
  Inherits AudioSink
    ...
End Class
```

In the MemoryStreamAudioSink, all the audio data is written to a MemoryStream object. This stream is then made available to the rest of the application through the AudioData property:

```
Private stream As MemoryStream
Public ReadOnly Property AudioData() As MemoryStream
    Get
        Return stream
    End Get
End Property
```

All audio sinks must provide an implementation of the AudioData property. They must also override several more MustInherit members. For example, the AudioFormat property track the format that the raw audio data is in, and the OnFormatChange() gives the audio sink the chance to react when the format changes:

```
Private _audioFormat As AudioFormat
Public ReadOnly Property AudioFormat() As AudioFormat
    Get
        Return _audioFormat
    End Get
End Property

Protected Overrides Sub OnFormatChange(ByVal audioFormat As AudioFormat)
    If _audioFormat Is Nothing Then
        _audioFormat = audioFormat
    Else
        ' Don't allow changes that could affect an existing capture.
        Throw New InvalidOperationException()
    End If
End Sub
```

Next, you need to override the OnCaptureStarted() method, which fires when recording is just about to begin. In the case of the MemoryStreamAudioSink, the OnCaptureStarted() method simply instantiates a new MemoryStream:

```
Protected Overrides Sub OnCaptureStarted()
    ' Prepare a new in-memory stream to store the captured audio.
    stream = New MemoryStream()
End Sub
```

The single most important method is OnSamples(), which is triggered repeatedly while a capture is underway, every time the audio sink receives a complete sample of audio data. This information is provided to your code as a byte array, which the MemoryStreamAudioSink simply writes to the memory array:

```
Protected Overrides Sub OnSamples(ByVal sampleTime As Long, _
  ByVal sampleDuration As Long, ByVal sampleData As Byte())
    ' Each time a sample is received, write it to the in-memory stream.
```

```
    ' (A more complex implementation might stream it over the network.)
    stream.Write(sampleData, 0, sampleData.Length)
End Sub
```

When the capture is stopped, the OnCaptureStopped() method is called. At this point, the MemoryStreamAudioSink has a bit of work to do. Currently, its internal memory stream has the raw wave data. But to use this as a true WAV file, with other applications that understand the WAV file format (for example, Windows Media Player), you need to add a proper header at the beginning. Ideally, the MemoryStreamAudioSink class would add this header before recording even begins, but this isn't possible, because the header includes information that isn't known until after the capture ends (such as the total length of the audio). To work around this, the OnCaptureStopped() method shown here writes the correct header and then copies all the data from the recorded memory stream to the new memory stream. It then replaces the old stream (which releases it for garbage collection). Although this technique works, it's memory-intensive, because you need enough memory for two copies of your recorded audio. A better (but more complicated) approach is to leave space for the header and attempt to patch it in later.

```
Protected Overrides Sub OnCaptureStopped()
    ' Genereate the header.
    Dim wavFileHeader As Byte() = WavFileHelper.GetWavFileHeader( _
        AudioData.Length, AudioFormat)

    ' Write the header to a new stream.
    Dim wavStream As New MemoryStream()
    wavStream.Write(wavFileHeader, 0, wavFileHeader.Length)

    ' Write the rest of the data, one chunk of 4096 bytes at a time.
    Dim buffer(4095) As Byte
    Dim read As Integer = 0
    stream.Seek(0, SeekOrigin.Begin)
    read = stream.Read(buffer, 0, buffer.Length)
    Do While read > 0
        wavStream.Write(buffer, 0, read)
        read = stream.Read(buffer, 0, buffer.Length)
    Loop

    ' Replace the raw stream with the new stream.
    stream = wavStream
    stream.Seek(0, SeekOrigin.Begin)
End Sub
```

■ **Note** Silverlight doesn't include any support for writing WAV file headers. You can write the header yourself, but the code is a bit tedious. This example uses the WavFileHelper, written by Mike Taulty, to do the job. He blogs about it at http://tinyurl.com/yk7rw7e.

Now, you can use the MemoryStreamAudioSink to create the example shown in Figure 11-13. First, create an instance of the audio sink and capture source as member fields:

```
Private audioSink As MemoryStreamAudioSink
Private capture As CaptureSource
```

To associate the two together, you simply need to set the AudioSink.CaptureSource property. Here's the code that starts the audio capture:

```
Private Sub cmdStartRecord_Click(ByVal sender As Object, _
  ByVal e As RoutedEventArgs)

    If CaptureDeviceConfiguration.AllowedDeviceAccess OrElse _
      CaptureDeviceConfiguration.RequestDeviceAccess() Then
        If audioSink Is Nothing Then
            capture = New CaptureSource()
            capture.AudioCaptureDevice = _
              CaptureDeviceConfiguration.GetDefaultAudioCaptureDevice()

            audioSink = New MemoryStreamAudioSink()
            audioSink.CaptureSource = capture
        Else
            audioSink.CaptureSource.Stop()
        End If

        audioSink.CaptureSource.Start()
        cmdStartRecord.IsEnabled = False

        ' Add a delay to make sure the recording is initialized.
        ' (Otherwise, an error may occur if the user stops it immediately.)

        System.Threading.Thread.Sleep(TimeSpan.FromSeconds(0.5))
        cmdStopRecord.IsEnabled = True

        lblStatus.Text = "Now recording ..."
    End If
End Sub
```

■ **Note** In this example, the user is allowed to record indefinitely. To avoid excessive memory usage, you should add a safeguard, like a timer that turns off recording after a certain maximum time limit.

As you can see, you can continue to use the CaptureSource object in the normal way, by calling Start() to initiate the capture and Stop() to end it:

```
Private Sub cmdStopRecord_Click(ByVal sender As Object, _
  ByVal args As RoutedEventArgs)
    audioSink.CaptureSource.Stop()

    cmdPlayClip.IsEnabled = True
    cmdStopRecord.IsEnabled = False
    cmdStartRecord.IsEnabled = True
    lblStatus.Text = "Finished recording. A clip is available to play."
End Sub
```

It's up to you to decide what to do with the audio data when the recording is finished. You could save it or send it to a web service or network server. This example simply plays it back using a MediaElement (named media). However, even this step isn't straightforward, because Silverlight includes no playback support for the WAV file format. Instead, you need to use a third-party library to convert the audio, like the WaveMediaStreamSource (available at http://code.msdn.microsoft.com/wavmss). Here's the code that uses it:

```
Private Sub cmdPlayClip_Click(ByVal sender As Object, _
  ByVal args As RoutedEventArgs)
    Dim wavMss As New WaveMSS.WaveMediaStreamSource(audioSink.AudioData)
    media.SetSource(wavMss)
End Sub
```

Although this example gives some perspective on raw video and audio capture with Silverlight, a realistic application needs much more. The key problem is that raw WAV data is very large. If you need to store multiple audio files or send frequent clips over a network, it's not practical. The problems with raw video are even worse—the samples are enormous, it requires reams of complex, handwritten code to convert it, and the conversion process is computationally expensive, which makes it extremely difficult to do in real time. Ideally, future versions of Silverlight or third-party libraries will include support that builds on CaptureSource and provides the infrastructure developers need to implement more useful scenarios with video and audio capture.

Deep Zoom

Now that you've explored the fine details of Silverlight's audio and video support, it's time to branch out to a very different type of multimedia: Silverlight's new Deep Zoom feature.

The idea behind Deep Zoom is to present a "zoomable" interface for huge images. The typical Deep Zoom image is far too large to be shown on the screen at once at its native resolution. Initially, the Deep Zoom image is shown at a greatly reduced size, so that the user gets a bird's-eye view of the entire picture. The user can then click to zoom in on a specific spot. As the user clicks, Silverlight zooms in more and more, eventually enlarging the selected area of the image to its native resolution (and beyond) and exposing the fine details that weren't initially visible.

Figure 11-14 shows the Deep Zoom process. At the top is the initial zoomed-out view of a beach scene. At the bottom is the wastebasket that you can see after zooming in on one small region at the right of the image.

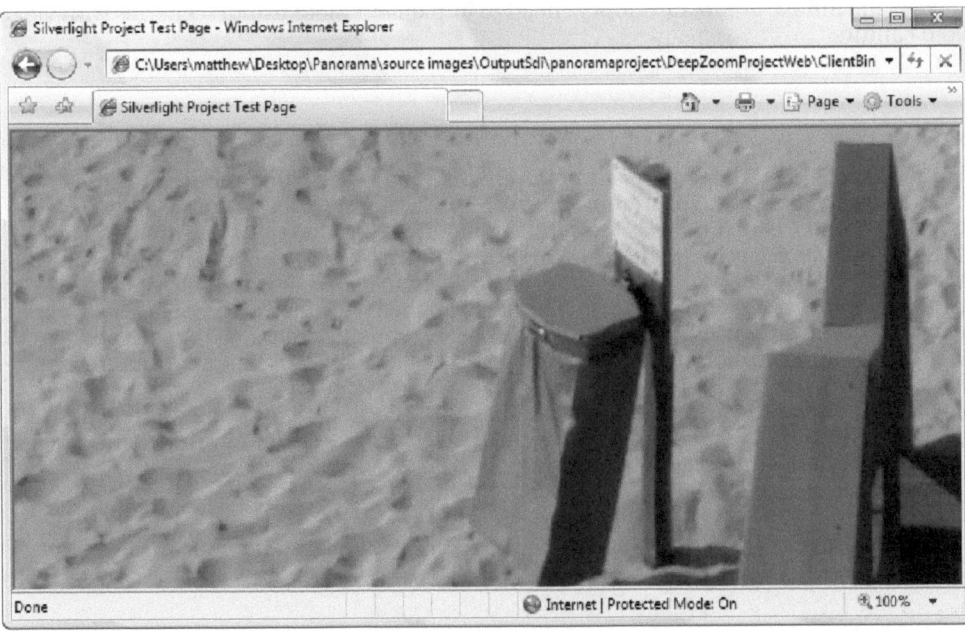

Figure 11-14. Using Deep Zoom to explore a panoramic image

Usually, Deep Zoom images are stitched together from dozens or hundreds of smaller images to create a seamless panorama. However, Deep Zoom can also work with a quilt of distinct images. One example is the Hard Rock Memorabilia website (http://memorabilia.hardrock.com), which uses Deep Zoom to allow visitors to examine different relics, which are tiled together into one huge picture.

■ **Note** Deep Zoom isn't a new idea. Many competitors already implement the same feature. One popular example is Zoomify, which is built using Adobe Flash. However, Deep Zoom feels surprisingly mature. It provides notably smooth zooming (rather than simply jumping between differently sized images) and fast performance that outdoes many more established competitors.

It's easy to create a Silverlight application that uses Deep Zoom, provided you have the right tools. The most important is the free Deep Zoom Composer tool. (To download it, surf to http://tinyurl.com/6xzp8v.) The Deep Zoom Composer allows you to convert a large image into the tiled groups of images that Deep Zoom needs for its zooming interface. It also lets you tile together smaller images to create a large image that's suitable for Deep Zoom, and it can even stitch overlapping images together automatically to create a panorama. (However, you may prefer to use more specialized stitching software such as AutoPano Pro, which can adjust geometry and lighting for a truly seamless compound image.)

■ **Tip** If you want to try the Deep Zoom feature, you have several options for getting the large image you need. Some dedicated photo stitchers post extremely large pictures to photo-sharing sites like Flickr. (Obviously, you need to ask for permission if you want to use the picture for anything other than a test on your local computer.) You can also grab huge satellite images from NASA's Visible Earth website (http://visibleearth.nasa.gov).

When you have the Deep Zoom Composer software and a suitable image (or images), you're ready to get to work.

Creating a Deep Zoom Image Set

To get started, load Deep Zoom Composer, and click New Project. You'll need to choose a project name and a project location. Deep Zoom Composer creates two folders in your initial project location. One folder, named Source Images, holds the original versions of all the pictures you import. The second folder, named Working Data, holds the dozens of image files that are generated when you lay these pictures out into a Deep Zoom image set.

■ **Note** Don't confuse the Deep Zoom project with a Silverlight project. A Deep Zoom project can be opened only in Deep Zoom composer. You must *export* the image set to generate a Silverlight project.

There are three steps to building a Deep Zoom image set with Deep Zoom Composer. First, you import the picture (or pictures) you plan to use. Next, you arrange the pictures. If you have a single picture, this won't take long. If you have multiple pictures, this is when you tile them together by hand. Finally, you export the Deep Zoom image set and create the Silverlight project.

You can switch from one step to another using the three tab buttons at the top of the Deep Zoom Composer window. Initially, you begin in the Import tab. Here's what to do:

1. To get the pictures you want, click the Import button in the panel at right, browse to the correct file, and click OK. Importing large pictures can be slow, so be prepared to wait.

2. Repeat step 1 until you've imported all the pictures you need.

3. Click the Compose button. Here, you start with a blank design surface where you can lay out your pictures (see Figure 11-15).

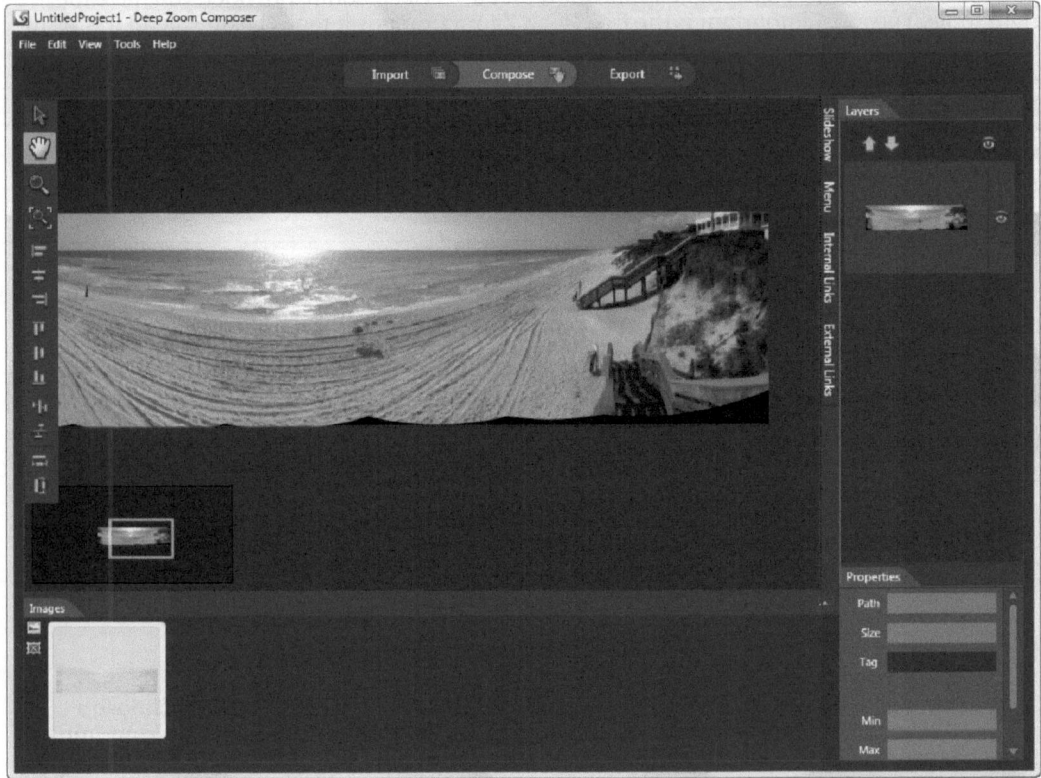

Figure 11-15. Laying out your images in Deep Zoom Composer

4. To add a picture to the design surface, drag it from the panel at the bottom. If you have several pictures, you must drag, position, and size each one. (Images can overlap.)

■ **Tip** Deep Zoom Composer provides a number of shortcuts to help you during the arranging process. For example, you can lay images into a regular grid. First, select the images you want (hold down Ctrl while clicking each one or press Ctrl+A to select them all). Then, right-click the selection, choose Arrange into Grid, fill in the appropriate options (row limits, column limits, and amount of padding), and click OK. This technique is useful if you're creating a Deep Zoom image set that's made up of distinctly separate images, like the tiled items in the Hard Rock Memorabilia display. If you want to create the illusion of a single huge picture, you can use Deep Zoom Composer to stitch overlapping images into a panorama. To do so, select the images, right-click the selection, and choose Create Panoramic Photo. The process may take some time as Deep Zoom Composer searches for matching segments of image data.

5. Click the Export button. Deep Zoom Composer gives you several export options (see Figure 11-16). The two most useful are to export your image set to DeepZoomPix (a Microsoft service for hosting Deep Zoom image sets online, with no code required) or to create a Silverlight project that you can edit, customize, and deploy to your own web server (which is the approach you'll take in the following steps).

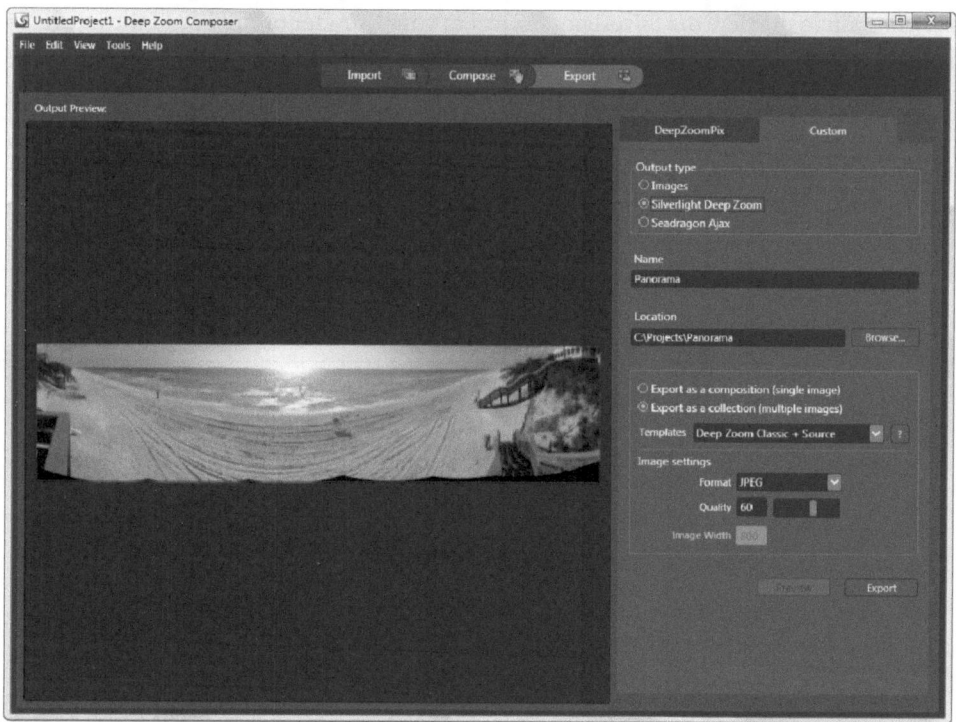

Figure 11-16. Exporting a Silverlight project from Deep Zoom Composer

431

6. To create a Silverlight project, click the Custom tab in the panel at the right. In the "Output type" box, choose Silverlight Deep Zoom.

7. In the Name text box, enter a name for your project. If you want to export the project to a different folder, change the path in the Location text box.

8. Choose "Export as a collection" to create a Deep Zoom image set. Underneath, the Templates box allows you to configure how the Silverlight project will be generated (and whether it will include source code). Although you can choose to export a Deep Zoom image set without project files, the exported project includes some genuinely useful code that allows the user to pan and zoom with the mouse. (If you create your project from scratch, you'll need to write your own code to make the page interactive.) The two most useful templates are "Deep Zoom Classic + Source" (which creates the standard panning and zooming interface you'll explore next), and "Blend 3 Behaviors + Source" (which creates essentially the same result but uses the new behavior feature discussed in Chapter 12 to implement interactivity).

9. In the "Image settings" box, choose either PNG or JPEG. PNG offers better quality through lossless compression. However, JPEG gives you the option to reduce the image quality, which decreases the size of your image files and thereby increases performance.

10. Click Export to create the image set and Silverlight project. This process may take some time. When it's finished, a window appears with several options (see Figure 11-17), allowing you to preview the Silverlight project in your browser or browse to the image folder or project folder.

■ **Tip** The export feature in Deep Zoom Composer is limited to C# code only. Unfortunately, that means your options for integrating Deep Zoom images in a VB-based Silverlight project are limited. You'll be forced to endure the rather awkward work of translating the C# project to VB. To get a head start, check out the downloadable code for this chapter, which includes a VB version of the Deep Zoom project. You can copy most of the code from this project to reduce your conversion work.

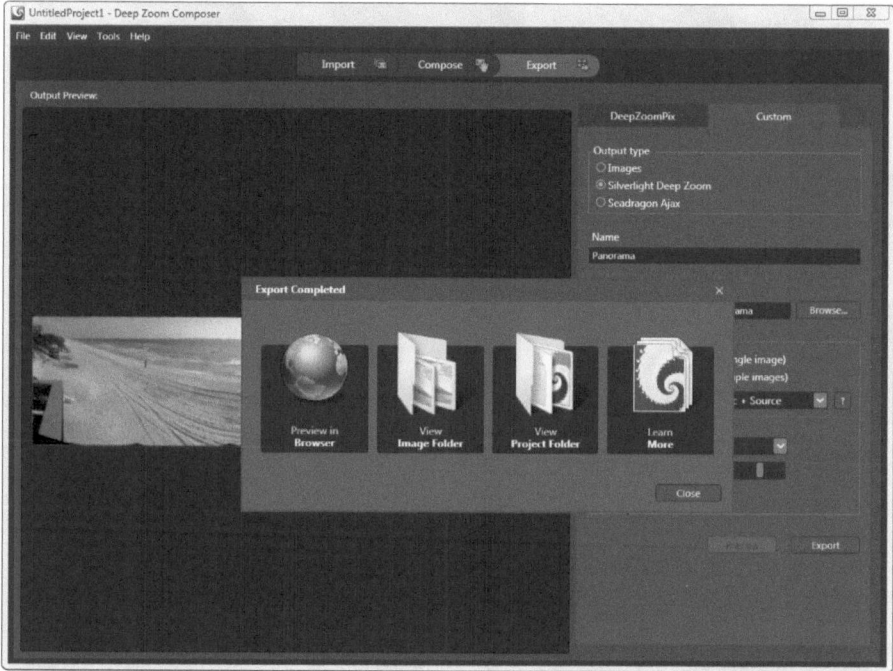

Figure 11-17. Completing an export

Using a Deep Zoom Image Set in Silverlight

When you export a Silverlight project, Deep Zoom Composer creates a Silverlight application named DeepZoomProject and a test website named DeepZoomProjectSite. The DeepZoomProject has all the Silverlight code for panning, scrolling, and zooming into your image. The DeepZoomProjectSite holds the compiled Silverlight project and the actual Deep Zoom image set—a set of XML files image tiles that represent small chunks of your picture at varying resolutions.

Figure 11-18 shows both pieces of the solution. As usual, when you run the project, Visual Studio compiles the Silverlight application into a XAP file and copies that to the ClientBin folder in the test website. However, you'll notice that the ClientBin folder has a subfolder named GeneratedImages. This holds the Deep Zoom image set.

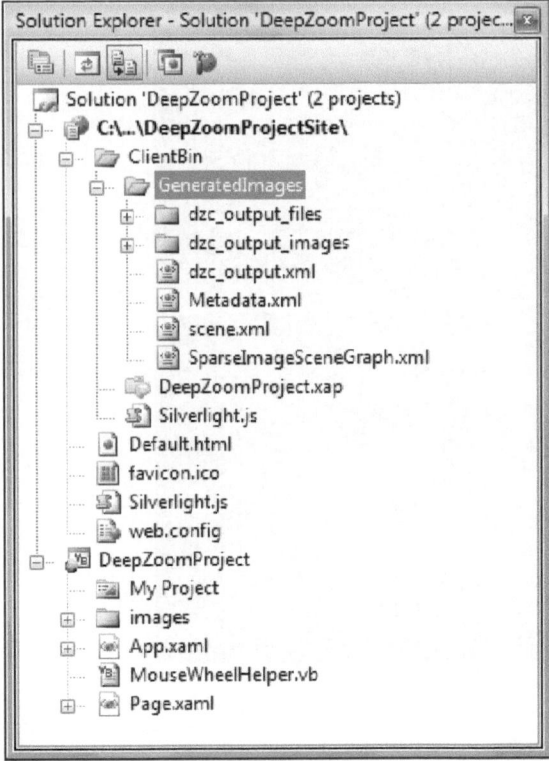

Figure 11-18. *The image set in a Deep Zoom solution*

Showing a Deep Zoom image in a Silverlight application is fairly easy. In fact, all you need is the MultiScaleImage element, as shown here:

```
<MultiScaleImage x:Name="msi" Height="600" Width="800"/>
```

In the automatically generated project, you'll find quite a bit more markup. However, almost all of it is for extra visual frills, including fancy animated buttons that zoom in, zoom out, restore the image to its initial size, and switch the application into full-screen mode (see Figure 11-19). If you decide to create a Deep Zoom project from scratch, you would start with nothing more than the MultiScaleImage.

Figure 11-19. *Image navigation buttons in a Deep Zoom project*

The source for the MultiScaleImage is an XML file that defines the Deep Zoom image set. Although you could set the source for the MultiScaleImage in markup, the automatically generated project uses code, as shown here:

```
msi.Source = New DeepZoomImageTileSource(uri)
```

The URI is passed in as a parameter from the test. By default, it's the XML file dzc_output.xml in the GeneratedImages folder.

The MultiScaleImage element has three key methods, which are detailed in Table 11-1.

Table 11-1. *Methods of the MultiScaleImage Element*

Method	Description
ElementToLogicalPoint()	Converts a physical on-screen point in the MultiScaleImage element to a logical point in the large, virtual image. This translation process lets you zoom into a specific area.
LogicalToElementPoint()	Converts a logical point in the virtual image to a physical location in the MultiScaleImage where that point is currently being displayed.
ZoomAboutLogicalPoint()	Zooms in or out using a logical center point you specify and a zoom factor. The zoom factor is a number greater than 0. Use 1 to fit the available space precisely. Numbers greater than 1 zoom in (for example, 3 zooms in to three-times magnification), and numbers less than 1 zoom out (for example, 0.5 zooms out to half magnification).

The automatically generated project uses these methods to control zooming. The lynchpin is a simple Zoom() method that translates a point in the element to logical coordinates and then zooms to that point. Although the code in an exported project is always written in C#, here's the VB equivalent:

```
Private Sub Zoom(ByVal newzoom As Double, ByVal p As Point)
    ' Don't allow the user to zoom the image out to less than half size.
    If newzoom < 0.5 Then
        newzoom = 0.5
    End If
    msi.ZoomAboutLogicalPoint(newzoom / zoom, p.X, p.Y)
    zoom = newzoom
End Sub
```

Using this method, you can programmatically zoom in on the center point, like this:

```
Zoom(1.2, New Point(Me.ActualWidth / 2, Me.ActualHeight / 2))
```

Or you can zoom out, like this:

```
Zoom(0.8, New Point(Me.ActualWidth / 2, Me.ActualHeight / 2))
```

The code in the automatically generated project goes a bit further. It allows the user to drag the image around the viewing area (using code that's similar to the dragging-circle example in Chapter 4). It also lets the user zoom in by clicking or turning the scroll wheel on the mouse. First, every time the mouse moves, its position is recorded:

```
Me.lastMousePos = e.GetPosition(Me.msi)
```

Then, when the user clicks, the image is zoomed in (or zoomed out if the Shift key is held down):

```
Dim newzoom As Double = zoom

If (Keyboard.Modifiers And ModifierKeys.Shift) = ModifierKeys.Shift Then
    newzoom /= 2
Else
    newzoom *= 2
End If

Zoom(newzoom, msi.ElementToLogicalPoint(Me.lastMousePos))
```

The scroll wheel has much the same effect, but the zoom amount is less:

```
Dim newzoom As Double = zoom

If e.Delta > 0 Then
    newzoom /= 1.3
Else
    newzoom *= 1.3
End If

Zoom(newzoom, msi.ElementToLogicalPoint(Me.lastMousePos))
```

Most Silverlight applications that use Deep Zoom include this code. However, you're free to extend it to suit your needs. For example, the Hard Rock Memorabilia website checks the clicked point to determine what item is in that location. It then zooms and displays a panel next to the image with information about the selected item.

The Last Word

In this chapter, you explored how to integrate sound and video into a Silverlight application. You also considered the best practices for dealing with video and ensuring optimum playback performance in the client and scalability on the server.

Microsoft has placed a great deal of emphasis on Silverlight's multimedia capabilities. In fact, multimedia is one area where Silverlight is gaining features that haven't appeared in the WPF world. For example, WPF has no VideoBrush (although it provides another way to accomplish the same effect with the VisualBrush). Furthermore, its version of the MediaElement lacks a few properties that Silverlight applications use to control buffering and interact with markers. Finally, WPF has no support for webcam and microphone input or Deep Zoom—so if you want a similar capability in a rich client application, you'll be forced to build it yourself.

■■■■

Styles and Behaviors

Silverlight applications would be a drab bunch if you were limited to the plain, gray look of ordinary buttons and other common controls. Fortunately, Silverlight has several features that allow you to inject some flair into basic elements and standardize the visual appearance of your application. In this chapter, you'll learn about two of the most important: styles and behaviors.

Styles are an essential tool for organizing and reusing for formatting choices. Rather than filling your XAML with repetitive markup to set details such as margins, padding, colors, and fonts, you can create a set of styles that encompass all these details. You can then apply the styles where you need them by setting a single property (or automatically by element type).

Behaviors are a more ambitious tool for reusing user interface code. The basic idea is that a behavior encapsulates a common bit of UI functionality (for example, the code that makes an element draggable). If you have the right behavior, you can attach it to any element with a line or two of XAML markup, saving you the effort of writing and debugging the code yourself. Currently, behaviors are still a new and developing feature, and many useful prebuilt behaviors are sure to be released in the coming months.

■ **What's New** Silverlight 4 finally adds support for *implicit styles*, a longstanding WPF feature that allows you to apply a style automatically to all the elements of a specific type (for example, all the buttons in a page). The section "Automatically Applying Styles by Type" has the details.

Styles

A *style* is a collection of property values that you can apply to an element in one step. In Silverlight, styles let you streamline your XAML markup by pulling repetitive formatting details out of your element tags.

The Silverlight style system plays a similar role to the Cascading Style Sheets (CSS) standard in HTML markup. Like CSS, Silverlight styles allow you to define a common set of formatting characteristics and apply them throughout your application to ensure consistency.

If you've used styles in WPF, you'll find that Silverlight styles have a few limitations. You can't share styles between different elements (for example, use a single style on a Button and a TextBlock), and you can't use triggers to change the style of a control when another property is modified. However, the most significant style shortcoming—the inability to apply styles automatically—has finally been removed in Silverlight 4.

Despite its limitations, the Silverlight style system is clearly a useful feature. You'll almost certainly use it to standardize and reuse formatting throughout your applications.

Defining a Style

Experienced Silverlight developers turn to styles when they expect to use a group of formatting characteristics more than once. For example, imagine that you need to standardize the font and foreground color that are used in all the buttons on a page. The first step is to define a Style object that wraps all the properties you want to set. You'll store this Style object as a resource, typically in the UserControl.Resources collection that holds resources for the entire page:

```
<UserControl.Resources>
  <Style x:Key="BigButtonStyle" TargetType="Button">
    ...
  </Style>
</UserControl.Resources>
```

Like all resources, the style has a key name so you can pull it out of the collection when needed. In this case, the key name is BigButtonStyle. (By convention, the key names for styles usually end with *Style*.) Additionally, every Silverlight style requires a TargetType, which is the type of element on which you apply the style. In this case, the TargetType property indicates that this style is designed to format buttons.

Every style holds a Setters collection with several Setter objects, one for each property you want to set. Each Setter object sets a single property in an element. The only limitation is that a setter can change only dependency properties—other properties can't be modified. (In practice, this isn't much of a limitation, because Silverlight elements consist almost entirely of dependency properties.) It's also important to note that property setters can act on any dependency property, even one that governs behavior rather than appearance. For example, if you're applying a style to a text box, you may choose to set the AcceptsReturn and IsReadOnly properties.

Here's a style that sets a combination of five properties, giving buttons large, light text with Georgia font:

```
<UserControl.Resources>
  <Style x:Key="BigButtonStyle" TargetType="Button">
    <Setter Property="FontFamily" Value="Georgia" />
    <Setter Property="FontSize" Value="40" />
    <Setter Property="Foreground" Value="SlateGray" />
    <Setter Property="Padding" Value="20" />
    <Setter Property="Margin" Value="10" />
  </Style>
</UserControl.Resources>
```

In some cases, you can't set the property value using a simple attribute string. For example, you can't create a complex brush like LinearGradientBrush or ImageBrush with a string. In this situation, you can use the familiar XAML trick of replacing the attribute with a nested element. Here's an example that sets the button background:

```
<Style x:Key="BigButtonStyle" TargetType="Button">
  <Setter Property="Background">
    <Setter.Value>
      <LinearGradientBrush StartPoint="0,0" EndPoint="1,0">
        <GradientStop Color="Blue"></GradientStop>
        <GradientStop Color="Yellow" Offset="1"></GradientStop>
      </LinearGradientBrush>
    </Setter.Value>
  </Setter>
```

```
  ...
</Style>
```

More powerfully, a style can set transforms, rendering effects, and control templates that completely revamp the appearance of an element.

Applying a Style

Every Silverlight element can use a single style (or no style). The style plugs into an element through the element's Style property (which is defined in the base FrameworkElement class). For example, to configure a button to use the style you created previously, you point the button to the style resource like this:

```
<Button Style="{StaticResource BigButtonStyle}"
  Content="A Customized Button"></Button>
```

Figure 12-1 shows a page with two buttons that use BigButtonStyle.

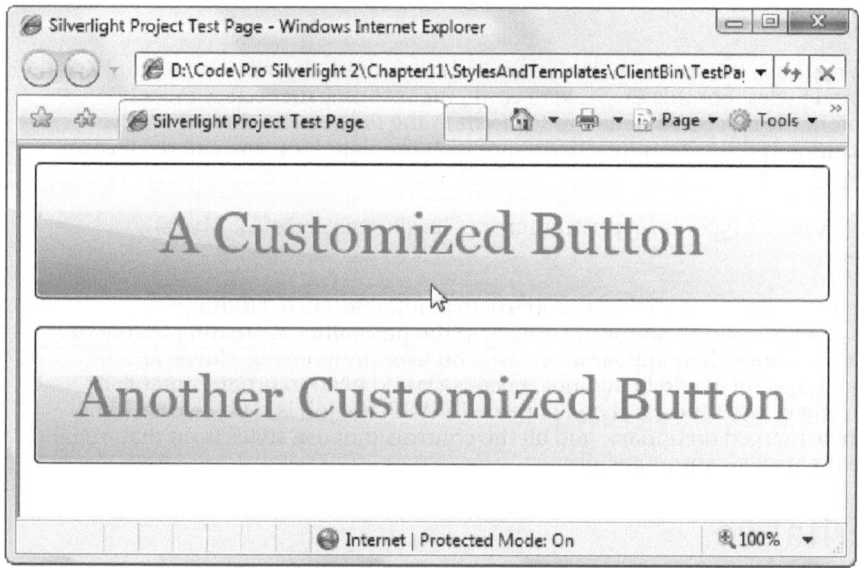

Figure 12-1. *Reusing button settings with a style*

The style system adds many benefits. Not only does it allow you to create groups of settings that are clearly related, but it also streamlines your markup by making it easier to apply these settings. Best of all, you can apply a style without worrying about what properties it sets. In the previous example, the font settings are organized into a style named BigButtonStyle. If you decide later that your big-font buttons also need more padding and margin space, you can add setters for the Padding and Margin properties as well. All the buttons that use the style automatically acquire the new style settings.

Styles set the initial appearance of an element, but you're free to override the characteristics they set. For example, if you apply the BigButtonStyle style and set the FontSize property explicitly, the FontSize setting in the button tag overrides the style. Ideally, you won't rely on this behavior. Instead, if you need to alter a style, you should create a new style that sets the appropriate properties. This gives you more flexibility to adjust your user interface in the future with minimum disruption.

Dynamic Styles

Although you'll usually want to link up styles in your markup, you can also set them programmatically. All you need to do is pull the style out of the appropriate Resources collection.

Here's the code you use to retrieve the BigButtonStyle from the page's Resources collection and apply it to a Button object named cmd:

```
cmd.Style = CType(this.Resources("AlternateBigButtonStyle"), Style)
```

You can even retrieve a style from a separate resource dictionary file in your project (or a referenced assembly). First, you need to create a ResourceDictionary object and supply the correct URI:

```
Dim resourceDictionary As New ResourceDictionary()
resourceDictionary.Source = New Uri("/Styles/AlternateStyles.xaml", _
  UriKind.Relative)
```

This example assumes the resource dictionary is a file named AlternateStyles.xaml in a project folder named Styles. For this example to work, the resource dictionary must be compiled as a content file (set Build Action to Content in the Properties window).

After you've configured the ResourceDictionary with the right URI, you can retrieve an object from inside:

```
Dim newStyle As Style = CType(resourceDictionary("BigButtonStyle"), Style)
cmd.Style = newStyle
```

Finally, to remove a style, set the Style property to a null reference (Nothing).

The ability to change styles dynamically opens up the possibility of creating *skinnable* applications that can change their appearance based on user preferences. However, implementing such a system would be tedious, because you'd need to programmatically retrieve each style object and set each styled control. In WPF, the job is much easier—you simply swap in a new merged dictionary, and all the controls that use styles from that merged dictionary update themselves automatically.

Style Inheritance

In some situations, you may want to use the properties from one style as the basis for another, more specialized style. You can use this sort of style inheritance by setting the BasedOn attribute of a style. For example, consider these two styles:

```
<UserControl.Resources>
  <Style x:Key="BigButtonStyle" TargetType="Button">
    <Setter Property="FontFamily" Value="Georgia" />
    <Setter Property="FontSize" Value="40" />
    <Setter Property="Padding" Value="20" />
    <Setter Property="Margin" Value="10" />
  </Style>

  <Style x:Key="EmphasizedBigButtonStyle" TargetType="Button"
    BasedOn="{StaticResource BigButtonStyle}">
```

```
    <Setter Property="BorderBrush" Value="Black" />
    <Setter Property="BorderThickness" Value="5" />
  </Style>
</UserControl.Resources>
```

The first style (BigButtonStyle) defines four properties. The second style (EmphasizedBigFontButtonStyle) acquires these aspects from BigFontButtonStyle and then supplements them with two more properties that apply a thick black outline around the button. This two-part design gives you the ability to apply just the font settings or the font and color combination. This design also allows you to create more styles that incorporate the font or color details you've defined (but not necessarily both).

You can use the BasedOn property to create an entire chain of inherited styles. The only rule is that if you set the same property twice, the last property setter (the one in the derived class farthest down the inheritance chain) overrides any earlier definitions.

■ **Tip** Surprisingly, an inherited style doesn't need to have the same TargetType as its parent. That means you can create a derived style that has the same formatting properties—but acts on a different element—than the parent. But you'll get an error if your style inherits a setter for a property that doesn't exist in its target element type. (For example, the TextBlock and Button classes both have a FontFamily property, so you can set FontFamily in a button style and create a TextBlock style that derives from it. However, only the button has the IsEnabled property, so if you set IsEnabled in a Button style, you can't derive a TextBlock style from it.)

STYLE INHERITANCE ADDS COMPLEXITY

Although style inheritance seems like a great convenience at first glance, it's usually not worth the trouble. That's because style inheritance is subject to the same problems as code inheritance: dependencies that make your application more fragile. For example, if you use the markup shown previously, you're forced to keep the same font characteristics for two styles. If you decide to change BigButtonStyle, EmphasizedBigButtonStyle changes as well—unless you explicitly add more setters that override the inherited values.

This problem is trivial enough in the two-style example, but it becomes a significant issue if you use style inheritance in a more realistic application. Usually, styles are categorized based on different types of content and the role the content plays. For example, a sales application may include styles such as ProductTitleStyle, ProductTextStyle, HighlightQuoteStyle, NavigationButtonStyle, and so on. If you base ProductTitleStyle on ProductTextStyle (perhaps because they both share the same font), you'll run into trouble if you apply settings to ProductTextStyle later that you don't want to apply to ProductTitleStyle (such as different margins). In this case, you'll be forced to define your settings in ProductTextStyle and explicitly override them in ProductTitleStyle. At the end, you'll be left with a more complicated model and very few style settings that are actually reused.

Unless you have a specific reason to base one style on another (for example, the second style is a special case of the first and changes just a few characteristics out of a large number of inherited settings), think carefully before using style inheritance.

Organizing Styles

In the previous examples, the Style object is defined at the page level and then reused in buttons inside that page. Although that's a common design, it's certainly not your only choice.

Strictly speaking, you don't need to use styles and resources together. For example, you can define the style of a particular button by filling its Style collection directly, as shown here:

```
<Button Content="A Customized Button">
  <Button.Style>
    <Style TargetType="Button">
      <Setter Property="FontFamily" Value="Georgia" />
      <Setter Property="FontSize" Value="40" />
      <Setter Property="Foreground" Value="White" />
    </Style>
  </Button.Style>
</Button>
```

This works, but it's obviously a lot less useful. Now there's no way to share this style with other elements.

More usefully, you may want to define styles in different resource collections. If you want to create more finely targeted styles, you can define them using the resources collection of their container, such as a StackPanel or a Grid. It's even possible for the same style to be defined at multiple levels (in a StackPanel containing a button and in the page that holds the StackPanel). In this situation, Silverlight follows the standard resource-resolution process you learned about in Chapter 2—it searches in the resources collection of the current element first, then the containing element, then its container, and so on, until it finds a style with the matching name.

If you want to reuse styles across an application, you should define them using the resources collection of your application (in the App.xaml file), which is the last place Silverlight checks. Or even better, put them in a separate resource dictionary that's merged into the resource collection of every page that needs it, as explained in Chapter 2.

Automatically Applying Styles by Type

So far, you've seen how to create named styles and refer to them in your markup. However, there's another approach. You can apply a style automatically to elements of a certain type.

Doing this is quite easy. You simply need to set the TargetType property to indicate the appropriate type (as described earlier) and leave out the key name altogether. Here's an example:

```
<Style TargetType="Button">
```

Now the style is automatically applied to any buttons all the way down the element tree. For example, if you define a style in this way on the UserControl.Resources collection, it applies to every button in that page (unless there's a style further downstream that replaces it).

Here's an example that sets the button styles automatically to get the same effect you saw in Figure 12-1:

```
<UserControl.Resources>
  <Style TargetType="Button">
    <Setter Property="FontFamily" Value="Georgia" />
    <Setter Property="FontSize" Value="40" />
    <Setter Property="Foreground" Value="SlateGray" />
    <Setter Property="Padding" Value="20" />
    <Setter Property="Margin" Value="10" />
```

```
    </Style>
  </UserControl.Resources>

<StackPanel x:Name="LayoutRoot" Background="White">
  <Button Content="A Customized Button"></Button>
  <Button Content="Another Customized Button"></Button>
</StackPanel>
```

If you want to prevent a style from being applied to an element automatically, you simply need to set the Style property to something else. If you don't want to use any style at all, simply set it to a null reference (Nothing), which effectively removes the style:

```
<Button Content="A Non-Customized Button" Style="{x:Null}"></Button>
```

Although automatic styles are convenient, they can complicate your design. Here are a few reasons why:

- In a complex page with many styles and multiple layers of styles, it becomes difficult to track down whether a given property is set through property value inheritance or a style (and if it's a style, which one). As a result, if you want to change a simple detail, you may need to wade through the markup of your entire page.

- The formatting in a page often starts out more general and becomes increasingly fine-tuned. If you apply automatic styles to the page early on, you'll probably need to override the styles in many places with explicit styles. This complicates the overall design. It's much more straightforward to create named styles for every combination of formatting characteristics you want and apply them by name.

- For example, if you create an automatic style for the TextBlock element, you'll wind up modifying other controls that use the TextBlock (such as a template-driven ListBox control).

To avoid problems, it's best to apply automatic styles judiciously. If you do decide to give your entire user interface a single, consistent look using automatic styles, try to limit your use of explicit styles to special cases.

Behaviors

Styles give you a practical way to reuse groups of property settings. They're a great first step that can help you build consistent, well-organized interfaces—but they're also broadly limited.

The problem is that property settings are only a small part of the user-interface infrastructure in a typical application. Even the most basic program usually needs reams of user-interface code that has nothing to do with the application's functionality. In many programs, the code that's used for UI tasks (such as driving animations, powering slick effects, maintaining user-interface state, and supporting user-interface features like dragging, zooming, and docking) far outweighs the business code in both size and complexity. Much of this code is generic, meaning you end up writing the same thing in every Silverlight project you create. Almost all of it is tedious.

Based on this reality, it's clear that a method for reusing user-interface code would be even more useful than a system for reusing property values through styles. So far, Silverlight hasn't met the challenge. But a new system of *behaviors* is beginning to fill the gap. The idea is simple: you (or another developer) create a behavior that encapsulates a common bit of user-interface functionality. This functionality can be basic (such as starting a storyboard or navigating to a hyperlink). Or, it can be complex (such as handling scrolling and the mouse wheel for a Deep Zoom image or modeling a collision with a real-time physics engine). Once built, this functionality can be added to another control in any application by hooking up the right behavior and setting its properties. In Expression Blend, using a behavior takes little more than a drag-and-drop operation.

■ **Note** Custom controls are another technique for reusing user-interface functionality in an application (or among multiple applications). However, a custom control must be developed as a tightly linked package of visuals and code. Although custom controls are extremely powerful, they don't address situations where you need to equip many different controls with similar functionality (for example, adding a mouse-over rendering effect to a group of different elements). For that reason, styles, behaviors, and custom controls are all complementary, and some of the most flexible applications use them in combination. You'll learn about behaviors in the rest of this chapter and custom controls in the next chapter.

Getting Support for Behaviors

There's one catch. The new infrastructure for reusing common blocks of user-interface code isn't part of the Silverlight SDK. Instead, it's bundled with Expression Blend. This is because behaviors began as a design-time feature for Expression Blend (and Expression Blend is currently the only tool that lets you add behaviors by dragging them onto the controls that need them). That doesn't mean behaviors are useful only in Expression Blend. You can create and use them in a Visual Studio application with only slightly more effort. (You simply need to write the markup by hand rather than using the Toolbox.)

To get the assemblies that support for behaviors, you have two options:

- You can install Expression Blend 4 (or the free preview that's available at www.microsoft.com/expression/try-it).
- You can install the Expression Blend SDK for Silverlight 4 (which is available at http://tinyurl.com/25sh2jt).

Either way, you'll find two important assemblies in a folder like c:\Program Files\Microsoft SDKs\Expression\Blend\Silverlight\v4.0\Librariest:

- *System.Windows.Interactivity.dll*: This assembly defines the base classes that support behaviors. It's the cornerstone of the behavior feature.
- *Microsoft.Expression.Interactions.dll*: This assembly adds some useful extensions, with optional action and trigger classes that are based on the core behavior classes.

In the following sections, you'll see how to develop a simple action and wire it up in an application, using nothing more than the System.Windows.Interactivity.dll assembly. Once you've completed this exercise and you understand the behavior model, you'll examine the Microsoft.Expression.Interactions.dll assembly and consider other ready-made behaviors.

Triggers and Actions

The behaviors feature consists of three smaller ingredients—trigger, action, and behavior classes—which, somewhat confusingly, are all called behaviors.

Triggers and actions work hand in hand. The trigger fires when something happens and invokes an action. Together, triggers and actions make up the simplest form of behavior.

Creating an Action

To get a grip on triggers and actions, it helps to design a simple action of your own. For example, imagine you want to play a sound when the user performs an operation (such as clicking a button). It's fairly easy to carry this out without behaviors—you add a MediaElement to the page, supply the URI of an audio file, and call the MediaElement.Play() method at the appropriate time. However, these details add unnecessary clutter. If you want to play a range of different sounds in response to different events, you need to manage a surprisingly large amount of code.

You can avoid the hassle with an action that takes care of the playback for you. And assuming no one has created just the action you need, you can create it yourself. First, create a Silverlight class library assembly. (In this example, it's called CustomBehaviorsLibrary.) Then, add a reference to the System.Windows.Interactivity.dll assembly. Finally, create an action class that derives from TriggerAction, as shown here:

```
Public Class PlaySoundAction
    Inherits TriggerAction(Of FrameworkElement)
    ...
End Class
```

All actions derive from the confusingly named TriggerAction class. That's because these are actions that are activated by a trigger. (In theory, the model could support actions that are wired up in different ways, although it currently doesn't and there's no reason it should.)

■ **Note** Ideally, you won't need to create an action yourself—instead, you'll use a ready-made action that someone else has created. Although the System.Windows.Interactivity.dll assembly doesn't include any action classes, you'll hunt some down later in this chapter.

As you can see in the previous example, the TriggerAction class uses generics. When you derive from TriggerAction, you supply the type of element that can use your action as a type parameter. Unless your action requires specialized functionality in the triggering element, it's common to use UIElement or FrameworkElement. In this example, PlaySoundAction supports any FrameworkElement. (It uses FrameworkElement rather than the somewhat more generic UIElement class, because it needs to use the VisualTreeHelper to search the visual tree, as you'll see shortly. But because all elements inherit from FrameworkElement, which in turn inherits from UIElement, there's really no difference.)

Like any other class, your action needs properties. Ideally, you'll use dependency properties to give yourself the best support for other Silverlight features. In this example, PlaySoundAction requires just one property—the URI source that points to the audio file:

```
Public Shared ReadOnly SourceProperty As DependencyProperty = _
  DependencyProperty.Register("Source", GetType(Uri), _
  GetType(PlaySoundAction), New PropertyMetadata(Nothing))

Public Property Source() As Uri
    Get
        Return CType(GetValue(PlaySoundAction.SourceProperty), Uri)
    End Get
    Set(ByVal value As Uri)
        SetValue(PlaySoundAction.SourceProperty, value)
    End Set
End Property
```

When a trigger fires, it activates your action by calling the Invoke() method. You override this method to supply all the code for your action. In this example, that means you need to supply the code that plays the audio.

As you know from Chapter 11, the media support in Silverlight has one clear restriction. To play anything through a MediaElement, you must place it in your hierarchy of elements. Even if your MediaElement is intended to play ordinary audio and has no visual presence, it needs to be in your page's visual tree.

You can handle this limitation several possible ways in the PlaySoundAction class. For example, you can include a property that takes the name of a MediaElement on the page. PlaySoundAction can then look up that MediaElement and use it. Another approach is to create the MediaElement you need and then remove it as soon as the playback is finished. This approach has several advantages: it saves you from needing to define the MediaElement yourself, and it allows you to play an unlimited number of sounds simultaneously. Here's the code that does the job, using a FindContainer() method that gets a container where the MediaElement can be inserted:

```
Private container As Panel
Private media As MediaElement

Protected Overrides Sub Invoke(ByVal args As Object)
    ' Find a place to insert the MediaElement.
    container = FindContainer()

    If container IsNot Nothing Then
        ' Create and configure the MediaElement.
        media = New MediaElement()
        media.Source = Me.Source

        AddHandler media.MediaEnded, AddressOf MediaEnded
        AddHandler media.MediaFailed, AddressOf MediaEnded

        ' Add the MediaElement and begin playback.
        media.AutoPlay = True
        container.Children.Add(media)
    End If
End Sub

Private Sub MediaEnded()
    container.Children.Remove(media)
End Sub
```

The FindContainer() method uses VisualTreeHelper to travel up the hierarchy of elements, starting at the current element. It stops as soon as it finds any sort of layout container (in other words, a class that derives from Panel) where the new MediaElement can be injected. If no panel is found, the MediaElement can't be added, and the media isn't played. This is possible in a window that contains a single control, but in no other situation. To get the current element (the element where the action is attached), you use the inherited TriggerAction.AssociatedObject property:

```
Private Function FindContainer() As Panel
    Dim element As FrameworkElement = Me.AssociatedObject

    ' Search for some sort of panel where the MediaElement can be inserted.
    Do While element IsNot Nothing
        If TypeOf element Is Panel Then
            Return CType(element, Panel)
        End If

        element = TryCast(VisualTreeHelper.GetParent(element), FrameworkElement)
    Loop
    Return Nothing
End Function
```

This completes the code for the action. As you can see, most action classes consist of nothing more than properties and the Invoke() method that does something. In the next section, you'll learn how to wire an element to your action.

Connecting an Action to an Element

To use an action, you need the help of a trigger. Technically, the trigger connects to your element, and the action connects to your trigger. That means the first step in using PlaySoundAction is choosing a suitable trigger.

All triggers derive from TriggerBase. The System.Windows.Interactivity.dll assembly includes a single trigger called EventTrigger, which fires when a specific event occurs. Although you can create your own triggers, the EventTrigger is flexible enough to handle a wide range of scenarios.

■ **Note** One reason that you might create your own trigger is to respond to a certain event and state *combination*. For example, a custom trigger can intercept an event, check a few other details, and then decide whether to fire. Building a trigger like this is easy: you derive from the TriggerBase class, override OnAttached() to wire up the appropriate event, and override OnDetaching() to disconnect the event handler. When the event you're watching occurs, your trigger handles it. If your trigger then decides to fire, it calls the inherited InvokeActions() method.

To test PlaySoundAction, begin by creating a new Silverlight project. Then, add a reference to the class library that defines the PlaySoundAction class (which you created in the previous section) and the System.Windows.Interactivity.dll assembly. Then, map both namespaces. Assuming the PlaySoundAction class is stored in a class library named CustomBehaviorsLibrary, you'll need markup like this:

```
<UserControl xmlns:i="clr-namespace:System.Windows.Interactivity;
assembly=System.Windows.Interactivity" xmlns:custom=
"clr-namespace:CustomBehaviorsLibrary;assembly=CustomBehaviorsLibrary" ... >
```

You can add a trigger to any element using the attached Interaction.Triggers collection that's defined in the System.Windows.Interactivity.dll assembly. Here's an example that adds an event trigger to a button:

```
<Button Content="Click to Play Sound">
  <i:Interaction.Triggers>
    <i:EventTrigger EventName="Click">
    </i:EventTrigger>
  </i:Interaction.Triggers>
</Button>
```

The EventTrigger.EventName property identifies the event that you want to respond to. In this example, the trigger fires when the Button.Click event occurs.

The final step is to add the behavior to the EventTrigger.Actions collection. You can do this in your markup by declaring the behavior inside the event trigger:

```
<Button Content="Click to Play Sound">
  <i:Interaction.Triggers>
    <i:EventTrigger EventName="Click">
      <custom:PlaySoundAction Source="test.mp3" />
    </i:EventTrigger>
  </i:Interaction.Triggers>
</Button>
```

Now you can run the page and test the button. When you click the button, playback begins for the test.mp3 file. From the point of view of the application developer, all it takes is the right behavior and a few straightforward lines of markup. You can use the same action to wire sound to any number of different elements and different events in the same page. You can even add a series of actions to your trigger, in which case they will be launched in quick succession.

But that's still not the whole story. If you're developing in Expression Blend, behaviors give you an even better design experience—one that can save you from writing any markup at all.

Design-Time Behavior Support in Blend

In Expression Blend, working with behaviors is a drag-drop-and-configure operation. First, you need to make sure your application has a reference to the assembly that has the behaviors you want to use. (In this case, that's the class library assembly where PlaySoundAction is defined.) Next, you need to ensure that it also has a reference to the System.Windows.Interactivity.dll assembly.

Expression Blend automatically searches all referenced assemblies for actions and displays them in the Asset Library (the same panel you use for choosing elements when designing a Silverlight page). It also adds the behaviors from the Microsoft.Expression.Interactions.dll assembly, even if they aren't yet referenced by your project.

To see the behaviors you have to choose from, start by drawing a button on the design surface of your page, click the Asset Library button, and click the Behaviors tab (Figure 12-2).

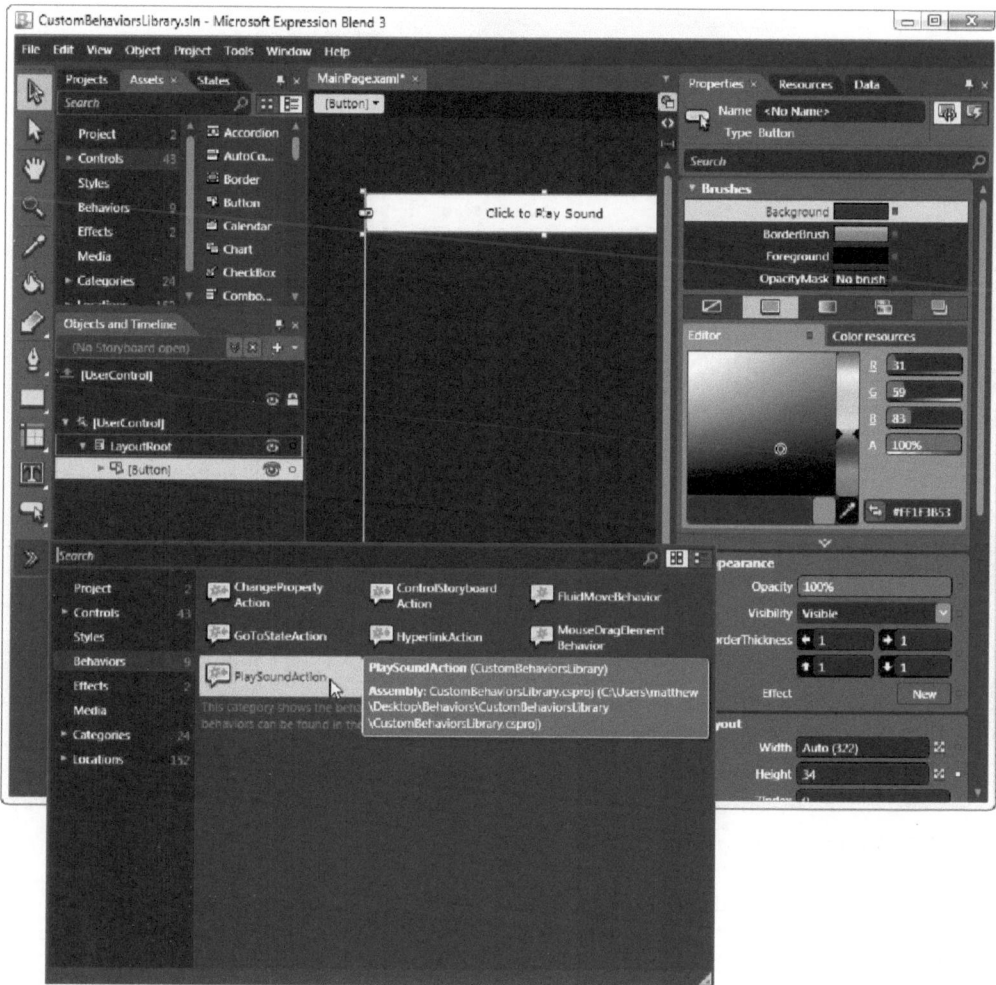

Figure 12-2. *Actions in the Asset Library*

To add an action to a control, drag it from the Asset Library, and drop it onto your control (in this case, the button). When you take this step, Expression Blend automatically creates a trigger for your element and creates the action inside. It's then up to you to configure both objects—typically, by selecting the action in the Objects and Timelines pane and then adjusting the options in the Properties window (Figure 12-3).

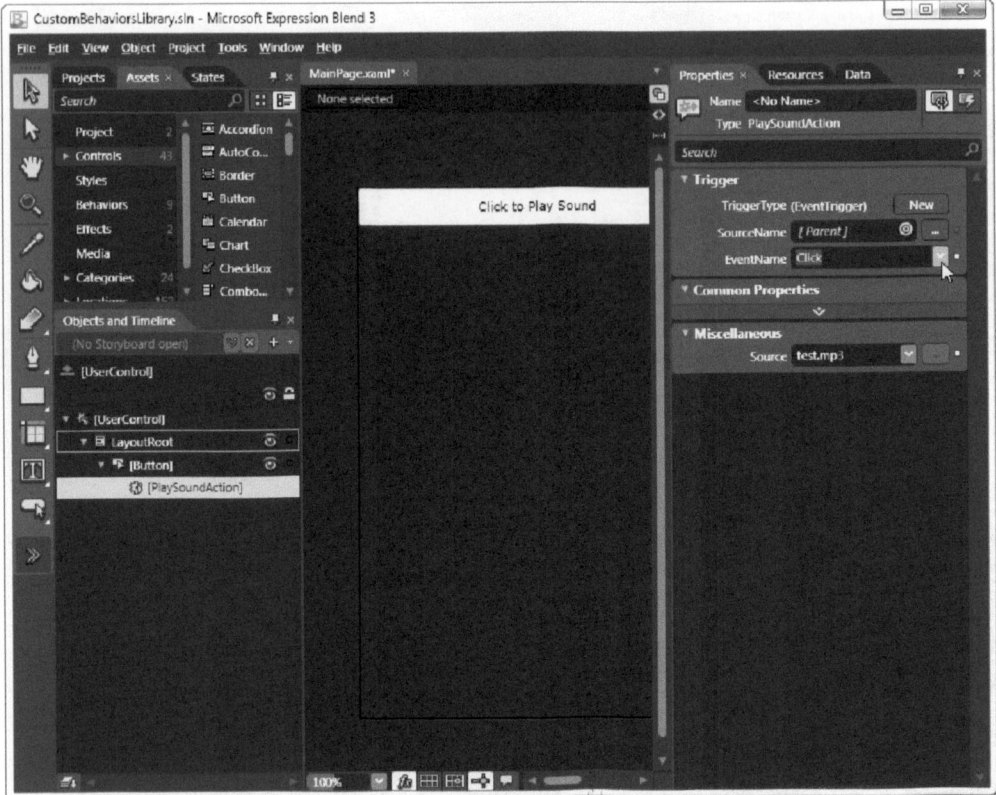

Figure 12-3. Configuring an action (and its trigger)

By default, when you add a new action to a button, Expression Blend creates an EventTrigger for the Click event. You can choose a new event (or a different trigger) in the Properties window (as shown in Figure 12-3). Other elements use different default events—for example, the ListBox uses the SelectionChanged event, and the Rectangle uses the Loaded event.

Incidentally, you can tell Expression Blend what trigger to use by default by adding the DefaultTrigger attribute to your action class. Here's an example that tells Expression Blend to use the ever-popular EventTrigger and set it to react to the MouseLeftButtonDown event:

```
<DefaultTrigger(GetType(UIElement), _
 GetType(EventTrigger), New Object() {"MouseLeftButtonDown"})> _
Public Class PlaySoundAction
    Inherits TriggerAction(Of FrameworkElement)
    ...
End Class
```

As you can see, the DefaultTrigger attribute takes several parameters, which specify the type of element that's being wired up, the type of trigger, and any additional information to pass to the trigger's constructor (such as the event name for the event trigger). You can get even more specific and choose a different trigger based on the type of element where the action is being applied. To do so, add multiple DefaultTrigger attributes, ordered from most to least specific. For example, the following combination ensures that when you wire the PlaySoundAction to a button, Expression Blend starts with an EventTrigger for the Click event. For Shape-derived elements like the Rectangle, it uses an EventTrigger set to the MouseEnter event. For all other elements, it uses an EventTrigger set to the MouseLeftButtonDown event:

```
<DefaultTrigger(GetType(ButtonBase), _
 GetType(EventTrigger), New Object() {"Click"})>, _
<DefaultTrigger(GetType(Shape), _
 GetType(EventTrigger), New Object() {"MouseEnter"})>, _
<DefaultTrigger(GetType(UIElement), _
 GetType(EventTrigger), New Object() {"MouseLeftButtonDown"})> _
Public Class PlaySoundAction
    Inherits TriggerAction(Of FrameworkElement)
    ...
End Class
```

Creating a Targeted Trigger

As you've seen, every action can access the element that it's attached to using the TriggerAction.AssociatedObject property. In the PlaySoundAction, this ability allowed the code to search up the element tree for a suitable layout container. Other actions can retrieve additional information from the source element or alter it in some way. But many actions need to go beyond the source element and perform work on another element. For example, clicking a button may cause a change in another control that's placed somewhere different. And although you could attempt to deal with this situation by adding additional properties, Silverlight has a more gracious solution—you can derive your action from the more specialized TargetedTriggerAction. The TargetedTriggerAction provides a Target property, which the application developer sets and your trigger code examines. You can then perform the necessary operations on that element. Aside from the Target property, the TargetedTriggerAction works exactly the same as the standard TriggerAction.

The following example shows a pair of actions that derive from TargetedTriggerAction. The first, FadeOutAction, runs an animation on the target that gradually fades the Opacity of the target element to 0. The second, FadeInAction, uses another animation to fade it back in. Here's the complete code for the FadeOutAction:

```
Public Class FadeOutAction
    Inherits TargetedTriggerAction(Of UIElement)

    ' The default fade out time is 2 seconds.
    Public Shared ReadOnly DurationProperty As DependencyProperty = _
      DependencyProperty.Register("Duration", GetType(TimeSpan), _
      GetType(FadeOutAction), New PropertyMetadata(TimeSpan.FromSeconds(2)))

    Public Property Duration() As TimeSpan
        Get
            Return CType(GetValue(FadeOutAction.DurationProperty), TimeSpan)
        End Get
        Set(ByVal value As TimeSpan)
            SetValue(FadeOutAction.DurationProperty, value)
        End Set
    End Property
```

```
        Private fadeStoryboard As New Storyboard()
        Private fadeAnimation As New DoubleAnimation()

        Public Sub New()
            fadeStoryboard.Children.Add(fadeAnimation)
        End Sub

        Protected Overrides Sub Invoke(ByVal args As Object)
            ' Make sure the storyboard isn't already running.
            fadeStoryboard.Stop()

            ' Set up the storyboard.
            Storyboard.SetTarget(fadeAnimation, Me.Target)
            Storyboard.SetTargetProperty(fadeAnimation, New PropertyPath("Opacity"))

            ' Set up the animation.
            ' It's important to do this at the last possible instant,
            ' in case the value for the Duration property changes.
            fadeAnimation.To = 0
            fadeAnimation.Duration = Duration

            fadeStoryboard.Begin()
        End Sub
    End Class
```

FadeInAction is almost identical. It animates the Opacity to 1, using a default time of just 0.5 seconds:

```
Public Class FadeInAction
    Inherits TargetedTriggerAction(Of UIElement)

    ' The default fade in is 0.5 seconds.
    Public Shared ReadOnly DurationProperty As DependencyProperty = _
      DependencyProperty.Register("Duration", GetType(TimeSpan), _
      GetType(FadeInAction), New PropertyMetadata(TimeSpan.FromSeconds(0.5)))

    Public Property Duration() As TimeSpan
        Get
            Return CType(GetValue(FadeInAction.DurationProperty), TimeSpan)
        End Get
        Set(ByVal value As TimeSpan)
            SetValue(FadeInAction.DurationProperty, value)
        End Set
    End Property

    Private fadeStoryboard As New Storyboard()
    Private fadeAnimation As New DoubleAnimation()

    Public Sub New()
        fadeStoryboard.Children.Add(fadeAnimation)
    End Sub
```

```
    Protected Overrides Sub Invoke(ByVal args As Object)
        ' Make sure the storyboard isn't already running.
        fadeStoryboard.Stop()

        ' Set up the storyboard.
        Storyboard.SetTarget(fadeAnimation, Me.Target)
        Storyboard.SetTargetProperty(fadeAnimation, New PropertyPath("Opacity"))

        ' Set up the animation.
        fadeAnimation.To = 1
        fadeAnimation.Duration = Duration

        fadeStoryboard.Begin()
    End Sub
End Class
```

And here's an example that uses that both actions. By clicking one of two buttons, the user can fade out or fade in a block of text.

```
<StackPanel Orientation="Horizontal" Margin="3,15">
  <Button Content="Click to Fade the TextBlock" Padding="5">
    <i:Interaction.Triggers>
      <i:EventTrigger EventName="Click">
        <custom:FadeOutAction TargetName="border" />
      </i:EventTrigger>
    </i:Interaction.Triggers>
  </Button>

  <Button Content="Click to Show the TextBlock" Padding="5">
    <i:Interaction.Triggers>
      <i:EventTrigger EventName="Click">
        <custom:FadeInAction TargetName="border" />
      </i:EventTrigger>
    </i:Interaction.Triggers>
  </Button>
</StackPanel>

<Border x:Name="border" Background="Orange" BorderBrush="Black" BorderThickness="1"
 Margin="3,0" >
  <TextBlock Margin="5" FontSize="17" TextWrapping="Wrap"
   Text="I'm the target of the FadeOutAction and FadeInAction."></TextBlock>
</Border>
```

Figure 12-4 shows the result.

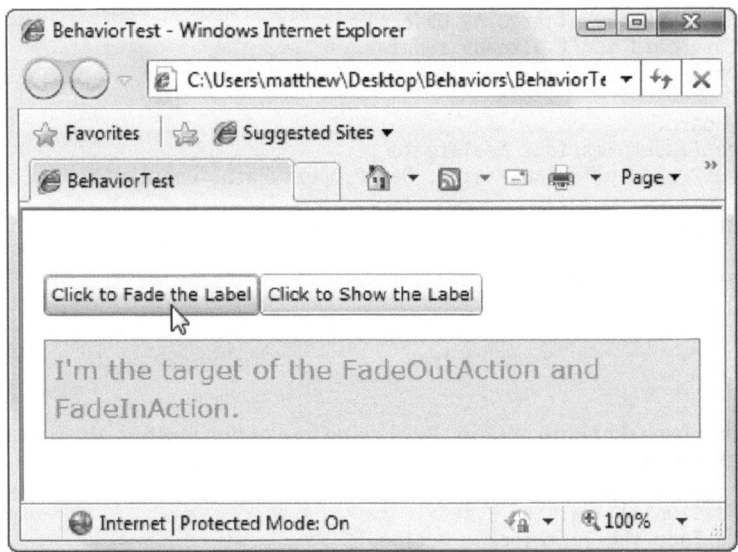

Figure 12-4. Using a targeted trigger

It's obviously possible to combine FadeOutAction and FadeInAction into a more general FadeAction that takes an additional property (say, TargetOpacity). However, behaviors are usually designed to be more specific so they can be dropped onto an element and used with a minimum amount of property setting. Similarly, you can run any animated effect using the ControlStoryboardAction (which is part of the Microsoft.Expression.Interactivity.dll assembly). However, it's a common design pattern for an action to encapsulate a programmatically created animation and expose any relevant details as higher-level properties. This simplifies your application and dramatically reduces the amount of markup that's needed.

Incidentally, targeted actions offer one way that you can rework the PlaySoundAction shown earlier if you want to avoid the overhead of creating and removing MediaElement objects (and you can live with the one-sound-at-a-time restriction that comes from using a single MediaElement for all your sounds). Rather than dynamically insert a MediaElement, ask the application developer to do the work, and then derive PlaySoundAction from TargetedTriggerAction. The trigger target would be the MediaElement.

SWAPPING THE SOURCE AND TARGET

There's a little-known fact about the source and target of an action: they're interchangeable. In fact, the application developer has the ability to set both the target and source element. You set the target using the TargetName property of the action, as you've already seen. Additionally, you can set the source using the SourceName property of the containing trigger. This means the source of an action doesn't necessarily need to be the element where it's placed.

This design allows two different scenarios. The most common scenario is the one you've already seen, where the trigger and action are nested inside the source element. This is called the *tell* model, because you tell the action to act on a specific target. But for greater flexibility, actions also support the *listen* model, where the action is defined in the target and you ask it to listen to a different source element. Either way, the action works in the same way and has the same convenient design-time support in Expression Blend. (It's equally possible to create an action in one element, set its source to a second element, and point its target to a third element, but it's hard to imagine a situation where this approach adds any benefit.)

Creating a Behavior

Although actions are often described as behaviors (and they appear in the Behaviors section of the Asset Library window in Expression Blend), there's a second sort of behavior that's represented by the Behavior class.

Behaviors have the same goal as actions: they aim to encapsulate bits of UI functionality so you can apply them to elements without writing the code yourself. The difference is that every action is a distinct piece of code that accomplishes a single task. Although you may create a set of related actions (like FadeOutAction and FadeInAction), each action is self-sufficient and can be used independently. Similarly, actions and triggers are complementary, and individual actions aren't linked to specific triggers. Any action can be used with any trigger.

On the other hand, behaviors group a combination of related operations. They encompass the job of triggers (listening for certain events or changes) and actions (performing the appropriate operations). They usually do several related things that can't be separated into smaller building blocks, because they're always used in combination. For example, the Deep Zoom functionality consists of several event handlers that allow you to pan and zoom a Deep Zoom image using the mouse buttons and mouse wheel. Together, this code establishes a basic level of support for navigating Deep Zoom images, and it's wrapped into a single behavior named DeepZoomBehavior.

■ **Tip** As a rule of thumb, if you find yourself trying to create actions that share data or interact or if you need a tight coupling between your action and a specific behavior, you should consider implementing your functionality as a behavior.

To gain a better understanding of behaviors, it's worth creating your own behavior. Imagine that you want to give any element the ability to be dragged around a Canvas with the mouse. You saw how to implement this feature for ellipses in Chapter 4, with the draggable-circle example. But with a bit more effort, you can turn that code into a reusable behavior that can give dragging support to any element on any Canvas.

The first step is to create a class that derives from the base Behavior class. As with the TriggerAction and TargetedTriggerAction classes, Behavior is a generic class that takes a type argument. You can use this type argument to restrict your behavior to specific elements, or you can use UIElement or FrameworkElement to include them all:

```
Public Class DragInCanvasBehavior
    Inherits Behavior(Of UIElement)
    ...
End Class
```

The first step in any behavior is to override the OnAttached() and OnDetaching() methods. When OnAttached() is called, you can access the element where the behavior is placed (through the AssociatedObject property) and attach event handlers. When OnDetaching() is called, you remove your event handlers.

Here's the code that the DragInCanvasBehavior uses to monitor the MouseLeftButtonDown, MouseMove, and MouseLeftButtonUp events:

```
Protected Overrides Sub OnAttached()
    MyBase.OnAttached()

    ' Hook up event handlers.
    AddHandler AssociatedObject.MouseLeftButtonDown, _
        AddressOf AssociatedObject_MouseLeftButtonDown
    AddHandler AssociatedObject.MouseMove, AddressOf AssociatedObject_MouseMove
    AddHandler AssociatedObject.MouseLeftButtonUp, _
        AddressOf AssociatedObject_MouseLeftButtonUp
End Sub

Protected Overrides Sub OnDetaching()
    MyBase.OnDetaching()

    ' Detach event handlers.
    RemoveHandler AssociatedObject.MouseLeftButtonDown, _
        AddressOf AssociatedObject_MouseLeftButtonDown
    RemoveHandler AssociatedObject.MouseMove, AddressOf AssociatedObject_MouseMove
    RemoveHandler AssociatedObject.MouseLeftButtonUp, _
        AddressOf AssociatedObject_MouseLeftButtonUp
End Sub
```

The final step is to run the appropriate code in the event handlers. For example, when the user clicks the left mouse button, DragInCanvasBehavior starts a dragging operation, records the offset between the upper-left corner of the element and the mouse pointer, and captures the mouse:

```
' Keep track of the Canvas where this element is placed.
Private canvas As Canvas

' Keep track of when the element is being dragged.
Private isDragging As Boolean = False

' When the element is clicked, record the exact position
' where the click is made.
```

```
Private mouseOffset As Point

Private Sub AssociatedObject_MouseLeftButtonDown(ByVal sender As Object, _
  ByVal e As MouseButtonEventArgs)
    ' Find the Canvas.
    If canvas Is Nothing Then
      canvas = CType(VisualTreeHelper.GetParent(Me.AssociatedObject), Canvas)
    End If

    ' Dragging mode begins.
    isDragging = True

    ' Get the position of the click relative to the element
    ' (so the top-left corner of the element is (0,0).
    mouseOffset = e.GetPosition(AssociatedObject)

    ' Capture the mouse. This way you'll keep receiving
    ' the MouseMove event even if the user jerks the mouse
    ' off the element.
    AssociatedObject.CaptureMouse()
End Sub
```

When the element is in dragging mode and the mouse moves, the element is repositioned:

```
Private Sub AssociatedObject_MouseMove(ByVal sender As Object, _
  ByVal e As MouseEventArgs)
    If isDragging Then
        ' Get the position of the element relative to the Canvas.
        Dim point As Point = e.GetPosition(canvas)

        ' Move the element.
        AssociatedObject.SetValue(Canvas.TopProperty, point.Y - mouseOffset.Y)
        AssociatedObject.SetValue(Canvas.LeftProperty, point.X - mouseOffset.X)
    End If
End Sub
```

And when the mouse button is released, dragging ends:

```
Private Sub AssociatedObject_MouseLeftButtonUp(ByVal sender As Object, _
  ByVal e As MouseButtonEventArgs)
    If isDragging Then
        AssociatedObject.ReleaseMouseCapture()
        isDragging = False
    End If
End Sub
```

To use this behavior, you simply need to add to any element inside a Canvas. The following markup creates a Canvas with three shapes. The two Ellipse elements use the DragInCanvasBehavior and can be dragged around the Canvas. The Rectangle element does not, and so cannot be moved.

```
<Canvas>
  <Rectangle Canvas.Left="10" Canvas.Top="10" Fill="Yellow" Width="40" Height="60">
  </Rectangle>
```

```
<Ellipse Canvas.Left="10" Canvas.Top="70" Fill="Blue" Width="80" Height="60">
  <i:Interaction.Behaviors>
    <custom:DragInCanvasBehavior></custom:DragInCanvasBehavior>
  </i:Interaction.Behaviors>
</Ellipse>

<Ellipse Canvas.Left="80" Canvas.Top="70" Fill="OrangeRed" Width="40" Height="70">
  <i:Interaction.Behaviors>
    <custom:DragInCanvasBehavior></custom:DragInCanvasBehavior>
  </i:Interaction.Behaviors>
</Ellipse>
</Canvas>
```

Figure 12-5 shows this example in action.

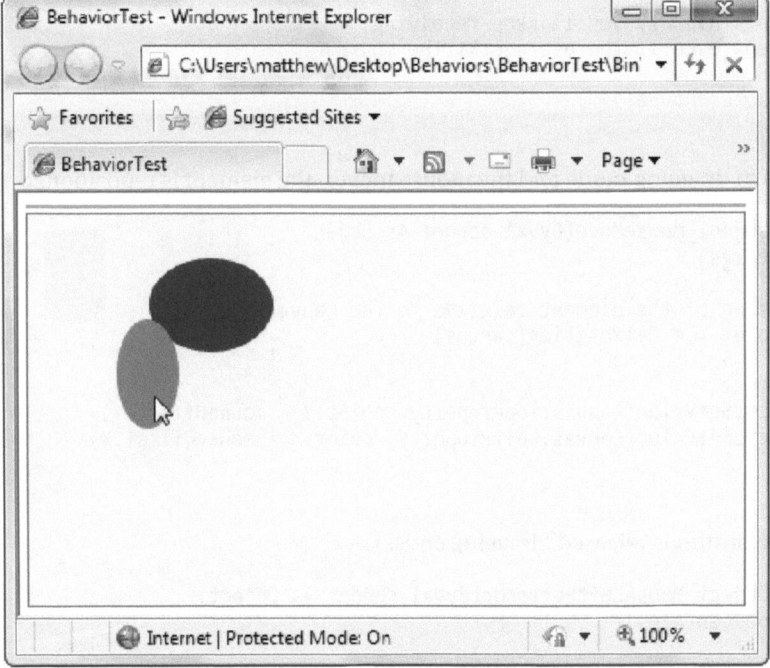

Figure 12-5. Making elements draggable with a behavior

Finding More Behaviors

Although you now know enough to create your own actions, triggers, and behaviors, most of the time you won't want to. Instead, you'll drop premade behavior classes into your applications. This allows you to quickly wire up a user-interface functionality without writing any code of your own. And if you're using Expression Blend, you won't need to type in the XAML markup either.

As you've already learned, the core System.Windows.Interactivity.dll assembly doesn't provide any actions or behaviors, and it includes just a single trigger class (the useful EventTrigger).

The Microsoft.Expression.Interactions.dll assembly is more practical. It includes a small set of essential actions (see Table 12-1), triggers (Table 12-2), and behaviors (Table 12-3).

Table 12-1. Action Classes in Microsoft.Expression.Interactions.dll

Class	Description
ChangePropertyAction	Changes any property to any value you specify, using reflection.
GoToStateAction	Switches a control to a specific visual state. (Visual states are a tool for building control templates. You'll explore them in Chapter 13.)
HyperlinkAction	Navigates to a new page using the URI you specify.
RemoveElementAction	Removes an element from the user interface.
PlaySoundAction	Works in essentially the same way as the PlaySoundAction demonstrated in this chapter. However, it adds the ability to set the media volume and displays the MediaElement in an invisible Popup control (ensuring that there's a place to put the MediaElement object even if your page doesn't include a single panel).
ControlStoryboardAction	Allows you to start, stop, pause, or resume an animation.

Table 12-2. Trigger Classes in Microsoft.Expression.Interactions.dll

Class	Description
KeyTrigger	Fires when a key is pressed
TimerTrigger	Fires at set intervals
StoryboardCompletedTrigger	Fires when an animation ends

Table 12-3. Behavior Classes in Microsoft.Expression.Interactions.dll

Class	Description
MouseDrag ElementBehavior	Allows the user to drag an element around the page. Its purpose is similar to the DragInCanvasBehavior demonstrated in this chapter, but it works with any layout container because it moves the element using a TranslateTransform.
FluidMoveBehavior	Watches an element (or a set of child elements) for layout changes. When these changes occur, the behavior moves the element smoothly to their new positions using an animation.

To get still more behaviors, you can try the Expression Blend Samples project at http://expressionblend.codeplex.com and the community behaviors gallery at http://gallery.expression.microsoft.com.

The Last Word

In this chapter, you saw how to use styles to reuse formatting settings with elements. You also considered how can use behaviors to develop tidy packages of user interface functionality, which can then be wired up to any element. Both tools give you a way to make more intelligent, maintainable user interfaces—ones that centralize formatting details and complex logic rather than forcing you to distribute it throughout your application and repeat it many times over.

In the next chapter, you'll continue learning how to make smart user interfaces with two still more powerful features: control templates and custom controls.

CHAPTER 13

■ ■ ■

Templates and Custom Controls

In the previous chapter, you learned how to use styles and behaviors to reuse user-interface property settings and code. In this chapter, you'll explore two more powerful tools: templates and custom controls.

Templates allow you to change the visual "face" of any common control. In other words, if you can't get the custom appearance you want by tweaking properties alone (and often you can't), you can almost certainly get it by applying a new template. And although creating custom templates is more work than just setting control properties, it's still far simpler and more flexible than developing an entirely new custom control, which many other programming frameworks force you to do.

Despite the power of styles and templates, you'll occasionally choose to create your own *custom control*. Usually, you'll take this step because you need functionality that's not offered by the core Silverlight controls. In this chapter, you'll learn how to create well-designed, extensible controls that use the template model. This way, you (and other developers) can change every aspect of the control's appearance without losing any part of its behavior.

Template Basics

In the previous chapter, you learned about styles, which allow you to change the appearance of an element. However, styles are limited to setting properties that are defined in the element class. For example, there are various visual details about a button that you can't change because they aren't exposed through properties. Examples include the shading in a button's background and the way it highlights itself when clicked.

But Silverlight has another, much more radical customization tool called *templates*. Although you can use styles with any Silverlight element, templates are limited to Silverlight controls—in other words, elements that inherit from the Control class in the System.Windows.Controls namespace. These elements acquire a property named Template, which you can set to apply a custom template, effectively overriding the control's standard visuals.

For example, by changing the template used by a Button object, you can create many exotic types of buttons that would be unthinkable with styles alone. You can create buttons that use round or shaped borders, and buttons that use eye-catching mouse-over effects (such as glowing, enlarging, or twinkling). All you need to do is draw on the drawing smarts you picked up in Chapter 8 and Chapter 9, as well as the animation techniques you learned in Chapter 10 when you build your custom template.

In the following sections, you'll peer into the templates used by common controls and see how to craft custom templates.

Creating a Template

Every control has a built-in recipe that determines how it should be rendered (as a group of more fundamental elements). That recipe is called a *control template*. It's defined using a block of XAML markup and applied to a control through the Template property.

For example, consider the basic button. Perhaps you want to get more control over the shading and animation effects that a button provides by creating a custom template. In this case, the first step is to try replacing the button's default template with one of your own devising.

To create a template for a basic button, you need to draw your own border and background and then place the content inside the button. There are several possible candidates for drawing the border, depending on the root element you choose:

- *Border*: This element does double duty—it holds a single element (say, a TextBlock with the button caption) and draws a border around it.

- *Grid*: By placing multiple elements in the same place, you can create a bordered button. Use a Silverlight shape element (such as a Rectangle or Path), and place a TextBlock in the same cell. Make sure the TextBlock is defined after the shape in XAML so it appears superimposed over the shape background. One advantage of the Grid is that it supports automatic sizing, so you can make sure your control is made only as large as its content requires.

- *Canvas*: The Canvas can place elements more precisely using coordinates. It's usually overkill, but it may be a good choice if you need to position a cluster of shapes in specific positions relative to each other as part of a more complex button graphic.

The following example uses the Border class to combine a rounded orange outline with an eye-catching red background and white text:

```
<Button Content="A Custom Button Template">
  <Button.Template>
    <ControlTemplate TargetType="Button" >
      <Border BorderBrush="Orange" BorderThickness="3" CornerRadius="10"
        Background="Red">
        <TextBlock Foreground="White" Text="A Custom Template"></TextBlock>
      </Border>
    </ControlTemplate>
  </Button.Template>
</Button>
```

Figure 13-1 shows the result.

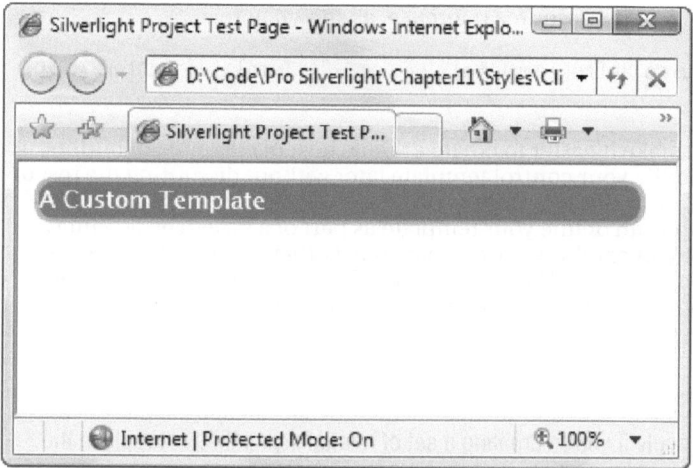

Figure 13-1. *A very basic new look for a button*

If you try this button, you'll find it's a pretty poor template. It loses many of the button features (such as changing appearance when the button is clicked). It also ignores virtually every property you set on the button, including the fundamentally important Content property. (Instead, it displays some hard-coded text.) However, this template is on its way to becoming a much better button template, and you'll begin refining it in the following sections.

■ **Note** At this point, you may be wondering why you've started building a custom button template without seeing the default button template. It's because default templates are extremely detailed. A simple button has a control template that's four printed pages long. But when you understand how a template is built, you'll be able to make your way through all the details in the default template.

Reusing Control Templates

In the previous example, the template definition is nested inside the element. But it's much more common to set the template of a control through a style. That's because you'll almost always want to reuse your template to skin multiple instances of the same control.

To accommodate this design, you need to define your control template as a resource:

```
<UserControl.Resources>
  <ControlTemplate x:Key="ButtonTemplate" TargetType="Button" >
    <Border BorderBrush="Orange" BorderThickness="3" CornerRadius="10"
      Background="Red">
      <TextBlock Foreground="White" Text="A Custom Template"></TextBlock>
    </Border>
  </ControlTemplate>
</UserControl.Resources>
```

You can then refer to it using a StaticResource reference, as shown here:

```
<Button Template="{StaticResource ButtonTemplate}" Content="A Templated Button"... >
</Button>
```

Not only does this approach make it easier to create a whole host of customized buttons, it also gives you the flexibility to modify your control template later without disrupting the rest of your application's user interface.

There's one more option—you can define your template as part of a style. The advantage to this approach is that your style can combine setters that adjust other properties, as well as a setter that applies the new control template. When you set the Style property of your button, all the setters come into action, giving your button a new template and adjusting any other related properties.

■ **Note** A few more considerations apply if you're creating a set of related styles that will replace the standard Silverlight controls to give your application a custom skinned look. In this situation, you should define all your styles in the App.xaml file, and you should place commonly used details in separate resources. For example, if all of your controls use the same highlighting effect when selected (which is a good idea for visual consistency), create a resource named HighlightBrush, and use that resource in your control templates.

The ContentPresenter

The previous example creates a rather unhelpful button that displays hard-coded text. What you really want to do is take the value of the Button.Content property and display it in your custom template. To pull this off, you need a specially designed placeholder called ContentPresenter.

The ContentPresenter is required for all content controls—it's the "insert content here" marker that tells Silverlight where to stuff the content. Here's how you can add it to the current example:

```
<ControlTemplate x:Key="ButtonTemplate" TargetType="Button">
  <Border BorderBrush="Orange" BorderThickness="3" CornerRadius="10"
    Background="Red">
    <ContentPresenter></ContentPresenter>
  </Border>
</ControlTemplate>
```

■ **Note** ContentPresenter isn't the only placeholder you'll use when developing custom templates, although it's the most common. Controls that represent lists and use ItemsControl will use an ItemsPresenter in their control templates, which indicates where the panel that contains the list of items will be placed. Scrollable content inside a ScrollViewer control is represented by a ScrollContentPresenter.

Template Bindings

Although the revised button template respects the content of the button, it ignores most other properties. For example, consider this instance that uses the template:

```
<Button Template="{StaticResource ButtonTemplate}" Content="A Templated Button"
 Margin="10" Padding="20"></Button>
```

This markup gives the button a Margin value of 10 and a Padding of 20. The element that holds the button is responsible for paying attention to the Margin property. However, the Padding property is ignored, leaving the contents of your button scrunched up against the sides. The problem here is the fact that the Padding property doesn't have any effect unless you specifically use it in your template. In other words, it's up to your template to retrieve the Padding value and use it to insert some extra space around your content.

Fortunately, Silverlight has a feature that's designed exactly for this purpose: *template bindings*. By using a template binding, your control template can pull out a value from the control to which you're applying the template. In this example, you can use a template binding to retrieve the value of the Padding property and use it to create a margin around the ContentPresenter:

```
<ControlTemplate x:Key="ButtonTemplate" TargetType="Button">
  <Border BorderBrush="Orange" BorderThickness="3" CornerRadius="10"
    Background="Red">
    <ContentPresenter Margin="{TemplateBinding Padding}">
    </ContentPresenter>
  </Border>
</ControlTemplate>
```

This achieves the desired effect of adding some space between the border and the content. Figure 13-2 shows your modest new button.

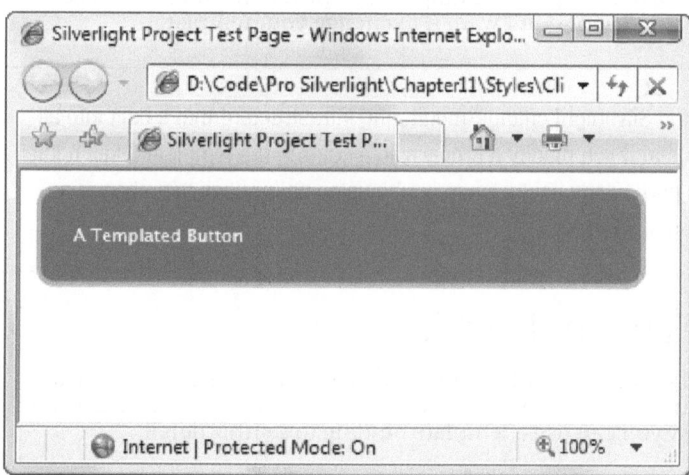

Figure 13-2. *A button with a customized control template*

■ **Note** Template bindings are similar to ordinary data bindings (which you'll consider in Chapter 16), but they're lighter weight because they're specifically designed for use in a control template. They only support one-way data binding (they can pass information from the control to the template but not the other way around).

It turns out that you need to set quite a few details in the ContentPresenter if you want to fully respect the properties of the Button class. For example, you need additional bindings if you want to get details such as text alignment, text wrapping, and so on. Buttons use a default control template that includes a ContentPresenter like this:

```
<ContentPresenter
  Content="{TemplateBinding Content}"
  ContentTemplate="{TemplateBinding ContentTemplate}"
  HorizontalContentAlignment="{TemplateBinding HorizontalContentAlignment}"
  Padding="{TemplateBinding Padding}"
  TextAlignment="{TemplateBinding TextAlignment}"
  TextDecorations="{TemplateBinding TextDecorations}"
  TextWrapping="{TemplateBinding TextWrapping}"
  VerticalContentAlignment="{TemplateBinding VerticalContentAlignment}"
  Margin="4,5,4,4">
</ContentPresenter>
```

The template binding for the Content property plays a key role: it extracts the content from the control and displays it in the ContentPresenter. However, this template binding is set implicitly. For that reason, you don't need to include it in your markup.

The only way you can anticipate what template bindings are needed is to check the default control template, as you'll see a bit later in this chapter (in the section "The Parts and States Model"). But in many cases, leaving out template bindings isn't a problem. You don't need to bind a property if you don't plan to use it or don't want it to change your template.

■ **Note** Template bindings support the Silverlight change-monitoring infrastructure that's built into all dependency properties. That means that if you modify a property in a control, the template takes it into account automatically. This detail is particularly useful when you're using animations that change a property value repeatedly in a short space of time.

Setting Templates Through Styles

Template bindings aren't limited to the ContentPresenter. You can use them anywhere in a control template. Consider the current button example, which hard-codes the red background in the Border element. Here's how you can use a template binding to set this detail:

```
<Border BorderBrush="Orange" BorderThickness="3" CornerRadius="10"
  Background="{TemplateBinding Background}">
```

This raises an obvious design question: is it better to hard-code the color to preserve the default appearance of your customized button or use a template binding to make it more flexible?

In this case, there's a compromise that lets you do both—you can combine templates with styles. The basic idea is to use style rules to set your template *and* set default values. Here's an example:

```xml
<Style x:Key="ButtonStyle" TargetType="Button">
  <Setter Property="Background" Value="Red"></Setter>
  <Setter Property="Template">
    <Setter.Value>
      <ControlTemplate TargetType="Button">
        <Border BorderBrush="Orange" BorderThickness="3" CornerRadius="10"
          Background="{TemplateBinding Background}">
          <ContentPresenter Margin="{TemplateBinding Padding}">
          </ContentPresenter>
        </Border>
      </ControlTemplate>
    </Setter.Value>
  </Setter>
</Style>
```

It's up to you whether you define the ControlTemplate inline (as in this example) or as a separate resource, as shown here:

```xml
<Style x:Key="ButtonStyle" TargetType="Button">
  <Setter Property="Background" Value="Red"></Setter>
  <Setter Property="Template" Value="{StaticResource ButtonTemplate}"></Setter>
</Style>
```

It's also useful to combine styles and templates if you need to set properties that aren't exposed by the ContentPresenter or the container elements in your control template. In the current example, you'll notice that there are no bindings that pass along the foreground color or font details of the button. That's because these properties (Foreground, FontFamily, FontSize, FontWeight, and so on) support *property inheritance*. When you set those values on a higher-level element (such as the Button class), they cascade down to contained elements (such as the TextBlock inside the button). The ContentPresenter doesn't provide any of these properties, because it doesn't need to do so. They flow from the control to the content inside, skipping right over the ContentPresenter.

In some cases, you'll want to change the inherited property values to better suit your custom control template. For instance, in the current example, it's important to set white as the foreground color, because white text stands out better against the button's colored background. But the standard font color is inherited from the containing Silverlight page, and it's black. Furthermore, you can't set the color through the ContentPresenter, because it doesn't offer the Foreground property. The solution is to combine the control template with a style setter that applies the white text:

```xml
<Style x:Key="ButtonStyle" TargetType="Button">
  <Setter Property="Foreground" Value="White"></Setter>
  <Setter Property="Background" Value="Red"></Setter>
  <Setter Property="Template" Value="{StaticResource ButtonTemplate}"></Setter>
</Style>
```

This approach gives you convenience and flexibility. If you take no extra steps, you automatically get the customized red background and white text. However, you also have the flexibility to create a new style that changes the color scheme but uses the existing control template, which can save a great deal of work.

Reusing Colors

As you've seen, flexible control templates can be influenced by control properties, which you can set through style rules. But Silverlight applications rarely change just a single control at a time. Most use an entire set of custom control templates to change the appearance of all Silverlight's common controls. In this situation, you need a way to share certain details (such as colors) between the controls.

The easiest way to implement this sharing is to pull hard-coded values out of styles and control templates and define them as separate resources, like this:

```
<SolidColorBrush x:Key="BackgroundBrush" Color="Red"></SolidColorBrush>
```

You can then use these resources in your styles and control templates:

```
<Style x:Key="ButtonStyle" TargetType="Button">
  <Setter Property="Foreground" Value="White"></Setter>
  <Setter Property="Background" Value="{StaticResource BackgroundBrush}"></Setter>
  <Setter Property="Template" Value="{StaticResource ButtonTemplate}"></Setter>
</Style>
```

This allows you to keep the same template but use a different border color simply by adding a resource with the right name. The drawback is that this approach can complicate your design.

For even greater flexibility, you can define your colors as separate resources and then use them in brush resources, as shown here:

```
<Color x:Key="BackgroundColor">#FF800000</Color>
<SolidColorBrush x:Key="ButtonBorderBrush"
 Color="{StaticResource BackgroundColor}"></SolidColorBrush>
```

This two-step approach let you reuse a color scheme in a variety of different ways (for example, in solid fills and in gradient brushes) without duplicating the color information in your markup. If you apply this pattern carefully, you can change the color scheme of your entire application by modifying a single set of color resources.

■ **Note** When you define a color as a resource, the content inside must be a color name or a hexadecimal HTML color code (as shown in the previous example). Unfortunately, you can't declare a color in XAML using the red, green, and blue components.

The Parts and States Model

If you try the button that you created in the previous section, you'll find it's a major disappointment. Essentially, it's nothing more than a rounded red rectangle—as you move the mouse over it or click it, there's no visual feedback. The button lies there, inert. (Of course, the Click event still fires when you click the button, but that's small consolation.) In WPF, you'd fix this problem with triggers. But Silverlight doesn't support triggers, and you need to include specially named elements and animations in your control template.

To understand how to make a template that can plug into the back-end code that a control uses, you need to study the Silverlight documentation. Online, you can view http://tinyurl.com/352brmx, which takes you to the Control Styles and Templates section. In this topic, you'll find a separate section that details the default templates for each control. There's one problem: the templates are intimidatingly huge.

To break a template into manageable pieces, you need to understand the parts and states model, which is how Silverlight templates are organized. *Parts* are the named elements that a control expects to find in a template. *States* are the named animations that are applied at specific times.

If your control template lacks a specific part or state, it usually won't cause an error. Instead, design best practices state that the control should degrade gracefully and ignore the missing information. However, if that part or state represents a key ingredient that's required for some part of the control's core functionality, the control may not work as expected (or at all). For example, this is why you lose the mouse-over behavior in the super-simple button template shown in the previous example.

The obvious question is this: how do you know what parts and states your control template needs to supply? There are two avenues. First, you can look at the documentation described in the previous section. Each control-specific page lists the parts and states that are required for that template, in two separate tables. Figure 13-3 shows an example for the Button control. Like many controls, Button requires certain states but no specific named parts, so you see just one table.

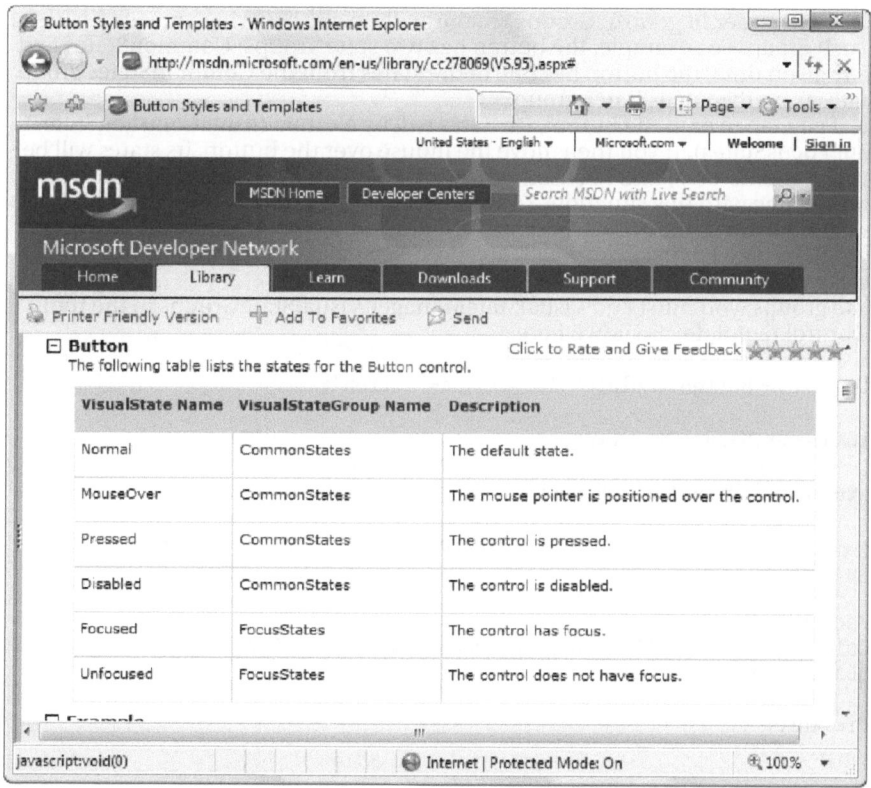

Figure 13-3. The named states for the Button class

Your other option is to use reflection in code to examine the control class. Each part is represented with a separate TemplatePart attribute applied to the class declaration. Each state is represented with a separate TemplateVisualState attribute. You'll take a closer look at these attributes in the following sections.

Understanding States with the Button Control

If you look at the declaration for the Button class (or the documentation shown in Figure 13-3), you'll discover that you need to supply six states to create a complete, well-rounded button:

```
<TemplateVisualState(Name:="Normal", GroupName:="CommonStates"), _
 TemplateVisualState(Name:="MouseOver", GroupName:="CommonStates"), _
 TemplateVisualState(Name:="Pressed", GroupName:="CommonStates"), _
 TemplateVisualState(Name:="Disabled", GroupName:="CommonStates"), _
 TemplateVisualState(Name:="Unfocused", GroupName:="FocusStates"), _
 TemplateVisualState(Name:="Focused", GroupName:="FocusStates")> _
Public Class Button
    Inherits ButtonBase
    ...
End Class
```

States are placed together in *groups*. Groups are mutually exclusive, which means a control has one state in each group. For example, the button has two state groups: CommonStates and FocusStates. At any given time, the button has one of the states from the CommonStates group *and* one of the states from the FocusStates group.

For example, if you tab over to the button, its states will be Normal (from CommonStates) and Focused (from FocusStates). If you then move the mouse over the button, its states will be MouseOver (from CommonStates) and Focused (from FocusStates). Without state groups, you'd have trouble dealing with this situation. You'd either be forced to make some states dominate others (so a button in the MouseOver state would lose its focus indicator) or need to create many more states (like FocusedNormal, UnfocusedNormal, FocusedMouseOver, UnfocusedMouseOver, and so on).

To define state groups, you must add VisualStateManager.VisualStateGroups in the root element of your control template, as shown here:

```
<ControlTemplate x:Key="ButtonTemplate" TargetType="Button">
  <Grid>
    <VisualStateManager.VisualStateGroups>
      ...
    </VisualStateManager.VisualStateGroups>

    <Border x:Name="ButtonBorder" BorderBrush="Orange" BorderThickness="3"
     CornerRadius="15">

      <Border.Background>
        <SolidColorBrush x:Name="ButtonBackgroundBrush" Color="Red" />
      </Border.Background>

      <ContentPresenter ... />
    </Border>
  </Grid>
</ControlTemplate>
```

To add the VisualStateManager element to your template, you need to use a layout panel. This layout panel holds both the visuals for your control and the VisualStateManager, which is invisible. Like the resources you first learned about in Chapter 2, the VisualStateManager defines objects—in this case, storyboards with animations—that the control can use at the appropriate time to alter its appearance.

Usually, you'll add a Grid at the root level of your template. In the button example, a Grid holds the VisualStateManager element and the Border element that renders the actual button.

Inside the VisualStateGroups element, you can create the state groups using appropriately named VisualStateGroup elements. In the case of the button, there are two state groups:

```
<VisualStateManager.VisualStateGroups>
  <VisualStateGroup x:Name="CommonStates">
    ...
  </VisualStateGroup>

  <VisualStateGroup x:Name="FocusStates">
    ...
  </VisualStateGroup>
</VisualStateManager.VisualStateGroups>
```

After you've added the VisualStateManager and the VisualStateGroup elements, you're ready to add a VisualState element for each state. You can add all the states that the control supports (as identified by the documentation and the TemplateVisualState attributes), or you can supply only those that you choose to use. For example, if you want to create a button that provides a mouse-over effect, you need to add the MouseOver state (which applies the effect) and the Normal state (which returns the button to its normal appearance). Here's an example that defines these two states:

```
<VisualStateManager.VisualStateGroups>
  <VisualStateGroup x:Name="CommonStates">
    <VisualState x:Name="MouseOver">
      ...
    </VisualState>

    <VisualState x:Name="Normal">
      ...
    </VisualState>
  </VisualStateGroup>

  <VisualStateGroup x:Name="FocusStates">
    ...
  </VisualStateGroup>
</VisualStateManager.VisualStateGroups>
```

Each state corresponds to a storyboard with one or more animations. If these storyboards exist, they're triggered at the appropriate times. For example, when the user moves the mouse over the button, you may want to use an animation to perform one of the following tasks:

- *Show a new visual*: To do this, you need to change the Opacity property of an element in the control template so it springs into view.
- *Change the shape or position*: You can use a TranslateTransform to tweak the positioning of an element (for example, offsetting it slightly to give the impression that the button's been pressed). You can use a ScaleTransform or a RotateTransform to twiddle the element's appearance slightly as the user moves the mouse over it.

- *Change the lighting or coloration*: To do this, you need an animation that acts on the brush that you use to paint the background. You can use a ColorAnimation to change colors in a SolidBrush, but more advanced effects are possible by animating more complex brushes. For example, you can change one of the colors in a LinearGradientBrush (which is what the default button control template does), or you can shift the center point of a RadialGradientBrush.

■ **Tip** Some advanced lighting effects use multiple layers of transparent elements. In this case, your animation modifies the opacity of one layer to let other layers show through.

Figure 13-4 shows an example of a button that uses customized state animations to change its background color when the user moves the mouse over it.

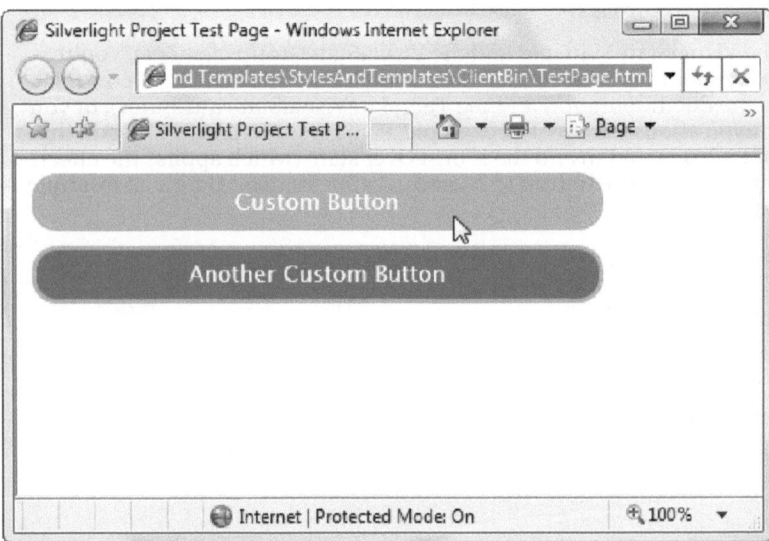

Figure 13-4. Animated effects in a custom button template

Here's the markup that does the trick:

```
<VisualStateManager.VisualStateGroups>
  <VisualStateGroup x:Name="CommonStates">
    <VisualState x:Name="MouseOver">
      <Storyboard>
        <ColorAnimation Duration="0:0:0"
          Storyboard.TargetName="ButtonBackgroundBrush"
          Storyboard.TargetProperty="Color" To="Orange" />
      </Storyboard>
    </VisualState>
  </VisualState>
```

```
<VisualState x:Name="Normal">
  <Storyboard>
    <ColorAnimation Duration="0:0:0"
      Storyboard.TargetName="ButtonBackgroundBrush"
      Storyboard.TargetProperty="Color" />
  </Storyboard>
</VisualState>
</VisualStateGroup>
</VisualStateManager.VisualStateGroups>
```

The MouseOver state applies a new, hard-coded color using a ColorAnimation. The Normal state uses a ColorAnimation with no set color, which means the animation reverts to the color that was set initially.

You can simplify this example by removing state settings that match the initial property settings of your template. That means you can remove the storyboard from the Normal state, because it reapplies the initial color. (However, you need to keep the VisualState element that defines the state.) Here's the result:

```
<VisualStateManager.VisualStateGroups>
  <VisualStateGroup x:Name="CommonStates">
    <VisualState x:Name="MouseOver">
      <Storyboard>
        <ColorAnimation Duration="0:0:0"
          Storyboard.TargetName="ButtonBackgroundBrush"
          Storyboard.TargetProperty="Color" To="Orange" />
      </Storyboard>
    </VisualState>

    <VisualState x:Name="Normal">
    </VisualState>
  </VisualStateGroup>
</VisualStateManager.VisualStateGroups>
```

This works because when you switch from the MouseOver state to the Normal state, Silverlight unwinds the MouseOver state and reverts the control to its initial property settings. By not explicitly specifying these details, you create cleaner markup.

HARD-CODING ANIMATION VALUES

You'll notice that this example has a hard-coded background color (Orange). It's also possible to pull details out of other properties and apply them to your animations using the TemplateBinding extension you saw earlier. However, this refactoring isn't necessary. As a general rule of thumb, it's acceptable for a customized control template to have hard-coded details like colors, fonts, and margins, because each template represents a specific, customized visual look.

When you create the default control template for a new custom control, it's much more important to make sure that the template is flexible. In this situation, control consumers should be able to customize the control's appearance by setting properties, and they shouldn't be forced to supply a new control template if only minor modifications are required. You'll learn more about creating a default control template later in this chapter, in the section "Creating Templates for Custom Controls."

Showing a Focus Cue

In the previous example, you used the Normal and MouseOver states from the CommonStates group to control how the button looks when the mouse moves overtop. You can also add the Pressed and Disabled states to customize your other two alternatives. These four states are mutually exclusive—if the button is pressed, the MouseOver state no longer applies, and if the button is disabled, all the other states are ignored no matter what the user does with the mouse. (There's a quirk here. If you don't supply a state animation, the previous animation keeps working. For example, if you don't supply a Pressed state animation, the MouseOver state animation stays active when the button is clicked.)

As you saw earlier, the button has two groups of states. Along with the four CommonStates are two FocusStates, which allows the button to be focused or unfocused. The CommonStates and FocusStates are independent, which means the buttons can be focused or unfocused no matter what's taking place with the mouse. Of course, there may be exceptions depending on the internal logic in the control. For example, a disabled button never gets the keyboard focus, so the Focused state will never apply when the common state is Disabled.

Many controls use a focus cue to indicate when they have focus. In the control template for the button, the focus cue is a Rectangle element with a dotted border. The focus cue is placed overtop the button surface using a Grid, which holds both the focus cue and the button border in the same cell. The animations in the FocusStates group show or hide the focus rectangle by adjusting its Opacity property:

```
<Grid>
  <VisualStateManager.VisualStateGroups>
    <VisualStateGroup x:Name="FocusStates">
      <VisualState x:Name="Focused">
        <Storyboard>
          <DoubleAnimation Duration="0" Storyboard.TargetName="FocusVisualElement"
            Storyboard.TargetProperty="Opacity" To="1" />
        </Storyboard>
      </VisualState>

      <VisualState x:Name="Unfocused">
        <!-- No storyboard is needed, because this state simply
             reverts to the initial Opacity for the rectangle (0). -->
      </VisualState>
    </VisualStateGroup>
    ...
  </VisualStateManager.VisualStateGroups>

  <Border x:Name="ButtonBorder" ... >
    <ContentPresenter ... />
  </Border>

  <Rectangle x:Name="FocusVisualElement" Stroke="Black" Margin="8" Opacity="0"
    StrokeThickness="1" StrokeDashArray="1 2"></Rectangle>
</Grid>
```

Now, the button shows the focus cue when it has the keyboard focus. Figure 13-5 shows an example with two buttons that use the same control template. The first button shows the focus cue.

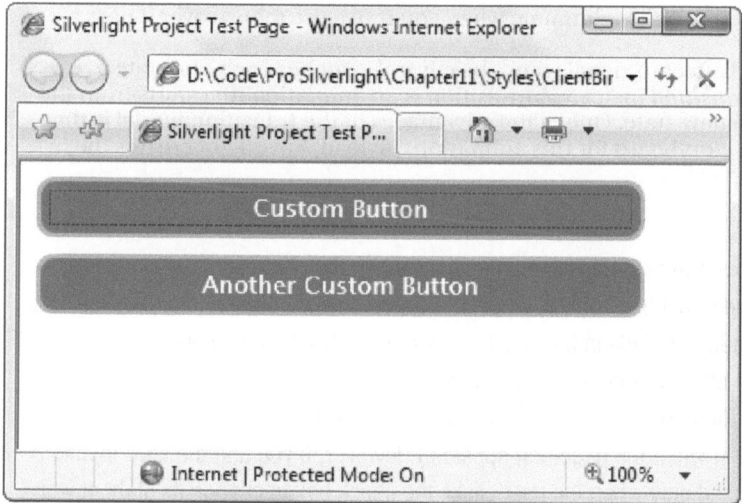

Figure 13-5. Focus in a custom button template

You should take care to avoid animating the same properties in different state groups. For example, if you animate the background color in the MouseOver state (which is in the CommonStates group), you shouldn't animate the background color in the Focused state (which is in the FocusStates group). If you do, the result will depend on the order that the control applies its states. For example, if the button applies the state from the FocusStates group first and then the state from the CommonStates group, your focused state animation will be active for just a split second before being replaced by the competing MouseOver state.

Transitions

The button shown in the previous example uses zero-length state animations. As a result, the color change happens instantly when the mouse moves over the button.

You can lengthen the duration to create a more gradual color-blending effect. Here's an example that fades in the new color over a snappy 0.2 seconds:

```
<VisualStateManager.VisualStateGroups>
  <VisualStateGroup x:Name="CommonStates">
    <VisualState x:Name="MouseOver">
      <Storyboard>
        <ColorAnimation Duration="0:0:0.2" ... />
      </Storyboard>
    </VisualState>
    ...
  </VisualStateGroup>
</VisualStateManager.VisualStateGroups>
```

Although this works, the concept isn't quite right. Technically, each visual state is meant to represent the appearance of the control while it's in that state (not including the transition used to get *into* that state). Ideally, a visual state animation should be either a zero-length animation like the ones shown earlier or a *steady-state* animation—an animation that repeats itself one or

477

more times. For example, a button that glimmers when you move the mouse over it uses a steady-state animation.

If you want an animated effect to signal when the control switches from one state to another, you should use a *transition* instead. A transition is an animation that starts from the current state and ends at the new state. One of the advantages of the transition model is that you don't need to create the storyboard for this animation. Instead, Silverlight creates the animation you need automatically.

■ **Note** Controls are smart enough to skip transition animations when the controls begin in a certain state. For example, consider the CheckBox control, which has an Unchecked state and a Checked state. You may decide to use an animation to fade in the check mark gracefully when the check box is selected. If you add the fade-in effect to the Checked state animation, it will apply when you show a checked check box for the first time. (For example, if you have a page with three checked check boxes, all three check marks will fade in when the page first appears.) However, if you add the fade-in effect through a transition, it will be used only when the user clicks the check box to change its state. It won't apply when the control is shown for the first time, which makes more sense.

The Default Transition

Transitions apply to state groups. When you define a transition, you must add it to the VisualStateGroup.Transitions collection. The simplest type of transition is a *default transition*, which applies to all the state changes for that group. To create the default transition, you need to add a VisualTransition element and set the GeneratedDuration property to set the length of the transition effect. Here's an example:

```
<VisualStateManager.VisualStateGroups>
  <VisualStateGroup x:Name="CommonStates">
    <VisualStateGroup.Transitions>
      <VisualTransition GeneratedDuration="0:0:0.2" />
    </VisualStateGroup.Transitions>

    <VisualState x:Name="MouseOver">
      <Storyboard>
        <ColorAnimation Duration="0:0:0"
          Storyboard.TargetName="ButtonBackgroundBrush"
          Storyboard.TargetProperty="Color" To="Orange" />
      </Storyboard>
    </VisualState>

    <VisualState x:Name="Normal">
    </VisualState>

  </VisualStateGroup>
</VisualStateManager.VisualStateGroups>
```

Now, whenever the button changes from one of the common states to another, the default 0.2-second transition kicks in. That means that when the user moves the mouse over the button and the button enters the MouseOver state, the new color fades in over 0.2 seconds, even though the MouseOver state animation has a zero length. Similarly, when the user moves the mouse off the button, the button blends back to its original color over 0.2 seconds.

Essentially, a transition is an animation that takes you from one state to another. VisualStateManager can create a transition animation as long as your state animations use one of the following types:

- ColorAnimation or ColorAnimationUsingKeyFrames
- PointAnimation or PointAnimationUsingKeyFrames
- DoubleAnimation or DoubleAnimationUsingKeyFrames

The button example works because the Normal and MouseOver states use a ColorAnimation, which is one of the supported types. If you use something else—say, an ObjectAnimationUsingKeyFrames—the transition won't have any effect. Instead, the old value will stay in place, the transition will run out its duration, and then the new value will snap in.

■ **Note** In some cases, a state uses several animations. In this situation, all the animations that use supported types are animated by the transition. Any unsupported types snap in at the end of the transition.

From and to Transitions

A default transition is convenient, but it's a one-size-fits-all solution that's not always suitable. For example, you may want a button to transition to the MouseOver state over 0.2 seconds but return instantly to the Normal state when the mouse moves away. To set this up, you need to define multiple transitions, and you need to set the From and To properties to specify when the transition will come into effect.

For example, if you have these transitions

```
<VisualStateGroup.Transitions>
  <VisualTransition To="MouseOver" GeneratedDuration="0:0:0.5" />
  <VisualTransition From="MouseOver" GeneratedDuration="0:0:0.1" />
</VisualStateGroup.Transitions>
```

the button will switch into the MouseOver state in 0.5 seconds, and it will leave the MouseOver state in 0.1 seconds. There is no default transition, so any other state changes will happen instantly.

This example shows transitions that apply when entering specific states and transitions that apply when leaving specific states. You can also use the To and From properties in conjunction to create even more specific transitions that apply only when moving between two specific states. When applying transitions, Silverlight looks through the collection of transitions to find the most specific one that applies, and it uses only that one. For example, when the mouse moves over a button, the VisualStateManager searches for states in this order, stopping when it finds a match:

1. A transition with From="Normal" and To="MouseOver"
2. A transition with To="MouseOver"
3. A transition with From="Normal"
4. The default transition

If there's no default transition, it switches between the two states immediately.

Transitioning to a Steady State

So far, you've seen how transitions work with *zero-length* state animations. However, it's equally possible to create a control template that uses transitions to move between *steady-state* animations. (Remember, a steady-state animation is a looping animation that repeats itself more than one time.)

To understand what happens in this situation, you need to realize that a transition to a steady-state animation moves from the current property value to the *starting* property value of the steady-state animation. For example, imagine you want to create a button that pulses steadily when the mouse is over it. As with all steady-state animations, you need to set the RepeatBehavior property to a number of repetitions you want or use Forever to loop indefinitely (as in this example). Depending on the data type, you may also need to set the AutoReverse property to True. For example, with a ColorAnimation, you need to use automatic reversal to return to the original color before repeating the animation. With a key-frame animation, this extra step isn't necessary because you can animate from the last key frame at the end of the animation to the first key frame of a new iteration.

Here's the steady-state animation for the pulsing button:

```
<VisualState x:Name="MouseOver">
  <Storyboard>
    <ColorAnimation Duration="0:0:0.4" Storyboard.TargetName="ButtonBackgroundBrush"
      Storyboard.TargetProperty="Color" From="DarkOrange" To="Orange"
      RepeatBehavior="Forever" AutoReverse="True" />
  </Storyboard>
</VisualState>
```

It's not necessary to use a transition with this button—after all, you may want the pulsing effect to kick in immediately. But if you do want to provide a transition, it will occur before the pulsing begins. Consider a standard transition like this one:

```
<VisualStateGroup.Transitions>
  <VisualTransition From="Normal" To="MouseOver" GeneratedDuration="0:0:1" />
</VisualStateGroup.Transitions>
```

This takes the button from its current color (Red) to the starting color of the steady-state animation (DarkOrange) using a one-second animation. After that, the pulsing begins.

Custom Transition

All the previous examples have used automatically generated transition animations. They change a property smoothly from its current value to the value set by the new state. However, you may want to define customized transitions that work differently. You may even choose to mix standard transitions with custom transitions that apply only to specific state changes.

■ **Tip** You may create a custom transition for several reasons. Here are some examples: to control the pace of the animation with a more sophisticated animation, to use an animation easing, to run several animations in succession (as in the FlipPanel example at the end of this chapter), or to play a sound at the same time as an animation.

To define a custom transition, you place a storyboard with one or more animations inside the VisualTransition element. Here's an example that creates an elastic compression effect when the user moves the mouse off a button:

```
<VisualStateGroup.Transitions>
  <VisualTransition To="Normal" From="MouseOver" GeneratedDuration="0:0:0.7">
    <Storyboard>
      <DoubleAnimationUsingKeyFrames Storyboard.TargetName="ScaleTransform"
        Storyboard.TargetProperty="ScaleX">
        <LinearDoubleKeyFrame KeyTime="0:0:0.5" Value="0" />
        <LinearDoubleKeyFrame KeyTime="0:0:0.7" Value="1" />
      </DoubleAnimationUsingKeyFrames>
    </Storyboard>
  </VisualTransition>
</VisualStateGroup.Transitions>
```

■ **Note** When you use a custom transition, you must still set the VisualTransition.GeneratedDuration property to match the duration of your animation. Without this detail, the VisualStateManager can't use your transition, and it will apply the new state immediately. (The actual time value you use still has no effect on your custom transition, because it applies only to automatically generated animations. See the end of this section to learn how you can mix and match a custom transition with automatically generated animations.)

This transition uses a key-frame animation. The first key frame compresses the button horizontally until it disappears from view, and the second key frame causes it to spring back into sight over a shorter interval of time. The transition animation works by adjusting the scale of this ScaleTransform object, which is defined in the control template:

```
<Grid RenderTransformOrigin="0.5,0.5">
  <Grid.RenderTransform>
    <ScaleTransform x:Name="ScaleTransform" ScaleX="1" />
  </Grid.RenderTransform>
  ...
</Grid>
```

When the transition is complete, the transition animation is stopped, and the animated properties return to their original values (or the values that are set by the current state animation). In this example, the animation returns the ScaleTransform to its initial ScaleX value of 1, so you don't notice any change when the transition animation ends.

It's logical to assume that a custom transition animation like this one replaces the automatically generated transition that the VisualStateManager would otherwise use. However, this isn't necessarily the case. Instead, it all depends on whether your custom transition animates the same properties as the VisualStateManager.

If your transition animates the same properties as the new state animation, your transition replaces the automatically generated transition. In the current example, the transition bridges the gap between the MouseOver state and the Normal state. The new state, Normal, uses a zero-length animation to change the button's background color. Thus, if you don't supply a custom animation for your transition, the VisualStateManager creates an animation that smoothly shifts the background color from the old state to the new state.

481

So what happens if you throw a custom transition into the mix? If you create a custom transition animation that targets the background color, the VisualStateManager will use your animation instead of its default transition animation. But that's not what happens in this example. Here, the custom transition doesn't modify the color—instead, it animates a transform. For that reason, the VisualStateManager still generates an automatic animation to change the background color. It uses its automatically generated animation in addition to your custom transition animation, and it runs them both at the same time, giving the generated transition the duration that's set by the VisualTransition.GeneratedDuration property. In this example, that means the new color fades in over 0.7 seconds, and at the same time the custom transition animation applies the compression effect.

Understanding Parts with the Slider Control

In the parts and states model, the states dominate. Many controls, like Button, use templates that define multiple state groups but no parts. But in other controls, like Slider, parts allow you to wire up elements in the control template to key pieces of control functionality.

To understand how parts work, you need to consider a control that uses them. Often, parts are found in controls that contain small working parts. For example, the DatePicker control uses parts to identify the drop-down button that opens the calendar display and the text box that shows the currently selected date. The ScrollBar control uses parts to delineate the draggable thumb, the track, and the scroll buttons. The Slider control uses much the same set of parts, although its scroll buttons are placed over the track, and they're invisible. This allows the user to move the slider thumb by clicking either side of the track.

A control indicates that it uses a specific part with the TemplatePart attribute. Here are the TemplatePart attributes that decorate the Slider control:

```
<TemplatePart(Name:="HorizontalTemplate", Type:=GetType(FrameworkElement)), _
 TemplatePart(Name:="HorizontalTrackLargeChangeIncreaseRepeatButton", _
  Type:=GetType(RepeatButton)), _
 TemplatePart(Name:="HorizontalTrackLargeChangeDecreaseRepeatButton", _
  Type:=GetType(RepeatButton)), _
 TemplatePart(Name:="HorizontalThumb", Type:=GetType(Thumb)), _
 TemplatePart(Name:="VerticalTemplate", Type:=GetType(FrameworkElement)), _
 TemplatePart(Name:="VerticalTrackLargeChangeIncreaseRepeatButton", _
  Type:=GetType(RepeatButton)), _
 TemplatePart(Name:="VerticalTrackLargeChangeDecreaseRepeatButton", _
  Type:=GetType(RepeatButton)), _
 TemplatePart(Name:="VerticalThumb", _
  Type:=GetType(Thumb)), _
 TemplateVisualState(Name:="Disabled", GroupName:="CommonStates"), _
 TemplateVisualState(Name:="Unfocused", GroupName:="FocusStates"), _
 TemplateVisualState(Name:="MouseOver", GroupName:="CommonStates"), _
 TemplateVisualState(Name:="Focused", GroupName:="FocusStates"), _
 TemplateVisualState(Name:="Normal", GroupName:="CommonStates")> _
Public Class Slider
    Inherits RangeBase
    ...
End Class
```

The Slider is complicated by the fact that it can be used in two different orientations, which require two separate templates that are coded side by side. Here's the basic structure:

```
<ControlTemplate TargetType="Slider">
  <!-- This Grid groups the two orientations together in the same template.-->
  <Grid>

    <!-- This Grid is used for the horizontal orientation. -->
    <Grid x:Name="HorizontalTemplate">
      ...
    </Grid>

    <!-- This Grid is used for the vertical orientation. -->
    <Grid x:Name="VerticalTemplate">
      ...
    </Grid>

  </Grid>
</ControlTemplate>
```

If Slider.Orientation is Horizontal, the Slider shows the HorizontalTemplate element and hides the VerticalTemplate element (if it exists). Usually, both of these elements are layout containers. In this example, each one is a Grid that contains the rest of the markup for that orientation.

When you understand that two distinct layouts are embedded in one control template, you'll realize that there are two sets of template parts to match. In this example, you'll consider a Slider that's always used in horizontal orientation and so provides only the corresponding horizontal parts: HorizontalTemplate, HorizontalTrackLargeChangeIncreaseRepeatButton, HorizontalTrackLargeChangeDecreaseRepeatButton, and HorizontalThumb.

Figure 13-6 shows how these parts work together. Essentially, the thumb sits in the middle, on the track. On the left and right are two invisible buttons that allow you to quickly scroll the thumb to a new value by clicking one side of the track and holding down the mouse button.

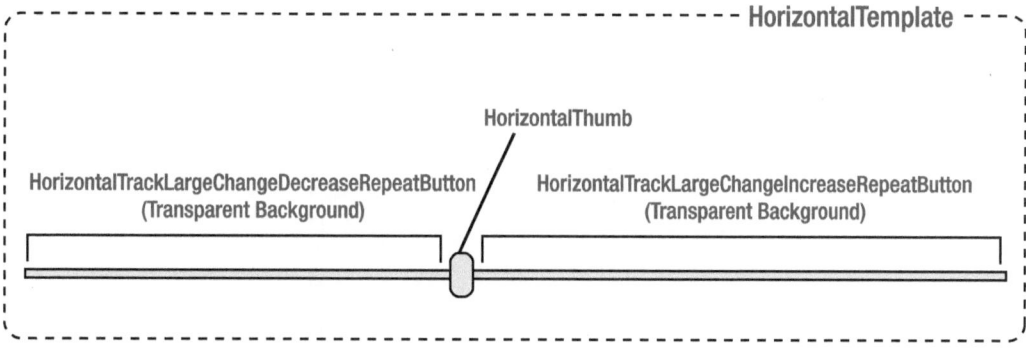

Figure 13-6. The named parts in the HorizontalTemplate part for the Slider

The TemplatePart attribute indicates the name the element must have, which is critical because the control code searches for that element by name. It also indicates the element type, which may be something very specific (such as Thumb, in the case of the HorizontalThumb part) or something much more general (for example, FrameworkElement, in the case of the HorizontalTemplate part, which allows you to use any element).

The fact that an element is used as a part in a control template tells you nothing about *how* that element is used. However, there are a few common patterns:

- *The control handles events from a part*: For example, the Slider code searches for the thumb when it's initialized and attaches event handlers that react when the thumb is clicked and dragged.

- *The control changes the visibility of a part*: For example, depending on the orientation, the Slider shows or hides the HorizontalTemplate and VerticalTemplate parts.

- *If a part isn't present, the control doesn't raise an exception*: Depending on the importance of the part, the control may continue to work (if at all possible), or an important part of its functionality may be missing. For example, when dealing with the Slider, you can safely omit HorizontalTrackLargeChangeIncreaseRepeatButton and HorizontalTrackLargeChangeDecreaseRepeatButton. Even without these parts, you can still set the Slider value by dragging the thumb. But if you omit the HorizontalThumb element, you'll end up with a much less useful Slider.

Figure 13-7 shows a customized Slider control. Here, a custom control template changes the appearance of the track (using a gently rounded Rectangle element) and the thumb (using a semitransparent circle).

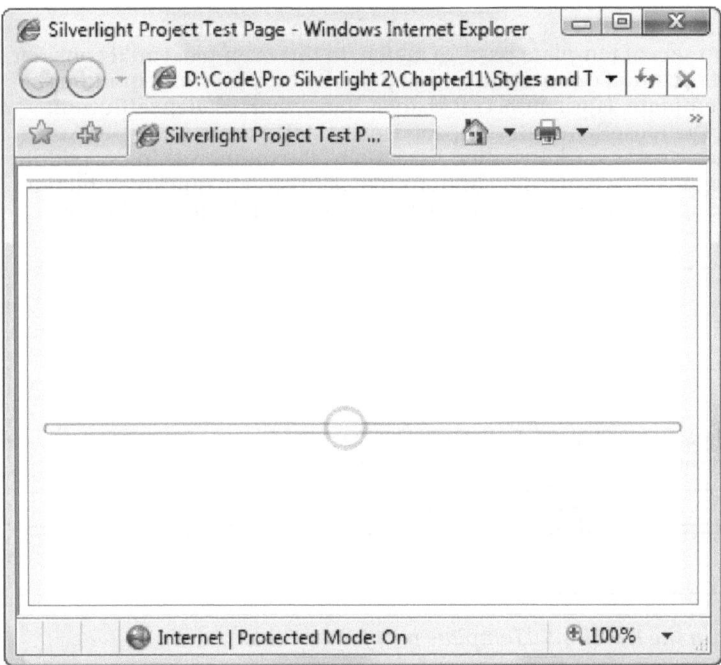

Figure 13-7. A customized Slider control

To create this effect, your custom template must supply a HorizontalTemplate part. In that HorizontalTemplate part, you must also include the HorizontalThumb part. The TemplatePart attribute makes it clear that you can't replace the Thumb control with another element. However, you can customize the control template of the Thumb to modify its visual appearance, as in this example.

Here's the complete custom control template:

```xml
<ControlTemplate TargetType="Slider">
  <Grid>
    <Grid x:Name="HorizontalTemplate">
      <Grid.ColumnDefinitions>
        <ColumnDefinition Width="Auto" />
        <ColumnDefinition Width="Auto" />
        <ColumnDefinition Width="*" />
      </Grid.ColumnDefinitions>

      <!-- The track -->
      <Rectangle Stroke="SteelBlue" StrokeThickness="1" Fill="AliceBlue"
        Grid.Column="0" Grid.ColumnSpan="3" Height="7" RadiusX="3" RadiusY="3" />

      <!-- The left RepeatButton -->
      <RepeatButton x:Name="HorizontalTrackLargeChangeDecreaseRepeatButton"
        Grid.Column="0" Background="Transparent" Opacity="0" IsTabStop="False" />

      <!-- The Thumb -->
      <Thumb x:Name="HorizontalThumb" Height="28" Width="28" Grid.Column="1">
        <Thumb.Template>
          <ControlTemplate TargetType="Thumb">
            <Ellipse x:Name="Thumb" Opacity="0.3" Fill="AliceBlue"
              Stroke="SteelBlue" StrokeThickness="3" Stretch="Fill"></Ellipse>
          </ControlTemplate>
        </Thumb.Template>
      </Thumb>

      <!-- The right RepeatButton -->
      <RepeatButton x:Name="HorizontalTrackLargeChangeIncreaseRepeatButton"
        Grid.Column="2" Background="Transparent" Opacity="0" IsTabStop="False" />

    </Grid>
    <!-- Add VerticalTemplate here if desired. -->
  </Grid>
</ControlTemplate>
```

CREATING SLICK CONTROL SKINS

The examples you've seen in this chapter demonstrate everything you need to know about the parts and states model. But they lack one thing: eye candy. For example, although you now understand the concepts you need to create customized Button and Slider controls, you haven't seen how to *design* the graphics that make a truly attractive control. And although the simple animated effects you've seen here—color changing, pulsing, and scaling—are respectable, they certainly aren't eye-catching. To get more dramatic results, you need to get creative with the graphics and animation skills you've picked up in earlier chapters.

To get an idea of what's possible, you should check out the Silverlight control examples that are available on the Web, including the many different glass and glow buttons that developers have created. You can also apply new templates using the expansive set of *themes* that are included with the Silverlight Toolkit (http://silverlight.codeplex.com). If you want to restyle your controls, you'll find that these themes give you a wide range of slick, professional choices. Best of all, themes work automatically thanks to a crafty tool called the ImplicitStyleManager. All you need to do is set the theme on some sort of container element (like a panel). The ImplicitStyleManager will automatically apply the correct styles to all the elements inside, complete with the matching control templates.

Creating Templates for Custom Controls

As you've seen, every Silverlight control is designed to be *lookless*, which means you can completely redefine its visuals (the *look*). What doesn't change is the control's behavior, which is hardwired into the control class. When you choose to use a control like Button, you choose it because you want button-like behavior—an element that presents content and can be clicked to trigger an action.

In some cases, you want different behavior, which means you need to create a custom control. As with all controls, your custom control will be lookless. Although it will provide a default control template, it won't force you to use that template. Instead, it will allow the control consumer to replace the default template with a fine-tuned custom template.

In the rest of this chapter, you'll learn how you can create a template-driven custom control. This custom control will let control consumers supply different visuals, just like the standard Silverlight controls you've used up to this point.

CONTROL CUSTOMIZATION

Custom control development is less common in Silverlight than in many other rich-client platforms. That's because Silverlight provides so many other avenues for customization, such as the following:

- *Content controls*: Any control that derives from ContentControl supports nested content. Using content controls, you can quickly create compound controls that aggregate other elements. (For example, you can transform a button into an image button or a list box into an image list.)

- *Styles and control templates:* You can use a style to painlessly reuse a combination of control properties. This means there's no reason to derive a custom control just to set a standard, built-in appearance. Templates go even further, giving you the ability to revamp every aspect of a control's visual appearance.

- *Control templates*: All Silverlight controls are *lookless*, which means they have hardwired functionality, but their appearance is defined separately through the control template. Replace the default template with something new, and you can revamp basic controls such as buttons, check boxes, radio buttons, and even windows.

- *Data templates*: Silverlight's list controls support data templates, which let you create a rich list representation of some type of data object. Using the right data template, you can display each item using a combination of text, images, and editable controls, all in a layout container of your choosing. You'll learn how in Chapter 16.

If possible, you should pursue these avenues before you decide to create a custom control or another type of custom element. These solutions are simpler, easier to implement, and often easier to reuse.

When *should* you create a custom element? Custom elements aren't the best choice when you want to fine-tune an element's appearance, but they make sense when you want to change its underlying functionality or design a control that has its own distinct set of properties, methods, and events.

Planning the FlipPanel Control

The following example develops a straightforward but useful control called FlipPanel. The basic idea behind the FlipPanel is that it provides two surfaces to host content, but only one is visible at a time. To see the other content, you "flip" between the sides. You can customize the flipping effect through the control template, but the default effect use a 3-D projection that looks like the panel is a sheet of paper being flipped around to reveal different content on its back (see Figure 13-8). Depending on your application, you could use the FlipPanel to combine a data-entry form with some helpful documentation, to provide a simple or a more complex view on the same data, or to fuse together a question and an answer in a trivia game.

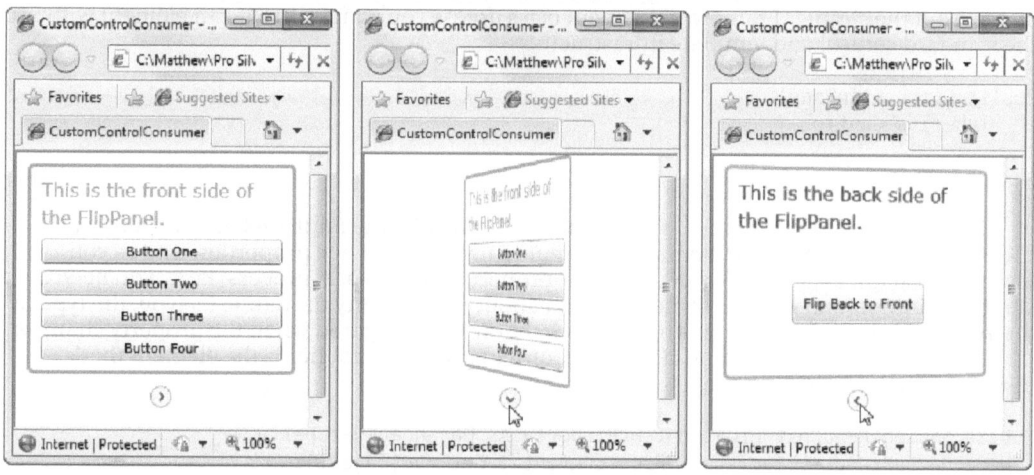

Figure 13-8. *Flipping the FlipPanel*

You can perform the flipping programmatically (by setting a property named IsFlipped), or the user can flip the panel using a convenient button (unless the control consumer removes it from the template).

Building the FlipPanel is refreshingly easy. You need to create a custom panel that adds an extra content region for the hidden surface, along with the animations that switch between the two sides. Ideally, you'll create a carefully structured control template that allows others to restyle the custom FlipPanel with different visuals.

Creating the Solution

Although you can develop a custom Silverlight control in the same assembly that holds your application, it's better to place it in a separate assembly. This approach allows you to refine, revise, and debug your control without affecting the application. It also gives you the option of using the same control with different Silverlight applications.

To add a Silverlight class library project to an existing solution that already holds a Silverlight application, choose File ➤ Add ➤ New Project. Then, choose the Silverlight Class Library project, choose the name and location, and click OK. Now, you're ready to begin designing your custom control.

Starting the FlipPanel Class

Stripped down to its bare bones, the FlipPanel is surprisingly simple. It consists of two content regions that the user can fill with a single element (most likely, a layout container that contains an assortment of elements). Technically, that means the FlipPanel isn't a true panel, because it doesn't use layout logic to organize a group of child elements. However, this isn't likely to pose a problem, because the structure of the FlipPanel is clear and intuitive. The FlipPanel also includes a flip button that lets the user switch between the two different content regions.

Although you can create a custom control by deriving from a control class such as ContentControl or Panel, the FlipPanel derives directly from the base Control class. If you don't need the functionality of a specialized control class, this is the best starting point. You shouldn't derive from the simpler FrameworkElement class unless you want to create an element without the standard control and template infrastructure:

```
Public Class FlipPanel
    Inherits Control
    ...
End Class
```

The first order of business is to create the properties for the FlipPanel. As with almost all the properties in a Silverlight element, you should use dependency properties. And as you learned in Chapter 4, defining a dependency property is a two-part process. First, you need a shared definition that records some metadata about the property: its name, its type, the type of the containing class, and an optional callback that will be triggered when the property changes.

Here's how FlipPanel defines the FrontContent property that holds the element that's displayed on the front surface:

```
Public Shared ReadOnly FrontContentProperty As DependencyProperty = _
    DependencyProperty.Register("FrontContent", GetType(Object), _
    GetType(FlipPanel), Nothing)
```

Next, you need to add a traditional .NET property procedure that calls the base GetValue() and SetValue() methods to change the dependency property. Here's the property procedure implementation for the FrontContent property:

```
Public Property FrontContent() As Object
    Get
        Return MyBase.GetValue(FrontContentProperty)
    End Get
    Set(ByVal value As Object)
        MyBase.SetValue(FrontContentProperty, value)
    End Set
End Property
```

The BackContent property is virtually identical:

```
Public Shared ReadOnly BackContentProperty As DependencyProperty = _
    DependencyProperty.Register("BackContent", GetType(Object), _
    GetType(FlipPanel), Nothing)

Public Property BackContent() As Object
    Get
        Return MyBase.GetValue(BackContentProperty)
    End Get
```

```
    Set(ByVal value As Object)
        MyBase.SetValue(BackContentProperty, value)
    End Set
End Property
```

You need to add just one more essential property: IsFlipped. This Boolean property keeps track of the current state of the FlipPanel (forward-facing or backward-facing) and lets the control consumer flip it programmatically:

```
Public Shared ReadOnly IsFlippedProperty As DependencyProperty = _
    DependencyProperty.Register("IsFlipped", GetType(Boolean), _
    GetType(FlipPanel), Nothing)

Public Property IsFlipped() As Boolean
    Get
        Return CBool(MyBase.GetValue(IsFlippedProperty))
    End Get
    Set(ByVal value As Boolean)
        MyBase.SetValue(IsFlippedProperty, value)
        ChangeVisualState(True)
    End Set
End Property
```

Keen eyes will notice that the IsFlipped property setter calls a custom method called ChangeVisualState(). This method makes sure the display is updated to match the current flip state (forward-facing or backward-facing). You'll consider the code that takes care of this task a bit later.

The FlipPanel doesn't need many more properties, because it inherits virtually everything it needs from the Control class. One exception is the CornerRadius property. Although the Control class includes BorderBrush and BorderThickness properties, which you can use to draw a border around the FlipPanel, it lacks the CornerRadius property for rounding square edges into a gentler curve, as the Border element does. Implementing the same effect in the FlipPanel is easy, provided you add the CornerRadius property and use it to configure a Border element in the FlipPanel's default control template:

```
Public Shared ReadOnly CornerRadiusProperty As DependencyProperty = _
    DependencyProperty.Register("CornerRadius", GetType(CornerRadius), _
    GetType(FlipPanel), Nothing)

Public Property CornerRadius() As CornerRadius
    Get
        Return CType(GetValue(CornerRadiusProperty), CornerRadius)
    End Get
    Set(ByVal value As CornerRadius)
        SetValue(CornerRadiusProperty, value)
    End Set
End Property
```

Adding the Default Style with Generic.xaml

Custom controls suffer from a chicken-and-egg dilemma. You can't write the code in the control class without thinking about the type of control template you'll use. But you can't create the control template until you know how your control works.

The solution is to build both the control class and the default control template at the same time. You can place the control class in any code file template in your Silverlight class library. The control template must be placed in a file named *generic.xaml*. If your class library contains multiple controls, all of their default templates must be placed in the same generic.xaml file. To add it, follow these steps:

1. Right-click the class library project in the Solution Explorer, and choose Add ➤ New Folder.
2. Name the new folder Themes.
3. Right-click the Themes folder, and choose Add ➤ New Item.
4. In the Add New Item dialog box, pick the XML file template, enter the name generic.xaml, and click Add.

The generic.xaml file holds a resource dictionary with styles for your custom controls. You must add one style for each custom control. And as you've probably guessed, the style must set the Template property of the corresponding control to apply the default control template.

■ **Note** You place the generic.xaml file in a folder named Themes for consistency with WPF, which takes the Windows theme settings into account. Silverlight keeps the Themes folder, even though it doesn't have a similar mechanism.

For example, consider the Silverlight project and class library combination shown in Figure 13-9. The CustomControl project is the class library with the custom control, and the CustomControlConsumer project is the Silverlight application that uses it.

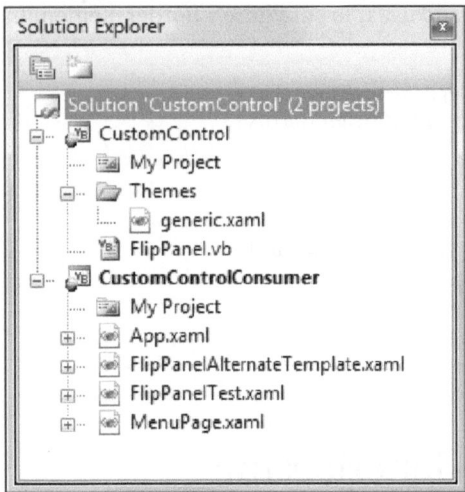

Figure 13-9. A Silverlight application and class library

In the generic.xaml file, you need to declare a resource dictionary. You then need to map the project namespace to an XML namespace prefix, so you can access your custom control in your markup (as you first saw in Chapter 2). In this example, the project namespace is FlipPanelControl, and the assembly is named FlipPanelControl.dll (as you would expect based on the project name):

```
<ResourceDictionary
 xmlns="http://schemas.microsoft.com/winfx/2006/xaml/presentation"
 xmlns:x="http://schemas.microsoft.com/winfx/2006/xaml"
 xmlns:local="clr-namespace:FlipPanelControl;assembly=FlipPanelControl">
 ...
</ResourceDictionary>
```

Notice that when you map the control namespace, you need to include both the project namespace *and* the project assembly name, which isn't the case when you use custom classes inside a Silverlight application. That's because the custom control will be used in other applications, and if you don't specify an assembly, Silverlight will assume that the application assembly is the one you want.

Inside the resource dictionary, you can define a style for your control. Here's an example:

```
<Style TargetType="local:FlipPanel">
  <Setter Property="Template">
    <Setter.Value>
      <ControlTemplate TargetType="local:FlipPanel">
        ...
      </ControlTemplate>
    </Setter.Value>
  </Setter>
</Style>
```

There's one last detail. To tell your control to pick up the default style from the generic.xaml file, you need to set the control's DefaultStyleKey property in the constructor:

```
Public Sub New()
    DefaultStyleKey = GetType(FlipPanel)
End Sub
```

DefaultStyleKey indicates the type that is used to look up the style. In this case, the style is defined with the TargetType of FlipPanel, so the DefaultStyleKey must also use the FlipPanel type. In most cases, this is the pattern you'll follow. The only exception is when you're deriving a more specialized control from an existing control class. In this case, you have the option of keeping the original constructor logic and inheriting the standard style from the base class. For example, if you create a customized Button-derived class with additional functionality, you can use the standard button style and save the trouble of creating a new style. On the other hand, if you do want a different style and a different default control template, you need to add the style using the TargetType of the new class and write a new constructor that sets the DefaultStyleKey property accordingly.

Choosing Parts and States

Now that you have the basic structure in place, you're ready to identify the parts and states that you'll use in the control template.

Clearly, the FlipPanel requires two states:

- *Normal*: This storyboard ensures that only the front content is visible. The back content is flipped, faded, or otherwise shuffled out of view.
- *Flipped*: This storyboard ensures that only the back content is visible. The front content is animated out of the way.

In addition, you need two parts:

- *FlipButton*: This is the button that, when clicked, changes the view from the front to the back (or vice versa). The FlipPanel provides this service by handling this button's events.
- *FlipButtonAlternate*: This is an optional element that works in the same way as the FlipButton. Its inclusion allows the control consumer to use two different approaches in a custom control template. One option is to use a single flip button outside the flippable content region. The other option is to place a separate flip button on both sides of the panel, in the flippable region.

You could also add parts for the front content and back content regions. However, the FlipPanel control doesn't need to manipulate these regions directly, as long as the template includes an animation that hides or shows them at the appropriate time. (Another option is to define these parts so you can explicitly change their visibility in code. That way, the panel can still change between the front and back content region even if no animations are defined, by hiding one section and showing the other. For simplicity's sake, the FlipPanel doesn't go to these lengths.)

To advertise the fact that the FlipPanel uses these parts and states, you should apply the TemplatePart attribute to your control class, as shown here:

```
<TemplateVisualState(Name := "Normal", GroupName := "ViewStates"), _
 TemplateVisualState(Name := "Flipped", GroupName := "ViewStates"), _
 TemplatePart(Name := "FlipButton", Type := GetType(ToggleButton)), _
 TemplatePart(Name := "FlipButtonAlternate", Type := GetType(ToggleButton))> _
Public Class FlipPanel
    Inherits Control
    ...
End Class
```

The FlipButton and FlipButtonAlternate parts are restricted—each one can only be a ToggleButton or an instance of a ToggleButton-derived class. (As you may remember from Chapter 5, the ToggleButton is a clickable button that can be in one of two states. In the case of the FlipPanel control, the ToggleButton states correspond to a normal front-forward view or a flipped back-forward view.)

■ **Tip** To ensure the best, most flexible template support, use the least-specialized element type that you can. For example, it's better to use FrameworkElement than ContentControl, unless you need some property or behavior that ContentControl provides.

NAMING CONVENTIONS FOR STATES, PARTS, AND STATE GROUPS

The naming conventions for parts and states are fairly straightforward. When you're naming a part or state, don't include a prefix or suffix—for example, use Flipped and FlipButton rather than FlippedState and FlipButtonPart. The exception is state groups, which should always end with the word *States*, as in ViewStates.

It also helps to look at similar controls in the Silverlight framework and use the same names. This is especially true if you need to use the states that are commonly defined in the CommonStates group (Normal, MouseOver, Pressed, and Disabled) or the FocusStates group (Focused and Unfocused). Remember, the control consumer must use the exact name. If you create a button-like control that breaks with convention and uses a Clicked state instead of a Pressed state and the control consumer inadvertently defines a Pressed state, its animation will be quietly ignored.

Starting the Default Control Template

Now, you can slot these pieces into the default control template. The root element is a two-row Grid that holds the content area (in the top row) and the flip button (in the bottom row). The content area is filled with two overlapping Border elements, representing the front and back content, but only one of the two is ever shown at a time.

To fill in the front and back content regions, the FlipPanel uses the ContentPresenter. This technique is virtually the same as in the custom button example, except you need two ContentPresenter elements, one for each side of the FlipPanel. The FlipPanel also includes a separate Border element wrapping each ContentPresenter. This lets the control consumer outline the flippable content region by setting a few straightforward properties on the FlipPanel (BorderBrush, BorderThickness, Background, and CornerRadius), rather than being forced to add a border by hand.

Here's the basic skeleton for the default control template:

```
<ControlTemplate TargetType="local:FlipPanel">
  <Grid>
    <VisualStateManager.VisualStateGroups>
      <!-- Place state animations here. -->
    </VisualStateManager.VisualStateGroups>

    <Grid.RowDefinitions>
      <RowDefinition Height="Auto"></RowDefinition>
      <RowDefinition Height="Auto"></RowDefinition>
    </Grid.RowDefinitions>

    <!-- This is the front content. -->
    <Border BorderBrush="{TemplateBinding BorderBrush}"
     BorderThickness="{TemplateBinding BorderThickness}"
     CornerRadius="{TemplateBinding CornerRadius}"
     Background="{TemplateBinding Background}">
      <ContentPresenter Content="{TemplateBinding FrontContent}">
      </ContentPresenter>
    </Border>

    <!-- This is the back content. -->
```

```
<Border BorderBrush="{TemplateBinding BorderBrush}"
 BorderThickness="{TemplateBinding BorderThickness}"
 CornerRadius="{TemplateBinding CornerRadius}"
 Background="{TemplateBinding Background}">
  <ContentPresenter Content="{TemplateBinding BackContent}">
  </ContentPresenter>
</Border>

<!-- This the flip button. -->
<ToggleButton Grid.Row="1" x:Name="FlipButton" Margin="0,10,0,0">
</ToggleButton>

  </Grid>
</ControlTemplate>
```

When you create a default control template, it's best to avoid hard-coding details that the control consumer may want to customize. Instead, you need to use template binding expressions. In this example, you set several properties using template-binding expressions: BorderBrush, BorderThickness, CornerRadius, Background, FrontContent, and BackContent. To set the default value for these properties (and thereby ensure that you get the right visual even if the control consumer doesn't set them), you must add additional setters to your control's default style.

The FlipButton Control

The control template shown in the previous example includes a ToggleButton. However, it uses the ToggleButton's default appearance, which makes the ToggleButton look like an ordinary button, complete with the traditional shaded background. This isn't suitable for the FlipPanel.

Although you can place any content you want inside the ToggleButton, the FlipPanel requires a bit more. It needs to do away with the standard background and change the appearance of the elements inside depending on the state of the ToggleButton. As you saw earlier in Figure 13-8, the ToggleButton points the way the content will be flipped (right initially, when the front faces forward, and left when the back faces forward). This makes the purpose of the button clearer.

To create this effect, you need to design a custom control template for the ToggleButton. This control template can include the shape elements that draw the arrow you need. In this example, the ToggleButton is drawn using an Ellipse element for the circle and a Path element for the arrow, both of which are placed in a single-cell Grid:

```
<ToggleButton Grid.Row="1" x:Name="FlipButton" RenderTransformOrigin="0.5,0.5"
 Margin="0,10,0,0">
  <ToggleButton.Template>
    <ControlTemplate>
      <Grid>
        <Ellipse Stroke="#FFA9A9A9" Fill="AliceBlue" Width="19"
         Height="19"></Ellipse>
        <Path RenderTransformOrigin="0.5,0.5" Data="M1,1.5L4.5,5 8,1.5"
         Stroke="#FF666666" StrokeThickness="2"
         HorizontalAlignment="Center" VerticalAlignment="Center"></Path>
      </Grid>
    </ControlTemplate>
  </ToggleButton.Template>
</ToggleButton>
```

Defining the State Animations

The state animations are the most interesting part of the control template. They're the ingredients that provide the flipping behavior. They're also the details that are most likely to be changed if a developer creates a custom template for the FlipPanel.

In the default control template, the animations use a 3-D projection to rotate the content regions. To hide a content region, it's turned until it's at a 90-degree angle, with the edge exactly facing the user. To show a content region, it's returned from this position to a flat 0-degree angle. To create the flipping effect, one animation turns and hides the first region (for example, the front), and a second animation picks up as the first one ends to show the second region (for example, the back).

To make this work, you first need to add a projection to the Border element that holds the front content:

```
<Border.Projection>
  <PlaneProjection x:Name="FrontContentProjection"></PlaneProjection>
</Border.Projection>
```

And you need to add a similar one to the Border element that holds the back content:

```
<Border.Projection>
  <PlaneProjection x:Name="BackContentProjection"></PlaneProjection>
</Border.Projection>
```

The content region isn't the only part of the FlipPanel that you need to animate. You must also add a RotateTransform to the ToggleButton so you can rotate the arrow to point to the other side when the content is flipped:

```
<ToggleButton.RenderTransform>
  <RotateTransform x:Name="FlipButtonTransform" Angle="-90"></RotateTransform>
</ToggleButton.RenderTransform>
```

Here are the animations that flip the front and back content regions and rotate the ToggleButton arrow:

```
<VisualStateGroup x:Name="ViewStates">
  <VisualState x:Name="Normal">
    <Storyboard>
      <DoubleAnimation Storyboard.TargetName="BackContentProjection"
        Storyboard.TargetProperty="RotationY" To="-90"
        Duration="0:0:0"></DoubleAnimation>
    </Storyboard>
  </VisualState>

  <VisualState x:Name="Flipped">
    <Storyboard>
      <DoubleAnimation Storyboard.TargetName="FrontContentProjection"
        Storyboard.TargetProperty="RotationY" To="90"
        Duration="0:0:0"></DoubleAnimation>

      <DoubleAnimation Storyboard.TargetName="FlipButtonTransform"
        Storyboard.TargetProperty="Angle" Duration="0:0:0" To="90"></DoubleAnimation>
    </Storyboard>
  </VisualState>
</VisualStateGroup>
```

Remember, the state animations only need to supply a storyboard for *changing* the initial values. That means the Normal state needs to indicate what to do with the back content region. The front content region is automatically restored to its initial state and rotated back into view. Similarly, the Flipped state needs to indicate what to do with the front content region and the arrow, while allowing the back content region to be rotated back into view.

Notice that all the animations are performed through transitions, which is the correct approach. For example, the Flipped state uses a zero-length animation to change the RotationY property of FrontContentProjection to 90 and rotate the arrow 90 degrees. However, there's a catch. To create the realistic flipping effect, you need to flip the visible content out of the way first and *then* flip the new content into view. The default transition can't handle this—instead, it rotates both content regions and the arrow with three simultaneous animations.

To fix the problem, you need to add the somewhat tedious custom transitions shown here. They explicitly use the Duration and BeginTime properties to ensure that the flipping animations happen in sequence:

```
<VisualStateManager.VisualStateGroups>
  <VisualStateGroup x:Name="ViewStates">
    <VisualStateGroup.Transitions>
      <VisualTransition To="Normal" From="Flipped" GeneratedDuration="0:0:0.7">
        <Storyboard>
          <DoubleAnimation Storyboard.TargetName="BackContentProjection"
           Storyboard.TargetProperty="RotationY" To="-90"
           Duration="0:0:0.5"></DoubleAnimation>
          <DoubleAnimation Storyboard.TargetName="FrontContentProjection"
           BeginTime="0:0:0.5" Storyboard.TargetProperty="RotationY" To="0"
           Duration="0:0:0.5"></DoubleAnimation>
        </Storyboard>
      </VisualTransition>

      <VisualTransition To="Flipped" From="Normal" GeneratedDuration="0:0:0.7">
        <Storyboard>
          <DoubleAnimation Storyboard.TargetName="FrontContentProjection"
           Storyboard.TargetProperty="RotationY" To="90"
           Duration="0:0:0.5"></DoubleAnimation>
          <DoubleAnimation Storyboard.TargetName="BackContentProjection"
           BeginTime="0:0:0.5" Storyboard.TargetProperty="RotationY" To="0"
           Duration="0:0:0.5"></DoubleAnimation>
        </Storyboard>
      </VisualTransition>
    </VisualStateGroup.Transitions>

    <VisualState x:Name="Normal">
      . . .
    </VisualState>

    <VisualState x:Name="Flipped">
      . . .
    </VisualState>
  </VisualStateGroup>
</VisualStateManager.VisualStateGroups>
```

The custom transition doesn't do anything to the ToggleButton arrow, because the automatically generated transition does a perfectly good job for it.

Wiring Up the Elements in the Template

Now that you've polished off a respectable control template, you need to fill in the plumbing in the FlipPanel control to make it work.

The trick is a protected method named OnApplyTemplate(), which is defined in the base Control class. This method is called when the control is being initialized. This is the point where the control needs to examine its template and fish out the elements it needs. The exact action a control performs with an element varies—it may set a property, attach an event handler, or store a reference for future use.

To use the template in a custom control, you override the OnApplyTemplate() method. To find an element with a specific name, you call the GetTemplateChild() method (which is inherited from FrameworkElement along with the OnApplyTemplate() method). If you don't find an element that you want to work with, the recommended pattern is to do nothing. Optionally, you can add code that checks that the element, if present, is the correct type and raises an exception if it isn't. (The thinking here is that a missing element represents a conscious opting out of a specific feature, whereas an incorrect element type represents a mistake.)

The OnApplyTemplate() method for the FlipPanel retrieves the ToggleButton for the FlipButton and FlipButtonAlternate parts and attaches event handlers to each, so it can react when the user clicks to flip the control. Finally, the OnApplyTemplate() method ends by calling a custom method named ChangeVisualState(), which ensures that the control's visuals match its current state:

```vb
Public Overrides Sub OnApplyTemplate()
    MyBase.OnApplyTemplate()

    ' Wire up the ToggleButton.Click event.
    Dim flipButton As ToggleButton = TryCast( _
      MyBase.GetTemplateChild("FlipButton"), ToggleButton)
    If flipButton IsNot Nothing Then
        AddHandler flipButton.Click, AddressOf flipButton_Click
    End If

    ' Allow for two flip buttons if needed (one for each side of the panel).
    Dim flipButtonAlternate As ToggleButton = TryCast( _
      MyBase.GetTemplateChild("FlipButtonAlternate"), ToggleButton)
    If flipButtonAlternate IsNot Nothing Then
        AddHandler flipButtonAlternate.Click, AddressOf flipButton_Click
    End If

    ' Make sure the visuals match the current state.
    ChangeVisualState(False)
End Sub
```

■ **Tip** When calling GetTemplateChild(), you need to indicate the string name of the element you want. To avoid possible errors, you can declare this string as a constant in your control. You can then use that constant in the TemplatePart attribute and when calling GetTemplateChild().

Here's the very simple event handler that allows the user to click the ToggleButton and flip the panel:

```
Private Sub flipButton_Click(ByVal sender As Object, ByVal e As RoutedEventArgs)
    IsFlipped = Not IsFlipped
    ChangeVisualState(True)
End Sub
```

Fortunately, you don't need to manually trigger the state animations. Nor do you need to create or trigger the transition animations. Instead, to change from one state to another, you call the shared VisualStateManager.GoToState() method. When you do, you pass in a reference to the control object that's changing state, the name of the new state, and a Boolean value that determines whether a transition is shown. This value should be True when it's a user-initiated change (for example, when the user clicks the ToggleButton) but False when it's a property setting (for example, if the markup for your page sets the initial value of the IsExpanded property).

Dealing with all the different states a control supports can become messy. To avoid scattering GoToState() calls throughout your control code, most controls add a custom method like the ChangeVisualState() method in the FlipPanel. This method has the responsibility of applying the correct state in each state group. The code inside uses one if block (or switch statement) to apply the current state in each state group. This approach works because it's completely acceptable to call GoToState() with the name of the current state. In this situation, when the current state and the requested state are the same, nothing happens.

Here's the code for the FlipPanel's version of the ChangeVisualState() method:

```
Private Sub ChangeVisualState(ByVal useTransitions As Boolean)
    If Not IsFlipped Then
        VisualStateManager.GoToState(Me, "Normal", useTransitions)
    Else
        VisualStateManager.GoToState(Me, "Flipped", useTransitions)
    End If
End Sub
```

Usually, you call the ChangeVisualState() method (or your equivalent) in the following places:

- After initializing the control at the end of the OnApplyTemplate() method.

- When reacting to an event that represents a state change, such as a mouse movement or a click of the ToggleButton.

- When reacting to a property change or a method that's triggered through code. For example, the IsFlipped property setter calls ChangeVisualState() and always supplies true, thereby showing the transition animations. If you want to give the control consumer the choice of not showing the transition, you can add a Flip() method that takes the same Boolean parameter you pass to ChangeVisualState().

As written, the FlipPanel control is remarkably flexible. For example, you can use it without a ToggleButton and flip it programmatically (perhaps when the user clicks a different control). Or, you can include one or two flip buttons in the control template and allow the user to take control.

Using the FlipPanel

Now that you've completed the control template and code for the FlipPanel, you're ready to use it in an application. Assuming you've added the necessary assembly reference, you can then map an XML prefix to the namespace that holds your custom control:

```
<UserControl x:Class="FlipPanelTest.Page"
  xmlns:lib="clr-namespace:FlipPanelControl;assembly=FlipPanelControl" ... >
```

Next, you can add instances of the FlipPanel to your page. Here's an example that creates the FlipPanel shown earlier in Figure 13-8, using a StackPanel full of elements for the front content region and a Grid for the back:

```
<lib:FlipPanel x:Name="panel" BorderBrush="DarkOrange"
 BorderThickness="3" CornerRadius="4" Margin="10">
  <lib:FlipPanel.FrontContent>
    <StackPanel Margin="6">
      <TextBlock TextWrapping="Wrap" Margin="3" FontSize="16"
       Foreground="DarkOrange">This is the front side of the FlipPanel.</TextBlock>
      <Button Margin="3" Padding="3" Content="Button One"></Button>
      <Button Margin="3" Padding="3" Content="Button Two"></Button>
      <Button Margin="3" Padding="3" Content="Button Three"></Button>
      <Button Margin="3" Padding="3" Content="Button Four"></Button>
    </StackPanel>
  </lib:FlipPanel.FrontContent>

  <lib:FlipPanel.BackContent>
    <Grid Margin="6">
      <Grid.RowDefinitions>
        <RowDefinition Height="Auto"></RowDefinition>
        <RowDefinition></RowDefinition>
      </Grid.RowDefinitions>
      <TextBlock TextWrapping="Wrap" Margin="3" FontSize="16"
       Foreground="DarkMagenta">This is the back side of the FlipPanel.</TextBlock>
      <Button Grid.Row="2" Margin="3" Padding="10" Content="Flip Back to Front"
       HorizontalAlignment="Center" VerticalAlignment="Center"
       Click="cmdFlip_Click"></Button>
    </Grid>
  </lib:FlipPanel.BackContent>
</lib:FlipPanel>
```

When clicked, the button on the back side of the FlipPanel programmatically flips the panel:

```
Private Sub cmdFlip_Click(ByVal sender As Object, ByVal e As RoutedEventArgs)
    panel.IsFlipped = Not panel.IsFlipped
End Sub
```

This has the same result as clicking the ToggleButton with the arrow, which is defined as part of the default control template.

Using a Different Control Template

Custom controls that have been designed properly are extremely flexible. In the case of the FlipPanel, you can supply a new template to change the appearance and placement of the ToggleButton and the animated effects that are used when flipping between the front and back content regions.

Figure 13-10 shows one such example. Here, the flip button is placed in a special bar that's at the bottom of the front side and the top of the back side. And when the panel flips, it doesn't turn its content like a sheet of paper. Instead, it squares the front content into nothingness at the top of the panel while simultaneously expanding the back content underneath. When the panel flips the other way, the back content squishes back down, and the front content expands from the top. For even more visual pizzazz, the content that's being squashed is also blurred with the help of the BlurEffect class.

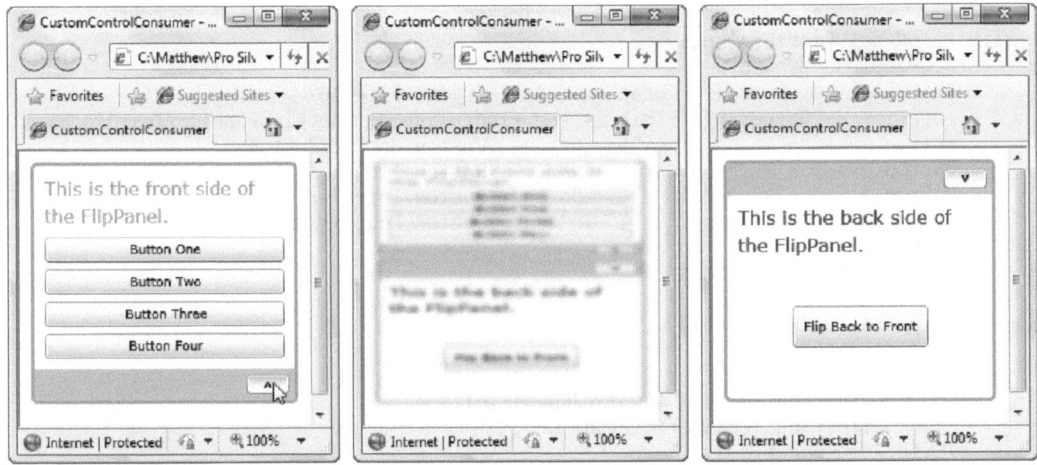

Figure 13-10. The FlipPanel with a different control template

Here's the portion of the template that defines the front content region:

```
<Border BorderBrush="{TemplateBinding BorderBrush}"
  BorderThickness="{TemplateBinding BorderThickness}"
  CornerRadius="{TemplateBinding CornerRadius}"
  Background="{TemplateBinding Background}">

  <Border.RenderTransform>
    <ScaleTransform x:Name="FrontContentTransform"></ScaleTransform>
  </Border.RenderTransform>
  <Border.Effect>
    <BlurEffect x:Name="FrontContentEffect" Radius="0"></BlurEffect>
  </Border.Effect>

  <Grid>
    <Grid.RowDefinitions>
      <RowDefinition></RowDefinition>
      <RowDefinition Height="Auto"></RowDefinition>
    </Grid.RowDefinitions>
```

```
    <ContentPresenter Content="{TemplateBinding FrontContent}"></ContentPresenter>
    <Rectangle Grid.Row="1" Stretch="Fill" Fill="LightSteelBlue"></Rectangle>
    <ToggleButton Grid.Row="1" x:Name="FlipButton" Margin="5" Padding="15,0"
     Content="^" FontWeight="Bold" FontSize="12" HorizontalAlignment="Right">
    </ToggleButton>
  </Grid>
</Border>
```

The back content region is almost the same. It consists of a Border that contains a ContentPresenter element, and it includes its own ToggleButton placed at the right edge of the shaded rectangle. It also defines the all-important ScaleTransform and BlurEffect on the Border, which is what the animations use to flip the panel.

Here are the animations that perform the flipping:

```
<VisualStateManager.VisualStateGroups>
  <VisualStateGroup x:Name="ViewStates">
    <VisualStateGroup.Transitions>
      <VisualTransition GeneratedDuration="0:0:0.7">
      </VisualTransition>
    </VisualStateGroup.Transitions>

    <VisualState x:Name="Normal">
      <Storyboard>
        <DoubleAnimation Storyboard.TargetName="BackContentTransform"
          Storyboard.TargetProperty="ScaleY" To="0"
          Duration="0:0:0"></DoubleAnimation>

        <DoubleAnimation Storyboard.TargetName="BackContentEffect"
          Storyboard.TargetProperty="Radius" To="40"
          Duration="0:0:0"></DoubleAnimation>
      </Storyboard>
    </VisualState>

    <VisualState x:Name="Flipped">
      <Storyboard>
        <DoubleAnimation Storyboard.TargetName="FrontContentTransform"
          Storyboard.TargetProperty="ScaleY" To="0"
          Duration="0:0:0"></DoubleAnimation>

        <DoubleAnimation Storyboard.TargetName="FrontContentEffect"
          Storyboard.TargetProperty="Radius" To="40"
          Duration="0:0:0"></DoubleAnimation>
      </Storyboard>
    </VisualState>
  </VisualStateGroup>
</VisualStateManager.VisualStateGroups>
```

Because the animation that changes the front content region runs at the same time as the animation that changes the back content region, you don't need a custom transition to manage them.

The Last Word

In the previous chapter, you saw how to use styles to reuse formatting. In this chapter, you learned how to use control templates to make more radical changes. You used the parts and states model to customize a Silverlight control and saw how you can create a respectable button without being forced to reimplement any core button functionality. These custom buttons support all the normal button behavior—you can tab from one to the next, you can click them to fire an event, and so on. Best of all, you can reuse your button template throughout your application and still replace it with a whole new design at a moment's notice.

What more do you need to know before you can skin all the Silverlight controls? In order to get the snazzy look you probably want, you may need to spend more time studying the details of Silverlight drawing and animation. Using the shapes, brushes, and transforms that you've already learned about, you can build sophisticated controls with glass-style blurs and soft glow effects. The secret is in combining multiple layers of shapes, each with a different gradient brush. The best way to get this sort of effect is to learn from the control template examples others have created. Two great starting points are the themes in the Silverlight Toolkit (http://silverlight.codeplex.com) and the Expression Blend community gallery (http://gallery.expression.microsoft.com).

CHAPTER 14

■ ■ ■

Browser Integration

Because Silverlight applications run in their own carefully designed environment, you're insulated from the quirks and cross-platform headaches that traditionally confront developers when they attempt to build rich browser-based applications. This is a tremendous advantage. It means you can work with an efficient mix of VB code and XAML markup rather than struggle through a quagmire of HTML, JavaScript, and browser-compatibility issues.

However, in some cases you'll need to create a web page that isn't just a thin shell around a Silverlight application. Instead, you may want to add Silverlight content to an existing page and allow the HTML and Silverlight portions of your page to interact.

There are several reasons you may choose to blend the classic browser world with the managed Silverlight environment. Here are some possibilities:

- *Compatibility*: You can't be sure your visitors will have the Silverlight plug-in installed. If you're building a core part of your website, your need to ensure broad compatibility (with HTML) may trump your desire to use the latest and greatest user interface frills (with Silverlight). In this situation, you may decide to include a Silverlight content region to show nonessential extras alongside the critical HTML content.

- *Legacy web pages:* If you have an existing web page that does exactly what you want, it may make more sense to extend it with a bit of Silverlight pizzazz than to replace it outright. Once again, the solution is to create a page that includes both HTML and Silverlight content.

- *Server-side features*: Some types of tasks require server-side code. For example, Silverlight is a poor fit for tasks that need to access server resources or require high security, which is why it makes far more sense to build a secure checkout process with a server-side programming framework like ASP.NET. But you can still use Silverlight to display advertisements, video content, product visualizations, and other value-added features in the same pages.

In this chapter, you'll consider how you can bridge the gap between Silverlight and the ordinary world of HTML. First, you'll see how Silverlight can reach out to other HTML elements on the page and manipulate them. Next, you'll learn how Silverlight can fire off JavaScript code and how JavaScript code can trigger a method in your Silverlight application. Finally, you'll look at a few more options for overlapping Silverlight content and ordinary HTML elements.

■ **Note** Thinking of building an out-of-browser application (as described in Chapter 21)? If so, you won't be able to use any of the features described in this chapter. However, you can use the new WebBrowser control to interact with HTML content (which is also described in Chapter 21).

Interacting with HTML Elements

Silverlight includes a set of managed classes that replicate the HTML document object model (DOM) in managed code. These classes let your Silverlight code interact with the HTML content on the same page. Depending on the scenario, this interaction may involve reading a control value, updating text, or adding new HTML elements to the page.

The classes you need to perform these feats are found in the System.Windows.Browser namespace and are listed in Table 14-1. You'll learn about them in the following sections.

Table 14-1. *The Key Classes in the System.Windows.Browser Namespace*

Class	Description
HtmlPage	Represents the current HTML page (where the Silverlight control is placed). The HtmlPage class is a jumping-off point for most of the HTML interaction features. It provides members for exploring the HTML elements on the page (the Document property), retrieving browser information (the BrowserInformation property), interacting with the current browser window (the Window property), and registering Silverlight methods that you want to make available to JavaScript (the RegisterCreatableType() and RegisterScriptableType() methods).
BrowserInformation	Provides some basic information about the browser that's being used to run your application, including the browser name, version, and operating system. You can retrieve an instance of the BrowserInformation class from the HtmlPage.BrowserInformation property.
HtmlDocument	Represents a complete HTML document. You can get an instance of HtmlDocument that represents the current HTML page from the HtmlPage.Document property. You can then use the HtmlDocument object to explore the structure and content of the page (as nested levels of HtmlElement objects).
HtmlElement	Represents any HTML element on the page. You can use methods like SetAttribute() and SetProperty() to manipulate that element. Usually, you look up HtmlElement objects in an HtmlDocument object.
HtmlWindow	Represents the browser window and provides methods for navigating to a new page or to a different anchor in the current page. You can get an instance of HtmlWindow that holds the current page from the HtmlPage.Window property.

Class	Description
HttpUtility	Provides shared methods for a few common HTML-related tasks, including HTML encoding and decoding (making text safe for display in a web page) and URL encoding and decoding (making text safe for use in a URL—for example, as a query string argument).
ScriptableType Attribute and ScriptableMember Attribute	Allows you to expose the classes and methods in your Silverlight application so they can be called from JavaScript code in the HTML page.
ScriptObject	Represents a JavaScript function that's defined in the page and allows you to invoke the function from your Silverlight application.

Getting Browser Information

Most of the time, you shouldn't worry about the specific browser that's being used to access your application. After all, one of the key advantages of Silverlight is that it saves you from the browser-compatibility hassles of ordinary web programming and lets you write code that behaves in the same way in every supported environment. However, in some scenarios you may choose to take a closer look at the browser—for example, when diagnosing an unusual error that can be browser related.

The browser information that's available in the BrowserInformation class is fairly modest. You're given four string properties that indicate the browser name, version, operating system, and user agent—a long string that includes technical details about the browser (for example, in Internet Explorer, it lists all the currently installed versions of the .NET Framework). You can also use the Boolean CookiesEnabled property to determine whether the current browser supports cookies and has them enabled (in which case it's True). You can then read or change cookies through the HtmlPage class.

■ **Note** The information you get from the BrowserInformation class depends on how the browser represents itself to the world, but it may not reflect the browser's true identity. Browsers can be configured to impersonate other browsers, and some browsers use this technique to ensure broader compatibility. If you write any browser-specific code, make sure you test it with a range of browsers to verify that you're detecting the correct conditions.

Here's some straightforward code that displays all the available browser information:

```
Dim b As BrowserInformation = HtmlPage.BrowserInformation
lblInfo.Text = "Name: " & b.Name
lblInfo.Text &= Environment.NewLine & "Browser Version: " & _
  b.BrowserVersion.ToString()
lblInfo.Text &= Environment.NewLine & "Platform: " & b.Platform
lblInfo.Text &= Environment.NewLine & "Cookies Enabled: " & b.CookiesEnabled
lblInfo.Text &= Environment.NewLine & "User Agent: " & b.UserAgent
```

Figure 14-1 shows the result.

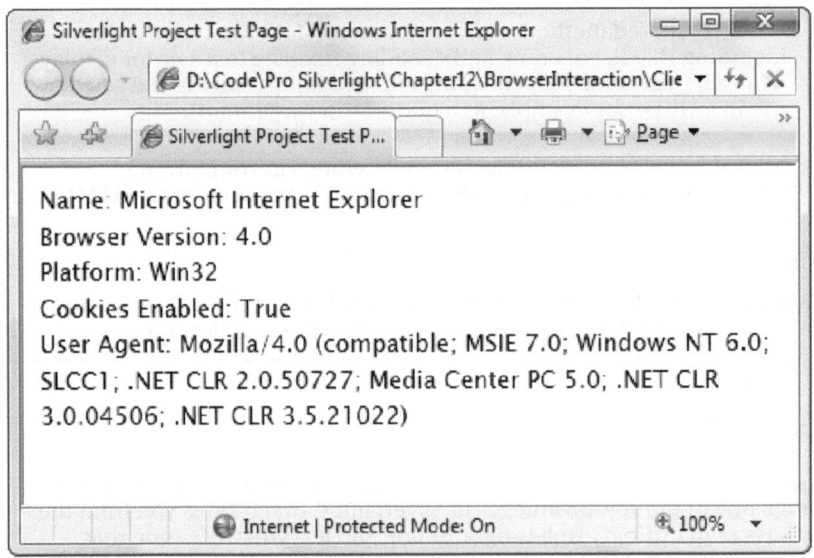

Figure 14-1. Profiling the browser

The HTML Window

Silverlight also gives you a limited ability to control the browser through the HtmlWindow class. It provides two methods that allow you to trigger navigation: Navigate() and NavigateToBookmark().

Navigate() sends the browser to another page. You can use an overloaded version of the Navigate() method to specify a target frame. When you use Navigate(), you abandon the current Silverlight application. It's the same as if the user had typed a new URL in the browser's address bar.

NavigateToBookmark() scrolls to a specific bookmark in the current page. A bookmark is an <a> element with an ID (or name) but no target:

```
<a id="myBookmark">...</a>
```

To navigate to a bookmark, you add the number sign (#) and bookmark name to the end of your URL:

```
<a href="page.html#myBookmark">Jump to bookmark</a>
```

You can retrieve the bookmark from the current browser URL at any time using the HtmlWindow.CurrentBookmark property, which is the only property the HtmlWindow class includes.

The NavigateToBookmark() method and CurrentBookmark property raise an interesting possibility. You can use a bookmark to store some state information. Because this state information is part of the URL, it's preserved in the browser history and (if you bookmark a page with Silverlight content) the browser's favorites list. This technique is the basis for the higher-level navigation framework you explored in Chapter 7.

Popup Windows

The HtmlPage class also provides a PopupWindow() method that allows you to open a pop-up window to show a new web page. The PopupWindow() method is intended for showing advertisements and content from other websites. It's not intended as a way to show different parts of the current Silverlight application. (If you want the ability to show a pop-up window inside a Silverlight application, you need the ChildWindow control described in Chapter 7.)

The PopupWindow() method is fairly reliable and dodges most pop-up blockers (depending on the user's settings). However, it also has a few quirks and should never be relied on for creating an integral part of your application. Instead, the pop-up window content should be an optional extra. Technically, the PopupWindow() method works by triggering a JavaScript window.open() call.

Here's an example that uses the PopupWindow() method. Note that this codes tests the IsPopupWindowAllowed property to avoid potential errors, because popup window are not supported in all scenarios:

```
If HtmlPage.IsPopupWindowAllowed Then
    ' Configure the popup window options.
    Dim options As New HtmlPopupWindowOptions()
    options.Resizeable = True

    ' Show the popup window.
    ' You pass in an absolute URI, an optional target frame, and the
    ' HtmlPopupWindowOptions.
    HtmlPage.PopupWindow(New Uri(uriForAdvertisement), Nothing, options)
End If
```

Here are the rules and restrictions of Silverlight popup windows:

- They don't work if the allowHtmlPopupWindow parameter is set to false in the HTML entry page. (See the "Securing HTML Interoperability" section at the end of this chapter.)
- If your HTML entry page and Silverlight application are deployed on different domains, popup windows are not allowed unless the HTML entry page includes the allowHtmlPopupWindow parameter and explicitly sets it to true.
- The PopupWindow() can be called only in response to a user-initiated click on a visible area of the Silverlight application.
- The PopupWindow() method can be called only once per event. This means you can't show more than one pop-up window at once.
- Popup window work with the default security settings in Internet Explorer and Firefox. However, they won't appear in Safari.
- You can configure the HtmlPopupWindowOptions object to determine whether the pop-up window should be resizable, how big it should be, where it should be placed, and so on, just as you can in JavaScript. However, these properties won't always be respected. For example, browsers refuse to show popup windows that are smaller than a certain size and, depending on settings, may show pop-up windows as separate tabs in the current window.
- When calling PopupWindow(), you must supply an absolute URI.

Inspecting the HTML Document

Retrieving browser information and performing navigation are two relatively straightforward tasks. Life gets a whole lot more interesting when you start peering into the structure of the page that hosts your Silverlight content.

To start your exploration, you use one of two shared properties from the HtmlPage class. The Plugin property provides a reference to the <object> element that represents the Silverlight control, as an HtmlElement object. The Document property provides something more interesting: an HtmlDocument object that represents the entire page, with the members set out in Table 14-2.

Table 14-2. Members of the HtmlDocument Class

Member	Description
DocumentUri	Returns the URL of the current document as a Uri object.
QueryString	Returns the query string portion of the URL as a single long string that you must parse.
DocumentElement	Provides an HtmlElement object that represents the top-level <html> element in the HTML page.
Body	Provides an HtmlElement object that represents the <body> element in the HTML page.
Cookies	Provides a collection of all the current HTTP cookies. You can read or set the values in these cookies. Cookies provide one easy, low-cost way to transfer information from server-side ASP.NET code to client-side Silverlight code. However, cookies aren't the best approach for storing small amounts of data on the client's computer—isolated storage, which is discussed in Chapter 18, provides a similar feature with better compatibility and programming support.
IsReady	Returns True if the browser is idle or False if it's still downloading the page.
CreateElement()	Creates a new HtmlElement object to represent a dynamically created HTML element, which you can then insert into the page.
AttachEvent() and DetachEvent()	Connect an event handler in your Silverlight application to a JavaScript event that's raised by the document.
Submit()	Submits the page by posting a form and its data back to the server. This is useful if you're hosting your Silverlight control in an ASP.NET page, because it triggers a postback that allows server-side code to run.

When you have the HtmlDocument object that represents the page, you can browse down through the element tree, starting at HtmlDocument.DocumentElement or HtmlDocument.Body. To step from one element to another, you use the Children property (to see the elements nested inside the current element) and the Parent property (to get the element that contains the current element).

Figure 14-2 shows an example—a Silverlight application that starts at the top-level <html> element and uses a recursive method to drill through the entire page. It displays the name and ID of each element.

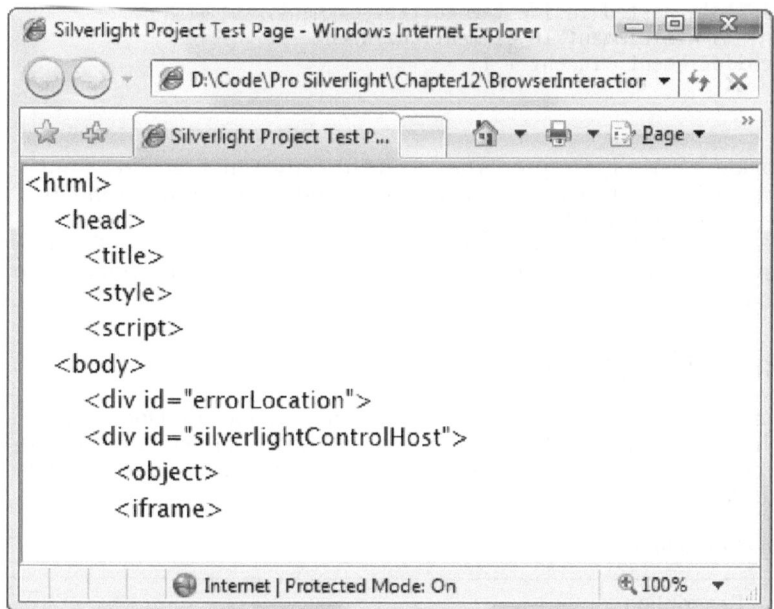

Figure 14-2. Dissecting the current page

Here's the code that creates this display when the page first loads:

```
Private Sub Page_Loaded(ByVal sender As Object, ByVal e As RoutedEventArgs)
    ' Start processing the top-level <html> element.
    Dim element As HtmlElement = HtmlPage.Document.DocumentElement
    ProcessElement(element, 0)
End Sub

Private Sub ProcessElement(ByVal element As HtmlElement, ByVal indent As Integer)
    ' Ignore comments.
    If element.TagName = "!" Then
        Return
    End If

    ' Indent the element to help show different levels of nesting.
    lblElementTree.Text += New String(" "c, indent * 4)
```

509

```
        ' Display the tag name.
        lblElementTree.Text &= "<" & element.TagName

        ' Only show the id attribute if it's set.
        If element.Id <> "" Then
            lblElementTree.Text &= " id=""" & element.Id & """"
        End If
        lblElementTree.Text &= ">" & Environment.NewLine

        ' Process all the elements nested inside the current element.
        For Each childElement As HtmlElement In element.Children
            ProcessElement(childElement, indent + 1)
        Next
    End Sub
```

The HtmlElement provides relatively few properties. Aside from the Children and Parent properties that allow you to navigate between elements, it also includes the TagName and Id demonstrated shown here, as well as a CssClass property that indicates the name of the Cascading Style Sheets (CSS) style that's set through the class attribute and used to configure the appearance of the current element. To get more information out of an element, you need to use one of the HtmlElement methods you'll learn about in the next section.

Manipulating an HTML Element

The Parent and Children properties aren't the only way to travel through an HtmlDocument object. You can also search for an element with a specific name using the GetElementByID() or GetElementsByTagName() method. When you have the element you want, you can manipulate it using one of the methods described in Table 14-3.

Table 14-3. Methods of the HtmlElement Class

Method	Description
AppendChild()	Inserts a new HTML element as the last nested element inside the current element. To create the element, you must first use the HtmlDocument.CreateElement() method.
RemoveChild()	Removes the specified HtmlElement object (which you supply as an argument). This HtmlElement must be one of the children that's nested in the current HtmlElement.
Focus()	Gives focus to the current element so it receives keyboard events.
GetAttribute(), SetAttribute(), and RemoveAttribute()	Let you retrieve the value of any attribute in the element, set the value (in which case the attribute is added if it doesn't already exist), or remove the attribute altogether, respectively.
GetStyleAttribute(), SetStyleAttribute(), RemoveStyleAttribute()	Let you retrieve a value of a CSS style property, set the value, or remove the style attribute altogether, respectively. (As you no doubt know, CSS properties are the modern way to format HTML elements, and they let you control details such as font, foreground and background color, spacing and positioning, and borders.)

Method	Description
GetProperty() and SetProperty()	Allow you to retrieve or set values that are defined as part of the HTML DOM. These are the values that are commonly manipulated in JavaScript code. For example, you can extract the text content from an element using the innerHTML property.
AttachEvent() and DetachEvent()	Connect and disconnect an event handler in your Silverlight application to a JavaScript event that's raised by an HTML element.

For example, imagine that you have a <p> element just underneath your Silverlight content region (and your Silverlight content region doesn't fill the entire browser window). You want to manipulate the paragraph with your Silverlight application, so you assign it a unique ID like this:

```
<p id="paragraph">...</p>
```

You can retrieve an HtmlElement object that represents this paragraph in any Silverlight event handler. The following code retrieves the paragraph and changes the text inside:

```
Dim element As HtmlElement = HtmlPage.Document.GetElementById("paragraph")
element.SetProperty("innerHTML", _
  "This HTML paragraph has been updated by Silverlight.")
```

This code works by calling the HtmlElement.SetProperty() method and setting the innerHTML property. Long-time JavaScript developers will recognize innerHTML as one of the fundamental ingredients in the DOM.

■ **Note** When you use methods like SetProperty() and SetStyleAttribute(), you leave the predictable Silverlight environment and enter the quirky world of the browser. As a result, cross-platform considerations may come into play. For example, if you use the innerText property (which is similar to innerHTML but performs automatic HTML escaping to ensure that special characters aren't interpreted as tags), you'll find that your code no longer works in Firefox, because Firefox doesn't support innerText.

Figure 14-3 shows a test page that demonstrates this code. At the top of the page is a Silverlight content region with a single button. When the button is clicked, the text is changed in the HTML element underneath (which is wrapped in a solid border to make it easy to spot).

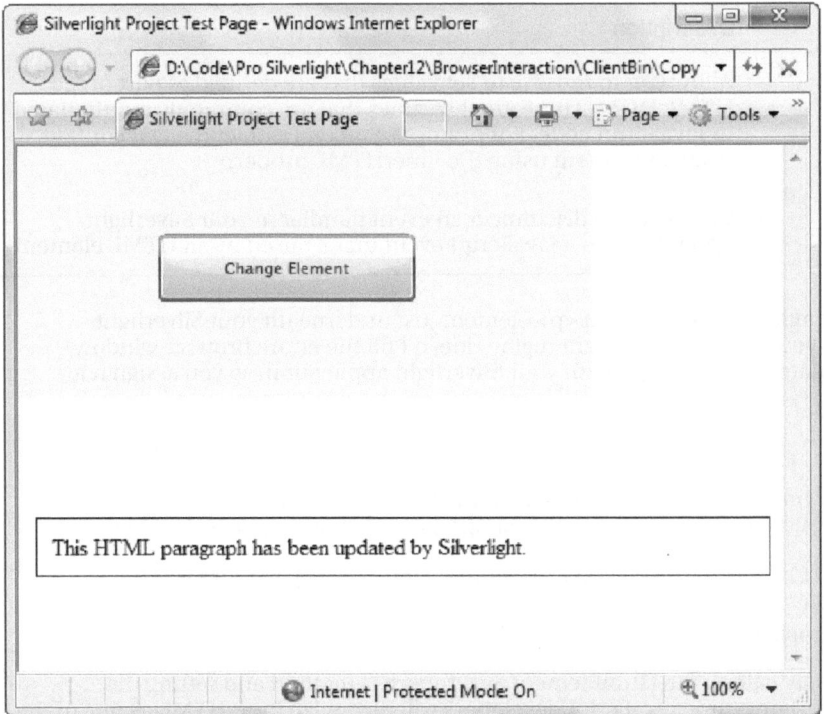

Figure 14-3. Changing HTML elements with Silverlight code

You'll notice that the transition between Silverlight and the HTML DOM isn't perfect. Silverlight doesn't include a full HTML DOM, just a lightweight version that standardizes on a basic HtmlElement class. To manipulate this element in a meaningful way, you often need to set an HTML DOM property (such as innerHTML in the previous example) using the SetProperty() method and supply the name of the property as a string. If you plan to do a lot of work with specific HTML elements, you may want to wrap them in higher-level custom classes (for example, by creating a custom Paragraph class) and replace their DOM properties or CSS style properties with strongly typed properties. Many developers use this approach to prevent minor typographic errors in property names that won't be caught at compile time.

ESCAPING SPECIAL CHARACTERS

When you set the innerHTML property, your text is interpreted as raw HTML. That means you're free to use nested elements, like this:

```
element.SetProperty("innerHTML", "This <b>word</b> is bold.")
```

If you want to use angle brackets that would otherwise be interpreted as special characters, you need to replace them with the < and > character entities, as shown here:

```
element.SetProperty("innerHTML", "To get bold text use the &lt;b&gt; element.")
```

If you have a string with many characters that need to be escaped or you don't want reduce the readability of your code with character entities, you can use the shared HttpUtility.HtmlEncode() method to do the work:

```
element.SetProperty("innerHTML", _
  HttpUtility.HtmlEncode("My favorite elements are <b>, <i>, <u>, and <p>."))
```

If you want to add extra spaces (rather than allow them to be collapsed to a single space character), you need to use the   character entity for a nonbreaking space.

Inserting and Removing Elements

The previous example modified an existing HTML element. It's just as easy to add elements to or remove them from an HTML page, using three methods: HtmlDocument.CreateElement(), HtmlElement.AppendChild(), and HtmlElement.RemoveChild().

For example, the following code assumes that the paragraph doesn't exist in the text page and creates it:

```
Dim element As HtmlElement = HtmlPage.Document.CreateElement("p")
element.Id = "paragraph"
element.SetProperty("innerHTML", _
  "This is a new element. Click to change its background color.")

HtmlPage.Document.Body.AppendChild(element)
```

In this example, the element is inserted as the last child of the <body> element, which means it's placed at the end of the document. If you have a place where you want to insert dynamic Silverlight content, it's easiest to define an empty <div> container with a unique ID. You can then retrieve the HtmlElement for that <div> and use AppendChild() to insert your new content.

■ **Note** You can execute this code more than once to add multiple paragraphs to the end of the HTML document. However, as it currently stands, each paragraph will be given the same ID, which isn't strictly correct. If you use the GetElementById() method on a document like this, you get only the first matching element.

Ordinarily, the AppendChild() method places the new element at the end of the collection of nested children. But it's possible to position an element more precisely by using an overloaded version of AppendChild() that accepts another HtmlElement object to act as a reference. When you use this approach, the element is inserted just *before* the referenced element:

```
' Get a reference to the first element in the <body>.
Dim referenceElement As HtmlElement = HtmlPage.Document.Body.Children(0)
```

```
' Make the new element the very first child in the <body> element,
' before all other nested elements.
HtmlPage.Document.Body.AppendChild(element, referenceElement)
```

Incidentally, it's even easier to remove an element. The only trick is that you need to use the RemoveChild() method of the *parent*, not the element you want to remove.

Here's the code that removes the paragraph element if it exists:

```
Dim element As HtmlElement = HtmlPage.Document.GetElementById("paragraph")
If element IsNot Nothing Then
  element.Parent.RemoveChild(element)
End If
```

Changing Style Properties

Setting style attributes is just as easy as setting DOM properties. You have essentially three options.

First, you can set the element to use an existing style class. To do this, you set the HtmlElement.CssClass property:

```
element.CssClass = "highlightedParagraph"
```

For this to work, the named style must be defined in the current HTML document or in a linked style sheet. Here's an example that defines the highlightedParagraph style in the <head> of the HTML page:

```
<html xmlns="http://www.w3.org/1999/xhtml">
  <head>
    <style type="text/css">
        .highlightedParagraph
        {
            color: White;
            border: solid 1px black;
            background-color: Lime;
        }
        ...
    </style>
    ...
  </head>
  <body>...</body>
</html>
```

This approach requires the least code and keeps the formatting details in your HTML markup. However, it's an all-or-nothing approach—if you want to fine-tune individual style properties, you must follow up with a different approach.

Another option is to set the element's style all at once. To do this, you use the HtmlElement.SetAttribute() method and set the style property. Here's an example:

```
element.SetAttribute("style", _
  "color: White; border: solid 1px black; background-color: Lime;")
```

But a neater approach is to set the style properties separately using the SetStyleAttribute() method several times:

```
element.SetStyleAttribute("color", "White")
element.SetStyleAttribute("border", "solid 1px black")
element.SetStyleAttribute("background", "Lime")
```

You can use the SetStyleAttribute() at any point to change a single style property, regardless of how you set the style initially (or even if you haven't set any other style properties).

■ **Tip** For a review of the CSS properties you can use to configure elements, refer to www.w3schools.com/css.

Handling JavaScript Events

Not only can you find, examine, and change HTML elements, you can also handle their events. Once again, you need to know the name of the HTML DOM event. In other words, you need to have your JavaScript skills handy in order to make the leap between Silverlight and HTML. Table 14-4 summarizes the most commonly used events.

Table 14-4. Common HTML DOM Events

Event	Description
onchange	Occurs when the user changes the value in an input control. In text controls, this event fires after the user changes focus to another control.
onclick	Occurs when the user clicks a control.
onmouseover	Occurs when the user moves the mouse pointer over a control.
onmouseout	Occurs when the user moves the mouse pointer away from a control.
onkeydown	Occurs when the user presses a key.
onkeyup	Occurs when the user releases a pressed key.
onselect	Occurs when the user selects a portion of text in an input control.
onfocus	Occurs when a control receives focus.
onblur	Occurs when focus leaves a control.
onabort	Occurs when the user cancels an image download.
onerror	Occurs when an image can't be downloaded (probably because of an incorrect URL).

Event	Description
onload	Occurs when a new page finishes downloading.
onunload	Occurs when a page is unloaded. (This typically occurs after a new URL has been entered or a link has been clicked. It fires just before the new page is downloaded.)

To attach your event handler, you use the HtmlElement.AttachEvent() method. You can call this method at any point and use it with existing or newly created elements. Here's an example that watches for the onclick event in the paragraph:

```
element.AttachEvent("onclick", AddressOf paragraph_Click)
```

■ **Tip** You can use HtmlElement.AttachEvent() to handle the events raised by any HTML element. You can also use HtmlWindow.AttachEvent() to deal with events raised by the browser window (the DOM window object) and HtmlDocument.AttachEvent() to handle the events raised by the top-level document (the DOM document object).

The event handler receives an HtmlEventArgs object that provides a fair bit of additional information. For mouse events, you can check the exact coordinates of the mouse (relative to the element that raised the event) and the state of different mouse buttons.

In this example, the event handler changes the paragraph's text and background color:

```
Private Sub paragraph_Click(ByVal sender As Object, ByVal e As HtmlEventArgs)
    Dim element As HtmlElement = CType(sender, HtmlElement)
    element.SetProperty("innerHTML", _
      "You clicked this HTML element, and Silverlight handled it.")
    element.SetStyleAttribute("background", "#00ff00")
End Sub
```

This technique achieves an impressive feat. Using Silverlight as an intermediary, you can script an HTML page with client-side VB code, instead of using the JavaScript that would normally be required.

Figure 14-4 shows this code in action.

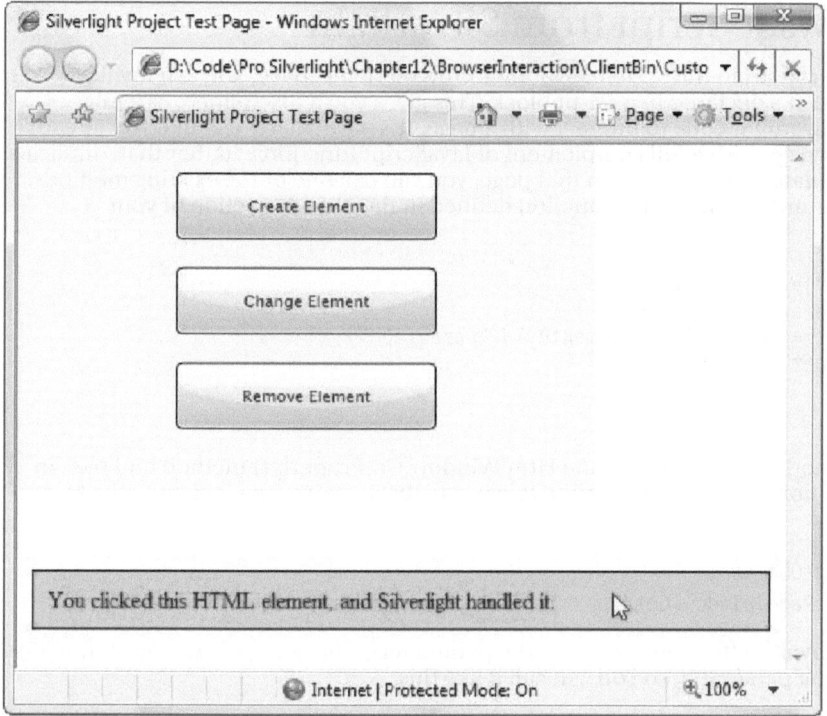

Figure 14-4. Silverlight and HTML interaction

Code Interaction

So far, you've seen how a Silverlight application can reach into the browser to perform navigation and manipulate HTML elements. The one weakness of this approach is that it creates tightly bound code—in other words, a Silverlight application that has hard-coded assumptions about the HTML elements on the current page and their unique IDs. Change these details in the HTML page, and the Silverlight code for interacting with them won't work anymore.

One alternative that addresses this issue is to allow interaction between *code*, not elements. For example, your Silverlight application can update the content of the HTML page by calling a JavaScript method that's in the page. Essentially, the JavaScript code creates an extra layer of flexibility in between the Silverlight code and HTML content. This way, if the HTML elements on the page are ever changed, the JavaScript method can be updated to match at the same time, and the Silverlight application won't need to be recompiled. The same interaction can work in the reverse direction—for example, you can create JavaScript code that calls a Silverlight method that's written in managed VB code. In the following sections, you'll see examples of both techniques.

Calling Browser Script from Silverlight

Using the Silverlight classes in the System.Windows.Browser namespace, you can invoke a JavaScript function that's declared in a script block. This gives you a disciplined, carefully controlled way for Silverlight code to interact with a page. It's particularly useful if you already have a self-sufficient page with a full complement of JavaScript functions. Rather than duplicate the code that manipulates the elements in that page, you can call one of the existing methods.

For example, assume you have this function defined in the <head> section of your HTML page:

```
<script type="text/javascript">
    function changeParagraph(newText) {
        var element = document.getElementById("paragraph");
        element.innerHTML = newText;
    }
</script>
```

To call this method, you need to use the HtmlWindow.GetProperty() method and pass in the name of the function. You receive a ScriptObject, which you can execute at any time by calling InvokeSelf().

```
Dim script As ScriptObject
script = CType(HtmlPage.Window.GetProperty("changeParagraph"), ScriptObject)
```

When you call InvokeSelf(), you pass in all the parameters. The changeParagraph() function requires a single string paragraph, so you can call it like this:

```
script.InvokeSelf("Changed through JavaScript.")
```

Calling Silverlight Methods from the Browser

Interestingly, Silverlight also has the complementary ability to let JavaScript code call a method written in managed code. This process is a bit more involved. To make it work, you need to take the following steps:

1. Create a public method in your Silverlight code that exposes the information or functionality you want the web page to use. You can place the method in your page class or in a separate class. You'll need to stick to simple data types, such as strings, Boolean values, and numbers, unless you want to go through the additional work of serializing your objects to a simpler form.

2. Add the ScriptableMember attribute to the declaration of the method that you want to call from JavaScript.

3. Add the ScriptableType attribute to the declaration of the class that includes the scriptable method.

4. To expose your Silverlight method to JavaScript, call the HtmlPage.RegisterScriptableObject() method.

Provided you take all these steps, your JavaScript code will be able to call your Silverlight method through the <object> element that represents the Silverlight content region. However, to make this task easier, it's important to give the <object> element a unique ID. By default, Visual Studio creates a test page that assigns a name to the <div> element that contains the <object> element (silverlightControlHost), but it doesn't give a name to the <object> element inside. Before continuing, you should create a test page that adds this detail, as shown here:

```
<div id="silverlightControlHost">
  <object data="data:application/x-silverlight,"
    type="application/x-silverlight-2-b1" width="400" height="300"
    id="silverlightControl">
    ...
  </object>
  <iframe style="visibility:hidden;height:0;width:0;border:0px"></iframe>
</div>
```

■ **Note** Remember, you can't modify the test page in a stand-alone Silverlight application, because it will be replaced when you rebuild your project. Instead, you need to create a new test page as described in Chapter 1. If you're using a solution that includes an ASP.NET test website, you can change the HTML test page directly. If you're using the server-side .aspx test page, you can change the ID of the server-side Silverlight control, which will be used when creating the client-side Silverlight control.

After you've named the Silverlight control, you're ready to create the scriptable Silverlight method. Consider the example shown in Figure 14-5. Here, a Silverlight region (the area with the gradient background) includes a single text block (left). Underneath is an HTML paragraph. When the user clicks the paragraph, a JavaScript event handler springs into action and calls a method in the Silverlight application that updates the text block (right).

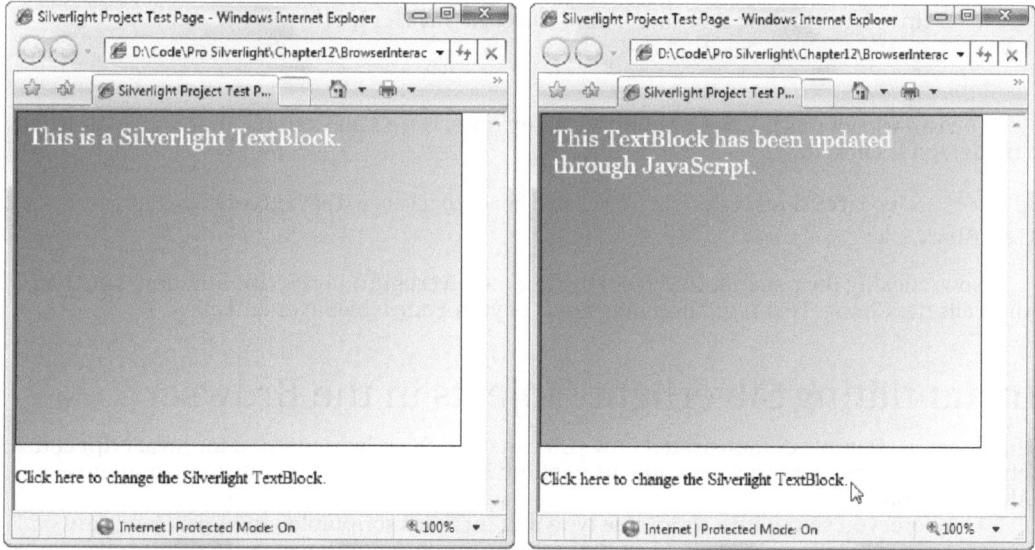

Figure 14-5. Calling Silverlight code from JavaScript

To create this example, you need the custom page class shown here. It includes a single scriptable method, which is registered when the page is first created:

```
<ScriptableType()> _
Public Partial Class ScriptableSilverlight
    Inherits UserControl

    Public Sub New()
        InitializeComponent()

        HtmlPage.RegisterScriptableObject("Page", Me)
    End Sub

    <ScriptableMember()> _
    Public Sub ChangeText(ByVal newText As String)
        lbl.Text = newText
    End Sub
End Class
```

When registering a scriptable type, you need to specify a JavaScript object name and pass a reference to the appropriate object. Here, an instance of the ScriptableSilverlight class is registered with the name Page. This tells Silverlight to create a property named Page in the Silverlight control on the JavaScript page. Thus, to call this method, the JavaScript code needs to use the find the Silverlight control, get its content, and then call its Page.ChangeText() method.

Here's an example of a function that does exactly that:

```
<script type="text/javascript">
  function updateSilverlightText()
  {
      var control = document.getElementById("silverlightControl");
      control.content.Page.ChangeText(
        "This TextBlock has been updated through JavaScript.");
  }
</script>
```

You can trigger this JavaScript method at any time. Here's an example that fires it off when a paragraph is clicked:

```
<p onclick="updateSilverlightText()">Click here to change the Silverlight
 TextBlock.</p>
```

Now, clicking the paragraph triggers the updateSilverlight() JavaScript function, which in turn calls the ChangeText () method that's part of your ScriptableSilverlight class.

Instantiating Silverlight Objects in the Browser

The previous example demonstrated how you can call a Silverlight method for JavaScript code. Silverlight has one more trick for code interaction: it allows JavaScript code to instantiate a Silverlight object.

As before, you start with a scriptable type that includes scriptable methods. Here's an example of a very basic Silverlight class that returns random numbers:

```
<ScriptableType()> _
Public Class RandomNumbers
    Private random As New Random()

    <ScriptableMember()> _
    Public Function GetRandomNumberInRange(ByVal fromValue As Integer, _
      ByVal toValue As Integer) As Integer
        Return random.Next(fromValue, toValue+1)
    End Function
End Class
```

As with the previous example, you need to register this class to make it available to JavaScript code. However, instead of using the RegisterScriptableObject() method, you use the RegisterCreateableType() method, as shown here:

```
HtmlPage.RegisterCreateableType("RandomNumbers", GetType(RandomNumbers))
```

To create an instance of a registered type, you need to find the Silverlight control and call its content.services.createObject() method. Here's an example with a JavaScript function that displays a random number from 1 to 6 using an instance of the Silverlight RandomNumbers class:

```
<script type="text/javascript">
  function getRandom1To6()
  {
      var control = document.getElementById("silverlightControl");
      var random = control.content.services.createObject("RandomNumbers");
      alert("Your number is: " + random.GetRandomNumberInRange(1, 6));
  }
</script>
```

The final detail is an HTML element that calls getRandom1To6():

```
<p onclick="getRandom1To6()">Click here to get a random number from 1 to 6.</p>
```

Figure 14-6 shows this code in action.

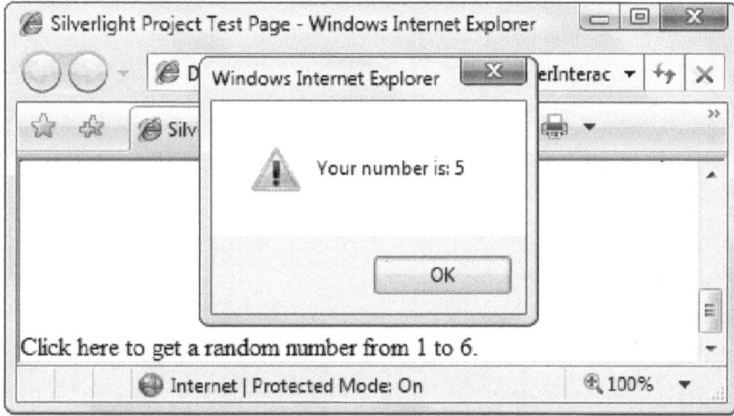

Figure 14-6. Creating a Silverlight object from JavaScript

Combining Silverlight and HTML Content

In Chapter 9, you learned how to create a windowless Silverlight content region. You can then use a transparent background to allow your Silverlight elements to "sit" directly on your HTML page. You can even use partial transparency to let the HTML content show through from underneath your Silverlight content.

This visual integration comes in handy when you use Silverlight code integration. For example, many developers have created custom-skinned media players using Silverlight's standard video window in conjunction with JavaScript-powered HTML elements. These controls can control playback by calling the scriptable methods in your Silverlight application.

When you combine HTML elements and Silverlight elements in the same visual space, it can take a bit of work to get the right layout. Usually, the trick is to fiddle around with CSS styles. For example, to constrain Silverlight content to a specific region of your page, you can place it in a <div> container. That <div> can even be placed with absolute coordinates. You can use other <div> containers to arrange blocks of HTML content alongside the Silverlight content. (You saw an example of this technique in Chapter 9, where a windowless Silverlight control was placed into a single column in a multicolumn layout.)

Occasionally, you'll want more layout control. For example, you may need to place or size your Silverlight control based on the current dimensions of the browser window or the location of other HTML elements. In the following sections, you'll see two examples that use Silverlight's HTML interoperability to place the Silverlight control dynamically.

Sizing the Silverlight Control to Fit Its Content

As you learned in Chapter 1, the default test page makes a Silverlight content region that fills the entire browser window. You can change this sizing, but you're still forced to assign an explicit size to your Silverlight control. If you don't, your Silverlight content is arranged according to the size of the page, but the page is truncated to fit a standard 200 by 200 pixel region, as shown in Figure 14-7.

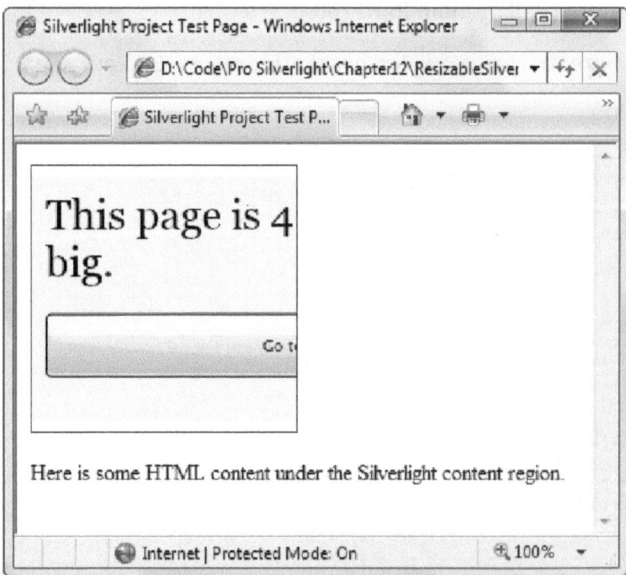

Figure 14-7. The default Silverlight control size

Sometimes, it would be nice to have a way to make the Silverlight content region size itself to match the dimensions of Silverlight page. Ordinarily, this doesn't happen. However, you can put it into practice with some simple code and Silverlight's HTML interoperability. It's easy. All you need to do is wait for your page to load, find the corresponding <object> element on the page, and resize it to match the dimensions of the page.

Here's an event handler that does the trick. It sizes the Silverlight control using the width and height style properties:

```
Private Sub Page_Loaded(ByVal sender As Object, ByVal e As RoutedEventArgs)
    Dim element As HtmlElement
    element = HtmlPage.Document.GetElementById("silverlightControl")
    element.SetStyleAttribute("width", Me.Width & "px")
    element.SetStyleAttribute("height", Me.Height & "px")
End Sub
```

You can use this code once, to size the Silverlight content region when the application is first loaded and the first page appears; or you can resize the content region to correspond to the content you're currently displaying by using the same code in several pages. Figure 14-8 shows the result of this approach, as the content changes inside a Silverlight application.

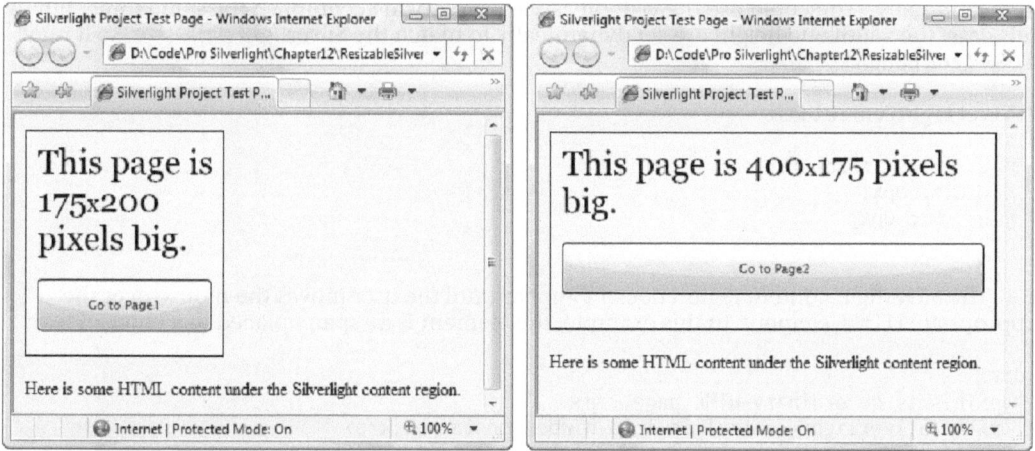

Figure 14-8. *Sizing the Silverlight control to fit the page*

Placing the Silverlight Control Next to an HTML Element

Much like you can resize the Silverlight control using style properties, you can also reposition it. The trick is to use a CSS style that specifies absolute positioning for the Silverlight control (or the <div> element that wraps it). You can then place the Silverlight control at the appropriate coordinates by setting the left and top style properties.

For example, in Figure 14-9, the goal is to pop up the Silverlight application in a floating window over of the page but next to a specific HTML element (which is highlighted in yellow). The specific position of the highlighted HTML element changes depending on the size of the browser window. Thus, to put the Silverlight content in the right place, you need to position it dynamically with code.

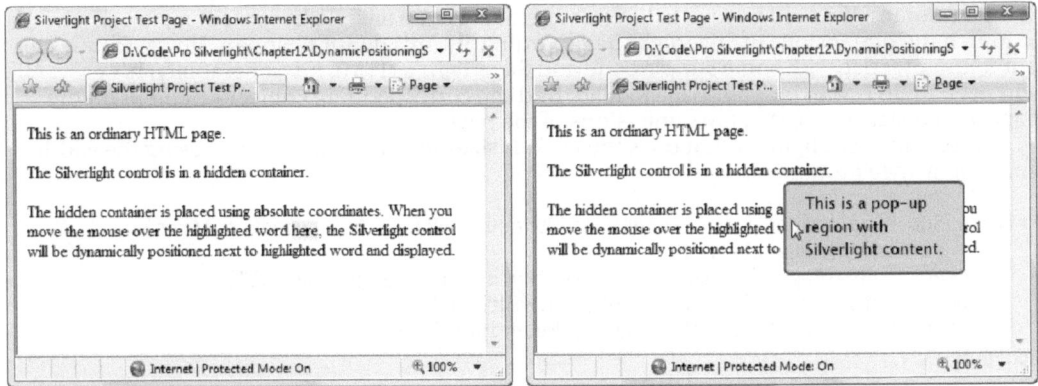

Figure 14-9. *Positioning Silverlight content next to an element*

To make this work, you must begin with a style that specifies absolute positioning for the Silverlight control. This style rule also sets the width and height to 0, so the control doesn't appear initially. (You could use the visibility style property to accomplish the same thing; but in this case, the width and height are set dynamically to match the Silverlight page size, so it may as well start at 0.)

```
#silverlightControlHost
{
    position: absolute;
    width: 0px;
    height: 0px;
}
```

The Silverlight content region doesn't appear until the user moves the mouse over the appropriate HTML element. In this example, the element is a <span> placed in a block of text:

```
<div>
  <p>This is an ordinary HTML page.</p>
  <p>The Silverlight control is in a hidden container.</p>
  <p>The hidden container is placed using absolute coordinates.
  When you move the mouse over the highlighted word <span id="target">here</span>,
  the Silverlight control will be dynamically positioned next to the highlighted
  word and displayed.
</div>
```

This span is given a yellow background through another style:

```
#target
{
    background-color: Yellow;
}
```

When the Silverlight page loads, the code finds the target <span> element and attaches an event handler to the JavaScript onmouseover event:

```
Private Sub Page_Loaded(ByVal sender As Object, ByVal e As RoutedEventArgs)
    Dim target As HtmlElement = HtmlPage.Document.GetElementById("target")
    target.AttachEvent("onmouseover", AddressOf element_MouseOver)
End Sub
```

When the user moves the mouse over the element, the event handler finds its current position using the HTML DOM properties offsetLeft and offsetTop. It then places the Silverlight container in a nearby location using the left and top style properties:

```
Private Sub element_MouseOver(ByVal sender As Object, ByVal e As HtmlEventArgs)
    ' Get the current position of the <span>.
    Dim target As HtmlElement = HtmlPage.Document.GetElementById("target")
    Dim targetLeft As Double = _
      Convert.ToDouble(target.GetProperty("offsetLeft")) - 20
    Dim targetTop As Double = _
      Convert.ToDouble(target.GetProperty("offsetTop")) - 20

    ' Get the Silverlight container, and position it.
    Dim silverlightControl As HtmlElement
    silverlightControl = HtmlPage.Document.GetElementById("silverlightControlHost")
    silverlightControl.SetStyleAttribute("left", targetLeft.ToString() & "px")
    silverlightControl.SetStyleAttribute("top", targetTop.ToString() & "px")

    ' Resize the Silverlight container to match the actual page size.
    ' This assumes the Silverlight user control has fixed values set for
    ' Width and Height (in this case, they're set in the XAML markup).
    silverlightControl.SetStyleAttribute("width", Me.Width & "px")
    silverlightControl.SetStyleAttribute("height", Me.Height & "px")
End Sub
```

The Silverlight content region is hidden using an ordinary Silverlight event handler that reacts to the MouseLeave event of the top-level user control:

```
Private Sub Page_MouseLeave(ByVal sender As Object, _
  ByVal e As MouseEventArgs)
    Dim silverlightControl As HtmlElement
    silverlightControl = HtmlPage.Document.GetElementById("silverlightControlHost")
    silverlightControl.SetStyleAttribute("width", "0px")
    silverlightControl.SetStyleAttribute("height", "0px")
End Sub
```

To give this example a bit more pizzazz, you can use an animation to fade the Silverlight content region into view. Here's an example that alternates the opacity of the top-level container from 0 to 1 over half a second:

```
<UserControl.Resources>
  <Storyboard x:Name="fadeUp">
    <DoubleAnimation Storyboard.TargetName="LayoutRoot"
     Storyboard.TargetProperty="Opacity"
     From="0" To="1" Duration="0:0:0.5" />
  </Storyboard>
</UserControl.Resources>
```

To use this animation, you need to add this statement to the end of the element_MouseOver() event handler:

```
fadeUp.Begin()
```

Securing HTML Interoperability

Silverlight's HTML interoperability features raise some new security considerations. This is particularly true if the Silverlight application and the hosting web page are developed by different parties. In this situation, there's a risk that malicious code in a Silverlight application could tamper with the HTML elsewhere on the page. Or, JavaScript code in the HTML page could call into the Silverlight application with malicious information, potentially tricking it into carrying out the wrong action.

If these issues are a concern, you can use a few options to clamp down on Silverlight's HTML interoperability. To prevent the Silverlight application from overstepping its bounds, you can set one of two parameters in the HTML entry page:

- *enableHtmlAccess*: When false, the Silverlight application won't be able to use most of the HTML interoperability features, including the Document, Window, Plugin, and BrowserInformation properties of the HtmlPage class. (However, you will still be allowed to call the HtmlPage.PopupWindow() method.) Ordinarily, enableHtmlAccess is set to true, and you must explicitly switch it off. However, if your Silverlight application is hosted on a different domain than your HTML entry page, enableHtmlAccess is set to false by default, and you can choose to explicitly switch it on to allow HTML interoperability.

- *allowHtmlPopupwindow*: When false, the Silverlight application can't use the HtmlPage.PopupWindow() method to show a pop-up window. By default, this parameter is true when the test page and Silverlight application are deployed together and false when the Silverlight application is hosted on a different domain.

Here's an example that sets enableHtmlAccess and allowHtmlPopupwindow:

```
<div id="silverlightControlHost">
  <object data="data:application/x-silverlight-2,"
    type="application/x-silverlight-2" width="100%" height="100%">
    <param name="enableHtmlAccess" value="false" />
    <param name="allowHtmlPopupwindow" value="false" />
    ...
  </object>
  <iframe style="visibility:hidden;height:0;width:0;border:0px"></iframe>
</div>
```

Silverlight also gives you the ability to protect your Silverlight application from JavaScript code. But first, it's important to remember that JavaScript code can't interact with your application unless you explicitly designate some classes and methods as *scriptable* (which you learned to do in the "Code Interaction" section of this chapter). Once you designate a method as scriptable, it will always be available to the HTML entry page, assuming both the HTML entry page and your Silverlight application are deployed together.

However, Silverlight's far stricter if the HTML entry page and Silverlight application are hosted on different domains. In this case, the HTML page will not be allowed to access to your scriptable classes and methods. Optionally, you can override this behavior and ensure that scriptable members are available to any HTML page by setting the ExternalCallersFromCrossDomain attribute in the application manifest file AppManifest.xml, as shown here:

```
<Deployment xmlns="http://schemas.microsoft.com/client/2007/deployment"
 xmlns:x="http://schemas.microsoft.com/winfx/2006/xaml"
 ExternalCallersFromsCrossDomain="ScriptableOnly" ...>
  <Deployment.Parts>
    ...
  </Deployment.Parts>
</Deployment>
```

Use this option with caution. It's entirely possible for an unknown individual to create an HTML page on another server that hosts your Silverlight application without your knowledge or consent. If you allow cross-domain access to your scriptable methods, anyone will be able to call these methods at any time, and with any information.

The Last Word

In this chapter, you saw how to build more advanced web pages by blending the boundaries between Silverlight and the containing HTML page. You learned how Silverlight can find and manipulate HTML elements directly and how it can call JavaScript code routines. You also learned how to use the reverse trick and let JavaScript call scriptable methods in your Silverlight application. Finally, you considered the security implications of breaking down the barriers between Silverlight code and the HTML world.

■ ■ ■

CHAPTER 15

ASP.NET Web Services

Some of the most interesting Silverlight applications have a hidden backbone of server-side code. They may call a web server to retrieve data from a database, perform authentication, store data in a central repository, submit a time-consuming task, or perform any number of other tasks that aren't possible with client-side code alone. The common ingredient in all these examples is that they are based on *web services*—libraries of server-side logic that any web-capable application can access.

In this chapter, you'll learn how to create ASP.NET web services and call them from a Silverlight application. You'll learn how to deal with different types of data, handle security, tap into ASP.NET services, monitor the client's network connection, and even build a two-way web service that calls your application when it has something to report.

Building Web Services for Silverlight

Without a doubt, the most effective way for a Silverlight application to tap into server-side code is through web services. The basic idea is simple: you include a web service with your ASP.NET website, and your Silverlight application calls the methods in that web service. Your web services can provide server-generated content that isn't available on the client (or would be too computationally expensive to calculate). Or, your web services can run queries and perform updates against a server-side database, as you'll see in Chapter 16. With a little extra work, it can even use ASP.NET services such as authentication, caching, and session state.

Silverlight applications can call traditional ASP.NET web services (.asmx services) as well as the WCF services, which are the newer standard. In the following sections, you'll learn how to build, call, and refine a WCF service. In Chapter 20, you'll consider how Silverlight applications can also call non-.NET web services, such as simpler REST services.

Creating a Web Service

To create a WCF service in Visual Studio, right-click your ASP.NET website in the Solution Explorer, and choose Add New Item. Choose the Silverlight-enabled WCF Service template, enter a file name, and click Add.

When you add a new WCF service, Visual Studio creates two files (see Figure 15-1):

- *The service endpoint*: The service endpoint has the extension .svc and is placed in your root website folder. For example, if you create a web service named TestService, you get a file named TestService.svc. When using the web service, the client requests a URL that points to the .svc file. But the .svc file doesn't contain any code—it includes one line of markup that tells ASP.NET where to find the corresponding web service code.
- *The service code*: The service code is placed in the App_Code folder of your website (if you're creating a projectless website) or in a separate code-behind file (if you're creating a web project). For example, if you create a web service named TestService, you get a code file named TestService.vb in a projectless website or TestService.svc.vb in a web project. Either way, the contents are the same: a code file with a class that implements the service interface and provides the actual code for your web service.

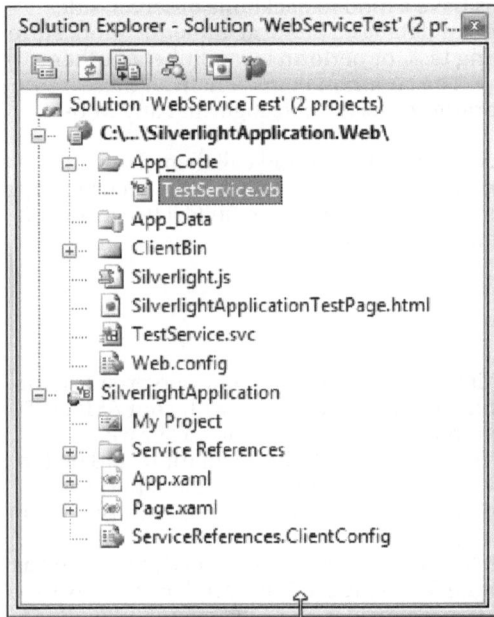

Figure 15-1. An ASP.NET website with a WCF service

The code file for your web service begins with two attributes. The ServiceContract attribute indicates that it defines a service contract—in other words, a set of methods that you plan to expose to remote callers as part of a service. The AspNetCompatibilityRequirements attribute indicates that it will have access to ASP.NET platform features such as session state:

```
<ServiceContract, AspNetCompatibilityRequirements( _
 RequirementsMode := AspNetCompatibilityRequirementsMode.Allowed)> _
Public Class TestService
    ...
End Class
```

To add a new web service method, you add a new method to the code file and make sure it's decorated with the OperationContract attribute. For example, if you want to add a method that returns the current time on the server, you can modify the interface like this:

```
<ServiceContract, AspNetCompatibilityRequirements( _
 RequirementsMode := AspNetCompatibilityRequirementsMode.Allowed)> _
Public Class TestService

    <OperationContract> _
    Public Function GetServerTime() As DateTime
        Return DateTime.Now
    End Function

End Class
```

Initially, a newly created service contains a single DoWork() web service method, which Visual Studio adds as an example. You're free to delete or customize this method.

■ **Note** When you add a WCF service, Visual Studio adds a significant amount of configuration information for it to the web.config file. For the most part, you don't need to pay much attention to these details. However, you will occasionally need to modify them to take advantage of a specialized feature (as with the duplexing service that you'll consider later in this chapter). It's also worth noting that, by default, the web.config file sets all WCF services to use binary encoding, instead of the ordinary text encoding that was the standard with Silverlight 2. This makes messages smaller, so they're quicker to transmit over the network and process on the server.

Adding a Service Reference

You consume a web service in a Silverlight application in much the same way that you consume one in a full-fledged.NET application. The first step is to create a proxy class by adding a Visual Studio service reference.

To add the service reference, follow these steps:

1. Right-click your Silverlight project in the Solution Explorer, and choose Add Service Reference. The Add Service Reference dialog box appears (see Figure 15-2).

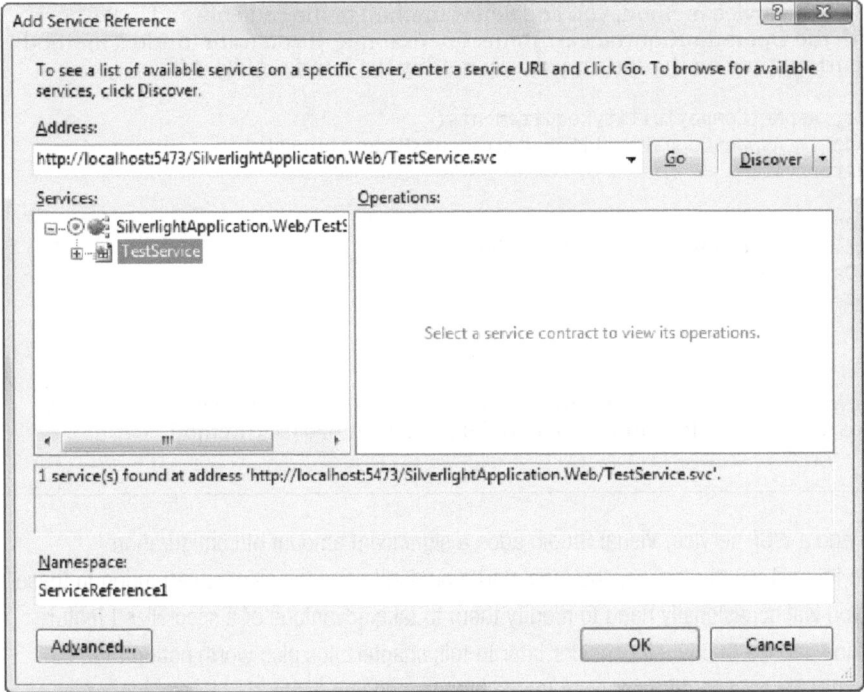

Figure 15-2. *Adding a service reference*

2. In the Address box, enter the URL that points to the web service, and click Go. However, you probably won't know the exact URL of your web service, because it incorporates the randomly chosen port used by Visual Studio's test web server (as in http://localhost:4198/ASPWebSite/TestService.svc). You could run your web application to find out, but an easier approach is to click the Discover button, which automatically finds all the web services that are in your current solution.

3. In the Namespace box, enter the *VB* namespace that Visual Studio should use for the automatically generated classes. This namespace is created inside your project namespace. So if your project is MyClient and you give the web service the namespace name WebServiceSite, the full namespace is MyClient.WebServiceSite.

4. Click OK. Visual Studio creates a proxy class that has the code for calling the web service. To see the file that contains this code, select the Silverlight project in the Solution Explorer, click the Show All Files button, expand the namespace node (which uses the name you picked in step 3), then expand the Service References node inside, then expand the Reference.svcmap node inside that, and open the Reference.vb file.

When you add a service reference, Visual Studio creates a *proxy class*—a class that you can interact with to call your web service. The proxy class is named after the original web service class with the word *Client* added at the end. For example, when adding a reference to the TestService shown earlier, Visual Studio creates a proxy class named TestServiceClient.

The proxy class contains methods that allow you to trigger the appropriate web service calls and all the events that allow you to receive the results. It takes care of the heavy lifting (creating the request message, sending it in an HTTP request, getting the response, and then notifying your code).

You can update your service reference at any time to taken into account web service changes (such as new methods or changes to the number of type of method parameters). To do so, recompile the web application, then right-click the service reference in the Solution Explorer, and finally choose Update Service Reference.

GENERATING PROXY CODE AT THE COMMAND LINE

Silverlight includes a command-line utility that does the same work as Visual Studio's service reference feature. This utility is named slsvcutil.exe (for Silverlight Service Model Proxy Generation Tool), and you can run it most easily from the Visual Studio Command Prompt. For example, the following command creates the proxy code for TestService example shown earlier (assuming the port number matches the port that the test web server is currently using):

```
slsvcutil http://localhost:4198/ASPWebSite/TestService.svc?WSDL
```

The ?WSDL that's appended to the service is an web service convention. It tells ASP.NET to provide the Web Service Description Language (WSDL) document that describes the web service. This document details the public interface of the web service (its methods and parameters) but doesn't expose any private details about its code or inner workings. The WSDL document has all the information Visual Studio or slsvcutil needs to generate the proxy code.

The most common reason for using slsvcutil is because you want to generate the proxy class code as part of an automated build process. To see a listing and description of all the parameters you can use, type in **slsvcutil** with no parameters.

Calling the Web Service

To use the proxy class, start by importing the namespace that you specified for the service reference in step 3. Assuming that you used the namespace MyWebServer and your project is named MySilverlightProject, you'd need this statement:

```
Imports MySilverlightProject.MyWebServer
```

In Silverlight, all web service calls must be asynchronous. That means you call a method to start the call (and send off the request). This method returns immediately. Your code can then carry on to perform other tasks, or the user can continue to interact with the application. When the response is received, the proxy class triggers a corresponding proxy class event, which is named in the form *MethodName*Completed. You must handle this event to process the results.

■ **Note** This two-part communication process means that it takes a bit more work to handle a web service call than to interact with an ordinary local object. However, it also ensures that developers create responsive Silverlight applications. After all, making an HTTP call to a web service can take as long as one minute (using the default timeout setting), so it's not safe to make the user wait. (And yes, Microsoft imposes this limitation to ensure that *your* code can't give *its* platform a bad name.)

Here's how to call the TestService.GetServerTime() method shown earlier:

```
' Create the proxy class.
Dim proxy As New TestServiceClient()

' Attach an event handler to the completed event.
AddHandler proxy.GetServerTimeCompleted, AddressOf GetServerTimeCompleted

' Start the web service call.
proxy.GetServerTimeAsync()
```

To get the results, you need to handle the completed event and examine the corresponding EventArgs object. When generating the proxy class, Visual Studio also creates a different EventArgs class for each method. The only difference is the Result property, which is typed to match the return value of the method. For example, the GetServerTime() method works in conjunction with a GetServerTimeCompletedEventArgs class that provides a DateTime object through its Result property.

When accessing the Result property for the first time, you need to use exception-handling code. That's because this is the point where an exception will be thrown if the web service call failed—for example, the server couldn't be found, the web service method returned an error, or the connection timed out. (As an alternative, you could check the Error property of the custom EventArgs object. For example, if GetServerTimeCompletedEventArgs.Error is null, no error occurred while processing the request, and it's safe to get the data from the Result property.)

Here's an event handler that reads the result (the current date and time on the server) and displays it in a TextBlock:

```
Private Sub GetServerTimeCompleted(ByVal sender As Object, _
  ByVal e As GetServerTimeCompletedEventArgs)
    Try
        lblTime.Text = e.Result.ToLongTimeString()
    Catch err As Exception
        lblTime.Text = "Error contacting web service"
    End Try
End Sub
```

■ **Tip** Even though web service calls are performed on a background thread, there's no need to worry about thread marshaling when the completed event fires. The proxy class ensures that the completed event fires on the main user-interface thread, allowing you to access the controls in your page without any problem.

By default, the proxy class waits for one minute before giving up if it doesn't receive a response. You can configure the timeout length by using code like this before you make the web service call:

```
proxy.InnerChannel.OperationTimeout = TimeSpan.FromSeconds(30)
```

WEB SERVICE EXCEPTIONS

You might think that when a web service method throws an exception, you can catch It in your Silverlight code. But life isn't that simple.

Although this chapter focuses on using web services for a single purpose—communicating between Silverlight and ASP.NET—thee standards that underpin web services are far broader and more general. They're designed to allow interaction between applications running on any web-enabled platform, and as such they don't incorporate any concepts that would tie them to a single, specific technology (like .NET exception classes).

There's another consideration: security. Web services can be consumed by any web-enabled application, and there's no way for your web service code to verify that it's your Silverlight application making the call. If web service methods returned specific, detailed exceptions, they would reveal far too much about their internal workings to potential attackers.

So, what happens when you call a web service method that goes wrong? First, the web server returns a generic *fault* message is returned to the client application. This message uses the HTTP status code 500, which signifies an internal error. Because of security restrictions in the browser, even if there were more information in the fault message, your Silverlight application wouldn't be allowed to access it because of the status code. Instead, Silverlight detects the fault message and immediately throws a CommunicationException with no useful information.

There is a way to work around this behavior and return more detailed exception information from the server, but because of the security concerns already mentioned, this feature is best for debugging. To get this error information, you need to take two somewhat tedious steps. First, you need to use a specialized WCF behavior that changes the HTTP status code of server-side fault messages from 500 to 200 before they're sent to the client. (The browser places no restriction on reading information from a response when the HTTP status code is 200.) Second, you need a mechanism to return the exception information. Silverlight includes a web service configuration option that, if enabled, inserts exception into the fault message. With these two details in place, you're ready to receive error information.

For more information, refer to http://msdn.microsoft.com/magazine/ee294456.aspx, which shows the WCF behavior that changes the status code, the configuration setting that inserts exception details into fault messages, and the client-side code that digs out the error information.

Configuring the Web Service URL

When you add a service reference, the automatically generated code includes the web service URL. As a result, you don't need to specify the URL when you create an instance of the proxy class.

But this raises a potential problem. All web service URLs are fully qualified—relative paths aren't allowed. If you're using the test web server in Visual Studio, that means you'll run into trouble if you try to run your application at a later point, when the test web server has chosen a different port number. Similarly, you'll need to update the URL when you deploy your final application to a production web server.

You can solve this problem by updating the service reference (and thereby regenerating all the proxy code), but there are two easier options.

Your first option is to configure Visual Studio to always use a specific port when running its test web server with your web application. This works only if you've created your web application as a web project (not a projectless website). In this case, you can configure the test web server for your project by double-clicking the My Project item in the Solution Explorer. Choose the Web tab. Then, in the Servers section, select "Specific port" and enter the port number you'd like to use. (You may as well choose the port number that the test server is already using for this session.) In the settings shown in Figure 15-3, that port number is 54752.

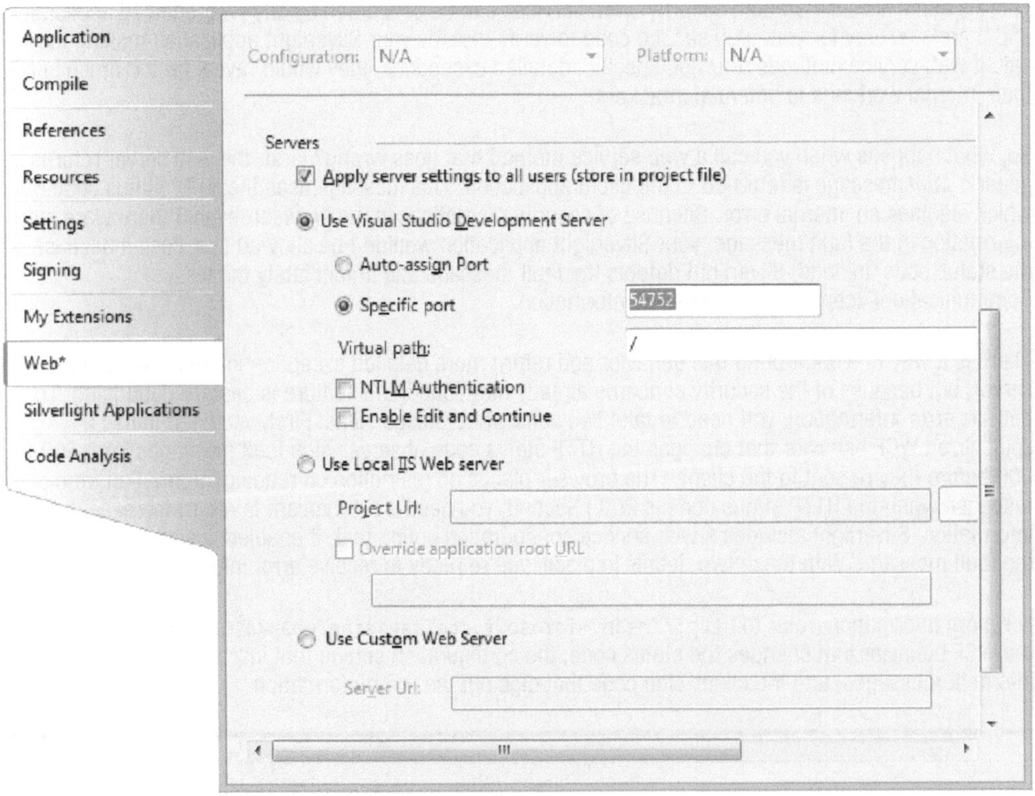

Figure 15-3. Setting the port for the test web server

Now you can modify the code that creates your proxy class. Instead of simply using this, which assumes the service is at the same port that it occupied when you added the reference:

```
Dim proxy As New TestServiceClient()
```

You can explicitly set the port with the EndpointAddress class:

```
' Create the proxy class.
Dim proxy As New TestServiceClient()

' Use the port that's hard-coded in the project properties.
Dim address As New EndpointAddress( _
  "http://localhost:54752/ASPWebSite/TestService.svc")

' Apply the new URI.
proxy.Endpoint.Address = address
```

Your second option is to change the address dynamically in your code so that it's synchronized with the port number that the test web server is currently using. To do so, you simply need to grab the URL of the Silverlight page and find its port number (because the Silverlight page is hosted on the same web server as the web service). Here's the code that does the trick:

```
' Create a new URL for the TestService.svc service using the current port number.
Dim address As New EndpointAddress( _
  "http://localhost:" & HtmlPage.Document.DocumentUri.Port & _
  "/SilverlightApplication.Web/TestService.svc")

' Use the new address with the proxy object.
Dim proxy As New TestServiceClient()
proxy.Endpoint.Address = address
```

You can use similar code to create a URL based on the current Silverlight page so that the web service continues to work no matter where you deploy it, as long as you keep the web service and Silverlight application together in the same web folder.

Using a Busy Indicator

Depending on exactly what your web method does, it may take a noticeable amount of time. If this delay isn't handled carefully, it can frustrate or confuse the user. For example, the user might assume there's a problem with the application, attempt to repeat the same action, or even restart the application.

One way to deal with slow services is to add some sort of indicator that informs the user that a web service call is underway. Although you can use any Silverlight element as an indicator (and even use the animation techniques you picked up in Chapter 10), the easiest solution is a dedicated control called the BusyIndicator, which is included in the Silverlight Toolkit (http://silverlight.codeplex.com).

The BusyIndicator has two display states: its ordinary state and its busy state. In its ordinary state, the BusyIndicator shows the content (if any) that's set in the Content property. In its busy state, the BusyIndicator fuses together an endlessly pulsating progress bar with whatever content you supply in the BusyContent property. Here's an example:

```
<toolkit:BusyIndicator x:Name="busy" BusyContent="Contacting Service..." />
```

To put the BusyIndicator into its busy state, simply set the IsBusy property. Here's the updated web service code:

```
Private Sub cmdCallSlowService_Click(ByVal sender As Object, _
  ByVal e As RoutedEventArgs)
    Dim proxy As New TestServiceClient()

    cmdCallSlowService.IsEnabled = False
    lblTime.Text = ""
    busy.IsBusy = True

    AddHandler proxy.GetServerTimeCompleted, AddressOf GetServerTimeCompleted
    proxy.GetServerTimeAsync()
End Sub

Private Sub GetServerTimeCompleted(ByVal sender As Object, _
  ByVal e As GetServerTimeCompletedEventArgs)
    Try
        lblTime.Text = e.Result.ToLongTimeString()
    Catch err As Exception
        lblTime.Text = "Error contacting service."
    Finally
        busy.IsBusy = False
        cmdCallSlowService.IsEnabled = True
    End Try
End Sub
```

And Figure 15-4 shows the result, while the web service call is underway.

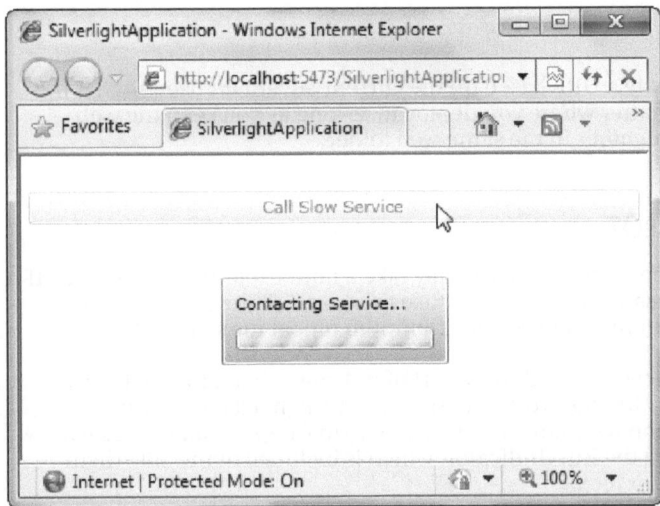

Figure 15-4. Showing the status of an in-progress call

Web Service Data Types

When you create a web service for use with Silverlight, you're limited to the core set of .NET data types. This includes strings, Boolean values, bytes, numeric data types, enumeration values, and DateTime objects. You can also use arrays, collections of any supported type, and—more interestingly—custom classes that are build with these same data types.

Custom Data Classes

To build a custom class that works with a web service, you need to meet a few basic requirements:

- Your class declaration must be decorated with the DataContract attribute.
- Your class must consist of public, writeable properties. Each property must use one of the previously discussed serializable data types, or another custom class.
- Each property must be decorated with the DataMember attribute to indicate that it should be serialized.
- Your class must include a zero-argument default constructor.
- Your class can include code, but it won't be accessible on the client. Instead, the client will get a stripped-down version of the class with no code.

Here's an example of a custom class that satisfies all these conditions:

```
<DataContract> _
Public Class Customer
    Private _firstName As String
    Private _lastName As String

    <DataMember> _
    Public Property FirstName() As String
        Get
            Return _firstName
        End Get
        Set(ByVal value As String)
            _firstName = value
        End Set
    End Property

    <DataMember> _
    Public Property LastName() As String
        Get
            Return _lastName
        End Get
        Set(ByVal value As String)
            _lastName = value
        End Set
    End Property
End Class
```

Now you can create a web service method that uses this class.

```
<AspNetCompatibilityRequirements( _
 RequirementsMode := AspNetCompatibilityRequirementsMode.Allowed)> _
Public Class TestService

    <OperationContract> _
    Public Function GetCustomer(ByVal customerID As Integer) As Customer
        Dim newCustomer As New Customer()
        ' (Look up and configure the Customer object here.)
        Return newCustomer
    End Function

End Class
```

The web method can use this class for a parameter or a return value (as in this example). Either way, when you add the service reference in your Silverlight project, Visual Studio generates a similar Customer class definition in your Silverlight application, alongside the proxy class. You can then interact with the Customer objects that the server sends back, or you can create Customer objects in the client and send them to the server.

```
Private Sub GetCustomerCompleted(ByVal sender As Object, _
  ByVal e As GetCustomerCompletedEventArgs)
    Try
        Dim newCustomer As Customer = e.Result
        ' (You can now display the customer information in the user interface).
    Catch err As Exception
        lblTime.Text = "Error contacting web service"
    End Try
End Sub
```

You'll see a much more in-depth example of a web service that uses custom classes later in this book. In Chapter 16, you'll build a web service that uses custom classes and collections to return data from a database.

Web Service Type Sharing

As you've seen so far, a layer of cross-platform standards separates your Silverlight client from the ASP.NET web services it uses. One consequence of this separation is that web services and clients can't share code. If a web service returns a data object like an instance of the Customer class shown in the previous section, the client gets a stripped-down version of that class with all the public data (as properties) and none of the code.

Usually, this design isn't a problem. As long as Visual Studio follows a few simple rules—for example, using property procedures instead of public fields and using ObservableCollection for any sort of collection of objects—the code-free data classes it generates will work fine with Silverlight data binding. However, there are some situations when you might want to work around this limitation, create a data class that includes code, and use that code on both the server and client sides of your solution. For example, you could use this technique to create a code-enriched version of the Customer class shown in the previous section and make that code accessible to both the web service and the Silverlight client.

To share a data class, you need to take two steps:

Share the code: You need to give a copy of the data class to both the web service and the Silverlight client. This is tricky, because the web service targets the full .NET Framework, whereas the client targets the scaled-down Silverlight libraries. Thus, even though they both can use the same data class *code*, they can't share a single data class *assembly*, because that assembly must be compiled for one platform or the other. The best solution is to create two class library assemblies—one for ASP.NET and one for Silverlight—and use Visual Studio linking to avoid duplicating your source files.

Identify the code: Once both projects have the data class, you still need a way to tell Visual Studio that they are one and the same. Otherwise, Visual Studio will still attempt to create a stripped-down client-side copy when you add the web reference. To identify your data class, you simply need to give it a unique XML namespace.
In the following sections, you'll tackle both of these steps, but in the reverse order.

■ **Note** Think carefully before you use web service type sharing. In most cases, it's best to avoid type sharing, because type sharing creates a tight dependency between your server-side code and your Silverlight client, which can complicate versioning, updating, and deployment. However, type sharing sometimes make sense when you're building smart data objects that have embedded details such as descriptive text and validation rules. In many cases, you aren't actually sharing code but the attributes that decorate it. You'll see an example that uses type sharing for this purpose in Chapter 17.

Identifying Your Data Classes

As you already know, the data classes you use in a WCF web service need to have the DataContract attribute. But the DataContract attribute doesn't just make your class usable in a web service. It also gives you the ability to uniquely identify your class by mapping it to an XML namespace of your choice. So far, the examples you've seen haven't used this ability, because it isn't required. But if you're deploying data-class code to the client, it's essential. That's because you must give the same XML namespace to both the Silverlight version and the ASP.NET version of each data class. Only then will Visual Studio understand that it represents the same entity.

Here's an example that maps the Customer class to the XML namespace http://www.prosetech.com/DataClasses/Customer:

```
<DataContract(Name := "Customer", _
 Namespace := "http://www.prosetech.com/DataClasses/Customer")> _
Public Class Customer
    Private _firstName As String
    Private _lastName As String

    <DataMember> _
    Public Property FirstName() As String
        Get
            Return _firstName
        End Get
        Set(ByVal value As String)
            _firstName = value
        End Set
    End Property

    <DataMember> _
    Public Property LastName() As String
        Get
```

```
            Return _lastName
        End Get
        Set(ByVal value As String)
            _lastName = value
        End Set
    End Property

    ' Ordinarily, this method would not be available on the client.
    Public Function GetFullName() As String
        Return firstName & " " & lastName
    End Function
End Class
```

Remember, XML namespaces don't need to point to web locations (even though they commonly use URIs). Instead, you can use a domain you control in your XML namespace to ensure that it's not inadvertently duplicated by another developer.

Sharing the Code

You may assume that after you've set up the DataContract attribute, you can copy the data-class code to your web service and Silverlight project. Unfortunately, life isn't this simple. Even though the DataContract attribute uniquely identifies the Customer class, Visual Studio still attempts to create new data classes when you create a reference to your web service (and it tries to regenerate them every time you refresh that reference). As it currently stands, that means your web service client code ends up using a stripped-down duplicate copy of the Customer class.

To fix this problem and get Visual Studio to use the same data classes in both projects, you need to create the correct project structure. At a bare minimum, you must place the client-side versions of the data classes in a separate assembly.

Here's the sequence of steps to follow:

1. Begin with a solution that includes your Silverlight project and the ASP.NET website with the data web service. (In the downloadable example for this chapter, that's a Silverlight application named TypeSharingClient and an ASP.NET website named TypeSharing.Web.)

■ **Note** This approach works equally well regardless of whether you create a web project or a projectless website in Visual Studio.

---

2. Add a new Silverlight class library application for the data classes. (In the downloadable example, it's called DataClasses.) This project needs to have a reference to the System.Runtime.Serialization.dll assembly in order to use the DataContract attribute.
3. Add a reference in your Silverlight project that points to the data-class project so the data classes are available in your application.
4. Add a new .NET class library for the server-side implementation of the data classes (for example, DataClasses.ServerSide). It will also need a reference to the

System.Runtime.Serialization.dll assembly, only now you're using the full .NET version
of that assembly. Figure 15-5 shows all the projects in the solution.

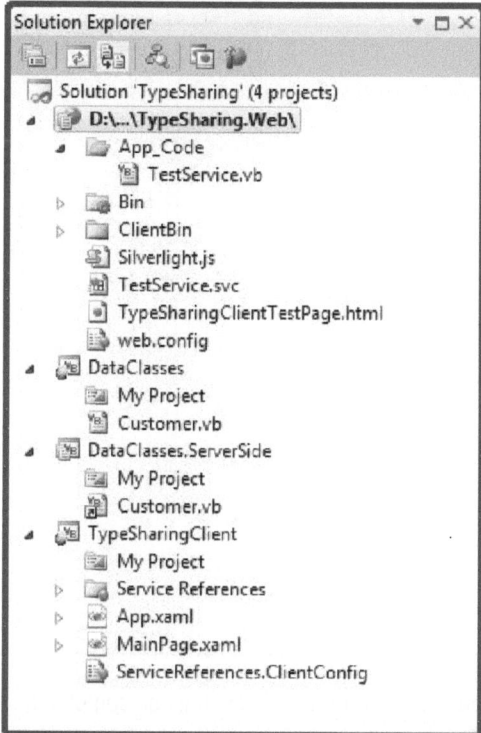

Figure 15-5. A solution that uses web service type sharing

5. Add a reference in your ASP.NET website (in this case, TypeSharing.Web) that points to
    the server-side class library (in this case, DataClasses.ServerSide).

6. The server-side class library needs the same code as the client-side class library. To
    accomplish this without duplication, you can use the Visual Studio *linking* feature. In
    your server-side class library, choose to add an existing item (right-click the project in
    the Solution Explorer, and choose Add ~TRA Existing Item). Browse to the code file for
    the Silverlight class library (for example, Customer.cs) and select it, but don't click Add.
    Instead, click the drop-down button to the right of the Add button, which pops open a
    small menu of addition options (see Figure 15-6), and choose Add As Link. This way,
    you'll have just one copy of the source code, which is shared between files. No matter
    which project you're editing, you'll update the same file, and your changes will be
    incorporated into both class libraries.

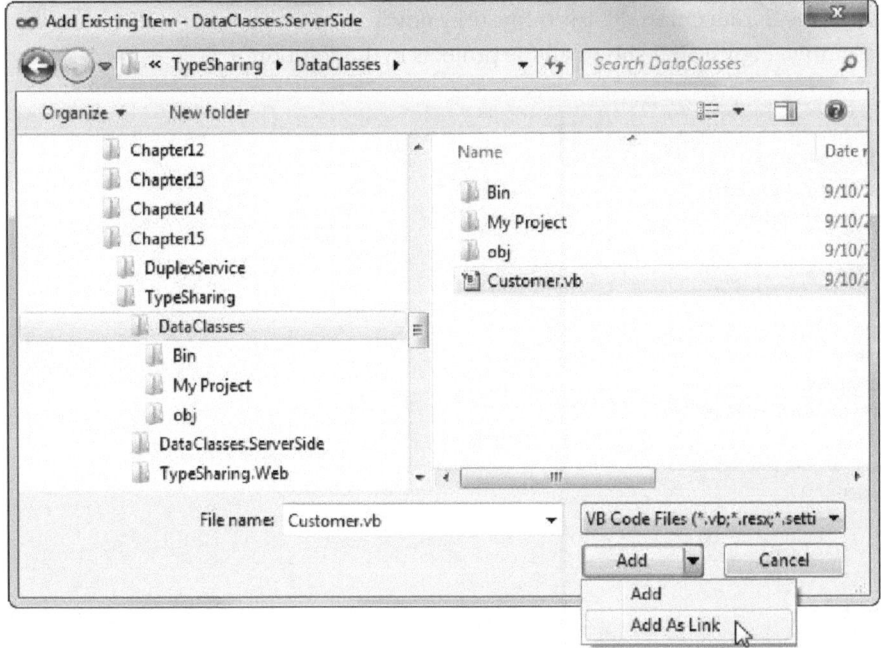

Figure 15-6. Linking to a source code file

■ **Note** Of course, it goes without saying that the code you write in your data class must be completely compatible with both the .NET *and* Silverlight platforms. Otherwise, an error will occur when you compile your assembly. If you need to write slightly different code for the Silverlight version of your data class, you may be forced to resort to the decidedly clumsy cut-and-paste approach.

---

7. Repeat step 6 if there are multiple source code files.

8. You've finished the type sharing setup. Now you can complete the usual steps—code your web service, compile your website, and add the service reference in your client.

Now, when you add a reference to the web service that uses the Customer class, Visual Studio won't generate a new copy. Instead, it will use the code-fortified copy from the Silverlight class library assembly. That means you can use the custom method you added to the Customer class, like this:

```
Private Sub GetCustomerCompleted(ByVal sender As Object, _
  ByVal e As GetCustomerCompletedEventArgs)
    Try
        Dim newCustomer As Customer = e.Result
        lblResut.Text = newCustomer.GetFullName()
    Catch err As Exception
```

```
        lblResult.Text = "Error contacting web service"
    End Try
End Sub
```

If this process seems convoluted—well, it is. But after you've set up a solution like this, you'll have no trouble creating more. If you're still in doubt, check out the downloadable code, which includes a data-class-sharing solution that's already set up.

PORTABLE ASSEMBLIES

Silverlight 4 introduced a new feature called *portable assemblies*. A portable assembly is a Silverlight class library assembly that you can share with both .NET applications and Silverlight applications, without being forced to recompile, cut-and-paste, link files, or resort to any other sort of clumsy workaround. At first glance this seems like the perfect solution for web service type sharing. Unfortunately, a portable assembly is only allowed to reference the following assemblies:

- Mscorlib.dll

- System.dll

- System.Core.dll

- System.ComponentModel.Composition.dll

- Microsoft.VisualBasic.dll

- Other portable assemblies that you've created

This means you can't create a portable assembly for a web service data class, because the data class needs to use nonportable assemblies such as System.Runtime.Serialization.dll (which provides the DataContract attribute).

If you still want to use portable assemblies for some other type of code sharing that doesn't involve web service data classes, it's fairly easy. Begin by creating a Silverlight class library, and make sure it uses only assemblies from the allowed list shown previously. Your Silverlight application can use this class library as normal. Your .NET application can use it also, but you must add a reference to the portable assembly's compiled .dll file. (You can't add a reference directly to the Silverlight class project.) As long as you follow these rules, you'll be able to share custom code routines without a worry.

More Advanced Web Services

You've now seen how to build and consume web services. In this section, you'll build on these basic skills with some more specialized techniques. First, you'll see how to give other websites access to your services. Next, you'll learn how to watch the current computer's network connection, so you know when it's safe to launch a web service call. And finally you'll see how you can use web services as a bridge to ASP.NET's server-side features, such as caching and authentication.

Cross-Domain Web Service Calls

Silverlight allows you to make web service calls to web services that are part of the same website with no restrictions. Additionally, Silverlight allows you to call web services on other web services *if* they explicitly allow it with a policy file.

In Chapter 20, you'll consider the implications this has when you're using third-party web services and downloading content on the Web. But now, it's worth understanding how you can configure your web service to allow cross-domain callers. To make this possible, you must create a file named clientaccesspolicy.xml and place it in the root of your website (for example, in the c:\inetpub\wwwroot directory of an IIS web server). The clientaccesspolicy.xml file indicates what domains are allowed to access your web service. Here's an example that allows any Silverlight application that's been downloaded from any web server to access your website:

```xml
<?xml version="1.0" encoding="utf-8"?>
<access-policy>
  <cross-domain-access>
    <policy>
      <allow-from>
        <domain uri="*"/>
      </allow-from>
      <grant-to>
        <resource path="/" include-subpaths="true"/>
      </grant-to>
    </policy>
  </cross-domain-access>
</access-policy>
```

When you take this step, third-party Silverlight applications can call your web services and make arbitrary HTTP requests (for example, download web pages). Ordinarily, neither task is allowed in a Silverlight application.

■ **Note** Desktop applications and server-side applications face no such restrictions—no matter what policy file you create, they can do everything an ordinary user can do, which means they can download any public content. Similarly, Silverlight applications that run with elevated trusted (a technique you'll consider in Chapter 21) can also make cross-domain web service requests, regardless of the existence of a policy file.

Alternatively, you can limit access to Silverlight applications that are running on web pages in specific domains. Here's an example that allows requests from Silverlight applications that are hosted at www.somecompany.com or www.someothercompany.com:

```xml
<?xml version="1.0" encoding="utf-8"?>
<access-policy>
  <cross-domain-access>
    <policy>
      <allow-from http-request-headers="*">
        <domain uri="http://www.somecompany.com" />
        <domain uri="http://www.someothercompany.com" />
      </allow-from>
      <grant-to>
```

```
        <resource path="/" include-subpaths="true"/>
      </grant-to>
    </policy>
  </cross-domain-access>
</access-policy>
```

You can use wildcards in the domain names to allow subdomains. For example, *.somecompany.com allows requests from mail.somecompany.com, admin.somecompany.com, and so on.

Furthermore, you can selectively allow access to part of your website. Here's an example that allows Silverlight applications to access the services folder in your root web domain, which is presumably where you'll place all your cross-domain web services:

```
<?xml version="1.0" encoding="utf-8"?>
<access-policy>
  <cross-domain-access>
    <policy>
      <allow-from>
        <domain uri="*"/>
      </allow-from>
      <grant-to>
        <resource path="/services/" include-subpaths="true"/>
      </grant-to>
    </policy>
  </cross-domain-access>
</access-policy>
```

■ **Note** Instead of using clientaccesspolicy.xml, you can create a crossdomain.xml file. This file has essentially the same purpose, but it uses a standard that was first developed for Flash applications. The only advantage to using it is if you want to give access to Silverlight and Flash applications in one step. Compared to crossdomain.xml, clientaccesspolicy.xml is slightly more sophisticated, because it lets you grant access to just a specific part of your website (both standards allow you to limit requests based on the caller's domain). For more information about crossdomain.xml, see Chapter 20.

Remember, if you make your web services publicly accessible, you must make sure they can't be easily abused. For example, you should never allow your web methods to return sensitive data or commit arbitrary changes. Even though you know your applications will use your web services properly, it's trivially easy for malicious users to create their own applications that don't. And even if your web service doesn't allow cross-domain access, it's a good idea to clamp down on your web methods as much as possible. Doing so prevents problems in a number of scenarios—for example, if a website configuration change inadvertently grants cross-domain access to your web services, if an attacker gain access to your website, or if an attacker fools your application into performing a damaging operation.

If you perform security checks in a cross-domain web service, remember that you can't trust cookies or any authentication information that's not part of the actual request message. That's because trusted users can visit malicious applications—and when they do, the malicious application gains access to their current cookies. To prevent issues like these, you can add checks and balances to your web service that look for improper usage—for example, users who access more data than they should in a short period of time or users who attempt to view or edit data that isn't directly related to them.

Monitoring the Network Connection

Chapter 21 describes how to create out-of-browser applications that can run outside the browser window, even when no network connection is available. Clearly, this raises the possibility that a user may run the application when the computer isn't online. In this case, attempts to call a web service are bound to fail.

Dealing with this problem is easy. As you've already seen, a failed web service call causes the completed event to fire. When you respond to this event and attempt to retrieve the result, an exception is thrown that notifies you of the problem. (If the network connection is present but the Internet comes and goes, the completed event may not fire until the call times out. On the other hand, if the computer is completely disconnected from the network, the completed event fires immediately. Either way, you need to catch the exception and either ignore the problem or inform the user.)

Exception-handling code gives your application a basic line of defense. However, if you have a client with intermittent connectivity, you may want to handle the issue more gracefully. For example, you may want to pay attention to the network status in your application and selectively disable certain features when they're not available, saving the user from potential confusion or frustration. This behavior is easy to implement using Silverlight's new network-monitoring support.

The network-monitoring feature consists of two extremely simple classes, both of which expose a single public member, and both of which are found in the System.Net.NetworkInformation namespace. First, you can use the GetIsNetworkAvailable() method of the NetworkInterface class to determine whether the user is online. Second, you can respond to the NetworkAddressChanged event of the NetworkChange class to determine when the network status (or IP address) changes. Usually, you'll work in that order—first use GetIsNetworkAvailable() to determine the network status and then handle NetworkAddressChanged to pick up any changes:

```
Public Sub New()
    InitializeComponent()

    ' Watch for network changes.
    AddHandler NetworkChange.NetworkAddressChanged, AddressOf NetworkChanged

    ' Set up the initial user interface
    CheckNetworkState()
End Sub

Private Sub NetworkChanged(ByVal sender As Object, ByVal e As EventArgs)
    ' Adjust the user interface to match the network state.
    CheckNetworkState()
End Sub

Private Sub CheckNetworkState()
    If NetworkInterface.GetIsNetworkAvailable() Then
        ' Currently online.
        cmdCallCachedService.IsEnabled = True
        cmdCallService.IsEnabled = True
    Else
        ' Currently offline.
        cmdCallCachedService.IsEnabled = False
        cmdCallService.IsEnabled = False
    End If
End Sub
```

It's important to remember that the network-monitoring feature was designed to help you build a more polished, responsive application. But it's no substitute for exception-handling code that catches network exceptions. Even if a network connection is present, there's no guarantee that it provides access to the Internet, that the requested website is online, and that the requested web service method will run without an error. For all these reasons, you need to treat your web service calls with caution.

Using ASP.NET Platform Services

Ordinarily, WCF services don't get access to ASP.NET platform features. Thus, even though ASP.NET is responsible for compiling your service and hosting it, your service can't use any of the following:

- Session state
- Data caching
- The authorization rules in the web.config file
- Provider-based features, such as authentication, membership, and profiles

In many cases, this makes sense, because WCF services are meant to be independent of the ASP.NET platform. It's dangerous to use ASP.NET-only features, because they limit your ability to move your service to other hosts, use other transport protocols, and so on. Although these considerations may not come into play with a Silverlight application, there's still a good philosophical basis for making your services as self-contained as possible.

Furthermore, some of the features don't make sense in a web service context. Currently, a number of workarounds are available to get session state to work with WCF services. However, the session-state feature fits awkwardly with the web service model, because the lifetime of the session isn't linked to the lifetime of the web service or proxy class. That means a session can unexpectedly time out between calls. Rather than introduce these headaches, it's better to store state information in a database.

But in some scenarios, ASP.NET features can legitimately save you a good deal of work. For example, you may want to build a service that uses in-memory caching if it's available. If it's not, the service can degrade gracefully and get its information from another source (like a database). But if the in-memory cache is working and has the information you need, it can save you the overhead of requerying it or re-creating it. Similarly, there's a case to be made for using some of the ASP.NET provider-based features to give you easy user-specific authentication, role-based security, and storage, without forcing you to reimplement a similar feature from scratch.

To access ASP.NET features in a web service, you use the shared Current property of the System.Web.HttpContext class. HttpContext represents the HTTP environment that hosts your service. It provides access to key ASP.NET objects through its properties, such as Session (the per-user session state), Application (the global application state), Cache (the data cache), Request (the HTTP request message, including HTTP headers, client browser details, cookies, the requested URL, and so on), User (the user making the request, if authenticated through ASP.NET), and so on. ASP.NET developers will be familiar with these details.

The following example uses HttpContext to get access to the data cache. It caches a collection of Product objects so the database doesn't have to be queried each time the web service method is called:

```
<OperationContract()> _
Public Function GetAllProducts() As Product()
    ' Check the cache.
    Dim context As HttpContext = HttpContext.Current
```

```
    If context.Cache("Products") IsNot Nothing Then
        ' Retrieve it from the cache
        Return CType(context.Cache("Products"), Product())
    Else
        ' Retrieve it from the database.
        Dim products As Product() = QueryProducts()

        ' Now store it in the cache for 10 minutes.
        context.Cache.Insert("Products", products, Nothing, _
          DateTime.Now.AddMinutes(10), TimeSpan.Zero)

        Return products
    End If
End Function

' This private method contains the database code.
Private Function QueryProducts() As Product()
    ...
End Function
```

The actual caching feature (and other ASP.NET features) is outside the scope of this book. However, this example shows how experienced ASP.NET developers can continue to use some of the features of ASP.NET when building a WCF service. To try an example of ASP.NET caching in a web service, check out the downloadable examples for this chapter.

■ **Note** For more information about ASP.NET platform services such as caching and authentication, refer to *Pro ASP.NET 4 in VB 2010* (Apress, 2010).

WCF RIA SERVICES

One of the hottest new developments in the Silverlight world is an add-on called WCF RIA Services. (RIA stands for rich Internet application.) Essentially, RIA Services is a model that breaks down the barriers between Silverlight and ASP.NET, making it easier for Silverlight to tap into server-side ASP.NET features through web services.

For example, RIA Services provides a framework for ASP.NET-backed authentication. Under this model, the user would supply some sort of login credentials (such as a user name and password), and the Silverlight application would pass these to the server by calling an RIA service. This service would then log the user in on the server and issue an ASP.NET authentication cookie, allowing the Silverlight application to make additional requests to other, secure web services. Similarly, RIA Services includes a tool for building data models, which can generate web methods for accessing and updating data based on the structure of your database tables.

RIA Services require a separate (but free download). They aren't covered in this book, although you can learn more about them from the book *Pro Business Applications with Silverlight 4* (Apress, 2010) or online at www.silverlight.net/getstarted/riaservices.

Duplex Services

Ordinarily, web services use a fairly straightforward and somewhat limiting form of interaction. The client (your Silverlight application) sends a request, waits for a response, and then processes it. This is a distinctly one-way type of communication—the client must initiate every conversation.

This model is no surprise, because it's based on the underlying HTTP protocol. Browsers request web resources, but websites can never initiate connections and transmit information to clients without first being asked. Although this model makes sense, it prevents you from building certain types of applications (such as chat servers) and implementing certain types of features (such as notification). Fortunately, there are several ways to work around these limitations in Silverlight:

- *Polling*: With polling, you create a client that connects to the server periodically and checks for new data. For example, if you want to create a chat application, you can create a chat client that checks the web server for new messages every second. The obvious problem with polling is that it's inefficient. On the client side, the overhead is fairly minimal, but the server can easily be swamped with work if a large number of clients keep bombarding it with requests.

- *Sockets*: The most powerful option is to use *sockets*—low-level network connections. Sockets avoid HTTP altogether, in favor of the leaner and more efficient TCP. However, using sockets is complex, and it requires you to worry about issues such as network timeouts, byte arrays, user concurrency, and firewalls. If you're still interested, Chapter 20 provides a complete example with a messaging application that uses sockets.

- *Duplex services*: Silverlight includes a feature for creating duplex services, which allow two-way communication (meaning the server can contact your client when needed). Behind the scenes, duplex services are based on polling, but they implement it in a more efficient manner. The client's network request is left open but in an inactive state that doesn't hassle the server. It stays open until it times out, 90 seconds later, at which point the client connects again.

Duplex services work best with small-scale use. If your application has a small, well-defined audience, duplex services offer an interesting technique for dealing with periodic updates or time-consuming operations. In the following sections, you'll see how to build a simple duplex service that handles a batch job. The client submits a job request, and the server completes the work asynchronously and then delivers the finished product back to the client.

Configuring the Service

To create a duplex service, you begin with the same steps you follow for an ordinary web service: you add a Silverlight-enabled WCF service to your project with the right name. In this example, the service is named AsyncTask.svc.

When you add a new web service, Visual Studio adds three familiar ingredients:

- *The .svc file*: This is the endpoint to your service. The client directs all its messages to this URL. In this example, the .svc file is named AsyncTask.svc, and you don't need to make any modifications to it.

- *The web service code*: This code isn't much help for a duplex service. In the following sections, you'll this service code with a more suitable version.
- *The web.config settings*: These are partially correct, but they need some tweaking to support duplex communication. This is the task you'll take on first.

The following are the changes you need to make to the automatically generated settings in the web.config file to transform an ordinary web service into a duplex service. You can see the full web.config file with the sample code for this chapter.

Before going any further, you need to add an assembly reference to the System.ServiceModel.PollingDuplex.dll assembly that has the duplexing support you need. You can find it in a folder like C:\Program Files\Microsoft SDKs\Silverlight\v4.0\Libraries\Server.

Once you've taken care of that, you're ready to make the first modification to the web.config file. Find the <system.serviceModel> element, and add this inside it:

```
<extensions>
  <bindingExtensions>
    <add name="pollingDuplexHttpBinding" type=
"System.ServiceModel.Configuration.PollingDuplexHttpBindingCollectionElement,
System.ServiceModel.PollingDuplex, Version=4.0.0.0, Culture=neutral,
PublicKeyToken=31bf3856ad364e35"/>
  </bindingExtensions>
</extensions>
```

This pulls the class you need out of the System.ServiceModel.PollingDuplex.dll assembly and uses it to set up a binding extension.

The next step is to find the <bindings> section. Remove the <customBinding> element that's already there and add this one instead, which uses the binding extension you just configured:

```
<pollingDuplexHttpBinding />
```

Finally, find the <services> section, which defines a single <service>. Remove the first <endpoint> element inside, and add this instead:

```
<endpoint address="" binding="pollingDuplexHttpBinding"
 contract="IAsyncTaskService"/>
```

For this to work, you must create the IAsyncTaskService interface, which is the task outlined in the next section. If you give your service interface a different name, you'll need to modify this configuration information to match.

This configures your service to use the duplex binding. Now you're ready to carry on and add the web service code.

The Interfaces

For a client application to have a two-way conversation with a web service, the client needs to know something about the web service, and the web service needs to know something about the client. Before you begin building any of the actual code, you need to formalize this arrangement by creating the interfaces that allow this interaction to happen. When calling the service, the client uses the service interface (which, in this example, is named IAsyncTaskService). When calling the client, the service uses the client interface (which is named IAsyncTaskClient).

In this example, the server interface consists of single method named SubmitTask(). The client calls this method to pass the task request to the server.

```
<ServiceContract(CallbackContract := GetType(IAsyncTaskClient))> _
Public Interface IAsyncTaskService
    <OperationContract(IsOneWay := True)> _
    Sub SubmitTask(ByVal task As TaskDescription)
End Interface
```

There are two important details to note here. First, the OperationContract that decorates the SubmitTask() method sets the IsOneWay property to True. This makes it a one-way method. When calling a one-way method, the client will disconnect after the request message is sent, without waiting for a response. This also makes the server-side programming model easier. Rather than starting a new thread or running a timer, the SubmitTask() can carry out its time-consuming work from start to finish, safe in the knowledge that the client isn't waiting.

The second important detail is found in the ServiceContract attribute that decorates the interface declaration. It sets the CallbackContract property to indicate the interface that the client will use. The client interface also consists of a single one-way method. This method is named ReturnResult(), and the server calls it to pass back the result to the client when the operation is complete.

```
<ServiceContract> _
Public Interface IAsyncTaskClient
    <OperationContract(IsOneWay := True)> _
    Sub ReturnResult(ByVal result As TaskResult)
End Interface
```

These interfaces require two data classes. The TaskDescription class encapsulates the information in the task request that the client sends to the server. The TaskResult class encapsulates the final, processed data that the server returns to the client.

```
<DataContract()> _
Public Class TaskDescription

    Private _dataToProcess As String
    <DataMember()> _
    Public Property DataToProcess() As String
        Get
            Return _dataToProcess
        End Get
        Set(ByVal value As String)
            _dataToProcess = value
        End Set
    End Property
End Class

<DataContract()> _
Public Class TaskResult

    Private _processedData As String
    <DataMember()> _
    Public Property ProcessedData() As String
        Get
            Return _processedData
        End Get
```

```
            Set(ByVal value As String)
                _processedData = value
            End Set
        End Property
    End Class
```

In this example, both classes wrap a single string, and the "processing" consists of reversing the characters in that string. A more sophisticated example might generate a made-to-order bitmap, look up an archived document, or perform a statistical analysis of a set of numbers.

The Service

The service implements the IAsyncTaskService and provides the code for the SubmitTask() method. It isn't decorated with the ServiceContract attribute (unlike the previous service examples), because that attribute is already present on the interface.

The actual code in the SubmitTask() method is refreshing simple. As in any other web service method, it carries out the operation and prepares the return value. The difference is that the return value is passed by explicitly calling the IAsyncTaskClient.ReturnResult() method.

```
<AspNetCompatibilityRequirements( _
 RequirementsMode:=AspNetCompatibilityRequirementsMode.Allowed)> _
Public Class AsyncTask
    Implements IAsyncTaskService

    Public Sub SubmitTask(ByVal taskDescription As TaskDescription) _
        Implements IAsyncTaskService.SubmitTask
        ' Simulate some work with a delay.
        Thread.Sleep(TimeSpan.FromSeconds(15))

        ' Reverse the letters in string.
        Dim data As Char() = taskDescription.DataToProcess.ToCharArray()
        Array.Reverse(data)

        ' Prepare the response.
        Dim result As New TaskResult()
        result.ProcessedData = New String(data)

        ' Send the response to the client.
        Try
            Dim client As IAsyncTaskClient = _
              OperationContext.Current.GetCallbackChannel(Of IAsyncTaskClient)()
            client.ReturnResult(result)
        Catch
            ' The client could not be contacted.
            ' Clean up any resources here before the thread ends.
        End Try
        ' Incidentally, you can call the client.ReturnResult() method mulitple times
        ' to return different pieces of data at different times.
        ' The connection remains available until the reference is released (when the
        ' method ends and the variable goes out of scope).
    End Sub
End Class
```

Incidentally, a web service method can call the client.ReturnResult() method multiple times to return different pieces of data at different times. The connection to the client remains available until the reference is released (when the method ends and the variable goes out of scope).

The Client

The client code is the easiest piece of the puzzle. First, you need a reference to the System.ServiceModel.PollingDuplex.dll assembly. However, you can't use the server-side version. Instead, you can find the Silverlight version in the folder C:\Program Files\Microsoft SDKs\Silverlight\v4.0\Libraries*Client*.

Now, when creating the proxy object, you need to explicitly create the duplex binding, as shown in the code here:

```
Private client As AsyncTaskServiceClient

Public Sub New()
    InitializeComponent()

    Dim address As New EndpointAddress("http://localhost:" & _
      HtmlPage.Document.DocumentUri.Port & "/DuplexService.Web/AsyncTask.svc")
    Dim binding As New PollingDuplexHttpBinding()
    client = New AsyncTaskServiceClient(binding, address)
    ...
```

When consuming an ordinary web service, you attach an event handler to the completed event. (You get one completed event for each web service method.) Using a duplex service is similar, but you get one event for each method in the client interface, and the word *Received* is added to the end instead of *Completed*. In the current example, the IAsyncTaskClient interface defines a single method named ReturnResult(), and so the proxy class includes an event named ReturnResultReceived().

```
    ...
    AddHandler client.ReturnResultReceived, AddressOf client_ReturnResultReceived
End Sub
```

Figure 15-7 shows a simple client that allows the user to enter a string of text. When the user clicks the button, this text is send to the web service, which then processes it asynchronously. When the server calls the client back, the new information is displayed in a TextBlock underneath.

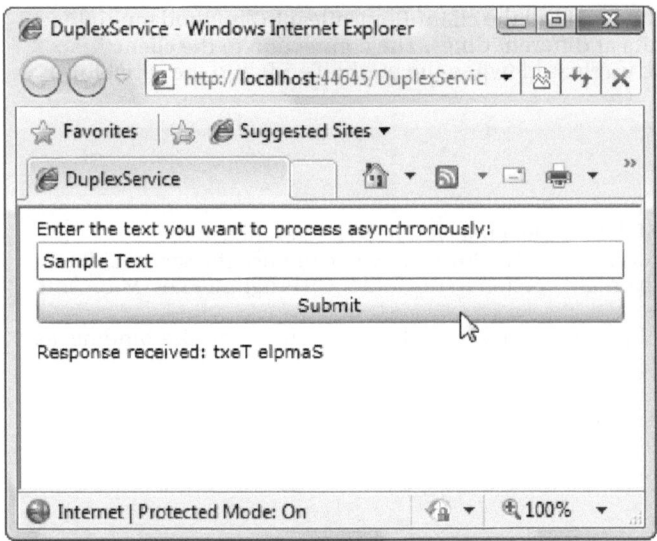

Figure 15-7. *Processing text with a duplex service*

Here's the code that makes it all happen:

```
Private Sub cmdSubmit_Click(ByVal sender As Object, ByVal e As RoutedEventArgs)
    Dim taskDescription As New TaskDescription()
    taskDescription.DataToProcess = txtTextToProcess.Text
    client.SubmitTaskAsync(taskDescription)
    lblStatus.Text = "Asynchronous request sent to server."
End Sub

Private Sub client_ReturnResultReceived(ByVal sender As Object, _
  ByVal e As ReturnResultReceivedEventArgs)
    lblStatus.Text = "Response received: " & e.result.ProcessedData
End Sub
```

From the client's point of view, the programming model seems quite similar. However, there are numerous differences:

- The client doesn't wait for the server's response but polls for it periodically.
- The server can hold onto the client reference for long periods of time and call the client multiple times before the method ends. The server could even keep the connection semipermanently or wire up a timer and send intermittent data refreshes to the client.
- The server can call different client methods from the same web service method. In fact, the service can call any method that's defined in the client interface.

■ **Note** Duplex services are not designed for huge numbers of users. By default, duplex services cap themselves at ten simultaneous connections, but you can override this by using code like what's shown at http://tinyurl.com/m9bdn4 (in C#). But as a general rule, duplex services will perform best with small numbers of simultaneously connected clients—think dozens, not hundreds.

The Last Word

In this chapter, you explored the interaction between ASP.NET web services and Silverlight. You saw how to build a basic and more advanced web service, how to monitor the network connection of the local computer, and how to support two-way web service communication. You'll build on these fundamentals in Chapter 16 and Chapter 17, as you explore how you can use a web service to provide your Silverlight application with information extracted from a server-side database.

CHAPTER 16

■ ■ ■

Data Binding

Data binding is the time-honored tradition of pulling information out of an object and displaying it in your application's user interface, without writing the tedious code that does all the work. Often, rich clients use *two-way* data binding, which adds the ability to push information from the user interface back into some object—again, with little or no code.

In Chapter 2, you learned how to use Silverlight data binding to link together two elements so that changing one affects the other. In this chapter, you'll learn how to use binding to pull data values out of an object, display them, format them, and let users edit them. You'll see how to get information from a server-side using a web service, how to format it with value converters, and how to shape it with data templates. You'll even take a look at validation techniques and the INotifyDataErrorInfo interface.

■ **What's New** Silverlight 4 introduces several refinements to the data-binding system. First, it adds binding properties that let you deal with null values and failed bindings more gracefully (see the section "Null Values and Failed Bindings"). More significantly, Silverlight introduces support for the IDataErrorInfo and INotifyDataErrorInfo interfaces, which allow you to add validation logic to your data objects (see "Creating Data Objects with Built-in Validation"). Finally, the new Binding.StringFormat property makes it possible to apply .NET format strings to bound data, saving you the trouble of building a custom value converter (see "String Formatting").

Binding to Data Objects

At its simplest, data binding is a process that tells Silverlight to extract a property value from a *source* object and use it to set a property in a *target* object. The source object can be just about anything, from an ordinary Silverlight element (as you saw in Chapter 2) to a custom data object (as you'll see in the examples in this chapter). The target object must be an instance of a class that derives from DependencyObject, and the target property must be a dependency property.

Usually, the target of a data binding is an element. This makes sense—after all, the ultimate goal of most Silverlight data binding is to display some information in your user interface. However, you can also use data binding to extract information from a source object and insert it in a brush, transform, bitmap effect, timeline, animation key frame, text element (in a RichTextBox), and more. For a complete list of classes that derive from DependencyObject (and are thus valid data binding targets), see the documentation at http://tinyurl.com/35gu6ph.

Building a Data Object

The best way to try Silverlight's data-binding features is to create a simple data object. Then, you can use data-binding expressions to display the data from your data object without writing tedious data-display code.

A *data object* is a package of related information. Any class will work as a data object, provided it consists of public properties. (A data object can also have fields and private properties, but you can't extract the information these members contain through data-binding expressions.) Furthermore, if you want the user to be able to modify a data object through data binding, its properties can't be read-only.

Here's a simple data object that encapsulates the information for a single product in a product catalog:

```
Public Class Product
    Private _modelNumber As String
    Public Property ModelNumber() As String
        Get
            Return _modelNumber
        End Get
        Set(ByVal value As String)
            _modelNumber = value
        End Set
    End Property

    Private _modelName As String
    Public Property ModelName() As String
        Get
            Return _modelName
        End Get
        Set(ByVal value As String)
            _modelName = value
        End Set
    End Property

    Private _unitCost As Double
    Public Property UnitCost() As Double
        Get
            Return _unitCost
        End Get
        Set(ByVal value As Double)
            _unitCost = value
        End Set
    End Property

    Private _description As String
    Public Property Description() As String
        Get
            Return _description
        End Get
        Set(ByVal value As String)
            _description = value
        End Set
    End Property

    Public Sub New(ByVal modelNumber As String, ByVal modelName As String, _
```

CHAPTER 16 ■ DATA BINDING

```
ByVal unitCost As Double, ByVal description As String)
    Me.ModelNumber = modelNumber
    Me.ModelName = modelName
    Me.UnitCost = unitCost
    Me.Description = description
End Sub

End Class
```

This class uses explicit properties, rather than shorter and slightly more convenient automatic properties. This design allows you to edit the property procedures later so that you can add support for change notification and validation.

Displaying a Data Object with DataContext

Consider the simple page shown in Figure 16-1. It shows the information for a single product using several text boxes in a Grid.

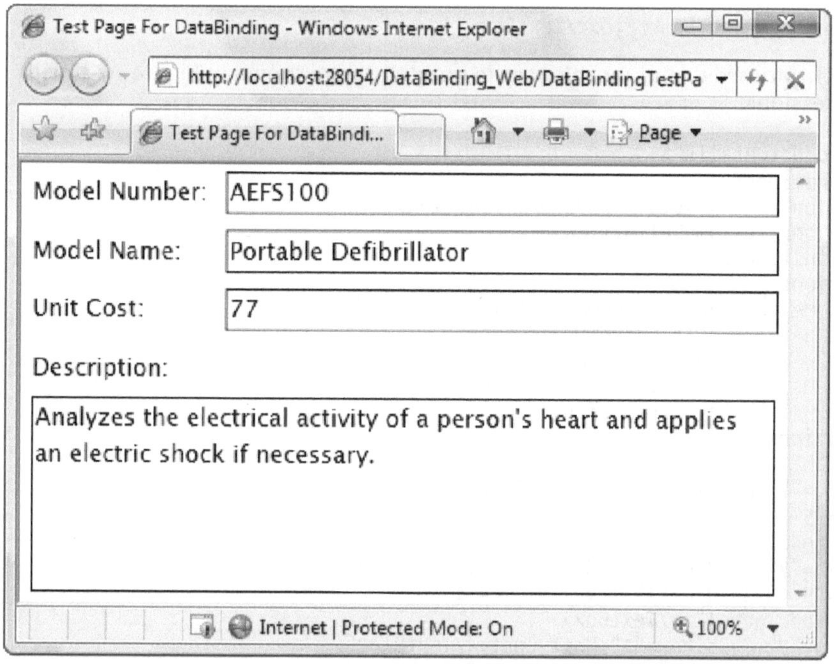

Figure 16-1. Displaying data from a Product object

To build this example, you need some code that creates the Product object you want to display. In this example, you'll use code to create a Product object using hard-coded details. Of course, in real life it's much more likely that you'll extract the data from another resource, such as a web service, an XML document, a file that's been downloaded from the Web (see Chapter 20), and so on. You'll explore a more realistic example that uses a full-fledged web service throughout this chapter, as you dig into data binding in more detail.

To display the information from a Product object, you can obviously resort to tedious data-copying code like this:

```
txtModelNumber = product.ModelNumber
```

This code is lengthy, error-prone, and brittle (for example, you'll probably need to rewrite it if you choose to use different display controls). Data binding allows you to move the responsibility for transferring the data from your VB code to your XAML markup.

To use data binding, you must set the target property using a *binding expression*. A binding expression is a markup extension (somewhat like the StaticResource extension you used in Chapter 2). It's delineated by curly braces and always starts with the word *Binding*. The simplest binding expression that you can create requires one more detail: the name of the property in the source object that has the data you want to extract.

For example, to access the Product.ModelNumber property, you use a binding expression like this:

```
{Binding ModelNumber}
```

And here's how you use it to set the Text property in a text box:

```
<TextBox Text="{Binding ModelNumber}"></TextBox>
```

Using this straightforward technique, it's easy to build the page shown in Figure 16-1, with its four binding expressions:

```xml
<Grid Name="gridProductDetails">
  <Grid.ColumnDefinitions>
    <ColumnDefinition Width="Auto"></ColumnDefinition>
    <ColumnDefinition></ColumnDefinition>
  </Grid.ColumnDefinitions>
  <Grid.RowDefinitions>
    <RowDefinition Height="Auto"></RowDefinition>
    <RowDefinition Height="Auto"></RowDefinition>
    <RowDefinition Height="Auto"></RowDefinition>
    <RowDefinition Height="Auto"></RowDefinition>
    <RowDefinition Height="*"></RowDefinition>
  </Grid.RowDefinitions>

  <TextBlock Margin="7">Model Number:</TextBlock>
  <TextBox Margin="5" Grid.Column="1"
   Text="{Binding ModelNumber}"></TextBox>
  <TextBlock Margin="7" Grid.Row="1">Model Name:</TextBlock>
  <TextBox Margin="5" Grid.Row="1" Grid.Column="1"
   Text="{Binding ModelName}"></TextBox>
  <TextBlock Margin="7" Grid.Row="2">Unit Cost:</TextBlock>
  <TextBox Margin="5" Grid.Row="2" Grid.Column="1"
   Text="{Binding UnitCost}"></TextBox>
  <TextBlock Margin="7,7,7,0" Grid.Row="3">Description:</TextBlock>
  <TextBox Margin="7" Grid.Row="4" Grid.Column="0" Grid.ColumnSpan="2"
   TextWrapping="Wrap" Text="{Binding Description}"></TextBox>
</Grid>
```

The binding expressions specify the name of the source property, but they don't indicate the source object. You can set the source object in one of two ways: by setting the DataContext property of an element or by setting the Source property of a binding.

In most situations, the most practical approach is to set the DataContext property, which every element includes. In the previous example, you could set the DataContext property of all four text boxes. However, there's an easier approach. If an element uses a binding expression and its DataContext property is set to Nothing (which is the default), the element continues its search up the element tree. This search continues until the element finds a data object or reaches the top-level container, which is the user control that represents the page. In the preceding example, that means you can save considerable effort by setting the Grid.DataContext property. All the text boxes then use the same data object.

Here's the code that creates the Product object and sets the Grid.DataContext property when the page first loads:

```
Private Sub Page_Loaded(ByVal sender As Object, ByVal e As RoutedEventArgs)
    Dim newProduct As New Product("AEFS100", "Portable Defibrillator", 77, _
        "Analyzes the electrical activity of a person's heart and applies " & _
        "an electric shock if necessary.")
    gridProductDetails.DataContext = newProduct
End Sub
```

If you don't run this code, no information will appear. Even though you've defined your bindings, no source object is available, so the elements in your page will remain blank.

■ **Tip** Usually, you'll place all your bound controls in the same container, and you'll be able to set the DataContext once on the container rather than for each bound element.

Storing a Data Object as a Resource

You have one other option for specifying a data object. You can define it as a resource in your XAML markup and then alter each binding expression by adding the Source property.

For example, you can create the Product object as a resource using markup like this:

```
<UserControl.Resources>
  <local:Product x:Key="resourceProduct"
  ModelNumber="AEFS100"
  ModelName="Portable Defibrillator" UnitCost="77"
  Description="Analyzes the electrical activity of a person's heart and applies
an electric shock if necessary.">
  </local:Product>
</UserControl.Resources>
```

This markup assumes you've mapped the project namespace to the XML namespace prefix local. For example, if the project is named DataBinding, you need to add this attribute to the UserControl start tag:

```
xmlns:local="clr-namespace:DataBinding"
```

To use this object in a binding expression, you need to specify the Source property. To set the Source property, you use a StaticResource expression that uses the resource's key name:

```
<TextBox
  Text="{Binding ModelNumber, Source={StaticResource resourceProduct} }">
</TextBox>
```

Unfortunately, you must specify the Source property in each data-binding expression. If you need to bind a significant number of elements to the same data object, it's easier to set the DataContext property of a container. In this situation, you can still use the StaticResource to set the DataContext property, which allows you to bind a group of nested elements to a single data object that's defined as a resource:

```
<Grid Name="gridProductDetails" DataContext="{StaticResource resourceProduct}">
```

Either way, when you define a data object as a resource, you give up a fair bit of freedom. Although you can still alter that object, you can't replace it. If you plan to retrieve the details for your data object from another source (such as a web service), it's far more natural to create the data object in code.

Incidentally, the Binding markup extension supports several other properties along with Source, including Mode (which lets you use two-way bindings to edit data objects) and Converter (which allows you to modify source values before they're displayed). You'll learn about Mode in the next section and Converter later in this chapter.

Null Values and Failed Bindings

Binding expressions also include two ways to fill in values when your binding doesn't work exactly the way you want it to work. First, you can use the TargetNullValue property of the binding to set a default value that should apply when the bound data is a null reference (Nothing). Here's an example:

```
<TextBox
  Text="{Binding ModelName, TargetNullValue='[No Model Name Set]' }">
</TextBox>
```

Now, if the Product.ModelName property is set to Nothing, you'll see the message "[No Model Name Set]" appear in the bound text box. Without the TargetNullValue setting, no text will appear.

The second tool for dealing with binding problems is the FallbackValue property. This sets a value that appears when the binding fails to load any data. Here's an example:

```
<TextBox
  Text="{Binding ModelName, TargetNullValue='[No Model Name Set]',
  FallbackValue='N/A' }">
</TextBox>
```

Now, the text "N/A" appears in the text box if the entire bound object is Nothing or if it doesn't contain a ModelName property. The "[No Model Name Set]" text appears if the object exists but the value of the ModelName property is Nothing.

Editing with Two-Way Bindings

At this point, you may wonder what happens if the user changes the bound values that appear in the text controls. For example, if the user types in a new description, is the in-memory Product object changed?

To investigate what happens, you can use code like this that grabs the current Product object from the DataContext and displays its properties in a TextBlock:

```
Dim product As Product = CType(gridProductDetails.DataContext, Product)

lblCheck.Text = "Model Name: " & product.ModelName + Environment.NewLine & _
  "Model Number: " & product.ModelNumber + Environment.NewLine & _
  "Unit Cost: " & product.UnitCost
```

If you run this code, you'll discover that changing the displayed values has no effect. The Product object remains in its original form.

This behavior results because binding expressions use one-way binding by default. However, Silverlight actually allows you to use one of three values from the System.Windows.Data.BindingMode enumeration when setting the Binding.Mode property. Table 16-1 has the full list.

Table 16-1. Values from the BindingMode Enumeration

Name	Description
OneWay	The target property is updated when the source property changes.
TwoWay	The target property is updated when the source property changes, and the source property is updated when the target property changes.
OneTime	The target property is set initially based on the source property value. However, changes are ignored from that point onward. Usually, you use this mode to reduce overhead if you know the source property won't change.

If you change one or more of your bindings to use two-way binding, the changes you make in the text box are committed to the in-memory object as soon as the focus leaves the text box (for example, as soon as you move to another control or click a button).

```
<TextBox Text="{Binding UnitCost, Mode=TwoWay}"></TextBox>
```

■ **Note** When you use two-way binding with a text box, the in-memory data object isn't modified until the text box loses focus. However, other elements perform their updates immediately. For example, when you make a selection in a list box, move the thumb in a slider, or change the state of a check box, the source object is modified immediately.

In some situations, you need to control exactly when the update is applied. For example, you may need to have a text box apply its changes as the user types, rather than wait for a focus change. In this situation, you need to do the job manually by calling the BindingExpression.UpdateSource() method in code. Here's the code that forces the text box to update the source data object every time the user enters or edits the text:

```
Private Sub txtUnitCost_TextChanged(ByVal sender As Object, _
  ByVal e As TextChangedEventArgs)
    Dim expression As BindingExpression
    expression = txtUnitCost.GetBindingExpression(TextBox.TextProperty)
    expression.UpdateSource()
End Sub
```

If you reach the point where all your updates are being made through code, you can disable Silverlight's automatic updating system using the UpdateSourceTrigger property of the Binding object, as shown here:

```
<TextBox Text=
  "{Binding UnitCost, Mode=TwoWay, UpdateSourceTrigger=Explicit}"></TextBox>
```

Silverlight supports only two values for UpdateSourceTrigger: Default and Explicit. It isn't possible to choose PropertyChanged (as it is in WPF). But with a little code and the UpdateSource() method, you can ensure that updates occur whenever you need.

■ **Note** Once you allow someone to edit bound data, you need to think about how you want to catch mistakes and deal with invalid data. Later in this chapter, in the "Validation" section, you'll learn about the many options Silverlight provides for preventing problems with data entry.

Change Notification

In some cases, you may want to modify a data object after it's been bound to one or more elements. For example, consider this code, which increases the current price by 10%:

```
Dim product As Product = CType(gridProductDetails.DataContext, Product)
product.UnitCost *= 1.1
```

■ **Note** If you plan to modify a bound object frequently, you don't need to retrieve it from the DataContext property each time. A better approach is to store it using a field in your page, which simplifies your code and requires less type casting.

This code won't have the effect you want. Although the in-memory Product object is modified, the change doesn't appear in the bound controls. That's because a vital piece of infrastructure is missing—quite simply, there's no way for the Product object to notify the bound elements.

To solve this problem, your data class needs to implement the System.ComponentModel.INotifyPropertyChanged interface. The INotifyPropertyChanged interface defines a single event, which is named PropertyChanged. When a property changes in your data object, you must raise the PropertyChanged event and supply the property name as a string.

Here's the definition for a revamped Product class that uses the INotifyPropertyChanged interface, with the code for the implementation of the PropertyChanged event:

```
Public Class Product
    Implements INotifyPropertyChanged

    Public Event PropertyChanged As PropertyChangedEventHandler _
      Implements INotifyPropertyChanged.PropertyChanged
```

```vbnet
    Public Sub OnPropertyChanged(ByVal e As PropertyChangedEventArgs)
        If PropertyChangedEvent IsNot Nothing Then
          RaiseEvent PropertyChanged(Me, e)
        End If
    End Sub

    ...
End Class
```

Now, you need to fire the PropertyChanged event in all your property setters:

```vbnet
Private _unitCost As Double
Public Property UnitCost() As Double
    Get
        Return _unitCost
    End Get
    Set(ByVal value As Double)
        _unitCost = value
        OnPropertyChanged(New PropertyChangedEventArgs("UnitCost"))
    End Set
End Property
```

If you use this version of the Product class in the previous example, you get the behavior you expect. When you change the current Product object, the new information appears in the bound text boxes immediately.

■ **Tip** If several values have changed, you can call OnPropertyChanged() and pass in an empty string. This tells Silverlight to reevaluate the binding expressions that are bound to any property in your class.

Building a Data Service

Although the examples you've seen so far have walked you through the basic details of Silverlight data binding, they haven't been entirely realistic. A more typical design is for your Silverlight application to retrieve the data objects it needs from an external source, such as a web service. In the examples you've seen so far, the difference is minimal. However, it's worth stepping up to a more practical example before you begin binding to collections. After all, it makes more sense to get your data from a database than to construct dozens or hundreds of Product objects in code.

In the examples in this chapter, you'll rely on a straightforward data service that returns Product objects. You've already learned to create a WCF service (and consume it) in Chapter 15. Building a data service is essentially the same.

The first step is to move the class definition for the data object to the ASP.NET website. (If you're creating a projectless website, you must place the code file in the App_Code folder. If you're creating a web project, you can place it anywhere.) The data object needs a few modifications: the addition of the DataContract and DataMember attributes to make it serializable and the addition of a public no-argument constructor that allows it to be serialized. Here's a partial listing of the code, which shows you the general outline you need:

```vb
<DataContract()> _
Public Class Product
    Implements INotifyPropertyChanged
    Private _modelNumber As String

    <DataMember()> _
    Public Property ModelNumber() As String
        Get
            Return _modelNumber
        End Get
        Set(ByVal value As String)
            _modelNumber = value
            OnPropertyChanged(New PropertyChangedEventArgs("ModelNumber"))
        End Set
    End Property

    Private _modelName As String

    <DataMember()> _
    Public Property ModelName() As String
        Get
            Return _modelName
        End Get
        Set(ByVal value As String)
            _modelName = value
            OnPropertyChanged(New PropertyChangedEventArgs("ModelName"))
        End Set
    End Property

    ...
    Public Sub New()
    End Sub

End Class
```

WEB SERVICES AND INOTIFYPROPERTYCHANGED

When you define data classes with a web service, you get Silverlight change notification for free. That's because when Visual Studio generates the client-side copy of a data class, it automatically implements the INotifyPropertyChanged interface, even if the server-side original doesn't.

For example, if you define a super-simple Product class like this:

```vb
<DataContract()> _
Public Class Product

    <DataMember()> _
    Public Property ModelNumber() As String

    <DataMember()> _
    Public Property ModelName() As String
```

```
<DataMember()> _
Public Property UnitCost() As Double

<DataMember()> _
Public Property Description() As String

End Class
```

the client-side copy still implements INotifyPropertyChanged, has full property procedures, and calls OnPropertyChanged() when to raise a notification event when any property changes.

You can inspect the client-side code for the data class by opening the Reference.vb file in your Silverlight project. (This file is hidden, so to see it, you need to first click the Show All Files button in the Solution Explorer. You'll find it in the Solution Explorer if you drill down. For example, the Product class in the current example is in Solution Explorer ➤ Service References ➤ DataService ➤ Reference.svcmap ➤ Reference.vb.)

This design raises a good question—namely, if INotifyPropertyChanged is implemented automatically in the client, should you even bother to implement it on the server? This decision is a matter of personal preference. Some developers feel that explicitly implementing the interface on the server side is clearer, while others don't bother because change notification is a client-side feature that's rarely used on the server.

Finally, it's worth noting that you can disable this feature and tell Visual Studio not to use INotifyPropertyChanged when it generates client-side data classes. To do so, you must edit the Reference.svcmap file in your Silverlight property. Just look for the <EnableDataBinding> element, and change its content from "true" to "false." However, there's really no good reason to take this step.

With the data object in place, you need a web service method that uses it. The web service class is exceedingly simple—it provides just a single method that allows the caller to retrieve one product record. Here's the basic outline:

```
<ServiceContract(Namespace := "")> _
Public Class StoreDb
    Private connectionString As String = _
      WebConfigurationManager.ConnectionStrings("StoreDb").ConnectionString

    <OperationContract()> _
    Public Function GetProduct(ByVal ID As Integer) As Product
        ...
    End Function
End Class
```

The query is performed through a stored procedure in the database named GetProduct. The connection string isn't hard-coded—instead, it's retrieved through a setting in the web.config file, which makes it easy to modify this detail later. Here's the section of the web.config file that defines the connection string:

```
<configuration>
  ...
  <connectionStrings>
    <add name="StoreDb" connectionString=
     "Data Source=localhost;Initial Catalog=Store;Integrated Security=True" />
  </connectionStrings>
  ...
</configuration>
```

The database component that's shown in the following example retrieves a table of product information from the Store database, which is a sample database for the fictional IBuySpy store included with some Microsoft case studies. You can get a script to install this database with the downloadable samples for this chapter (or you can use an alternative version that grabs the same information from an XML file).

In this book, we're primarily interested in how data objects can be bound to Silverlight elements. The actual process that deals with creating and filling these data objects (as well as other implementation details, such as whether StoreDb caches the data over several method calls, whether it uses stored procedures instead of inline queries, and so on) isn't our focus. However, just to get an understanding of what's taking place, here's the complete code for the data service:

```
<ServiceContract(Namespace := ""), _
 AspNetCompatibilityRequirements( _
   RequirementsMode := AspNetCompatibilityRequirementsMode.Allowed)> _
Public Class StoreDb
    Private connectionString As String = _
      WebConfigurationManager.ConnectionStrings("StoreDb").ConnectionString

    <OperationContract()> _
    Public Function GetProduct(ByVal ID As Integer) As Product
        Dim con As New SqlConnection(connectionString)
        Dim cmd As New SqlCommand("GetProductByID", con)
        cmd.CommandType = CommandType.StoredProcedure
        cmd.Parameters.AddWithValue("@ProductID", ID)

        Try
            con.Open()
            Dim reader As SqlDataReader
            reader = cmd.ExecuteReader(CommandBehavior.SingleRow)

            If reader.Read() Then
                ' Create a Product object that wraps the
                ' current record.
                Dim product As New Product(CStr(reader("ModelNumber")), _
                  CStr(reader("ModelName")), Convert.ToDouble(reader("UnitCost")), _
                  CStr(reader("Description")))
                Return product
            Else
                Return Nothing
            End If
        Finally
            con.Close()
        End Try
    End Function
End Class
```

■ **Note** Currently, the GetProduct() method doesn't include any exception-handling code, so exceptions will bubble up the calling code. This is a reasonable design choice, but you may want to catch the exception in GetProduct(), perform cleanup or logging as required, and then rethrow the exception to notify the calling code of the problem. This design pattern is called *caller inform*.

Using the ADO.NET objects directly (as in this example) is a simple, clean way to write the code for a data service. Generally, you won't use ADO.NET's disconnected data objects, such as the DataSet, because Silverlight doesn't include these classes and so can't manipulate them.

Calling a Data Service

To use a data service, you need to begin by adding a web reference in your Silverlight project, a basic step that's covered in Chapter 15. Once that's taken care of, you're ready to use the automatically generated web service code in your application. In this case, it's a class named StoreDbClient.

Figure 16-2 shows a Silverlight page that lets the user retrieve the details about any product.

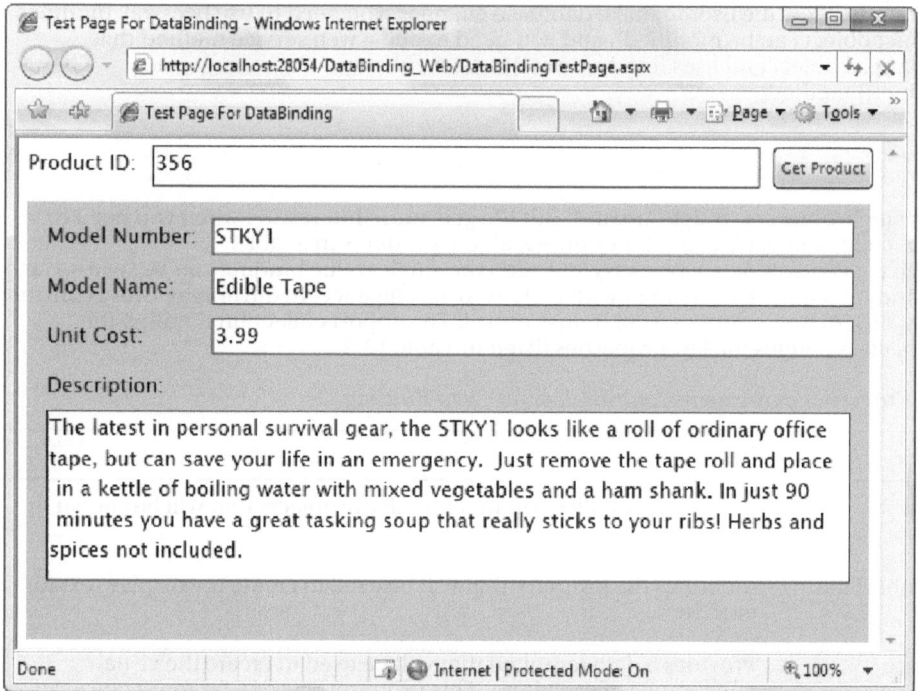

Figure 16-2. Retrieving product data from a web service

When the user clicks Get Product, this code runs:

```
Private Sub cmdGetProduct_Click(ByVal sender As Object, ByVal e As RoutedEventArgs)
    ' Set the URL, taking the port of the test web server into account.
    Dim client As New StoreDbClient()

    ' Call the service to get the Product object.
    AddHandler client.GetProductCompleted, AddressOf client_GetProductCompleted
    client.GetProductAsync(356)
End Sub
```

When the web service returns its data, you need to set the DataContext property of the container, as in previous examples:

```
Private Sub client_GetProductCompleted(ByVal sender As Object, _
  ByVal e As GetProductCompletedEventArgs)
    Try
        gridProductDetails.DataContext = e.Result
    Catch err As Exception
        lblError.Text = "Failed to contact service."
    End Try
End Sub
```

If you want to allow the user to make database changes, you need to use two-way bindings (so the Product object can be modified), and you need to add a web service method that accepts a changed object and uses it to commit databases changes (for example, an UpdateProduct() method).

Binding to a Collection of Objects

Binding to a single object is straightforward. But life gets more interesting when you need to bind to some collection of objects—for example, all the products in a table.

Although every dependency property supports the single-value binding you've seen so far, collection binding requires an element with a bit more intelligence. In Silverlight, every control that displays a list of items derives from ItemsControl. To support collection binding, the ItemsControl class defines the key properties listed in Table 16-2.

Table 16-2. Properties in the ItemsControl Class for Data Binding

Name	Description
ItemsSource	Points to the collection that has all the objects that will be shown in the list.
DisplayMemberPath	Identifies the property that will be used to create the display text for each item.
ItemTemplate	Provides a data template that will be used to create the visual appearance of each item. This property acts as a far more powerful replacement for DisplayMemberPath.
ItemsPanel	Provides a template that will be used to create the layout container that holds all the items in the list.

At this point, you're probably wondering what types of collections you can stuff in the ItemsSource property. Happily, you can use just about anything. All you need is support for the IEnumerable interface, which is provided by arrays, all types of collections, and many more specialized objects that wrap groups of items. However, the support you get from a basic IEnumerable interface is limited to read-only binding. If you want to edit the collection (for example, you want to allow inserts and deletions), you need a bit more infrastructure, as you'll see shortly.

Displaying and Editing Collection Items

Consider the page shown in Figure 16-3, which displays a list of products. When you choose a product, the information for that product appears in the bottom section of the page, where you can edit it. (In this example, a GridSplitter control lets you adjust the space given to the top and bottom portions of the page.)

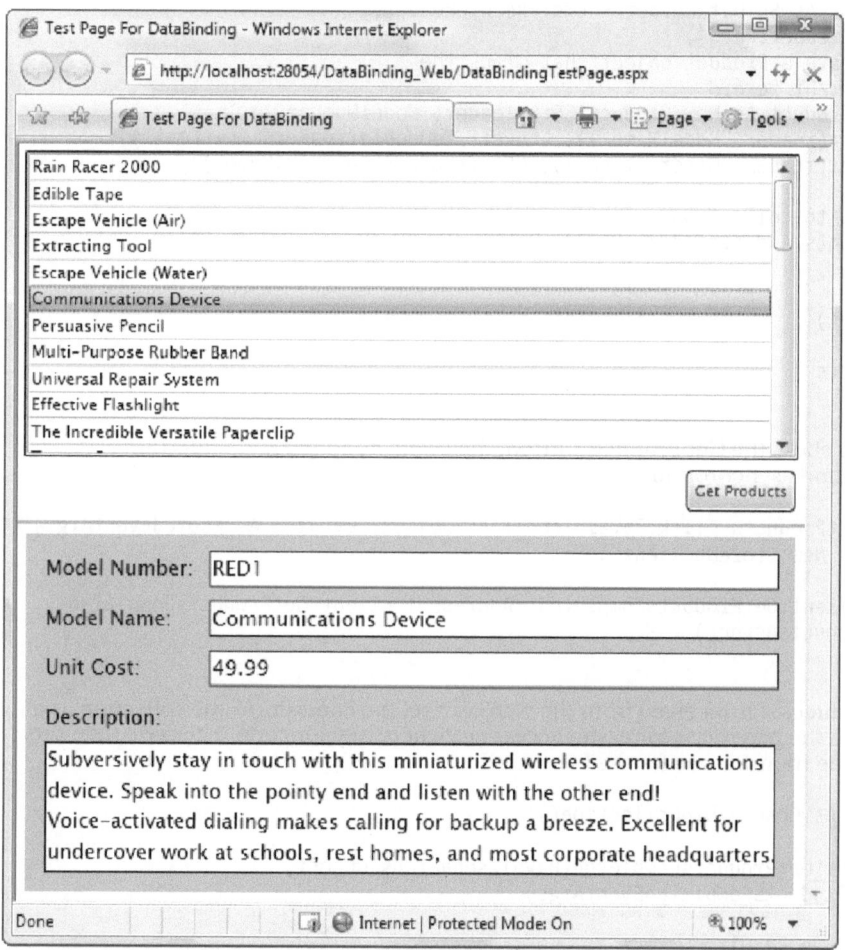

Figure 16-3. A list of products

To create this example, you need to begin by building your data-access logic. In this case, the StoreDb.GetProducts() method retrieves the list of all the products in the database using the GetProducts stored procedure. A Product object is created for each record and added to a generic List collection. (You can use any collection here—for example, an array or a weakly typed ArrayList would work equivalently.)

Here's the GetProducts() code:

```
<OperationContract()> _
Public Function GetProducts() As List(Of Product)
    Dim con As New SqlConnection(connectionString)
    Dim cmd As New SqlCommand("GetProducts", con)
    cmd.CommandType = CommandType.StoredProcedure

    Dim products As New List(Of Product)()
    Try
        con.Open()
        Dim reader As SqlDataReader = cmd.ExecuteReader()
        Do While reader.Read()
            ' Create a Product object that wraps the
            ' current record.
            Dim product As New Product(CStr(reader("ModelNumber")), _
                CStr(reader("ModelName")), Convert.ToDouble(reader("UnitCost")), _
                CStr(reader("Description")), CStr(reader("CategoryName")))

            ' Add to collection
            products.Add(product)
        Loop
    Finally
        con.Close()
    End Try
    Return products
End Function
```

When the user clicks the Get Products button, the event-handling code calls the GetProducts() method asynchronously:

```
Private Sub cmdGetProducts_Click(ByVal sender As Object, ByVal e As RoutedEventArgs)
    Dim client As New StoreDbClient()

    AddHandler client.GetProductsCompleted, AddressOf client_GetProductsCompleted
    client.GetProductsAsync()
End Sub
```

When the product list is received from the web service, the code stores the collection as a member variable in the page class for easier access elsewhere in your code. The code then sets that collection as the ItemsSource for the list:

```
Private products As ObservableCollection()

Private Sub client_GetProductsCompleted(ByVal sender As Object, _
  ByVal e As GetProductsCompletedEventArgs)
    Try
        products = e.Result
        lstProducts.ItemsSource = products
    Catch err As Exception
```

```
        lblError.Text = "Failed to contact service."
    End Try
End Sub
```

■ **Note** Keen eyes will notice one unusual detail in this example. Although the web service returned an array of Product objects, the client applications receives them in a different sort of package: the ObservableCollection. You'll learn why Silverlight performs this sleight of hand in the next section.

This code successfully fills the list with Product objects. However, the list doesn't know how to display a Product object, so it calls the ToString() method. Because this method hasn't been overridden in the Product class, this has the unimpressive result of showing the fully qualified class name for every item (see Figure 16-4).

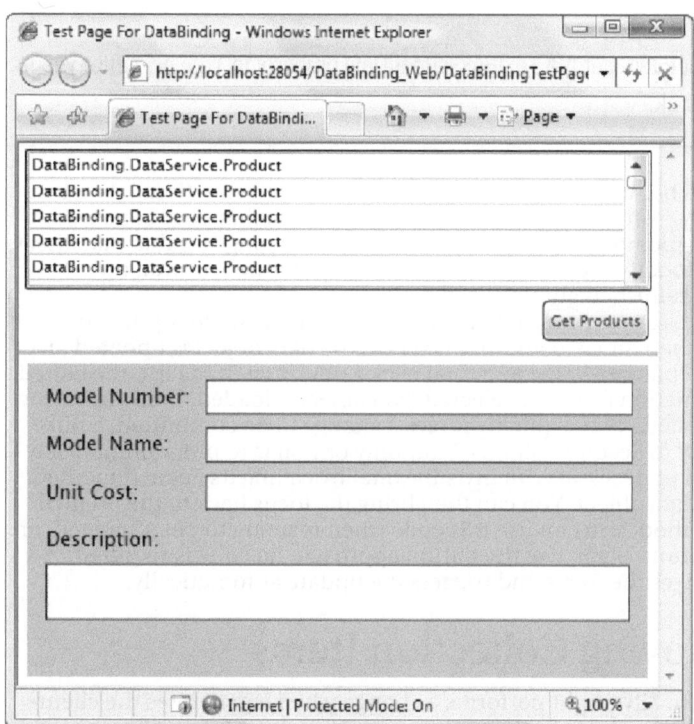

Figure 16-4. An unhelpful bound list

You have three options to solve this problem:

- *Set the list's DisplayMemberPath property*: For example, set it to ModelName to get the result shown in Figure 16-3.
- *Override the Product.ToString() method to return more useful information*: For example, you can return a string with the model number and model name of each item. This

approach gives you a way to show more than one property in the list (for example, it's great for combining the FirstName and LastName properties in a Customer class).

- *Supply a data template*: This way, you can show any arrangement of property values (and along with fixed text). You'll learn how to use this trick later in this chapter.

When you've decided how to display information in the list, you're ready to move on to the second challenge: displaying the details for the currently selected item in the grid that appears below the list. To make this work, you need to respond to the SelectionChanged event and change the DataContext of the Grid that contains the product details. Here's the code that does it:

```
Private Sub lstProducts_SelectionChanged(ByVal sender As Object, _
  ByVal e As SelectionChangedEventArgs)
    gridProductDetails.DataContext = lstProducts.SelectedItem
End Sub
```

■ **Tip** To prevent a field from being edited, set the TextBox.IsReadOnly property to True or, better yet, use a read-only control like a TextBlock.

If you try this example, you'll be surprised to see that it's already fully functional. You can edit product items, navigate away (using the list), and then return to see that your edits were successfully committed to the in-memory data objects. You can even change a value that affects the display text in the list. If you modify the model name and tab to another control, the corresponding entry in the list is refreshed automatically.

But there's one quirk. Changes are committed only when a control loses focus. If you change a value in a text box and then move to another text box, the data object is updated just as you'd expect. However, if you change a value and then click a new item in the list, the edited value is discarded, and the information from the selected data object is loaded. If this behavior isn't what you want, you can add code that explicitly forces a change to be committed. Unlike WPF, Silverlight has no direct way to accomplish this. Your only option is to programmatically send the focus to another control (if necessary, an invisible one) by calling its Focus() method. This commits the change to the data object. You can then bring the focus back to the original text box by calling its Focus() method. You can use this code when reacting to TextChanged, or you can add a Save or Update button. If you use the button approach, no code is required, because clicking the button changes the focus and triggers the update automatically.

Inserting and Removing Collection Items

As you saw in the previous section, Silverlight performs a change when it generates the client-side code for communicating with a web service. Your web service may return an array or List collection, but the client-side code places the objects into an ObservableCollection. The same translation step happens if you return an object with a collection property.

This shift takes place because the client doesn't really know what type of collection the web server is returning. Silverlight assumes that it should use an ObservableCollection to be safe, because an ObservableCollection is more fully featured than an array or an ordinary List collection.

So, what does the ObservableCollection add that arrays and List objects lack? First, like the List, the ObservableCollection has support for adding and removing items. For example, you try deleting an item with a Delete button that executes this code:

```
Private Sub cmdDeleteProduct_Click(ByVal sender As Object, _
  ByVal e As RoutedEventArgs)
    products.Remove(CType(lstProducts.SelectedItem, Product))
End Sub
```

This obviously doesn't work with an array. It does work with a List collection, but there's a problem: although the deleted item is removed from the collection, it remains stubbornly visible in the bound list.

To enable collection change tracking, you need to use a collection that implements the INotifyCollectionChanged interface. In Silverlight, the only collection that meets this bar is the ObservableCollection class. When you execute the previous code with an ObservableCollection like the collection of products returned from the web service, you'll see the bound list is refreshed immediately. Of course, it's still up to you to create the data-access code that can commit changes like these permanently—for example, the web service methods that insert and remove products from the back-end database.

Binding to a LINQ Expression

One of Silverlight's many surprises is its support for Language Integrated Query, which is an all-purpose query syntax that was introduced in .NET 3.5.

LINQ works with any data source that has a LINQ provider. Using the support that's included with Silverlight, you can use similarly structured LINQ queries to retrieve data from an in-memory collection or an XML file. And as with other query languages, LINQ lets you apply filtering, sorting, grouping, and transformations to the data you retrieve.

Although LINQ is somewhat outside the scope of this chapter, you can learn a lot from a simple example. For example, imagine you have a collection of Product objects named *products*, and you want to create a second collection that contains only those products that exceed $100 in cost. Using procedural code, you can write something like this:

```
' Get the full list of products.
Dim products As List(Of Product) = App.StoreDb.GetProducts()

' Create a second collection with matching products.
Dim matches As New List(Of Product)()

For Each product As Product In products
    If product.UnitCost >= 100 Then
        matches.Add(product)
    End If
Next
```

Using LINQ, you can use the following *expression*, which is far more concise:

```
' Get the full list of products.
Dim products As List(Of Product) = App.StoreDb.GetProducts()

' Create a second collection with matching products.
Dim matches As IEnumerable(Of Product) = _
  From product In products _
  Where product.UnitCost >= 100 _
  Select product
```

This example uses LINQ to Objects, which means it uses a LINQ expression to query the data in an in-memory collection. LINQ expressions use a set of new language keywords, including From, In, Where, and Select. These LINQ keywords are a genuine part of the VB language.

■ **Note** A full discussion of LINQ is beyond the scope of this book. For a detailed treatment, you can refer to the book *Pro LINQ: Language Integrated Query in C# 2010*, the LINQ developer center at `http://msdn.microsoft.com/netframework/aa904594.aspx`, or the huge catalog of LINQ examples at `http://msdn.microsoft.com/vbasic/bb737913.aspx`.

LINQ revolves around the IEnumerable(Of T) interface. No matter what data source you use, every LINQ expression returns some object that implements IEnumerable(Of T). Because IEnumerable(Of T) extends IEnumerable, you can bind it in a Silverlight page just as you bind an ordinary collection (see Figure 16-5):

```
lstProducts.ItemsSource = matches
```

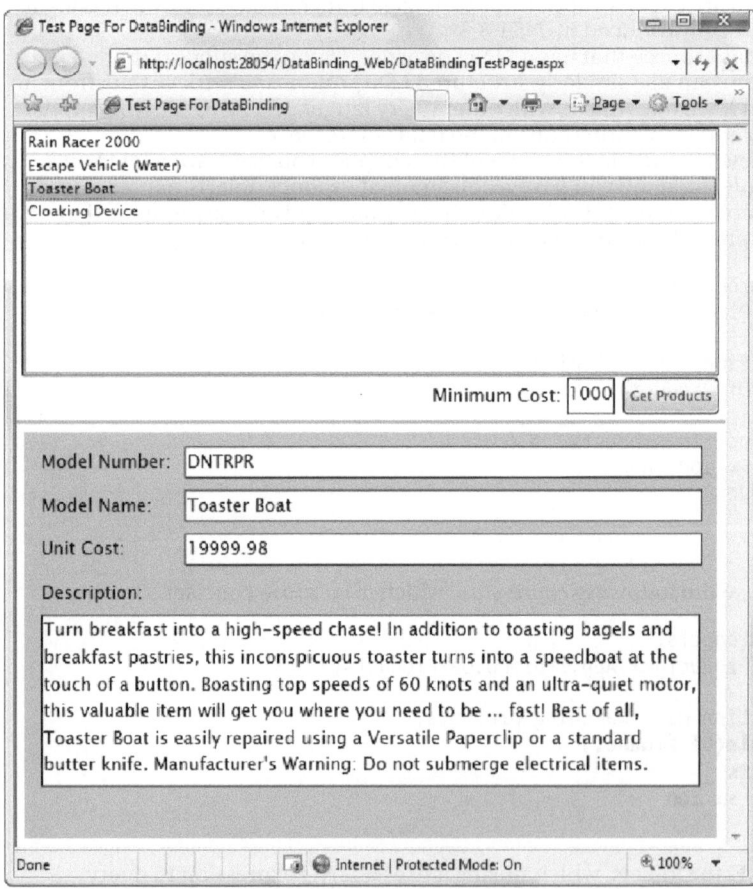

Figure 16-5. Filtering a collection with LINQ

Unlike the List and ObservableCollection classes, the IEnumerable(Of T) interface doesn't provide a way to add or remove items. If you need this capability, you must first convert your IEnumerable(Of T) object into an array or List collection using the ToArray() or ToList() method.

Here's an example that uses ToList() to convert the result of a LINQ query (shown previously) into a strongly typed List collection of Product objects:

```
Dim productMatches As List(Of Product) = matches.ToList()
```

■ **Note** ToList() is an extension method, which means it's defined in a different class from the one in which it's used. Technically, ToList() is defined in the System.Linq.Enumerable helper class, and it's available to all IEnumerable(Of T) objects. However, it isn't available if the Enumerable class isn't in scope, which means the code shown here won't work if you haven't imported the System.Linq namespace.

The ToList() method causes the LINQ expression to be evaluated immediately. The end result is an ordinary List collection, which you can deal with in all the usual ways. If you want to make the collection editable so that changes show up in bound controls immediately, you'll need to copy the contents of the List to a new ObservableCollection.

Master-Details Display

As you've seen, you can bind other elements to the SelectedItem property of your list to show more details about the currently selected item. Interestingly, you can use a similar technique to build a master-details display of your data. For example, you can create a page that shows a list of categories and a list of products. When the user chooses a category in the first list, you can show just the products that belong to that category in the second list. Figure 16-6 shows this example.

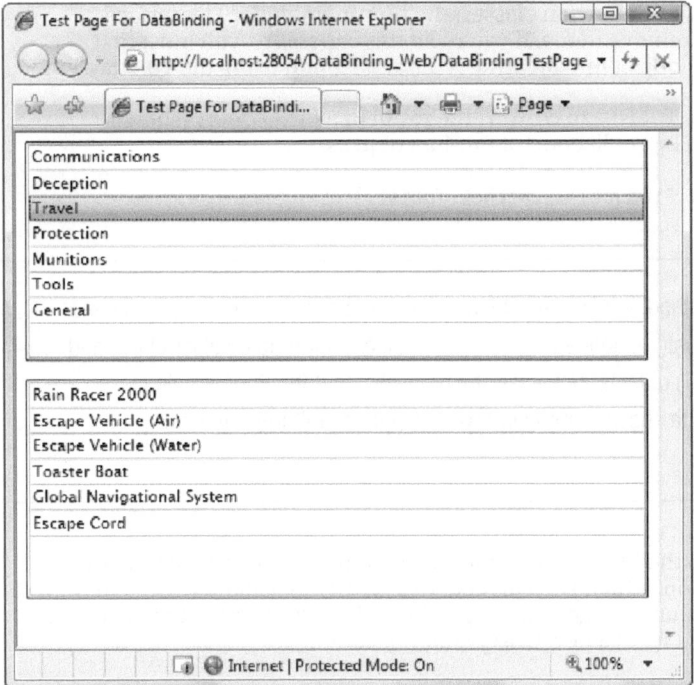

Figure 16-6. A master-details list

To pull this off, you need a *parent* data object that provides a collection of related *child* data objects through a property. For example, you can build a Category class that provides a property named Category.Products with the products that belong to that category. Like the Product class, the Category class can implement the INotifyPropertyChanged to provide change notifications. Here's the complete code:

```
Public Class Category
    Implements INotifyPropertyChanged
    Private _categoryName As String
    Public Property CategoryName() As String
        Get
            Return _categoryName
        End Get
        Set(ByVal value As String)
            _categoryName = value
            OnPropertyChanged(New PropertyChangedEventArgs("CategoryName"))
        End Set
    End Property

    Private _products As List(Of Product)
    Public Property Products() As List(Of Product)
        Get
            Return _products
        End Get
        Set(ByVal value As List(Of Product))
```

```
            _products = value
            OnPropertyChanged(New PropertyChangedEventArgs("Products"))
        End Set
    End Property

    Public Event PropertyChanged As PropertyChangedEventHandler
    Public Sub OnPropertyChanged(ByVal e As PropertyChangedEventArgs)
        If PropertyChangedEvent IsNot Nothing Then
            RaiseEvent PropertyChanged(Me, e)
        End If
    End Sub

    Public Sub New(ByVal categoryName As String, _
      ByVal products As List(Of Product))
        Me.CategoryName = categoryName
        Me.Products = products
    End Sub

    Public Sub New()
    End Sub
End Class
```

To use the Category class, you also need to modify the data-access code that you saw earlier. Now, you query the information about products and categories from the database. The example in Figure 16-6 uses a web service method named GetCategoriesWithProducts(), which returns a collection of Category objects, each of which has a nested collection of Product objects:

```
<OperationContract()> _
Public Function GetCategoriesWithProducts() As List(Of Category)
    ' Perform the query for products using the GetProducts stored procedure.
    Dim con As New SqlConnection(connectionString)
    Dim cmd As New SqlCommand("GetProducts", con)
    cmd.CommandType = CommandType.StoredProcedure

    ' Store the results (temporarily) in a DataSet.
    Dim adapter As New SqlDataAdapter(cmd)
    Dim ds As New DataSet()
    adapter.Fill(ds, "Products")

    ' Perform the query for categories using the GetCategories stored procedure.
    cmd.CommandText = "GetCategories"
    adapter.Fill(ds, "Categories")

    ' Set up a relation between these tables.
    ' This makes it easier to discover the products in each category.
    Dim relCategoryProduct As New DataRelation("CategoryProduct", _
      ds.Tables("Categories").Columns("CategoryID"), _
      ds.Tables("Products").Columns("CategoryID"))
    ds.Relations.Add(relCategoryProduct)

    ' Build the collection of Category objects.
    Dim categories As List(Of Category) = New List(Of Category)()

    For Each categoryRow As DataRow In ds.Tables("Categories").Rows
```

```
' Add the nested collection of Product objects for this category.
Dim products As List(Of Product) = New List(Of Product)()
For Each productRow As DataRow In _
  categoryRow.GetChildRows(relCategoryProduct)
    products.Add(New Product(productRow("ModelNumber").ToString(), _
      productRow("ModelName").ToString(), _
      Convert.ToDouble(productRow("UnitCost")), _
      productRow("Description").ToString()))
Next

categories.Add( _
  New Category(categoryRow("CategoryName").ToString(), products))
Next
Return categories
End Function
```

To display this data, you need the two lists shown here:

```
<ListBox x:Name="lstCategories" DisplayMemberPath="CategoryName"
  SelectionChanged="lstCategories_SelectionChanged"></ListBox>
<ListBox x:Name="lstProducts" Grid.Row="1" DisplayMemberPath="ModelName">
</ListBox>
```

After you receive the collection from the GetCategoriesWithProducts() method, you can set the ItemsSource of the topmost list to show the categories:

```
lstCategories.ItemsSource = e.Result
```

To show the related products, you must react when an item is clicked in the first list and then set the ItemsSource property of the second list to the Category.Products property of the selected Category object:

```
lstProducts.ItemsSource = (CType(lstCategories.SelectedItem, Category)).Products
```

Alternatively, you can perform this step using another binding expression in XAML, rather than code. The trick is to use the SelectedValuePath property, which is provided by the ListBox and ComboBox.

Essentially, the SelectedValuePath tells the list control to expose a specific property of the bound object through the SelectedValue property of the list. If you've bound a collection of Category objects, you can tell the ListBox to provide the Category.Products collection through the ListBox.SelectedValue property like this:

```
<ListBox x:Name="lstCategories" DisplayMemberPath="CategoryName"
  SelectedValuePath="Products"></ListBox>
```

Then you can use the SelectedValuePath to write a binding expression for the second list box:

```
<ListBox x:Name="lstProducts" Grid.Row="1" DisplayMemberPath="ModelName"
  ItemsSource="{Binding ElementName=lstCategories, Path=SelectedValue}">
</ListBox>
```

Now, selected a category in the first list box causes the list of products to appear in the second list box, with no code required.

■ **Note** In this example, the full category and product information is downloaded from the web service at once. But if your application uses a much larger product catalog and most users need to view only one or two categories, you might be better off downloading just the subset of products you want to view. In this case, you would react to the SelectionChanged event of the category list box and retrieve the CategoryID of the currently selected category (using the SelectedItem or SelectedValue property). Then, you would perform a new web service call to get the products in that category. But be warned—this scenario can multiply the work the database server needs to perform and slow down your application. You'll need to weight the trade-offs carefully and (at the very least) consider caching product information in the client after it's downloaded for the first time.

Validation

When the Silverlight data-binding system encounters invalid data, it usually ignores it. For example, consider the following list, which details the three types of errors that can occur when you're editing a two-way field:

- *Incorrect data type*: For example, a numeric property like UnitCost can't accommodate letters or special characters. Similarly, it can't hold extremely large numbers (numbers larger than 1.79769313486231570E+308).
- *Property setter exception*: For example, a property like UnitCost may use a range check and throw an exception if you attempt to set a negative number.
- *Read-only property*: This can't be set at all.

If you run into these errors, you're likely to miss them, because the Silverlight data-binding system doesn't give you any visual feedback. The incorrect value remains in the bound control, but it's never applied to the bound object.

To avoid confusion, it's a good idea to alert users to their mistakes as soon as possible. In the following sections, you'll learn how to respond to these issues and design data objects that have built-in validation.

Error Notifications

The first step to implementing error notifications is to set the ValidatesOnExceptions property of the binding to True. This tells the data-binding system to pay attention to all errors, whether they occur in the type converter or the property setter. But when ValidatesOnException is set to False (the default), the data-binding system fails silently when it hits these conditions. The data object isn't updated, but the offending value remains in the bound control.

Here's an example that applies this property to the binding for UnitCost:

```
<TextBox Margin="5" Grid.Row="2" Grid.Column="1" x:Name="txtUnitCost"
  Text="{Binding UnitCost, Mode=TwoWay, ValidatesOnExceptions=True}"></TextBox>
```

This simple change gives your application the ability to catch and display errors, provided you're using two-way data binding with a control that supports the ValidationState group of control states. These are some of the controls that support this feature (with no extra work required):

- TextBox
- PasswordBox
- CheckBox
- RadioButton
- ListBox
- ComboBox

In Chapter 13, you learned that *control states* are animations that change the way a control looks at certain times. In the case of validation, a control must support three states: Valid, InvalidUnfocused, and InvalidFocused. Together, these states make up the ValidationState group, and they allow a control to vary its appearance when it contains invalid data.

To understand how this works, it helps to consider the simple example of a text box with invalid data. First, consider a version of the Product class that uses this code to catch negative prices and raise an exception:

```
Private _unitCost As Double
Public Property UnitCost() As Double
    Get
        Return _unitCost
    End Get
    Set(ByVal value As Double)
        If value < 0 Then
            Throw New ArgumentException("Can't be less than 0.")
        End If

        _unitCost = value
    End Set
End Property
```

Now, consider what happens if the user enters a negative number. In this case, the property setter will throw an ArgumentException. BecauseValidatesOnException is set to True, this exception is caught by the data-binding system, which then switches the ValidationState of the text box from Valid to InvalidFocused (if the text box currently has focus) or InvalidUnfocused (if the text box doesn't).

■ **Tip** If you have Visual Studio set to break an all exceptions, Visual Studio will notify you when the ArgumentException is thrown and switch into break mode. To carry on and see what happens when the exception reaches the data-binding system, choose Debug ➤ Continue or just press the shortcut key F5.

In the unfocused state, the text box gets a dark red border with an error-notification icon (a tiny red triangle) in the upper-right corner. In its focused state, or when the user moves the mouse over the error icon, the exception message text appears in a pop-up red alert balloon. Figure 16-7 shows both states.

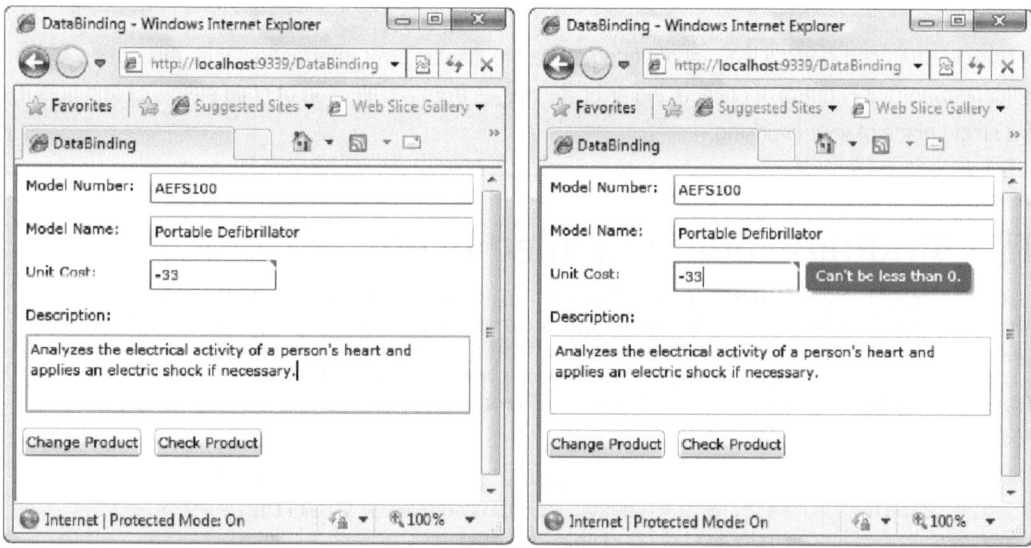

Figure 16-7. The InvalidUnfocused state (left) and InvalidFocused state (right) of a text box

---

■ **Note** For the red pop-up balloon to appear properly, sufficient space must be available between the text box and the edges of the browser window. If there is space on the right side of the text box, the balloon appears there. If not, it appears on the left. The balloon appears on top of any other elements that are in the same place, such as buttons or labels. However, it can't stretch out of the browser window. In the example shown in Figure 16-7, the width of the UnitPrice text box is limited to make sure there is room on the right side. Finally, if the message is too long to fit in the available space, part of it is chopped off.

At first glance, the error pop-ups seem easy and incredibly useful. Because the control takes care of the visual details, you simply need to worry about reporting helpful error messages. But there is a disadvantage to wiring the validation display into the control template: if you want to change the way a control displays error messages (or disable error display altogether), you need to replace the entire control template, making sure to include all the other unrelated states and markup details. And as you already know, the average control template is quite lengthy, so this process is tedious and potentially limiting. (For example, it may prevent you from using someone else's customized template to get more attractive visuals if you're already relying on your own custom template to tweak the error-display behavior.)

■ **Note** In Chapter 17, you'll learn about another way to display error information, with the ValidationSummary control. It collects the error messages from a collection of child elements and lists it in a single place of your choosing.

The BindingValidationFailed Event

Error notifications are a great way to inform the user about potential problems. But there are a whole host of situations in which you'll want more control over what happens next. In these cases, you need to intercept the BindingValidationFailed event and use custom code.

To enable the BindingValidationFailed event, you must set ValidatesOnExceptions to True (to detect errors) and NotifyOnValidationError to True (to fire the validation events). Here's an example:

```
<TextBox Margin="5" Grid.Row="2" Grid.Column="1" x:Name="txtUnitCost"
  Text="{Binding UnitCost, Mode=TwoWay, ValidatesOnExceptions=True,
NotifyOnValidationError=True}"></TextBox>
```

BindingValidationError is a bubbling event, which means you can handle it where it occurs (in the text box) or at a higher level (such as the containing Grid). Handling errors where they occur gives you the opportunity to write targeted error-handling logic that deals separately with errors in different fields. Handling them at a higher level (as shown here) allows you to reuse the same logic for many different types of errors:

```
<Grid Name="gridProductDetails"
 BindingValidationError="Grid_BindingValidationError">
```

The final step is to do something when the problem occurs. You may choose to display a message or change the appearance of some part of your application, but the real power of the BindingValidationError event is that it lets you perform other actions, such as changing focus, resetting the incorrect value, trying to correct it, or offering more detailed, targeted help based on the specific mistake that was made.

The following example displays an error message and indicates the current value (see Figure 16-8). It also transfers focus back to the offending text box, which is a heavy-handed (but occasionally useful) technique. It has the side effect of making sure the control remains in the InvalidFocused state rather than the InvalidUnfocused state, so the pop-up error message also remains visible:

```
Private Sub Grid_BindingValidationError(ByVal sender As Object, _
  ByVal e As ValidationErrorEventArgs)
    ' Display the error.
    lblInfo.Text = e.Error.Exception.Message
    lblInfo.Text &= Environment.NewLine & "The stored value is still: " & _
      (CType(gridProductDetails.DataContext, Product)).UnitCost

    ' Suggest the user try again.
    txtUnitCost.Focus()
End Sub
```

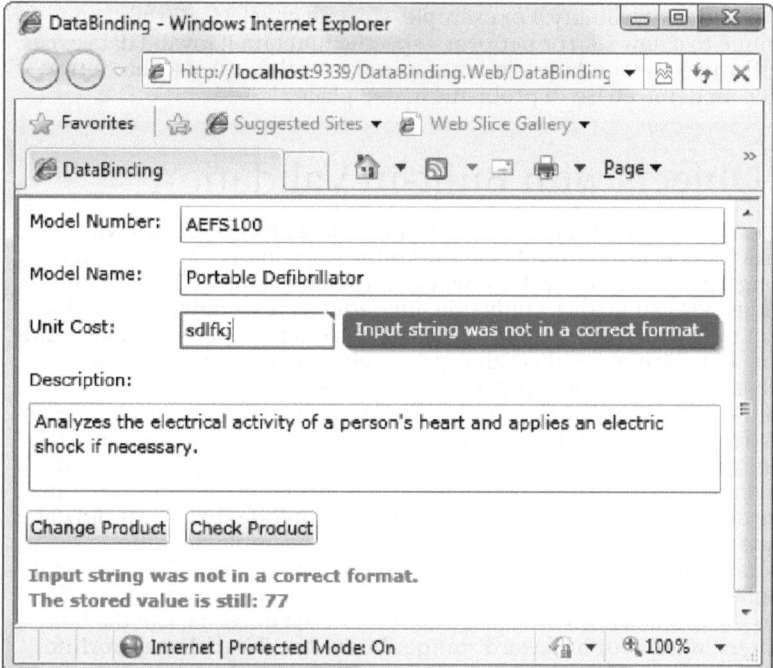

Figure 16-8. *Pointing out a validation error*

The BindingValidationError event happens only when the value is changed and the edit is committed. In the case of the text box, this doesn't happen until the text box loses focus. If you want errors to be caught more quickly, you can use the BindingExpression.UpdateSource() method to force immediate updates as the user types, as described in the previous section.

■ **Tip** If you don't reset the value in the text box, the incorrect value remains on display, even though it isn't stored in the bound data object. You might choose to allow this behavior so that users have another chance to edit invalid values.

Whatever steps you take in this event handler happen in addition to the control state change. Unfortunately, you can't selectively disable control error reporting *and* choose to receive the BindingValidationError event.

The Validation Class

You don't need to wait for the BindingValidationError event to detect invalid data. You can check a bound control at any time using the static methods of the Validation class. Validation.GetHasErrors() returns True if the control has failed validation, and Validation.GetErrors() returns the appropriate collection of one of more exception objects.

587

These methods give you added flexibility. For example, you can check HasErrors() and refuse to let the user continue to a new step or perform a specific function if invalid data exists. Similarly, you can use GetErrors() to round up a series of mistakes at the end of a data-entry process, so you can provide an itemized list of problems in one place.

Creating Data Objects with Built-in Validation

Earlier, you saw the simplest way to build validation logic into a data object—by throwing exceptions for bad data (such as negative numbers in the UnitCost property). However, raising exceptions is a slightly dangerous and often reckless way to implement validation. One problem is that the exception-raising code can inadvertently rule out perfectly reasonable uses of the data object. For example, it might not make sense for the UnitsInStock property to hold a value of –10, but if the underlying database stores this value, you might still want to create the corresponding Product object so you can manipulate it in your application.

Most purists prefer not to raise exceptions to indicate user input errors. There are several reasons that underlie this design choice—for example, a user input error isn't an exceptional condition, error conditions may depend on the interaction between multiple property values, and it's sometimes worthwhile to hold on to incorrect values for further processing rather than reject them outright. Instead, cautious developers rely on other techniques to flag invalid data.

In Silverlight, your data objects can provide built-in validation by implement the IDataErrorInfo or INotifyDataErrorInfo interface. Both of these interfaces have the same goal—they replace aggressive unhandled exceptions with a politer system of error notification. The IDataErrorInfo interface is the original error-tracking interface, which dates back to the first version of .NET. Silverlight includes it for backward compatibility. The INotifyDataErrorInfo interface is a similar but richer interface that was introduced with Silverlight 4. It has support for additional features, such as multiple errors per property and rich error objects.

No matter which interface you use, there are three basic steps to follow:

- Implement the IDataErrorInfo or INotifyDataErrorInfo interface in your data object.
- Tell the data binding infrastructure to check the data object for errors and show error notifications. If you're using the IDataErrorInfo interface, you set ValidatesOnDataErrors to True. If you're using the INotifyDataErrorInfo interface, you set ValidatesOnNotifyDataErrors to True.
- Optionally, set NotifyOnValidationError to True if you want to receive the BindingValidationFailed event.
- Optionally, set the ValidatesOnExceptions property to True if you want to catch other types of errors, such as data conversion problems.

■ **Note** If you're using a web service, remember that Visual Studio can't transfer the code from a server-defined data object to the automatically generated client-side copy. This isn't a problem for basic data objects, but it is a serious obstacle if you want to create a rich data object that implements features such as validation. In this situation, you need to use the type sharing feature described in Chapter 15, which lets you manually add a synchronized copy of the data class to both projects.

The following example shows how to use the INotifyDataErrorInfo interface to detect problems with the Product object. The first step is to implement the interface:

```
Public Class Product
    Implements INotifyPropertyChanged, INotifyDataErrorInfo
    ...
End Class
```

The INotifyDataErrorInfo interface requires just three members. The ErrorsChanged event fires when an error is added or removed. The HasErrors property returns True or False to indicate whether the data object has errors. Finally, the GetErrors() method provides the full error information.

Before you can implement these methods, you need a way to track the errors in your code. The best bet is a private collection, like this:

```
Private errors As New Dictionary(Of String, List(Of String))()
```

At first glance, this collection looks a little odd. To understand why, you need to know to facts. First, the INotifyPropertyChanged interface expects you to link your errors to a specific property. Second, each property can have one or more errors. The easiest way to track this error information is with a Dictionary(Of T, Of K) collection that's indexed by property name:

```
Private errors As New Dictionary(Of String, List(Of String))()
```

Each entry in the dictionary is itself a collection of errors. This example uses a simple List<Of T> List<T> of strings:

```
Private errors As New Dictionary(Of String, List(Of String))()
```

However, you could use a full-fledged error object to bundle together multiple pieces of information about the error, including details such as a text message, error code, severity level, and so on.

Once you have this collection in place, you simply need to add to it when an error occurs (and remove the error information if the error is corrected). The make this process easier, the Product class in this example adds a pair of private methods named SetErrors() and ClearErrors():

```
Public Event ErrorsChanged As EventHandler(Of DataErrorsChangedEventArgs) _
  Implements INotifyDataErrorInfo.ErrorsChanged

Private Sub SetErrors(ByVal propertyName As String, _
  ByVal propertyErrors As List(Of String))
    ' Clear any errors that already exist for this property.
    errors.Remove(propertyName)

    ' Add the list collection for the specified property.
    errors.Add(propertyName, propertyErrors)

    ' Raise the error-notification event.
    RaiseEvent ErrorsChanged(Me, New DataErrorsChangedEventArgs(propertyName))
End Sub

Private Sub ClearErrors(ByVal propertyName As String)
    ' Remove the error list for this property.
```

```
        errors.Remove(propertyName)

        ' Raise the error-notification event.
        RaiseEvent ErrorsChanged(Me, New DataErrorsChangedEventArgs(propertyName))
End Sub
```

And here's the error-handling logic that ensures that the Product.ModelNumber property is restricted to a string of alphanumeric characters. (Punctuation, spaces, and other special characters are not allowed.)

```
Private _modelNumber As String
Public Property ModelNumber() As String
    Get
        Return _modelNumber
    End Get
    Set(ByVal value As String)
        _modelNumber = value

        Dim valid As Boolean = True
        For Each c As Char In _modelNumber
            If (Not Char.IsLetterOrDigit(c)) Then
                valid = False
                Exit For
            End If
        Next
        If (Not valid) Then
            Dim errors As List(Of String) = New List(Of String)()
            errors.Add("The ModelNumber can only contain letters and numbers.")
            SetErrors("ModelNumber", errors)
        Else
            ClearErrors("ModelNumber")
        End If

        OnPropertyChanged(New PropertyChangedEventArgs("ModelNumber"))
    End Set
End Property
```

The final step is to implement the GetErrors() and HasErrors() method. The GetErrors() method returns the list of errors for a specific property (or all the errors for all the properties). The HasErrors() property returns True if the Product class has one or more errors.

```
Public Function GetErrors(ByVal propertyName As String) As IEnumerable _
    Implements System.ComponentModel.INotifyDataErrorInfo.GetErrors

    If propertyName = "" Then
        ' Provide all the error collections.
        Return (errors.Values)
    Else
        ' Provice the error collection for the requested property
        ' (if it has errors).
        If errors.ContainsKey(propertyName) Then
            Return (errors(propertyName))
        Else
            Return Nothing
        End If
```

```
        End If
End Function

Public ReadOnly Property HasErrors As Boolean _
    Implements INotifyDataErrorInfo.HasErrors

        Get
            ' Indicate whether this entire object is error-free.
            Return (errors.Count > 0)
        End Get
End Property
```

To tell Silverlight to use the INotifyDataErrorInfo interface and use it to check for errors when a property is modified, the ValidatesOnNotifyDataErrors property of the binding must be True:

```
<TextBox Margin="5" Grid.Row="2" Grid.Column="1" x:Name="txtModelNumber"
    Text="{Binding ModelNumber, Mode=TwoWay, ValidatesOnNotifyDataErrors=True,
NotifyOnValidationError=True}"></TextBox>
```

Technically, you don't need to explicitly set ValidatesOnNotifyDataErrors, because it's True by default (unlike the similar ValidatesOnDataErrors property that's used with the IDataErrorInfo interface). However, it's still a good idea to set it explicitly to make your intention to use it clear in the markup.

If you also use the NotifyOnValidationError event (as shown here), the BindingValidationError event will fire whenever the error collection changes. However, there won't be an exception object provided in the ValidationErrorEventArgs.Exception property, because no exception occurred. Instead, you can get the error information that you return from your data object through the ValidationErrorEventArgs.Error.ErrorContent property. In the current example, this is the simple string in the error list, but you could return any object you want. And if you return a list with multiple error items for a property, the BindingValidationError event fires once for each item.

```
Private Sub Grid_BindingValidationError(ByVal sender As Object, _
    ByVal e As ValidationErrorEventArgs)
    If e.Error.Exception IsNot Nothing Then
        ' Validation failed due to an exception.
        ...
    Else
        ' Validation error reported through an interface.
        lblInfo.Text = e.Error.ErrorContent.ToString()
        txtModelNumber.Focus()
    End If
End Sub
```

Incidentally, you can combine both approaches by creating a data object that throws exceptions for some types of errors and uses IDataErrorInfo or INotifyDataErrorInfo to report others. However, keep in mind that these two approaches have a broad difference. When an exception is triggered, the property is not updated in the data object. But when you use the IDataErrorInfo or INotifyDataErrorInfo interface, invalid values are allowed but flagged. The data object is updated, but you can use notifications and the BindingValidationFailed event to inform the user.

■ **Note** In Chapter 17, you'll see another way to build validation into a data object—using declarative annotations. This approach is less powerful but allows you to separate the validation logic from the rest of your code. It also integrates nicely with the data controls that support these attributes.

Data Formatting and Conversion

In a basic binding, the information travels from the source to the target without any change. This seems logical, but it's not always the behavior you want. Your data source may use a low-level representation that you don't want to display directly in your user interface. For example, you may have numeric codes you want to replace with human-readable strings, numbers that need to be cut down to size, dates that need to be displayed in a long format, and so on. If so, you need a way to convert these values into the correct display form. And if you're using a two-way binding, you also need to do the converse—take user-supplied data and convert it to a representation suitable for storage in the appropriate data object.

Fortunately, Silverlight has two tools that can help you:

- *String formatting*: This feature allows you to convert data that's represented as text—for example, strings that contain dates and numbers—by setting the Binding.StringFormat property. It's a convenient technique that works for at least half of all formatting tasks.
- *Value converters*: This is a far more powerful (and somewhat more complicated) feature that lets you convert any type of source data into any type of object representation, which you can then pass on to the linked control.

In the following sections, you'll consider both approaches.

String Formatting

String formatting is the perfect tool for formatting numbers that need to be displayed as text. For example, consider the UnitCost property of the Product class used in this chapter. UnitCost is stored as a decimal. As a result, when it's displayed in a text box, you see values like 3.9900. Not only does this display format show more decimal places than you'd probably like, but it also leaves out the currency symbol. A more intuitive representation is the currency-formatted value $49.99, as shown in Figure 16-9.

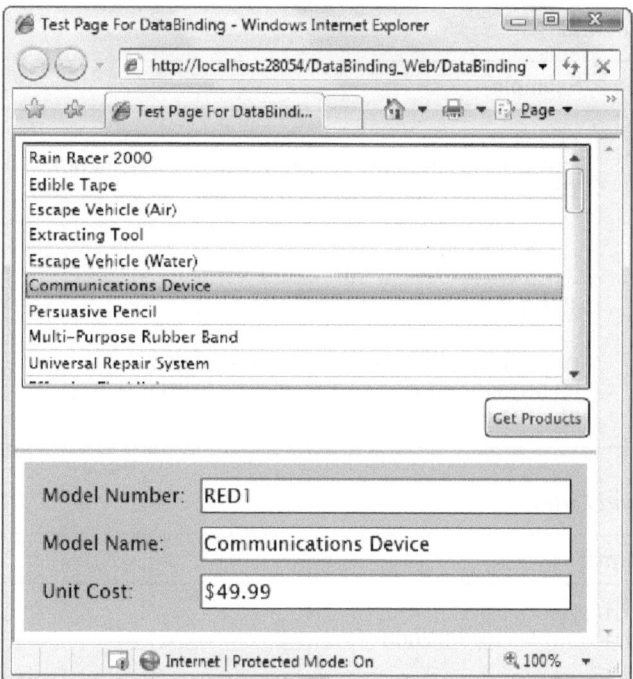

Figure 16-9. Displaying formatted currency values

The easiest solution to fix this problem is to set the Binding.StringFormat property. Silverlight will use the format string to convert the raw text to its display value, just before it appears in the control. Just as importantly, Silverlight will (in most cases) use this string to perform the reverse conversion, taking any edited data and using it to update the bound property.

When setting the Binding.StringFormat property, you use standard .NET format strings. For example, you can use the format string *C* to apply the locale-specific currency format, which translates 3.99 to $3.99 on a U.S. computer. You wrap the currency string in single quotation marks. Here's an example that applies the currency format string to the UnitCost field:

```
<TextBox Margin="5" Grid.Row="2" Grid.Column="1"
 Text="{Binding UnitCost, StringFormat='C' }">
</TextBox>
```

To get the results you want with the StringFormat property, you need the right format string. You can learn about all the format strings that are available in the Visual Studio help. However, Table 16-3 and Table 16-4 show some of the most common options you'll use for numeric and date values, respectively. And here's a binding expression that uses a custom date format string to format the OrderDate property:

```
<TextBlock Text="{Binding Date, StringFormat='0:MM/dd/yyyy' }"></TextBlock>
```

Table 16-3. Format Strings for Numeric Data

Type	Format String	Example
Currency	C	$1,234.50. Parentheses indicate negative values: ($1,234.50). The currency sign is locale-specific.
Scientific (Exponential)	E	1.234.50E+004.
Percentage	P	45.6%.
Fixed Decimal	F?	Depends on the number of decimal places you set. F3 formats values like 123.400. F0 formats values like 123.

Table 16-4. Format Strings for Times and Dates

Type	Format String	Format
Short Date	d	M/d/yyyy For example: 10/30/2008
Long Date	D	dddd, MMMM dd, yyyy For example: Wednesday, January 30, 2008
Long Date and Short Time	f	dddd, MMMM dd, yyyy HH:mm aa For example: Wednesday, January 30, 2008 10:00 AM
Long Date and Long Time	F	dddd, MMMM dd, yyyy HH:mm:ss aa For example: Wednesday, January 30, 2008 10:00:23 AM
ISO Sortable Standard	s	yyyy-MM-dd HH:mm:ss For example: 2008-01-30 10:00:23
Month and Day	M	MMMM dd For example: January 30
General	G	M/d/yyyy HH:mm:ss aa (depends on locale-specific settings) For example: 10/30/2008 10:00:23 AM

Value Converters

The Binding.StringFormat property is created for simple, standard formatting with numbers and dates. But many data binding scenarios need a more powerful tool, called a *value converter* class.

A value converter plays a straightforward role. It's responsible for converting the source data just before it's displayed in the target and (in the case of a two-way binding) converting the new target value just before it's applied back to the source.

Value converters are an extremely useful piece of the Silverlight data-binding puzzle. You can use them several ways:

- *To format data to a string representation*: For example, you can convert a number to a currency string. This is the most obvious use of value converters, but it's certainly not the only one.
- *To create a specific type of Silverlight object*: For example, you can read a block of binary data and create a BitmapImage object that can be bound to an Image element.
- *To conditionally alter a property in an element based on the bound data*: For example, you may create a value converter that changes the background color of an element to highlight values in a specific range.

In the following sections, you'll consider an example of each of these approaches.

Formatting Strings with a Value Converter

To get a basic idea of how a value converter works, it's worth revisiting the currency-formatting example you looked at in the previous section. Although that example used the Binding.StringFormat property, you can accomplish the same thing—and more—with a value converter. For example, you could round or truncate values (changing 3.99 to 4), use number names (changing 1,000,000 into 1 million), or even add a dealer markup (multiplying 3.99 by 15%). You can also tailor the way that the reverse conversion works to change user-supplied values into the right data values in the bound object.

To create a value converter, you need to take three steps:

1. Create a class that implements IValueConverter (from the System.Windows.Data namespace). You place this class in your Silverlight project, which is where the conversion takes place—not in the web service.
2. Implement a Convert() method that changes data from its original format to its display format.
3. Implement a ConvertBack() method that does the reverse and changes a value from display format to its native format.

Figure 16-10 shows how it works.

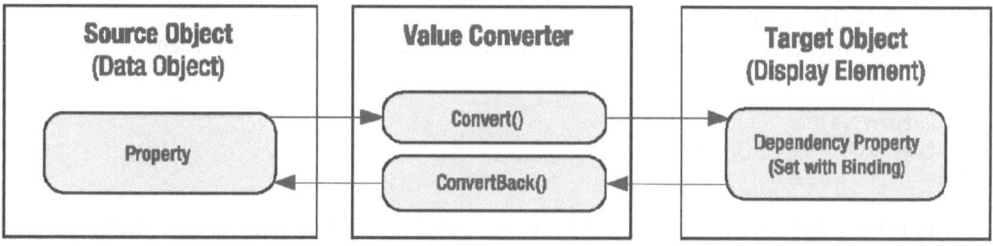

Figure 16-10. Converting bound data

In the case of the decimal-to-currency conversion, you can use the Decimal.ToString() method to get the formatted string representation you want. You need to specify the currency format string "C", as shown here:

```
Dim currencyText As String = decimalPrice.ToString("C")
```

This code uses the culture settings that apply to the current thread. A computer that's configured for the English (United States) region runs with a locale of en-US and displays currencies with the dollar sign ($). A computer that's configured for another locale may display a different currency symbol. (This is the same way that Binding,StringFormat property works when set to *C*.) If this isn't the result you want (for example, you always want the dollar sign to appear), you can specify a culture using the overload of the ToString() method shown here:

```
Dim culture As New CultureInfo("en-US")
Dim currencyText As String = decimalPrice.ToString("C", culture)
```

Converting from the display format back to the number you want is a little trickier. The Parse() and TryParse() methods of the Double type are logical choices to do the work, but ordinarily they can't handle strings that include currency symbols. The solution is to use an overloaded version of the Parse() or TryParse() method that accepts a System.Globalization.NumberStyles value. If you supply NumberStyles.Any, you can successfully strip out the currency symbol, if it exists.

Here's the complete code for the value converter that deals with price values such as the Product.UnitCost property:

```
Public Class PriceConverter
    Implements IValueConverter

    Public Function Convert(ByVal value As Object, ByVal targetType As Type, _
      ByVal parameter As Object, ByVal culture As CultureInfo) As Object _
      Implements IValueConverter.Convert

        Dim price As Double = CDbl(value)
        Return price.ToString("C", culture)
    End Function

    Public Function ConvertBack(ByVal value As Object, ByVal targetType As Type, _
      ByVal parameter As Object, ByVal culture As CultureInfo) As Object _
      Implements IValueConverter.ConvertBack

        Dim price As String = value.ToString()
        Dim result As Double
        If Double.TryParse(price, NumberStyles.Any, culture, result) Then
            Return result
        End If
        Return value
    End Function
End Class
```

To put this converter into action, you need to begin by mapping your project namespace to an XML namespace prefix you can use in your markup. Here's an example that uses the namespace prefix local and assumes your value converter is in the namespace DataBinding:

```
xmlns:local="clr-namespace:DataBinding"
```

Typically, you'll add this attribute to the <UserControl> start tag at the top of your markup.

Now, you need to create an instance of the PriceConverter class in your page's Resources collection, as shown here:

```
<UserControl.Resources>
  <local:PriceConverter x:Key="PriceConverter"></local:PriceConverter>
</UserControl.Resources>
```

Then, you can point to it in your binding using a StaticResource reference:

```
<TextBox Margin="5" Grid.Row="2" Grid.Column="1"
 Text="{Binding UnitCost, Mode=TwoWay, Converter={StaticResource PriceConverter}}">
</TextBox>
```

■ **Note** Unlike WPF, Silverlight lacks the IMultiValueConverter interface. As a result, you're limited to converting individual values, and you can't combine values (for example, join together a FirstName and a LastName field) or perform calculations (for example, multiply UnitPrice by UnitsInStock).

Creating Objects with a Value Converter

Value converters are indispensable when you need to bridge the gap between the way data is stored in your classes and the way it's displayed in a page. For example, imagine you have picture data stored as a byte array in a field in a database. You can convert the binary data into a System.Windows.Media.Imaging.BitmapImage object and store that as part of your data object. However, this design may not be appropriate.

For example, you may need the flexibility to create more than one object representation of your image, possibly because your data library is used in both Silverlight applications and Windows Forms applications (which use the System.Drawing.Bitmap class instead). In this case, it makes sense to store the raw binary data in your data object and convert it to a BitmapImage object using a value converter.

■ **Tip** To convert a block of binary data into an image, you must first create a BitmapImage object and read the image data into a MemoryStream. Then, you can call the BitmapImage.SetSource() method to pass the image data in the stream to the BitmapImage.

The Products table from the Store database doesn't include binary picture data, but it does include a ProductImage field that stores the file name of an associated product image. In this case, you have even more reason to delay creating the image object. First, the image may not be available, depending on where the application is running. Second, there's no point in incurring the extra memory overhead from storing the image unless it's going to be displayed.

The ProductImage field includes the file name but not the full URI of an image file. This gives you the flexibility to pull the image files from any location. The value converter has the task of creating a URI that points to the image file based on the ProductImage field and the website you want to use. The root URI is stored using a custom property named RootUri, which defaults to the same URI where the current web page is located.

Here's the complete code for the ImagePathConverter that performs the conversion:

```vb
Public Class ImagePathConverter
    Implements IValueConverter

    Private _rootUri As String
    Public Property RootUri() As String
        Get
            Return _rootUri
        End Get
        Set(ByVal value As String)
            _rootUri = value
        End Set
    End Property

    Public Sub New()
        Dim uri As String = HtmlPage.Document.DocumentUri.ToString()

        ' Remove the web page from the current URI to get the root URI.
        RootUri = uri.Remove(uri.LastIndexOf("/"c), _
          uri.Length - uri.LastIndexOf("/"c))
    End Sub

    Public Function Convert(ByVal value As Object, ByVal targetType As Type, _
      ByVal parameter As Object, ByVal culture As CultureInfo) As Object _
      Implements IValueConverter.Convert

        Dim imagePath As String = RootUri & "/" & CStr(value)
        Return New BitmapImage(New Uri(imagePath))
    End Function

    Public Function ConvertBack(ByVal value As Object, ByVal targetType As Type, _
      ByVal parameter As Object, ByVal culture As CultureInfo) As Object _
      Implements IValueConverter.ConvertBack
        ' Images aren't editable, so there's no need to support ConvertBack.
        Throw New NotSupportedException()
    End Function
End Class
```

To use this converter, begin by adding it to Resources. Although you can set the RootUri property on the ImagePathConverter element, this example doesn't. As a result, the ImagePathConverter uses the default value that points to the current application website.

```xml
<UserControl.Resources>
    <local:ImagePathConverter x:Key="ImagePathConverter"></local:ImagePathConverter>
</UserControl.Resources>
```

Now it's easy to create a binding expression that uses this value converter:

```xml
<Image Margin="5" Grid.Row="2" Grid.Column="1" Stretch="None"
 HorizontalAlignment="Left" Source=
 "{Binding ProductImagePath, Converter={StaticResource ImagePathConverter}}">
</Image>
```

This works because the Image.Source property expects an ImageSource object, and the BitmapImage class derives from ImageSource.

Figure 16-11 shows the result.

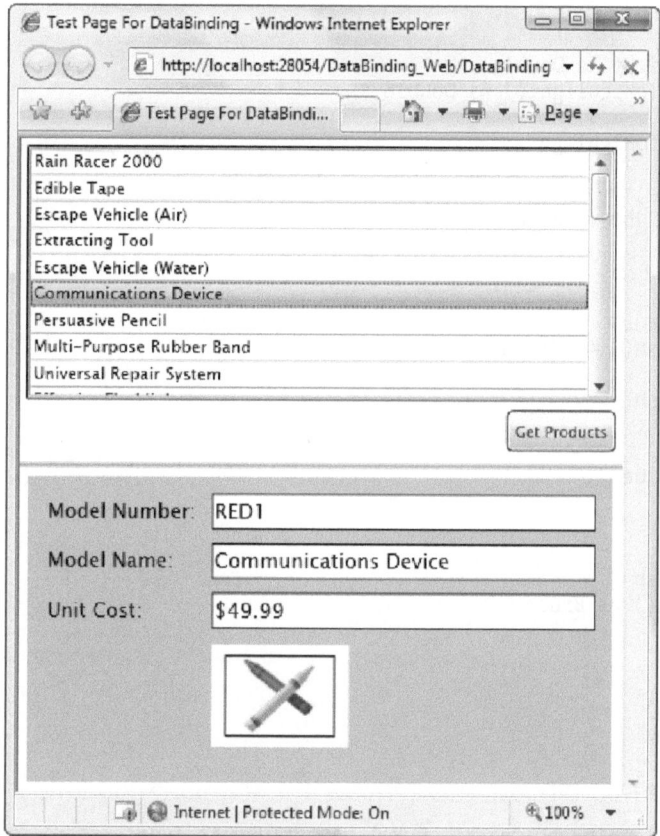

Figure 16-11. Displaying bound images

You can improve this example in a couple of ways. First, attempting to create a BitmapImage that points to a nonexistent file causes an exception, which you'll receive when setting the DataContext, ItemsSource, or Source property. Alternatively, you can add properties to the ImagePathConverter class that let you configure this behavior. For example, you may introduce a Boolean SuppressExceptions property. If it's set to True, you can catch exceptions in the Convert() method and return an empty string. Or, you can add a DefaultImage property that takes a placeholder BitmapImage. ImagePathConverter can then return the default image if an exception occurs.

Applying Conditional Formatting

Some of the most interesting value converters aren't designed to format data for presentation. Instead, they're intended to format some other appearance-related aspect of an element based on a data rule.

For example, imagine that you want to flag high-priced items by giving them a different background color. You can easily encapsulate this logic with the following value converter:

```vb
Public Class PriceToBackgroundConverter
    Implements IValueConverter

    Private _minimumPriceToHighlight As Double
    Public Property MinimumPriceToHighlight() As Double
        Get
            Return _minimumPriceToHighlight
        End Get
        Set(ByVal value As Double)
            _minimumPriceToHighlight = value
        End Set
    End Property

    Private _highlightBrush As Brush
    Public Property HighlightBrush() As Brush
        Get
            Return _highlightBrush
        End Get
        Set(ByVal value As Brush)
            _highlightBrush = value
        End Set
    End Property

    Private _defaultBrush As Brush
    Public Property DefaultBrush() As Brush
        Get
            Return _defaultBrush
        End Get
        Set(ByVal value As Brush)
            _defaultBrush = value
        End Set
    End Property

    Public Function Convert(ByVal value As Object, ByVal targetType As Type, _
        ByVal parameter As Object, ByVal culture As CultureInfo) As Object _
        Implements IValueConverter.Convert

        Dim price As Double = CDbl(value)
        If price >= MinimumPriceToHighlight Then
            Return HighlightBrush
        Else
            Return DefaultBrush
        End If
    End Function

    Public Function ConvertBack(ByVal value As Object, ByVal targetType As Type, _
        ByVal parameter As Object, ByVal culture As CultureInfo) As Object _
        Implements IValueConverter.ConvertBack

        Throw New NotSupportedException()
    End Function
End Class
```

■ **Tip** If you decide you can't perform the conversion, you can return the value Binding.UnsetValue to tell Silverlight to ignore your binding. The bound property (in this case, Background) will keep its default value.

Once again, the value converter is carefully designed with reusability in mind. Rather than hard-coding the color highlights in the converter, they're specified in the XAML by the code that *uses* the converter:

```
<local:PriceToBackgroundConverter x:Key="PriceToBackgroundConverter"
  DefaultBrush="{x:Null}" HighlightBrush="Orange" MinimumPriceToHighlight="50">
</local:PriceToBackgroundConverter>
```

Brushes are used instead of colors so that you can create more advanced highlight effects using gradients and background images. And if you want to keep the standard, transparent background (so the background of the parent elements is used), set the DefaultBrush or HighlightBrush property to Nothing, as shown here.

All that's left is to use this converter to set the background of an element, such as the border that contains all the other elements:

```
<Border Background=
  "{Binding UnitCost, Converter={StaticResource PriceToBackgroundConverter}}"
  ... >
```

In many cases, you'll need to pass information to a converter beyond the data you want to convert. In this example, PriceToBackgroundConverter needs to know the highlight color and minimum price details, and this information is passed along through properties. However, you have one other alternative. You can pass a single object (of any type) to a converter through the binding expression, by setting the ConverterParameter property. Here's an example that uses this approach to supply the minimum price:

```
<Border Background=
  "{Binding UnitCost, Converter={StaticResource PriceToBackgroundConverter},
ConverterParameter=50}"
  ... >
```

The parameter is passed as an argument to the Convert() method. Here's how you can rewrite the earlier example to use it:

```
Public Function Convert(ByVal value As Object, ByVal targetType As Type, _
  ByVal parameter As Object, ByVal culture As CultureInfo) As Object _
  Implements IValueConverter.Convert

    Dim price As Double = CDbl(value)
    If price >= Double.Parse(parameter) Then
        Return HighlightBrush
    Else
        Return DefaultBrush
    End If
End Function
```

In general, the property-based approach is preferred. It's clearer, more flexible, and strongly typed. (When set in the markup extension, ConverterParameter is always treated as a string.) But in some situations, you may want to reuse a single value converter for multiple elements, and you may need to vary a single detail for each element. In this situation, it's more efficient to use ConverterParameter than to create multiple copies of the value converter.

Data Templates

A *data template* is a chunk of XAML markup that defines how a bound data object should be displayed. Two types of controls support data templates:

- Content controls support data templates through the ContentTemplate property. The content template is used to display whatever you've placed in the Content property.
- List controls (controls that derive from ItemsControl) support data templates through the ItemTemplate property. This template is used to display each item from the collection (or each row from a DataTable) that you've supplied as the ItemsSource.

The list-based template feature is based on content control templates: each item in a list is wrapped by a content control, such as ListBoxItem for the ListBox, ComboBoxItem for the ComboBox, and so on. Whatever template you specify for the ItemTemplate property of the list is used as the ContentTemplate of each item in the list.

What can you put inside a data template? It's simple. A data template is an ordinary block of XAML markup. Like any other block of XAML markup, the template can include any combination of elements. It should also include one or more data-binding expressions that pull out the information that you want to display. (After all, if you don't include any data-binding expressions, each item in the list will appear the same, which isn't very helpful.)

The best way to see how a data template works is to start with a basic list that doesn't use a template. For example, consider this list box, which was shown previously:

```
<ListBox Name="lstProducts" DisplayMemberPath="ModelName"></ListBox>
```

You can get the same effect with this list box that uses a data template:

```
<ListBox Name="lstProducts">
  <ListBox.ItemTemplate>
    <DataTemplate>
      <TextBlock Text="{Binding ModelName}"></TextBlock>
    </DataTemplate>
  </ListBox.ItemTemplate>
</ListBox>
```

When you bind the list to the collection of products (by setting the ItemsSource property), a single ListBoxItem is created for each Product object. The ListBoxItem.Content property is set to the appropriate Product object, and the ListBoxItem.ContentTemplate is set to the data template shown earlier, which extracts the value from the Product.ModelName property and displays it in a TextBlock.

So far, the results are underwhelming. But now that you've switched to a data template, there's no limit to how you can creatively present your data. Here's an example that wraps each item in a rounded border, shows two pieces of information, and uses bold formatting to highlight the model number:

```
<ListBox Name="lstProducts" HorizontalContentAlignment="Stretch"
 SelectionChanged="lstProducts_SelectionChanged">
  <ListBox.ItemTemplate>
    <DataTemplate>
      <Border Margin="5" BorderThickness="1" BorderBrush="SteelBlue"
       CornerRadius="4">
        <Grid Margin="3">
          <Grid.RowDefinitions>
            <RowDefinition></RowDefinition>
            <RowDefinition></RowDefinition>
          </Grid.RowDefinitions>
          <TextBlock FontWeight="Bold"
           Text="{Binding ModelNumber}"></TextBlock>
          <TextBlock Grid.Row="1"
           Text="{Binding ModelName}"></TextBlock>
        </Grid>
      </Border>
    </DataTemplate>
  </ListBox.ItemTemplate>
</ListBox>
```

When this list is bound, a separate Border object is created for each product. Inside the Border element is a Grid with two pieces of information, as shown in Figure 16-12.

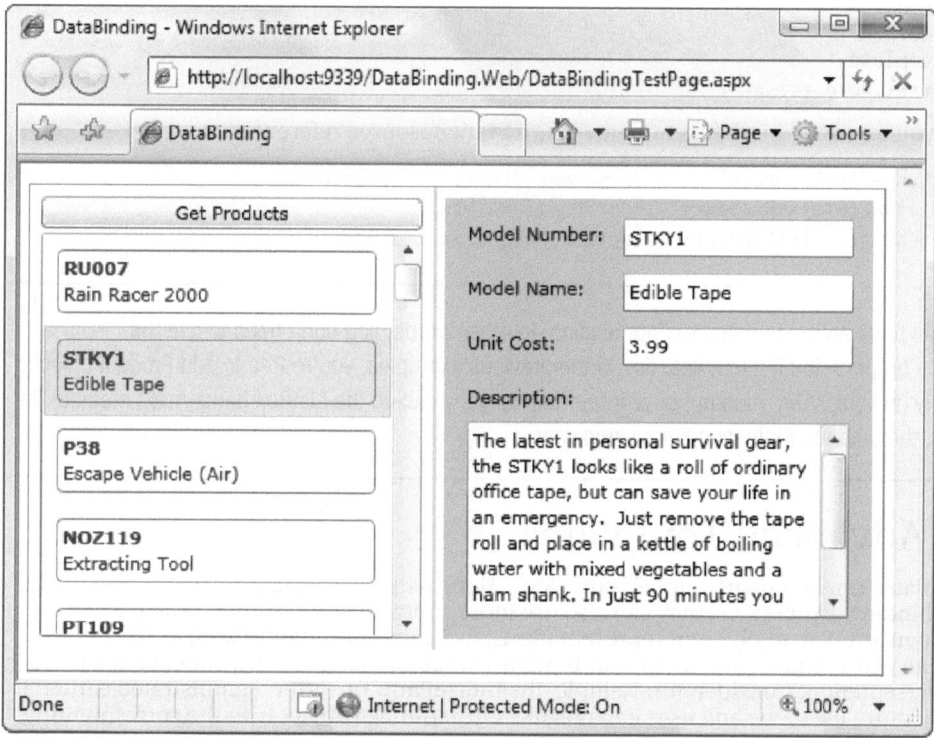

Figure 16-12. A list that uses a data template

Separating and Reusing Templates

Like styles, templates are often declared as a page or application resource rather than defined in the list where you use them. This separation is often clearer, especially if you use long, complex templates or multiple templates in the same control (as described in the next section). It also gives you the ability to reuse your templates in more than one list or content control if you want to present your data the same way in different places in your user interface.

To make this work, all you need to do is to define your data template in a resources collection and give it a key name. Here's an example that extracts the template shown in the previous example:

```xml
<UserControl.Resources>
  <DataTemplate x:Key="ProductDataTemplate">
    <Border Margin="5" BorderThickness="1" BorderBrush="SteelBlue"
     CornerRadius="4">
      <Grid Margin="3">
        <Grid.RowDefinitions>
          <RowDefinition></RowDefinition>
          <RowDefinition></RowDefinition>
        </Grid.RowDefinitions>
        <TextBlock FontWeight="Bold"
         Text="{Binding ModelNumber}"></TextBlock>
        <TextBlock Grid.Row="1"
         Text="{Binding ModelName}"></TextBlock>
      </Grid>
    </Border>
  </DataTemplate>
</UserControl.Resources>
```

Now you can use your data template using a StaticResource reference:

```xml
<ListBox Name="lstProducts" HorizontalContentAlignment="Stretch"
 ItemTemplate="{StaticResource ProductDataTemplate}"
 SelectionChanged="lstProducts_SelectionChanged"></ListBox>
```

■ **Note** Data templates don't require data binding. In other words, you don't need to use the ItemsSource property to fill a template list. In the previous examples, you're free to add Product objects declaratively (in your XAML markup) or programmatically (by calling the ListBox.Items.Add() method). In both cases, the data template works the same way.

More Advanced Templates

Data templates can be remarkably self-sufficient. Along with basic elements such as TextBlock and data-binding expressions, they can also use more sophisticated controls, attach event handlers, convert data to different representations, use animations, and so on.

You can use a value converter in your binding expressions to convert your data to a more useful representation. Consider, for example, the ImagePathConverter demonstrated earlier. It accepts a picture file name and uses it to create a BitmapImage object with the corresponding image content. This BitmapImage object can then be bound directly to the Image element.

You can use the ImagePathConverter to build the following data template that displays the image for each product:

```
<UserControl.Resources>
  <local:ImagePathConverter x:Key="ImagePathConverter"></local:ImagePathConverter>
  <DataTemplate x:Key="ProductDataTemplate">
    <Border Margin="5" BorderThickness="1" BorderBrush="SteelBlue"
     CornerRadius="4">
      <Grid Margin="3">
        <Grid.RowDefinitions>
          <RowDefinition></RowDefinition>
          <RowDefinition></RowDefinition>
          <RowDefinition></RowDefinition>
        </Grid.RowDefinitions>
        <TextBlock FontWeight="Bold" Text="{Binding Path=ModelNumber}"></TextBlock>
        <TextBlock Grid.Row="1" Text="{Binding Path=ModelName}"></TextBlock>
        <Image Grid.Row="2" Grid.RowSpan="2" Source=
"{Binding Path=ProductImagePath, Converter={StaticResource ImagePathConverter}}">
        </Image>
      </Grid>
    </Border>
  </DataTemplate>
</UserControl.Resources>
```

Although this markup doesn't involve anything exotic, the result is a much more interesting list (see Figure 16-13).

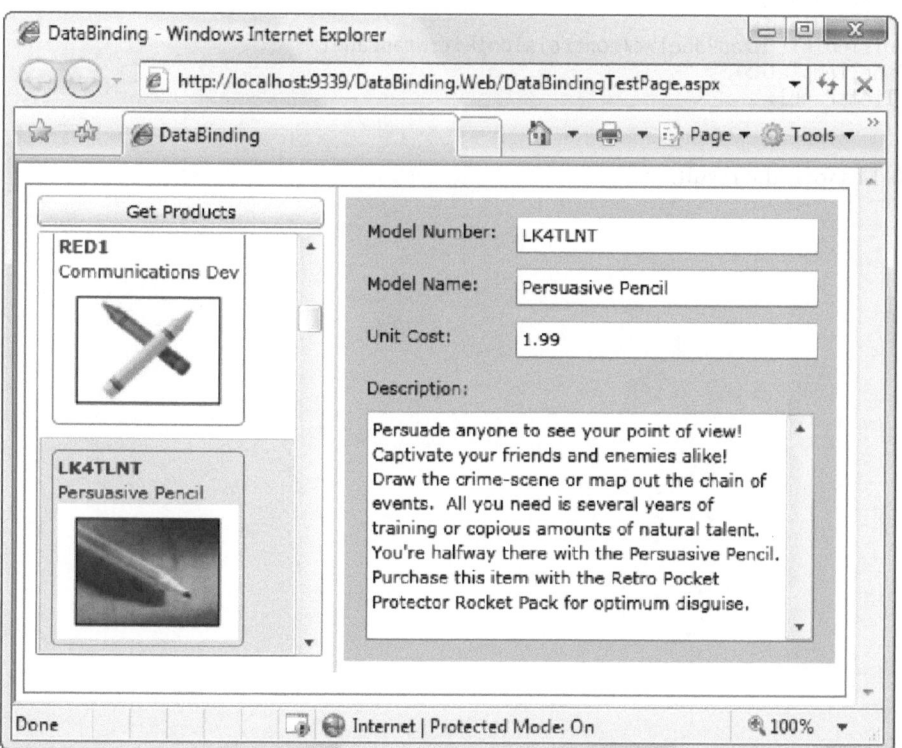

Figure 16-13. A list with image content

■ **Note** If there is an error in your template, you don't receive an exception. Instead, the control is unable to display your data and remains blank.

Changing Item Layout

Data templates give you remarkable control over every aspect of item presentation. However, they don't allow you to change how the items are organized with respect to each other. No matter what templates and styles you use, the list box puts each item into a separate horizontal row and stacks each row to create the list.

You can change this layout by replacing the container that the list uses to lay out its children. To do so, you set the ItemsPanel property with a block of XAML that defines the panel you want to use. This panel can be any class that derives from System.Windows.Controls.Panel, including a custom layout container that implements your own specialized layout logic.

The following uses the WrapPanel from the Silverlight Toolkit (http://silverlight.codeplex.com), which was described in Chapter 3. It arranges items from left to right over multiple rows:

```
<ListBox Margin="7,3,7,10" Name="lstProducts"
 ItemTemplate="{StaticResource ProductDataTemplate}">
  <ListBox.ItemsPanel>
    <ItemsPanelTemplate>
      <controlsToolkit:WrapPanel></controlsToolkit:WrapPanel>
    </ItemsPanelTemplate>
  </ListBox.ItemsPanel>
</ListBox>
```

Figure 16-14 shows the result.

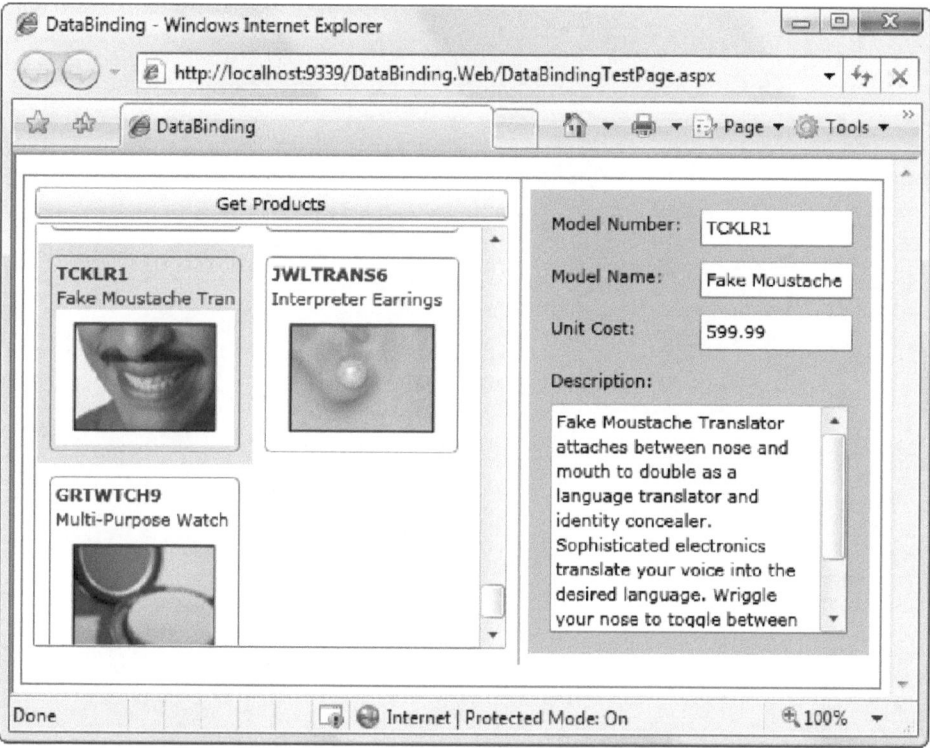

Figure 16-14. Tiling a list

The Last Word

This chapter took a thorough look at data binding. You learned how to create data-binding expressions that draw information from custom objects, use change notification and validation, bind entire collections of data, and get your records from a web service. You also explored a range of techniques you can use to customize the display of your data, from data conversion and conditional formatting with IValueConverter to data templates and custom layout.

In the next chapter, you'll build on these concepts as you take a deeper look into validation and consider rich data controls like the DataGrid, DataForm, and TreeView.

CHAPTER 17

■ ■ ■

Data Controls

So far, you've learned how to use data binding to pull information out of data objects, format it, and make it available for editing. However, although data binding is a flexible and powerful system, getting the result you want can still take a lot of work. For example, a typical data form needs to bind a number of different properties to different controls, arrange them in a meaningful way, and use the appropriate converters, templates, and validation logic. Creating these ingredients is as time-consuming as any other type of UI design.

Silverlight offers several features that can help offload some of the work:

- *The Label and DescriptionViewer controls*: They pull metadata out of your data objects and display it in your pages—automatically.
- *Data annotations*: Originally introduced with ASP.NET Dynamic Data, they let you embed validation rules in your data classes. Pair data annotations with the ValidationSummary control for an easy way to list all the validation errors in a page.
- *The DataGrid control*: It's the centerpiece of Silverlight's rich data support—a highly customizable table of rows and columns with support for sorting, editing, grouping, and (with the help of the DataPager) paging.
- *The TreeView control*: Silverlight's hierarchical tree isn't limited to data binding and doesn't support editing. However, it's a true time-saver when dealing with hierarchical data, such as a list of categories with nested lists of products.

In this chapter, you'll learn how to extend the data-binding basics you picked up in the previous chapter. You'll also learn how to pass your smart data objects across the great web service divide so that the same metadata and validation logic is available to your server-side ASP.NET code and your client-side Silverlight applications.

■ **What's New** Silverlight 4 keeps the same rich data controls that were introduced in Silverlight 3. The DataGrid sports a few minor refinements—for example, it now lets users scroll with the mouse wheel and copy the selected row of the data to the clipboard (so it can be pasted into another program). More significantly, the DataGrid adds a new star-sizing feature that automatically resizes columns to fit the available space, with no awkward horizontal scrollbar. You'll learn more in the "Controlling Column Width" section.

Better Data Forms

In the previous chapter, you learned how to use data binding to build basic data forms. These forms—essentially, ordinary pages made up of text boxes and other bound controls—allow users to enter, edit, and review data. Best of all, they require relatively little code.

But the reality isn't as perfect as it seems. To build a data form, you need a fair bit of handwritten XAML markup, which must include hard-coded details such as caption text, prompts, and error messages. Managing all these details can be a significant headache, especially if the data model changes frequently. Change the database, and you'll be forced to alter your data classes *and* your user interface—and Visual Studio's compile-time error checking can't catch invalid bindings or outdated validation rules.

For these reasons, the creators of Silverlight are hard at work building higher-level data controls, helper classes, and even a whole server-based data-management framework (the forthcoming RIA Services). And although these features are still evolving rapidly, some components have already trickled down into the Silverlight platform. In the following sections, you'll see how three of them—the Label, DescriptionViewer, and ValidationSummary—can make it easier to build rich data forms right now, particularly when you combine them with the powerful *data annotations* feature.

■ **Note** To get access to the Label, DescriptionViewer, and ValidationSummary controls, you must add a reference to the System.Windows.Controls.Data.Input.dll assembly. If you add one of these controls from the Toolbox, Visual Studio will add the assembly reference and map the namespace like this:

```
xmlns:dataInput="clr-namespace:System.Windows.Controls;assembly=System.Windows.
Controls.Data.Input"
```

The Goal: Data Class Markup

Although you can use Label, DescriptionViewer, and ValidationSummary on their own, their greatest advantages appear when you use them with *smart* data classes—classes that use a small set of attributes to embed extra information. These attributes allow you to move data-related details (such as property descriptions and validation rules) into your data classes rather than force you to include them in the page markup.

The attribute-based approach has several benefits. First, it's an impressive time-saver that lets you build data forms faster. Second, and more important, it makes your application far more maintainable because you can keep data details properly *synchronized*. For example, if the underlying data model changes and you need to revise your data classes, you simply need to tweak the attributes. This is quicker and more reliable than attempting to track down the descriptive text and validation logic that's scattered through multiple pages—especially considering that the data class code is usually compiled in a separate project (and possibly managed by a different developer).

■ **Note** The attribute-based approach also has a drawback when you're using web services. Because all the attributes are placed on the data class, they'll be stripped out of the automatically generated client-side code. To avoid this problem, you need to use web service type sharing, which lets you share the same class definition between the server-side ASP.NET code and the client-side Silverlight code. Chapter 15 describes how to use this technique, and the downloadable code for this chapter puts it into practice.

In the following sections, you'll see how the data attributes works. Using them, you'll learn how you can embed captions, descriptions, and validation rules directly in your data objects.

The Label

The Label takes the place of the TextBlock that captions your data controls. For example, consider this markup, which displays the text "Model Number" followed by the text box that holds the model number:

```
<TextBlock Margin="7">Model Number:</TextBlock>
<TextBox Margin="5" Grid.Column="1" x:Name="txtModelNumber"
 Text="{Binding ModelNumber, Mode=TwoWay}"></TextBox>
```

You can replace the TextBlock using a label like this:

```
<dataInput:Label Margin="7" Content="Model Number:"></dataInput:Label>
<TextBox Margin="5" Grid.Column="1" x:Name="txtModelNumber"
 Text="{Binding ModelNumber, Mode=TwoWay}"></TextBox>
```

Used in this way, the label confers no advantage. Its real benefits appear when you use binding to latch it onto the control you're captioning using the Target property, like this:

```
<dataInput:Label Margin="7" Target="{Binding ElementName=txtModelNumber}">
</dataInput:Label>
<TextBox Margin="5" Grid.Column="1" x:Name="txtModelNumber"
 Text="{Binding ModelNumber, Mode=TwoWay}"></TextBox>
```

When used this way, the label does something interesting. Rather than rely on you to supply it with a fixed piece of text, it examines the referenced element, finds the bound property, and looks for a Display attribute like this one:

```
<Display(Name := "Model Number")> _
Public Property ModelNumber() As String
End Property
```

The label then displays that text—in this case, the cleanly separated two-word caption "Model Number."

■ **Note** Before you can add the Display attribute to your data class, you need to add a reference to the System.ComponentModel.DataAnnotations.dll assembly. (You also need to import the System.ComponentModel.DataAnnotations namespace where the Display attribute is defined.)

From a pure markup perspective, the label doesn't save any keystrokes. You may feel that it takes more effort to write the binding expression that connects the label than it does to fill the TextBlock with the same text. However, the label approach has several advantages. Most obviously, it's highly maintainable—if you change the data class at any time, the new caption will flow seamlessly into any data forms that use the data class, with no need to edit a line of markup.

The label isn't limited to displaying a basic caption. It also varies its appearance to flag required properties and validation errors. To designate a required property (a property that must be supplied in order for the data object to be valid), add the Required attribute:

```
<Required>, _
<Display(Name := "Model Number")> _
Public Property ModelNumber() As String
End Property
```

By default, the label responds by bolding the caption text. But you can change this formatting—or even add an animated effect—by modifying the Required and NotRequired visual states in the label's control template. (To review control templates and visual states, refer to Chapter 13.)

Similarly, the label pays attentions to errors that occur when the user edits the data. For this to work, your binding must opt in to validation using the ValidatesOnExceptions and NotifyOnValidationError properties, as shown here with the UnitCost property:

```
<dataInput:Label Margin="7" Grid.Row="2"
  Target="{Binding ElementName=txtUnitCost}"></dataInput:Label>
<TextBox Margin="5" Grid.Row="2" Grid.Column="1" x:Name="txtUnitCost" Width="100"
  HorizontalAlignment="Left" Text="{Binding UnitCost, Mode=TwoWay,
  ValidatesOnExceptions=true, NotifyOnValidationError=true}"></TextBox>
```

To test this, type non-numeric characters into the UnitCost field, and tab away. The caption text in the label shifts from black to red (see Figure 17-1). If you want something more interesting, you can change the control template for the label—this time, you need to modify the Valid and Invalid visual states.

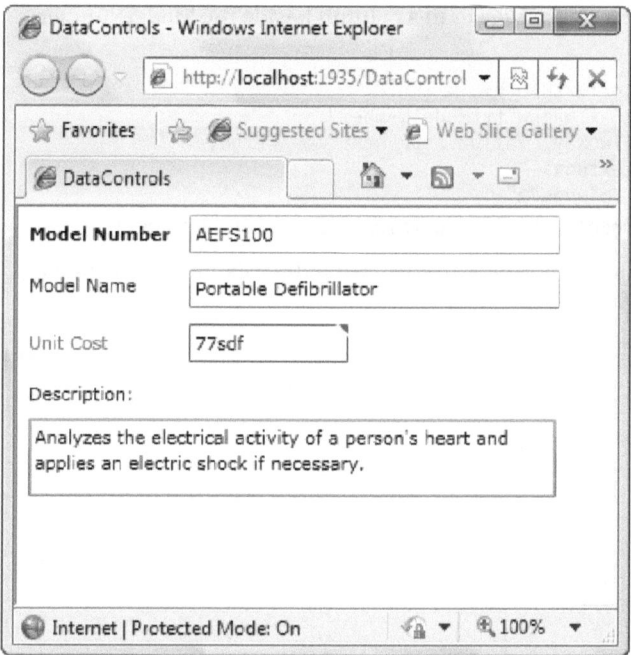

Figure 17-1. A required ModelNumber and invalid UnitCost

■ **Note** The error notification in the label is in *addition* to the standard error indicator in the input control. For example, in the page shown on Figure 17-1, the UnitCost text box shows a red outline and red triangle in the upper-right corner to indicate that the data it contains is invalid. In addition, when the UnitCost text box gets the focus, a red balloon pops up with the error description.

The DescriptionViewer

The Label control takes care of displaying caption text, and it adds the ability to highlight required properties and invalid data. However, when users are filling out complex forms, they sometimes need a little more. A few words of descriptive text can work wonders, and the DescriptionViewer control gives you a way to easily incorporate this sort of guidance into your user interface.

It all starts with the Display attribute you saw in the previous section. Along with the Name property, it accepts a Description property that's intended for a sentence or two or more detailed information:

```
<Display(Name := "Model Number", _
 Description := "This is the alphanumeric product tag used in the warehouse.")> _
Public Property ModelNumber() As String
End Property
```

Here's the markup that adds a DescriptionViewer to a column beside the ModelNumber text box:

```
<TextBlock Margin="7">Model Number</TextBlock>
<TextBox Margin="5" Grid.Column="1" x:Name="txtModelNumber"
 Text="{Binding ModelNumber, Mode=TwoWay, ValidatesOnExceptions=true,
NotifyOnValidationError=true}"></TextBox>
<dataInput:DescriptionViewer Grid.Column="2"
 Target="{Binding ElementName=txtModelNumber}"></dataInput:DescriptionViewer>
```

The DescriptionViewer shows a small information icon. When the user moves the mouse over it, the description text appears in a tooltip (Figure 17-2).

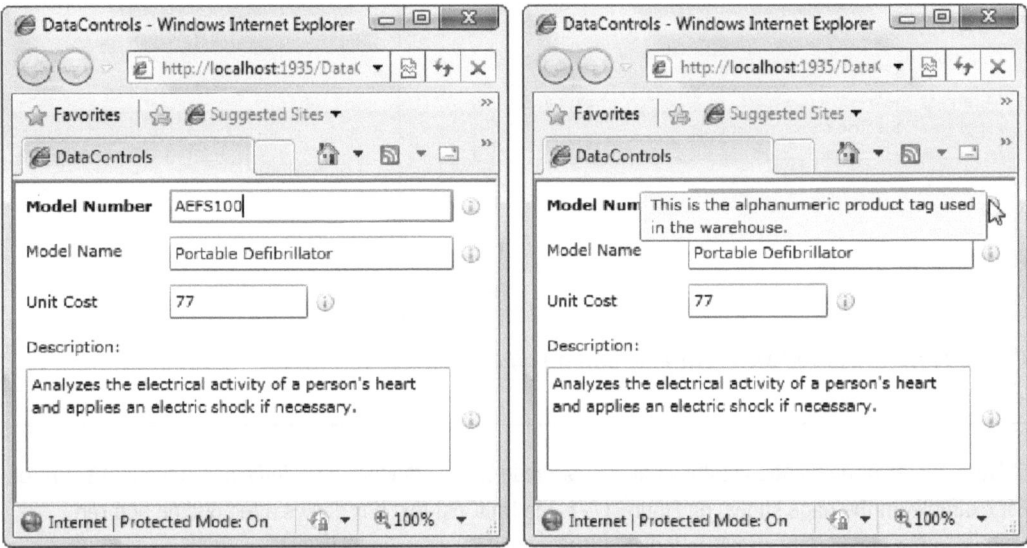

Figure 17-2. The DescriptionViewer

You can replace the icon with something different by setting the GlyphTemplate property, which determines the display content of the DescriptionViewer. Here's an example that swaps in a new icon:

```
<dataInput:DescriptionViewer Grid.Row="1" Grid.Column="2"
 Target="{Binding ElementName=ModelName}">
  <dataInput:DescriptionViewer.GlyphTemplate>
    <ControlTemplate>
      <Image Source="info.jpg" Stretch="None"></Image>
    </ControlTemplate>
  </dataInput:DescriptionViewer.GlyphTemplate>
</dataInput:DescriptionViewer>
```

The DescriptionViewer doesn't change its appearance when the bound data has a validation error. However, it does include an IsValid property, and it does support the four basic visual states for validation (ValidFocused, ValidUnfocused, InvalidFocused, and InvalidUnfocused). That means you can change the DescriptionViewer template and add some sort of differentiator that changes its appearance to highlight errors or applies a steady-state animation.

The ValidationSummary

You've now seen several ways that Silverlight helps you flag invalid data. First, as you learned in the previous chapter, most input controls change their appearance when something is amiss—for example, changing their border to a red outline. Second, these input controls also show a pop-up error message when the control has focus. Third, if you're using the Label control, it turns its caption text red. And fourth, if you're using the DescriptionViewer control, you can replace the default control template with one that reacts to invalid data (much as you can change the way a label and input controls display their error notifications by giving them custom control templates).

All these techniques are designed to give *in-situ* error notifications—messages that appear next to or near the offending input. But in long forms, it's often useful to show an error list that summarizes the problems in a group of controls. You can implement a list like this by reacting to the BindingValidationError described in the previous chapter. But Silverlight has an even easier option that does the job with no need for code: the ValidationSummary control.

The ValidationSummary monitors a container for error events. For example, if you have Grid with input controls, you can point the ValidationSummary to that Grid using the Target property. It will then detect the errors that occur in any of the contained input controls. (Technically, you could point the ValidationSummary at a single input control, but that wouldn't have much point.) Most of the time, you don't need to set the Target property. If you don't, the ValidationSummary retrieves a reference to its container and monitors all the controls inside. To create the summary shown in Figure 17-3, you need to add the ValidationSummary somewhere inside the Grid that holds the product text boxes:

```
<dataInput:ValidationSummary Grid.Row="6" Grid.ColumnSpan="3" Margin="7" />
```

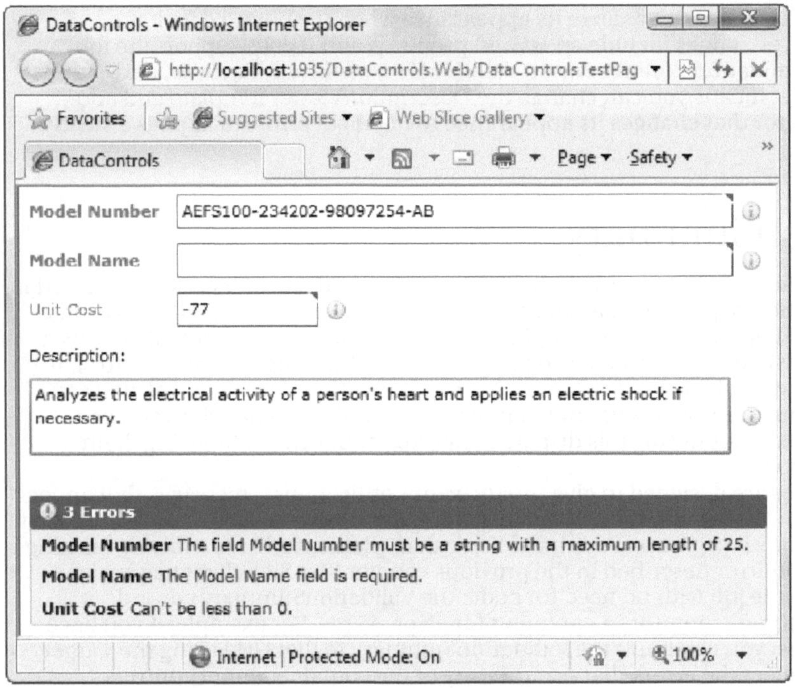

Figure 17-3. A validation summary with three errors

■ **Note** Remember, to catch errors with the ValidationSummary, your bindings must have Mode set to TwoWay, and they must set ValidatesOnExceptions and NotifyOnValidationError to true.

When no errors are present, the ValidationSummary is invisible and collapsed so that it takes no space. When there are one or more errors, you see the display shown in Figure 17-3. It consists of a header (which displays an error icon and the number of errors) and a list of errors that details the offending property and the exception message. If the user clicks one of the error message, the ValidationSummary fires the FocusingInvalidControl event and transfers focus to the input control with the data (unless you've explicitly set the FocusControlsOnClick property to False).

If you want to prevent a control from adding its errors to the ValidationSummary, you can set the attached ShowErrorsInSummary property, as shown here:

```
<TextBox Margin="5" x:Name="txtUnitCost" Width="100" HorizontalAlignment="Left"
  dataInput:ValidationSummary.ShowErrorsInSummary="False"
  Text="{Binding UnitCost, Mode=TwoWay, ValidatesOnExceptions=true,
NotifyOnValidationError=true}"></TextBox>
```

The ValidationSummary also provides several properties you can use to customize its appearance. Use HeaderTemplate to supply a data template that changes how the title is presented, use Header to supply your own custom header text, and use SummaryListBoxStyle to change how the error list is formatted. Programmatically, you may want to check the HasErrors property to determine whether the form is valid and the Errors collection to examine all the problems that were detected.

Data Annotations

Now that you've seen how to improve the error reporting in your forms, it's worth considering a complementary feature that makes it easier to implement the validation rules that check your data. Currently, Silverlight validation responds to unhandled exceptions that occur when you attempt to set a property. If you want to implement custom validation, you're forced to write code in the property setter that tests the new value and throws an exception when warranted. The validation code you need is repetitive to write and tedious to maintain. And if you need to check several different error conditions, your code can grow into a tangled mess that inadvertently lets certain errors slip past.

Silverlight offers a solution with its new support for *data annotations*, which allow you to apply validation rules by attaching one or more attributes to the properties in your data class. Done right, data annotations let you move data validation out of your code and into the declarative metadata that decorates it, which improves the clarity of your code and the maintainability of your classes.

■ **Note** The data annotations system was originally developed for ASP.NET Dynamic Data, but Silverlight borrows the same model. Technically, you've already seen annotations at work, as the Display and Required attributes demonstrated in the previous section are both data annotations.

Raising Annotation Errors

Before you can use data annotations, you need to add a reference to the System.ComponentModel.DataAnnotations.dll assembly, which is the same assembly you used to access the Display and Required attributes in the previous section. You'll find all the data-annotation classes in the matching namespace, System.ComponentModel.DataAnnotations.

Data annotations work through a small set of attributes that you apply to the property definitions in your data class. Here's an example that uses the StringLength attribute to cap the maximum length of the ModelName field at 25 characters:

```
<StringLength(25)>, _
<Display(Name := "Model Name", Description := "This is the retail product name.")> _
Public Property ModelName() As String
    Get
        Return modelName
    End Get
    Set(ByVal value As String)
        modelName = value
        OnPropertyChanged(New PropertyChangedEventArgs("ModelName"))
    End Set
End Property
```

This setup looks perfect: the validation rule is clearly visible, easy to isolate, and completely separate from the property setting code. However, it's not enough for Silverlight's data-binding system. Even with data annotations, all of Silverlight's standard controls require an exception before they recognize the presence of invalid data.

Fortunately, there's an easy way to throw the exception you need, when you need it. The trick is the Validator class, which provides several shared helper methods that can test your data annotations and check your properties for bad data. The ValidateProperty() method throws an exception if a specific value is invalid for a specific property. The ValidateObject() method examines an entire object for problems and throws an exception if any property is out of whack. The TryValidateProperty() and TryValidateObject() methods perform much the same tasks, but they provide a ValidationResult object that explains potential problems rather than throwing a ValidationException.

The following example shows the three lines of code you use to check a property value with the ValidateProperty() method. When called, this code examines all the validation attributes attached to the property and throws a ValidationException as soon as it finds one that's been violated:

```
<StringLength(25)>, _
<Display(Name := "Model Name", Description := "This is the retail product name.")> _
Public Property ModelName() As String
    Get
        Return modelName
    End Get
    Set(ByVal value As String)
```

```
    ' Explicitly raise an exception if a data annotation attribute
    ' fails validation.
    Dim context As New ValidationContext(Me, Nothing, Nothing)
    context.MemberName = "ModelName"
    Validator.ValidateProperty(value, context)

    modelName = value
    OnPropertyChanged(New PropertyChangedEventArgs("ModelName"))
  End Set
End Property
```

By adding code like this to all your property setters, you can enjoy the best of the data-annotation system—straightforward attributes that encode your validation logic—and still plug your validation into the Silverlight data-binding system.

■ **Note** Data annotations are powerful, but they aren't perfect for every scenario. In particular, they still force your data class to throw exceptions to indicate error conditions. This design pattern isn't always appropriate (for example, it runs into problems if you need an object that's temporarily in an invalid state or you want to impose restrictions only for user edits, not programmatic changes). It's also a bit dangerous, because making the wrong change to a data object in your code has the potential to throw an unexpected exception and derail your application. (To get around this, you can create an AllowInvalid property in your data classes that, when True, tells them to bypass the validation-checking code. But it's still awkward at best.) For this reason, many developers prefer to use the IDataError or INotifyDataError interface with their data objects, as described in Chapter 16.

The Annotation Attributes

To use validation with data annotations, you need to add the right attributes to your data classes. The following sections list the attributes you can use, all of which derive from the base ValidationAttribute class and are found in the System.ComponentModel.DataAnnotations namespace. All of these attributes inherit the ValidationAttribute.ErrorMessage property, which you can set to add custom error message text. This text is featured in the pop-up error balloon and shown in the ValidationSummary control (if you're using it).

■ **Tip** You can add multiple restrictions to a property by stacking several different data annotation attributes.

Required

This attribute specifies that the field must be present—if it's left blank, the user receives an error. This works for zero-length strings, but it's relatively useless for numeric values, which start out as perfectly acceptable 0 values:

```
<Required()> _
Public Property ModelNumber() As String
    ...
End Property
```

Here's an example that adds an error message:

```
<Required(ErrorMessage := "You must valid ACME Industries ModelNumber.")> _
Public Property ModelNumber() As String
    ...
End Property
```

StringLength

This attribute sets the maximum length of a string. You can also (optionally) set a minimum length by setting the MinimumLength property, as shown here:

```
<StringLength(25, MinimumLength := 5)> _
Public Property ModelName() As String
    ...
End Property
```

When you're using an attribute that has important parameters, like StringLength, you can add these details to your error message using numbered placeholders, where {0} is the name of the field that's being edited, {1} is the constructor argument, {2} is the first attribute property, {3} is the second, and so on. Here's an example:

```
<StringLength(25, MinimumLength := 5, _
 ErrorMessage := "Model names must have between {2} and {1} characters.")> _
Public Property ModelName() As String
    ...
End Property
```

When the StringLength causes a validation failure, it sets the error message to this text: "Model names must have between 5 and 25 characters."

Range

This attribute forces a value to fall within a range between a minimum value and a maximum value, as shown here:

```
<Range(0,1000)> _
Public Property UnitsInStock() As Integer
    ...
End Property
```

The Range attribute is generally used for numeric data types, but you can use any type that implements IComparable—just use the overloaded version of the constructor that takes the data type as a Type argument, and supply your values in string form:

```
<Range(GetType(DateTime), "1/1/2005", "1/1/2010")> _
Public Property ExpiryDate() As DateTime
    ...
End Property
```

RegularExpression

This attribute tests a text value against a *regular expression*—a formula written in a specialized pattern-matching language.

Here's an example that allows one or more alphanumeric characters (capital letters from *A–Z*, lowercase letters from *a–z*, and numbers from 0–9, but nothing else):

```
^[A-Za-z0-9]+$
```

The first character (^) indicates the beginning of the string. The portion in square brackets identifies a range of options for a single character—in essence, it says that the character can fall between *A* to *Z*, or *a* to *z*, or 0 to 9. The + that immediately follows extends this range to match a sequence of one or more characters. Finally, the last character ($) represents the end of the string.

To apply this to a property, you use the RegularExpression attribute like this:

```
<RegularExpression("^[A-Za-z0-9]+$")> _
Public Property ModelNumber() As String
    ...
End Property
```

In this example, the characters ^, [], +, and $ are all metacharacters that have a special meaning in the regular-expression language. Table 17-1 gives a quick outline of all the metacharacters you're likely to use.

Table 17-1. Regular-Expression Metacharacters

Character	Rule
*	Represents zero or more occurrences of the previous character or subexpression. For example, a*b matches *aab* or just *b*.
+	Represents one or more occurrences of the previous character or subexpression. For example, a+b matches *aab* but not *a*.
( )	Groups a subexpression that is treated as a single element. For example, (ab)+ matches *ab* and *ababab*.
{*m*}	Requires *m* repetitions of the preceding character or group. For example, a{3} matches *aaa*.
{*m, n*}	Requires *n* to *m* repetitions of the preceding character or group. For example, a{2,3} matches *aa* and *aaa* but not *aaaa*.

Character	Rule
|	Represents either of two matches. For example, a|b matches *a* or *b*.
[ ]	Matches one character in a range of valid characters. For example, [A-C] matches *A*, *B*, or *C*.
[^]	Matches a character that isn't in the given range. For example, [^A-C] matches any character except *A*, *B*, and *C*.
.	Represents any character except newline.
\s	Represents any whitespace character (like a tab or space).
\S	Represents any nonwhitespace character.
\d	Represents any digit character.
\D	Represents any character that isn't a digit.
\w	Represents any alphanumeric character (letter, number, or underscore).
^	Represents the start of the string. For example, ^ab can find a match only if the string begins with *ab*.
$	Represents the end of the string. For example, ab$ can find a match only if the string ends with *ab*.
\	Indicates that the following character is a literal (even though it may ordinarily be interpreted as a metacharacter). For example, use \\ for the literal \ and use \+ for the literal +.

Regular expressions are all about patterned text. In many cases, you won't devise a regular expression yourself—instead, you'll look for the correct premade expression that validates postal codes, email addresses, and so on. For a detailed exploration of the regular-expression language, check out a dedicated book like the excellent *Mastering Regular Expressions* (O'Reilly, Jeffrey Friedl).

REGULAR EXPRESSION BASICS

All regular expressions are made up of two kinds of characters: literals and metacharacters. *Literals* represent a specific defined character. *Metacharacters* are wildcards that can represent a range of values. Regular expressions gain their power from the rich set of metacharacters that they support (see Table 17-1).

Two examples of regular-expression metacharacters include the ^ and $ characters you've already seen, which designate the beginning and ending of the string. Two more common metacharacters are \s (which represents any whitespace character) and \d (which represents any digit). Using these characters, you can construct the following expression, which will successfully match any string that starts with the numbers 333, followed by a single whitespace character and any three numbers. Valid matches include 333 333, 333 945, but not 334 333 or 3334 945:

```
^333\s\d\d\d$
```

You can also use the plus (+) sign to represent a repeated character. For example, 5+7 means "any number of *5* characters, followed by a single *7*." The number 57 matches, as does 555557. In addition, you can use the brackets to group together a subexpression. For example, (52)+7 matches any string that starts with a sequence of 52. Matches include 527, 52527, 52552527, and so on.

You can also delimit a range of characters using square brackets. [a-f] matches any single character from *a* to *f* (lowercase only). The following expression matches any word that starts with a letter from *a* to *f*, contains one or more letters, and ends with *ing*—possible matches include *acting* and *developing*:

```
^[a-f][a-z]+ing$
```

This discussion just scratches the surface of regular expressions, which constitute an entire language of their own. However, you don't need to learn everything there is to know about regular expressions before you start using them. Many programmers look for useful prebuilt regular expressions on the Web. Without much trouble, you can find examples for e-mails, phone numbers, postal codes, and more, all of which you can drop straight into your applications.

CustomValidation

The most interesting validation attribute is CustomValidation. It allows you to write your own validation logic in a separate class and then attach that logic to a single property or use it to validate an entire data object.

Writing a custom validator is remarkably easy. All you need to do is write a shared method—in any class—that accepts the property value you want to validate (and, optionally, the ValidationContext) and returns a ValidationResult. If the value is valid, you return ValidationResult.Success. If the value isn't valid, you create a new ValidationResult, pass in a description of the problem, and return that object. You then connect that custom validation class to the field you want to validate with the CustomValidation attribute.

Here's an example of a custom validation class named ProductValidation. It examines the UnitCost property and only allows prices that end in 75, 95, or 99:

```
Public Class ProductValidation
    Public Shared Function ValidateUnitCost(ByVal value As Double, _
      ByVal context As ValidationContext) As ValidationResult
        ' Get the cents portion.
        Dim valueString As String = value.ToString()
```

```
            Dim cents As String = ""
            Dim decimalPosition As Integer = valueString.IndexOf(".")
            If decimalPosition <> -1 Then
                cents = valueString.Substring(decimalPosition)
            End If

            ' Perform the validation test.
            If (cents <> ".75") AndAlso (cents <> ".99") AndAlso (cents <> ".95") Then
                Return New ValidationResult( _
                    "Retail prices must end with .75, .95, or .99 to be valid.")
            Else
                Return ValidationResult.Success
            End If
        End Function
End Class
```

To enforce this validation, use the CustomValidation attribute to attach it the appropriate property. You must specify two arguments: the type of your custom validation class and the name of the static method that does the validation. Here's an example that points to the ValidateUnitCost() method in the ProductValidation class:

```
<CustomValidation(GetType(ProductValidation), "ValidateUnitCost")> _
Public Property UnitCost() As Double
    ...
End Property
```

Figure 17-4 shows this rule in action.

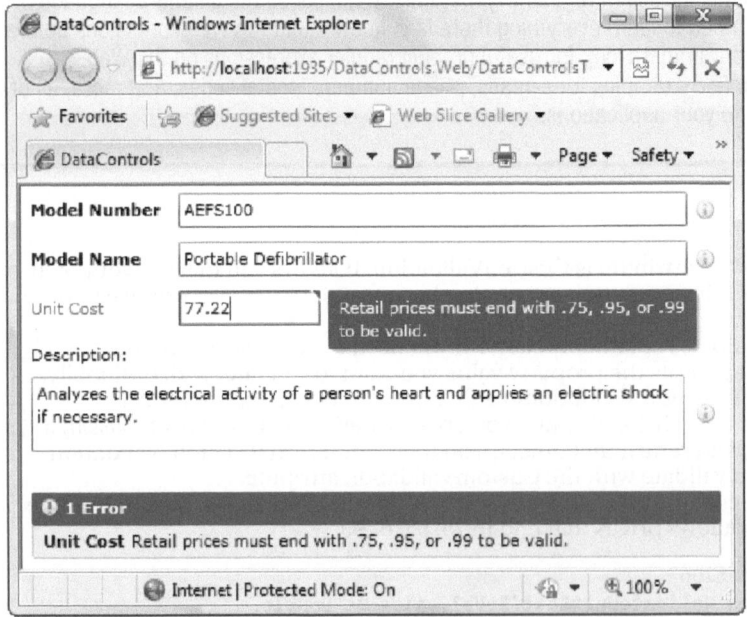

Figure 17-4. Violating a custom validation rule

You can also use the CustomValidation attribute to attach a class-wide validation rule. This is useful if you need to perform validation that compares properties (for example, making sure one property is less than another). Here's a validation method that checks to make sure the ModelNumber and ModelName properties have different values:

```
Public Shared Function ValidateProduct(ByVal product As Product, _
  ByVal context As ValidationContext) As ValidationResult
    If product.ModelName = product.ModelNumber Then
        Return New ValidationResult( _
          "You can't use the same model number as the model name.")
    Else
        Return ValidationResult.Success
    End If
End Function
```

And here's the CustomValidation attribute that attaches it to the Product class:

```
<CustomValidation(GetType(ProductValidation), "ValidateProduct")> _
Public Class Product
    Implements INotifyPropertyChanged
    ...
End Class
```

Class-wide validation rules have a significant drawback. Like property-validation rules, it's up to you to enforce them by calling the Validator.ValidateObject() method. Unfortunately, it doesn't make sense to do this in any of your property setters, because class-wide validation should be performed after the entire editing process is complete for the current object. Two of Silverlight's rich data controls—DataGrid and DataForm—solve the problem by triggering the Validator.ValidateObject() method themselves as soon as the user moves to a different record. But if you're not using either of these controls, custom validators may not be worth the trouble.

The DataGrid

DataGrid is, at its name suggests, a data-display control that takes the information from a collection of objects and renders it in a grid of rows and cells. Each row corresponds to a separate object, and each column corresponds to a property in that object.

The DataGrid adds much-needed versatility for dealing with data in Silverlight. Its advantages include the following:

- *Flexibility*: You use a column-based model to define exactly the columns you want to use and supply the binding expressions that extract the data from the bound objects. The DataGrid also supports a few important tools you learned about in Chapter 16, such as templates and value converters.

- *Customizability*: You can radically alter the appearance of the DataGrid using properties, along with headers and styles that format individual components of the grid. And if you're truly daring, you can give the entire DataGrid a new control template, complete with custom-drawn visuals and animations.

- *Editing*: The DataGrid gives you the ability to monitor the editing process and roll back invalid changes. It supports the exception-based validation you learned about in

Chapter 16, and it also supports *data annotations*—flexible validation rules that you add to data classes using attributes instead of code.

* *Performance*: The DataGrid boasts excellent performance with large sets of data because it uses *virtualization*. That means the DataGrid only retains in-memory objects for the data that's currently visible, not for the entire set of data that's loaded. This reduces the memory overhead dramatically and lets the control hold tens of thousands of rows without a serious slowdown.

Creating a Simple Grid

The DataGrid is defined in the familiar System.Windows.Controls namespace, but it's deployed in a different assembly from other Silverlight elements: the System.Windows.Controls.Data.dll assembly. By default, your Silverlight project doesn't have a reference to this assembly. However, as soon as you add a DataGrid from the Toolbox, Visual Studio adds the reference and inserts a new namespace mapping like the one shown here:

```
<UserControl xmlns:data=
 "clr-namespace:System.Windows.Controls;assembly=System.Windows.Controls.Data" ... >
```

This maps the DataGrid and its related classes to the namespace prefix *data*.

To create a quick-and-dirty DataGrid, you can use automatic column generation. To do so, you need to set the AutoGenerateColumns property to True (which is the default value):

```
<data:DataGrid x:Name="gridProducts" AutoGenerateColumns="True">
</data:DataGrid>
```

Now, you can fill the DataGrid as you fill a list control, by setting the ItemsSource property:

```
gridProducts.DataSource = products
```

Figure 17-5 shows a DataGrid that uses automatic column generation with the collection of Product objects you used throughout Chapter 16. When using automatic column generation, the DataGrid uses reflection to find every public property in the bound data object. It creates a column for each property. To display nonstring properties, the DataGrid calls ToString(), which works well for numbers, dates, and other simple data types but won't work as well if your objects include a more complex data object. (In this case, you may want to explicitly define your columns, which gives you the chance to bind to a subproperty, use a value converter, or apply a template to get the right display content.)

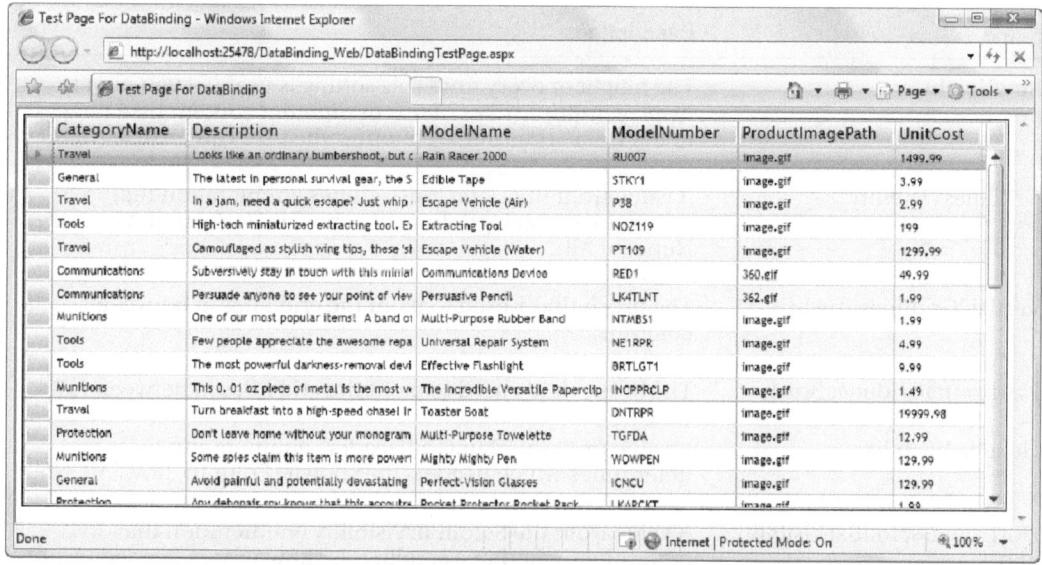

Figure 17-5. A DataGrid with automatically generated columns

Table 17-2 lists some of the properties you can use to customize a DataGrid's basic display. In the following sections, you'll see how to get fine-grained formatting control with styles and templates. You'll also see how the DataGrid deals with sorting and selection, and you'll consider many more properties that underlie these features.

Table 17-2. Basic Display Properties for the DataGrid

Name	Description
RowBackground and AlternatingRowBackground	RowBackground sets the brush that's used to paint the background behind every row. If you set AlternatingRowBackground, alternate rows are painted with a different background color, making it easier to distinguish rows at a glance. By default, the DataGrid gives odd-number rows a white background and gives the alternating, even-numbered rows a light gray background.
ColumnHeaderHeight	The height (in pixels) of the row that has the column headers at the top of the DataGrid.
RowHeaderWidth	The width (in pixels) of the column that has the row headers. This is the column at the far left of the grid, which shows no data but indicates the currently selected row (with an arrow) and indicates when the row is being edited (with an arrow in a circle).
ColumnWidth	The default width of every column. If you define columns explicitly, you can override this width to size individual columns. By default, columns are 100 pixels wide.

Name	Description
RowHeight	The height of every row. This setting is useful if you plan to display multiple lines of text or different content (like images) in the DataGrid. Unlike columns, the user can't resize rows.
GridlinesVisibility	A value from the DataGridGridlines enumeration that determines which gridlines are shown (Horizontal, Vertical, None, or All).
VerticalGridlinesBrush	The brush that's used to paint the grid lines in between columns.
HorizontalGridlinesBrush	The brush that's used to paint the grid lines in between rows.
HeadersVisibility	A value from the DataGridHeaders enumeration that determines which headers are shown (Column, Row, All, None).
HorizontalScrollBarVisibility and VerticalScrollBarVisibility	A value from the ScrollBarVisibility enumeration that determines whether a scrollbar is shown when needed (Auto), always (Visible), or never (Hidden). The default for both properties is Auto. No matter what this property value is, the horizontal scroll bar is not shown if you include one or more star-sized columns.

How Columns Are Resized and Rearranged

When displaying automatically generated columns, the DataGrid attempts to size the width of each column intelligently. Initially, it makes each column just wide enough to show the largest value that's currently in view (or the header, if that's wider).

The DataGrid attempts to preserve this intelligent sizing approach when the user starts scrolling through the data. As soon as you come across a row with longer data, the DataGrid widens the appropriate columns to fit it. This automatic sizing is one-way only, so columns don't shrink when you leave large data behind.

The automatic sizing of the DataGrid columns is interesting and often useful, but it's not always what you want. Consider the example shown in Figure 17-5, which contains a Description column that holds a long string of text. Initially, the Description column is made extremely wide to fit this data, crowding the other columns out of the way. (In Figure 17-5, the user has manually resized the Description column to a more sensible size. All the other columns are left at their initial widths.) After a column has been resized, it doesn't exhibit the automatic enlarging behavior when the user scrolls through the data.

■ **Note** Obviously, you don't want to force your users to grapple with ridiculously wide columns. To size columns correctly from the start, you need to define your columns explicitly, as described in the next section.

Ordinarily, users can resize columns by dragging the column edge to either size. You can prevent the user from resizing the columns in your DataGrid by setting the CanUserResizeColumns property to False. If you want to be more specific, you can prevent the user from resizing an individual column by setting the CanUserResize property of that column to False. You can also prevent the user from making the column extremely narrow by setting the column's MinWidth property.

The DataGrid has another surprise frill that lets users customize the column display. Not only can columns be resized, but they can also be dragged from one position to another. If you don't want users to have this reordering ability, set the CanUserReorderColumns property of the DataGrid or the CanUserReorder property of a specific column to False.

Defining Columns

Using automatically generated columns, you can quickly create a DataGrid that shows all your data. However, you give up a fair bit of control. For example, you can't control how columns are ordered, how wide they are, how the values inside are formatted, and what header text is placed at the top.

A far more powerful approach is to turn off automatic column generation by setting AutoGenerateColumns to False. You can then explicitly define the columns you want, with the settings you want, and in the order you want. To do this, you need to fill the DataGrid.Columns collection with the right column objects.

Currently, the DataGrid supports three types of columns, which are represented by three different classes that derive from DataGridColumn:

- *DataGridTextColumn*: This column is the standard choice for most data types. The value is converted to text and displayed in a TextBlock. When you edit the row, the TextBlock is replaced with a standard text box.

- *DataGridCheckBoxColumn:* This column shows a check box. This column type is used automatically for Boolean (or nullable Boolean) values. Ordinarily, the check box is read-only; but when you edit the row, it becomes a normal check box.

- *DataGridTemplateColumn*: This column is by far the most powerful option. It allows you to define a data template for displaying column values, with all the flexibility and power you have when using templates in a list control. For example, you can use a DataGridTemplateColumn to display image data or use a specialized Silverlight control (like a drop-down list with valid values or a DatePicker for date values).

For example, here's a revised DataGrid that creates a two-column display with product names and prices. It also applies clearer column captions and widens the Product column to fit its data:

```
<data:DataGrid x:Name="gridProducts" Margin="5" AutoGenerateColumns="False">
  <data:DataGrid.Columns>
    <data:DataGridTextColumn Header="Product" Width="175"
     Binding="{Binding ModelName}"></data:DataGridTextColumn>
    <data:DataGridTextColumn Header="Price"
     Binding="{Binding UnitCost}"></data:DataGridTextColumn>
  </data:DataGrid.Columns>
</data:DataGrid>
```

When you define a column, you almost always set three details: the header text that appears at the top of the column, the width of the column, and—most importantly—the binding that gets the data.

The DataGrid's approach to data binding is different from the approach of most other list controls. List controls include a DisplayMemberPath property instead of a Binding property. The Binding approach is more flexible—it allows you to incorporate a value converter without needing to step up to a full template column. For example, here's how you can format the UnitCost column as a currency value (see Figure 17-6) using the StringFormat property of the binding:

```
<data:DataGridTextColumn Header="Price" Binding=
 "{Binding UnitCost, StringFormat='C'}">
</data:DataGridTextColumn>
```

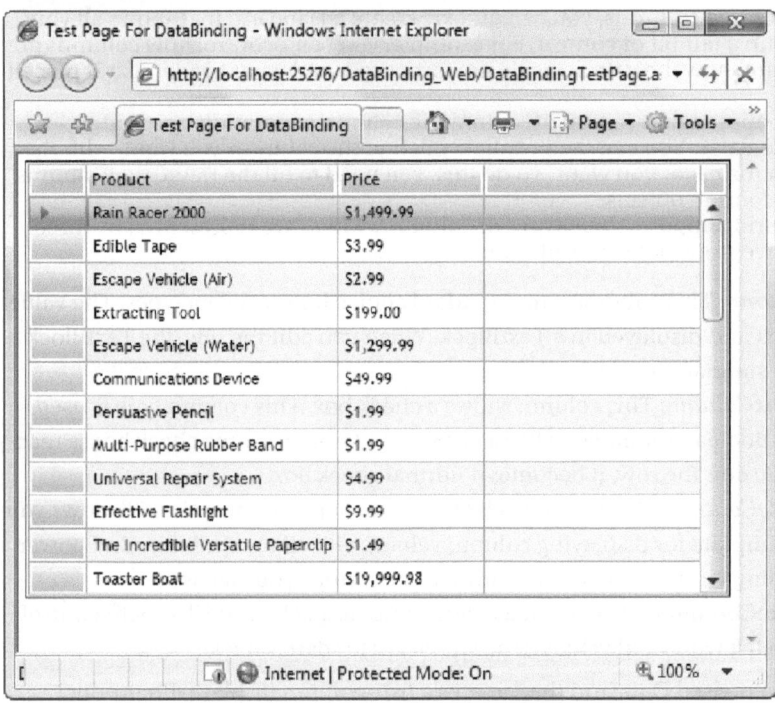

Figure 17-6. Setting the header text and formatting column values

---

■ **Tip** You can dynamically show and hide columns by modifying the Visibility property of the corresponding column object. Additionally, you can move columns at any time by changing their DisplayIndex values.

The DataGridCheckBoxColumn

The Product class doesn't include any Boolean properties. If it did, the DataGridCheckBoxColumn would be a useful option.

As with DataGridTextColumn, the Binding property extracts the data—in this case, the True or False value that's used to set the IsChecked property of the CheckBox element inside. The DataGridCheckBoxColumn also adds a property named Content that lets you show optional content alongside the check box. Finally, the DataGridCheckBoxColumn includes an IsThreeState property that determines whether the check box supports the undetermined state as well as the more obvious checked and unchecked states. If you're using the DataGridCheckBoxColumn to show the information from a nullable Boolean value, you can set IsThreeState property to True. That way, the user can click back to the undetermined state (which shows a lightly shaded check box) to return the bound value to Nothing.

The DataGridTemplateColumn

The DataGridTemplateColumn uses a data template, which works in the same way as the data-template features you explored with list controls earlier. The only difference in the DataGridTemplateColumn is that it allows you to define two templates: one for data display (the CellTemplate) and one for data editing (the CellEditingTemplate), which you'll consider shortly. Here's an example that uses the template data column to place a thumbnail image of each product in the grid (see Figure 17-7):

```
<data:DataGridTemplateColumn>
  <data:DataGridTemplateColumn.CellTemplate>
    <DataTemplate>
      <Image Stretch="None" Source=
      "{Binding ProductImagePath, Converter={StaticResource ImagePathConverter}}">
      </Image>
    </DataTemplate>
  </data:DataGridTemplateColumn.CellTemplate>
</data:DataGridTemplateColumn>
```

This example assumes you've added the ImagePathConverter value converter to the UserControl.Resources collection:

```
<UserControl.Resources>
  <local:ImagePathConverter x:Key="ImagePathConverter"></local:ImagePathConverter>
</UserControl.Resources>
```

The full ImagePathConverter code is shown in Chapter 16.

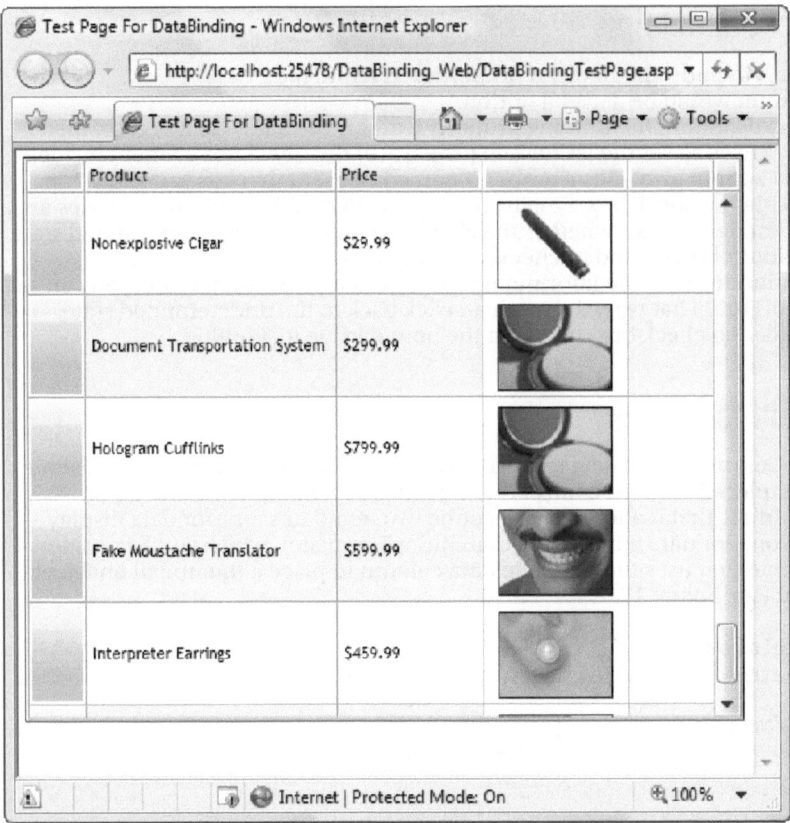

Figure 17-7. A DataGrid with image content

Formatting and Styling Columns

You can format a DataGridTextColumn in the same way that you format a TextBlock element, by setting the Foreground, FontFamily, FontSize, FontStyle, and FontWeight properties. However, the DataGridTextColumn doesn't expose all the properties of the TextBlock. For example, there's no way to set the often-used Wrapping property if you want to create a column that shows multiple lines of text. In this case, you need to use the ElementStyle property instead.

Essentially, the ElementStyle property lets you create a style that is applied to the element inside the DataGrid cell. In the case of a simple DataGridTextColumn, that's a TextBlock. (In a DataGridCheckBoxColumn, it's a check box; and in a DataGridTemplateColumn, it's whatever element you've created in the data template.)

Here's a simple style that allows the text in a column to wrap:

```
<data:DataGridTextColumn Header="Description" Width="400"
  Binding="{Binding Description}">
  <data:DataGridTextColumn.ElementStyle>
    <Style TargetType="TextBlock">
      <Setter Property="TextWrapping" Value="Wrap"></Setter>
```

```
        </Style>
      </data:DataGridTextColumn.ElementStyle>
  </data:DataGridTextColumn>
```

To see the wrapped text, you must expand the row height. Unfortunately, the DataGrid can't size itself as flexibly as Silverlight layout containers can. Instead, you're forced to set a fixed row height using the DataGrid.RowHeight property. This height applies to all rows, regardless of the amount of content they contain. Figure 17-8 shows an example with the row height set to 70 pixels.

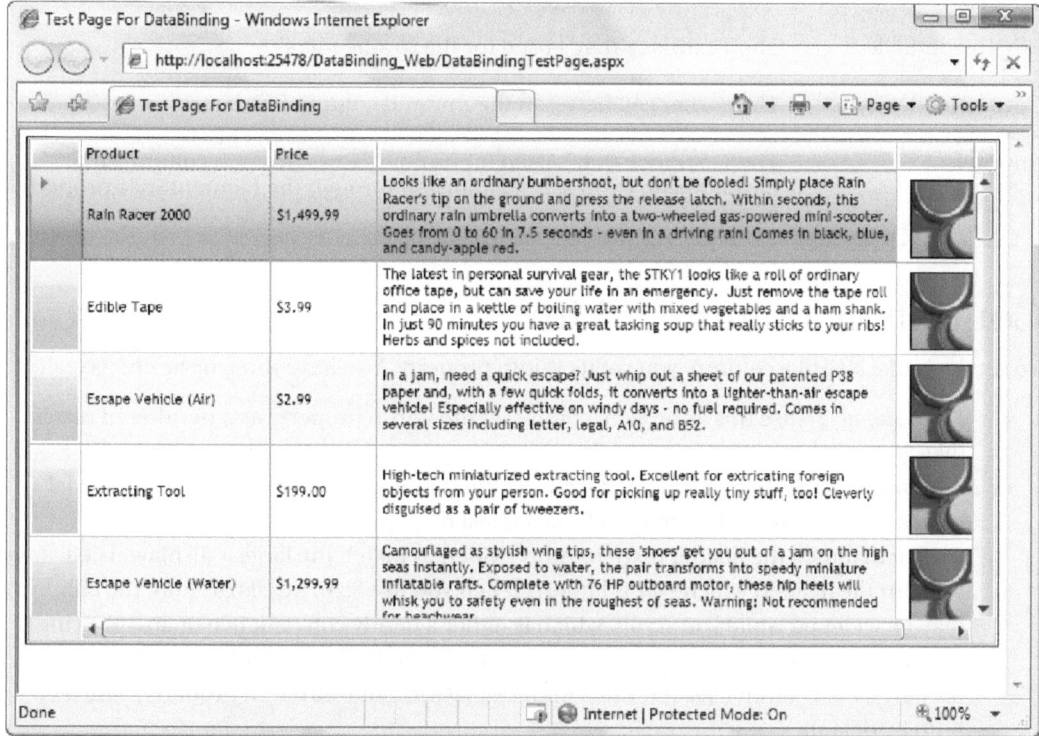

Figure 17-8. A DataGrid with wrapped text

■ **Tip** If you want to apply the same style to multiple columns (for example, to deal with wrappable text in several places), you can define the style in the Resources collection and then refer to it in each column using a StaticResource.

You can use EditingElementStyle to style the element that's used when you're editing a column. In the case of DataGridTextColumn, the editing element is the TextBox control.

The ElementStyle, ElementEditingStyle, and column properties give you a way to format all the cells in a specific column. However, in some cases you might want to apply formatting settings to every cell in every column. The simplest way to do so is to configure a style for the DataGrid.RowStyle property. The DataGrid also exposes a small set of additional properties that allow you to format other parts of the grid, such as the column headers and row headers. Table 17-3 has the full story.

Table 17-3. Style-Based DataGrid Properties

Property	Style Applies To...
ColumnHeaderStyle	The TextBlock that's used for the column headers at the top of the grid
RowHeaderStyle	The TextBlock that's used for the row headers
CornerHeaderStyle	The corner cell between the row and column headers
RowStyle	The TextBlock that's used for ordinary rows (rows in columns that haven't been expressly customized through the ElementStyle property of the column)

Controlling Column Width

To set the initial size of a column, you set its Width property. You have three basic choices:

- *Fixed sizing*: To use this approach, simply set the Width property as a number of pixels (like 200).
- *Automatic sizing*: To use this approach, set the Width property to one of three special values: SizeToCells (widen to match the largest displayed cell value), SizeToHeader (widen to match the header text), or Auto (widen to match the largest displayed cell value or the header, whichever is larger). When you use SizeToCells or Auto, the column may be widened while you scroll, which is either a handy convenience or an annoying distraction, depending on your perspective. Incidentally, if you don't set the Width property, the column uses the value of the DataGrid.ColumnWidth property, which is Auto by default.
- *Proportional sizing*: To use this approach (which is also called *star-sizing*), set the Width property to an asterisk (*). This approach works like the proportional sizing of the Grid layout container. Once all fixed-sized and automatic-sized columns are place, the remaining space is shared out among the proportional columns. If you want to give some columns a larger share of the free space, set a width that uses a number followed by an asterisk. For example, column with a width of 2* gets twice the space of a column with a width of just *.

Figure 17-9 shows an example that uses a fixed width for one column (ModelNumber), automatic sizing for another (Price), and proportional sizing for the remaining two (ModelName and Description):

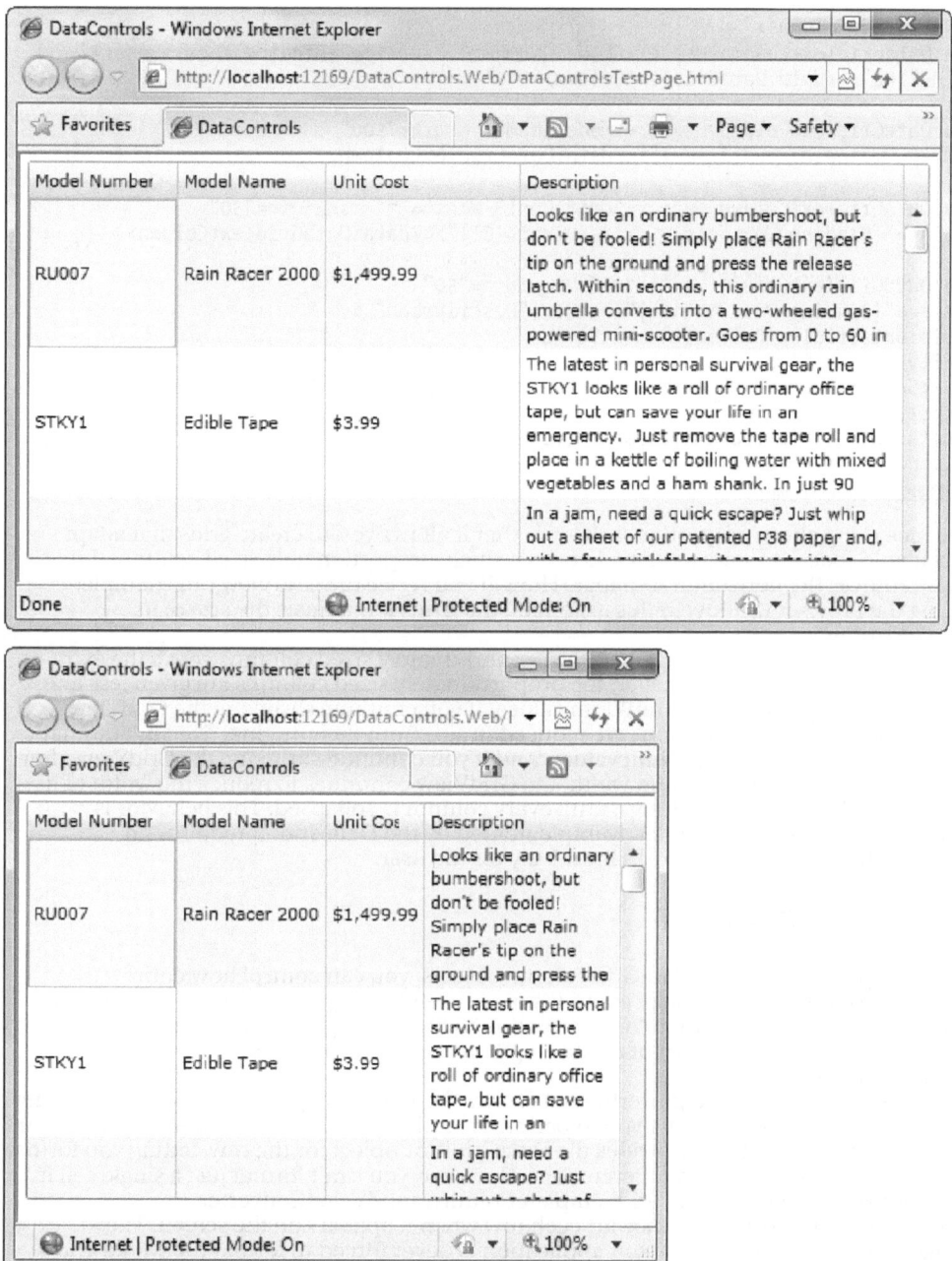

Figure 17-9. *Resizing a grid with proportional columns*

And here's the markup that creates the grid shown in Figure 17-9:

```
<data:DataGrid.Columns>
  <data:DataGridTextColumn Header="Model Number" Binding="{Binding ModelNumber}"
  Width="100"></data:DataGridTextColumn>

  <data:DataGridTextColumn Header="Model Name" Width="100"
   Binding="{Binding ModelName}"></data:DataGridTextColumn>

  <data:DataGridTextColumn Header="Unit Cost" Width="*" MinWidth="50"
   Binding="{Binding UnitCost, StringFormat='C'}"></data:DataGridTextColumn>

  <data:DataGridTextColumn Width="2*" MinWidth="50"
   Binding="{Binding Description}" Header="Description" >
    <data:DataGridTextColumn.ElementStyle>
      <Style TargetType="TextBlock">
        <Setter Property="TextWrapping" Value="Wrap"></Setter>
      </Style>
    </data:DataGridTextColumn.ElementStyle>
  </data:DataGridTextColumn>
</data:DataGrid.Columns>
```

The nice thing about proportional sizing is that it allows you to create grids that adapt to the available space. As soon as you add even a single proportionately sized column, the GridView removes the horizontal scrollbar. Then, if you resize the GridView (for example, by resizing the browser window in this example), the GridView adjusts the size of its columns accordingly.

If you size the GridView to be larger, the proportionately sized columns get all the extra space. If you size the GridView smaller, the proportionately sized columns are given less and less space, until they reach their MinWidth values. If you continue shrinking the GridView beyond this point, the *other* columns are reduced in size from their original fixed or automatic sizes, until they reach their MinWidth values. And if you continue shrinking the GridView after all its columns reach their minimum width, the GridView continues to reduce the width of its columns below the MinWidth settings, until every column is collapsed. This behavior is somewhat complex, but it's completely implemented by the DataGrid. It requires no application intervention, and it works intuitively for the user.

Formatting Rows

By setting the properties of the DataGrid column objects, you can control how entire columns are formatted. But in many cases, it's more useful to flag rows that contain specific data. For example, you may want to draw attention to high-priced products or expired shipments. You can apply this sort of formatting programmatically by handling the DataGrid.LoadingRow event.

The LoadingRow event is a powerful tool for row formatting. It gives you access to the data object for the current row, allowing you to perform simple range checks, comparison, and more complex manipulations. It also provides the DataGridRow object for the row, letting you format the row with different colors or a different font. However, you can't format just a single cell in that row—for that, you need DataGridTemplateColumn and IValueConverter.

The LoadingRow event fires once for each row when it appears on the screen. The advantage of this approach is that your application is never forced to format the whole grid—instead, the LoadingRow fires only for the rows that are currently visible. But there's also a downside. As the user scrolls through the grid, the LoadingRow event is triggered continuously. As a result, you can't place time-consuming code in the LoadingRow method unless you want scrolling to grind to a halt.

There's also another consideration: virtualization. To lower its memory overhead, the DataGrid reuses the same DataGrid objects to show new data as you scroll through the data.

(That's why the event is called LoadingRow rather than CreatingRow.) If you're not careful, the DataGrid can load data into an already-formatted DataGridRow. To prevent this from happening, you must explicitly restore each row to its initial state.

In the following example, high-priced items are given a bright orange background (see Figure 17-10). Regular-price items are given the standard white background:

```
' Reuse brush objects for efficiency in large data displays.
Private highlightBrush As New SolidColorBrush(Colors.Orange)
Private normalBrush As New SolidColorBrush(Colors.White)

Private Sub gridProducts_LoadingRow(ByVal sender As Object, _
  ByVal e As DataGridRowEventArgs)
    ' Check the data object for this row.
    Dim product As Product = CType(e.Row.DataContext, Product)

    ' Apply the conditional formatting.
    If product.UnitCost > 100 Then
        e.Row.Background = highlightBrush
    Else
        ' Restore the default white background. This ensures that used,
        ' formatted DataGrid object are reset to their original appearance.
        e.Row.Background = normalBrush
    End If
End Sub
```

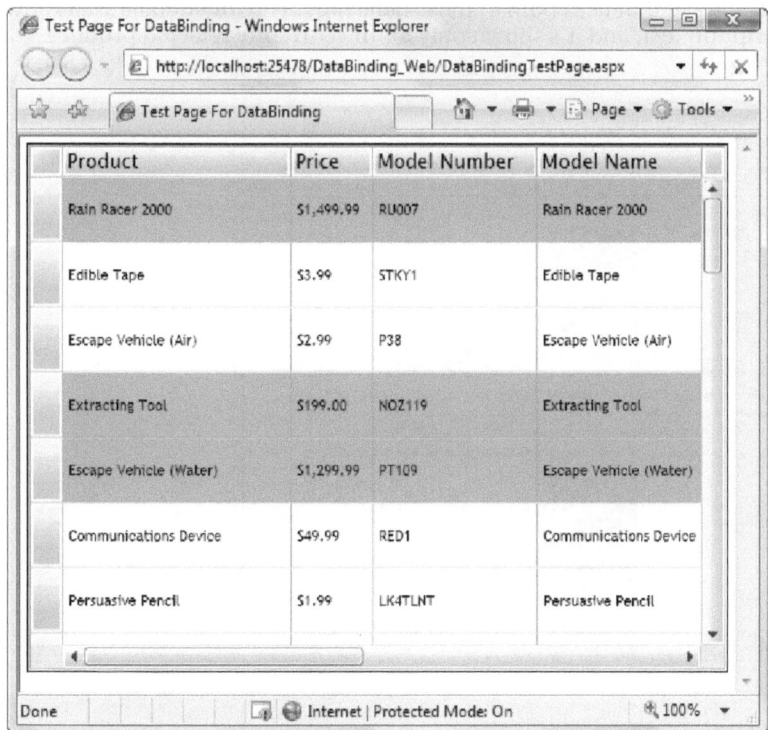

Figure 17-10. Highlighting rows

637

Remember, you have another option for performing value-based formatting: you can use an IValueConverter that examines bound data and converts it to something else. This technique is especially powerful when combined with a DataGridTemplateColumn. For example, you can create a template-based column that contains a TextBlock and bind the TextBlock.Background property to an IValueConverter that sets the color based on the price. Unlike the LoadingRow approach shown previously, this technique allows you to format just the cell that contains the price, not the whole row. For more information about this technique, refer to the "Applying Conditional Formatting" section in Chapter 16.

■ **Tip** The formatting you apply in the LoadingRow event handler applies only when the row is loaded. If you edit a row, this LoadingRow code doesn't fire (at least, not until you scroll the row out of view and then back into sight).

Row Details

The DataGrid also supports *row details*—an optional, separate display area that appears just under the column values for a row. The row-details area adds two things that you can't get from columns alone. First, it spans the full width of the DataGrid and isn't carved into separate columns, which gives you more space to work with. Second, you can configure the row-details area so that it appears only for the selected row, allowing you to tuck the extra details out of the way when they're not needed.

Figure 17-11 shows a DataGrid that uses both of these behaviors. The row-details area shows the wrapped product description text, and it's shown only for the currently selected product.

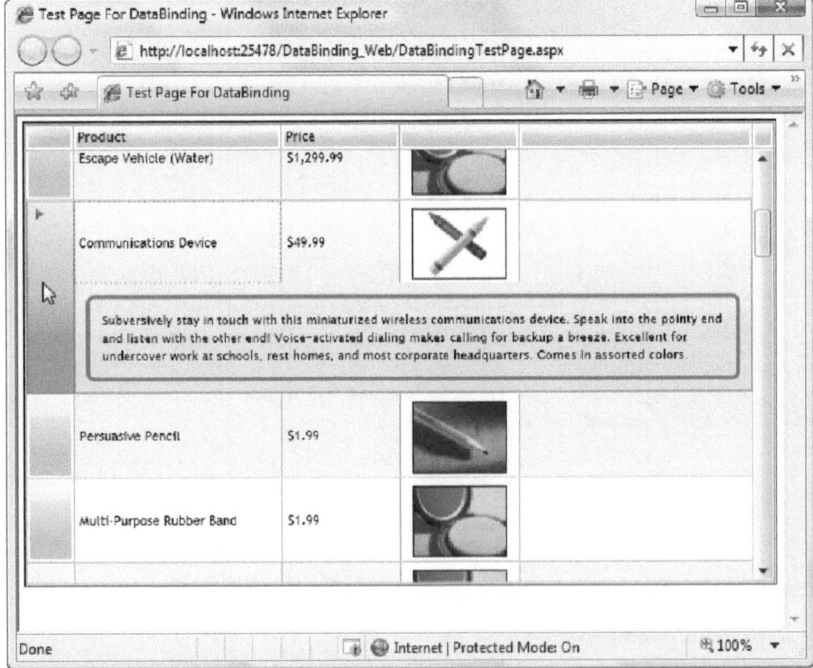

Figure 17-11. Using the row-details area

To create this example, you need to first define the content that's shown in the row-details area by setting the DataGrid.RowDetailsTemplate property. In this case, the row-details area uses a basic template that includes a TextBlock that shows the full product text and adds a border around it:

```
<data:DataGrid.RowDetailsTemplate>
  <DataTemplate>
    <Border>
      <Border Margin="10" Padding="10" BorderBrush="SteelBlue" BorderThickness="3"
        CornerRadius="5">
        <TextBlock Text="{Binding Description}" TextWrapping="Wrap" FontSize="10">
        </TextBlock>
      </Border>
    </Border>
  </DataTemplate>
</data:DataGrid.RowDetailsTemplate>
```

Other options include adding controls that allow you to perform various tasks (for example, getting more information about a product, adding it to a shopping list, editing it, and so on).

■ **Note** There's a quick with the way the DataGrid sizes the row-details area: it doesn't take the margin of the root element into account. As a result, if you set the Margin property on the root element, part of your content will be cut off at the bottom and right edges. To correct this problem, you can add an extra container, as in this example. Here, the root-level element doesn't include a margin, but the nested Border element inside does, which dodges the sizing problem.

You can configure the display behavior of the row-details area by setting the DataGrid.RowDetailsVisibilityMode property. By default, this property is set to VisibleWhenSelected, which means the row-details area is shown when the row is selected. Alternatively, you can set it to Visible, which means the row detail area of every row will be shown at once. Or, you can use Collapsed, which means the row detail area won't be shown for any row, at least not until you change the RowDetailsVisibilityMode in code (for example, when the user selects a certain type of row).

Freezing Columns

A *frozen* column stays in place at the left size of the DataGrid, even as you scroll to the right. Figure 17-12 shows how a frozen Product column remains visible during scrolling. Notice how the horizontal scrollbar extends only under the scrollable columns, not the frozen columns.

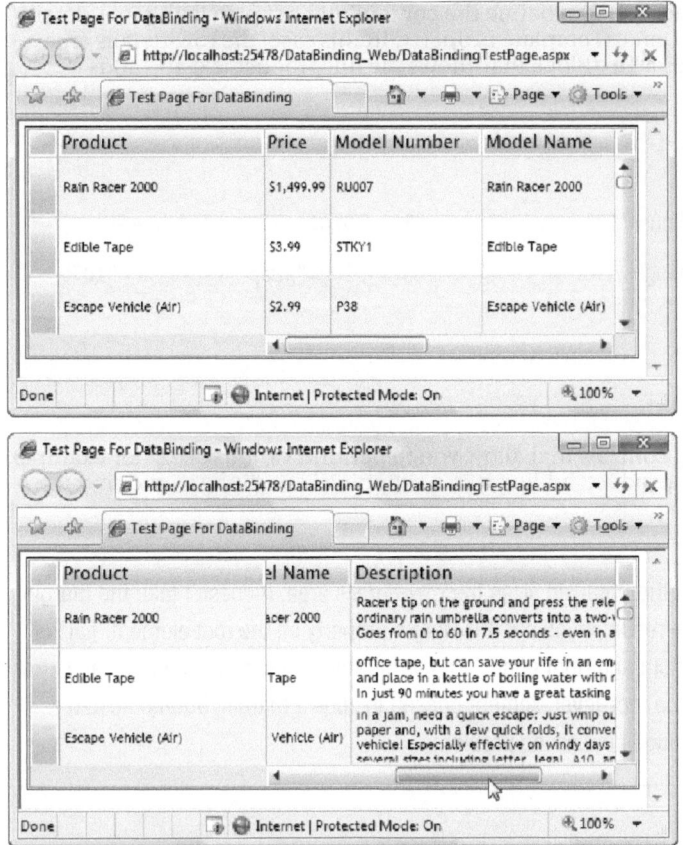

Figure 17-12. *Freezing the Product column*

Column freezing is a useful feature for very wide grids, especially when you want to make sure certain information (such as the product name or a unique identifier) is always visible. To use it, you set the IsFrozen property of the column to True:

```
<data:DataGridTextColumn Header="Product" Width="175" IsFrozen="True"
 Binding="{Binding ModelName}"></data:DataGridTextColumn>
```

There's one catch: frozen columns must always be on the left side of the grid. If you freeze one column, it must be the leftmost column; if you free two columns, they must be the first two on the left; and so on.

Selection

Like an ordinary list control, the DataGrid lets the user select individual items. You can react to the SelectionChanged event when this happens. To find out what data object is currently selected, you can use the SelectedItem property. If you want the user to be able to select multiple rows, set the SelectionMode property to Extended. (Single is the only other option and the default.) To select multiple rows, the user must hold down the Shift or Ctrl key. You can retrieve the collection of selected items from the SelectedItems property.

■ **Tip** You can set the selection programmatically using the SelectedItem property. If you're setting the selection to an item that's not currently in view, it's a good idea to follow up with a call to the DataGrid.ScrollIntoView() method, which forces the DataGrid to scroll forward or backward until the item you've indicated is visible.

Sorting

The DataGrid features built-in sorting as long as you're binding a collection that implements IList (such as the List(Of T) and ObservableCollection(Of T) collections). If you meet this requirement, your DataGrid gets basic sorting for free.

To use the sorting, the user needs to click a column header. Clicking once sorts the column in ascending order based on its data type (for example, numbers are sorted from 0 up, and letters are sorted alphabetically). Click the column again, and the sort order is reversed. An arrow appears at the far-right side of the column header, indicating that the DataGrid is sorted based on the values in this column. The arrow points up for an ascending sort and down for a descending sort. (When you click a column more than once, the arrow flips with a quick animation effect.)

Users can sort based on multiple columns by holding down Shift while they click. For example, if you hold down Shift and click the Category column followed by the Price column, products are sorted into alphabetical category groups, and the items in each category group are ordered by price.

It's possible to exercise some control over the DataGrid sorting process, depending on how much effort you're willing to make (and how much code you're willing to live with). Here are your options:

- *The SortMemberPath property*: Every column provides the SortMemberPath property, which allows you to specify the property in the bound data object that's used for sorting. If SortMemberPath isn't set, the column is sorted using the bound data, which makes sense. However, if you have a DataGridTemplateColumn, you need to use SortMemberPath because there's no Binding property to provide the bound data. If you don't, your column won't support sorting.

- The *PagedCollectionView class*: The PagedCollectionView wraps an ordinary collection and gives you added abilities to sort, filter, group, and page its contents. (You'll use PagedCollectionView later in this chapter for DataGrid grouping and sorting.)

- *A custom template*: If you don't like the arrows that indicate when a sort order has been applied (or you want to add glitzier animation), you need to use the DataGrid.ColumnHeaderStyle property to apply a new template. It has three key states: Unsorted State (when no sorting is applied), SortedAscending State (when the column is first sorted), and SortedDescending State (when the column header is clicked twice, and the sort order is reversed). Customize these to plug in your own visuals.

You can also disable sorting by setting the CanUserSortColumns property to False (or turn it off for specific columns by setting the column's CanUserSort property).

DataGrid Editing

One of the DataGrid's greatest conveniences is its support for editing. A DataGrid cell switches into edit mode when the user double-clicks it. But the DataGrid lets you restrict this editing ability in several ways:

- *DataGrid.IsReadOnly*: When this property is True, users can't edit anything.
- *DataGridColumn.IsReadOnly*: When this property is True, users can't edit any of the values in that column.
- *Read-only properties*: If your data object has a property with no property setter, the DataGrid is intelligent enough to notice this detail and disable column editing just as if you had set DataGridColumn.IsReadOnly to True. Similarly, if your property isn't a simple text, numeric, or date type, the DataGrid makes it read-only (although you can remedy this situation by switching to the DataGridTemplateColumn, as described shortly).

What happens when a cell switches into edit mode depends on the column type. A DataGridTextColumn shows a text box (although it's a seamless-looking text box that fills the entire cell and has no visible border). A DataGridCheckBox column shows a check box that you can check or uncheck. But the DataGridTemplateColumn is by far the most interesting. It allows you to replace the standard editing text box with a more specialized input control, like a DatePicker or ComboBox.

Editing with Templates

You've already seen how to supply a CellTemplate for the DataGridTemplateColumn. But the DataGridTemplateColumn supports two templates. The CellTemplate determines how the cell looks when it's not being edited. The CellEditingTemplate specifies the controls that should be shown in editing mode, using a two-way binding expression to connect to the appropriate field. It's up to you whether you use the same controls in both templates.

For example, the following column shows a date. When the user double-clicks to edit that value, it turns into a drop-down DatePicker (see Figure 17-13) with the current value preselected:

```
<data:DataGridTemplateColumn Header="Date Added">
  <data:DataGridTemplateColumn.CellTemplate>
    <DataTemplate>
      <TextBlock Margin="4" Text=
"{Binding DateAdded, Converter={StaticResource DateOnlyConverter}}"></TextBlock>
    </DataTemplate>
  </data:DataGridTemplateColumn.CellTemplate>
  <data:DataGridTemplateColumn.CellEditingTemplate>
    <DataTemplate>
      <controls:DatePicker SelectedDate="{Binding DateAdded, Mode=TwoWay}">
      </controls:DatePicker>
    </DataTemplate>
  </data:DataGridTemplateColumn.CellEditingTemplate>
</data:DataGridTemplateColumn>
```

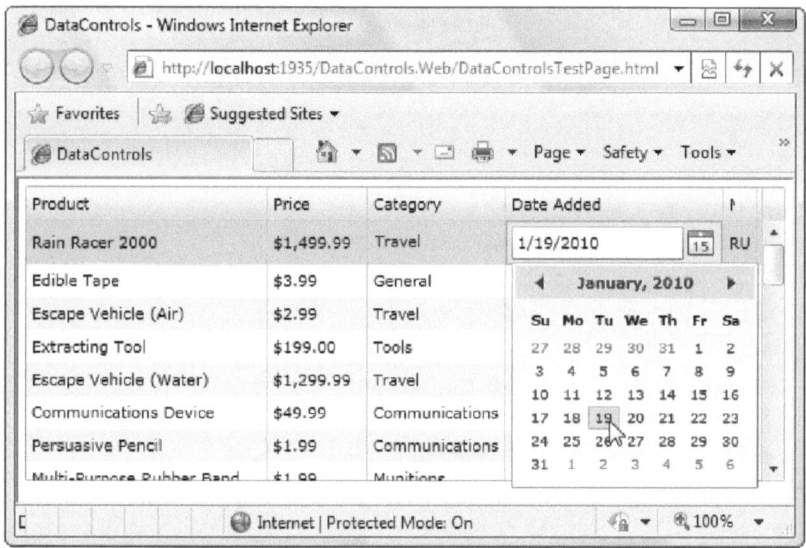

Figure 17-13. *Editing dates with the DatePicker*

You can even use a template column to supply a lookup list of options for data entry. For example, you may want to constrain the Category choice to a list of predefined categories. The easiest way to do this is to create a combo box in the CellEditingTemplate. Then, bind the ComboBox.SelectedItem property to the Product.CategoryName using a two-way binding, and bind the ComboBox.ItemsSource property to a collection that contains the allowed values. In the following example, that's a collection that's exposed by the Product.CategoryChoices property:

```
<data:DataGridTemplateColumn Header="Category">
  <data:DataGridTemplateColumn.CellTemplate>
    <DataTemplate>
      <TextBlock Margin="4" Text="{Binding CategoryName}"></TextBlock>
    </DataTemplate>
  </data:DataGridTemplateColumn.CellTemplate>

  <data:DataGridTemplateColumn.CellEditingTemplate>
    <DataTemplate>
      <ComboBox Margin="4" ItemsSource="{Binding CategoryChoices}"
        SelectedItem="{Binding CategoryName, Mode=TwoWay}">
      </ComboBox>
    </DataTemplate>
  </data:DataGridTemplateColumn.CellEditingTemplate>
</data:DataGridTemplateColumn>
```

Validation and Editing Events

The DataGrid automatically supports the same basic validation system you learned about in the previous chapter, which reacts to problems in the data-binding system (such as the inability to convert supplied text to the appropriate data type) or exceptions thrown by the property setter. The error message appears as a red pop-up next to the offending column (Figure 17-14).

643

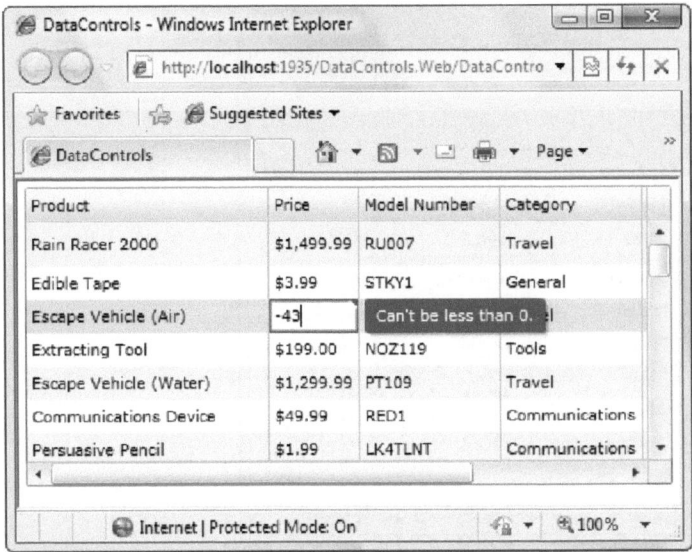

Figure 17-14. *A property setting exception*

The DataGridTextColumn automatically uses binding expressions that support validation. However, if you're using a DataGridTemplateColumn, you must add both the ValidatesOnExceptions and NotifyOnValidationError properties to the binding expression in the CellEditingTemplate, as shown here:

```
<data:DataGridTemplateColumn Header="Price">
  <data:DataGridTemplateColumn.CellTemplate>
    <DataTemplate>
      <TextBlock Margin="4"
       Text="{Binding UnitCost, StringFormat='C'}">
      </TextBlock>
    </DataTemplate>
  </data:DataGridTemplateColumn.CellTemplate>
  <data:DataGridTemplateColumn.CellEditingTemplate>
    <DataTemplate>
      <TextBox Margin="4"
       Text="{Binding UnitCost, Mode=TwoWay, ValidatesOnExceptions=true,
NotifyOnValidationError=true}">
      </TextBox>
    </DataTemplate>
  </data:DataGridTemplateColumn.CellEditingTemplate>
</data:DataGridTemplateColumn>
```

You can implement validation a couple of other ways with a DataGrid. One option is to use the DataGrid's editing events, which are listed in Table 17-4. The order of rows matches the order that the events fire in the DataGrid.

Table 17-4. DataGrid Editing Events

Name	Description
BeginningEdit	Occurs when the cell is about to be put in edit mode. You can examine the column and row that are currently being edited, check the cell value, and cancel this operation using the DataGridBeginningEditEventArgs.Cancel property.
PreparingCellForEdit	Used for template columns. At this point, you can perform any last-minute initialization that's required for the editing controls. Use DataGridPreparingCellForEditEventArgs.EditingElement to access the element in the CellEditingTemplate.
CellEditEnding	Occurs when the cell is about to exit edit mode. DataGridCellEditEndingEventArgs.EditAction tells you whether the user is attempting to accept the edit (for example, by pressing Enter or clicking another cell) or cancel it (by pressing the Escape key). You can examine the new data and set the Cancel property to roll back an attempted change.
CellEditEnded	Occurs after the cell has returned to normal. You can use this point to update other controls or display a message that notes the change.
RowEditEnding	Occurs when the user navigates to a new row after editing the current row. As with CellEditEnding, you can use this point to perform validation and cancel the change. Typically, you'll perform validation that involves several columns—for example, ensuring that the value in one column isn't greater than the value in another.
RowEditEnded	Occurs after the user has moved on from an edited row. You can use this point to update other controls or display a message noting the change.

If you need a place to perform validation logic that is specific to your page (and so can't be baked into the data objects), you can write custom validation logic that responds to the CellEditEnding and RowEditEnding events. Check column rules in the CellEditEnding event handler, and validate the consistency of the entire row in the RowEditEnding event. And remember, if you cancel an edit, you should provide an explanation of the problem (usually in a TextBlock elsewhere on the page, although you can also use the ChildWindow control or a message box).

Finally, it's worth noting that the DataGrid supports data annotations in a different way than the ordinary input controls you've used so far. If your property setters use the Validator.ValidateProperty() method to check for invalid values and throw a ValidationException (as shown earlier), the DataGrid responds in the typical way, by immediately recognizing the error and displaying the error message in a red pop-up. But if you don't use the validator, the DataGrid *still* validates all the properties you've set and validates the entire object. The difference is that it doesn't perform this validation until the user attempts to move to another row. Furthermore, if a validation error is detected at this point, the DataGrid handles it in a different way. It returns the user to the invalid row, keeping it in edit mode, and then shows the error message in a shaded bar that appears over the bottom of the DataGrid. Figure 17-15 shows an example where an edit violates the custom validation routine from the ProductValidation class shown earlier.

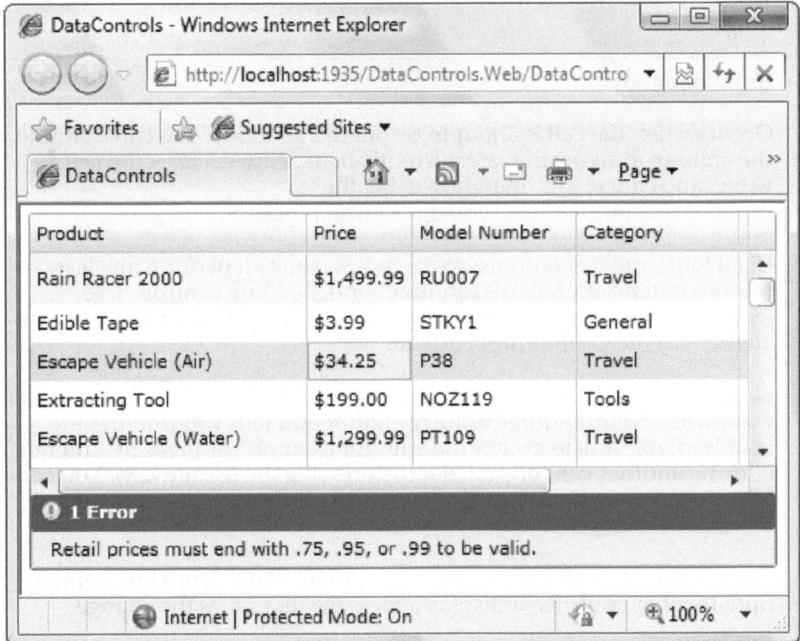

Figure 17-15. *Violating a data annotation*

Here's another way to think about it. Any exceptions you raise in the property setter are handled when the CellEditEnding event fires. And data annotations you apply but don't explicitly enforce with code are checked by the DataGrid when the RowEditEnding event fires.

It's up to you whether you use this ability. If you intend to perform editing in ordinary controls as well as the DataGrid, you need to keep using the Validator.ValidateProperty() method to defend against invalid data. But if you intend to use the DataGrid exclusively for your editing needs, it may make sense to omit the property-setting code and let the DataGrid perform the validation. Just remember that when used in this way, the data annotation rules won't kick in when you set values programmatically. (This also raises the possibility of a particularly odd error condition: if the DataGrid is loaded with invalid data and the user attempts to edit that data, the user will be trapped in edit mode until the value is changed. The edit can't be cancelled, because the original value is invalid.)

The PagedCollectionView

The DataGrid has a few more features that require the support of the PagedCollectionView, which is found in the System.Windows.Data namespace. The PagedCollectionView wraps a collection and gives you a different way to look at it. Conceptually, the PagedCollectionView is a window onto your data, and that window can apply sorting, filtering, grouping, and paging before your data appears in a bound control like the DataGrid.

To use the PagedCollectionView, you need to explicitly create it in your code. You supply the source collection with your data as a constructor argument. You then bind the PagedCollectionView to the appropriate control instead of your original collection.

To implement this approach with the current example, you'd change this code, which reacts when the web service returns the collection of products:

```
gridProducts.ItemsSource = e.Result
```

to this:

```
Dim view As New PagedCollectionView(e.Result)
gridProducts.ItemsSource = view
```

To change the way your data appears in the bound control, you tweak the settings of the PagedCollectionView. In the following sections, you'll see examples that use the most commonly modified PagedCollectionView properties.

Sorting

You can apply a multilevel sort order by adding SortDescription objects (from the System.ComponentModel namespace) to the PagedCollectionView.SortDescriptions collection. Each SortDescription object applies a single level of sorting, based on a single property. The SortDescription objects are applied in the order you add them, which means the following code orders products by category and then sorts each category group from lowest to highest price:

```
Dim view As New PagedCollectionView(e.Result)

' Sort by category and price.
view.SortDescriptions.Add(New SortDescription("CategoryName", _
  ListSortDirection.Ascending))
view.SortDescriptions.Add(New SortDescription("UnitCost", _
  ListSortDirection.Ascending))

gridProducts.ItemsSource = view
```

This approach integrates perfectly with the built-in DataGrid sorting you considered earlier in this chapter. The DataGrid displays the up or down sort arrow in the header of whatever columns the PagedCollectionView uses for sorting. And if the user clicks a column header, the old sort order is abandoned, and the rows are re-ordered appropriately.

Filtering

You can use the PagedCollectionView.Filter property to set a filtering callback: a routine that tests whether each row should be shown or hidden. To show a row, the callback returns True. To hide it, the callback returns False.

For example, if you create a PagedCollectionView and set it to a filtering method named SelectTravelProducts, like this:

```
Dim view As New PagedCollectionView(e.Result)
view.Filter = AddressOf SelectTravelProducts

gridProducts.ItemsSource = view
```

You can then add filtering logic to select all the products in the Travel category:

```
' Show only travel items.
' The full list of products remains in the source collection, but the
' non-travel items are not visible through the PagedCollectionView.
```

```
Private Function SelectTravelProducts(ByVal filterObject As Object) As Object
    Dim product As Product = CType(filterObject, Product)
    Return (product.CategoryName = "Travel")
End Function
```

Grouping

The DataGrid also has support for grouping, which allows you to organize rows together into logical categories. The basic idea is that you pick a property to use for grouping (such as CategoryName). Objects that have the same value for that property (for example, products with the same CategoryName) are placed into a single group, which can be collapsed in the DataGrid display, as shown in Figure 17-16.

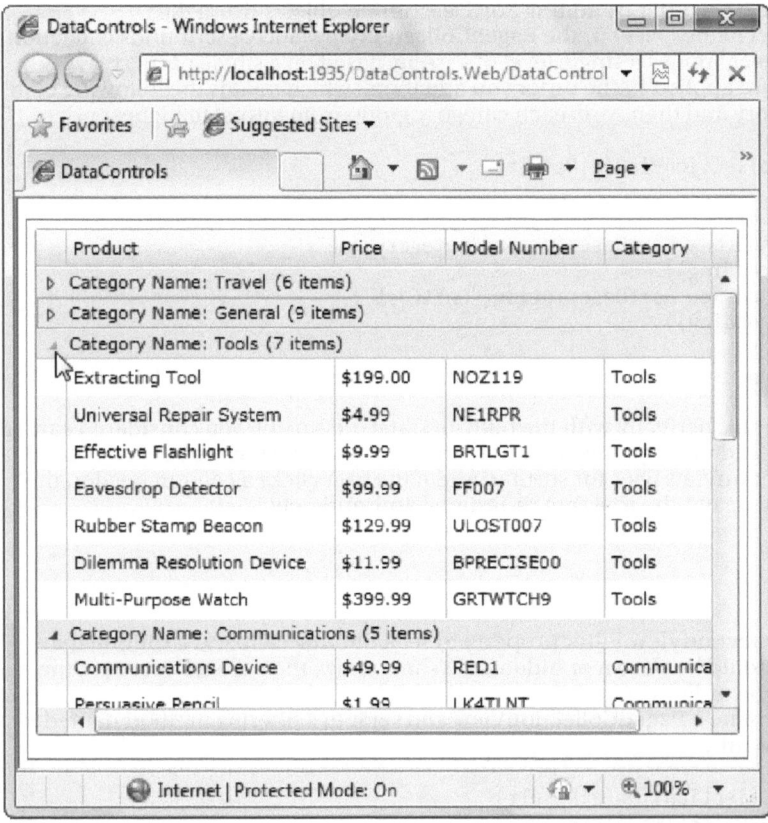

Figure 17-16. Products grouped by CategoryName

To implement grouping, you pick the field on which you want to group. You then add a PropertyGroupDescription object (from the System.ComponentModel namespace) to the PagedCollectionView.GroupDescriptions collection. Here's an example that creates the category groups shown in Figure 17-16:

```
Dim view As New PagedCollectionView(e.Result)
view.GroupDescriptions.Add(New PropertyGroupDescription("CategoryName"))
gridProducts.ItemsSource = view
```

If you want to perform grouping and subgrouping, you add more than one PropertyGroupDescription. The following code splits the products by category and then further divides each category by product status:

```
Dim view As New PagedCollectionView(e.Result)

view.GroupDescriptions.Add(New PropertyGroupDescription("CategoryName"))
view.GroupDescriptions.Add(New PropertyGroupDescription("Status"))

gridProducts.ItemsSource = view
```

The DataGrid allows the user to collapse and expand each group. Initially, all groups are expanded. However, the DataGrid gives you the ability to programmatically expand and collapse groups through its ExpandRowGroup() and CollapseRowGroup() methods. All you need to do is find the group you want in the PagedCollectionView.Groups collection. For example, the following code collapses the Travel group:

```
For Each group As CollectionViewGroup In view.Groups
    If group.Name = "Travel" Then
      gridProducts.CollapseRowGroup(group, True)
    End If
Next
```

The CollapseRowGroup() and ExpandRowGroup() take two parameters: first, the group you want to collapse or expand; and second, a Boolean value that indicates whether you want to collapse or expand all the subgroups inside.

■ **Note** In theory, the PagedCollectionView can support any bound ItemsControl, including the modest ListBox. This is true of sorting, filtering, and paging, but it isn't the case for grouping. With grouping, the bound control needs a way to show the appropriate headers for each group. Currently, only the DataGrid has this ability.

The standard PagedCollectionView grouping is simple—it works by matching values exactly. In some cases, you might want to create broader groups—for example, you might want to group all the products that have names starting with a certain letter or that prices within a set range. To accomplish this, you need to use the PropertyGroupDescription.Converter property. This takes an IValueConverter (just like the ones you created in Chapter 16), which changes the source value into the value you want to use for grouping. For example, to implement first-letter grouping, the IValueConverter would simply extract the first letter from the supplied string.

Another challenge is changing the appearance of the headers that precede each group. The DataGrid helps out a bit with its RowGroupHeaderStyles property, which allows you to create a style object that will pass its property settings down to the group header. Here's an example that changes the background and foreground colors:

```
<data:DataGrid.RowGroupHeaderStyles>
  <Style TargetType="data:DataGridRowGroupHeader">
    <Setter Property="Background" Value="#FF112255" />
    <Setter Property="Foreground" Value="#FFEEEEEE" />
  </Style>
</data:DataGrid.RowGroupHeaderStyles>
```

■ **Note** Your style can change any of the properties of the DataGridRowGroupHeader class. But changing the text is a bit of work—you need to supply a new control template for the DataGridRowGroupHeader.Template property.

The RowGroupHeaderStyles property is a collection, which means you can supply as many Style objects as you want. This allows you to apply customized formatting to the headers in a DataGrid that uses multiple grouping levels. If you supply more than one style, the first one will apply to the top-level group, the second one will apply to the subgroups inside, the third one will apply to the subgroups in the subgroups, and so on.

Paging

Paging is the ability of the PagedCollectionView to split your data into pages, each of which has a fixed number of rows. The user can then browse from one page to another. Paging is useful when you have a huge amount of data, because it allows the user to review it in more manageable chunks.

The PagedCollectionView provides two properties that configure paging:

- *PageSize*: This property sets the maximum number of records that's allowed on a page. By default, its set to 0, which means the PagedCollectionView does not use paging and all the records are kept together.

- *PageIndex*: This property indicates the user's current page, where 0 is the first page, 1 is the second, and so on. You can't set the PageIndex property programmatically, but the PagedCollectionView provides several methods for changing pages, including MoveToFirstPage(), MoveToLastPage(), MoveToPreviousPage(), MoveToNextPage(), and MoveToPage().

Paging would be a bit of a chore if you had to create the controls that allow the user to move from one page to another. Fortunately, Silverlight has a DataPager control that's dedicated to exactly this task. You simply need to add the DataPager to your page (typically, you'll place it under the DataGrid), set a few properties to configure its appearance, and then wire it up to the PagedCollectionView.

Here's the markup that creates the DataPager shown in Figure 17-7:

```
<data:DataPager Margin="5,0,5,5" Grid.Row="1" x:Name="pager"
  PageSize="5" DisplayMode="FirstLastPreviousNextNumeric"
  NumericButtonCount="3" IsTotalItemCountFixed="True"></data:DataPager>
```

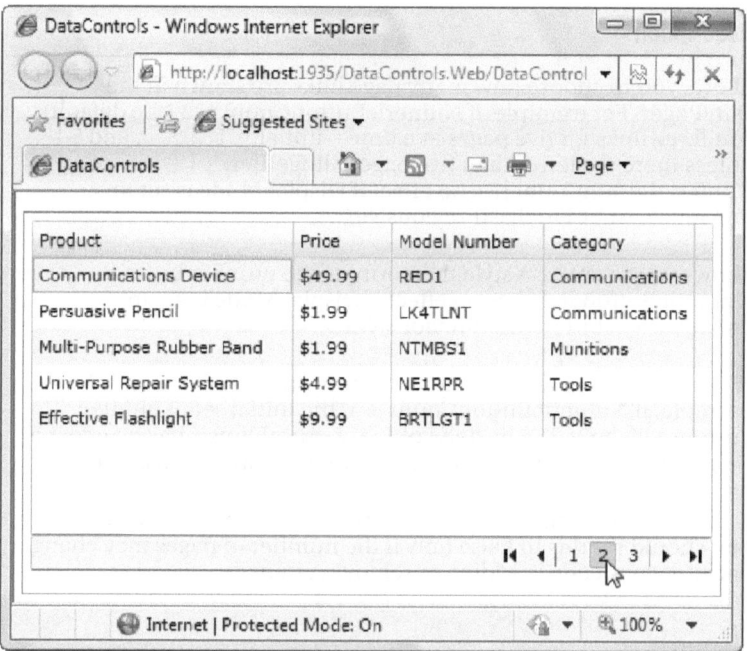

Figure 17-17. Using the DataPager to move through paged data

To make the DataPager operational, you need to add one line of code after you create your view, which connects the DataPager to the PagedCollectionView:

```
pager.Source = view;
```

The DataPager is a fairly intuitive control. Table 17-5 lists its key properties.

Table 17-5. DataPager Properties

Property	Description
PageCount	Gets or sets the PagedCollectionView.PageCount property. This allows you to set the number of records per page through the DataPager, rather than forcing you to go through the PagedCollectionView.
Source	Gets or sets the PagedCollectionView that wraps the source data and implements the paging.
DisplayMode	Allows you to choose one of six common arrangements for pager buttons, using the PagerDisplayMode enumeration. Your options are shown in Figure 17-18. If you want to customize the pager display beyond this, you can either create your own paging controls that interact directly with the PagedCollecitonView or supply a custom control template for the DataPager.

Property	Description
NumericButtonCount	Allows you to choose how many page links are shown in the DataPager. For example, if NumericButtonCount is 5 (the default), you'll see links for five pages at a time—initially, 1, 2, 3, 4, and 5—unless there are fewer than five pages altogether. NumericButtonCount has no effect if DisplayMode is set to PreviousNext or FirstLastPreviousNext.
NumericButtonStyle	Allows you to create a style that formats the number buttons. NumericButtonStyle has no effect if DisplayMode is set to PreviousNext or FirstLastPreviousNext.
AutoEllipsis	If True, replaces the last number button with an ellipsis (…). For example, if NumericButtonCount is 3, the initial set of number buttons will be 1, 2, … instead of 1, 2, 3. AutoEllipsis has no effect if DisplayMode is set to PreviousNext or FirstLastPreviousNext.
IsTotalItemCountFixed	If True, the Next button is disabled when the user is on the last page. You should set this to False only if the number of pages may change because your code is adding or removing items.

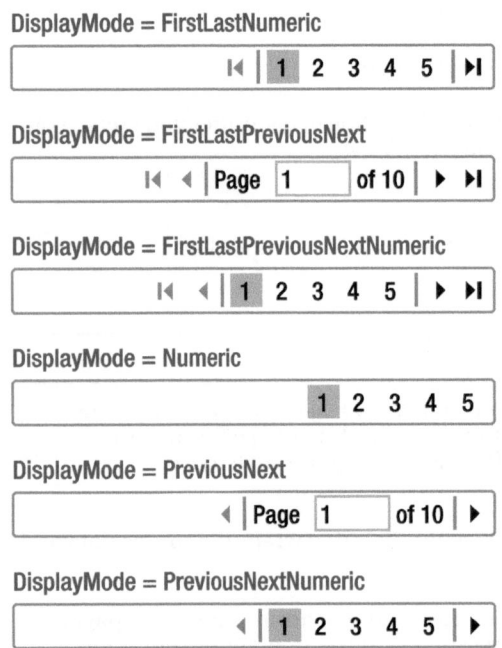

Figure 17-18. PagerDisplayMode options

The TreeView

The TreeView control allows you to display items in a collapsible, hierarchical tree, just like the kind that are a staple of the Windows world, appearing in everything from the Windows Explorer file browser to the .NET help library.

At its heart, the TreeView is a specialized ItemsControl that hosts TreeViewItem objects. The trick is that each TreeViewItem is its own distinct ItemsControl, with the ability to hold more TreeViewItem objects. This flexibility allows you to create a deeply layered data display.

Filling a TreeView

Here's the skeleton of a very basic TreeView, which is declared entirely in markup:

```
<controls:TreeView>
  <controls:TreeViewItem Header="Fruit">
    <controls:TreeViewItem Header="Orange"/>
    <controls:TreeViewItem Header="Banana"/>
    <controls:TreeViewItem Header="Grapefruit"/>
  </controls:TreeViewItem>
 <controls:TreeViewItem Header="Vegetables">
    <controls:TreeViewItem Header="Aubergine"/>
    <controls:TreeViewItem Header="Squash"/>
    <controls:TreeViewItem Header="Spinach"/>
  </controls:TreeViewItem>
</controls:TreeView>
```

It's not necessary to construct a TreeView out of TreeViewItem objects. In fact, you have the ability to add virtually any element to a TreeView, including buttons, panels, and images. However, if you want to display nontext content, the best approach is to use a TreeViewItem wrapper and supply your content through the TreeViewItem.Header property, like this:

```
<controls:TreeViewItem>
  <controls:TreeViewItem.Header>
    <Button Content="There's a Button in this TreeView"></Button>
  </controls:TreeViewItem.Header>
</controls:TreeViewItem>
```

This gives you the same effect as adding non-TreeViewItem elements directly to your TreeView but gives you access to the rich set of TreeViewItem properties, such as properties that tell you whether a node is selected or collapsed (IsSelected and IsExpanded) and events that can inform you when it happens (Selected, Unselected, Expanded, and Collapsed).

You can also display an ordinary data object in a TreeViewItem, like a Product object. You do this in much the same way that you showed data objects in the ListBox in Chapter 16. Just use the Header property to supply the data object, and use the HeaderTemplate property to supply a data template that formats it.

A Data-Bound TreeView

Usually, you won't fill a TreeView with fixed information that's hard-coded in your markup. Instead, you'll construct the TreeViewItem objects you need programmatically, or you'll use data binding to display a collection of objects.

Filling a TreeView with data is easy enough—as with any ItemsControl, you simply set the ItemsSource property. However, this technique fills only the first level of the TreeView. A more interesting use of the TreeView incorporates *hierarchical data* that has some sort of nested structure.

For example, consider the TreeView shown in Figure 17-19. The first level consists of Category objects, while the second level shows the Product objects that fall into each category.

Figure 17-19. A TreeView of categories and products

The TreeView makes hierarchical data display easy. You simply need to specify the right data templates. Your templates indicate the relationship between the different levels of the data.

For example, imagine you want to build the example shown in Figure 17-19. In Chapter 16, you saw how you could call a web method named GetCategoriesWithProducts() to retrieve a collection of Category objects, each of which holds a collection of Product objects. You can use the same method with this example—you simply need to bind the Category collection to the tree so that it appears in the first level. Here's the page code that queries the web service and displays the result:

```
Private Sub Page_Loaded(ByVal sender As Object, ByVal e As RoutedEventArgs)
    Dim client As New StoreDbClient()
    AddHandler client.GetCategoriesWithProductsCompleted, _
       AddressOf client_GetCategoriesWithProductsCompleted
    client.GetCategoriesWithProductsAsync()

    lblStatus.Text = "Contacting service ..."
End Sub
```

```
Private Sub client_GetCategoriesWithProductsCompleted(ByVal sender As Object, _
  ByVal e As GetCategoriesWithProductsCompletedEventArgs)
    Try
        treeCategoriesProducts.ItemsSource = e.Result
        lblStatus.Text = "Received results from web service."
    Catch err As Exception
        lblStatus.Text = "An error occured: " & err.Message
    End Try
End Sub
```

To display the categories, you need to supply a TreeView.ItemTemplate that can process the bound objects. In this example, you need to display the CategoryName property of each Category object, in bold. Here's the data template that does it, as a resource in the UserControls.Resources collection:

```
<UserControl.Resources>
  <common:HierarchicalDataTemplate x:Key="CategoryTemplate">
    <TextBlock Text="{Binding CategoryName}" FontWeight="Bold" />
  </common:HierarchicalDataTemplate>
</UserControl.Resources>
```

The only unusual detail here is that the TreeView.ItemTemplate is set using a HierarchicalDataTemplate object instead of a DataTemplate. The HierarchicalDataTemplate has the added advantage that it can wrap a second template. The HierarchicalDataTemplate can then pull a collection of items from the first level and provide that to the second-level template. You simply set the ItemsSource property to identify the property that has the child items (in this case, it's the Category.Products collection), and you set the ItemTemplate property to indicate how each object should be formatted. In this example, the child product objects are formatted using a second HierarchicalDataTemplate, which simply displays the ModelName in italics. Here are the two templates that do it:

```
<UserControl.Resources>
  <common:HierarchicalDataTemplate x:Key="CategoryTemplate"
   ItemsSource="{Binding Products}" ItemTemplate="{StaticResource ProductTemplate}">
    <TextBlock Text="{Binding CategoryName}" FontWeight="Bold" />
  </common:HierarchicalDataTemplate>

  <common:HierarchicalDataTemplate x:Key="ProductTemplate">
    <TextBlock FontStyle="Italic" Text="{Binding ModelName}" />
  </common:HierarchicalDataTemplate>
</UserControl.Resources>
```

Essentially, you now have two templates, one for each level of the tree. The second template uses the selected item from the first template as its data source.

Finally, here's the TreeView, which specifies that the root-level items (the categories) should be formatted with the CategoryTemplate:

```
<controls:TreeView x:Name="treeCategories" Margin="5"
  ItemTemplate="{StaticResource CategoryTemplate}">
</controls:TreeView>
```

This is all you need to get the category and product tree shown in Figure 17-19.

The Last Word

In this chapter, you delved deeper into data binding, one of the key pillars of Silverlight. You learned how to build smart data forms that require less code and draw the information they need from attributes. On the way, you mastered the Label, DescriptionViewer, and ValidationSummary controls, and you learned how to implement validation with the new data annotation model.

Most of your time in this chapter was spent exploring Silverlight's rich data controls, particularly the remarkably flexible DataGrid, with its fine-grained support for formatting and editing. You saw how to implement advanced filtering, grouping, and paging with the PagedCollectionView. You ended with a look at the TreeView, which lets you bind nested layers of data objects, with no extra code required.

This brings you to the end of your data binding odyssey. In the next chapter, you'll start with a new topic, learning how to store data on the local computer with isolated storage.

■ ■ ■

CHAPTER 18

File Access

Silverlight includes a surprisingly capable set of features for saving, reading, and managing files. The key fact you need to understand is that ordinary Silverlight applications aren't allowed to write to (or read from) arbitrary locations on the file system. Instead, Silverlight applications that need to store data permanently must use a feature called *isolated storage*.

Isolated storage gives you access to a small segment of hard-disk space, with certain limitations. For instance, you don't know exactly where your files are being stored. You also can't read the files left by another Silverlight application or recorded for another user. In essence, isolated storage provides carefully restricted, tamperproof file access for applications that need to store permanent information on the local computer—usually so this information can be retrieved the next time the user runs the application.

In recent releases, Silverlight has added three back doors that allow you to bypass the rules of isolated storage in specific scenarios:

- *OpenFileDialog and SaveFileDialog*: These classes allow you to access a file directly, provided the user explicitly select it in a dialog box. For example, if the user picks a file in the OpenFileDialog, your application is granted permission to read that file's contents. If the user picks a file in the SaveFileDialog, your application is allowed to write to it. However, in neither case are you given the specifics about that file (such as its exact location on the hard drive), nor are you allowed to access any other files in the same location.

- *Drag and drop*: The second way to get around the limits of isolated storage is using Silverlight's drag-and-drop feature. Once again, it's up to the user to explicitly pass the file to your code—in this case, by dragging it over your application and dropping it on one of your elements. You can then handle the Drop event and read the file content.

- *Elevated trust*: Elevated-trust applications are applications that must be explicitly installed by the user and show a security prompt requesting extra privileges. In the case of file access, elevated trust applications can read and write to the files in the set of documents folder for the current user (which is sometimes called "My Documents").

In this chapter, you'll learn how to create files in isolated storage and write and read data. You'll see how to store miscellaneous data, application settings, and entire objects. You'll also learn how to request more isolated storage space for your application and upload files to a web server with the help of web services. Finally, you'll look at Silverlight's file dialog classes and the drag-and-drop file transfer feature.

657

■ **What's New** Silverlight 4 adds the ability for your application to receive files that the user drags and drops onto an element (see the section "Dragging and Dropping Files"). Silverlight 4 also allows out-of-browser applications with elevated trust to access the files in the current user's documents folder (see Chapter 21).

Isolated Storage

Isolated storage provides a virtual file system that lets you write data to a small, user-specific and application-specific slot of space. The actual location on the hard drive is obfuscated (so there's no way to know beforehand exactly where the data will be written), and the default space limit is a mere 1 MB (although you can request that the user grant you more).

Essentially, isolated storage is the Silverlight equivalent of persistent cookies in an ordinary web page. It allows small bits of information to be stored in a dedicated location that has specific controls in place to prevent malicious attacks (such as code that attempts to fill the hard drive or replace a system file).

The Scope of Isolated Storage

With isolated storage, a unique storage location is created for every combination of user and application. In other words, the same computer can have multiple isolated storage locations for the same application, assuming each one is for a different user. Similarly, the same user can have multiple isolated storage locations, one for each Silverlight application. Isolated storage isn't affected by browser, so a Windows user switching from Internet Explorer to Firefox will get the same isolated storage location in both browsers.

■ **Note** Data in one user's isolated store is restricted from other users (unless they're Windows administrators).

The critical factor that gives a Silverlight application its identity is the URL of the XAP file. That means the following:

- Different XAP files on the same web server and in the same folder have different isolated stores.
- If you host the website on different domains, each instance gets its own isolated store.
- If you create different test pages that use the same application at the same location, they share the same isolated storage.
- If you rename the XAP file (or the folder that it's in), you get a new isolated store.
- If you change the GUID, version, or other assembly metadata for your Silverlight application, you keep the same isolated store.
- If you replace a Silverlight application with another application that has the same XAP file name, it acquires the previous application's isolated store.

What to Put in Isolated Storage

Isolated storage is a great way to store small amounts of nonessential information. Good choices include user-specific details, user preferences, and information about recent user actions. Isolated storage is also great temporary storage. For example, imagine you create a Silverlight application that allows a user to fill out a multipart form (over several pages) and then send it to a web service, where it will be stored permanently. Each time the user moves from one part of the form to the next, you can save the current data to isolated storage. Then, when the user completes the operation and successfully submits the data to the web service, you can delete it. This commonsense approach prevents the user from losing data if the application can't contact the web service (because the network isn't working) or the user accidentally restarts the application (for example, by clicking the browser's Back button). Your application can check for the temporary data on startup and give the user the option of reloading that data.

Isolated storage is persistent—unlike the browser cache, it never expires, and it's not removed if the user chooses to explicitly delete temporary Internet files. However, isolated storage isn't a good storage place for important documents, because they're never backed up, are easily deleted, and can be even more easily lost (for example, if the user changes accounts or computers). Furthermore, isolated storage generally isn't a good place to cache resources (for example, external bitmaps and media files). It may seem tempting, but isolated storage is intended to be a limited-size storage location for data, not a handcrafted replacement for HTTP caching.

Using Isolated Storage

Isolated storage is easy to use because it exposes the same stream-based model that's used in ordinary .NET file access. You use the types in the System.IO.IsolatedStorage namespace, which are a core part of the Silverlight runtime.

Opening an Isolated Store

Silverlight creates isolated stores automatically. To interact with an isolated store, you use the IsolatedStorageFile class. You get the IsolatedStorageFile object for the current user and application by calling the shared IsolatedStorageFile.GetUserStoreForApplication() method, as shown here:

```
Dim store As IsolatedStorageFile = IsolatedStorageFile.GetUserStoreForApplication()
```

Usually, this gives you exactly what you want: an application-specific, user-specific location where you can store data. However, the IsolatedStorageFile class also includes a similar but slightly different shared method named GetUserStoreForSite(). This method provides a storage site that's accessible to all the Silverlight applications on the same website domain. However, these settings are still user-specific. You may choose to use a domain-wide isolated store if you're developing a group of Silverlight applications together and you want all of them to share some personalization information.

Either way, once you've opened the isolated store and have a live IsolatedStorageFile object, you're ready to start creating files.

File Management

The IsolatedStorageFile class name is somewhat misleading, because it doesn't represent a single file. Instead, it provides access to the collection of files in the isolated store. The methods that the IsolatedStorageFile class provides are similar to the file-management methods you can use through the File and Directory classes in a full-fledged .NET application. Table 18-1 lists the methods you can use.

Table 18-1. File-Management Methods for IsolatedStorageFile

Method	Description
CreateDirectory()	Creates a new folder in the isolated store, with the name you specify.
DeleteDirectory()	Deletes the specified folder from the isolated store.
CreateFile()	Creates a new file with the name you supply and returns an IsolatedStorageFileStream object that you can use to write data to the file.
DeleteFile()	Deletes the specified file from the isolated store.
Remove()	Removes the isolated store, along with all its files and directories.
OpenFile()	Opens a file in the isolated store and returns an IsolatedStorageFileStream object that you can use to manipulate the file. Usually, you'll use this method to open an existing file for reading, but you can supply different FileMode and FileAccess values to create a new file or overwrite an existing file.
FileExists()	Returns True or False, depending on whether the specified file exists in the isolated store. You can use an overloaded version of this method to look in a specific subfolder or match a file with a search expression (using the wildcards ? and *).
DirectoryExists()	Returns True or False, depending on whether the specified folder exists in the isolated storage location.
GetFileNames()	Returns an array of strings, one for each file in the root of the isolated store. Optionally, you can use an overloaded version of this method that accepts a single string argument. This argument lets you specify a subfolder you want to search or a search expression (using the wildcards ? and *).
GetDirectoryNames()	Returns an array of strings, one for each subfolder in the root of the isolated store. Optionally, you can use an overloaded version of this method that accepts a single string argument. This argument lets you get subfolders in a specific directory or specify a search expression (using the wildcards ? and *).

Writing and Reading Data

Using the methods in Table 18-1, you can create files and use streams to write and read data. Of course, you're unlikely to deal with the IsolatedStorageFileStream class directly, unless you want to read and write your data one byte at a time. Instead, you'll use one of the more capable classes from the System.IO namespace that wrap streams:

- *StreamWriter and StreamReader*: Use these classes if you want to write and read data as ordinary text strings. You can write the data in several pieces and retrieve it line by line or in one large block using StreamReader.ReadToEnd().

- *BinaryWriter and BinaryReader*: Use these classes if you want to write data more strictly (and somewhat more compactly). When retrieving data, you need to use the data type. (For example, you must use the BinaryReader.ReadInt32() method to retrieve a 32-bit integer from the file, BinaryReader.ReadString() to read a string, and so on.)

The following example gets the current isolated store, creates a new file named date.txt, and writes the current date to that file as a piece of text:

```
' Write to isolated storage.
Try
    Using store As IsolatedStorageFile = _
      IsolatedStorageFile.GetUserStoreForApplication()

        Using stream As IsolatedStorageFileStream = store.CreateFile("date.txt")
            Dim writer As New StreamWriter(stream)
            writer.Write(DateTime.Now)
            writer.Close()
        End Using

        lblStatus.Text = "Data written to date.txt"
    End Using
Catch err As Exception
    lblStatus.Text = err.Message
End Try
```

Retrieving information is just as easy. You simply need to open the IsolatedStorageFileStream object in read mode:

```
' Read from isolated storage.
Try
    Using store As IsolatedStorageFile = _
      IsolatedStorageFile.GetUserStoreForApplication()

        Using stream As IsolatedStorageFileStream = store.OpenFile("date.txt", _
          FileMode.Open)
            Dim reader As New StreamReader(stream)
            lblData.Text = reader.ReadLine()
            reader.Close()
        End Using
    End Using
Catch err As Exception
    ' An exception will occur if you attempt to open a file that doesn't exist.
    lblStatus.Text = err.Message
End Try
```

In this example, on a Windows Vista or Windows 7 computer, you'll find the date.txt file in a path in this form:

`C:\Users\[UserName]\AppData\LocalLow\Microsoft\Silverlight\is\[Unique_Identifier]`

Several automatically generated folder names are tacked onto the end of this path. Here's an example of a dynamically created path that Silverlight may use for isolated storage:

`C:\Users\matthew\AppData\LocalLow\Microsoft\Silverlight\is\sid3dsxe.u1y\lstesiyg.ezx`
`\s\atkj2fb5vjnabwjsx2nfj3htrsq1ku1h\f\date.txt`

If you're curious, you can get the path for the current isolated store using the Visual Studio debugger. To do so, hover over the IsolatedStorageFile object while in break mode, and look for the nonpublic RootDirectory property, as shown in Figure 18-1.

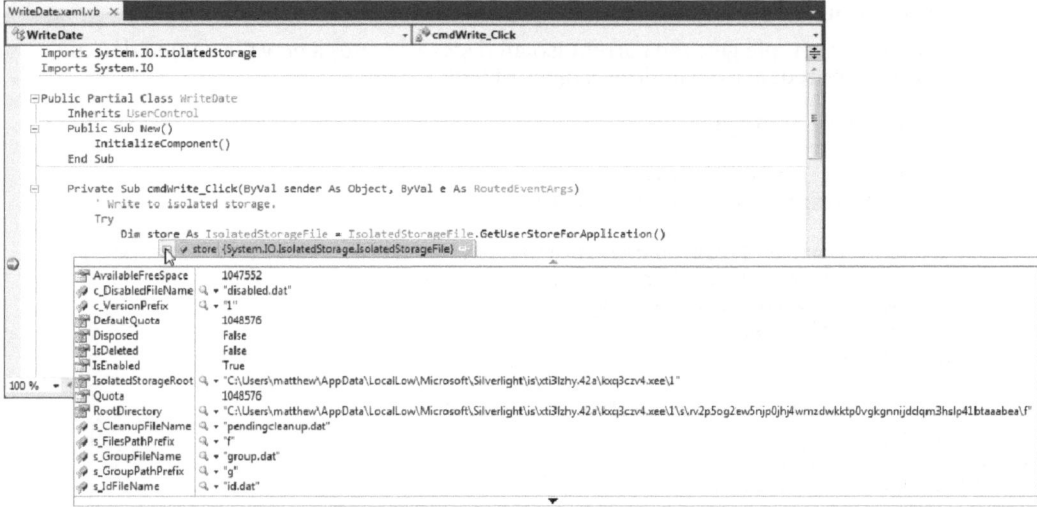

Figure 18-1. *Finding the isolated storage location*

Fortunately, you don't need to worry about the directory structure that's used for isolated storage. You can check for files and retrieve from isolated storage using the methods of IsolatedStorageFile, such as GetFileNames() and OpenFile().

■ **Note** Notably, Silverlight doesn't obfuscate the names of files in isolated storage. That means if the user knows the file name, the user can perform a file search to find the file.

Requesting More Space

Initially, each Silverlight application gets 1 MB of space in its isolated store. You can examine the IsolatedStorageFile.AvailableFreeSpace property to find out how much free space remains.

■ **Note** There's one exception to the initial 1 MB quota. If the application has been installed as an out-of-browser application (see Chapter 21), its quota automatically increases to 25 MB. This increased quota applies regardless of whether you launch the application in a stand-alone window or run it through the browser—either way, the application uses the same isolated store.

If your application needs more space, you can use an option: the IsolatedStorageFile IncreaseQuotaTo() method. When you call this method, you request the number of bytes you want. Silverlight then shows a message box with the current number of bytes the application is using in isolated storage (*not* the current quota limit) and the new requested amount of space. The dialog box also shows the URL of the Silverlight application (or file:// if you're running it locally).

Figure 18-2 shows an example where the application currently has no files stored in isolated storage and is attempting to increase the limit to 1 MB. If the user clicks Yes to accept the request, the quota is increased, and the IncreaseQuotaTo() method returns True. If the user clicks No, the request is denied, and IncreaseQuotaTo() returns False.

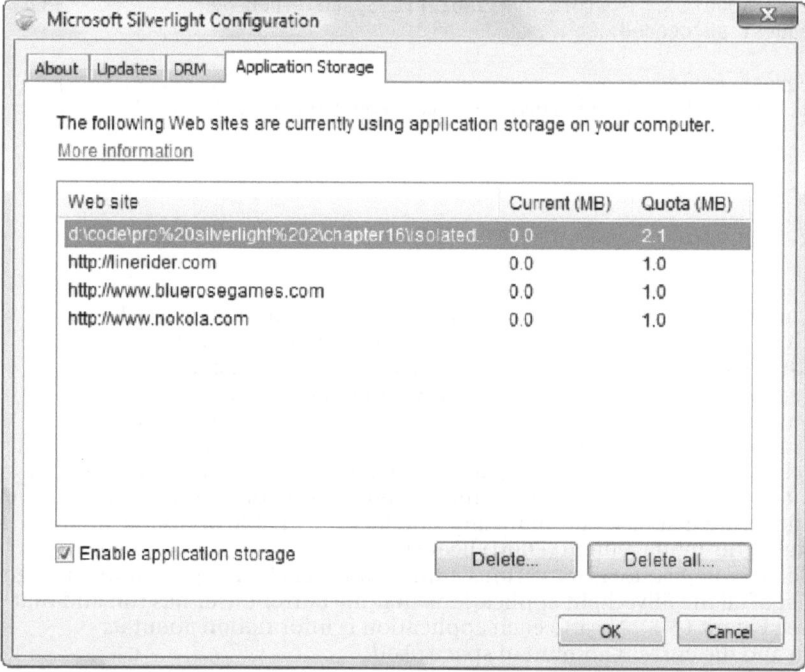

Figure 18-2. Asking to increase the isolated store quota

Two considerations limit how you can use IncreaseQuotaTo():

- You must use it in an event handler that reacts to a user action (for example, a button click). If you attempt to use it elsewhere—say, when a page loads—the call will be ignored. This is a security measure designed to prevent users from inadvertently accepting large quotas if the confirmation dialog suddenly steals the focus.
- You must request a value that's higher than the current quota. Otherwise, you'll receive an exception. That means you can't use the IncreaseQuotaTo() method to ensure that there's a certain level of free space—instead, you need to explicitly check whether you have the required amount of space.

You can determine the current quota size by checking the IsolatedStorageFile.Quota property. You can find the amount of space that remains in the isolated store using the IsolatedStorageFile.AvailableFreeSpace property. (It therefore follows that you can calculate the amount of space you're using in isolated storage by calculating IsolatedStorageFile.Quota – IsolatedStorageFile.AvailableFreeSpace.)

Here's an example of the IncreaseQuotaTo() method in action:

```
Using store As IsolatedStorageFile = _
  IsolatedStorageFile.GetUserStoreForApplication()

    ' In an application that writes 1000 KB files, you need to ask for an increase
    ' if there is less than 1000 KB free.
    If store.AvailableFreeSpace < 1000*1024 Then
        If store.IncreaseQuotaTo( _
          store.Quota + 1000*1024 - store.AvailableFreeSpace) Then
            ' The request succeeded.
        Else
            ' The request failed.
            lblError.Text = "Not enough room to save temporary file."
            Return
        End If
    End If

    ' (Write the big file here.)
End Using
```

The preceding example uses a calculation to request an exact amount of space. The potential problem with this approach is that every time you need a bit more space, you'll need to present the user with a new request. To avoid these constant requests, it makes sense to request an amount of space that's comfortably above your immediate needs.

There's an easy way to find out how much isolated space storage has been allocated to every Silverlight application you've ever used. To do so, you must first browse to a page with Silverlight content. Right-click the Silverlight content region, and choose Silverlight. A tabbed dialog box appears that displays information about the current version of Silverlight, allows you to control whether updates are installed automatically, and lets you enable or disable media content that uses digital rights management (DRM) licensing.

To review the isolated storage quotas for various applications, click the Application Storage tab. There, you see a list of all the Silverlight applications that the current user has run and that use isolated storage (see Figure 18-3). Next to each application is information about its maximum space quota and the current amount of space used.

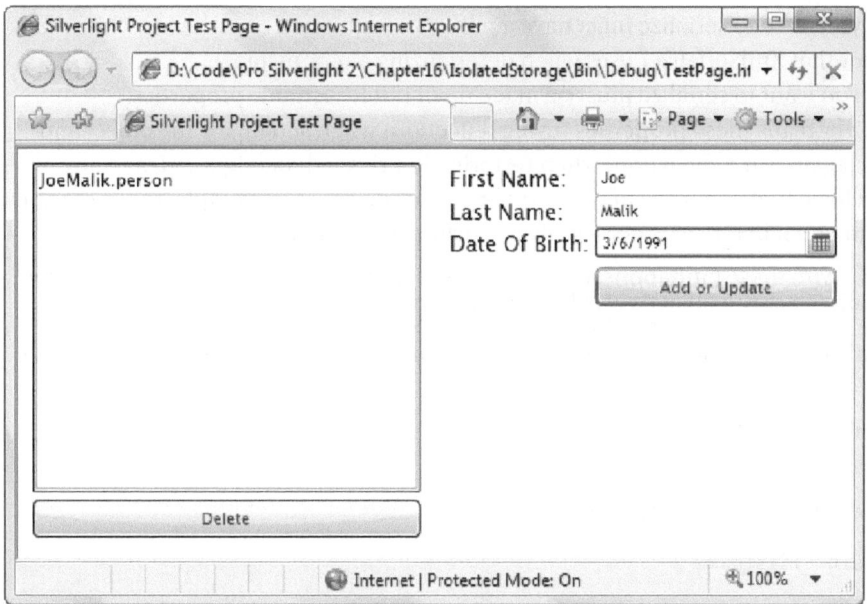

Figure 18-3. Reviewing the isolated stores of different applications

The Application Storage tab gives you the only way to remove isolated storage. Simply select the application, and click Delete. When you do so, two things happen: all the files in isolated storage for that application are removed, and the quota is reset to the standard 1 MB.

■ **Note** There's no way to lower an application's isolated storage quota without removing the current contents of its isolated store. You can also do this programmatically using the IsolatedStorageFile.Remove() method.

Storing Objects with XmlSerializer

As you've already seen, you can write to files in isolated storage using the same classes you use for ordinary file access in a .NET application, such as StreamWriter and BinaryWriter. To read from them, you use the corresponding StreamReader and BinaryReader classes. Although this approach gives you the most direct control over your files, it's not the only option.

The XmlSerializer class provides a higher-level alternative that allows you to serialize and deserialize objects rather than write and read individual pieces of data. XmlSerializer works by converting a live object into a stream of bytes, which you can push out to any stream. XmlSerializer can also perform the reverse trick and convert a stream of bytes into an object instance. To use XmlSerializer, you need to add a reference to the System.Xml.Serialization.dll assembly, which will be included in the XAP file for your compiled application.

XmlSerializer can't work with every class. It has two non-negotiable requirements:

- The class you want to serialize must have a public no-argument constructor. This is the constructor that XmlSerializer uses when deserializing a new instance.
- The class you want to serialize must be made up of public settable properties. XmlSerializer reads these properties (using reflection) when serializing the object and sets them (again using reflection) when restoring it. Private data is ignored, and any validation logic that you place in your property procedures—for example, requiring one property to be set before another—is likely to cause a problem.

If you can live with these limitations, the advantage is that XmlSerializer gives you a clean, concise way to store an entire object's worth of information.

Ideally, the classes you use to store information with XmlSerializer will be simple data packages with little or no functionality built in. Here's a simple Person class that's serialized in the next example you'll consider:

```
Public Class Person

    Private _firstName As String
    Public Property FirstName() As String
        Get
            Return _firstName
        End Get
        Set(ByVal value As String)
            _firstName = value
        End Set
    End Property

    Private _lastName As String
    Public Property LastName() As String
        Get
            Return _lastName
        End Get
        Set(ByVal value As String)
            _lastName = value
        End Set
    End Property

    Private _dateOfBirth As DateTime
    Public Property DateOfBirth() As DateTime
        Get
            Return _dateOfBirth
        End Get
        Set(ByVal value As DateTime)
            _dateOfBirth = value
        End Set
    End Property

    Public Sub New(ByVal firstName As String, ByVal lastName As String, _
      ByVal dateOfBirth As Nullable(Of DateTime))
        Me.FirstName = firstName
        Me.LastName = lastName
        Me.DateOfBirth = dateOfBirth
    End Sub
```

```
       ' Required for serialization support.
       Public Sub New()
       End Sub
End Class
```

Figure 18-4 shows a test page that uses XmlSerializer and the Person class. It lets the user specify the three pieces of information that make up a Person object and then stores that data in isolated storage. Person files are named using the first name, last name, and extension *.person*, as in JoeMalik.person. The list on the left of the page shows all the .person files in isolated storage and allows the user to select one to view or update its data.

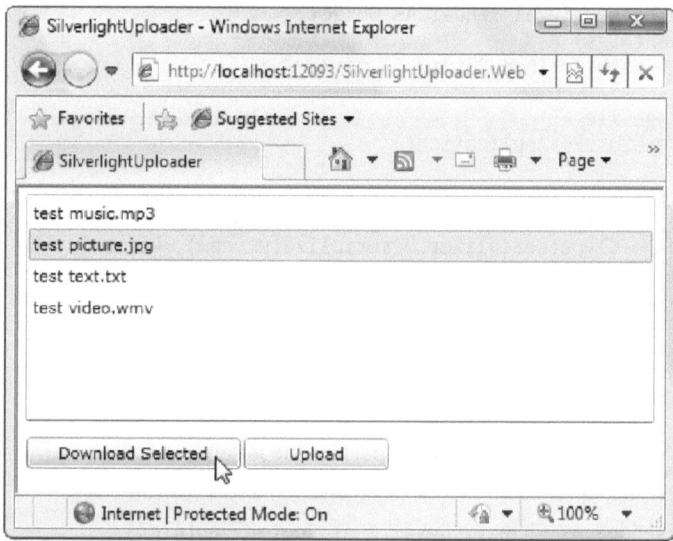

Figure 18-4. Storing person objects

Building this example is easy. First, you need an instance of XmlSerializer that's customized to use the Person class and is available to all your event-handling code:

```
Private serializer As New XmlSerializer(GetType(Person))
```

When the user clicks the Add button, the current information in the two text boxes and the DatePicker control is used to build a Person object, and that Person object is serialized to isolated storage:

```
Private Sub cmdAdd_Click(ByVal sender As Object, ByVal e As RoutedEventArgs)
    Dim person As New Person(txtFirstName.Text, txtLastName.Text, _
      dpDateOfBirth.SelectedDate)

    Using store As IsolatedStorageFile = _
      IsolatedStorageFile.GetUserStoreForApplication()

        ' The CreateFile() method creates a new file or overwrites an existing one.
        Using stream As FileStream = store.CreateFile( _
          person.FirstName + person.LastName & ".person")
            ' Store the person details in the file.
```

```
                    serializer.Serialize(stream, person)
                End Using

                ' Update the list.
                lstPeople.ItemsSource = store.GetFileNames("*.person")
        End Using
End Sub
```

When the user clicks one of the person files in the list, the data is retrieved from isolated storage:

```
Private Sub lstPeople_SelectionChanged(ByVal sender As Object, _
    ByVal e As SelectionChangedEventArgs)
        If lstPeople.SelectedItem Is Nothing Then Return

        Using store As IsolatedStorageFile = _
          IsolatedStorageFile.GetUserStoreForApplication()

            Using stream As FileStream = _
              store.OpenFile(lstPeople.SelectedItem.ToString(), FileMode.Open)
                Dim person As Person = CType(serializer.Deserialize(stream), Person)
                txtFirstName.Text = person.FirstName
                txtLastName.Text = person.LastName
                dpDateOfBirth.SelectedDate = person.DateOfBirth
            End Using
        End Using
End Sub
```

And finally, if the user clicks Delete button, the selected person file is removed from the isolated store:

```
Private Sub Delete_Click(ByVal sender As Object, ByVal e As RoutedEventArgs)
        If lstPeople.SelectedItem Is Nothing Then Return

        Using store As IsolatedStorageFile = _
          IsolatedStorageFile.GetUserStoreForApplication()
            store.DeleteFile(lstPeople.SelectedItem.ToString())
            lstPeople.ItemsSource = store.GetFileNames("*.person")
        End Using
End Sub
```

Storing Application Settings

A common pattern with isolated storage is to load it when the application starts (or as needed) and then save it automatically when the application ends and the Application.Exit event fires. Silverlight has a higher-level class that allows you to implement this pattern to store miscellaneous pieces of information (typically, application settings). This class is IsolatedStorageSettings.

The IsolatedStorageSettings class provides two shared properties, both of which hold collections of information that you want to store. The most commonly used collection is IsolatedStorageSettings.ApplicationSettings, which is a name-value collection that can hold any items you like. Behind the scenes, the ApplicationSettings class uses XmlSerializer to store the information you add.

To add an item, you need to assign it a new string key name. Here's an example that stores the date under the key name LastRunDate:

```
IsolatedStorageSettings.ApplicationSettings("LastRunDate") = DateTime.Now
```

And here's an example that stores a Person object under the key name CurrentUser:

```
IsolatedStorageSettings.ApplicationSettings("CurrentUser") = New Person(...)
```

Retrieving it is just as easy, although you need to cast the object to the right type:

```
Dim runDate As DateTime
runDate = CDate(IsolatedStorageSettings.ApplicationSettings("LastRunDate"))
Dim person As Person
person = CType(IsolatedStorageSettings.ApplicationSettings("CurrentUser"), Person)
```

You can also use the Contains() method to check whether a key exists in the ApplicationSettings collection, and you can use the Remove() method to delete an existing piece of information.

The ApplicationSettings class stores all the information it contains automatically when the Silverlight application shuts down (for example, when the user navigates to a new page). Thus, the information will be present in the ApplicationSettings collection the next time the user runs the application. The IsolatedStorageSettings class also provides a SiteSettings collection that works in much the same way—it's an untyped collection that can hold any type of serializable data—but is scoped to the current website domain. That means any Silverlight applications running at that domain have access to these settings.

The ApplicationSettings collection and SiteSettings collection are really just niceties that simplify what you can already do directly with isolated storage. However, they're a convenient place to store small scraps of configuration information without needing to build a more complex data model for your application.

Accessing Files Outside of Isolated Storage

As you've seen, Silverlight applications aren't allowed to browse the file system. But you can use several back doors to read and write individual files—provided the user selects them first. These back doors are the OpenFileDialog and SaveFileDialog classes and the drag-and-drop feature.

The OpenFileDialog and SaveFileDialog classes allow you to show a standard Open and Save dialog box in response to a user-initiated action (like a button click). The user then selects a file in the dialog box, which is returned to your code as a stream. If you show the Open dialog box, you get given a read-only stream for accessing the file. If you show the Save dialog box, you get a writeable stream. Either way, the OpenFileDialog and SaveFileDialog classes give you access to a single specific file, while walling off everything else.

■ **Note** For security reasons, Silverlight does not support the OpenFileDialog and SaveFileDialog classes in full-screen mode. Although Silverlight's standard behavior is to switch out of full-screen mode as soon as you show either one, it's better for your code to set explicitly set Application.Current.Host.Content.IsFullScreen to False to avoid any possible problems on different browsers and operating systems.

Silverlight's drag-and-drop feature is similar to the OpenFileDialog. It allows your application to access one or more files in read-only mode, provided the user explicitly drags these files onto an element in your application. The advantage of the drag-and-drop feature is that the file selection requires fewer user steps. It's particularly convenient if the user needs to select a large number of files at once.

In the following sections, you'll learn how to use all these approaches. You'll also consider a realistic application that uploads user-selected files to a web service.

Reading Files with OpenFileDialog

OpenFileDialog allows you to show the ordinary Open File dialog box. After the user chooses a file, it's made available to your application for reading only. No restrictions are placed on the OpenFileDialog, so it's possible for the user to choose any file. However, there's no way for you to access any file without the user explicitly choosing it and clicking Open, which is considered to be a high enough bar for security.

To use OpenFileDialog, you first create a new instance and then set the Filter and FilterIndex properties to configure what file types the user sees. The Filter property determines what appears in the file-type list.

You need to indicate the text that should appear in the file-type list and the corresponding expression that the OpenFileDialog box will use to filter files. For example, if you want to allow the user to open text files, you can show the text "Text Files (*.txt)" and use the filter expression *.txt to find all files with the .txt extension. Here's how you then set the Filter property:

```
Dim dialog As New OpenFileDialog()
dialog.Filter = "Text Files (*.txt)|*.txt"
```

You use the pipe (|) character to separate the display text from the filter expression in the filter string. If you have multiple file types, string them one after the other, separated by additional pipe characters. For example, if you want to let the user see different types of images, you can write a filter string like this:

```
dialog.Filter = "Bitmaps (*.bmp)|*.bmp|JPEGs (*.jpg)|*.jpg|All files (*.*)|*.*"
```

You can also create a filter expression that matches several file types, by separating them with semicolons:

```
dialog.Filter = "Image Files (*.bmp;*.jpg;*.gif)|*.bmp;*.jpg;*.gif"
```

After you've configured the OpenFileDialog, you then show the dialog box by calling ShowDialog(). The ShowDialog() method returns a DialogResult value that indicates what the user selected. If the result is True, the user picked a file, and you can go ahead and open it:

```
If dialog.ShowDialog() = True Then
    ...
End If
```

The file is exposed through the OpenFileDialog.File property, which is a FileInfo object. The FileInfo is a relatively simple class that exposes a small set of useful members, including a Name property that returns the file name, an OpenRead() method that returns a FileStream object in read-only mode, and an OpenText() method that creates the FileStream and returns a StreamReader for it:

```
If dialog.ShowDialog() = True Then
    Using reader As StreamReader = dlg.File.OpenText()
```

```
        Dim data As String = reader.ReadToEnd()
    End Using
End If
```

Obviously, the OpenText() method is a good shortcut if you're dealing with text data, and OpenRead() is a better choice if you need to create a BinaryReader or use the FileStream.Read() method directly to pull out a block of bytes.

■ **Tip** The OpenFileDialog also supports multiple selection. Set OpenFileDialog.Multiselect to True before you call ShowDialog(). Then, retrieve all the selected files through the OpenFileDialog.Files property.

One interesting way to use OpenFileDialog is to copy a selected file from the local hard drive to isolated storage so the application can manipulate it later. Here's an example that performs this trick:

```
Dim dialog As New OpenFileDialog()
dialog.Filter = "All files (*.*)|*.*"
dialog.Multiselect = True

' Show the dialog box.
If dialog.ShowDialog() = True Then

    ' Copy all the selected files to isolated storage.
    Dim store As IsolatedStorageFile = _
      IsolatedStorageFile.GetUserStoreForApplication()

    For Each file As FileInfo In dialog.Files
        Using fileStream As Stream = file.OpenRead()
            ' Check for free space.
            If fileStream.Length > store.AvailableFreeSpace Then
                ' (Cancel the operation or use IncreaseQuotaTo().)
            End If

            Using storeStream As IsolatedStorageFileStream = _
              store.CreateFile(file.Name)
                ' Write 1 KB block at a time.
                Dim buffer(1023) As Byte
                Dim count As Integer = 0
                Do
                    count = fileStream.Read(buffer, 0, buffer.Length)
                    If count > 0 Then
                        storeStream.Write(buffer, 0, count)
                    End If
                Loop While count > 0
            End Using
        End Using
    Next
End If
```

Writing Files with SaveFileDialog

When you've mastered OpenFileDialog, the SaveFileDialog class will seem straightforward. Like OpenFileDialog, it allows the user to hunt around the hard drive and choose a file that is exposed to your application. You can't retrieve any details about where this file is located or what other files exist in the same folder. Instead, SaveFileDialog gives you a stream into which you can write data.

To use the SaveFileDialog class, begin by creating an instance and setting the file-type filter. Then, show the dialog box (using the familiar ShowDialog() method), and grab the stream for the selected file (using the OpenFile() method). Here's a simple code routine that demonstrates these steps by copying text out of a text box into a user-designated file:

```
Dim saveDialog As New SaveFileDialog()
saveDialog.Filter = "Text Files (*.txt)|*.txt"

If saveDialog.ShowDialog() = True Then
    Using stream As Stream = saveDialog.OpenFile()
        Using writer As New StreamWriter(stream)
            writer.Write(txtData.Text)
        End Using
    End Using
End If
```

For security reasons, you can't set a default filename for SaveFileDialog, although you can set a default file extension using the DefaultExt property:

```
saveDialog.DefaultExt = "txt"
```

SaveFileDialog adds the default extension to the file name the user types in, unless the file name already includes the same extension. If the user includes a different extension (for example, the DefaultExt is txt and the user enters myfile.test), the default extension is still added to the end (for example, making the file myfile.test.txt).

If the user picks a file that already exists, a confirmation message appears asking whether the user wants to overwrite the existing file. This will happen when the user selects an existing file from the displayed list or if the user types in a file name that, with the addition of the default extension, matches an existing file. Either way, the user must confirm the operation to close the dialog box and continue.

Finally, after ShowDialog() returns, you can retrieve the file name the user selected, without any folder or path information, from the SafeFileName property.

Transmitting Files with a Web Service

With the combination of OpenFileDialog and SaveFileDialog, it's possible to build an application that copies server content to the local computer or uploads local files to a location on the web server. In fact, building an application like this is easy—all you need is a back-end web service that manages the files.

Figure 18-5 shows a simple example that demonstrates the concept. When this application first starts, it requests from the server a list of available files. The user can then choose to upload new files or download one of the existing ones. In a more sophisticated example, the web server could require some form of authentication and give each user access to a different collection of files.

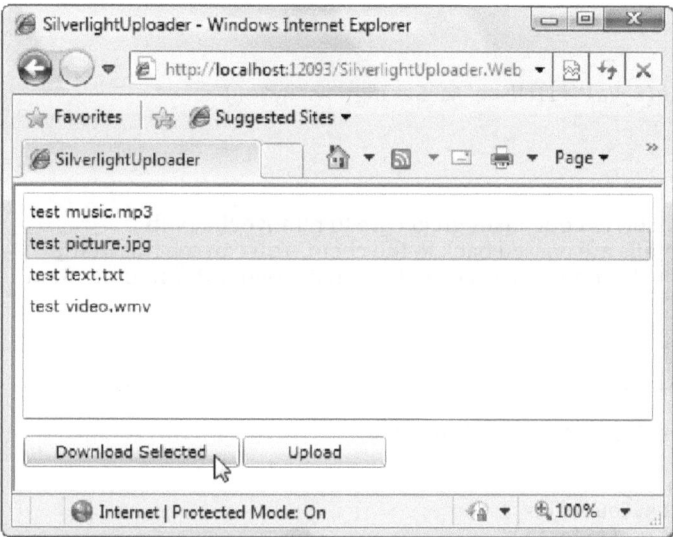

Figure 18-5. *A Silverlight-based file uploader*

You can try this complete example with the downloadable code for this chapter. In the following sections, you'll walk through all the essential code.

The File Service

The backbone of this example is a set of server-side methods that allows the Silverlight application to do three things: retrieve a list of files, download an existing file, and upload a new file. In this example, a single service named FileService takes care of all three tasks.

FileService provides access to the files in a predefined location. In this case, files are stored in a subfolder on the server named Files. Here's the basic outline of the web service code:

```
<ServiceContract(Namespace := "")> _
<AspNetCompatibilityRequirements( _
 RequirementsMode := AspNetCompatibilityRequirementsMode.Allowed)> _

Public Class FileService
    Private filePath As String

    Public Sub New()
        filePath = HttpContext.Current.Server.MapPath("Files")
    End Sub

    <OperationContract> _
    Public Function GetFileList() As String()
        ...
    End Function

    <OperationContract> _
    Public Sub UploadFile(ByVal fileName As String, ByVal data As Byte())
        ...
```

```
        End Sub

        <OperationContract> _
        Public Function DownloadFile(ByVal fileName As String) As Byte()
            ...
        End Function
    End Class
```

When handling file names, the server code takes great care to remove the path information so that no sensitive details are passed back to the client. You can see this in the GetFileList() method, which uses the System.IO.Path class to strip the path information out of each of the file names:

```
<OperationContract> _
Public Function GetFileList() As String()
    ' Scan the folder for files.
    Dim files As String() = Directory.GetFiles(filePath)

    ' Trim out path information.
    For i As Integer = 0 To files.Count() - 1
        files(i) = Path.GetFileName(files(i))
    Next i

    ' Return the file list.
    Return files
End Function
```

The DownloadFile() method needs to take similar care but for a different reason. It makes sure to strip any path information out of the caller-supplied file name. This prevents malicious callers from passing in relative paths such as ../../../Windows/System/somefile.dll, which could otherwise trick the application into returning a sensitive file.

Once the DownloadFile() code has safely filtered out the file name, it opens the file, copies its contents to a byte array, and returns the data:

```
<OperationContract> _
Public Function DownloadFile(ByVal fileName As String) As Byte()
    ' Make sure the filename has no path information.
    Dim file As String = Path.Combine(filePath, Path.GetFileName(fileName))

    ' Open the file, copy its raw data into a byte array, and return that.
    Using fs As New FileStream(file, FileMode.Open)
        Dim data(fs.Length - 1) As Byte
        fs.Read(data, 0, fs.Length)
        Return data
    End Using
End Function
```

■ **Note** The file transfer technique used in DownloadFile() requires loading the contents of the file into memory. Thus, this approach isn't suitable for extremely large files, and it's a good idea to add a safeguard that checks the file length before attempting to create the byte array. When dealing with larger files, you'll probably want to pass a URI to the client and let the client download the file from that URI. To keep your files fairly private, you can use a randomly generated file name that incorporates a globally unique identifier (GUID) using the System.Guid class.

Finally, the web service allows the user to submit a block of data that will be blindly written to the Files folder. The user gets to choose the file name, and once again, any path information is stripped out before the file is created.

```
<OperationContract> _
Public Sub UploadFile(ByVal fileName As String, ByVal data As Byte())
    ' Make sure the filename has no path information.
    Dim file As String = Path.Combine(filePath, Path.GetFileName(fileName))

    Using fs As New FileStream(file, FileMode.Create)
        fs.Write(data, 0, data.Length)
    End Using
End Sub
```

You might think the UploadFile() method is a logical place to check the size of the byte[] array so that a malicious user can't pass extremely large files that will consume the hard drive. However, WCF already clamps down on this ability by restricting the maximum message size it accepts and the maximum size of transmitted arrays in that message. These limits are meant to stop *denial-of-service attacks* by making it impossible for an attacker to tie the server up with huge or complex messages that are time-consuming to process.

If you actually do want to build a web service that accepts large amounts of data, you'll need to perform a fair bit of tweaking in both the web.config file on the web server and the ServiceReferences.ClientConfig in the client. Although these configuration changes are outside the scope of this book, you can get the full details at http://tinyurl.com/nc8xkn. You can also see them at work with the downloadable code for this chapter, which is configured to allow large file uploads and downloads.

The Silverlight Client

The code for the client is fairly straightforward. All web service calls go through a single FileServiceClient instance, which is stored as a field in the page class. When the page first loads, the code attaches all the event handlers it will use for the various completed events and then calls the GetFileListAsync() method to fetch the list of files for the list box.

```
Private client As New FileServiceClient()

Private Sub Page_Loaded(ByVal sender As Object, ByVal e As RoutedEventArgs)
    ' Attach these event handlers for uploads and downloads.
    AddHandler client.DownloadFileCompleted, AddressOf client_DownloadFileCompleted
    AddHandler client.UploadFileCompleted, AddressOf client_UploadFileCompleted
```

```
    ' Get the initial file list.
    AddHandler client.GetFileListCompleted, AddressOf client_GetFileListCompleted
    client.GetFileListAsync()
End Sub

Private Sub client_GetFileListCompleted(ByVal sender As Object, _
  ByVal e As GetFileListCompletedEventArgs)
    Try
        lstFiles.ItemsSource = e.Result
    Catch
        lblStatus.Text = "Error contacting web service."
    End Try
End Sub
```

When the user selects a file and clicks the Download button, the application shows the SaveFileDialog so the user can pick the downloaded location. You can't show SaveFileDialog after, when the DownloadFileCompleted event occurs, because this event isn't user-initiated. (If you try, you'll receive a SecurityException.)

However, even though the code starts by showing the SaveFileDialog, it doesn't attempt to open the FileStream right away. Doing so would leave the file open while the download is underway. Instead, the code passes the SaveFileDialog object to DownloadFileCompleted event as a state object, using the optional second argument that's available with all web service methods.

```
Private Sub cmdDownload_Click(ByVal sender As Object, ByVal e As RoutedEventArgs)
    If lstFiles.SelectedIndex <> -1 Then
        Dim saveDialog As New SaveFileDialog()
        If saveDialog.ShowDialog() = True Then
            client.DownloadFileAsync(lstFiles.SelectedItem.ToString(), saveDialog)
            lblStatus.Text = "Download started."
        End If
    End If
End Sub
```

The DownloadFileCompleted event retrieves the SaveFileDialog object and uses it to create the FileStream. It then copies the data from the byte array into this file.

```
Private Sub client_DownloadFileCompleted(ByVal sender As Object, _
  ByVal e As DownloadFileCompletedEventArgs)
    If e.Error Is Nothing Then
        lblStatus.Text = "Download completed."

        ' Get the SaveFileDialog that was passed in with the call.
        Dim saveDialog As SaveFileDialog = CType(e.UserState, SaveFileDialog)

        Using stream As Stream = saveDialog.OpenFile()
            stream.Write(e.Result, 0, e.Result.Length)
        End Using

        lblStatus.Text = "File saved to " & saveDialog.SafeFileName
    Else
        lblStatus.Text = "Download failed."
    End If
End Sub
```

A nice side effect of this approach is that this code allows the user to start multiple simultaneous downloads. Each one has its own SaveFileDialog object, and so each one can be saved to the appropriate file when the download is complete.

The uploading code is similar, but it shows the OpenFileDialog and retrieves the data from the file as soon as the user selects the file. The data is placed in a byte array and passed to the UploadFileAsync() method. The code the Silverlight client uses to accomplish this task is almost the same as the code the web service uses to open a file in the DownloadFile() method.

```
Private Sub cmdUpload_Click(ByVal sender As Object, ByVal e As RoutedEventArgs)
    Dim openDialog As New OpenFileDialog()

    If openDialog.ShowDialog() = True Then
        Try
            Using stream As Stream = openDialog.File.OpenRead()
                ' Don't allow really big files (more than 5 MB).
                If stream.Length < 5120000 Then
                    Dim data(stream.Length - 1) As Byte
                    stream.Read(data, 0, stream.Length)

                    client.UploadFileAsync(openDialog.File.Name, data)
                    lblStatus.Text = "Upload started."
                Else
                    lblStatus.Text = "Files must be less than 5 MB."
                End If
            End Using
        Catch
            lblStatus.Text = "Error reading file."
        End Try
    End If
End Sub

Private Sub client_UploadFileCompleted(ByVal sender As Object, _
  ByVal e As System.ComponentModel.AsyncCompletedEventArgs)
    If e.Error Is Nothing Then
        lblStatus.Text = "Upload succeeded."

        ' Refresh the file list.
        client.GetFileListAsync()
    Else
        lblStatus.Text = "Upload failed."
    End If
End Sub
```

This completes the example and gives you a fully functional client that can transfer content to and from the web server.

Dragging and Dropping Files

The OpenFileDialog class provides the most practical way for a Silverlight application to access a file that isn't in isolated storage. However, Silverlight offers one more approach. If a selects one or more files (for example, in Windows Explorer or on the desktop), drags these files over your Silverlight application, and then drops them onto an element, your application can read these files. Web-based uploading tools (for example, with SharePoint and Microsoft's SkyDrive service) often include this sort of feature to make it easier to upload an entire batch of files at once.

Figure 18-6 shows a Silverlight application that uses this feature. First, the user drags one or more image files and drops them on the large shaded rectangle. Then, the application reads the image data from all the files and adds a thumbnail for each one in the list on the right.

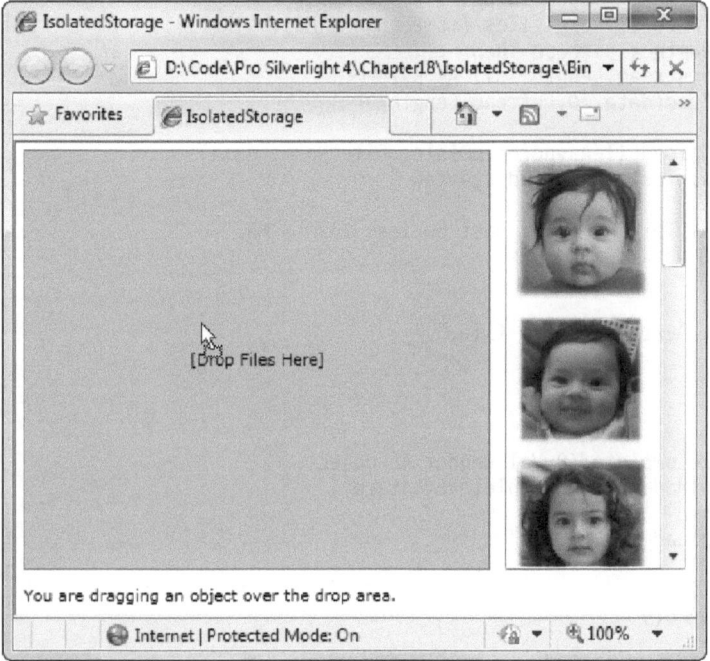

Figure 18-6. Dragging and dropping image files

■ **Note** The most practical applications of the drag and drop usually involve uploading content to a web server. For example, you could easily duplicate the multiple-document uploading feature in SharePoint or SkyDrive using a combination of the Silverlight drag-and-drop feature and a file-uploading web service (like the one described earlier in the section "Transmitting Files with a Web Service.")

DRAG-AND-DROP FEATURE ON A MAC

The drag-and-drop feature works seamlessly on a Windows computer but not on a Mac. The problem is that Mac web browsers, like Safari, have a slightly different plug-in model. The key difference is that they don't provide HTML DOM events to the plug-ins that are hosted in a web page. As a result, when you drop a file onto a Silverlight content region on a Mac, the web page receives the event, but the Silverlight application does not.

Currently, there is a Mac workaround that solves this issue on Safari web browsers but not on Firefox. It involves adding a small section of JavaScript to the hosting web page, which will then manually forward drag-and-drop events to the Silverlight application. For the full details, refer to the documentation at http://msdn.microsoft.com/library/ee670998.aspx. Unfortunately, at the time of this writing, Mac users running Firefox won't be able to drag files onto Silverlight applications, even using the workaround.

It's fairly easy to build the application shown in Figure 18-6. The first step is to add the element that will receive the dropped data. Often, this is displayed a large rectangular area, so it makes sense to use a Border or a simple Rectangle (although any element supports the drag-and-drop feature). You must then set the AllowDrop property to True, as shown here:

```
<Rectangle x:Name="rectDropSurface" Margin="5" Fill="LightSteelBlue"
 Stroke="SteelBlue" StrokeThickness="1" AllowDrop="True"
</Rectangle>
```

Now you simply need to handle the right events. Optionally, you can respond to the DragEnter, DragLeave, and DragOver events to provide some sort of status information as the drag is in progress. More importantly, you must handle the Drop event to receive the dropped data:

```
<Rectangle x:Name="rectDropSurface" Margin="5" Fill="LightSteelBlue"
 Stroke="SteelBlue" StrokeThickness="1" AllowDrop="True"
 DragEnter="rectDropSurface_DragEnter" DragLeave="rectDropSurface_DragLeave"
 Drop="rectDropSurface_Drop">
</Rectangle>
```

In the example shown in Figure 18-6, the DragEnter and DragLeave event handlers simply update a label in the display:

```
Private Sub rectDropSurface_DragEnter(ByVal sender As Object, _
  ByVal e As DragEventArgs)
    lblResults.Text = "You are dragging an object over the drop area."
End Sub

Private Sub rectDropSurface_DragLeave(ByVal sender As Object, _
  ByVal e As DragEventArgs)
    lblResults.Text = ""
End Sub
```

The Drop event is more interesting. It checks what data has been dropped using the DrageEventArgs.GetDataPresent() method. When calling this method, you need to specify the type of data you want to read using a value from the DataFormats enumeration. Currently,

Silverlight supports just one value: DataFormats.FileDrop. That means it isn't possible to drag a text selection from Microsoft Word or a chart object from Microsoft Excel and drop it into a Silverlight application. However, it's still considered good form to explicitly test the type of data, because future versions of Silverlight may add support for other data types.

```
Private Sub rectDropSurface_Drop(ByVal sender As Object, ByVal e As DragEventArgs)
    ' Check if data is present and in the correct format.
    If (e.Data IsNot Nothing) AndAlso _
       (e.Data.GetDataPresent(DataFormats.FileDrop)) Then
        ...
```

Once you've performed this check, you can call the DragEventArgs.GetData() method to actually retrieve the data. Silverlight provides the dropped-in files as an array of FileInfo objects:

```
        ...
        Dim files As FileInfo()
        files = CType(e.Data.GetData(DataFormats.FileDrop), FileInfo())
        ...
```

It's important to note that you can't use FileInfo properties such as DirectoryName and FullPath to get the full path information for dropped files. Doing so causes a SecurityException to be thrown. (The exception is if you're running an elevated trust application, as described in Chapter 21). However, you can use the Name property to get the file name with no path information, and you can use the Length property to get the file size in bytes.

The following code uses the Name property to extract the file extension. If the file is a recognized image type, it carries ahead, opening a FileStream, reading the data, and loading the image content into an ImageSource object. The ImageSource is then added to a list.

```
        ...
        For Each file As FileInfo In files
            Dim ext As String = System.IO.Path.GetExtension(file.Name)

            ' Check if it's an image.
            Select Case ext.ToLower()
                Case ".jpg", ".gif", ".png", ".bmp"
                    Try
                        ' Read the image, wrap it in a BitmapImage, and
                        ' add it to the list.
                        Using fs As FileStream = file.OpenRead()
                            Dim source As New BitmapImage()
                            source.SetSource(fs)
                            lstImages.Items.Add(source)
                        End Using
                    Catch err As Exception
                        lblResults.Text = "Error reading file: " & err.Message
                    End Try

                Case Else
                    lblResults.Text = _
                    "The dropped file was not recognized as a supported image type."
            End Select

            lblResults.Text = files.Count().ToString() & _
              " files successfully dropped."
        Next
```

```
    End If
End Sub
```

Finally, the list box converts the ImageSource objects into a list of images, using this straightforward template:

```
<ListBox Grid.Column="1" Margin="5" x:Name="lstImages">
  <ListBox.ItemTemplate>
    <DataTemplate>
      <Image Margin="1"
        Source="{Binding}" Width="95" />
    </DataTemplate>
  </ListBox.ItemTemplate>
</ListBox>
```

The Last Word

In this chapter, you saw how Silverlight allows you to access the local hard drive but with careful restrictions in place. First, you took a thorough look at isolated storage, the obfuscated, space-limited storage location that you can use to store miscellaneous data, serialized objects, and application settings. Then, you saw how you can use the OpenFileDialog class to retrieve information from a user-selected file anywhere on the hard drive and how to use SaveFileDialog to perform the reverse feat and write to a user-selected file. Finally, you learned how to access file content through a drag-and-drop operation.

Taken together, these features give Silverlight applications an impressive balance of safety and performance. They ensure that malicious applications can't tamper with local files or read sensitive data but that legitimate software can store details from one user session to the next.

■ ■ ■

CHAPTER 19

Multithreading

One of Silverlight's least expected surprises is its support for *multithreading*—the fine art of executing more than one piece of code at the same time. It's a key part of the full .NET Framework and a commonly used feature in rich client applications built with WPF and Windows Forms. However, multithreading hasn't appeared in the toolkit of most browser-based developers, and it's notably absent from both JavaScript and Flash.

The second surprise is how similar Silverlight's threading tools are to those in the full .NET Framework. As with ordinary .NET programming, Silverlight developers can create new threads with the Thread class, manage a long-running operation with the BackgroundWorker, and even submit tasks to a pool of worker threads with the ThreadPool. All of these ingredients are closely modeled after their counterparts in the full .NET Framework, so developers who have written multithreaded client applications will quickly find themselves at home with Silverlight. And although there are some clear limitations—for example, you can't control thread priorities with Silverlight code—these issues don't stop Silverlight threading from being remarkably powerful.

In this chapter, you'll begin by taking a look at the lower-level Thread class, which gives you the most flexible way to create new threads at will. Along the way, you'll explore the Silverlight threading model and the rules it imposes. Finally, you'll examine the higher-level BackgroundWorker class, which gives you a conveniently streamlined, practical way to deal with background tasks.

Understanding Multithreading

When you program with threads, you write your code as though each thread is running independently. Behind the scenes, the Windows operating system gives each thread a brief unit of time (called a *time slice*) to perform some work, and then it freezes the thread in a state of suspended animation. A little later (perhaps only a few milliseconds), the operating system unfreezes the thread and allows it to perform a little more work.

This model of constant interruption is known as *preemptive multitasking*. It takes place completely outside the control of your program. Your application acts (for the most part) as though all the threads it has are running simultaneously, and each thread carries on as though it's an independent program performing some task.

> ■ **Note** If you have multiple CPUs or a dual-core CPU, it's possible that two threads will execute at once, but it's not necessarily likely—after all, the Silverlight plug-in, other applications and services, and the client's operating system can also compete for the CPU's attention. Furthermore, the high-level tasks you perform with a programming platform such as Silverlight will be translated into many more low-level instructions. In some cases, a dual-core CPU can execute more than one instruction at the same time, meaning a single thread can keep more than one CPU core busy.

The Goals of Multithreading

Multithreading increases complexity. If you decide to use multithreading, you need to code carefully to avoid minor mistakes that can lead to mysterious errors later. Before you split your application into separate threads, you should carefully consider whether the additional work is warranted.

There are essentially three reasons for using multiple threads in a program:

- *Making the client more responsive*: If you run a time-consuming task on a separate thread, the user can still interact with your application's user interface to perform other tasks. You can even give the user the ability to cancel the background work before it's complete. By comparison, a single-threaded application locks up the user interface when it performs time-consuming work on the main thread.

- *Completing several tasks at once*: On its own, multithreading doesn't improve performance for the typical single-CPU computer. (In fact, the additional overhead needed to track the new threads actually decreases performance slightly.) But certain tasks may involve a high degree of latency, such as fetching data from an external source (a web page, a database, or a file on a network) or communicating with a remote component. While these tasks are underway, the CPU is essentially idle. Although you can't reduce the wait time, you can use the time to perform other work. For example, you can send requests to three web services at the same time to reduce the total time taken, or you can perform CPU-intensive work while waiting for a call to complete.

- *Making a server application scalable*: A server-side application needs to be able to handle an arbitrary number of clients. Depending on the technology you're using, this may be handled for you (as it is if you're creating an ASP.NET web application). In other cases, you may need to create this infrastructure on your own—for example, if you're building a socket-based application with the .NET networking classes, as demonstrated in Chapter 20. This type of design usually applies to .NET-based server applications, not Silverlight applications.

In this chapter, you'll explore an example where multithreading makes good sense: dealing with a time-consuming operation in the background. You'll see how to keep the application responsive, avoid threading errors, and add support for progress notification and cancellation.

■ **Tip** The CPU is rarely the limiting factor for the performance of a Silverlight application. Network latency, slow web services, and disk access are more common limiting factors. As a result, multithreading rarely improves overall performance, even on a dual-core CPU. However, by improving responsiveness, it can make an application feel much more performant to the user.

The DispatcherTimer

In some cases, you can avoid threading concerns altogether using the DispatcherTimer class from the System.Windows.Threading namespace. DispatcherTimer was used in Chapter 10 to power the bomb-dropping animations in a simple arcade game.

The DispatcherTimer doesn't offer true multithreaded execution. Instead, it triggers a periodic Tick event on the main application thread. This event interrupts whatever else is taking place in your application, giving you a chance to perform some work. But if you need to frequently perform small amounts of work (for example, starting a new set of bomb-dropping animations every fraction of a second), the DispatcherTimer works as seamlessly as actual multithreading.

The advantage of the DispatcherTimer is that the Tick event always executes on the main application thread, thereby sidestepping synchronization problems and the other headaches you'll consider in this chapter. However, this behavior also introduces a number of limitations. For example, if your timer event-handling code performs a time-consuming task, the user interface locks up until it's finished. Thus, the timer doesn't help you make a user interface more responsive, and it doesn't allow you to collapse the waiting time for high-latency operations. To get this functionality, you need the real multithreading discussed in this chapter.

However, clever use of the DispatcherTimer can achieve the effect you need in some situations. For example, it's a great way to periodically check a web service for new data. As you learned in Chapter 15, all web service calls are asynchronous and are carried out on a background thread. Thus, you can use the DispatcherTimer to create an application that periodically downloads data from a slow web service. For example, it might fire every five minutes and then launch the web service call asynchronously, allowing the time-consuming download to take place on a background thread.

■ **Note** The name of the DispatcherTimer refers to the *dispatcher*, which controls the main application thread in a Silverlight application. You'll learn more about the Dispatcher in this chapter.

The Thread Class

The most straightforward way to create a multithreaded Silverlight application is to use the Thread class from the System.Threading namespace. Each Thread object represents a separate thread of execution.

To use the Thread class, you begin by creating a new Thread object, at which point you supply a delegate to the method you want to invoke asynchronously. A Thread object can point to only a single method. This signature of this method is limited in several ways. It can't have a return value, and it must have either no parameters (in which case it matches the ThreadStart delegate) or a single object parameter (in which case it matches the ParameterizedThreadStart delegate).

For example, if you have a method like this:

```
Private Sub DoSomething()
    ...
End Sub
```

you can create a thread that uses it like this:

```
Dim thread As New Thread(AddressOf DoSomething)
```

After you've created the Thread object, you can start it on its way by calling the Thread.Start() method. If your thread accepts an object parameter, you pass it in at this point.

```
thread.Start()
```

The Start() method returns immediately, and your code begins executing asynchronously on a new thread. When the method ends, the thread is destroyed and can't be reused. In between, you can use a small set of properties and methods to control the thread's execution. Table 19-1 lists the most significant.

Table 19-1. Members of the Thread Class

Member	Description
IsAlive	Returns True unless the thread is stopped, aborted, or not yet started.
ManagedThreadId	Provides an integer that uniquely identifies this thread.
Name	Enables you to set a string name that identifies the thread. This is primarily useful during debugging, but it can also be used to distinguish different threads. Once set, the Name property can't be set again.
ThreadState	A combination of ThreadState values that indicate whether the thread is started, running, finished, and so on. The ThreadState property should be used only for debugging. If you want to determine whether a thread has completed its work, you need to track that information manually.
Start()	Starts a thread executing for the first time. You can't use Start() to restart a thread after it ends.
Join()	Waits until the thread terminates (or a specified timeout elapses).
Sleep()	Pauses the current thread for a specified number of milliseconds. This method is shared.

■ **Note** Seasoned .NET programmers will notice that the Silverlight version of the Thread class leaves out a few details. In Silverlight, all threads are background threads, you can't set thread priorities, and you have no ability to temporarily pause and then resume a thread. Similarly, although the Thread class includes an Abort() method that kills a thread with an unhandled exception, this method is marked with the SecurityCritical attribute and so can be called only by the Silverlight plug-in, not by your application code.

The challenge of multithreaded programming is communicating between the background thread and the main application thread. It's easy enough to pass information to the thread when it starts (using parameters). But trying to communicate with the thread while it's running and trying to return data when it's complete are two more difficult tasks. You may need to use locking to ensure that the same data isn't accessed on two threads at once (a cardinal sin of multithreaded programming) and marshaling to make sure you don't access a user interface element from a background thread (an equally bad mistake). Even worse, threading mistakes don't result in compile-time warnings and don't necessarily lead to clear, showstopper bugs. They may cause subtler problems that appear only under occasional, difficult-to-diagnose circumstances. In the following sections, you'll learn how to use a background thread safely.

Marshaling Code to the User Interface Thread

Much like .NET client applications (for example, WPF applications and Windows Forms applications), Silverlight supports a *single-threaded apartment* model. In this model, a single thread runs your entire application and owns all the objects that represent user-interface elements. Furthermore, all these elements have *thread affinity*. The thread that creates them owns them, and other threads can't interact with them directly. If you violate this rule—for example, by trying to access a user-interface object from a background thread—you're certain to cause an immediate exception, a lock-up, or a subtler problem.

To keep your application on an even keel, Silverlight uses a *dispatcher*. The dispatcher owns the main application thread and manages a queue of work items. As your application runs, the dispatcher accepts new work requests and executes one at a time.

■ **Note** The dispatcher is an instance of the System.Windows.Threading.Dispatcher class, which was introduced with WPF.

You can retrieve the dispatcher from any element through the Dispatcher property. The Dispatcher class includes just two members: a CheckAccess() method that allows you to determine whether you're on the correct thread to interact with your application's user interface, and a BeginInvoke() method that lets you marshal code to the main application thread that the dispatcher controls.

■ **Tip** The Dispatcher.CheckAccess() method is hidden from Visual Studio IntelliSense. You can use it in code; you just won't see it in the pop-up list of members.

For example, the following code responds to a button click by creating a new System.Threading.Thread object. It then uses that thread to launch a small bit of code that changes a text box in the current page:

```
Private Sub cmdBreakRules_Click(ByVal sender As Object, ByVal e As RoutedEventArgs)
    Dim thread As New Thread(AddressOf UpdateTextWrong)
    thread.Start()
End Sub

Private Sub UpdateTextWrong()
    ' Simulate some work taking place with a five-second delay.
    Thread.Sleep(TimeSpan.FromSeconds(5))

    txt.Text = "Here is some new text."
End Sub
```

This code is destined to fail. The UpdateTextWrong() method will be executed on a new thread, and that thread isn't allowed to access Silverlight objects. The result is an UnauthorizedAccessException that derails the code.

To correct this code, you need to get a reference to the dispatcher that owns the TextBox object (which is the same dispatcher that owns the page and all the other Silverlight objects in the application). When you have access to that dispatcher, you can call Dispatcher.BeginInvoke() to marshal some code to the dispatcher thread. Essentially, BeginInvoke() schedules your code as a task for the dispatcher. The dispatcher then executes that code.

Here's the corrected code:

```
Private Sub cmdFollowRules_Click(ByVal sender As Object, ByVal e As RoutedEventArgs)
    Dim thread As New Thread(AddressOf UpdateTextRight)
    thread.Start()
End Sub

Private Sub UpdateTextRight()
    ' Simulate some work taking place with a five-second delay.
    Thread.Sleep(TimeSpan.FromSeconds(5))

    ' Get the dispatcher from the current page, and use it to invoke
    ' the update code.
    Me.Dispatcher.BeginInvoke(AddressOf SetText)
End Sub

Private Sub SetText()
    txt.Text = "Here is some new text."
End Sub
```

The Dispatcher.BeginInvoke() method takes a single parameter: a delegate that points to the method with the code you want to execute.

■ **Note** The BeginInvoke() method also has a return value, which isn't used in the earlier example. BeginInvoke() returns a DispatcherOperation object, which allows you to follow the status of your marshaling operation and determine when your code has been executed. However, the DispatcherOperation is rarely useful, because the code you pass to BeginInvoke() should take very little time.

Remember, if you're performing a time-consuming background operation, you need to perform this operation on a separate thread and *then* marshal its result to the dispatcher thread (at which point you'll update the user interface or change a shared object). It makes no sense to perform your time-consuming code in the method that you pass to BeginInvoke(). For example, this slightly rearranged code still works but is impractical:

```
Private Sub UpdateTextRight()
    ' Get the dispatcher from the current page, and use it to invoke
    ' the update code.
    Me.Dispatcher.BeginInvoke(AddressOf SetText)
End Sub

Private Sub SetText()
    ' Simulate some work taking place with a five-second delay.

    Thread.Sleep(TimeSpan.FromSeconds(5))
    txt.Text = "Here is some new text."
End Sub
```

The problem here is that all the work takes place on the dispatcher thread. That means this code ties up the dispatcher in the same way a nonmultithreaded application would.

Creating a Thread Wrapper

The previous example shows how you can update the user interface directly from a background thread. However, this approach isn't ideal. It creates complex, tightly coupled applications that mingle the code for performing a task with the code for displaying data. The result is an application that's more complex, less flexible, and difficult to change. For example, if you change the name of the text box in the previous example or replace it with a different control, you'll also need to revise your threading code.

A better approach is to create a thread that passes information back to the main application and lets the application take care of the display details. To make it easier to use this approach, it's common to wrap the threading code and the data into a separate class. You can then add properties to that class for the input and output information. This custom class is often called a *thread wrapper*.

Before you create your thread wrapper, it makes sense to factor out all the threading essentials into a base class. That way, you can use the same pattern to create multiple background tasks without repeating the same code each time.

You'll examine the ThreadWrapperBase class piece by piece. First, you declare the ThreadWrapperBase with the MustInherit keyword so it can't be instantiated on its own. Instead, you need to create a derived class.

```vb
Public MustInherit Class ThreadWrapperBase
    ...
End Class
```

The ThreadWrapperBase defines one public property, named Status, which returns one of three values from an enumeration (Unstarted, InProgress, or Completed):

```vb
' Track the status of the task.
Private _status As StatusState = StatusState.Unstarted
Public ReadOnly Property Status() As StatusState
    Get
        Return _status
    End Get
End Property
```

The ThreadWrapperBase wraps a Thread object. It exposes a public Start() method that, when called, creates the thread and starts it:

```vb
' This is the thread where the task is carried out.
Private thread As Thread

' Start the new operation.
Public Sub Start()
    If Status = StatusState.InProgress Then
        Throw New InvalidOperationException("Already in progress.")
    Else
        ' Initialize the new task.
        _status = StatusState.InProgress

        ' Create the thread.
        thread = New Thread(AddressOf StartTaskAsync)

        ' Start the thread.
        thread.Start()
    End If
End Sub
```

The thread executes a private method named StartTaskAsync(). This method farms out the work to two other methods: DoTask() and OnCompleted(). DoTask() performs the actual work (calculating prime numbers). OnCompleted() fires a completion event or triggers a callback to notify the client. Both of these details are specific to the particular task at hand, so they're implemented as MustOverride methods that the derived class will override:

```vb
' Start the new operation.
Private Sub StartTaskAsync()
    DoTask()
    _status = StatusState.Completed
    OnCompleted()
End Sub

' Override this class to supply the task logic.
Protected MustOverride Sub DoTask()

' Override this class to supply the callback logic.
Protected MustOverride Sub OnCompleted()
```

This completes the ThreadWrapperBase class. Now, you need to create a derived class that uses it. The following section presents a practical example with an algorithm for finding prime numbers.

Creating the Worker Class

The basic ingredient for any test of multithreading is a time-consuming process. The following example uses a common algorithm called the *sieve of Eratosthenes* for finding prime numbers in a given range, which was invented by Eratosthenes in about 240 BC. With this algorithm, you begin by making a list of all the integers in a range of numbers. You then strike out the multiples of all primes less than or equal to the square root of the maximum number. The numbers that are left are the primes.

In this example, you won't consider the theory that proves the sieve of Eratosthenes works or the fairly trivial code that performs it. (Similarly, don't worry about optimizing it or comparing it against other techniques.) However, you will see how to perform the sieve of Eratosthenes algorithm on a background thread.

The full code for the FindPrimesThreadWrapper class is available with the online examples for this chapter. Like any class that derives from ThreadWrapperBase, it needs to supply four things:

- *Fields or properties that store the initial data*: In this example, those are the from and to numbers that delineate the search range.

- *Fields or properties that store the final data*: In this example, that's the final prime list, which is stored in an array.

- *An overridden DoTask() method that performs the actual operation*: It uses the initial data and sets the final result.

- *An overridden OnCompleted() method that raises the completion event*: Typically, this completion event uses a custom EventArgs object that supplies the final data. In this example, the FindPrimesCompletedEventArgs class wraps the from and to numbers and the prime list array.

Here's the code for the FindPrimesThreadWrapper:

```
Public Class FindPrimesThreadWrapper
    Inherits ThreadWrapperBase

    ' Store the input and output information.
    Private fromNumber, toNumber As Integer
    Private primeList As Integer()

    Public Sub New(ByVal fromNumber As Integer, ByVal toNumber As Integer)
        Me.fromNumber = fromNumber
        Me.toNumber = toNumber
    End Sub

    Protected Overrides Sub DoTask()
        ' Find the primes between fromNumber and toNumber,
        ' and return them as an array of integers.
        ' (See the code in the downloadable examples.)
    End Sub
```

```
    Public Event Completed As EventHandler(Of FindPrimesCompletedEventArgs)
    Protected Overrides Sub OnCompleted()
        ' Signal that the operation is complete.
        If CompletedEvent IsNot Nothing Then
            RaiseEvent Completed(Me, _
                New FindPrimesCompletedEventArgs(fromNumber, toNumber, primeList))
        End If
    End Sub
End Class
```

It's important to note that the data the FindPrimesThreadWrapper class uses—the from and to numbers and the prime list—aren't exposed publically. This prevents the main application thread from accessing that information while it's being used by the background thread, which is a potentially risky scenario that can lead to data errors. One way to make the prime list available is to add a public property. This property can then check the ThreadWrapperBase.Status property and return the prime list only if the thread has completed its processing.

An even better approach is to notify the user with a callback or event, as with the completion event demonstrated in the thread wrapper. However, it's important to remember that events fired from a background thread continue to execute on that thread, no matter where the code is defined. Thus, when you handle the Completed event, you still need to use marshaling code to transfer execution to the main application thread before you attempt to update the user interface or any data in the current page.

■ **Note** If you really need to expose the same object to two threads that may use it at the same time, you must safeguard the access to that object with locking. As in a full-fledged .NET application, you can use the SyncLock keyword to obtain exclusive access to an in-memory object. However, locking complicates application design and raises other potential problems. It can slow performance, because other threads must wait to access a locked object, and it can lead to deadlocks if two threads try to achieve locks on the same objects.

Using the Thread Wrapper

The last ingredient is a Silverlight sample application that uses the FindPrimesThreadWrapper. Figure 19-1 shows one such example. This page lets the user choose the range of numbers to search. When the user clicks Find Primes, the search begins, but it takes place in the background. When the search is finished, the list of prime numbers appears in a list.

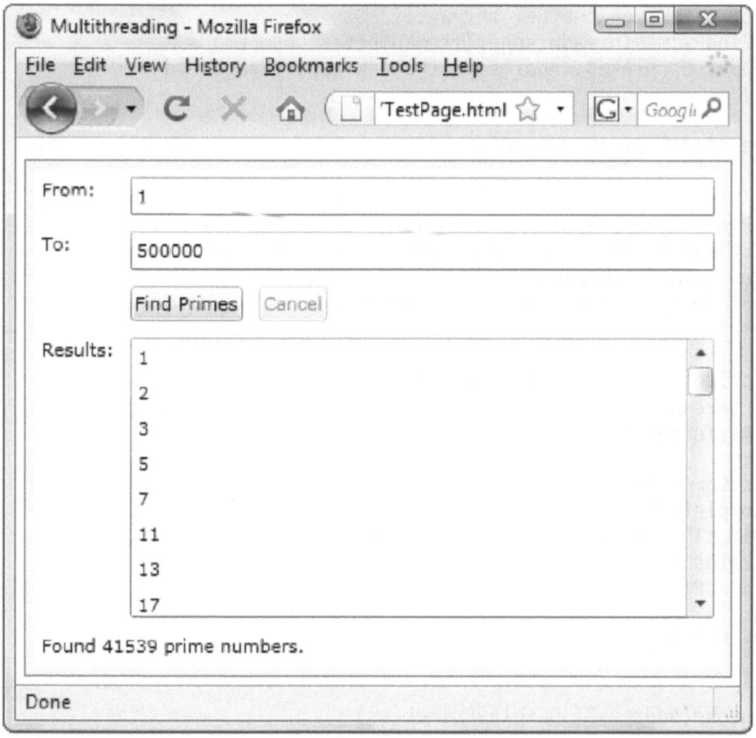

Figure 19-1. A completed prime-number search

The code that underpins this page is straightforward. When the user clicks the Find Primes button, the application disables the button (preventing multiple concurrent searches, which are possible but potentially confusing to the user) and determines the search range. Then, it creates the FindPrimesThreadWrapper object, hooks up an event handler to the Completed event, and calls Start() to begin processing:

```
Private threadWrapper As FindPrimesThreadWrapper

Private Sub cmdFind_Click(ByVal sender As Object, ByVal e As RoutedEventArgs)
    ' Disable the button and clear previous results.
    cmdFind.IsEnabled = False
    lstPrimes.ItemsSource = Nothing

    ' Get the search range.
    Dim fromNumber, toNumber As Integer
    If Not Int32.TryParse(txtFrom.Text, fromNumber) Then
        lblStatus.Text = "Invalid From value."
        Return
    End If
    If Not Int32.TryParse(txtTo.Text, toNumber) Then
        lblStatus.Text = "Invalid To value."
        Return
    End If
```

```
' Start the search for primes on another thread.
threadWrapper = New FindPrimesThreadWrapper(fromNumber, toNumber)
AddHandler threadWrapper.Completed, AddressOf threadWrapper_Completed
threadWrapper.Start()

    lblStatus.Text = "The search is in progress..."
End Sub
```

When the task is in process, the application remains remarkably responsive. The user can click other controls, type in the text boxes, and so on, without having any indication that the CPU is doing additional work in the background.

When the job is finished, the Completed event fires, and the prime list is retrieved and displayed:

```
' Temporarily store the prime list here while the call is
' marshaled to the right thread.
Private recentPrimeList As Integer()

Private Sub threadWrapper_Completed(ByVal sender As Object, _
  ByVal e As FindPrimesCompletedEventArgs)
    Dim thread As FindPrimesThreadWrapper = CType(sender, FindPrimesThreadWrapper)
    If thread.Status = StatusState.Completed Then
        recentPrimeList = e.PrimeList
    Else
        recentPrimeList = Nothing
    End If

    Me.Dispatcher.BeginInvoke(AddressOf DisplayPrimeList)
End Sub

Private Sub DisplayPrimeList()
    If recentPrimeList IsNot Nothing Then
        lblStatus.Text = "Found " & recentPrimeList.Length & " prime numbers."
        lstPrimes.ItemsSource = recentPrimeList
    End If

    cmdFind.IsEnabled = True
End Sub
```

Cancellation Support

Now that you have the basic infrastructure in place, it takes just a bit more work to add additional features such as cancellation and progress notification.

For example, to make cancellation work, your thread wrapper needs a field that, when True, indicates that it's time to stop processing. Your worker code can check this field periodically. Here's the code you can add to the ThreadWrapperBase to make this a standard feature:

```
' Flag that indicates a stop is requested.
Private _cancelRequested As Boolean = False
Protected ReadOnly Property CancelRequested() As Boolean
    Get
        Return _cancelRequested
```

```
        End Get
    End Property

    ' Call this to request a cancel.
    Public Sub RequestCancel()
        _cancelRequested = True
    End Sub

    ' When cancelling, the worker should call the OnCancelled() method
    ' to raise the Cancelled event.
    Public Event Cancelled As EventHandler
    Protected Sub OnCancelled()
        If CancelledEvent IsNot Nothing Then
          RaiseEvent Cancelled(Me, EventArgs.Empty)
        End If
    End Sub
```

And here's a modified bit of worker code in the FindPrimesThreadWrapper.DoWork() method that makes periodic checks (about 100 of them over the course of the entire operation) to see whether a cancellation has been requested:

```
Dim iteration As Integer = list.Length / 100

If i Mod iteration = 0 Then
    If CancelRequested Then
        Return
    End If
End If
```

You also need to modify the ThreadWrapperBase.StartTaskAsync() method so it recognizes the two possible ways an operation can end—by completing gracefully or by being interrupted with a cancellation request:

```
Private Sub StartTaskAsync()
    DoTask()
    If CancelRequested Then
        _status = StatusState.Cancelled
        OnCancelled()
    Else
        _status = StatusState.Completed
        OnCompleted()
    End If
End Sub
```

To use this cancellation feature in the example shown in Figure 19-1, you simply need to hook up an event handler to the Cancelled event and add a new Cancel button. The following code initiates a cancel request for the current task:

```
Private Sub cmdCancel_Click(ByVal sender As Object, ByVal e As RoutedEventArgs)
    threadWrapper.RequestCancel()
End Sub
```

And here's the event handler that runs when the cancellation is finished:

```
Private Sub threadWrapper_Cancelled(ByVal sender As Object, ByVal e As EventArgs)
    Me.Dispatcher.BeginInvoke(UpdateDisplay)
End Sub

Private Sub UpdateDisplay()
    lblStatus.Text = "Search cancelled."
    cmdFind.IsEnabled = True
    cmdCancel.IsEnabled = False
End Sub
```

Remember, Silverlight threads can't be halted with the Abort() method, so you have no choice but to request a polite stop that the worker code is free to honor or ignore.

The BackgroundWorker

So far, you've seen the no-frills approach to multithreading—creating a new System.Threading.Thread object by hand, supplying your asynchronous code, and launching it with the Thread.Start() method. This approach is powerful, because the Thread object doesn't hold anything back. You can create dozens of threads at will, pass information to them at any time, temporarily delay them with Thread.Sleep(), and so on. However, this approach is also a bit dangerous. If you access shared data, you need to use locking to prevent subtle errors. If you create threads frequently or in large numbers, you'll generate additional, unnecessary overhead.

One of the simplest and safest approaches to multithreading is provided by the System.ComponentModel.BackgroundWorker component, which was first introduced with .NET 2.0 to simplify threading considerations in Windows Forms applications. Fortunately, the BackgroundWorker is equally at home in Silverlight. The BackgroundWorker component gives you a nearly foolproof way to run a time-consuming task on a separate thread. It uses the dispatcher behind the scenes and abstracts away the marshaling issues with an event-based model.

As you'll see, the BackgroundWorker also supports two frills: progress events and cancel messages. In both cases, the threading details are hidden, making for easy coding. In fact, the BackgroundWorker ranks as the single most practical tool for Silverlight multithreading.

■ **Note** BackgroundWorker is perfect if you have a single asynchronous task that runs in the background from start to finish (with optional support for progress reporting and cancellation). If you have something else in mind—for example, an asynchronous task that runs throughout the entire life of your application or an asynchronous task that communicates with your application while it does its work—you must design a customized solution that uses the threading features you've already seen.

Creating the BackgroundWorker

To use the BackgroundWorker, you begin by creating an instance in your code and attaching the event handlers programmatically. The BackgroundWorker's core events are DoWork, ProgressChanged, and RunWorkerCompleted. You'll consider each of them in the following example.

■ **Tip** If you need to perform multiple asynchronous tasks, you can create your BackgroundWorker objects when needed and store them in some sort of collection for tracking. The example described here uses just one BackgroundWorker, and it's created in code when the page is first instantiated.

Here's the initialization code that enables support for progress notification and cancellation and attaches the event handlers:

```
Private backgroundWorker As New BackgroundWorker()

Public Sub New()
    InitializeComponent()

    backgroundWorker.WorkerReportsProgress = True
    backgroundWorker.WorkerSupportsCancellation = True
    AddHandler backgroundWorker.DoWork, AddressOf backgroundWorker_DoWork
    AddHandler backgroundWorker.ProgressChanged, _
      AddressOf backgroundWorker_ProgressChanged
    AddHandler backgroundWorker.RunWorkerCompleted, _
      AddressOf backgroundWorker_RunWorkerCompleted
End Sub
```

Running the BackgroundWorker

The first step to using the BackgroundWorker with the prime-number search example is to create a custom class that allows you to transmit the input parameters to the BackgroundWorker. When you call BackgroundWorker.RunWorkerAsync(), you can supply any object, which is delivered to the DoWork event. However, you can supply only a single object, so you need to wrap the to and from numbers into one class:

```
Public Class FindPrimesInput
    Private _toNumber As Integer
    Public Property ToNumber() As Integer
        Get
            Return _toNumber
        End Get
        Set(ByVal value As Integer)
            _toNumber = value
        End Set
    End Property

    Private _fromNumber As Integer
    Public Property FromNumber() As Integer
        Get
            Return _fromNumber
        End Get
        Set(ByVal value As Integer)
            _fromNumber = value
        End Set
    End Property
```

```
        Public Sub New(ByVal fromNumber As Integer, ByVal toNumber As Integer)
            Me.ToNumber = toNumber
            Me.FromNumber = fromNumber
        End Sub
    End Class
```

To start the BackgroundWorker on its way, you need to call the
BackgroundWorker.RunWorkerAsync() method and pass in the FindPrimesInput object. Here's
the code that does this when the user clicks the Find Primes button:

```
Private Sub cmdFind_Click(ByVal sender As Object, ByVal e As RoutedEventArgs)
    ' Disable this button and clear previous results.
    cmdFind.IsEnabled = False
    cmdCancel.IsEnabled = True
    lstPrimes.Items.Clear()

    ' Get the search range.
    Dim fromNumber, toNumber As Integer
    If Not Int32.TryParse(txtFrom.Text, fromNumber) Then
        MessageBox.Show("Invalid From value.")
        Return
    End If
    If Not Int32.TryParse(txtTo.Text, toNumber) Then
        MessageBox.Show("Invalid To value.")
        Return
    End If

    ' Start the search for primes on another thread.
    Dim input As New FindPrimesInput(fromNumber, toNumber)
    backgroundWorker.RunWorkerAsync(input)
End Sub
```

When the BackgroundWorker begins executing, it fires the DoWork event on a separate
thread. Rather than create this thread (which incurs some overhead), the BackgroundWorker
borrows a thread from the runtime thread pool. When the task is complete, the
BackgroundWorker returns this thread to the thread pool so it can be reused for another task.
The thread-pool threads are also used for the asynchronous operations you've seen in other
chapters, such as receiving a web service response, downloading a web page, and accepting a
socket connection.

■ **Note** Although the thread pool has a set of workers at the ready, it can run out if a large number of
asynchronous tasks are underway at once, in which case the later ones are queued until a thread is free.
This prevents the computer from being swamped (say, with hundreds of separate threads), at which
point the overhead of managing the threads would impede the CPU from performing other work.

You handle the DoWork event and begin your time-consuming task. However, you need to be careful not to access shared data (such as fields in your page class) or user-interface objects. After the work is complete, the BackgroundWorker fires the RunWorkerCompleted event to notify your application. This event fires on the dispatcher thread, which allows you to access shared data and your user interface without incurring any problems.

When the BackgroundWorker acquires the thread, it fires the DoWork event. You can handle this event to call the Worker.FindPrimes() method. The DoWork event provides a DoWorkEventArgs object, which is the key ingredient for retrieving and returning information. You retrieve the input object through the DoWorkEventArgs.Argument property and return the result by setting the DoWorkEventArgs.Result property:

```vb
Private Sub backgroundWorker_DoWork(ByVal sender As Object, _
  ByVal e As DoWorkEventArgs)
    ' Get the input values.
    Dim input As FindPrimesInput = CType(e.Argument, FindPrimesInput)

    ' Start the search for primes and wait.
    ' This is the time-consuming part, but it won't freeze the
    ' user interface because it takes place on another thread.
    Dim primes As Integer() = Worker.FindPrimes(input.FromNumber, input.ToNumber)

    ' Return the result.
    e.Result = primes
End Sub
```

When the method completes, the BackgroundWorker fires the RunWorkerCompleted event on the dispatcher thread. At this point, you can retrieve the result from the RunWorkerCompletedEventArgs.Result property. You can then update the interface and access page-level variables without worry:

```vb
Private Sub backgroundWorker_RunWorkerCompleted(ByVal sender As Object, _
  ByVal e As RunWorkerCompletedEventArgs)
    If e.Error IsNot Nothing Then
        ' An error was thrown by the DoWork event handler.
        MessageBox.Show(e.Error.Message)
    Else
        Dim primes As Integer() = CType(e.Result, Integer())
        For Each prime As Integer In primes
            lstPrimes.Items.Add(prime)
        Next
    End If

    cmdFind.IsEnabled = True
    cmdCancel.IsEnabled = False
    progressBar.Width = 0
End Sub
```

Notice that you don't need any locking code, and you don't need to use the Dispatcher.BeginInvoke() method. The BackgroundWorker takes care of these issues for you.

Tracking Progress

The BackgroundWorker also provides built-in support for tracking progress, which is useful for keeping the client informed about how much work has been completed in a long-running task.

To add support for progress, you need to first set the BackgroundWorker.WorkerReportsProgress property to True. Actually, providing and displaying the progress information is a two-step affair. First, the DoWork event-handling code needs to call the BackgroundWorker.ReportProgress() method and provide an estimated percent complete (from 0% to 100%). You can do this as little or as often as you like. Every time you call ReportProgress(), the BackgroundWorker fires the ProgressChanged event. You can react to this event to read the new progress percentage and update the user interface. Because the ProgressChanged event fires from the user interface thread, there's no need to use Dispatcher.BeginInvoke().

The FindPrimes() method reports progress in 1% increments, using code like this:

```
Dim iteration As Integer = list.Length / 100
For i As Integer = 0 To list.Length - 1
    ...

    ' Report progress only if there is a change of 1%.
    ' Also, don't bother performing the calculation if there
    ' isn't a BackgroundWorker or if it doesn't support
    ' progress notifications.
    If (i Mod iteration = 0) AndAlso (backgroundWorker IsNot Nothing) Then

        If backgroundWorker.WorkerReportsProgress Then
            backgroundWorker.ReportProgress(i \ iteration)
        End If
    End If
Next
```

■ **Note** To set this system up, the worker code needs access to the BackgroundWorker object so it can call the ReportProgress() method. In this example, the FindPrimesWorker class has a constructor that accepts a reference to a BackgroundWorker object. If supplied, the FindPrimesWorker uses the BackgroundWorker for progress notification and cancellation. To see the complete worker code, refer to the downloadable examples for this chapter.

After you've set the BackgroundWorker.WorkerReportsProgress property, you can respond to these progress notifications by handling the ProgressChanged event. However, Silverlight doesn't include a progress bar control, so it's up to you to decide how you want to display the progress information. You can display the progress percentage in a TextBlock, but it's fairly easy to build a basic progress bar out of common Silverlight elements. Here's one that uses two rectangles (one for the background and one for the progress meter) and a TextBlock that shows the percentage in the center. All three elements are placed in the same cell of a Grid, so they overlap.

```
<Rectangle x:Name="progressBarBackground" Fill="AliceBlue" Stroke="SlateBlue"
 Grid.Row="4" Grid.ColumnSpan="2" Margin="5" Height="30" />
<Rectangle x:Name="progressBar" Width="0" HorizontalAlignment="Left"
 Grid.Row="4" Grid.ColumnSpan="2" Margin="5" Fill="SlateBlue" Height="30" />
<TextBlock x:Name="lblProgress" HorizontalAlignment="Center" Foreground="White"
 VerticalAlignment="Center" Grid.Row="4" Grid.ColumnSpan="2" />
```

To make sure the progress bar looks right even if the user resizes the browser window, the following code reacts to the SizeChanged event and stretches the progress bar to fit the current page:

```
Private maxWidth As Double

Private Sub UserControl_SizeChanged(ByVal sender As Object, _
  ByVal e As SizeChangedEventArgs)
    maxWidth = progressBarBackground.ActualWidth
End Sub
```

Now, you simply need to handle the BackgroundWorker.ProgressChanged event, resize the progress meter, and display the current progress percentage:

```
Private Sub backgroundWorker_ProgressChanged(ByVal sender As Object, _
  ByVal e As ProgressChangedEventArgs)
    progressBar.Width = CDbl(e.ProgressPercentage)/100 * maxWidth
    lblProgress.Text = (CDbl(e.ProgressPercentage)/100).ToString("P0")
End Sub
```

It's possible to pass information in addition to the progress percentage. The ReportProgress() method also provides an overloaded version that accepts two parameters. The first parameter is the percent done, and the second parameter is any custom object you want to use to pass additional information. In the prime-number search example, you may want to use the second parameter to pass information about how many numbers have been searched so far or how many prime numbers have been found. Here's how to change the worker code so it returns the most recently discovered prime number with its progress information:

```
backgroundWorker.ReportProgress(i / iteration, i)
```

You can then check for this data in the ProgressChanged event handler and display it if it's present:

```
If e.UserState IsNot Nothing Then
    lblStatus.Text = "Found prime: " & e.UserState.ToString() & "..."
End If
```

Figure 19-2 shows the progress meter while the task is in progress.

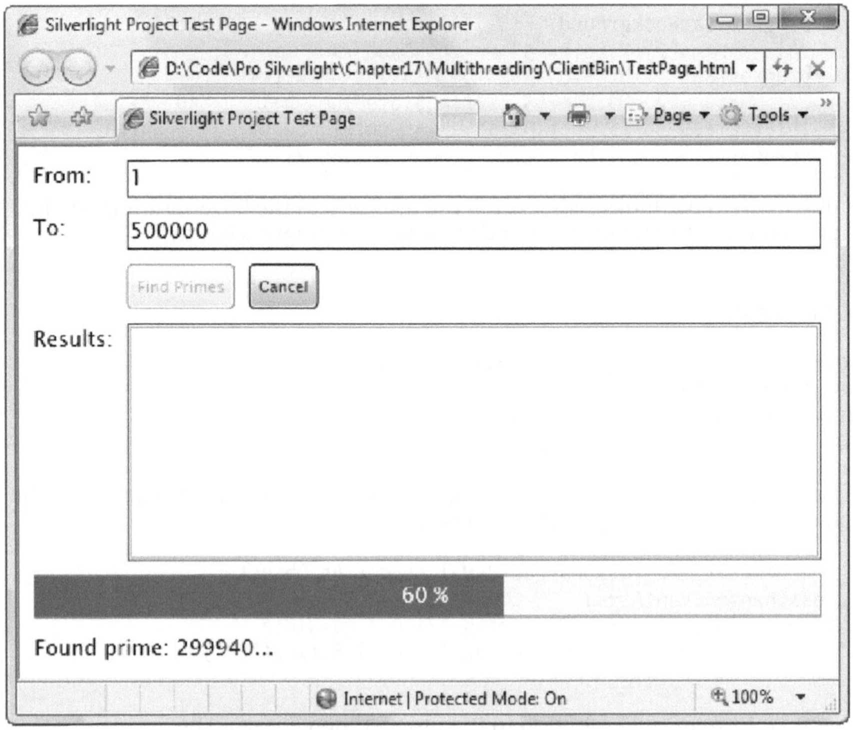

Figure 19-2. Tracking progress for an asynchronous task

Supporting Cancellation

It's just as easy to add support for canceling a long-running task with the BackgroundWorker. The first step is to set the BackgroundWorker.WorkerSupportsCancellation property to True.

To request a cancellation, your code needs to call the BackgroundWorker.CancelAsync() method. In this example, the cancellation is requested when the user clicks the Cancel button:

```
Private Sub cmdCancel_Click(ByVal sender As Object, ByVal e As RoutedEventArgs)
    backgroundWorker.CancelAsync()
End Sub
```

Nothing happens automatically when you call CancelAsync(). Instead, the code that's performing the task needs to explicitly check for the cancel request, perform any required cleanup, and return. Here's the code in the FindPrimes() method that checks for cancellation requests just before it reports progress:

```
For i As Integer = 0 To list.Length - 1
    ...
    If (i Mod iteration) AndAlso (backgroundWorker IsNot Nothing) Then
        If backgroundWorker.CancellationPending Then
            ' Return without doing any more work.
            Return
```

```
        End If

        If backgroundWorker.WorkerReportsProgress Then
            backgroundWorker.ReportProgress(i / iteration)
        End If
    End If
Next
```

The code in your DoWork event handler also needs to explicitly set the DoWorkEventArgs.Cancel property to True to complete the cancellation. You can then return from that method without attempting to build up the string of primes:

```
Private Sub backgroundWorker_DoWork(ByVal sender As Object, _
  ByVal e As DoWorkEventArgs)
    Dim input As FindPrimesInput = CType(e.Argument, FindPrimesInput)
    Dim primes As Integer() = Worker.FindPrimes(input.FromNumber, input.ToNumber, _
      backgroundWorker)

    If backgroundWorker.CancellationPending Then
        e.Cancel = True
        Return
    End If

    ' Return the result.
    e.Result = primes
End Sub
```

Even when you cancel an operation, the RunWorkerCompleted event still fires. At this point, you can check whether the task was cancelled and handle it accordingly:

```
Private Sub backgroundWorker_RunWorkerCompleted(ByVal sender As Object, _
  ByVal e As RunWorkerCompletedEventArgs)

    If e.Cancelled Then
        MessageBox.Show("Search cancelled.")
    ElseIf e.Error IsNot Nothing Then
        ' An error was thrown by the DoWork event handler.
        MessageBox.Show(e.Error.Message)
    Else
        Dim primes As Integer() = CType(e.Result, Integer())
        For Each prime As Integer In primes
            lstPrimes.Items.Add(prime)
        Next
    End If
    cmdFind.IsEnabled = True
    cmdCancel.IsEnabled = False
    progressBar.Value = 0
End Sub
```

Now, the BackgroundWorker component allows you to start a search and end it prematurely.

The Last Word

In this chapter, you saw two powerful ways to incorporate multithreading into a Silverlight application. Of course, just because you can write a multithreaded Silverlight application doesn't mean you should. Before you delve too deeply into the intricacies of multithreaded programming, it's worth considering the advice of Microsoft architects. Because of the inherent complexity of deeply multithreaded code, especially when combined with dramatically different operating systems and hardware, Microsoft's official guidance is to use multithreading sparingly. Certainly, you should use it to move work to the background, avoid long delays, and create more responsive applications. However, when possible, it's better to use the straightforward BackgroundWorker component than the lower-level Thread class. And when you need to use the Thread class, it's better to stick to just one or two background threads. It's also a good idea to set up your threads to work with distinct islands of information and thereby avoid locking complications and synchronization headaches.

■ ■ ■

CHAPTER 20

Networking

Like most software, Silverlight applications need to interact with the outside world to get relevant, current information. You've already seen one tremendously useful way to pull information into a Silverlight application: using WCF services, which allow Silverlight applications to retrieve data from the web server by calling a carefully encapsulated piece of .NET code. However, WCF services won't provide all the data you need to use. In many situations you'll want to retrieve information from other non-.NET repositories, such as representational state transfer (REST) web services, RSS feeds, and ordinary HTML web pages.

In this chapter, you'll learn about this other side of the Silverlight networking picture. You'll pick up the techniques you need to download data from a variety of different non-.NET sources and convert it to the form you need. On the way, you'll also learn how to process XML data with the remarkable XDocument class and LINQ to XML. But the most ambitious task you'll consider in this chapter is using Silverlight's socket support to build a basic messaging application.

■ **What's New** Silverlight 4 makes a number of tweaks to the networking stack. In Chapter 15, you saw its built-in support for HTTP authentication. In this chapter, you'll learn how Silverlight 4 allows you to pass login credentials to the server (see the section "Using Network Credentials") and lets socket-based applications download policy files by HTTP (see the sidebar "Serving the Policy File Over HTTP"). And in Chapter 21, you'll learn how out-of-browser applications can bypass two of the standard network security restrictions if they run with elevated trust. (They can download content from any website, including those that don't explicitly permit cross-domain access, and they can use any port number for socket connections.)

Finally, Silverlight 4 adds low-level support for UDP socket connections, in addition to the TCP socket connections you'll learn about in this chapter. UDP is a lightweight protocol that's sometimes used when speed is more important than reliability—for example, when streaming video. Although UDP isn't covered in this book, you can learn more from the official documentation at http://tinyurl.com/34tn68y.

Interacting with the Web

In Chapter 6, you saw how you can use the WebClient class to download a file from the Web. This technique allows you to grab a resource or even a Silverlight assembly at the exact point in time when an application needs it.

The WebClient isn't just for downloading binary files. It also opens some possibilities for accessing HTML pages and web services. And using its bigger brother, WebRequest, you gain the ability to post values to a web page. In the following sections, you'll see a variety of approaches that use these classes to pull information from the Web. But before you begin, you need to reconsider the security limitations that Silverlight applies to any code that uses HTTP.

■ **Note** The networking examples in this chapter assume you're using a solution with an ASP.NET test website, as described in Chapter 1. You need to use a test website both to build simple web services and to use Silverlight's downloading features, which aren't available when you launch a Silverlight application directly from your hard drive.

Cross-Domain Access

If you've ever created a web page using Ajax techniques, you've no doubt used the XMLHttpRequest object, which lets you perform web requests in the background. However, the XMLHttpRequest object imposes a significant limitation: the web page can only access web resources (HTML documents, web services, files, and so on) that are on the same web server. There's no direct way to perform a cross-domain call to fetch information from another website.

Silverlight imposes almost exactly the same restrictions in its WebClient and WebRequest classes. The issue is security. If a Silverlight application could call other websites without informing the user, it would open up the possibility for phishing attacks. For example, if a user was logged on to a service like Hotmail, a malicious Silverlight application could quietly retrieve pages that provide the user's Hotmail data. There are some possible changes that could stave off these attacks—for example, linking user credentials to their source URLs—but these would require a fairly significant change to the way browsers work.

However, Silverlight isn't completely restrictive. It borrows a trick from Flash to let websites opt in to cross-domain access through an XML policy file. When you attempt to download data from a website, Silverlight looks on that website for a file named clientaccesspolicy.xml (which you learned to create in Chapter 15). If this file isn't present, Silverlight looks for a file named crossdomain.xml. This file plays the same role but was originally developed for Flash applications. The end result is that websites that can be accessed by Flash applications can also be accessed by Silverlight applications.

The clientaccesspolicy.xml or crossdomain.xml file must be stored in the web root. So, if you attempt to access web content with the URL www.somesite.com/~luther/services/CalendarService.ashx, Silverlight checks for www.somesite.com/clientaccesspolicy.xml and then (if the former isn't found) www.somesite.com/crossdomain.xml. If neither of these files exists, or if the one that exists doesn't grant access to your Silverlight application's domain, your application won't be allowed to access any content on that website. Often, companies that provide public web services place them on a separate domain to better control this type of access. For example, the photo-sharing website Flickr won't allow you to access URLs on the domain www.flickr.com, but it will let you access those on api.flickr.com (thanks to the `http://api.flickr.com/crossdomain.xml` file).

■ **Tip** Before you attempt to use the examples in this chapter with different websites, you should verify that they support cross-domain access. To do so, try requesting the clientaccesspolicy.xml and crossdomain.xml files in the root website.

In Chapter 15, you learned what the clientaccesspolicy.xml file looks like. The crossdomain.xml file is similar. For example, here's a crossdomain.xml file that allows all access. It's similar to what you'll find on the Flickr website `http://api.flickr.com`:

```
<?xml version="1.0"?>
<cross-domain-policy>
  <allow-access-from domain="*" />
</cross-domain-policy>
```

On the other hand, the Twitter social-networking website uses its clientaccesspolicy.xml file to allow access to just a few domains, which means your Silverlight code can't retrieve any of its content:

```
<?xml version="1.0"?>
<cross-domain-policy>
  <allow-access-from domain="twitter.com" />
  <allow-access-from domain="api.twitter.com" />
  <allow-access-from domain="search.twitter.com" />
  <allow-access-from domain="static.twitter.com" />
  ...
</cross-domain-policy>
```

If you need to access web content from a website that doesn't allow cross-domain access, you have two options. One approach is to create an out-of-browser application that runs with elevated trust, which you'll learn to do in Chapter 21. In this situation, users must explicitly install your application and accept a security warning—but once they do, your application will be able to download content from any website.

Your other option is to build a server-side proxy. To implement this design, you must create an ASP.NET website that includes a web service, as you learned to do in Chapter 15. Your web page will be allowed to call that service, because it's on the same website (and even if it isn't, you'll need to add your own clientaccesspolicy.xml file alongside the web service). Your web service can then access the website you want and return the data to your page. This works because the web service is allowed to call any website, regardless of the cross-domain access rules. That's because web services run on the server, not the browser, and so they don't face the same security considerations. Figure 20-1 compares this arrangement to the more straightforward direct-downloading approach.

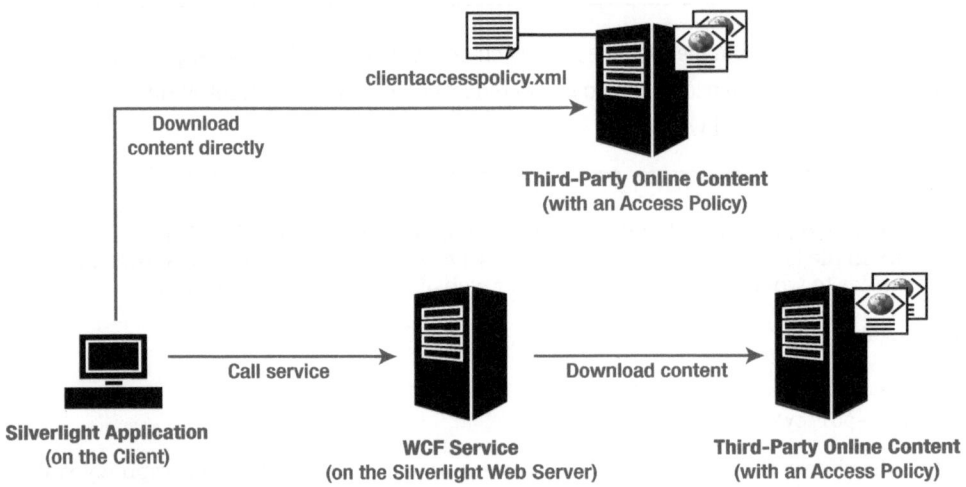

Figure 20-1. *Downloading web content in Silverlight*

Creating a server-side proxy requires a bit more work, but it's an acceptable solution if you need to retrieve small amounts of information infrequently. However, if you need to make frequent calls to your web service (for example, you're trying to read the news items in an RSS feed on a server that doesn't allow cross-domain access), the overhead can add up quickly. The web server ends up doing a significant amount of extra work, and the Silverlight application waits longer to get its information because every call goes through two delays: first, the web page's request to the web service; and second, the web service's request to the third-party website.

Now that you understand the rules that govern what websites you can access, you're ready to start downloading content. In this chapter, you'll learn how to manipulate several different types of content, but you'll start out with the absolute basic—ordinary HTML files.

HTML Scraping

One of the crudest ways to get information from the Web is to dig through the raw markup in an HTML page. This approach is fragile, because the assumptions your code makes about the structure of a page can be violated easily if the page is modified. But in some circumstances, HTML scraping is the only option. In the past, before websites like Amazon and eBay provided web services, developers often used screen-scraping techniques to get price details, sales rank, product images, and so on.

In the following example, you'll see how HTML screen scraping allows you to pull information from the table shown in Figure 20-2. This table lists the world's population at different points in history, and it's based on information drawn from Wikipedia.

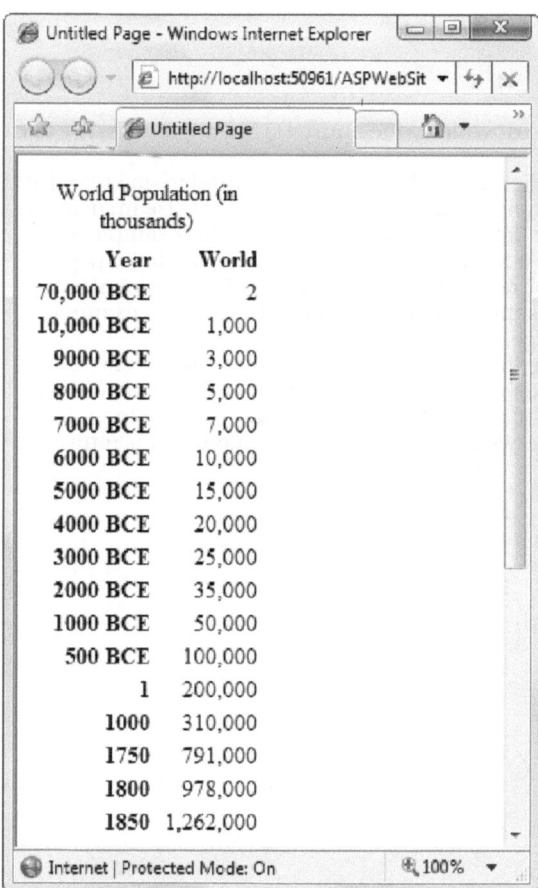

Figure 20-2. A plain HTML page

The information in the table has a structure in this format:

```
<table>
  <tr>
    <th>Year</th>
    <th width="70">World</th>
  </tr>
  <tr>
    <th>70,000 BCE</th>
    <td>2</td>
  </tr>
  <tr>
    <th>10,000 BCE</th>
    <td>1,000</td>
  </tr>
  <tr>
    <th>9000 BCE</th>
   <td>3,000</td>
```

```
    </tr>
    ...
</table>
```

The WebClient class gives you the ability to download the entire HTML document. It's then up to you to parse the data.

In Chapter 6, you learned to use the WebClient.OpenReadAsync() method to download a file from the Web as a stream of bytes. You then have the flexibility to read that stream using a StreamReader (for text data) or a BinaryReader (for binary information). In this example, you can use the OpenAsync() method and then use a StreamReader to browse through the page. However, the WebClient provides a shortcut for relatively small amounts of text content—the DownloadStringAsync() method, which returns the results as a single string. In this example, that string includes the HTML for the entire page.

Figure 20-3 shows a simple Silverlight page that lets you query the table from Figure 20-2 for information. The user enters a year. The code then searches the web page for a matching cell and returns the population number from the next column. No attempt is made to interpolate values—if the indicated year falls between values in the table, no result is returned.

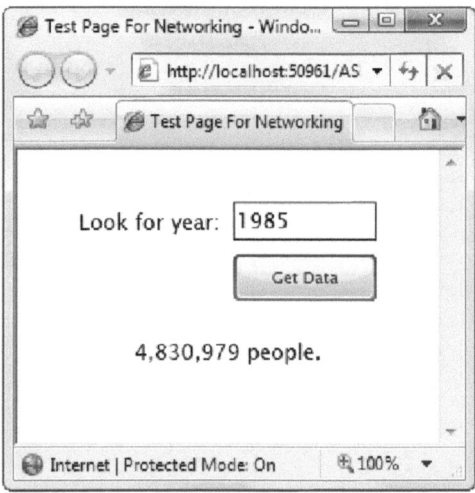

Figure 20-3. *Querying an HTML page with WebClient*

When the user clicks the Get Data button, a new WebClient object is created. The DownloadStringAsync() method is called with the appropriate website address:

```
Private Sub cmdGetData_Click(ByVal sender As Object, ByVal e As RoutedEventArgs)
    Dim client As New WebClient()
    Dim address As New Uri("http://localhost:" & _
      HtmlPage.Document.DocumentUri.Port & "/ASPWebSite/PopulationTable.html")

    AddHandler client.DownloadStringCompleted, _
      AddressOf client_DownloadStringCompleted
    client.DownloadStringAsync(address)
End Sub
```

■ **Tip** When you begin an asynchronous operation like this one, it's a good time to update the user interface with some sort of status message. For example, you can display the text "Contacting web service" in a TextBlock.

Here's the code that receives the results:

```
Private Sub client_DownloadStringCompleted(ByVal sender As Object, _
  ByVal e As DownloadStringCompletedEventArgs)
    Dim pageHtml As String = ""

    Try
        pageHtml = e.Result
    Catch
        lblResult.Text = "Error contacting service."
        Return
    End Try

    ...
```

When you read the Result property, an exception is thrown if the web request failed—for example, if the specified web page can't be found, or it doesn't allow cross-domain access. For this reason, exception-handling code is required.

It takes a bit more work to coax the information you want out of the HTML string. Although you can manually step through the string, examining each character, it's far easier to use *regular expressions*. Regular expressions are a pattern-matching language that's often used to search text or validate input. Using the ordinary methods of the String class, you can search for a series of specific characters (for example, the word *hello*) in a string. Using a regular expression, however, you can find any word in a string that is five letters long and begins with an *h*. You first learned about regular expressions in Chapter 17, where you used them to perform validation in a data object.

In this example, you need to find scraps of HTML in this form:

```
<th>500 BCE</th><td>100,000</td>
```

Here, the year in the <th> element is the lookup value, which is provided by the user. The number in the following <td> element is the result you want to retrieve.

There are several ways to construct a regular expression that does the trick, but the cleanest approach is to use a *named group*. A named group is a placeholder that represents some information you want to retrieve. You assign the group a name and then retrieve its value when you need it. Named groups use this syntax:

```
(?<NamedGroupName>MatchExpression)
```

Here's the named group used in this example:

```
(?<population>.*)
```

This named group is named population. It uses .* as its expression, which is just about as simple as a regular expression can get. The period (.) matches any character except a newline. The asterisk (*) indicates that there can be zero, one, or more occurrences of this pattern—in other words, the population value can have any number of characters.

What makes this named group useful is its position inside a larger regular expression. Here's an example that's very similar to the final expression used in this example:

```
<th>1985</th>\s*<td>(?<population>.*)</td>
```

If you break down this expression piece by piece, it's relatively straightforward. First, this regular expression looks for the column with the year value 1985:

```
<th>1985</th>
```

That can be followed by zero or more whitespace characters (spaces, lines, hard returns, and so on), which are represented by the \s metacharacter:

```
<th>1985</th>\s*
```

Then, the <td> tag for the next column appears, followed by the value you want to capture (the population number), in a named group:

```
<th>1985</th>\s*<td>(?<population>.*)
```

Finally, the closing </td> tag represents the end of column and the end of the expression.

The only difference in the final version of this expression that the code uses is that the year isn't hard-coded. Instead, the user enters it in a text box, and this value is inserted into the expression string:

```
Dim pattern As String = "<th>" & txtYear.Text & "</th>" & "\s*" & _
   "<td>" & "(?<population>.*)" & "</td>"
```

When you have the regular expression in place, the rest of the code is easy. You need to create a Regex object that uses the expression and pass in the search string to the Regex.Match() method. You can then look up your group by name and extract the value:

```
    ...
    Dim regex As New Regex(pattern)
    Dim match As Match = regex.Match(pageHtml)
    Dim people As String = match.Groups("population").Value
    If people = "" Then
        lblResult.Text = "Year not found."
    Else
        lblResult.Text = match.Groups("population").Value & " people."
    End If
End Sub
```

This isn't the most elegant way to get information from the Web, but it demonstrates how WebClient can work as a straightforward tool for reading HTML and other text sources on the Web. This behavior becomes even more useful when you begin to dabble in web services that use representational state transfer (REST), as described in the following sections.

REST and Other Simple Web Services

Recently, there's been a resurgence of simple web services—web services that avoid the detailed SOAP protocol and the complexity of the WS-* standards. Simple web services will never replace SOAP-based web services, because they don't provide solutions for the real challenges of distributed processing, such as routing, transactions, and security. However, their clean, stripped-down structure makes them an ideal choice for building public web services that need to be compatible with the broadest range of clients possible. Many top-notch websites (like Amazon, eBay, and Google) provide REST-based and SOAP-based interfaces for their web services.

SOAP VS. REST

At this point, you may be wondering what the differences are between SOAP, REST, and other web service standards. All web services pass messages over HTTP. But there are differences in the way information is presented, both when it's passed to the web service and when it's returned from the web service.

Full-fledged SOAP web services place their data into a specific XML structure: a SOAP document. SOAP can be verbose, which means it's more work to construct a SOAP message on a platform that doesn't have built-in SOAP support. (Silverlight is an example of a platform that does have built-in SOAP support, which is why you need to add a web reference to a SOAP service in order to use it, rather than construct the XML you need by hand.) SOAP also provides some significant advantages—it uses strongly typed data, and it's highly extensible thanks to SOAP *headers* (separate pieces of information that can be passed along with a message but aren't placed in the message body). SOAP headers are a key extensibility point that other SOAP-based standards use.

Non-SOAP web services have simpler ways to pass in information. Input values can be supplied in the URL (in which cased they're tacked on to the end as query string parameters) or supplied as a combination of name-value pairs in the message body. Either way, there's less overhead, but no real type checking. The web service response may use plain string data or XML.

Simple web services that return HTML documents are often described as using XML over HTTP. Simple web services are often also described as REST services, but in truth REST is a philosophical idea rather than a concrete standard. The fundamental idea behind REST is that every URL represents a unique object rather than a mere method call. The different HTTP verbs represent what you want to do with the object (for example, you use an HTTP GET to retrieve the object and HTTP POST to update it). Most web services that describe themselves as REST-based don't completely adhere to this idea and are actually just simple non-SOAP web services.

In this section, you'll see how to consume a simple web service that returns plain-text data. Later in this chapter, you'll go a little further and consider a web service that returns XML.

Earlier, you looked at a page that included a table with world population numbers throughout history. If you want to convert this to a web service, you can write a bit of web code that receives a year and writes out the relevant population figure. The requested year can be supplied through a query string argument (in an HTTP GET request) or posted to your page (with an HTTP POST request). The strategy you choose determines whether the client must use WebClient or the somewhat more complex WebRequest class. WebClient is enough for an ordinary HTTP GET request, but only WebRequest allows your Silverlight code to post a value.

You can build your web service using ASP.NET, but you need to avoid the full web form model. After all, you don't want to return a complete page to the user, with unnecessary elements like <html>, <head>, and <body>. Instead, you need to create what ASP.NET calls an *HTTP handler*.

To do so, right-click your ASP.NET website in the Solution Explorer, and choose Add New Item. Then, choose the Generic Handler template, supply a name, and click Add. By default, HTTP handlers have the extension .ashx. In this example, the handler is called PopulationService.ashx.

All HTTP handlers are code as classes that implement IHttpHandler, and they must provide a ProcessRequest() method and an IsReusable property getter. The IsReusable property indicates whether your HTTP handler can, after it's created, be reused to handle more than one request, which is slightly more efficient than creating it each time. If you don't store any state information in the fields of your class, you can safely return True:

```
Public ReadOnly Property IsReusable() As Boolean Implements IHttpHandler.IsReusable
    Get
        Return True
    End Get
End Property
```

The ProcessRequest() method does the actual work. It receives an HttpContext object through which it can access the current request details and write the response. In this example, ProcessRequest() checks for a posted value named year. It then checks if the year string includes the letters, and gets the corresponding population statistic using a custom method called GetPopulation (which isn't shown). The result is written to the page as plain text:

```
Public Sub ProcessRequest(ByVal context As HttpContext) _
    Implements IHttpHandler.ProcessRequest

    ' Get the posted year.
    Dim year As String = context.Request.Form("year")

    ' Remove any commas in the number, and excess spaces at the ends.
    year = year.Replace(",", "")
    year = year.Trim()

    ' Check if this year is BC.
    Dim isBc As Boolean = False
    If year.EndsWith("BC", StringComparison.OrdinalIgnoreCase) Then
        isBc = True
        year = year.Remove(year.IndexOf("BC", StringComparison.OrdinalIgnoreCase))
        year = year.Trim()
    End If

    ' Get the population.
    Dim yearNumber As Integer = Int32.Parse(year)
    Dim population As Integer = GetPopulation(yearNumber, isBc)

    ' Write the response.
    context.Response.ContentType = "text/plain"
    context.Response.Write(population)
End Sub
```

On the client side, you need to use the WebRequest class from the System.Net namespace. To make this class available, you must add a reference to the System.Net.dll assembly, which isn't included by default.

WebRequest requires that you do all your work asynchronously. Whereas WebClient has one asynchronous step (downloading the response data), WebRequest has two: creating the request stream and then downloading the response.

To use WebRequest, you first need to create a WebRequest object, configure it with the correct URI, and then call BeginGetRequestStream(). When you call BeginGetRequestStream(), you supply a callback that will write the request to the request stream when it's ready. In this example, that task falls to another method named CreateRequest():

```
Private searchYear As String

Private Sub cmdGetData_Click(ByVal sender As Object, ByVal e As RoutedEventArgs)
    Dim address As New Uri("http://localhost:" & _
      HtmlPage.Document.DocumentUri.Port & _
      "/ASPWebSite/PopulationService.ashx")

    ' Create the request object.
    Dim request As WebRequest = WebRequest.Create(address)
    request.Method = "POST"
    request.ContentType = "application/x-www-form-urlencoded"

    ' Store the year you want to use.
    searchYear = txtYear.Text

    ' Prepare the request asynchronously.
    request.BeginGetRequestStream(AddressOf CreateRequest, request)
End Sub
```

This code contains one other detail. Before calling BeginGetRequestStream(), the code copies the search year from the text box into a private field named searchYear. This technique serves two purposes. First, it ensures that the CreateRequest() callback can access the original search value, even if the user is somehow able to edit the text box before the CreateRequest() code runs. More important, this technique avoids threading headaches. Because the CreateRequest() callback runs on a background thread (not the main application thread), it can't directly access the elements in the page. As you saw in Chapter 19, you can work around this problem using Dispatcher.BeginInvoke(). However, copying the search year sidesteps the problem.

Typically, Silverlight calls your CreateRequest() method a fraction of a second after you call BeginGetRequestStream(). At this point, you need to write the posted values as part of your request. Often, web services use the same standard for posted values as HTML forms. That means each value is supplied as a name-value pair, separated by an equal sign; multiple values are chained together with ampersands (&), as in FirstName=Matthew&LastName=MacDonald. To write the data, you use a StreamWriter:

```
Private Sub CreateRequest(ByVal asyncResult As IAsyncResult)
    Dim request As WebRequest = CType(asyncResult.AsyncState, WebRequest)

    ' Write the year information in the name-value format "year=1985".
    Dim requestStream As Stream = request.EndGetRequestStream(asyncResult)
    Dim writer As New StreamWriter(requestStream)
    writer.Write("year=" & searchYear)

    ' Clean up (required).
    writer.Close()
    requestStream.Close()

    ' Read the response asynchronously.
    request.BeginGetResponse(AddressOf ReadResponse, request)
End Sub
```

After you've written the request, you need to close the StreamWriter (to ensure all the data is written) and then close the request stream. Next, you must call BeginGetResponse() to supply the callback that will process the response stream when it's available. In this example, a method named ReadResponse() does the job.

To read the response, you use a StreamReader. You also need error-handling code at this point, to deal with the exceptions that are thrown if the service can't be found. If the response uses XML, it's also up to you to parse that XML now:

```
Private Sub ReadResponse(ByVal asyncResult As IAsyncResult)
    Dim result As String
    Dim request As WebRequest = CType(asyncResult.AsyncState, WebRequest)

    ' Get the response stream.
    Dim response As WebResponse = request.EndGetResponse(asyncResult)
    Dim responseStream As Stream = response.GetResponseStream()

    Try
        ' Read the returned text.
        Dim reader As New StreamReader(responseStream)
        Dim population As String = reader.ReadToEnd()
        result = population & " people."
    Catch
        result = "Error contacting service."
    Finally
        response.Close()
    End Try
    ...
```

As with the callback for BeginGetRequestStream(), the callback for BeginGetResponse() runs on a background thread. If you want to interact with an element, you need to use Dispatcher.BeginInvoke() to marshal the call to the foreground thread. But first, you need a separate method that can perform the updating. The code in this method is simple—in this example, it merely copies the returned text information into a label:

```
Private Sub UpdateLabel(text As String)
    lblResult.Text = text
End Sub
```

Now, all you need to do is add the code that calls this method at the end of the ReadResponse() method. To tell Dispatcher.BeginInvoke() which method to invoke, you must either pass a delegate that points to the method or create an Action object. If you use the delegate approach, you need to define the delegate with the appropriate method signature (in this case, a subroutine that accepts a single string parameter) elsewhere in your code. If you create an Action object, you specify the method signature by supplying type arguments. This approach is more streamlined, so it's the one this code uses:

```
    ...
    ' Create an action that represents a method that takes
    ' a single string parameter.
    ' Point the action to UpdateLabel().
    Dim updateAction As New Action(Of String)(AddressOf UpdateLabel)

    ' Call the action and pass in the new text.
    Dispatcher.BeginInvoke(updateAction, result)
End Sub
```

Ironically, calling a simple web service is more work in Silverlight than calling a SOAP-based web service, because Silverlight can't generate any code for you. This is part of the drawback with simple web services—although they're easier to call, they aren't self-describing. That means they lack the low-level documentation details that allow development tools like Visual Studio to generate some of the code you need.

Using Network Credentials

Both the WebClient and WebRequest classes allow you to pass network credentials—specifically, a user name and password—with your web request. This allows you to access secure web servers that require authentication.

To use network credentials, you must first call the shared WebRequest.RegisterPrefix() method to make sure you're using the Client HTTP stack, as shown here:

```
WebRequest.RegisterPrefix("http://", _
    System.Net.Browser.WebRequestCreator.ClientHttp)
```

You then set the UseDefaultCredentials property and the Credentials property of your WebClient or WebRequest object. Here's an example with the WebRequest class:

```
Dim request As WebRequest = WebRequest.Create(address)
request.UseDefaultCredentials = False
request.Credentials = New NetworkCredential("username", "password")
```

Once you've finished setting these properties, you can use the WebRequest or WebClient object in the normal way to contact the web server.

Processing Different Types of Data

So far, you've seen how to retrieve ordinary text data from the Web, whether it's from a static file or dynamically generated by a web service. You've also seen how to search through that text if it contains HTML markup. But both plain text and HTML are limited from a programming point of view, because they're difficult to parse. More often, you'll deal with more complex structured data. Web services that return structured data usually adopt a standardized format, such as ordinary XML, SOAP messages, or JSON. Silverlight supports all three formats, and you'll see how to use them in the following sections.

XML

Many simple web services return their data in XML. When consuming this sort of service, you need to decide how to process the XML.

Silverlight includes several options for dealing with XML:

- *XmlWriter and XmlReader:* These classes offer a bare-bones approach for dealing with XML, with the fewest features. Using them, you can write or read XML content one element at a time.

- *XmlSerializer:* This class allows you to convert a live object into an XML representation, and vice versa. The only limitation is that the XML is forced to adhere to the structure of the class. To use XmlSerializer, you need a reference to the System.Xml.Serialization.dll assembly.

- *XDocument:* This class is the foundation of LINQ to XML. It lets you transform XML objects (and back), but it gives you far more flexibility than XmlSerializer. Using the right LINQ expression, you can filter out just the details you want and change the structure of your content. To use XDocument, you need a reference to the System.Xml.Linq.dll assembly.

So which is the best approach to use? XmlReader and XmlWriter offer the lowest-level approach. For example, to read an XML document with XmlReader, you need to loop through all the nodes, keeping track of the structure on your own and ignoring comments and whitespace. You're limited to travelling in one direction (forward). If you want just a single node, you're still forced to read through every node that occurs before it in the document. Similarly, when writing a document, you need to write all the elements sequentially, relying on the order of your statements to generate the right structure. You also need to explicitly write the start and end tag for each element that contains nested elements.

Generally, most Silverlight applications are better off using the higher-level XmlSerializer and XDocument classes. The only exception is if you need to deal with a huge amount of XML and you want to avoid the overhead of loading it all into memory at once. In this case, the bit-by-bit processing of XmlWriter and XmlReader may be required.

Between XmlSerializer and XDocument, XmlSerializer is a reasonable option if you're in complete control of the data format—in other words, you've created the class you want to serialize and you don't need to conform to a specific XML format. However, XDocument provides much more flexibility, giving you the ability to look at XML as a collection of elements or transform it into a collection of objects. It's particularly useful when you're consuming someone else's XML—for example, when you're retrieving data from a web service.

■ **Note** Silverlight doesn't include a class that uses the XML DOM model (such as the XmlDocument class you can use in .NET). If you want to perform in-memory XML processing, you're better off with the more streamlined and efficient XDocument.

In the next section, you'll see how to use XDocument to parse the data that's returned from a web service and create an XML document to send to a web service. If you have a specialized scenario that requires XmlWriter, XmlReader, or XmlSerializer, you'll find that they work much the same way as in the full .NET Framework.

Services That Return XML Data

Flickr is an image-sharing website that provides REST-like services. You supply your parameters by tacking query string arguments onto the end of the URL. The Flickr web service returns a response in XML.

Figure 20-4 shows an example that lets the user supply a search keyword and then displays a list of images that are described with that keyword on Flickr.

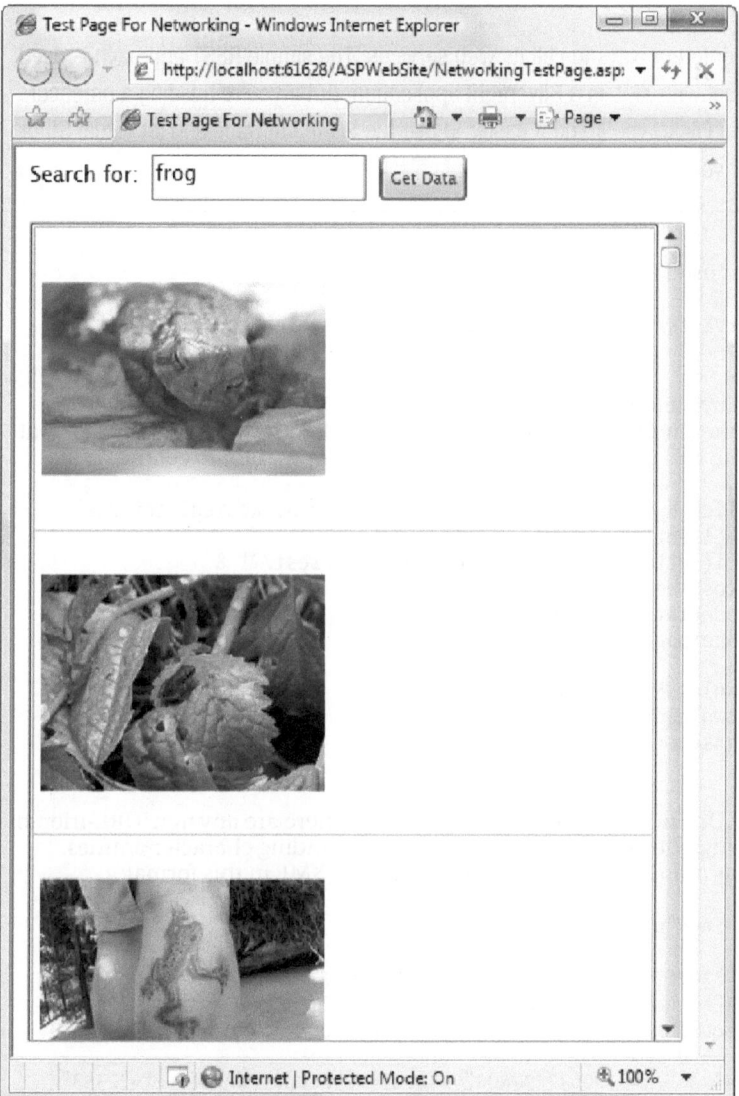

Figure 20-4. *Retrieving pictures from Flickr*

In this example, the Flickr request includes the following query string arguments: method (indicates the type of operation being performed), tags (the search keywords), perpage (the number of results you want to retrieve), and api_key (a unique ID that allows you to get access to Flickr's service). You can use many more arguments to fetch multiple pages of results, apply sorting, filter by dates, and so on. To get more information and obtain your own free API key, visit www.flickr.com/services/api.

■ **Tip** Flickr provides several different ways to call its web services. The simple REST approach is used here to demonstrate how to deal with XML in a Silverlight application. But if you're building a practical Silverlight application that uses Flickr, you'll find it easier to use the SOAP interface and let Visual Studio generate some of the code for you.

Here's what the request used in Figure 20-4 looks like:

```
http://api.flickr.com/services/rest/?method=flickr.photos.search&tags=frog&
api_key=...&perpage=10
```

Because all the input parameters are passed in through the URL, there's no need to post anything, and you can use the simpler WebClient instead of WebRequest. Here's the code that builds the Flickr request URL and then triggers an asynchronous operation to get the result:

```
Private Sub cmdGetData_Click(ByVal sender As Object, ByVal e As RoutedEventArgs)
    Dim client As New WebClient()
    Dim address As New Uri("http://api.flickr.com/services/rest/?" & _
      "method=flickr.photos.search" & "&tags=" & _
      HttpUtility.UrlEncode(txtSearchKeyword.Text) & _
      "&api_key=..." & "&perpage=10")

    AddHandler client.DownloadStringCompleted, _
      AddressOf client_DownloadStringCompleted
    client.DownloadStringAsync(address)
End Sub
```

The shared HttpUtility.UrlEncode() method ensures that if there are any non-URL-friendly characters in the search string, they're replaced with the corresponding character entities.

The result is retrieved as a single long string, which contains XML in this format:

```
<?xml version="1.0" encoding="utf-8" ?>
<rsp stat="ok">
  <photos page="1" pages="1026" perpage="100" total="102577">
    <photo id="2519140273" owner="85463968@N00" secret="9d215a1b8b" server="2132"
    farm="3" title="He could hop in, but he couldn't hop out" ispublic="1"
    isfriend="0" isfamily="0" />
    <photo id="2519866774" owner="72063229@N00" secret="05bccd89cd" server="2353"
    farm="3" title="Small Frog on a Leaf" ispublic="1" isfriend="0" isfamily="0" />
    ...
  </photos>
</rsp>
```

To parse this information, the first step is to load the entire document into a new XDocument object. The XDocument class provides two shared methods to help you: a Load() method for pulling content out of an XmlReader, and a Parse() method for pulling content out a string. When the WebClient.DownloadStringCompleted event fires, you use the Parse() method:

```
Dim document As XDocument = XDocument.Parse(e.Result)
```

When you have the XDocument object, you can use one of two strategies to extract the information you need. You can move through the collections of elements and attributes in the XDocument, which are represented as XElement and XAttribute objects. Or, you can use a LINQ expression to retrieve the XML content you want and convert it into the most suitable object representation. The following sections demonstrate both approaches.

Navigating Over an XDocument

Every XDocument holds a collection of XNode objects. The XNode is a MustInherit base class. Other more specific classes, like XElement, XComment, and XText, derive from it and are used to represent elements, comments, and text content. Attributes are an exception—they're treated not as separate nodes but as simple name-value pairs that are attached to an element.

Once you have a live XDocument with your content, you can dig into the tree of nodes using a few key properties and methods of the XElement class. Table 20-1 lists the most useful methods.

Table 20-1. Essential Methods of the XElement Class

Method	Description
Attributes()	Gets the collection of XAttribute objects for this element.
Attribute()	Gets the XAttribute with a specific name (or Nothing if there's no match).
Elements()	Gets the collection of XElement objects that are contained by this element. (This is the top level only—these elements may in turn contain more elements.) Optionally, you can specify an element name, and only those elements will be retrieved.
Element()	Gets the XElement contained by this element that has a specific name (or Nothing if there's no match). If there is more than matching element, this returns the first.
Nodes()	Gets all the XNode objects contained by this elements. This includes elements and other content, like comments.

There's a critically important detail here: XDocument exposes nested elements through methods, not properties. This gives you added flexibility to filter out just the elements that interest you. For example, when using the XDocument.Elements method, you have two overloads to choose from. You can get all the child elements (in which case you supply no parameters) or get only those child elements that have a specific element name (in which case you specify the element name as a string).

In the Flickr example, the top-level element is named <rsp>. Thus, you can access it like this:

```
Dim element As XElement = document.Element("rsp")
```

Of course, what you're really interested in is the <photos> element inside the <rsp> element. You can get this in two steps:

```
Dim rspElement As XElement = document.Element("rsp")
Dim photosElement As XElement = rspElement.Element("photos")
```

or even more efficiently in one:

```
Dim photosElement As XElement = document.Element("rsp").Element("photos")
```

To get the <photo> elements inside the <photos> element, you use the Elements() method (because there are multiple matching elements). You don't need to specify the name <photos>, because there isn't any other type of element inside:

```
Dim elements As IEnumerable(Of XElement)
elements = document.Element("rsp").Element("photos").Elements()
```

All the information you need is in the attributes of each <photo> element. To get the Flickr image (which you can then display using the Image element), you need to construct the right URL, which involves combining several pieces of information in the correct format:

```
Private Sub client_DownloadStringCompleted(ByVal sender As Object, _
  ByVal e As DownloadStringCompletedEventArgs)
    Dim document As XDocument = XDocument.Parse(e.Result)

    ' Clear the list.
    images.Items.Clear()

    ' Examine each <photo> element.
    For Each element As XElement In _
      document.Element("rsp").Element("photos").Elements()
        ' Get the attribute values and combine them to build the image URL.
        Dim imageUrl As String = String.Format(
          "http://farm{0}.static.flickr.com/{1}/{2}_{3}_m.jpg", _
          CStr(element.Attribute("farm")), CStr(element.Attribute("server")), _
          CStr(element.Attribute("id")), CStr(element.Attribute("secret")))

        ' Create an Image object that shows the image.
        Dim img As New Image()
        img.Stretch = Stretch.Uniform
        img.Width = 200
        img.Height = 200
        img.Margin = New Thickness(10)
        img.Source = New BitmapImage(New Uri(imageUrl))

        ' Add the Image element to the list.
        images.Items.Add(img)
    Next
End Sub
```

■ **Tip** The easiest way to get the actual value out of an XAttribute or XElement object is to cast it to the desired type. In the previous example, all the attributes are treated as string values.

You've already seen how to use the Element() and Elements() methods to filter out elements that have a specific name. However, both these methods go only one level deep. The XDocument and XElement classes also include two methods that search more deeply:

Ancestors() and Descendants(). The Descendants() method finds all XElement objects contained by the current element, at any depth. The Ancestors() method finds all the XElement objects that contain the current element, again at any level. Using Descendants(), you can rewrite this statement from the earlier code block

```
For Each element As XElement In _
  document.Element("rsp").Element("photos").Elements()
```

like this:

```
For Each element As XElement In document.Descendants("photo")
```

The XDocument and XElement classes are small miracles of efficiency. If you take a closer look at them, you'll find many more members for navigation. For example, they have properties for quickly stepping from one node to the next (FirstNode, LastNode, NextNode, PreviousNode, and Parent) and methods for retrieving sibling nodes at the same level as the current node (ElementsAfterSelf() and ElementsBeforeSelf()). You'll also find methods for manipulating the document structure, which you'll consider later in this chapter.

Querying an XDocument with LINQ

As you've seen, it's easy to use methods like Element(), Elements(), and Ancestors() to reach into an XDocument and get the content you want. However, in some situations you want to transform the content to a different structure. For example, you may want to extract the information from various elements and flatten it into a simple structure. This technique is easy if you use the XDocument in conjunction with a LINQ expression.

As you learned in Chapter 16, LINQ expressions work with objects that implement IEnumerable(Of T). The XDocument and XElement classes include several ways to get IEnumerable(Of T) collections of elements, including the Elements() and Descendants() methods you've just considered.

After you place your collection of elements in a LINQ expression, you can use all the standard LINQ operators. That means you can use sorting, filtering, grouping, and projections to get the data you want.

Here's an example that selects all the <photo> elements in an XML document (using the Descendants() method), extracts the most important attribute values, and sets these as the properties of an object:

```
Dim photos = From results In document.Descendants("photo") _
            Select New With { _
                .Id = CStr(results.Attribute("id")), _
                .Farm = CStr(results.Attribute("farm")), _
                .Server = CStr(results.Attribute("server")), _
                .Secret = CStr(results.Attribute("secret"))}
```

This technique uses the standard LINQ feature of *anonymous types*. Essentially, this expression generates a collection of a dynamically defined type that includes the properties you specify.

The VB compiler creates the class definition you need. Elsewhere in your code, you can loop over the photos collection and interact with the properties of the dynamically generated type to build Image elements, as you saw earlier:

```
For Each photo In photos
    imageUrl = String.Format( _
      "http://farm{0}.static.flickr.com/{1}/{2}_{3}_m.jpg", _
      photo.Farm, photo.Server, photo.Id, photo.Secret)
    ...
Next
```

This technique of mapping a portion of an XML document to new class is called *projection*. Often, projection is combined with anonymous types for one-off tasks, when you don't need to use the same grouping of data elsewhere in your application. However, it's just as easy to use a projection to create instances of a custom class. You'll need to use this approach if you plan to perform data binding with the newly generated objects.

To see how this works, it helps to consider an alternative way to build the example that's shown in Figure 20-4. Instead of manually constructing each Image element, you can define a data template that takes bound objects, extracts the URL information, and uses it in an Image element:

```
<ListBox x:Name="images">
  <ListBox.ItemTemplate>
    <DataTemplate>
      <Image Stretch="Uniform" Width="200" Height="200"
        Margin="5" Source="{Binding ImageUrl}"></Image>
    </DataTemplate>
  </ListBox.ItemTemplate>
</ListBox>
```

To make this work, you need a custom class that provides an ImageUrl property (and may include other details). Here's the simplest possibility:

```
Public Class FlickrImage
    Private _imageUrl As String
    Public Property ImageUrl() As String
        Get
            Return _imageUrl
        End Get
        Set(ByVal value As String)
            _imageUrl = value
        End Set
    End Property
End Class
```

Now, you can use a LINQ expression to create a collection of FlickrImage objects:

```
Dim photos = From results In document.Descendants("photo") _
             Select New FlickrImage With {.ImageUrl = _
             String.Format( _
               "http://farm{0}.static.flickr.com/{1}/{2}_{3}_m.jpg", _
               results.Attribute("farm").Value.ToString(), _
               results.Attribute("server").Value.ToString(), _
               results.Attribute("id").Value.ToString(), _
               results.Attribute("secret").Value.ToString())}
images.ItemsSource = photos
```

This approach requires the least amount of code and provides the most streamlined solution.

Services That Require XML Data

Simple web services often allow you to supply all the input parameters through query string arguments. However, query string arguments are limited by the rules of web browser URIs. They can only be so long, and they're hard-pressed to represent structured data.

For that reason, web services that need more detailed data usually accept some form of XML. SOAP (described next) is one example. Non-SOAP web services often use a basic standard called XML-RPC. For example, Flickr provides an XML-RPC interface for its image search. To use it, you post an XML request in this format:

```
<methodCall>
  <methodName>flickr.photos.search</methodName>
  <params>
    <param>
      <value>
        <struct>
          <member>
            <name>tags</name>
            <value><string>value</string></value>
          </member>
          <member>
            <name>api_key</name>
```

```
            <value><string>...</string></value>
          </member>
        </struct>
      </value>
    </param>
  </params>
</methodCall>
```

You can add additional parameters by adding more <member> elements. For example, you can add the optional perpage parameter, as in the previous examples.

To use an XML-RPC service (or any web service that requires an XML request message), you need to send the XML document in the body of an HTTP POST. That means you need the higher-powered WebRequest class rather than WebClient.

To construct the XML message, you can construct a new XDocument, using classes like XElement, XAttribute, XComment, XDeclaration, and so on. However, VB has a tidier approach with its *XML literals* feature.

Essentially, XML literals give you an easy way to declare XML content directly in your code. Here's an example:

```
Dim element As XElement
element = <photo src="http://www.someplace.com/someimage.jpg">
            <tag>horse</tag>
            <tag>plow</tag>
          </photo>
```

When using XML literals, you're free to split your content over multiple lines without the line-break underscore. Visual Studio also indents your XML content automatically.

The most practical part of XML literals is the ability to embed variables and snippets of code inside the static XML content. For example, if you want to draw one of the keywords in the previous example from a variable named pictureKeyword, here's what you need to do:

```
element = <photo src="http://www.someplace.com/someimage.jpg">
            <tag><%= pictureKeyword %></tag>
            <tag>plow</tag>
          </photo>
```

Here's a complete example that constructs the request message for an XML-RPC request for a Flickr image search, using the search keywords from a text box, and writes it to the request stream:

```
Dim writer As New StreamWriter(requestStream)
writer.Write(<methodCall>
                <methodName>flickr.photos.search</methodName>
                <params>
                    <param>
                        <value>
                            <struct>
                                <member>
                                    <name>tags</name>
                                    <value>
                                        <string><%= searchKeyword %></string>
                                    </value>
                                </member>
                                <member>
                                    <name>api_key</name>
```

```
                      <value><string>...</string></value>
                  </member>
                  <member>
                      <name>perpage</name>
                      <value><string>10</string></value>
                  </member>
              </struct>
          </value>
      </param>
  </params>
</methodCall>)
```

When you call the Flickr image search through XML-RPC, you also get an XML-RPC response. To get the photo information you used earlier, you need to call HttpUtility.HtmlDecode() on the message and then use LINQ to XML to filter out the <photo> elements. For the complete code, see the downloadable examples for this chapter.

■ **Note** You've now learned how to read and create XML with XDocument and LINQ to XML. These techniques are useful when you're dealing with XML-based web services, but they also come in handy if you need to work with XML in other scenarios (for example, if you have a locally stored file in isolated storage that has XML content). If you dig into the XDocument and XElement classes, you'll find they have many more elements that make it easy to modify XML documents after you've created them. Not only can you set the value of any element or attribute, but you can also use methods for inserting, removing, and otherwise manipulating the XML tree of nodes, such as Add(), AddAfterSelf(), AddBeforeSelf(), RemoveNodes(), Remove(), ReplaceWith(), and so on.

Services That Return SOAP Data

As you learned in Chapter 15, Silverlight works seamlessly with .NET web services. These web services send SOAP-encoded data. SOAP is a form of XML, so it's technically possible to use Silverlight's XML processing (for example, the XDocument class) to create request messages and parse response messages, as in the previous sections. However, it's far easier to add a service reference in Visual Studio.

What you may not know is that the same technique applies to any SOAP-based web service. In other words, you can add references to SOAP-based services that aren't built in .NET. Silverlight has no way of distinguishing between the two and no way of knowing what code powers the service it's calling.

When you add a web reference to any SOAP-based web service, Visual Studio creates the proxy class you need, complete with asynchronous methods and events for each web method in the web service. For more information, see Chapter 15.

Services That Return JSON Data

JavaScript Object Notation (JSON) is an object-notation syntax that's sometimes used as a lightweight alternative to XML. You need to use the JSON serializer is when you're consuming a web service that returns JSON data and provides no SOAP alternative. (If the web service returns JSON or simple XML, it's up to you whether you prefer the JSON approach or XDocument.) To make matters even more interesting, Silverlight provides two distinct ways to parse JSON data: you can deserialize it with the JSON deserializer, as the next example demonstrates; or you can use LINQ to JSON, which works much the same way as LINQ to XML. Although this chapter doesn't discuss LINQ to JSON, you can get more information in the Silverlight SDK documentation (or read a quick review at http://blogs.msdn.com/mikeormond/archive/2008/08/21/linq-to-json.aspx).

Before you can deal with JSON data, you need to add references to three additional assemblies: System.Runtime.Serialization.dll, System.ServiceModel.dll, and System.ServiceModel.Web.dll.

Deserializing JSON is a lot like deserializing XML with the XmlSerializer class. The first requirement is to have a suitable class that matches the structure of your JSON data. You can then use the DataContractJsonSerializer class to convert instances of this class into JSON data and vice versa.

For example, Yahoo! provides a JSON interface for its image-search service (described at http://developer.yahoo.com/search/image/V1/imageSearch.html). It returns data that looks like this:

```
{"ResultSet":{
  "totalResultsAvailable":"957841",
  "totalResultsReturned":10,
  "firstResultPosition":1,
  "Result":[
    {
      "Title":"tree_frog.jpg",
      "Summary":"Red-Eyed Tree Frog",
      "Url":"http:\/\/www.thekidscollege.com\/images\/animals\/redeyetree_frog.jpg",
      ...
    },
    {
      "Title":"tree_frog_large-thumb.jpg",
      "Summary":"Before I came back in though I got another shot of the frog.",
      "Url":"http:\/\/www.silveriafamily.com\/blog\/john\/treefrog.jpg",
      ...
    }
  ]
}}
```

The data is in name-value pairs and is grouped into classes using curly braces {} and into arrays using square brackets []. To model the data shown here with classes, you need a class for each individual search result (named Result in the JSON), a class for the entire result set (named ResultSet in the JSON), and a top-level class that holds the search result set. You can give these classes any name you want, but the property names must match the names in the JSON representation exactly, including case. Your classes don't need to include properties for details you don't want to retrieve—they can be safely ignored.

Here are the classes you need (with the property procedures omitted to save space). The property names (which are based on the JSON representation) are highlighted:

```
Public Class SearchResults
    Public ResultSet As SearchResultSet
End Class

Public Class SearchResultSet
    Private _totalResultsAvailable As Integer
    Public Property totalResultsAvailable() As Integer
        ...
    End Property

    Private _totalResultsReturned As Integer
    Public Property totalResultsReturned() As Integer
        ...
    End Property

    Private _result As SearchResult()
    Public Property Result() As SearchResult()
        ...
    End Property
End Class

Public Class SearchResult
    Private _title As String
    Public Property Title() As String
        ...
    End Property

    Private _summary As String
    Public Property Summary() As String
        ...
    End Property

    Private _url As String
    Public Property Url() As String
        ...
    End Property
End Class
```

Now you can use these classes to deserialize the results of a search. It's a two-step affair. First, you create an instance of the DataContractJsonSerializer, specifying the type you want to serialize or deserialize as a constructor argument:

```
Dim serializer As New DataContractJsonSerializer(GetType(SearchResults))
```

Then, you can use ReadObject() to deserialize JSON data or WriteObject() to create it:

```
Dim results As SearchResults = CType(serializer.ReadObject(jsonData), SearchResults)
```

Figure 20-5 shows a sample Silverlight page that searches for images by keyword.

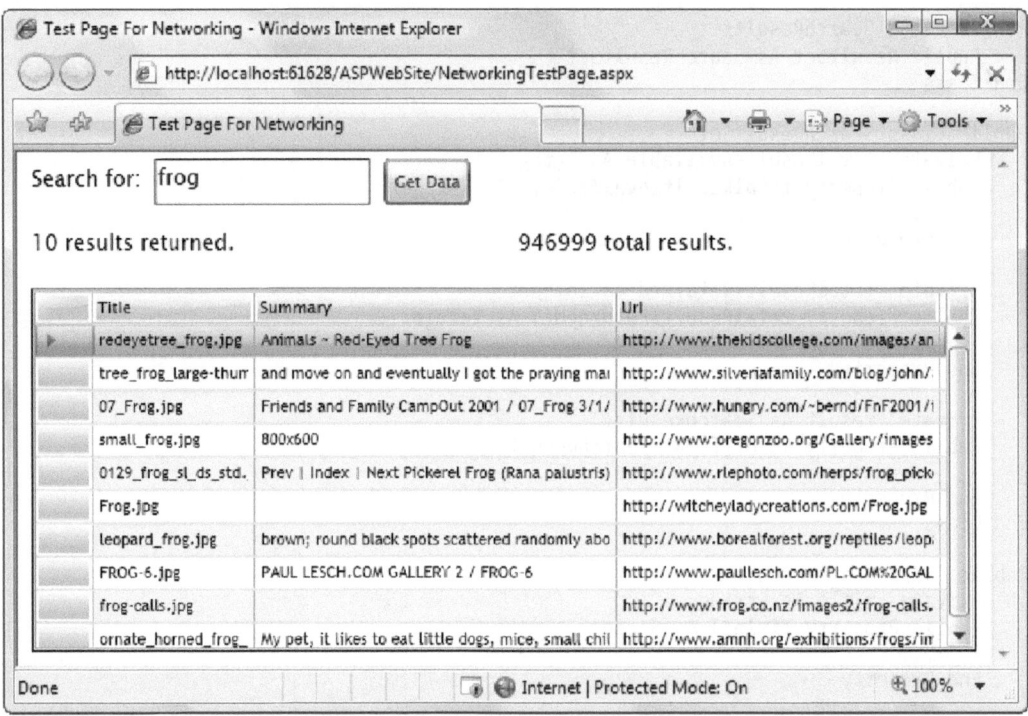

Figure 20-5. *Searching for images with Yahoo!*

Here's the code that underpins this page:

```
Private Sub cmdGetData_Click(ByVal sender As Object, ByVal e As RoutedEventArgs)
    Dim client As New WebClient()
    Dim address As New Uri( _
      "http://search.yahooapis.com/ImageSearchService/V1/imageSearch?" & _
      "appid=YahooDemo&query=" & HttpUtility.UrlEncode(txtSearchKeyword.Text) & _
      "&output=json")

    AddHandler client.OpenReadCompleted, AddressOf client_OpenReadCompleted
    client.OpenReadAsync(address)
End Sub

Private Sub client_OpenReadCompleted(ByVal sender As Object, _
  ByVal e As OpenReadCompletedEventArgs)
    Dim serializer As New DataContractJsonSerializer(GetType(SearchResults))
    Dim results As SearchResults = CType(serializer.ReadObject(e.Result), _
      SearchResults)

    lblResultsTotal.Text = results.ResultSet.totalResultsAvailable & " _
      total results."
    lblResultsReturned.Text = results.ResultSet.totalResultsReturned & " _
      results returned."
```

```
        gridResults.ItemsSource = results.ResultSet.Result
End Sub
```

RSS

Really Simple Syndication (RSS) is an XML-based format for publishing summaries of frequently updated content, such as blog entries or news stories. These documents are called *feeds*. Client applications called *RSS readers* can check RSS feeds periodically and notify you about newly added items.

.NET 3.5 introduced classes that support the RSS 2.0 or Atom 1.0 formats. Silverlight borrows these same classes, allowing you to read feed information without tedious XML-parsing code. These classes are defined in the System.ServiceModel.Syndication namespace, and to get access to them you need to add a reference to the System.ServiceModel.Syndication.dll assembly.

When you use RSS, it's important to remember that you're limited by the cross-domain rules explained at the beginning of this chapter. Obviously, if you try to access a feed on a web server that doesn't allow cross-domain access, you'll get an error. However, feeds also contain links. For example, a typical feed item contains a summary and a link that points to the full page for the corresponding blog entry or news item. If you attempt to download the page at this location, you must also be sure it's on a web server that allows cross-domain access.

You need to consider one other issue. The items in an RSS feed usually point to full-fledged HTML pages. But even if you download this HTML content, there's no way to display it in its properly formatted form in the Silverlight content region. A better approach is to show it on another part of the current HTML page—for example, just below the Silverlight control. Figure 20-6 shows an example that combines a Silverlight page that displays feed items (on top) with an ordinary HTML <iframe> element, which shows the page that corresponds to the currently selected item.

Figure 20-6. Browsing an RSS feed with news items

Creating this example is surprisingly straightforward. First, you need a feed URI. This example uses the URI http://feeds.feedburner.com/ZDNetBlogs, which points to blogged news items on the high-tech website ZDNet. Feeds are XML documents, and you can download them easily using the familiar DownloadStringAsycn() and OpenReadAsync() methods. The latter is more efficient, because the entire XML document doesn't need to be held in memory at once as a string:

```
Private Sub cmdGetData_Click(ByVal sender As Object, ByVal e As RoutedEventArgs)
    Dim client As New WebClient()
    Dim address As New Uri("http://feeds.feedburner.com/ZDNetBlogs")
    AddHandler client.OpenReadCompleted, AddressOf client_OpenReadCompleted
    client.OpenReadAsync(address)
End Sub
```

When you read the response, you can load the XML content into a SyndicationFeed object. The SyndicationFeed class includes various properties that describe details about the feed, such as its author, its last update, a summary of what the feed is about, and so on. The most important detail is the Items property, which holds a collection of SyndicationItem objects. The SyndicationItem objects are shown in the Grid in Figure 20-6:

```vb
Private Sub client_OpenReadCompleted(ByVal sender As Object, _
  ByVal e As OpenReadCompletedEventArgs)
    Try
        Dim reader As XmlReader = XmlReader.Create(e.Result)
        Dim feed As SyndicationFeed = SyndicationFeed.Load(reader)
        gridFeed.ItemsSource = feed.Items
        reader.Close()
    Catch
        lblError.Text = "Error downloading feed."
    End Try
End Sub
```

To display the information from each SyndicationItem object, you need to pull out the right information with custom binding expressions. Useful properties include Authors, Title, Summary, and PublishDate, each of which returns a different type of syndication object (all of which are defined in the System.ServiceModel.Syndication namespace). The example in Figure 20-6 uses the title and summary information:

```xml
<data:DataGrid>
  <data:DataGrid.Columns>
    <data:DataGridTextColumn Binding="{Binding Title.Text}"
      ElementStyle="{StaticResource DataGridWrapStyle}" />
    <data:DataGridTextColumn Width="400"
      Binding="{Binding Summary.Text, Converter={StaticResource HtmlCleanUp}}"
      ElementStyle="{StaticResource DataGridWrapStyle}" />
  </data:DataGrid.Columns>
</data:DataGrid>
```

The DataGrid also uses a custom style for text wrapping (as described in Chapter 17) and a custom value converter to remove the HTML tags from the summary and shorten it if it exceeds a certain maximum number of characters. (To see the custom value converter, refer to the downloadable code examples for this chapter.)

When an item is clicked in the DataGrid, the following event handler grabs the corresponding SyndicationItem object and examines the Links property to find the URI that points to the web page with the full story. It then uses a dash of HTML interoperability (as described in Chapter 14) to point an <iframe> to that page:

```vb
Private Sub gridFeed_SelectionChanged(ByVal sender As Object, ByVal e As EventArgs)
    ' Get the selected syndication item.
    Dim selectedItem As SyndicationItem = CType(gridFeed.SelectedItem, _
      SyndicationItem)

    ' Find the <iframe> element on the page.
    Dim element As HtmlElement = HtmlPage.Document.GetElementById("rssFrame")
    ' Point the <iframe> to the full page for the selected feed item.
    element.SetAttribute("src", selectedItem.Links(0).Uri.ToString())
End Sub
```

Sockets

So far, you've focused exclusively on retrieving information over HTTP. Even though HTTP was developed for downloading simple HTML documents in the earliest days of the Internet, it also works surprisingly well as a transport mechanism for XML documents and the request and response messages used to interact with web services.

That said, HTTP isn't without a few significant drawbacks. First, HTTP is a high-level standard that's built on Transmission Control Protocol (TCP). It will never be as fast as a raw network connection. Second, HTTP uses a request model that forces the client to ask for data. There's no way for the server to call back to the client if new information arrives. This limitation means that HTTP is a poor choice for everything from real-time Internet games to stock monitoring. If you need to go beyond these limitations in order to build a certain type of application, you'll need to step up to a rich client platform (like WPF) or use Silverlight's support for *sockets*.

Understanding Sockets and TCP

Strictly speaking, sockets are nothing more than endpoints on a network. They consist of two numbers:

- *IP address:* The IP address identifies your computer on a network or the Internet.
- *Port:* The port number corresponds to a specific application or service that's communicating over the network.

The combination of two sockets—one on the client that's running the Silverlight application, and one on a web server that's running a server application—defines a connection, as shown in Figure 20-7.

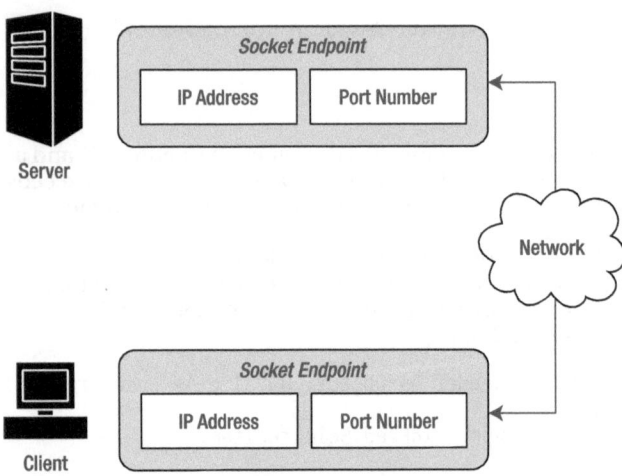

Figure 20-7. A socket-based connection

■ **Note** Port numbers don't correspond to anything physical—they're a method for separating different application endpoints on the same computer. For example, if you're running a web server, your computer will respond to requests on port 80. Another application may use port 8000. Essentially, ports map the network communication on a single computer to the appropriate applications. Silverlight lets you open connections using any port in the range 4502–4534.

Silverlight supports *stream sockets*, which are sockets that use TCP communication. TCP is a connection-oriented protocol that has built-in flow control, error correction, and sequencing. Thanks to these features, you don't need to worry about resolving any one of the numerous possible network problems that can occur as information is segmented into packets and then transported and reassembled in its proper sequence at another computer. Instead, you can write data to a stream on one side of the connection and read it from the stream on the other side.

To create a TCP connection, your application must perform a three-stage handshaking process:

1. The server enters listening mode by performing a passive open. At this point, the server is idle, waiting for an incoming request.

2. A client uses the IP address and port number to perform an active open. The server responds with an acknowledgment message in a predetermined format that incorporates the client sequence number.

3. The client responds to the acknowledgment. The connection is now ready to transmit data in either direction.

In the following sections, you'll use Silverlight to build a socket client and .NET to build a socket server. The result is a simple chat application that allows multiple users to log in at the same time and send messages back and forth. Figure 20-8 shows two of these instances of the client engaged in conversation.

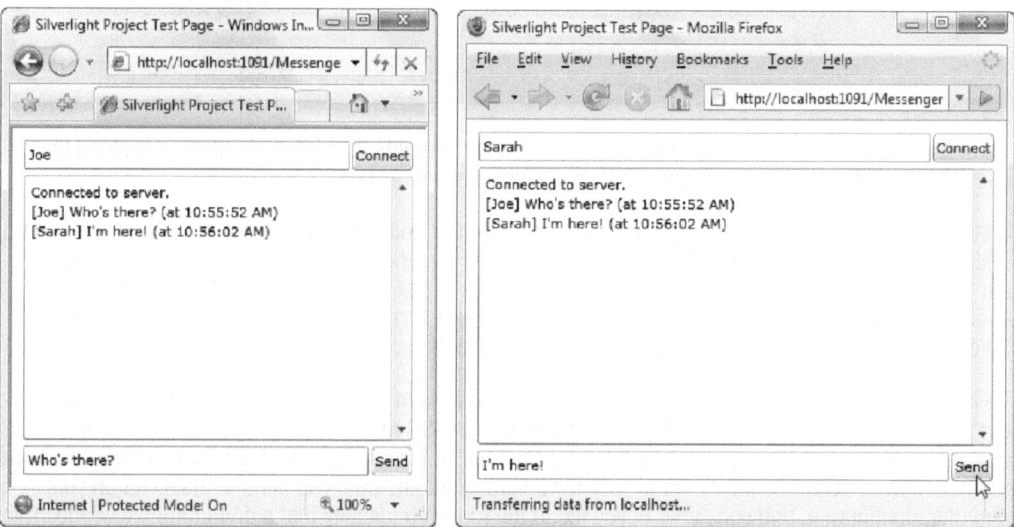

Figure 20-8. *A socket-based chat client*

Implementing this solution takes a fair bit of networking code. However, the result is well worth it and takes you far beyond the bounds of ordinary HTML pages.

The Policy Server

Before you can even think about designing a socket server, you need to develop something else: a policy server that tells Silverlight what clients are allowed to connect to your socket server.

As you saw earlier in this chapter, Silverlight doesn't let you download content or call a web service if the domain doesn't have a clientaccesspolicy.xml or crossdomain.xml file that explicitly allows it. A similar restriction applies to your socket server. Unless it provides a way for the client to download a clientaccesspolicy.xml file that allows remote access, Silverlight will refuse to make a connection.

Unfortunately, providing the clientaccesspolicy.xml file for a socket-based application takes more work than providing it with a website. With a website, the web server software can hand out the clientaccesspolicy.xml file for you, as long as you remember to include it. With a socket-based application, you must open a socket that clients can call with their policy requests, and you need to manually write the code that serves it. To perform these functions, you must create a policy server.

As you'll see, the policy server works in much the same way as the messaging server—it just has a simpler range of interactions. Although you can create the policy server and messaging server separately, you can also combine them both in one application, as long as they're listening for requests on different threads. In this example, you'll begin with a simple policy server and then enhance it to be a messaging server.

To create a policy server, you begin by creating a .NET application. Although you can use any type of .NET application to serve as a policy server, a simple command-line Console application is the most common choice. (After you've perfected your server, you might choose to move the code to a Windows service so it can run quietly in the background at all times.)

SERVING THE POLICY FILE OVER HTTP

One of the enhancements that Silverlight 4 introduces is the ability for socket clients to download the policy file via HTTP, rather than request it through a socket connection.

To take advantage of this feature, you first need to copy your policy file (as shown in the next section) to an IIS web server. However, there's a catch—IIS must be running on the same computer that's hosting your socket server. For example, if you're connecting to a server with the address 102.168.212.226, Silverlight will attempt to retrieve the policy file from http://102.168.212.226/clientaccesspolicy.xml. (Silverlight always uses port 80 to download the policy file, because this is the standard for HTTP traffic.)

The HTTP download feature works well for a deployed Silverlight application, because you'll usually place your Silverlight content on an IIS web server, often with an ASP.NET website. However, it's not as convenient when testing an application, because the built-in Visual Studio test web server doesn't use port 80. Thus, it can't serve the policy file to your Silverlight client. Instead, you need to install IIS on your computer, configure it, and copy your policy file to the root web (usually that's c:\inetpub\wwwroot). This setup task is outside the scope of this book, and for that reason the following sections assume you're using the policy server approach, which works in a deployed application and when testing in Visual Studio.

Finally, it's important to note that if you do have a deployed Silverlight application on an IIS web server with a policy file, you still need to explicitly indicate that you want to use the HTTP download feature. You'll learn how to do this "Connecting to the Server" section, later in this chapter. Unfortunately, you need to choose one approach or the other—if you tell Silverlight to use the HTTP download feature, it won't attempt to contact a policy server if the download fails.

The Policy File

Here's the policy file that the policy server provides:

```xml
<?xml version="1.0" encoding="utf-8" ?>
<access-policy>
  <cross-domain-access>
    <policy>
      <allow-from>
        <domain uri="*"/>
      </allow-from>
      <grant-to>
        <socket-resource port="4502-4534" protocol="tcp"/>
      </grant-to>
    </policy>
  </cross-domain-access>
</access-policy>
```

This policy file establishes three rules:

- It allows access on all ports from 4502 to 4534, which is the full range supported by Silverlight. To change this detail, modify the port attribute in the <socket-resource> element.
- It allows TCP access through the protocol attribute in the <socket-resource> element.
- It allows callers from any domain. In other words, the Silverlight application that's making the connection can be hosted on any website. To change this detail, modify the uri attribute in the <domain> element.

To make life easy, this policy is placed in a file named clientaccesspolicy.xml, and added to the policy-server project. In Visual Studio, the file's Copy to Output Directory setting is set to "Copy always." That way, the policy file is copied to the same folder as the policy server application, and the policy server simply needs to find the file, open it, and return its contents to the client.

The PolicyServer Classes

The policy server's functionality resides in two key classes. The first class, PolicyServer, is responsible for waiting and listening for connections. When a connection is received, it's handed off to a new instance of the second class, PolicyConnection, which then sends the policy file. This two-part design is common in network programming, and you'll see it again with the messaging server.

When the PolicyServer class is created, it loads the policy file from the hard drive and stores it in a field as an array of bytes:

```vbnet
Public Class PolicyServer
    Private policy As Byte()

    Public Sub New(ByVal policyFile As String)
        ' Load the policy file.
        Dim policyStream As New FileStream(policyFile, FileMode.Open)
        policy = New Byte(policyStream.Length - 1){}
        policyStream.Read(policy, 0, policy.Length)
```

```
        policyStream.Close()
    End Sub
    ...
```

To start listening, the server application must call PolicyServer.Start(). This creates a
TcpListener object, which waits for requests. The TcpListener is configured to listen on port
943, because Silverlight reserves this port for policy servers. (As you'll see, when Silverlight
applications make policy files requests, they automatically send them to this port.)

```
    ...
    Private listener As TcpListener

    Public Sub Start()
        ' Create the listener.
        listener = New TcpListener(IPAddress.Any, 943)

        ' Begin listening. This method returns immediately.
        listener.Start()

        ' Wait for a connection. This method returns immediately.
        ' The waiting happens on a separate thread.
        listener.BeginAcceptTcpClient(OnAcceptTcpClient, Nothing)
    End Sub
    ...
```

To accept any pending connections, the policy server calls BeginAcceptTcpClient(). Like all
the Begin*Xxx*() methods in .NET, this method returns immediately and starts the real work on a
separate thread. This is an important detail for a networking application, because it allows you
to handle multiple policy file requests at the same time.

■ **Note** Newcomers to network programming often wonder how they can handle more than one
simultaneous request, and they sometimes assume that multiple server ports are required. This isn't the
case—if it were, a small set of applications could quickly exhaust the available ports. Instead, server
applications handle multiple requests with the same port. This process is almost completely transparent
because the underlying TCP architecture in Windows automatically identifies messages and routes them
to the appropriate object in your code. Connections are uniquely identified based on four pieces of
information: the IP address and server port, and the IP address and client port.

Each time a request is made, the OnAcceptTcpClient() callback is triggered. That callback
then calls BeginAcceptTcpClient() again to start waiting for the next request on *another* thread,
and then gets to the real work of dealing with the current request:

```
    ...
    Public Sub OnAcceptTcpClient(ByVal ar As IAsyncResult)
        If isStopped Then Return
        Console.WriteLine("Received policy request.")

        ' Wait for the next connection.
```

```
        listener.BeginAcceptTcpClient(AddressOf OnAcceptTcpClient, Nothing)

        ' Handle this connection.
        Try
            Dim client As TcpClient = listener.EndAcceptTcpClient(ar)
            Dim policyConnection As New PolicyConnection(client, policy)
            policyConnection.HandleRequest()
        Catch err As Exception
            Console.WriteLine(err.Message)
        End Try
    End Sub
    ...
```

Each time a new connection is received, a new PolicyConnection object is created to deal with it. The task of serving the policy file is handled by the PolicyConnection class, which you'll consider in the next section.

The final ingredient in the PolicyServer class is a Stop() method that stops waiting for requests. The application can call this if it's shutting down:

```
    ...
    Private isStopped As Boolean
    Public Sub [Stop]()
        isStopped = True
        Try
            listener.Stop()
        Catch err As Exception
            Console.WriteLine(err.Message)
        End Try
    End Sub

End Class
```

To start the policy server, the Main() method of the server application uses the following code, which is placed in a file named Module1.vb:

```
Shared Sub Main(ByVal args As String())
    Dim policyServer As New PolicyServer("clientaccesspolicy.xml")
    policyServer.Start()
    Console.WriteLine("Policy server started.")

    Console.WriteLine("Press Enter to exit.")
    ' Wait for an Enter key. You could also wait for a specific input
    ' string (like "quit") or a single key using Console.ReadKey().
    Console.ReadLine()

    policyServer.Stop()
    Console.WriteLine("Policy server shut down.")
End Sub
```

The PolicyConnection Classes

The PolicyConnection class has a simple task. When created, it stores a reference to the policy file data. Then, when the HandleRequest() method is called, the code accesses the network stream for the new connection and attempts to read from it. If all is well, the client will have sent a string that contains the text "<policy-file-request/>". After reading that string, the client writes the policy data to that stream and closes the connection.

Here's the complete code:

```
Public Class PolicyConnection
    Private client As TcpClient
    Private policy As Byte()

    Public Sub New(ByVal client As TcpClient, ByVal policy As Byte())
        Me.client = client
        Me.policy = policy
    End Sub

    ' The request that the client sends.
    Private Shared policyRequestString As String = "<policy-file-request/>"

    Public Sub HandleRequest()
        Dim s As Stream = client.GetStream()

        ' Read the policy request string.
        ' This code doesn't actually check the content of the request string.
        ' Instead, it returns the policy for every request.
        Dim buffer As Byte() = New Byte(policyRequestString.Length - 1){}

        ' Only wait 5 seconds. That way, if you attempt to read the request string
        ' and it isn't there or it's incomplete, the client only waits for 5
        ' seconds before timing out.
        client.ReceiveTimeout = 5000
        s.Read(buffer, 0, buffer.Length)

        ' Send the policy.
        s.Write(policy, 0, policy.Length)

        ' Close the connection.
        client.Close()

        Console.WriteLine("Served policy file.")
    End Sub
End Class
```

You now have a complete, fully functioning policy server. Unfortunately, you can't test it yet. That's because Silverlight doesn't allow you to explicitly request policy files. Instead, it automatically requests them when you attempt to use a socket-based application. And before you build a client for that socket-based application, you need to build the server.

The Messaging Server

Although you can create the messaging server as a separate application, it's tidier to place it in the same application as the policy server. Because the policy server does its listening and request-handling work on separate threads, the messaging server can do its work at the same time.

Like the policy server, the messaging server is broken into two classes: MessengerServer, which listens for requests and tracks clients, and MessengerConnection, which handles the interaction of a single client. To see the full code, refer to the downloadable examples for this chapter. In this section, you'll explore the differences between the policy server and messaging server.

First, the messaging server performs its listening on a different port. As described earlier, Silverlight allows socket-based applications to use any port in a limited band from 4502 to 4534. The messaging server uses port 4530:

```
listener = New TcpListener(IPAddress.Any, 4530)
```

When the messaging server receives a connection request, it performs an extra step. As with the policy server, it creates an instance of a new class (in this case, MessengerConnection) to handle the communication. Additionally, it adds the client to a collection so it can keep track of all the currently connected users. This is the only way you can allow interaction between these clients—for example, allowing messages to be sent from one user to another. Here's the collection that performs the tracking, and a field that helps the server give each new client a different identifying number:

```
Private clientNum As Integer
Private clients As New List(Of MessengerConnection)
```

When the client connects, this code creates the MessengerConnection and adds the client to the clients collection:

```
clientNum += 1
Console.WriteLine("Messenger client #" & clientNum.ToString() & " connected.")

' Create a new object to handle this connection.
Dim clientHandler As New MessengerConnection(client, "Client " & _
  clientNum.ToString(), Me)
clientHandler.Start()

SyncLock clients
    clients.Add(clientHandler)
End SyncLock
```

Because the possibility exists that several clients will be connected at the same time, this code locks the clients collection before adding the client. Otherwise, subtle threading errors could occur when two threads in the messaging server attempt to access the clients collection simultaneously.

When the messaging server is stopped, it steps through this complete collection and makes sure every client is disconnected:

```
For Each client As MessengerConnection In clients
    client.Close()
Next
```

You've now seen how the basic framework for the messaging server is designed. However, it still lacks the message-delivery feature—the ability for one client to submit a message that's then delivered to all clients.

To implement this feature, you need two ingredients. First, you must handle the message submission in the MessengerConnection class. Then, you need to handle the message delivery in the MessengerServer class.

When a MessengerConnection object is created and has its Start() method called, it begins listening for any data:

```
Public Sub Start()
    Try
        ' Listen for messages.
        client.Client.BeginReceive(message, 0, message.Length, SocketFlags.None, _
          New AsyncCallback(AddressOf OnDataReceived), Nothing)
    Catch se As SocketException
        Console.WriteLine(se.Message)
    End Try
End Sub
```

The OnDataReceived() callback is triggered when the client sends some data. It reads one byte at a time until it has all the information the client has sent. It then passes the data along to the MessengerServer.Deliver() method and begins listening for the next message:

```
Public Sub OnDataReceived(ByVal asyn As IAsyncResult)
    Try
        Dim bytesRead As Integer = client.Client.EndReceive(asyn)

        If bytesRead > 0 Then
            ' Ask the server to send the message to all the clients.
            server.DeliverMessage(message, bytesRead)

            ' Listen for more messages.
            client.Client.BeginReceive(message, 0, message.Length, _
              SocketFlags.None, New AsyncCallback(AddressOf OnDataReceived), _
              Nothing)
        End If
    Catch err As Exception
        Console.WriteLine(err.Message)
    End Try
End Sub
```

■ **Note** When a message is received, the messenger assumes that message is made up of text that needs to be delivered to other recipients. A more sophisticated application would allow more complex messages. For example, you might serialize and send a Message object that indicates the message text, sender, and intended recipient. Or, you might use a library of string constants that identify different commands—for example, for sending messages, sending files, querying for a list of currently connected users, logging off, and so on. The design of your messaging application would be the same, but you would need much more code to analyze the message and decide what action to take.

The MessengerServer.DeliverMessage() method walks through the collection of clients and calls each one's ReceiveMessage() method to pass the communication along. Once again, threading issues are a concern, and an error could occur if the messenger is adding a new client to the clients collection while another thread is iterating over the clients collection to perform a message delivery. But locking the entire collection isn't ideal, because the delivery process can take some time, particularly if a client isn't responding. To avoid any slowdowns, the DeliverMessage() code begins by creating a snapshot copy of the collection. It then uses that copy to deliver its message:

```
Public Sub DeliverMessage(ByVal message As Byte(), ByVal bytesRead As Integer)
    Console.WriteLine("Delivering message.")

    ' Duplicate the collection to prevent threading issues.
    Dim connectedClients As MessengerConnection()
    SyncLock clients
        connectedClients = clients.ToArray()
    End SyncLock

    For Each client As MessengerConnection In connectedClients
        Try
            client.ReceiveMessage(message, bytesRead)
        Catch
            ' Client is disconnected.
            ' Remove the client to avoid future attempts.
            SyncLock clients
                clients.Remove(client)
            End SyncLock

            client.Close()
        End Try
    Next
End Sub
```

The MessengerConnection.ReceiveMessage() method writes the message data back into the network stream so the client can receive it:

```
Public Sub ReceiveMessage(ByVal data As Byte(), ByVal bytesRead As Integer)
    client.GetStream().Write(data, 0, bytesRead)
End Sub
```

The final change you need to make is to modify the startup code so the application creates and starts both the policy server and the messaging server. Here's the code, with additions in bold:

```
Shared Sub Main(ByVal args As String())
    Dim policyServer As New PolicyServer("clientaccesspolicy.xml")
    policyServer.Start()
    Console.WriteLine("Policy server started.")

    Dim messengerServer As New MessengerServer()
    messengerServer.Start()
    Console.WriteLine("Messenger server started.")
    Console.WriteLine("Press Enter to exit.")

    ' Wait for an Enter key. You could also wait for a specific input
    ' string (like "quit") or a single key using Console.ReadKey().
    Console.ReadLine()

    policyServer.Stop()
    Console.WriteLine("Policy server shut down.")

    messengerServer.Stop()
    Console.WriteLine("Messenger server shut down.")
End Sub
```

Figure 20-8 showed what happens when two clients begin talking to each other through the socket server. Figure 20-9 shows the back end of the same process—the messages that appear in the Console window of the socket server while the clients are connecting and then interacting.

Figure 20-9. *The policy and messaging server*

The Messenger Client

So far, you've focused exclusively on the server-side .NET application that powers the messaging server. Although this is the most complex piece of the puzzle, the Silverlight socket client also requires its fair share of code.

The messaging client has three basic tasks: to connect to the server, to send messages, and to receive and display them. The code is similar to the socket server, but it requires slightly more work. That's because Silverlight doesn't have a TcpClient class but forces you to use the lower-level Socket class instead.

To use the Socket class, you use three asynchronous methods: ConnectAsync() to make a connection, SendAsync() to send an outgoing message, and ReceiveAsync() to listen for an incoming. All three of these methods require a SocketAsyncEventArgs object.

The SocketAsyncEventArgs plays two crucial roles:

- It acts as a package that holds any additional data you want to transmit.
- It notifies you when the asynchronous operation is complete with the Completed event.

To perform any task with a socket in Silverlight, you must create and configure a SocketAsyncEventArgs object, and then pass it to one of the asynchronous methods in the Socket class.

Connecting to the Server

The first task in the messaging client is to establish a connection when the user clicks the Connect button. To do so, the client needs to create a new Socket object and a new SocketAsyncEventArgs object. Here's what happens:

```
' The socket for the underlying connection.
Private socket As Socket

Private Sub cmdConnect_Click(ByVal sender As Object, ByVal e As RoutedEventArgs)
    Try
```

```vb
        If (socket IsNot Nothing) AndAlso (socket.Connected) Then
            socket.Close()
        End If
    Catch err As Exception
        AddMessage("ERROR: " & err.Message)
    End Try

    Dim endPoint As New DnsEndPoint( _
      Application.Current.Host.Source.DnsSafeHost, 4530)
    socket = New Socket(AddressFamily.InterNetwork, _
      SocketType.Stream, ProtocolType.Tcp)

    Dim args As New SocketAsyncEventArgs()

    ' To configure a SocketAsyncEventArgs object, you need to set
    ' the corresponding Socket object (in the UserToken property)
    ' and the location of the remote server (in the RemoteEndPoint property).
    args.UserToken = socket
    args.RemoteEndPoint = endPoint

    ' You must also attach an event handler for the Completed event.
    AddHandler args.Completed, AddressOf OnSocketConnectCompleted

    ' Start the asynchronous connect process.
    socket.ConnectAsync(args)
End Sub
```

Most of these details are straightforward. If the socket is already opened, it's closed. Then, a DnsEndPoint object is created to identify the location of the remote host. In this case, the location of the remove host is the web server that hosts the Silverlight page, and the port number is 4530. Finally, the code creates the SocketAsyncEventArgs object and attaches the OnSocketConnectCompleted() event to the Completed event.

■ **Note** Remember, unless you specify otherwise, the client's port is chosen dynamically from the set of available ports when the connection is created. That means you can have multiple clients open connections to the same server. On the server side, each connection is dealt with uniquely, because each connection has a different client port number.

If you want to download the policy file from an IIS web server over HTTP (rather than a dedicated policy server), you need to make one change to the previous code. Before you call ConnectAsync(), you must set the SocketAsyncEventArgs. SocketClientAccessPolicyProtocol property, as shown here:

```
args.SocketClientAccessPolicyProtocol = SocketClientAccessPolicyProtocol.Http
```

But be aware that when you use this approach, Silverlight will not attempt to contact a policy server if it fails to download the policy file (or if the policy file is incorrect). Instead, your application will be unable to connect to the server.

The code in the cmdConnect_Click() event handler uses a custom method named AddMessage() to add information to the message list. This method takes the extra step of making sure it's running on the user-interface thread. This is important, because AddMessage() may be called during one of the client's asynchronous operations:

```
Private Sub AddMessage(ByVal message As String)
    If Me.CheckAccess() Then
        ' This is the right thread. Go ahead and make the update.
        lblMessages.Text &= message & Environment.NewLine

        ' Scroll down to the bottom of the list, so the new message is visible.
        scrollViewer.ScrollToVerticalOffset(scrollViewer.ScrollableHeight)
    Else
        ' Call this method on the right thread.
        Dispatcher.BeginInvoke( _
          New Action(Of String)(AddressOf AddMessage), _
          message)
    End If
End Sub
```

When the client's connection attempt finishes, the OnSocketConnectCompleted() event handler runs. It updates the display and reconfigures the SocketAsyncEventArgs object so it can be used to receive messages, wiring the Completed event to a new event handler. The client then begins listening for messages:

```
Private Sub OnSocketConnectCompleted(ByVal sender As Object, _
  ByVal e As SocketAsyncEventArgs)
    If Not socket.Connected Then
        AddMessage("Connection failed.")
        Return
    End If

    AddMessage("Connected to server.")

    ' Messages can be a maximum of 1024 bytes.
    Dim response(1023) As Byte
    e.SetBuffer(response, 0, response.Length)
    RemoveHandler e.Completed, AddressOf OnSocketConnectCompleted
    AddHandler e.Completed, AddressOf OnSocketReceive

    ' Listen for messages.
    socket.ReceiveAsync(e)
End Sub
```

To listen for a message, you must create a buffer that will hold the received data (or at least a single chunk of that data). The messaging client creates a 1024-byte buffer and doesn't attempt to read more than one chunk. It assumes that messages won't be greater than 1024 bytes. To prevent potential errors, the messaging application should enforce this restriction as well. One good safety measure is to set a MaxLength property of the text box where the user enters new messages.

Sending Messages

The messages in the chat application are slightly more detailed than simple strings. Each message includes three details: the text, the sender's chosen name, and the sender's time when the message was submitted. These three details are encapsulated in a custom Message class:

```
Public Class Message
    Private _messageText As String
    Public Property MessageText() As String

        ...
    End Property

    Private _sender As String
    Public Property Sender() As String

        ...
    End Property

    Private _sendTime As DateTime
    Public Property SendTime() As DateTime

        ...
    End Property

    Public Sub New(ByVal messageText As String, ByVal sender As String)
        Me.MessageText = messageText
        Me.Sender = sender
        SendTime = DateTime.Now
    End Sub

    ' A no-argument constructor allows instances of this class to be serialized.
    Public Sub New()
    End Sub
End Class
```

To send a message, the user enters some text and clicks the Send button. At this point, you must create a new SocketAsyncEventArgs object. (Remember, the first one is still in use, waiting to receive new messages on a background thread.) The new SocketAsyncEventArgs object needs to store the buffer of message data. To create it, you begin by constructing a Message object. You then serialize that message object to a stream with the XmlSerializer, convert it to a simple byte array, and add it to the SocketAsyncEventArgs object using the BufferList property, as shown here:

```
Private Sub cmdSend_Click(ByVal sender As Object, ByVal e As RoutedEventArgs)
    If (socket Is Nothing) OrElse (socket.Connected = False) Then
        AddMessage("ERROR: Not connected.")
        Return
    End If

    ' Create the MemoryStream where the serialized data will be placed.
    Dim ms As New MemoryStream()

    ' Use the XmlSerializer to serialize the data.
    Dim serializer As New XmlSerializer(GetType(Message))
    serializer.Serialize(ms, New Message(txtMessage.Text, txtName.Text))

    ' Convert the serialized data in the MemoryStream to a byte array.
```

```
        Dim messageData As Byte() = ms.ToArray()

        ' Place the byte array in the SocketAsyncEventArgs object,
        ' so it can be sent to the server.
        Dim args As New SocketAsyncEventArgs()
        Dim bufferList As New List(Of ArraySegment(Of Byte))()
        bufferList.Add(New ArraySegment(Of Byte)(messageData))
        args.BufferList = bufferList

        ' Send the message.
        socket.SendAsync(args)
End Sub
```

Unfortunately, because the Socket class in Silverlight works at a lower level than the TcpClient class in .NET, you don't have the straightforward stream-based access to the network connection that you have on the server side.

■ **Tip** You can write any type of data you want to the server, in any form. You certainly don't need to use XmlSerializer. However, serialization gives you a simple way to pass along a bundle of information as an instance of some class.

Receiving Messages

When a message is sent to the client, the other SocketAsyncEventArgs object fires its Completed event, which triggers the OnSocketReceive() event handler. You now need to deserialize the message, display it, and wait for the next one:

```
Private Sub OnSocketReceive(ByVal sender As Object, ByVal e As SocketAsyncEventArgs)
    If e.BytesTransferred = 0 Then
        AddMessage("Server disconnected.")
        Try
            socket.Close()
        Catch
        End Try
        Return
    End If

    Try
        ' Retrieve and display the message.
        Dim serializer As New XmlSerializer(GetType(Message))
        Dim ms As New MemoryStream()
        ms.Write(e.Buffer, 0, e.BytesTransferred)
        ms.Position = 0
        Dim message As Message = CType(serializer.Deserialize(ms), Message)

        AddMessage("[" & message.Sender & "] " & _
          message.MessageText & _
          " (at " & message.SendTime.ToLongTimeString() & ")")
    Catch err As Exception
        AddMessage("ERROR: " & err.Message)
```

```
    End Try

    ' Listen for more messages.
    socket.ReceiveAsync(e)
End Sub
```

This completes the messaging client. To experiment with the complete solution, try the downloadable code for this chapter.

■ **Note** You can make a number of refinements to polish the messaging application. You've already considered how you can replace the simple message-passing mechanism on the server side with more complex logic that recognizes different types of messages and performs various operations. Other changes you may want to implement include managing user-interface state (for example, disabling or enabling controls based on whether a connection is available), intercepting the application shutdown event and politely disconnecting from the server, allowing users to deliver to specific people, adding authentication, and informing newly connected clients about how many other people are currently online. With all that in mind, the messaging application is still an impressive first start that shows how far a Silverlight application can go with direct network communication.

Local Connections

As you've seen, Silverlight gives you considerable power to communicate with sockets, but the cost is complexity. Interestingly, Silverlight has another communication mechanism that's far simpler: the *local connection* model.

Whereas socket support allows a Silverlight application to communicate with virtually any networked program, on any computer (provided there's no firewall blocking the way), the local connection feature is far more modest. It provides a simple way for two Silverlight applications running on the same computer to interact. This feature is particularly suited to out-of-browser applications, because users are more likely to run them in combination and leave them loaded for long periods of time.

In the following sections, you'll take a quick look at the local connection model, and see how it allows you to build the application shown in Figure 20-10.

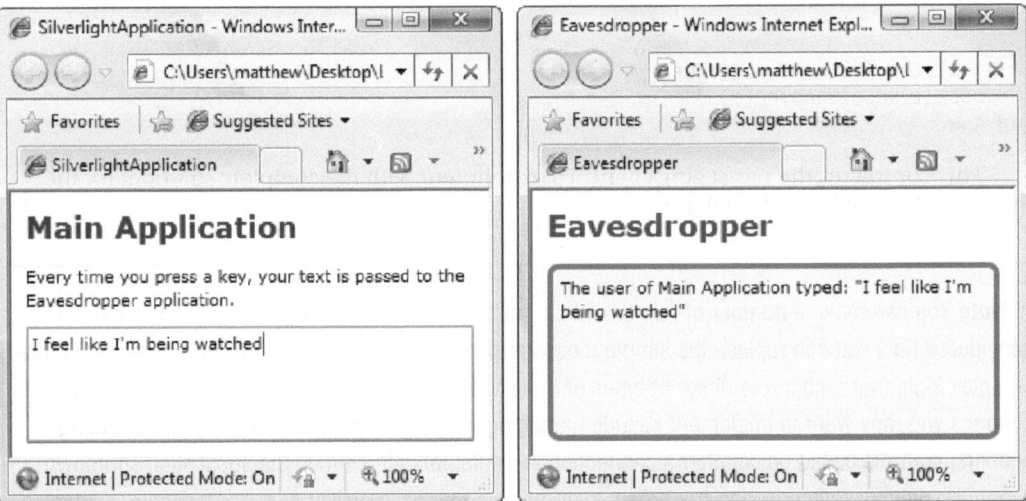

Figure 20-10. *Sending messages between local applications*

The local connection model works through two classes in the System.Windows.Messaging namespace: LocalMessageSender and LocalMessageReceiver. Together, these two classes allow one way communication. The *sender* application transmits a message to the *receiver*. If you want both applications to have the chance to send and receive messages, you simply need to use the LocalMessageSender and LocalMessageReceiver in combination.

It's also worth noting that communication doesn't need to be one-to-one. The message that one application sends can be received by any number of other Silverlight applications that are running at the same time. Or, it can be received by none. Unlike sockets, local connection messaging is a one-way, fire-and-forget system. The message sender has no idea whether its message has been received, unless the responder sends a confirmation message back.

Sending a Message

To send a message, you begin by creating an instance of the LocalMessageSender class. It's usually convenient to create a single instance, and keep it around as a member variable in your page:

```
Private messageSender As New LocalMessageSender("EavesdropperReceiver")
```

When you create the LocalMessageSender object, you need to supply a receiver name. This is the name that the receiver will use to listen for messages. The actual name isn't important, but the sender and receiver must use the same name to communicate.

Ordinarily, local connections only work with applications that are running from the same web domain. However, you have the option of specifying a different domain as a second constructor argument:

```
Private messageSender As New LocalMessageSender( _
  "EavesdropperReceiver", "anotherWebDomain.com")
```

Alternatively, you can use the syntax shown here to create a *global* message sender:

```
Private messageSender As New LocalMessageSender( _
   "EavesdropperReceiver", LocalMessageSender.Global)
```

Now, any application from any domain can receive the messages you send. If you use this approach, it's a good idea to take extra care to choose a receiver name that's likely to be unique. Don't use a common naming shorthand like "receiver", "MessageReceiver", and so on, as other Silverlight applications use the same names.

Once you've created the LocalMessageSender, sending a message is easy. You simply need to call SendAsync() method and pass in your message as a string. Optionally, you can handle the LocalMessageSender.Completed event, which fires when the message has been sent (but won't tell you if it's been received).

Here's the code used to send messages as the user types, as shown in the application in Figure 20-10:

```
Private Sub txt_KeyUp(ByVal sender As Object, ByVal e As KeyEventArgs)
   messageSender.SendAsync(txt.Text)
End Sub
```

Receiving a Message

As you probably expect, you receive messages by creating a LocalMessageReceiver object. When you do, you must specify the same receiver name that was used when creating the LocalMessageSender:

```
Private receiver As New LocalMessageReceiver("EavesdropperReceiver")
```

This LocalMessageReceiver will receive message from Silveright applications that are running from the same domain. Alternatively, you can pass in an array that specifies one or more domains that you want to allow. Here's an example that listens to messages sent to the Eavesdropper receiver, by applications running on anotherWebDomain.com:

```
Dim messageReceiver As New LocalMessageReceiver( _
   "Eavesdropper", ReceiverNameScope.Domain, New String(){"anotherWebDomain.com"})
```

Finally, you can choose to accept messages from all domains with this syntax:

```
Dim messageReceiver As New LocalMessageReceiver( _
   "Eavesdropper", ReceiverNameScope.Global, LocalMessageReceiver.AnyDomain)
```

Once you've created the LocalMessageReceiver, you need to attach an event handler to the MessageReceived event, and call the Listen() method to start listening.

```
Private Sub Page_Loaded(ByVal sender As Object, ByVal e As RoutedEventArgs)
   AddHandler receiver.MessageReceived, AddressOf receiver_MessageReceived
   receiver.Listen()
End Sub
```

Listen() is an asynchronous method, so all the message listening takes place on a separate thread while your application continues its normal operations. When a message is received, the listening thread fires the MessageReceived event on the user interface thread and resumes listening. Here's the code that's used in the example in Figure 20-10 to display the received message in a TextBlock:

```
Private Sub receiver_MessageReceived(ByVal sender As Object, _
  ByVal e As MessageReceivedEventArgs)
    lblDisplay.Text = "The user of Main Application typed: """ & e.Message & """"
End Sub
```

The Last Word

In this chapter, you saw a wide range of Silverlight networking features. You learned how to use them to do everything from directly downloading HTML files to calling simple XML-based web services to building an entire messaging system based on socket communication. Along the way, you considered several techniques for parsing different types of information, including regular expressions (to search HTML), LINQ to XML (to process XML), and serialization (to save or restore the contents of an in-memory object). These techniques can come in handy in a variety of situations—for example, they're just as useful when you need to manage information that's stored on the client computer in isolated storage.

CHAPTER 21

■■■

Out-of-Browser Applications

As you already know, the code for every Silverlight application is contained in a XAP file. The Silverlight browser plug-in downloads this file from the web server and executes it on the client. After this point, there's no requirement for the web server to get involved again—all the code runs on the local computer.

This design raises an interesting possibility. Although Silverlight applications depend on the Silverlight browser plug-in, there's no technical reason that they need to be embedded in a live web page. In fact, as long as there's a reliable way to run the Silverlight plug-in outside of a browser, it's also possible to run a Silverlight application on its own.

This is the essential idea behind Silverlight out-of-browser applications. With an out-of-browser application, a user visits your website to download and install your Silverlight application. After the installation process, the application can be launched in a stand-alone window, straight from the Start menu. Aside from being in a separate window, the out-of-browser application functions in almost exactly the same way as an ordinary Silverlight application that's hosted in a web page. You can even download and automatically apply new updates on the client's computer.

Although out-of-browser applications are still standard Silverlight applications, they open two new possibilities. First, because they are cached on the client's computer, they can run without a web connection. This is a useful tool for user with intermittent network connections—for example, traveling employees with laptops. Second, out-of-browser applications they can be installed with elevated trust, which gives them a number of new capabilities. These include customizing the application window, accessing the current user's documents, and running already-installed COM components.

■ **What's New** Silverlight 4 adds two new features to ordinary out-of-browser applications: the WebBrowser control (for displaying HTML content) and notification windows. Both are described in the "Out-of-Browser Application Features" section. Silverlight 4 also adds the option for *elevated trust* applications. If you use this approach, you gain a number of new features, such as custom window chrome, direct file access, and COM support, all of which are described in the "Elevated Trust" section.

Understanding Out-of-Browser Support

The first detail you should understand about Silverlight out-of-browser applications is that despite their name, they don't run without the browser. Instead, out-of-browser applications *conceal* the browser's presence. When you run one, a specialized Silverlight tool named sllauncher.exe (which you can find in a directory like c:\Program Files\Microsoft Silverlight\ on a Windows computer) creates a stand-alone window that hosts a browser control inside. This browser window is stripped down to little more than a bare frame and includes none of the standard browser user interface (which means no toolbar, favorites, navigation buttons, and so on).

■ **Note** Out-of-browser applications work on all supported Silverlight platforms, including Windows and Mac computers.

Figure 21-1 shows the EightBall application from Chapter 2, running as an out-of-browser application.

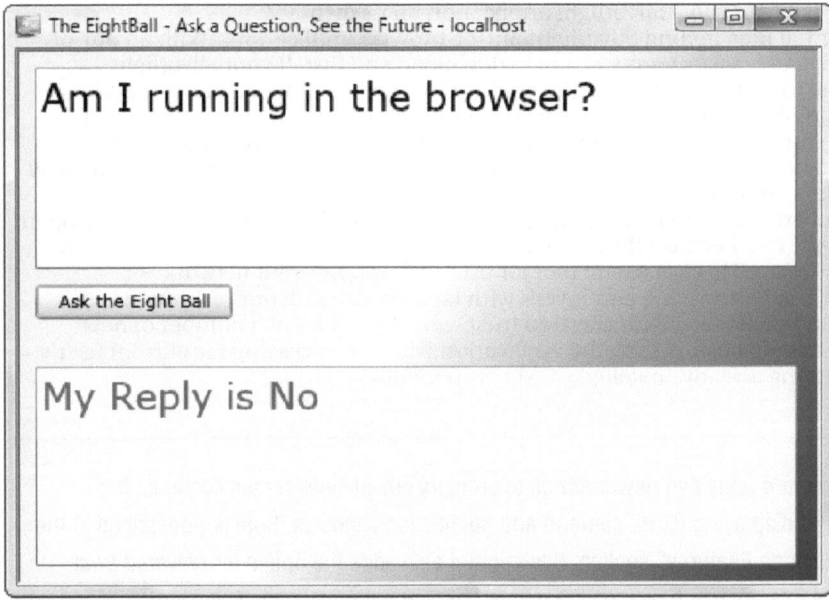

Figure 21-1. *The EightBall application outside of the browser*

Given that out-of-browser applications are really just slick illusions (and given your limited control over the stand-alone window, they're not even that slick), why would you use them? There are several good reasons:

- *To get a desktop presence*: An out-of-browser application must be "installed" through a lightweight process that downloads the XAP file (as usual) and adds desktop and Start menu shortcuts. If you want to give your users the ability to launch a Silverlight application this way, rather than forcing them to load a browser and navigate to the appropriate URL, an out-of-browser application makes sense.

- *To allow the application to run when offline*: Ordinarily, Silverlight applications are accessed through a web page, and that web page is on a public or private network. As a result, clients can't run the application unless they have a connection. But after going through the install process for an out-of-browser application, a copy is cached locally and permanently (unless the user explicitly removes the application).

- *To support intermittent connectivity*: This is similar to the previous point but represents an even more common scenario. Many clients—particularly those who use laptops and access the Internet through a wireless connection—have periods of connectivity interrupted with periodic disconnections. Using an out-of-browser application (and the network detection features described in Chapter 15), you can create an application that deals with both scenarios. When it's connected, it can call web services to retrieve updated data and perform server actions. When disconnected, it remains self-sufficient and allows the user to keep working.

- *To gain the features of elevated trust*: Elevated trust applications are a specific type of out-of-browser application. In other words, if you want to take advantage of the elevated trust features described later in this chapter, you must create an out-of-browser application—a normal Silverlight application will not work.

■ **Note** Although out-of-browser applications have the same feature set as in-browser applications, there is one difference: their initial allotment of file space. As explained in Chapter 18, every Silverlight application gets its own carefully walled-off area of disk space where it can create and store files. Ordinary in-browser applications get a mere 1MB of disk space (although they can request more by prompting the user). But out-of-browser applications start with a significantly increased quota of 25MB, which means that in many cases they won't need to ask the user for additional space. To learn more about isolated storage, refer to Chapter 18.

Creating an Out-of-Browser Application

To run a Silverlight application outside of the browser window, the user must first install it locally. But before that's possible, you must specifically allow this feature. To do so, follow these steps:

1. Double-click the My Project item in the Solution Explorer to show the application configuration options.
2. Click the Silverlight tab.
3. Select the "Enable running application out of the browser" setting.

4. Optionally, click the Out-of-Browser Settings button to show a window where you can set additional options (see Figure 21-2).

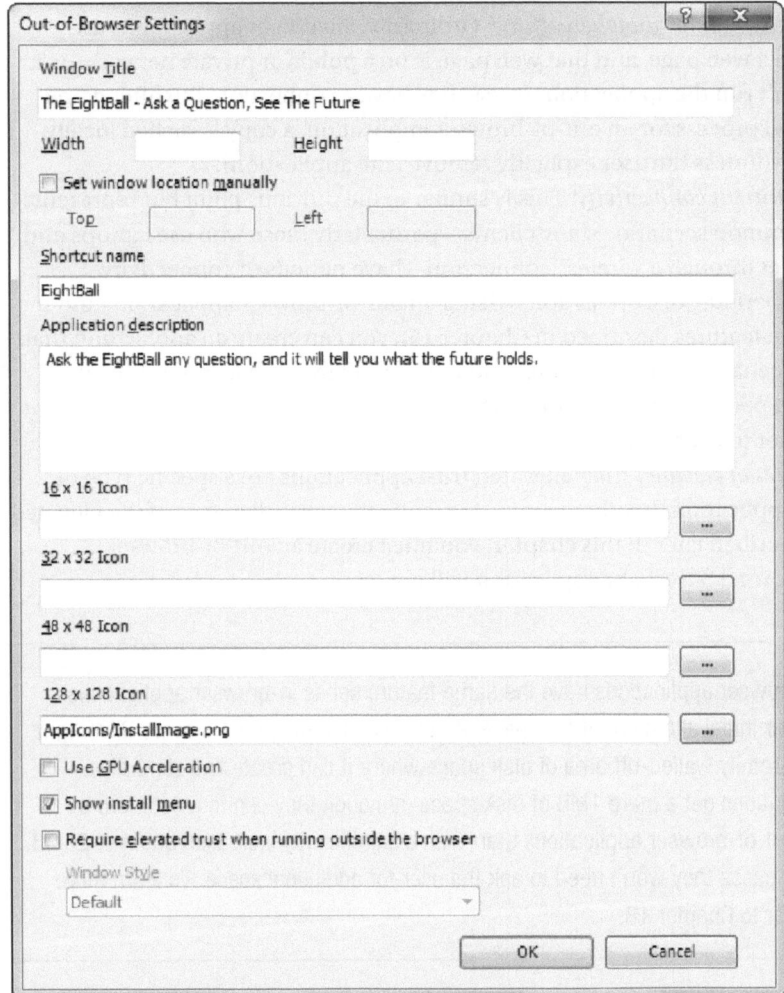

Figure 21-2. Configuring out-of-browser settings

The additional options that you can set in the Out-of-Browser Settings window include the following:

- *Window Title*: Name used in the title bar of the window when the application is running outside of the browser.
- *Width and Height*: The initial size of the window for the out-of-browser application. The user can resize the window after the application launches. If you don't supply width and height values, Silverlight creates a window that's 800 pixels wide and 600 pixels tall.

- *Set window location manually*: If checked, allows you to fill in the coordinates (in pixels) for the top-left corner of the window. Usually, you'll leave this unset and let Silverlight place the window in the center of the screen.

- *Shortcut name*: Name used in the installation window and in any shortcuts that are created.

- *Application description*: Descriptive text that describes the application. It appears in a tooltip when you hover over the application shortcut.

- *Icon settings*: Allow you to customize the images used for the installation window and the shortcuts, as described a bit later in the "Customizing Icons" section.

- *Enable GPU Acceleration*: Determines whether the out-of-browser will support the video-card caching that boosts performance in some scenarios. Selecting this check box simply gives you the option to use hardware acceleration—it's still up to your elements to opt in when appropriate, as described in Chapter 10.

- *Show install menu*: If checked, shows an option for installing your application when the user right-clicks the Silverlight content region. If you don't use this option, it's up to you to start the installation process programmatically. Both approaches are described in the next section.

- *Require elevated trust*: This configures your application to be an elevated trust application, which grants it additional privileges (but requires that the user accept a security warning when first installing the application). For more information, see the "Elevated Trust" section later in this chapter.

The values you enter are placed in a file named OutOfBrowserSettings.xml and saved with your project.

Once you've performed these steps, your application gains the ability to be installed locally and then launched outside of the browser. However, it doesn't need to be installed—the user can continue running it as a standard Silverlight application in the browser.

Installing an Out-of-Browser Application

There are two ways to install an application that has out-of-browser capability. The first option is for the user to perform this step explicitly. To do so, the user must request the entry page for the application, right-click the Silverlight content region, and choose "Install [*ApplicationShortName*] on this computer," as shown in Figure 21-3.

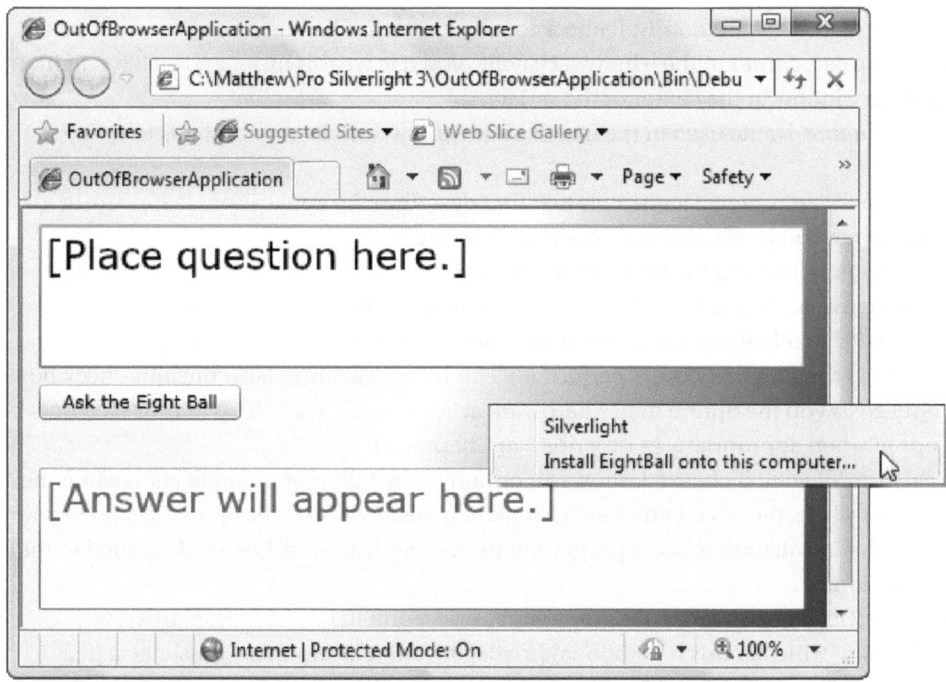

Figure 21-3. *Explicitly installing an out-of-browser application*

The other option is to start the installation process programmatically by calling the Application.Install() method. You must call this method in an event handler that responds to a user-initiated action (like clicking a button). The Install() method returns True if the user accepts the install prompt and continues or False if the user declines to install the application.

The Install() method has one potential problem: it throws an exception if the application is already installed on the local computer, even if the user is currently running the application in the browser. To avoid this error, you should check the Application.InstallState value before calling Install(). Here's the complete process, launched in response to a button click:

```
Private Sub cmdInstall_Click(ByVal sender As Object, ByVal e As RoutedEventArgs)
    ' Make sure that the application is not already installed.
    If Application.Current.InstallState <> InstallState.Installed Then
        ' Attempt to install it.
        Dim installAccepted As Boolean = Application.Current.Install()

        If Not installAccepted Then
            lblMessage.Text = "You declined the install. " & _
              "Click Install to try again."
        Else
            cmdInstall.IsEnabled = False
            lblMessage.Text = "The application is installing... "
        End If
    End If
End Sub
```

■ **Tip** Optionally, you can remove the install option from the Silverlight menu and force a user to install an application through your code. To do so, simply clear the "Show install menu" check box in the Out-of-Browser Settings window. The only reason that you would take this approach is if you need to perform some other tasks before your installation, such as collecting user information or calling a web service.

When an application is installed, either through the user's choice or the Install() method, several things happen. First, an installation window appears (see Figure 21-4) that asks the user for confirmation.

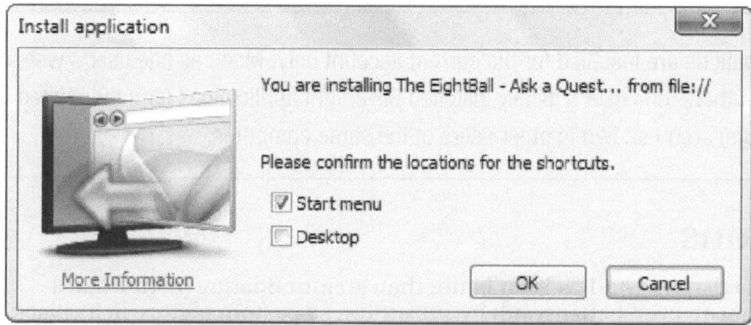

Figure 21-4. *Installing the EightBall application*

The installation window performs several services:

- It provides the name of the application and indicates the domain where it's hosted (or shows file:// for an application that's executed locally, without an ASP.NET test website).
- It provides a More Information link, which, if clicked, launches another browser window and navigates to a page on the Silverlight website that describes the out-of-browser feature.
- It allows the user to choose whether the installation should create a Start menu shortcut (which is checked by default) and a desktop shortcut (which isn't). If the user clears both check boxes, the OK button becomes disabled, and the install can't continue. It's also worth noting that the Start menu shortcut appears in the first level of the Start menu (not in a subgroup).

■ **Note** The install window looks slightly different on a Mac, to follow the conventions of that platform. For example, it doesn't include check boxes for creating shortcuts. Instead, Mac users are expected to drag the installed application bundle into the location of their choice after the install completes, much as they would do when installing any other application.

The most important feature of the installation window is that it explains what's about to happen in clear, nonthreatening terms. It doesn't require an administrator account or privilege escalation, and it isn't followed or preceded by any additional security warnings. Compare this to the prompts you get when installing a standard .NET application through ClickOnce, and you'll see that the experience is much friendlier with a Silverlight out-of-browser application. As a result, the user is much more likely to go through with the operation and install the application, rather than be scared off by obscurely worded warnings.

The installation process places the application in a randomly generated folder in the current user's profile. When the installation process finishes, it launches the newly installed application in a stand-alone window. But the existing browser window remains open, which means there are now two instances of the application running. You can deal with this situation by handling the InstallStateChanged event, as described in the "Tracking Application State" section.

■ **Note** Out-of-browser applications are installed for the current account only. Much as one user's web bookmarks aren't available to others, one user's locally installed Silverlight applications (and the related desktop or Start menu shortcuts) aren't shown to other users of the same computer.

Customizing Icons

The default image in the installation window is far better than an intimidating picture like a warning icon. But you can get an even better result by substituting a custom picture in its place. Silverlight lets you supply small image files to customize several details:

- Use a 16×16 image for the application icon in the title bar and in the Start menu.
- Use a 32×32 image for the application icon on the desktop (and in other large icon views).
- Use a 48×48 image for the application icon in tile mode.
- Use a 128×128 image for the installation window.

To customize these details, begin by adding the images to your project. Each image file must be a PNG file with the correct dimensions. For better organization, put the images in a project subfolder, like AppIcons. Then, select each one in the Solution Explorer, and set Build Action to Content (not Resource) so the images are packaged in the XAP as separate files. Finally, return to the Out-of-Browser Settings window shown in Figure 21-2 to identify the icons you want to use.

Figure 21-5 shows the installation window with a customized image.

Figure 21-5. Customizing the EightBall installation

■ **Note** The image is the only detail you can change in the installation window. You can't alter the options it provides or the descriptive text it uses.

Tracking Application State

As you've seen, out-of-browser applications are ordinary Silverlight applications that have a bit of additional information in the application manifest. This gives users the *option* of installing them locally, but it doesn't prevent them from running in the browser. This flexibility can be helpful, but it many situations you'll want to pay attention to the application's execution state—in other words, whether it's running in the browser or in a stand-alone window. You may want to provide less functionality in the browser or even prevent the user from running the application.

The tool that allows you to implement this design is the Application. IsRunningOutOfBrowser property. It returns True if the application was launched as a stand-alone application or False if it's running in the browser. To differentiate your application in these two states, you need to check this property and adjust the user interface accordingly.

For example, if you want to create an application that supports offline use only, you can use code like this in the Startup event handler:

```
Private Sub Application_Startup(ByVal o As Object, ByVal e As StartupEventArgs) _
    Handles Me.Startup

    If Application.Current.IsRunningOutOfBrowser Then
        ' Show the full user interface.
        Me.RootVisual = New MainPage()
    Else
        ' Show a window with an installation message and an Install button.
        Me.RootVisual = New InstallPage()
    End If
End Sub
```

■ **Tip** It's also important to check the IsRunningOutOfBrowser property before attempting to use any features that aren't supported in out-of-browser applications—namely, the browser interaction features described in Chapter 14.

There's one possible quirk with this code. It tests the IsRunningOutOfBrowser property and uses that to decide whether to launch the install process. This makes sense, because if the application is currently running in the browser (meaning IsRunningOutOfBrowser is True), the application is obviously installed. However, the reverse isn't necessarily correct. If the application isn't running in the browser (meaning IsRunningOutOfBrowser is False), it *may* still be installed. The only way to know for sure is to check the Application.InstallState property, which takes one of the values from the InstallState enumeration (as described in Table 21-1).

Table 21-1. Values of the InstallState Enumeration

Value	Description
NotInstalled	The application is running inside the browser and hasn't been installed as an out-of-browser application.
Installing	The application is in the process of being installed as an out-of-browser application.
InstallFailed	The out-of-browser install process failed.
Installed	The application is installed as an out-of-browser application. This doesn't necessarily mean it's currently running outside of the browser—to determine this fact, you need to check the IsRunningOutOfBrowser property.

The InstallPage takes this detail into account. When it loads, it uses the following code in its constructor to determine whether the application is already installed. If the application hasn't been installed, it enables an Install button. Otherwise, it disables the Install button and shows a message asking the user to run the application from the previously installed shortcut.

```
Public Sub New
    InitializeComponent()

    If Application.Current.InstallState = InstallState.Installed Then
        lblMessage.Text = "This application is already installed. " & _
          "You cannot use the browser to run it. " & _
          "Instead, use the shortcut on your computer."
        cmdInstall.IsEnabled = False
    Else
        lblMessage.Text = "You need to install this application to run it."
        cmdInstall.IsEnabled = True
    End If
End Sub
```

■ **Tip** This behavior isn't mandatory. It's perfectly acceptable to have an application that supports in-browser and out-of-browser use. In this case, you may choose to check InstallState and show some sort of install button that gives the user the option to install the application locally. However, you won't redirect the user to an installation page if the application isn't installed.

Once the application has been successfully installed, it makes sense to inform the user. However, you can't assume that the application has been successfully installed just because the Install() method returns True. This simply indicates that the user has clicked the OK button in the installation window to start the install. The actual install process happens asynchronously.

As the install progresses, Silverlight adjusts the Applicaton.InstallState property and triggers the Application.InstallStateChanged event to notify you. When InstallStateChanged fires and InstallState is Installed, your application has just finished being installed as an out-of-browser application. At this point, you should notify the user. The following example does exactly that using some extra code in the App class. It reacts as the application is being installed and notifies the InstallPage:

```
Private Sub Application_InstallStateChanged(ByVal o As Object, _
  ByVal e As EventArgs) Handles Me.InstallStateChanged

    Dim page As InstallPage = TryCast(Me.RootVisual, InstallPage)
    If page IsNot Nothing Then
        ' Tell the root visual to show a message by calling a method
        ' in InstallPage that updates the display.
        Select Case Me.InstallState
            Case InstallState.InstallFailed
                page.DisplayFailed()
            Case InstallState.Installed
                page.DisplayInstalled()
        End Select
    End If
End Sub
```

Finally, you need to add the following methods to the InstallPage class to show the updated status text:

```
Public Sub DisplayInstalled()
    lblMessage.Text = _
        "The application installed and launched. You can close this page."
End Sub

Public Sub DisplayFailed()
    lblMessage.Text = "The application failed to install."
    cmdInstall.IsEnabled = True
End Sub
```

In this example, the application displays some text in the first, browser-based application instance, which informs users that they can now close the browser window (see Figure 21-6). To be fancier, you could use a bit of JavaScript and the browser-interaction features described in Chapter 14 to forcibly close the browser window.

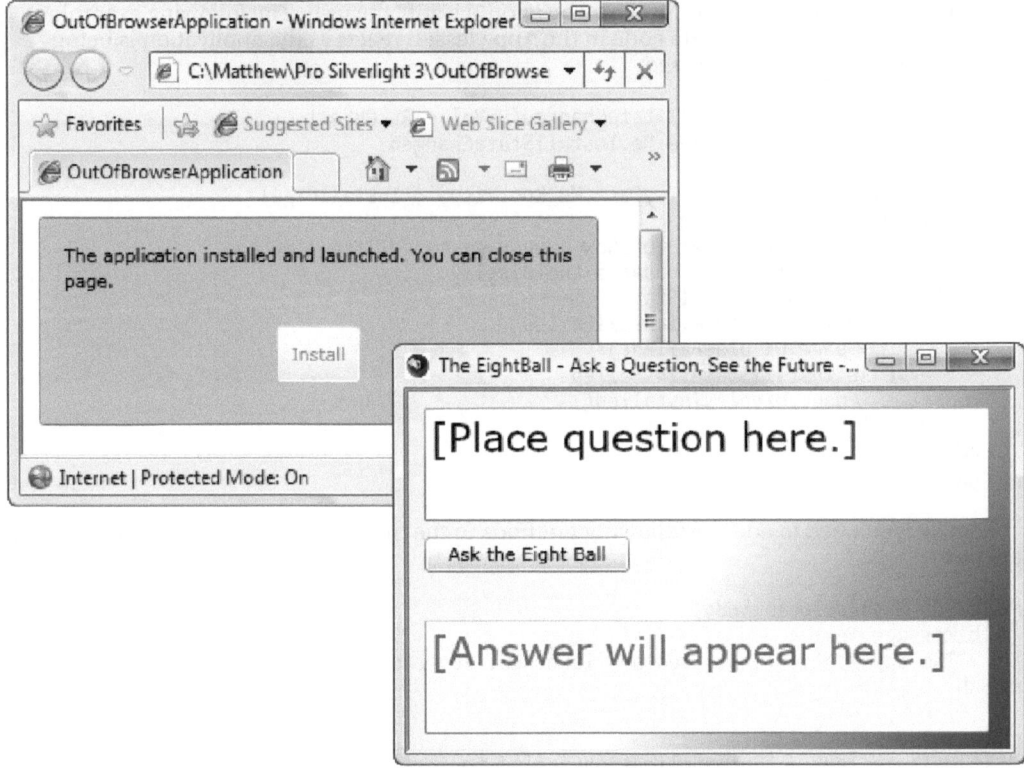

Figure 21-6. Using the browser-based application for installation

Although Silverlight notifies you when the installation process is completed (successfully or unsuccessfully), it doesn't fire the InstallStateChanged even if the application is uninstalled, as described in the next section.

Removing and Updating an Application

Now that you've explored the installation process of an out-of-browser application in detail, it's time to ask about two other common tasks for installed applications: updates and removal.

The removal or uninstallation process is easy: the user right-clicks the running application (either in a browser or in a stand-alone window) and chooses "Remove this application." A confirmation dialog box appears; if the user clicks OK, the application is quickly uninstalled, and its shortcuts are removed. All removals must be initiated by the user, because there is no corresponding Application class method.

Application updates are more interesting. Silverlight has the built-in ability to check for an updated version of your application. In fact, it requires just a single line of code that calls the Application.CheckAndDownloadUpdateAsync() method. This method launches an asynchronous process that checks the web server to see whether a newer XAP file is available. (That's a XAP file with a more recent file date. The actual version number that you used when you compiled the application has no effect.)

Here's an example that checks for updates when the application starts:

```vb
Private Sub Application_Startup(ByVal o As Object, ByVal e As StartupEventArgs) _
  Handles Me.Startup

    If Application.Current.IsRunningOutOfBrowser Then
        ' Check for updates.
        Application.Current.CheckAndDownloadUpdateAsync()

        Me.RootVisual = New MainPage()
    Else
        Me.RootVisual = New InstallPage()
    End If
End Sub
```

■ **Note** Although Microsoft recommends that you call CheckAndDownloadUpdateAsync() in response to a user-initiated action (like clicking an Update button), it doesn't enforce this rule, and you're free to check for updates on application startup.

If a network connection is present, the web server can be contacted, and an updated XAP file is available, the application downloads it automatically and then fires the Application. CheckAndDownloadUpdateCompleted event.

For simplicity's sake, application updates are mandatory once you call the CheckAndDownloadUpdateAsync() method. The user has no way to decline an update, and your application has no way to check whether an update is present without downloading and installing it. However, the update doesn't kick in until the application is restarted.

If you want the user to switch to the new version immediately, you can handle the CheckAndDownloadUpdateCompleted event to display an informative message:

```vb
Private Sub Application_CheckAndDownloadUpdateCompleted(ByVal o As Object, _
  ByVal e As CheckAndDownloadUpdateCompletedEventArgs) _
  Handles Me.CheckAndDownloadUpdateCompleted

    If e.UpdateAvailable Then
        MessageBox.Show("A new version has been installed. " & _
          "Please restart the application.")
        ' (You could add code here to call a custom method in MainPage
        '  that disables the user interface.)
    ElseIf e.Error IsNot Nothing Then
        If TypeOf e.Error Is PlatformNotSupportedException Then
            MessageBox.Show("An application update is available, " & _
              "but it requires a new version of Silverlight. " & _
              "Visit http://silverlight.net to upgrade.")
        Else
            MessageBox.Show("An application update is available, " & _
              "but it cannot be installed. Please remove the current version " & _
              "before installing the new version.")
        End If
    End If
End Sub
```

To try the application update feature, you'll need to create an ASP.NET test website (as described in Chapter 1). That's because Silverlight only supports downloading from a web location, not the file system. However, there's still a catch. As you know, Visual Studio chooses a random port number when it starts the test web server. If you close and restart Visual Studio, it will pick a new port number for its test web server, but any previously installed out-of-browser applications will continue using the old port number to check for updates. Their attempts will fail silently, and your application won't be updated until you manually remove and reinstall it. To avoid this problem altogether, you can deploy your Silverlight application to an IIS test server on your computer or the local network.

Silent Installation

There's one other, less traveled, road to installing an out-of-browser application. Using the sllauncher.exe tool, you can install the application automatically and silently on a Windows computer, using the command line. This functionality is primarily useful if you want to install a Silverlight application through a batch file, from a setup CD, or as some sort of automated installation sequence. (In all these cases, it's also worth considering if a full-fledged WPF application would suit the problem better than a Silverlight application.)

As you learned earlier, sllauncher.exe is part of the Silverlight runtime and the tool that's used to launch out-of-browser applications. To use it install a Silverlight application, the destination computer must already have an installed copy of Silverlight 4. Assuming this detail is in place, you can use a command like this to install your application:

```
"%ProgramFiles%\Microsoft Silverlight\sllauncher.exe" /install:"MyApplication.xap"
```

You run this command from the folder where your XAP file is placed (for example, a setup CD).

This line installs the application for the first time, with no support for updating. It's more common to add the /origin parameter (to specify the server the application will use for updates when you call the Application.CheckAndDownloadUpdateAsync() method), the /overwrite parameter (to ensure that the application installs over any old, already installed versions of the same application), and the /shortcut parameter (to explicitly specify whether you want a desktop shortcut, a Start menu shortcut, or both).

```
"%ProgramFiles%\Microsoft Silverlight\sllauncher.exe" /install:"MyApplication.xap"
/origin:http://www.mysite.net/MyApplication/MyApplication.xap /overwrite
/shortcut:desktop+startmenu
```

■ **Tip** It's important to use the variable %ProgramFiles% instead of hard-coding the Program Files directory, because this directory varies (it's typically c:\Program Files\ on a 32-bit operating system and c:\Program Files (x86)\ on a 64-bit operating system).

Out-of-Browser Application Features

Although an out-of-browser application looks superficially different from an ordinary Silverlight application, what happens inside the stand-alone window is virtually the same. However, there are exceptions, and in this section you'll learn about the three that apply to all out-of-browser applications, regardless of their trust level. First, you'll see how the WebBrowser control allows you to embed HTML content inside your Silverlight application. Next, you'll see learn how to display notifications in separate pop-up windows. Finally, you'll consider how you can interact with the main application window to change its size, position, or state.

The WebBrowser Control

Ordinary Silverlight applications can use Silverlight's browser integration features to interact with the HTML and JavaScript of the hosting page. As you learned in Chapter 14, that means they can change HTML content and trigger JavaScript functions. None of this is possible with an out-of-browser application, because it doesn't have a host web page that you can configure.

To make up for this potential shortcoming, out-of-browser applications can use the WebBrowser control, which hosts HTML content (see Figure 21-7). However, just as the browser interaction features are limited to in-browser applications, the WebBrowser control is limited to out-of-browser applications. If you use the WebBrowser in an ordinary Silverlight application, you'll get a mostly blank rectangle with a message stating "HTML is enabled only in Out-of-Browser mode."

Figure 21-7. *The WebBrowser in action*

Figure 21-7 combines a TextBox, Button, and WebBrowser control in a Grid. Here's the markup:

```
<Grid x:Name="LayoutRoot" Background="White">
  <Grid.RowDefinitions>
    <RowDefinition Height="Auto"></RowDefinition>
    <RowDefinition></RowDefinition>
  </Grid.RowDefinitions>
  <Grid.ColumnDefinitions>
    <ColumnDefinition></ColumnDefinition>
    <ColumnDefinition Width="Auto"></ColumnDefinition>
  </Grid.ColumnDefinitions>

  <TextBox x:Name="txtUrl" Margin="5,5,1,5"
   Text="http://www.prosetech.com"></TextBox>
  <Button Grid.Column="1" Click="cmdGo_Click" Content="Go"
   Margin="1,5,5,5"></Button>
  <Border Grid.Row="1" Grid.ColumnSpan="2" Margin="5"
   BorderBrush="Black" BorderThickness="1">
    <WebBrowser x:Name="browser"></WebBrowser>
  </Border>
</Grid>
```

The WebBrowser provides an extremely simple programming model, with just a handful of members. The following sections show you how it works.

Showing HTML Content

When using the WebBrowser, you can specify the content in three ways.

The most direct approach is to call the NavigateToString() method and pass in a string with the HTML content you want to show. This allows you to use the WebBrowser to show static content you have on hand or content you've grabbed from a web service. Usually, you'll supply a complete HTML document, but an HTML snippet like this works just as well:

```
browser.NavigateToString("<h1>Welcome to the WebBrowser Test</h1>")
```

Your second approach is to call the Navigate() method and pass in a fully qualified or relative URI that points to the content you want to show. For example, this code reacts to a button click and directs the user to the URL that was typed in the text box:

```
Private Sub cmdGo_Click(ByVal sender As Object, ByVal e As RoutedEventArgs)
    Try
        browser.Navigate(New Uri(txtUrl.Text))
    Catch
        ' Possible errors include a UriFormatException (if the text can't be
        ' converted to a valid URI for the Uri class) and a SecurityException
        ' (if the URL uses any scheme other than http://, including ftp://,
        ' file://, and https://).
    End Try
End Sub
```

The third possible approach is to set the Source property of the WebBrowser to a fully qualified or relative URI. This is effectively the same as calling the Navigate() method.

```
browser.Source = New Uri(txtUrl.Text))
```

You can't use the Navigate() method or the Source property with an application URI. In other words, it's not possible for the WebBrowser to load an HTML file that's and embedded in your XAP file (unless you manually access that resource, load it into a string, and call the NavigateToString() method).

■ **Note** To test a Silverlight application that loads web pages into the WebBrowser control, you must include an ASP.NET website in your solution. This shouldn't come as a surprise, because this is a requirement for many of Silverlight's web-based features, including web services and networking.

The WebBrowser has some inevitable limitations. Unlike the HTML interaction features described in Chapter 14, the WebBrowser doesn't expose the HTML document as a collection of HtmlElement objects, so there's no way to programmatically explore its structure. However, you can grab the complete HTML document as a string at any time by calling the WebBrowser.SaveToString() method.

The WebBrowserBrush

The WebBrowser doesn't integrate with the Silverlight rendering system. For example, you can't apply pixel shader effects and transforms to change the appearance of a web page. However, Silverlight does provide a crafty workaround for developers who truly need this ability. You can display the visual content from a WebBrowser window on another element (for example, a Rectangle) by painting it with the WebBrowserBrush. Consider this example, which places both a WebBrowser and Rectangle in the same cell of a Grid, with the Rectangle superimposed on top but showing the content from the WebBrowser underneath:

```
<Grid>
  <WebBrowser x:Name="browser" Source="Hello.html" Height="150" Width="150" />
  <Rectangle>
    <Rectangle.Fill>
      <WebBrowserBrush SourceName="browser" />
    </Rectangle.Fill>
  </Rectangle>
</Grid>
```

This is essentially the same technique as the one you used to paint video content with the VideoBrush in Chapter 11. The only limitation is that the copy you paint won't be user-interactive. For example, if you paint a Rectangle with a WebBrowserBrush, the user can't scroll the page or click a link.

For this reason, you'll probably use the WebBrowserBrush briefly, to apply specific effects. For example, you could create an animation that tweaks a rotate transform and scale transform to make a Rectangle that's painted with the WebBrowserBrush "pop" onto the screen. Once the animation finishes, you would hide the Rectangle and show the real WebBrowser with the source content.

Interacting with JavaScript Code

Although it's not terribly common, you can enable basic interaction between your Silverlight application and JavaScript code using the WebBrowser control. To invoke a JavaScript method in the currently loaded HTML document, you use the InvokeScript() method, as shown here:

```
browser.InvokeScript("MyJavaScriptFunction")
```

The InvokeScript() accepts an optional array of string values as a second parameter, which it will pass to the function as arguments. It also provides the return value from the JavaScript function as its own return value:

```
Dim result As Object = browser.InvokeScript("AddTwoNumbers", _
  New String() { "1", "2" })
```

To react to a JavaScript method, you can handle the WebBrowser.ScriptNotify event. This event fires when JavaScript code calls window.external.notify. For example, if you use this statement in a JavaScript block:

```
window.external.notify("This is a notification from JavaScript")
```

you can retrieve the string in an event handler like this:

```
Private Sub Browser_ScriptNotify(ByVal sender As Object, _
  ByVal e As NotifyEventArgs)
    MessageBox.Show("Received message: " & e.Value)
End Sub
```

However, the WebBrowser control doesn't support the scriptable type system that's described in Chapter 14 for JavaScript-to-Silverlight interaction.

Notification Windows

A notification window is one that pops up in the lower-right corner of the screen, similar to the ones that Outlook uses to notify users about incoming e-mails or Messenger uses to alert the user to a new instant message. (On the Macintosh, notification windows appear at the top of the screen rather than the bottom.)

In Silverlight, a basic notification window is a completely unremarkable blank box, 400 pixels wide and 100 pixels tall. You can resize it to have smaller dimensions but not larger ones. To create a notification window, you create an instance of the NotificationWindow class. To place a notification into the notification window, you simply set its Content property.

If you want your notification window to contain something more than a plain TextBlock, you should create a custom user control for its content. You can then create an instance of that user control and use it to set the NotificationWindow.Content property. The following example use a two-row Grid for the notification window content. At the top of the Grid is a title bar. Underneath is a TextBlock (for the application-supplied message) that's placed over a multicolor gradient fill:

```
<UserControl x:Class="ElevatedTrust.CustomNotification" ...>
  <Grid>
    <Grid.RowDefinitions>
      <RowDefinition Height="Auto"></RowDefinition>
      <RowDefinition></RowDefinition>
```

```
      </Grid.RowDefinitions>

      <Border Background="LightGray" Height="20">
        <TextBlock Margin="3" FontSize="10"
        Text="MyApplication Notification"></TextBlock>
      </Border>

      <Border Grid.Row="1">
        <Border.Background>
          <LinearGradientBrush>
            ...
          </LinearGradientBrush>
        </Border.Background>
        <TextBlock x:Name="lblMessage" Margin="10" FontWeight="Bold" FontSize="16"
          Foreground="White" TextWrapping="Wrap" HorizontalAlignment="Center"
          VerticalAlignment="Center">Notification goes here.</TextBlock>
      </Border>
    </Grid>
  </UserControl>
```

The code that for the CustomNotification user control exposes a Message property that sets the TextBlock:

```
Public Property Message() As String
    Get
        Return lblMessage.Text
    End Get
    Set(ByVal value As String)
        lblMessage.Text = value
    End Set
End Property
```

To show the notification window, you simply need to create an instance of CustomNotification, place it in an instance of the framework-supplied NotificationWindow class, and call NotificationWindow.Show() to display your notification:

```
If Application.Current.IsRunningOutOfBrowser Then
    Dim notification As New CustomNotification()
    notification.Message = "You have just been notified. The time is " & _
      DateTime.Now.ToLongTimeString() & "."

    Dim window As New NotificationWindow()
    window.Content = notification

    ' Specify the number of milliseconds before the window closes.
    ' This example sets 5 seconds.
    window.Show(5000)
Else
    ' (Implement a different notification strategy here.)
End If
```

Figure 21-8 shows the notification message.

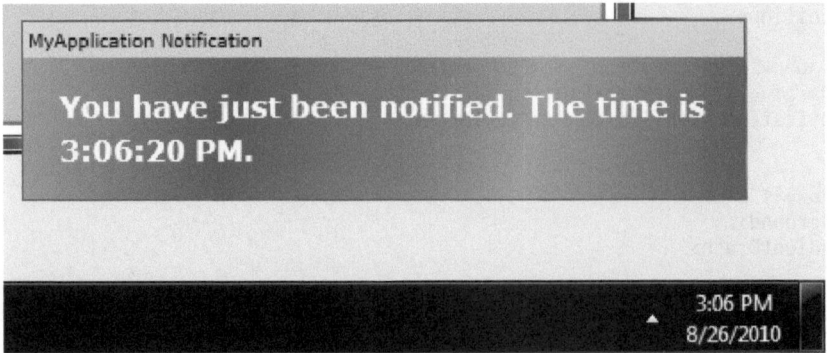

Figure 21-8. A custom notification

The user can't move a notification window. However, you could add a button inside the notification window that calls NotificationWindow.Close(), which would allow the user to close the notification before it times out.

■ **Tip** This example assumes you want to use the standard notification window width (400 by 100 pixels). If you want to resize the window smaller, you should hard-code a Width and Height in the custom user control. Then, when you create the NotificationWindow, take the Width and Height of the custom user control and assign it to the NotificationWindow. This way, you get a true sense of what your notification window looks like when you edit the user control, and you can change the size whenever you want by modifying just one file (the user control XAML).

Notification windows aren't designed with any sort of queuing mechanism. If you create a second instance of NotifyWindow and call its Show() method while the first notification window is still visible, nothing will happen (and the new notification message will be lost). A better approach is to store the NotifyWindow you want to use as a class-level field in your code:

```
Private window As New NotificationWindow()
```

Then, you can explicitly call Close() to hide the first notification just before you call Show() to show the second. Best of all, you don't even need to check the current state of the notification window, because calling Close() on an already closed window has no effect.

```
Dim notification As New CustomNotification()
notification.Message = "You have just been notified. The time is " & _
  DateTime.Now.ToLongTimeString() & "."
window.Content = notification

window.Close()
window.Show(5000)
```

In some applications, you may expect to have multiple notifications occurring in close proximity. In this case, the Close() and Show() approach may not be suitable, because it may hide a notification before the user has time to read it. To avoid this problem, you can implement a queuing system. The basic idea is to maintain an application-wide Queue(Of T) collection of notification messages and to handle the NotificationWindow.Closed event. Every time the Closed event occurs, you can respond by dequeuing the next message and calling Show() to show the notification window again. Tim Heuer shows one possible implementation at http://tinyurl.com/yfefkud.

Controlling the Main Window

An ordinary Silverlight application runs inside a web browser window, and the only way to interact with that window is through JavaScript. But in an out-of-browser application, you get a stand-alone window, which you can interact with as a Window object through the shared Application.MainWindow property.

The Window class provides a small set of properties and methods, as listed in Table 21-2. You can read these properties at any time, and you can change most of them in response to user-initiated events and in an event handler for the Application.Startup event (the exception being the Top and Left properties, which *must* be set in the Application.Startup event handler).

Table 21-2. Properties of the Window Class

Properties	Description
Top and Left	Top is the number of pixels between the top edge of the window and the top edge of the screen. Left is the number of pixels between the left edge of the window and the left edge of the screen. You can only set these properties in an event handler for the Application.Startup event and only if you have specified initial startup coordinates in the Out-of-Browser Settings window.
Height and Width	Provides the height and width of the main window, in pixels. You can only set these properties in an event handler for the Application.Startup event or in response to a user-initiated event.
WindowState	Indicates whether the window is Normal, Minimized, or Maximized, as a value from the WindowState enumeration. You can only set this property in an event handler for the Application.Startup event or in response to a user-initiated event.
TopMost	If True, the window is "pinned" so it appears above all other application windows. You can only set this property in an event handler for the Application.Startup event or in response to a user-initiated event.

The best way to set the initial position and size of the main window is using the Out-of-Browser Settings window. But the Window properties are useful if you need to change the size dynamically at runtime—for example, if you need to accommodate changing amounts of content.

The Windows properties also allow you to create an application that remembers its window position. To do this, you simply save the relevant details when the application shuts down and restore them when it starts up. Here's an example that performs this feat using isolated storage:

```
    Private Sub Application_Startup(ByVal sender As Object, ByVal e As StartupEventArgs)
        If Application.Current.IsRunningOutOfBrowser Then
            ' (You can check for updates here, if you want.)

            ' Restore the window state.
            Try
                Dim store As IsolatedStorageFile = _
                  IsolatedStorageFile.GetUserStoreForApplication()

                If store.FileExists("window.Settings") Then
                    Using fs As IsolatedStorageFileStream = _
                      store.OpenFile("window.Settings", FileMode.Open)
                        Dim r As New BinaryReader(fs)
                        Application.Current.MainWindow.Top = r.ReadDouble()
                        Application.Current.MainWindow.Left = r.ReadDouble()
                        Application.Current.MainWindow.Width = r.ReadDouble()
                        Application.Current.MainWindow.Height = r.ReadDouble()
                        r.Close()
                    End Using
                End If
            Catch err As Exception
                ' Can't restore the window details. No need to report the error.
            End Try
        End If
        Me.RootVisual = New MainPage()
    End Sub

    Private Sub Application_Exit(ByVal sender As Object, ByVal e As EventArgs)
        If Application.Current.IsRunningOutOfBrowser Then
            ' Store window state.
            Try
                Dim store As IsolatedStorageFile = _
                  IsolatedStorageFile.GetUserStoreForApplication()

                Using fs As IsolatedStorageFileStream = _
                  store.CreateFile("window.Settings")
                    Dim w As New BinaryWriter(fs)
                    w.Write(Application.Current.MainWindow.Top)
                    w.Write(Application.Current.MainWindow.Left)
                    w.Write(Application.Current.MainWindow.Width)
                    w.Write(Application.Current.MainWindow.Height)
                    w.Close()
                End Using
            Catch err As Exception
                ' Can't save the window details. No need to report the error.
            End Try
        End If
    End Sub
```

Remember, to set the Top and Left properties of the main window, you must have checked the "Set window location manually" setting in the Out-of-Browser Settings window and provided a fixed default value for the Top and Left settings. Unfortunately, the approach used here isn't perfect, because you'll see the window appear briefly in the default position before it is relocated to its previous position and restored to its previous size.

The Window class also provides a small set of methods. The most useful of these is Close(), which closes the main window (in response to a user-initiated action). There are a few more methods that are designed for elevated trust applications, like DragMove(), DragResize(), and Activate(). These are particularly useful when creating a customized window frame, and they're discussed later in the "Window Customization" section.

Elevated Trust

An elevated trust application is a special type of out-of-browser application—one that has an impressive set of additional privileges. These privileges don't match the capabilities of a traditional desktop application, but they do allow a range of potentially dangerous actions, from activating third-party programs to directly accessing the file system. For this reason, elevated trust applications make the most sense in controlled environments, such as a corporate network where users already know the application and trust the publisher. In other scenarios, users may be reluctant to grant the additional privileges that an elevated trust application needs, and they may choose not to install it at all. (System administrators can also configure computers so that users are prevented from installing or running out-of-browser applications that require elevated trust.)

■ **Note** Before you begin developing a Silverlight application that requires elevated trust, consider whether a full-fledged Windows Presentation Foundation (WPF) application will better suit your needs. WPF applications are .NET-powered desktop applications that support a wide range of rich features, including some that Silverlight can't emulate (like 3-D drawing and document display). Like Silverlight, WPF applications can be installed from the Web. Unlike Silverlight, WPF applications don't run in the browser and so guarantee the best possible performance. The only drawback is that WPF applications are designed exclusively for the Windows operating system.

Installing an Elevated Trust Application

To indicate that an application requires elevated trust, you simply need to open the Out-of-Browser Settings window (as described at the beginning of this chapter) and check the "Require elevated trust" setting.

Now, when the installation process starts (either through the right-click Silverlight menu or through a call to the Application.Install() method), the browser presents a more intimidating security warning, like the one shown for the application named ElevatedTrust in Figure 21-9.

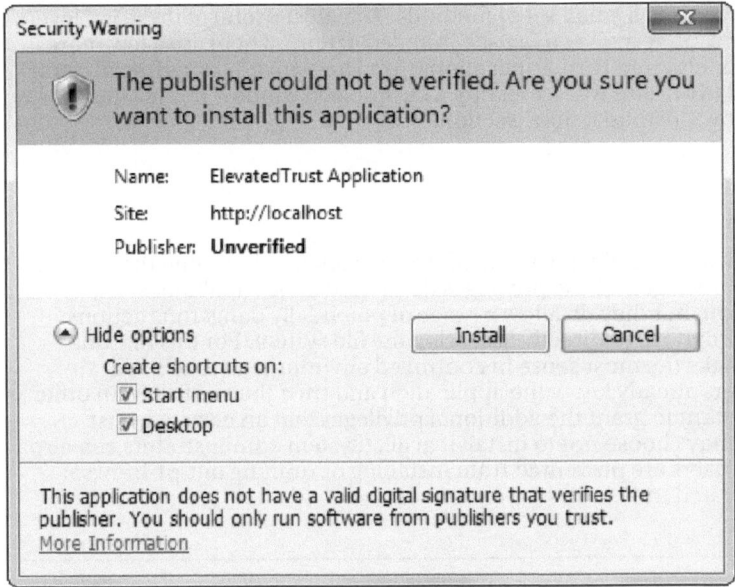

Figure 21-9. Installing an elevated trust application

The situation improves a bit if you've signed the application with an X.509 certificate from a certificate authority (CA). In this case, the warning includes the publisher name and the exclamation icon is replaced with a less threatening question mark, as shown in Figure 21-10.

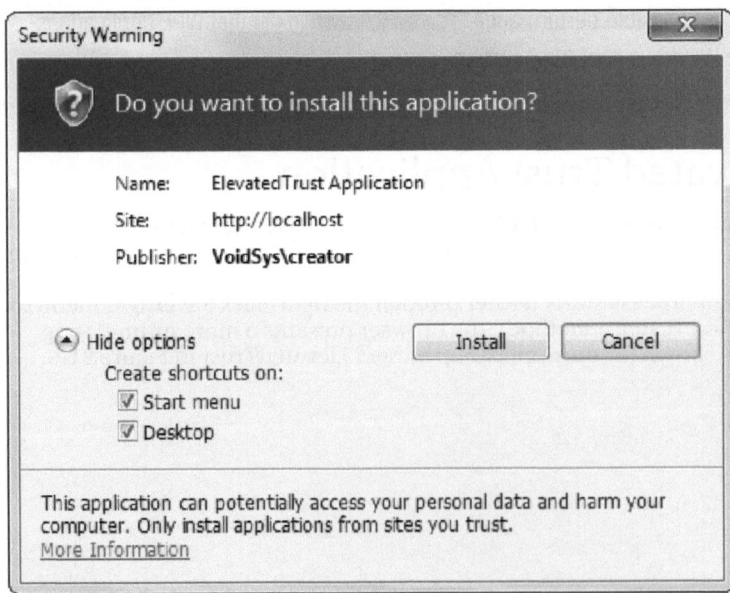

Figure 21-10. Installing a signed elevated trust application

USING A TEST CERTIFICATE

You'll sign a professional Silverlight application with a certificate from a known certificate authority (such as VeriSign). If you sign your application with a test certificate that you've created yourself, users won't get the friendly warning shown in Figure 21-10. The problem here is that when you create a test certificate, you essentially assume the role of the certificate authority. And although other people's computers are already configured to trust well-known certificate authorities, they aren't be configured to trust you. That means that if you use a test certificate, there won't be a chain of trust to vouch for your identity as the application publisher.

However, it's sometimes worth signing an application with a test certificate for the purposes of testing. For example, you might do this if you want to test an elevated trust application on a test web server in a company network. In this situation, you won't be able to perform automatic updates unless you sign your application first.

To create a test certificate, follow these steps:

1. Double-click the My Project node in the Solution Explorer, and click the Signing tab.

2. Place a check mark next to the option named Sign the XAP File.

3. If you've already created a certificate, you can select it by clicking the Select From Store or Select From File button. Otherwise, click the Create Test Certificate button to create a new certificate. Visual Studio will ask you for a password to use to protect the certificate, and it will then add the certificate as a .pfx file in your project. From now on, when you compile your application, Visual Studio will sign it automatically.

4. To establish the chain of trust, you must install your test certificate in the certificate store on your computer. To do this, browse to the .pfx certificate file in Windows Explorer and double-click it. This starts the Certificate Import Wizard. As you walk through the wizard, you'll need to supply two details: the certificate password and the certificate store where the test certificate should be placed. To ensure your application is trusted (so that you get the nicer installation messages and the automatic update feature), you must place your certificate in the Trusted Root Certification Authorities store.

When you're finished testing your application, it's a good idea to remove the certificate from the certificate store on your computer, simply to reduce the possibility of an attack that exploits this opening.

There's another reason to prefer signed applications when using elevated trust. Silverlight does not allow automatic updates with unsigned applications that require elevated trust. In other words, calling Application.CheckAndDownloadUpdateAsync() has no effect. This restriction is designed to prevent an innocent application from being replaced with a more dangerous successor.

The only way to update an unsigned, elevated trust application is to manually remove it and then reinstall it. But if your application is signed, the Application.CheckAndDownloadUpdateAsync() method continues to work in its normal way. That means Silverlight will check for a new version of your application and install it automatically—provided this new version is signed with the same certificate as the original application.

■ **Note** There's one exception to Silverlight's automatic update restrictions. If your application is configured to download updates from the local computer (in other words, the network address 127.0.0.1), Silverlight will allow automatic updates, even if the application is not signed. This behavior is designed to make testing easier. It means you can run an elevated trust application from inside Visual Studio and take advantage of automatic updating in the same way as you would with a lower-trust application.

Once the user chooses to install the application, a shortcut is created, and the application is started, just as with any other Silverlight application. From this point on, the user will not receive any further security prompts. However, when the application is launched, it will automatically run with elevated trust.

From a threat-modeling point of view, you should assume that an elevated trust application has the same privileges as the user who runs it. However, an elevated trust application will not receive administrator privileges in Windows, even if the current user is an administrator, thanks to the permission elevation feature in the Windows operating system.

■ **Note** An application that requires elevated trust cannot be installed as a normal, low-trust application. If the user chooses not to continue with the installation because of the security warning, the application will not be installed. However, an elevated trust application can still be run inside the browser, unless you take steps to prevent it (such as checking the install state and disabling the application user interface, as demonstrated earlier in this chapter). When running in the browser, the application does not receive elevated trust, and it will not be able to use any of the special features described in this section.

The Abilities of an Elevated Trust Application

So, what can a trusted application do that an ordinary Silverlight application can't? Here's a quick list:

Window customization: If you don't like the standard window border and controls that are provided by the operating system, you can remove them from your elevated trust application and then draw your own custom chrome with the help of Silverlight's standard elements.

File system access: An elevated trust application can access the files in the current user's document folders, including MyDocuments, MyMusic, MyPictures, and MyVideos.

COM interop: An elevated trust application can make use of the vast libraries of functionality that are exposed on Windows computers through COM. For example, you can interact with Outlook, Office applications, and built-in components of the Windows operating system, such as Windows Script Host and Windows Management Instrumentation (WMI).

Unrestricted full-screen support: Elevated trust applications do not show the "Press ESC to exit full-screen mode" message when they switch into full-screen mode (nor does the Escape key return the application to normal mode, unless you explicitly add this functionality). More significantly, full-screen applications that have elevated trust can continue to receive keyboard input.

No cross-domain access restrictions: An elevated trust application is allowed to download content from any website, call a web service on any website, and open a socket connection to any server, just like an ordinary desktop application.

Fewer user-consent restrictions: There are a host of situations where an ordinary Silverlight application must explicitly ask for user consent, such as when accessing the clipboard, storing files in isolated storage, configuring a full-screen application to stay in full-screen mode when it loses focus, and so on. These restrictions are removed in an elevated trust application, except when accessing audio and video capture devices. Similarly, there are a number of actions that can be performed only in response to a user-initiated action, such as switching to full-screen mode, using the clipboard, manipulating the main window, and so on. In an elevated trust application, these tasks can be performed at any time.

The following sections expand on the three new features of elevated trust applications: window customization, file access, and COM interop.

■ **Tip** To determine whether your application is currently running with elevated trust, check the Application.HasElevatedPermissions property. If it's True, you can use the features described next. If not, you need to either disable the application (if you don't want to allow running in partial trust) or provide alternate code.

Window Customization

An ordinary out-of-browser application runs in the standard window frame provided by the operating system. It has the familiar buttons (which you can use to minimize, maximize, and close the window), and its color, shading, and transparency are controlled by the operating system. Although this works perfectly well, it might not suit your design. If you're creating a slick, highly graphical application, you might prefer to pair your visuals with a custom window frame and hand-crafted window buttons.

There are two steps to customizing the window frame. The first step is to completely remove the standard window frame. To do this, double-click the My Project node in the Solution Explorer. Click the Silverlight tab and then the Out-of-Browser Settings button, and choose an option from the Window Style list. The first option, Default, gives you the standard window frame. The second option, No Border, removes the standard window frame and leaves a floating rectangle with your user control content. The third option, Borderless Round Corners, removes the standard window frame and gently rounds the corners of the application window. The difference between these last two options is purely cosmetic—either way, the standard window frame is taken away. Figure 21-11 shows a borderless window with a light yellow background, superimposed over a Notepad window.

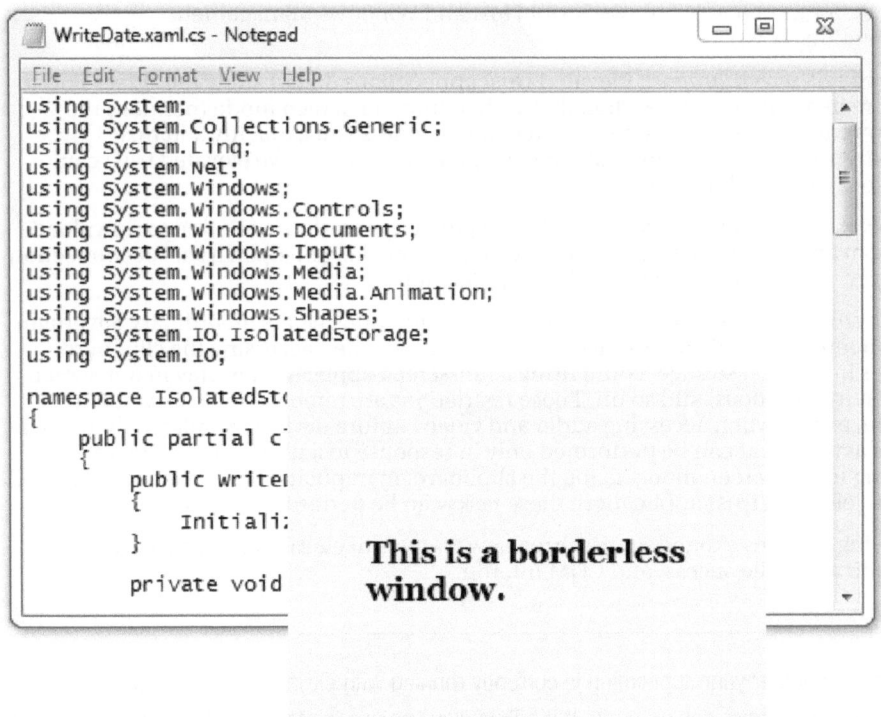

Figure 21-11. A Silverlight window with no frame

The second step is to draw the frame you want and add basic controls for closing the window, dragging the window, and changing the window state. Because of cross-platform considerations, Silverlight doesn't give you any way to modify the standard operating system window frame. Instead, you must draw the window frame yourself, using the appropriate image content or Silverlight elements. For example, you could place the main content in a single-cell Grid, superimposed over another Silverlight element, like a Rectangle or an Image. Or, you could wrap the root user control in a Border element, like this:

```xaml
<UserControl x:Class="ElevatedTrust.CustomWindow" ... >
  <Border x:Name="windowFrame" BorderBrush="DarkSlateBlue" BorderThickness="5"
    CornerRadius="2" Margin="0,0,1,1">
    <Grid x:Name="LayoutRoot" Background="LightSteelBlue" Margin="5">
      ...
    </Grid>
  </Border>
</UserControl>
```

Here, the application is using the Borderless Round Corners option. The 1-pixel margin on the right and bottom edges ensures the frame appears in the correct position, and the CornerRadius of 2 lines up with rounded corners of the window region. Figure 21-12 shows the result.

Figure 21-12. A frameless window that uses the Border element

---

■ **Note** Silverlight does not support shaped or irregular windows. Thus, when you draw the window frame, you must limit yourself to the rectangle (or rounded rectangle) that defines the main window. Essentially, Silverlight gives you a rectangle to work with, and you simply paint it with the appropriate window frame and content. Similarly, you can't create transparent or partially transparent regions that allow other application windows to show through.

Although the added border in Figure 21-12 looks a bit better, there's still a serious problem. Without the frame that's provided by the operating system, the user has no way to resize the window, move it to a different position, minimize it, maximize it, or close it. In fact, it's up to you to build this functionality into your customized main window. Fortunately, the members of the Window class make this work fairly easy.

The first step is to create a title bar. The title bar has three purposes: it holds a text caption, it provides a place that the user can click and drag to move the window, and it holds the minimize, maximize, and close buttons (usually at the far right). Here's the markup you could use to create a basic title bar. The shapes in the minimize, maximize, and close buttons are drawn using Path and Rectangle elements.

```
<Border x:Name="titleBar" Background="LightSteelBlue"
 MouseLeftButtonDown="titleBar_MouseLeftButtonDown">
  <Grid>
    <Grid.ColumnDefinitions>
      <ColumnDefinition></ColumnDefinition>
      <ColumnDefinition Width="Auto"></ColumnDefinition>
      <ColumnDefinition Width="Auto"></ColumnDefinition>
      <ColumnDefinition Width="Auto"></ColumnDefinition>
    </Grid.ColumnDefinitions>

    <TextBlock Margin="5">Title Bar</TextBlock>

    <Button Grid.Column="1" x:Name="cmdMinimize" Width="24"
     Click="cmdMinimize_Click">
      <Path Stroke="Black" StrokeThickness="4" Data="M 1,10 L 13,10" />
    </Button>

    <Button Grid.Column="2" x:Name="cmdMaximize" Width="24"
     Click="cmdMaximize_Click">
      <Rectangle Stroke="Black" StrokeThickness="3" Height="12" Width="12" />
    </Button>

    <Button Grid.Column="3" x:Name="cmdClose" Width="24" Click="cmdClose_Click">
      <Path Stroke="Black" StrokeThickness="3" Data="M 2,2 L 12,12 M 12,2 L 2,12" />
    </Button>
  </Grid>
</Border>
```

You can then place this title bar in a Grid with two rows. The first row holds the title bar, and the second row holds the rest of the window content. The entire Grid fits into the Border that outlines the window. Figure 21-13 shows the final result.

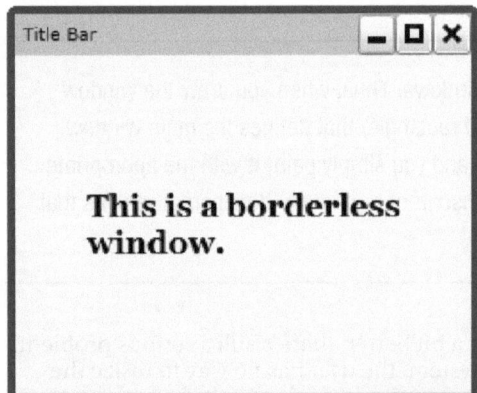

Figure 21-13. *A window with a custom title bar*

Of course, to make this window do what it's supposed to do, you need to add code. Changing the window state and closing the window is easy, thanks to the WindowState property and the Close() method. Surprisingly, it's just as easy to allow the user to drag the window. All you need to do is react when the left mouse button is pressed over the title bar and call the Window.DragMove() method. From that point on, the window will move in tandem with the mouse, until the user releases the mouse button.

Here's the complete code:

```
Private Sub titleBar_MouseLeftButtonDown(ByVal sender As System.Object, _
  ByVal e As System.Windows.Input.MouseButtonEventArgs)
    Application.Current.MainWindow.DragMove()
End Sub

Private Sub cmdMinimize_Click(ByVal sender As System.Object, _
  ByVal e As System.Windows.RoutedEventArgs)
    Application.Current.MainWindow.WindowState = WindowState.Minimized
End Sub

Private Sub cmdMaximize_Click(ByVal sender As System.Object, _
  ByVal e As System.Windows.RoutedEventArgs)
    If Application.Current.MainWindow.WindowState = WindowState.Normal Then
        Application.Current.MainWindow.WindowState = WindowState.Maximized
    Else
        Application.Current.MainWindow.WindowState = WindowState.Normal
    End If
End Sub

Private Sub cmdClose_Click(ByVal sender As System.Object, _
  ByVal e As System.Windows.RoutedEventArgs)
    Application.Current.MainWindow.Close()
End Sub
```

■ **Tip** In an elevated trust application, you can use the Close() method without restriction. That means you can call it without waiting for a user-initiated action to occur. This allows you to forcibly close the window after detecting an update in the event handler for the CheckAndDownloadUpdateCompleted event. If you choose to use this approach, make sure you show a message explaining that an update has occurred and the application must be restarted before you call Close().

There's still one missing feature: a way to resize the window by clicking and dragging the edge of the window border. The easiest way to implement this feature is to wrap the entire window in a Grid that uses invisible Rectangle elements. Altogether, you'll need eight Rectangle elements—one on each edge and one at each corner. Each one is 5 pixels wide (or high). Figure 21-14 shows where each rectangle is placed around the cells of the Grid with three rows and three columns.

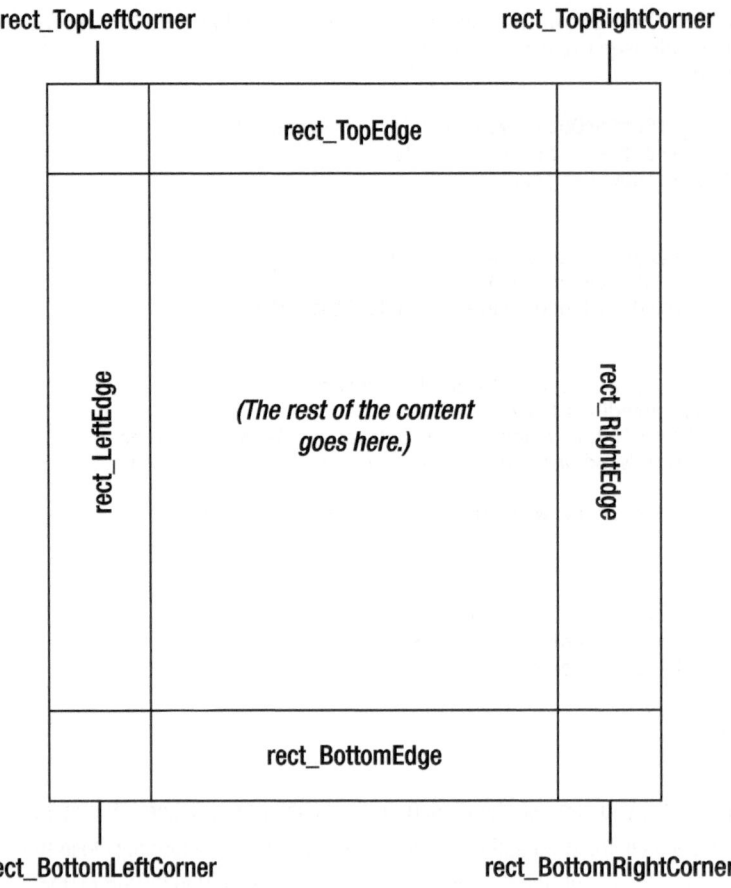

Figure 21-14. Using invisible resize rectangles

Here's the Grid that defines this structure, with the invisible rectangles that allow resizing:

```
<Border x:Name="windowFrame" ...>
  <Grid x:Name="resizeContainer">
    <Grid.ColumnDefinitions>
      <ColumnDefinition Width="5"></ColumnDefinition>
      <ColumnDefinition Width="*"></ColumnDefinition>
      <ColumnDefinition Width="5"></ColumnDefinition>
    </Grid.ColumnDefinitions>

    <Grid.RowDefinitions>
      <RowDefinition Height="5"></RowDefinition>
      <RowDefinition Height="*"></RowDefinition>
      <RowDefinition Height="5"></RowDefinition>
    </Grid.RowDefinitions>

    <Rectangle x:Name="rect_TopLeftCorner" Grid.Row="0" Grid.Column="0"
     Cursor="SizeNWSE" Fill="Transparent" MouseLeftButtonDown="rect_Resize" />
```

```
   <Rectangle x:Name="rect_TopEdge" Grid.Row="0" Grid.Column="1"
    Cursor="SizeNS" Fill="Transparent" MouseLeftButtonDown="rect_Resize" />
   <Rectangle x:Name="rect_TopRightCorner" Grid.Row="0" Grid.Column="2"
    Cursor="SizeNESW" Fill="Transparent" MouseLeftButtonDown="rect_Resize" />
   <Rectangle x:Name="rect_LeftEdge" Grid.Row="1" Grid.Column="0"
    Cursor="SizeWE" Fill="Transparent" MouseLeftButtonDown="rect_Resize" />
   <Rectangle x:Name="rect_RightEdge" Grid.Row="1" Grid.Column="2"
    Cursor="SizeWE" Fill="Transparent" MouseLeftButtonDown="rect_Resize" />
   <Rectangle x:Name="rect_BottomLeftCorner" Grid.Row="2" Grid.Column="0"
    Cursor="SizeNESW" Fill="Transparent" MouseLeftButtonDown="rect_Resize" />
   <Rectangle x:Name="rect_BottomEdge" Grid.Row="2" Grid.Column="1"
    Cursor="SizeNS" Fill="Transparent" MouseLeftButtonDown="rect_Resize" />
   <Rectangle x:Name="rect_BottomRightCorner" Grid.Row="2" Grid.Column="2"
    Cursor="SizeNWSE" Fill="Transparent" MouseLeftButtonDown="rect_Resize" />

   <Grid Grid.Row="1" Grid.Column="1">
     <!-- The rest of the content goes here, including the title bar. -->
   </Grid>
  </Grid>
</Border>
```

Each Rectangle has the job of showing the appropriate resize mouse cursor. When clicked with the left mouse button, the Rectangle calls Window.DragResize() with the correct argument to initiate the resize operation:

```
Private Sub rect_Resize(ByVal sender As System.Object, _
  ByVal e As System.Windows.Input.MouseButtonEventArgs)
    If sender Is rect_TopLeftCorner Then
        Application.Current.MainWindow.DragResize(WindowResizeEdge.TopLeft)
    ElseIf sender Is rect_TopEdge Then
        Application.Current.MainWindow.DragResize(WindowResizeEdge.Top)
    ElseIf sender Is rect_TopRightCorner Then
        Application.Current.MainWindow.DragResize(WindowResizeEdge.TopRight)
    ElseIf sender Is rect_LeftEdge Then
        Application.Current.MainWindow.DragResize(WindowResizeEdge.Left)
    ElseIf sender Is rect_RightEdge Then
        Application.Current.MainWindow.DragResize(WindowResizeEdge.Right)
    ElseIf sender Is rect_BottomLeftCorner Then
        Application.Current.MainWindow.DragResize(WindowResizeEdge.BottomLeft)
    ElseIf sender Is rect_BottomEdge Then
        Application.Current.MainWindow.DragResize(WindowResizeEdge.Bottom)
    ElseIf sender Is rect_BottomRightCorner Then
        Application.Current.MainWindow.DragResize(WindowResizeEdge.BottomRight)
    End If
End Sub
```

Much like the DragMove() method, the DragResize() method takes care of the rest, automatically resizing the window as the user moves the mouse pointer until the mouse button is released (see Figure 21-15).

Figure 21-15. *Resizing a custom window*

This completes the infrastructure you need to create a respectable custom window. Once you have these details in place (a title bar, window buttons, dragging, and resizing), you're ready to customize the visual appearance with images and elements to get the exact effect you want.

File Access

Applications with elevated trust can access files in several user-specific folders on the Windows operating system. They include the following:

- My Documents
- My Music
- My Pictures
- My Videos

You get the physical path to these locations using the Environment.GetFolderPath() method and passing in one of the values from the Environment.SpecialFolder enumeration. Here's an example:

```
Dim documentPath As String = _
  Environment.GetFolderPath(Environment.SpecialFolder.MyDocuments)
```

■ **Note** The Environment.SpecialFolder enumeration provides values for additional paths besides the "My" folders listed earlier. For example, it includes a path for the desktop, program files directory, Internet cache, Start menu folder, and so on. However, if you attempt to use any of these other folders, you'll receive a SecurityException.

Once you've retrieve the path to a "My" folder, you can create, modify, and delete files and directories in it. The most convenient way to perform file and directory process is to use the shared methods of the File and Directory classes, which are transplants from the full .NET Framework. For example, you can write an entire file in one step using methods such as WriteAllText(), WriteAllBytes(), or WriteAllLines(), and you can read a file in one step using ReadAllText(), ReadText(), or ReadLines().

Here's an example that write a file in the My Documents folder:

```
Dim documentPath As String = _
  Environment.GetFolderPath(Environment.SpecialFolder.MyDocuments)
Dim fileName As String = System.IO.Path.Combine(documentPath, "TestFile.txt")

File.WriteAllText(fileName, "This is a test.")
```

And here's the complementary code that reads the file:

```
Dim documentPath As String = _
  Environment.GetFolderPath(Environment.SpecialFolder.MyDocuments)
Dim fileName As String = System.IO.Path.Combine(documentPath, "TestFile.txt")

If File.Exists(fileName) Then
    Dim contents As String = File.ReadAllText(fileName)
    lblResults.Text = contents
End If
```

If you don't want to write or read the contents of a file in one chunk, you can open a FileStream for the file and then wrap it in a StreamReader or StreamWriter. Here's an example that writes to a file through a StreamWriter, in much the same way as if you called File.WriteAllText():

```
Dim documentPath As String = _
  Environment.GetFolderPath(Environment.SpecialFolder.MyDocuments)
Dim fileName As String = System.IO.Path.Combine(documentPath, "TestFile.txt")

Using fs As New FileStream(fileName, FileMode.Create)
    Dim writer As New StreamWriter(fs)
    writer.Write("This is a test with FileStream.")
    writer.Close()
End Using
```

No matter which approach you'll use, you should wrap your file access code in an exception-handling block to deal with potential errors, as you did in Chapter 18.

COM

One of the most surprising features that's granted to elevated trust applications is the ability to create COM objects and use them to do pretty much anything.

For those who need a history lesson, Component Object Model (COM) is the original standard for component reuse and application integration. It's the predecessor to .NET, a key part of past and present versions of Windows and still a cornerstone technology for certain tasks (for example, interacting with Office applications such as Word and Excel). Because other applications can provide hooks into their functionality through COM libraries and because these libraries can then be used to "drive" (or *automate*) the applications, this feature is sometimes called *COM automation*.

■ **Note** COM is an unashamedly Windows-specific technology. Thus, this feature won't work on non-Windows operating systems, like Mac OS.

In the past, before the widespread adoption of .NET, developers often created their own COM components to share and deploy functionality. However, the COM support in Silverlight is not intended for the use of custom COM components. Although this is technically possible, deploying COM components runs into a wide range of configuration and versioning headaches, so it's not recommended. Instead, the COM support is designed to give your applications a pass into preinstalled libraries, such as those provided by the Windows operating system and other already-installed applications (such as Microsoft Office), and those provided to access the functionality in other components (such as scanners, cameras, and so on).

The AutomationFactory class is the entry point to Silverlight's COM support. It's found in the System.Runtime.InteropServices.Automation namespace. To use Silverlight's COM support, several details need to be in place. Your application needs to be running out-of-browser, with elevated trust, on a Windows computer. If all these requirements are met and COM support is available, the AutomationFactory.IsAvailable property will return True.

Here's an example method that checks for COM supports and shows an explanatory message if it isn't available:

```
Private Function TestForComSupport() As Boolean
    If Application.Current.InstallState <> InstallState.Installed Then
        MessageBox.Show( _
            "This feature is not available because the application is not installed.")
    ElseIf Not Application.Current.IsRunningOutOfBrowser Then
        MessageBox.Show( _
            "This feature is not available because you are running in the browser.")
    ElseIf Not Application.Current.HasElevatedPermissions Then
        MessageBox.Show( _
            "This feature is not available because the application " & _
            "does not have elevated trust.")
    ElseIf Not AutomationFactory.IsAvailable Then
        MessageBox.Show("This feature is not available because the operating " & _
            "system does not appear to support COM.")
    Else
        Return True
    End If
    Return False
End Function
```

To create a COM object, you can call the AutomationFactory.CreateObject() method, with the full COM type name. Here's an example:

```
Dim speech As Object = AutomationFactory.CreateObject("Sapi.SpVoice")
```

To use a COM object, you need the help of *late binding*. When using late binding, you define a variable that points to the COM object using the basic Object type. But here's the trick—you can call all the methods that the COM object supports, as though they were methods of the Object type. Here's an example that calls the Speak() method of the Sapi.SpVoice COM object:

```
Dim speech As Object = AutomationFactory.CreateObject("Sapi.SpVoice")
speech.Speak("This is a test")
```

This works even though the .NET Object type doesn't include a method named Speak().

When you use late binding, you first must make sure that your project or code file is not using strict type checking. One way is to add the following statement to the top of your code file:

```
Option Strict Off
```

Another approach is to turn off strict type checking for your entire project using the project properties. To do that, just double-click My Project in the Solution Explorer, choose the Compile tab, and change the Option Strict setting to Off. By default, new Visual Basic projects have strict type checking off, so you don't need to make any changes or add the Option Strict Off statement to your code.

There's a drawback to Silverlight's use of late binding with COM—it means Visual Studio can't provide IntelliSense for any COM object, and the VB compiler won't catch possible errors when you build your application. For example, it won't notice if you attempt to use a method that the COM object doesn't provide. For this reason, it's always a good idea to wrap your COM code in an exception-handling block. This way, you can trap the errors that occur if you attempt to instantiate a COM object that doesn't exist or if you use it in a way that isn't supported at runtime.

The following code puts all these details together. It starts by checking whether COM support is available. It then creates the COM component for text-to-speech, which is included with the Windows operating system. The result is a synthesized voice that speaks the sentence "This is a test."

```
If TestForComSupport() Then
    Try
        Using speech As Object = AutomationFactory.CreateObject("Sapi.SpVoice")
            speech.Volume = 100
            speech.Speak("This is a test")
        End Using
    Catch err As Exception
        ' An exception occurs if the COM library doesn't exist, or if the
        ' properties and methods you attempted to use aren't part of it.
    End Try
End If
```

As this illustrates, using COM takes surprisingly little effort. The real work is finding the component you want to use, learning its object model (which, depending on the application, may be complex and highly idiosyncratic), and catching typos and other potential usage errors.

Here's an example that uses a different component—the Windows Script Host, which provides functionality for retrieving system information and environment variables, working with the registry, and managing shortcuts. Here's the code Windows Script Host is used to launch a completely separate application (the Windows Calculator) using the Run() method:

```
Using shell As Object = AutomationFactory.CreateObject("WScript.Shell")
    shell.Run("calc.exe")
End Using
```

And here's an example that uses two components in concert. First, it uses the Windows Script Host to retrieve the location of the current user's desktop folder from the registry. Then, it uses the Windows FileSystemObject (FSO) to create a file in that location.

```
Using shell As Object = AutomationFactory.CreateObject("WScript.Shell")
    ' Get the desktop path from the registry.
    Dim desktopPath As String = shell.RegRead( _
      "HKCU\Software\Microsoft\Windows\CurrentVersion\" & _
      "Explorer\User Shell Folders\Desktop")

    Using fso As Object = _
      AutomationFactory.CreateObject("Scripting.FileSystemObject")
        Dim filePath As String = System.IO.Path.Combine(desktopPath, "TestFile.txt")

        ' Create the file.
        Dim file As Object = fso.CreateTextFile(filePath, True)
        file.WriteLine("An elevated trust Silverlight application can write " & _
          "anywhere that doesn't require adminsitrative privileges.")
        file.Close()

        ' Read the file.
        file = fso.OpenTextFile(filePath, 1, True)
        MessageBox.Show(file.ReadAll())
        file.Close()
    End Using
End Using
```

This is a step beyond what an elevated trust application can do without COM. As you learned in the previous section, elevated trust applications get direct file access to various "My" folders, but nothing else. But the FSO provides a slew of methods for reading, writing, and managing files and folders anywhere, so long as the location you choose doesn't require administrator privileges. (That's because Silverlight won't use the Windows permission elevation feature to request administrator privileges, like some other applications.) To learn more about the FSO model, refer to http://tinyurl.com/325a2jt.

COM interop isn't limited to built-in Windows components. It also works with many other well-known applications, like the Microsoft Office applications (Word, Excel, PowerPoint, Outlook and so on). All of these applications have complex object models with dozens of different classes.

For example, here's a snippet that starts Word, inserts two short paragraphs of text, and then makes the Word window visible so the user can continue working with it:

```
Using word As Object = AutomationFactory.CreateObject("Word.Application")
    Dim document As Object = word.Documents.Add()

    Dim paragraph As Object = document.Content.Paragraphs.Add
    paragraph.Range.Text = "Heading 1"
    paragraph.Range.Font.Bold = True
    paragraph.Format.SpaceAfter = 18
```

```
    paragraph.Range.InsertParagraphAfter()

    paragraph = document.Content.Paragraphs.Add
    paragraph.Range.Font.Bold = False
    paragraph.Range.Text = "This is some more text"

    word.Visible = True
End Using
```

Figure 21-16 shows the result of this code.

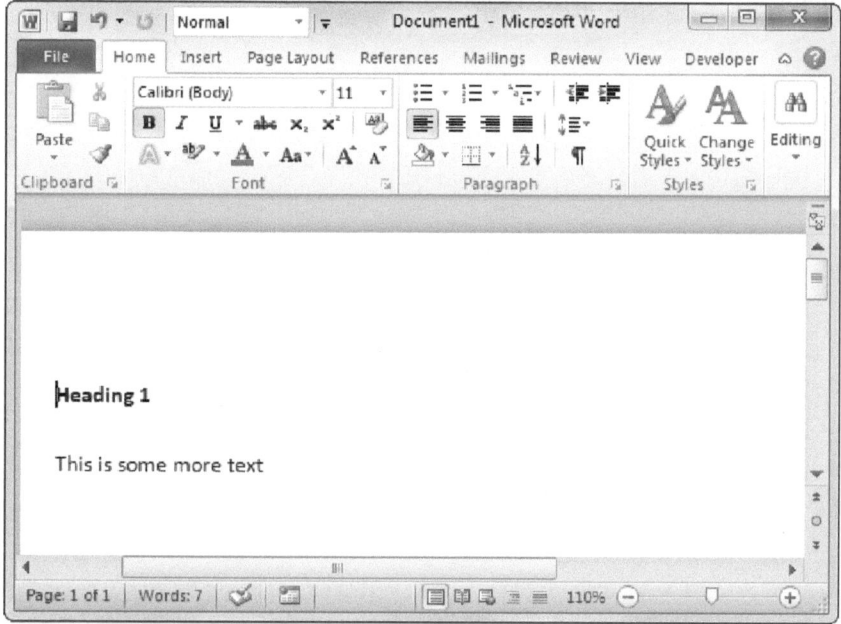

Figure 21-16. Interacting with Word programmatically

There's really no limit to what a crafty use of COM interop can do. For even more mind-stretching examples, check out Justin Angel's block post at http://justinangel.net/CuttingEdgeSilverlight4ComFeatures, where he uses various COM libraries to mimic user input, pin applications to the Windows 7 taskbar, configure applications to run automatically at startup, connect to a camera over USB, watch the file system for changes, execute SQL on a local database, and more. As with the previous examples, the challenge is in ferreting out the details of the COM library you want to use, because the Silverlight side of the bargain is pure simplicity.

The Last Word

In this chapter, you looked at how Silverlight applications can go beyond the browser. At its simplest, Silverlight's out-of-browser feature is simply a way to let the user launch your applications from the Start menu and run them when the computer is offline. The actual functioning of ordinary out-of-browser applications is mostly the same as normal Silverlight applications, except for the update process, the WebBrowser control, and the small added frill of notification windows. But out-of-browser applications also open the way for something far more interesting—elevated trust applications. And as you saw in this chapter, elevated trust applications have a set of new capabilities, including window customization, document file access, and COM interop.

Index

◼◼◼

■ **T**

■ X, Y